Precalculus

Ninth Edition

Hybrid

Ron Larson

The Pennsylvania State University
The Behrend College

With the assistance of David C. Falvo

The Pennsylvania State University
The Behrend College

BROOKS/COLE
CENGAGE Learning™

Australia • Brazil • Japan • Korea • Mexico • Singapore • Spain • United Kingdom • United States

Precalculus
Ninth Edition, Hybrid

Ron Larson

Publisher: Liz Covello

Acquisitions Editor: Gary Whalen

Senior Development Editor: Stacy Green

Assistant Editor: Cynthia Ashton

Editorial Assistant: Samantha Lugtu

Media Editor: Lynh Pham

Senior Content Project Manager: Jessica Rasile

Art Director: Linda May

Rights Acquisition Specialist: Shalice Shah-Caldwell

Manufacturing Planner: Doug Bertke

Text/Cover Designer: Larson Texts, Inc.

Compositor: Larson Texts, Inc.

Cover Image: diez artwork/Shutterstock.com

For product information and technology assistance, contact us at
Cengage Learning Customer & Sales Support, 1-800-354-9706.

For permission to use material from this text or product, submit all requests online at **www.cengage.com/permissions.** Further permissions questions can be emailed to **permissionrequest@cengage.com.**

Library of Congress Control Number: 2012948319

Student Edition:
ISBN-13: 978-1-133-95054-7
ISBN-10: 1-133-95054-X

Brooks/Cole
20 Channel Center Street
Boston, MA 02210
USA

Cengage Learning is a leading provider of customized learning solutions with office locations around the globe, including Singapore, the United Kingdom, Australia, Mexico, Brazil, and Japan. Locate your local office at: **international.cengage.com/region**

Cengage Learning products are represented in Canada by Nelson Education, Ltd.

For your course and learning solutions, visit **www.cengage.com.**

Purchase any of our products at your local college store or at our preferred online store **www.cengagebrain.com.**

Instructors: Please visit **login.cengage.com** and log in to access instructor-specific resources.

Printed in the United States of America
1 2 3 4 5 6 7 16 15 14 13

Contents

Technology

The technology feature gives suggestions for effectively using tools such as calculators, graphing calculators, and spreadsheet programs to help deepen your understanding of concepts, ease lengthy calculations, and provide alternate solution methods for verifying answers obtained by hand.

Historical Notes

These notes provide helpful information regarding famous mathematicians and their work.

Algebra of Calculus

Throughout the text, special emphasis is given to the algebraic techniques used in calculus. Algebra of Calculus examples and exercises are integrated throughout the text and are identified by the symbol ∫ .

Project

The projects at the end of selected sections involve in-depth applied exercises in which you will work with large, real-life data sets, often creating or analyzing models. These projects are offered online at LarsonPecalculus.com.

Project: Department of Defense The table shows the total numbers of military personnel P (in thousands) on active duty from 1980 through 2010. *(Source: U.S. Department of Defense)*

Year	Personnel, P	Year	Personnel, P
1980	2051	1995	1518
1981	2083	1996	1472
1982	2109	1997	1439
1983	2123	1998	1407
1984	2138	1999	1386
1985	2151	2000	1384
1986	2169	2001	1385
1987	2174	2002	1414
1988	2138	2003	1434
1989	2130	2004	1427
1990	2044	2005	1389
1991	1986	2006	1385
1992	1807	2007	1380
1993	1705	2008	1402
1994	1610	2009	1419
		2010	1431

(a) Use a graphing utility to plot the data. Let t represent the year, with $t = 0$ corresponding to 1980.

(b) A model that approximates the data is given by

$$P = \frac{9.6518t^2 - 244.743t + 2044.77}{0.0059t^2 - 0.131t + 1}$$

where P is the total number of personnel (in thousands) and t is the year, with $t = 0$ corresponding to 1980. Construct a table showing the actual values of P and the values of P obtained using the model.

▷ **TECHNOLOGY** You can use a graphing utility to check that a solution is reasonable. One way to do this is to graph the left side of the equation, then graph the right side of the equation, and determine the point of intersection. For instance, in Example 2, if you graph the equations

$y_1 = 6(x - 1) + 4$ The left side

$y_2 = 3(7x + 1)$ The right side

in the same viewing window, they should intersect at $x = -\frac{1}{3}$, as shown in the graph below.

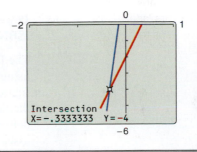

Chapter Summaries

The Chapter Summary now includes explanations and examples of the objectives taught in each chapter.

ENHANCED
WebAssign

Enhanced WebAssign combines exceptional Precalculus content that you know and love with the most powerful online homework solution, WebAssign. Enhanced WebAssign engages you with immediate feedback, rich tutorial content and interactive, fully customizable eBooks (YouBook) helping you to develop a deeper conceptual understanding of the subject matter.

Instructor Resources

Print

Annotated Instructor's Edition
ISBN-13: 978-1-133-94902-2

This AIE is the complete student text plus point-of-use annotations for you, including extra projects, classroom activities, teaching strategies, and additional examples. Answers to even-numbered text exercises, Vocabulary Checks, and Explorations are also provided.

Complete Solutions Manual
ISBN-13: 978-1-133-95442-2

This manual contains solutions to all exercises from the text, including Chapter Review Exercises, and Chapter Tests.

Media

PowerLecture with ExamView™
ISBN-13: 978-1-133-95440-8

The DVD provides you with dynamic media tools for teaching Precalculus while using an interactive white board. PowerPoint® lecture slides and art slides of the figures from the text, together with electronic files for the test bank and a link to the Solution Builder, are available. The algorithmic ExamView allows you to create, deliver, and customize tests (both print and online) in minutes with this easy-to-use assessment system. The DVD also provides you with a tutorial on integrating our instructor materials into your interactive whiteboard platform. Enhance how your students interact with you, your lecture, and each other.

Solution Builder
(*www.cengage.com/solutionbuilder*)
This online instructor database offers complete worked-out solutions to all exercises in the text, allowing you to create customized, secure solutions printouts (in PDF format) matched exactly to the problems you assign in class.

ENHANCED WebAssign

www.webassign.net
Printed Access Card: 978-0-538-73810-1
Online Access Code: 978-1-285-18181-3

Exclusively from Cengage Learning, Enhanced WebAssign combines the exceptional mathematics content that you know and love with the most powerful online homework solution, WebAssign. Enhanced WebAssign engages students with immediate feedback, rich tutorial content, and interactive, fully customizable eBooks (YouBook), helping students to develop a deeper conceptual understanding of their subject matter. Online assignments can be built by selecting from thousands of text-specific problems or supplemented with problems from any Cengage Learning textbook.

Student Resources

Print

Student Study and Solutions Manual
ISBN-13: 978-1-133-95441-5

This guide offers step-by-step solutions for all odd-numbered text exercises, Chapter and Cumulative Tests, and Practice Tests with solutions.

Text-Specific DVD
ISBN-13: 978-1-133-96287-8

Keyed to the text by section, these DVDs provide comprehensive coverage of the course—along with additional explanations of concepts, sample problems, and application—to help you review essential topics.

Note Taking Guide
ISBN-13: 978-1-133-94904-6

This innovative study aid, in the form of a notebook organizer, helps you develop a section-by-section summary of key concepts.

Media

ENHANCED WebAssign
www.webassign.net
Printed Access Card: 978-0-538-73810-1
Online Access Code: 978-1-285-18181-3

Enhanced WebAssign (assigned by the instructor) provides you with instant feedback on homework assignments. This online homework system is easy to use and includes helpful links to textbook sections, video examples, and problem-specific tutorials.

CengageBrain.com

Visit *www.cengagebrain.com* to access additional course materials and companion resources. At the CengageBrain.com home page, search for the ISBN of your title (from the back cover of your book) using the search box at the top of the page. This will take you to the product page where free companion resources can be found.

Acknowledgements

I would like to thank the many people who have helped me prepare the text and the supplements package. Their encouragement, criticisms, and suggestions have been invaluable.

Thank you to all of the instructors who took the time to review the changes in this edition and to provide suggestions for improving it. Without your help, this book would not be possible.

Reviewers

Timothy Andrew Brown, *South Georgia College*
Blair E. Caboot, *Keystone College*
Shannon Cornell, *Amarillo College*
Gayla Dance, *Millsaps College*
Paul Finster, *El Paso Community College*
Paul A. Flasch, *Pima Community College West Campus*
Vadas Gintautas, *Chatham University*
Lorraine A. Hughes, *Mississippi State University*
Shu-Jen Huang, *University of Florida*
Renyetta Johnson, *East Mississippi Community College*
George Keihany, *Fort Valley State University*
Mulatu Lemma, *Savannah State University*
William Mays Jr., *Salem Community College*
Marcella Melby, *University of Minnesota*
Jonathan Prewett, *University of Wyoming*
Denise Reid, *Valdosta State University*
David L. Sonnier, *Lyon College*
David H. Tseng, *Miami Dade College – Kendall Campus*
Kimberly Walters, *Mississippi State University*
Richard Weil, *Brown College*
Solomon Willis, *Cleveland Community College*
Bradley R. Young, *Darton College*

My thanks to Robert Hostetler, The Behrend College, The Pennsylvania State University, and David Heyd, The Behrend College, The Pennsylvania State University, for their significant contributions to previous editions of this text.

I would also like to thank the staff at Larson Texts, Inc. who assisted with proofreading the manuscript, preparing and proofreading the art package, and checking and typesetting the supplements.

On a personal level, I am grateful to my spouse, Deanna Gilbert Larson, for her love, patience, and support. Also, a special thanks goes to R. Scott O'Neil. If you have suggestions for improving this text, please feel free to write to me. Over the past two decades I have received many useful comments from both instructors and students, and I value these comments very highly.

Ron Larson, Ph.D.
Professor of Mathematics
Penn State University
www.RonLarson.com

1 Functions and Their Graphs

Snowstorm

Bacteria

Average Speed

Alternative-Fueled Vehicles

Americans with Disabilities Act *(page 28)*

1.1 Rectangular Coordinates

- ■ Plot points in the Cartesian plane.
- ■ Use the Distance Formula to find the distance between two points.
- ■ Use the Midpoint Formula to find the midpoint of a line segment.
- ■ Use a coordinate plane to model and solve real-life problems.

The Cartesian Plane

Just as you can represent real numbers by points on a real number line, you can represent ordered pairs of real numbers by points in a plane called the **rectangular coordinate system,** or the **Cartesian plane,** named after the French mathematician René Descartes (1596–1650).

Two real number lines intersecting at right angles form the Cartesian plane, as shown in Figure 1.1. The horizontal real number line is usually called the **x-axis,** and the vertical real number line is usually called the **y-axis.** The point of intersection of these two axes is the **origin,** and the two axes divide the plane into four parts called **quadrants.**

The Cartesian plane can help you visualize relationships between two variables. For instance, given how far north and west one city is from another, plotting points to represent the cities can help you visualize these distances and determine the flying distance between the cities.

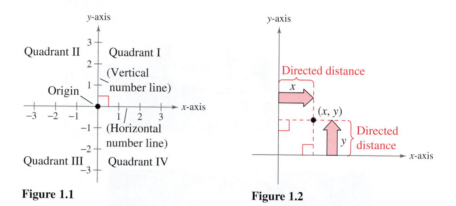

Figure 1.1

Figure 1.2

Each point in the plane corresponds to an **ordered pair** (x, y) of real numbers x and y, called **coordinates** of the point. The **x-coordinate** represents the directed distance from the y-axis to the point, and the **y-coordinate** represents the directed distance from the x-axis to the point, as shown in Figure 1.2.

The notation (x, y) denotes both a point in the plane and an open interval on the real number line. The context will tell you which meaning is intended.

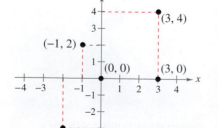

Figure 1.3

EXAMPLE 1 Plotting Points in the Cartesian Plane

Plot the points $(-1, 2)$, $(3, 4)$, $(0, 0)$, $(3, 0)$, and $(-2, -3)$.

Solution To plot the point $(-1, 2)$, imagine a vertical line through -1 on the x-axis and a horizontal line through 2 on the y-axis. The intersection of these two lines is the point $(-1, 2)$. Plot the other four points in a similar way, as shown in Figure 1.3.

✓ **Checkpoint** ◀))) Audio-video solution in English & Spanish at LarsonPrecalculus.com.

Plot the points $(-3, 2)$, $(4, -2)$, $(3, 1)$, $(0, -2)$, and $(-1, -2)$.

The beauty of a rectangular coordinate system is that it allows you to *see* relationships between two variables. It would be difficult to overestimate the importance of Descartes's introduction of coordinates in the plane. Today, his ideas are in common use in virtually every scientific and business-related field.

EXAMPLE 2   **Sketching a Scatter Plot**

The table shows the numbers *N* (in millions) of subscribers to a cellular telecommunication service in the United States from 2001 through 2010, where *t* represents the year. Sketch a scatter plot of the data.   *(Source: CTIA-The Wireless Association)*

Solution   To sketch a *scatter plot* of the data shown in the table, represent each pair of values by an ordered pair (*t*, *N*) and plot the resulting points, as shown below. For instance, the ordered pair (2001, 128.4) represents the first pair of values. Note that the break in the *t*-axis indicates omission of the years before 2001.

DATA	Year, *t*	Subscribers, *N*
	2001	128.4
	2002	140.8
	2003	158.7
	2004	182.1
	2005	207.9
	2006	233.0
	2007	255.4
	2008	270.3
	2009	290.9
	2010	311.0

Spreadsheet at LarsonPrecalculus.com

Subscribers to a Cellular
Telecommunication Service

✓ **Checkpoint** ◀))) *Audio-video solution in English & Spanish at LarsonPrecalculus.com.*

The table shows the numbers *N* (in thousands) of cellular telecommunication service employees in the United States from 2001 through 2010, where *t* represents the year. Sketch a scatter plot of the data.   *(Source: CTIA-The Wireless Association)*

DATA	*t*	*N*
	2001	203.6
	2002	192.4
	2003	205.6
	2004	226.0
	2005	233.1
	2006	253.8
	2007	266.8
	2008	268.5
	2009	249.2
	2010	250.4

Spreadsheet at LarsonPrecalculus.com

▷ **TECHNOLOGY** The scatter plot in Example 2 is only one way to represent the data graphically. You could also represent the data using a bar graph or a line graph. Try using a graphing utility to represent the data given in Example 2 graphically.

In Example 2, you could have let *t* = 1 represent the year 2001. In that case, there would not have been a break in the horizontal axis, and the labels 1 through 10 (instead of 2001 through 2010) would have been on the tick marks.

The Pythagorean Theorem and the Distance Formula

The following famous theorem is used extensively throughout this course.

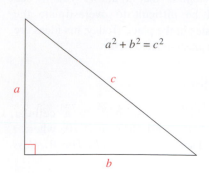

$$a^2 + b^2 = c^2$$

Figure 1.4

Pythagorean Theorem

For a right triangle with hypotenuse of length c and sides of lengths a and b, you have $a^2 + b^2 = c^2$, as shown in Figure 1.4. (The converse is also true. That is, if $a^2 + b^2 = c^2$, then the triangle is a right triangle.)

Suppose you want to determine the distance d between two points (x_1, y_1) and (x_2, y_2) in the plane. These two points can form a right triangle, as shown in Figure 1.5. The length of the vertical side of the triangle is $|y_2 - y_1|$ and the length of the horizontal side is $|x_2 - x_1|$.

By the Pythagorean Theorem,

$$d^2 = |x_2 - x_1|^2 + |y_2 - y_1|^2$$
$$d = \sqrt{|x_2 - x_1|^2 + |y_2 - y_1|^2} = \sqrt{(x_2 - x_1)^2 + (y_2 - y_1)^2}.$$

This result is the **Distance Formula.**

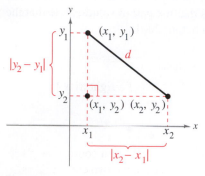

Figure 1.5

The Distance Formula

The distance d between the points (x_1, y_1) and (x_2, y_2) in the plane is

$$d = \sqrt{(x_2 - x_1)^2 + (y_2 - y_1)^2}.$$

EXAMPLE 3 **Finding a Distance**

Find the distance between the points $(-2, 1)$ and $(3, 4)$.

Algebraic Solution

Let

$$(x_1, y_1) = (-2, 1) \quad \text{and} \quad (x_2, y_2) = (3, 4).$$

Then apply the Distance Formula.

$$d = \sqrt{(x_2 - x_1)^2 + (y_2 - y_1)^2} \qquad \text{Distance Formula}$$
$$= \sqrt{[3 - (-2)]^2 + (4 - 1)^2} \qquad \text{Substitute for } x_1, y_1, x_2, \text{ and } y_2.$$
$$= \sqrt{(5)^2 + (3)^2} \qquad \text{Simplify.}$$
$$= \sqrt{34} \qquad \text{Simplify.}$$
$$\approx 5.83 \qquad \text{Use a calculator.}$$

So, the distance between the points is about 5.83 units. Use the Pythagorean Theorem to check that the distance is correct.

$$d^2 \overset{?}{=} 5^2 + 3^2 \qquad \text{Pythagorean Theorem}$$
$$\left(\sqrt{34}\right)^2 \overset{?}{=} 5^2 + 3^2 \qquad \text{Substitute for } d.$$
$$34 = 34 \qquad \text{Distance checks. } \checkmark$$

Graphical Solution

Use centimeter graph paper to plot the points $A(-2, 1)$ and $B(3, 4)$. Carefully sketch the line segment from A to B. Then use a centimeter ruler to measure the length of the segment.

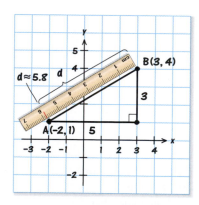

The line segment measures about 5.8 centimeters. So, the distance between the points is about 5.8 units.

✓ Checkpoint 🔊)) *Audio-video solution in English & Spanish at LarsonPrecalculus.com.*

Find the distance between the points $(3, 1)$ and $(-3, 0)$.

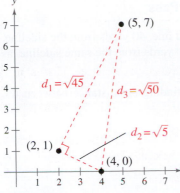

Figure 1.6

ALGEBRA HELP You can review the techniques for evaluating a radical in Appendix A.2.

EXAMPLE 4 Verifying a Right Triangle

Show that the points

$(2, 1)$, $(4, 0)$, and $(5, 7)$

are vertices of a right triangle.

Solution The three points are plotted in Figure 1.6. Using the Distance Formula, the lengths of the three sides are as follows.

$$d_1 = \sqrt{(5-2)^2 + (7-1)^2} = \sqrt{9+36} = \sqrt{45}$$

$$d_2 = \sqrt{(4-2)^2 + (0-1)^2} = \sqrt{4+1} = \sqrt{5}$$

$$d_3 = \sqrt{(5-4)^2 + (7-0)^2} = \sqrt{1+49} = \sqrt{50}$$

Because $(d_1)^2 + (d_2)^2 = 45 + 5 = 50 = (d_3)^2$, you can conclude by the Pythagorean Theorem that the triangle must be a right triangle.

✓ *Checkpoint* Audio-video solution in English & Spanish at LarsonPrecalculus.com.

Show that the points $(2, -1)$, $(5, 5)$, and $(6, -3)$ are vertices of a right triangle.

The Midpoint Formula

To find the **midpoint** of the line segment that joins two points in a coordinate plane, you can find the average values of the respective coordinates of the two endpoints using the **Midpoint Formula**.

The Midpoint Formula

The midpoint of the line segment joining the points (x_1, y_1) and (x_2, y_2) is given by the Midpoint Formula

$$\text{Midpoint} = \left(\frac{x_1 + x_2}{2}, \frac{y_1 + y_2}{2} \right).$$

For a proof of the Midpoint Formula, see Proofs in Mathematics on page 75.

EXAMPLE 5 Finding a Line Segment's Midpoint

Find the midpoint of the line segment joining the points

$(-5, -3)$ and $(9, 3)$.

Solution Let $(x_1, y_1) = (-5, -3)$ and $(x_2, y_2) = (9, 3)$.

$$\text{Midpoint} = \left(\frac{x_1 + x_2}{2}, \frac{y_1 + y_2}{2} \right) \qquad \textcolor{red}{\text{Midpoint Formula}}$$

$$= \left(\frac{-5 + 9}{2}, \frac{-3 + 3}{2} \right) \qquad \textcolor{red}{\text{Substitute for } x_1, y_1, x_2, \text{ and } y_2.}$$

$$= (2, 0) \qquad \textcolor{red}{\text{Simplify.}}$$

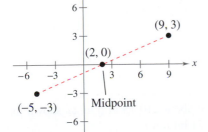

Figure 1.7

The midpoint of the line segment is $(2, 0)$, as shown in Figure 1.7.

✓ *Checkpoint* Audio-video solution in English & Spanish at LarsonPrecalculus.com.

Find the midpoint of the line segment joining the points $(-2, 8)$ and $(4, -10)$.

Applications

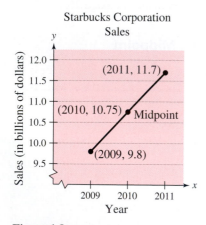

Football Pass

(40, 28)

(20, 5)

Figure 1.8

EXAMPLE 6   **Finding the Length of a Pass**

A football quarterback throws a pass from the 28-yard line, 40 yards from the sideline. A wide receiver catches the pass on the 5-yard line, 20 yards from the same sideline, as shown in Figure 1.8. How long is the pass?

Solution   You can find the length of the pass by finding the distance between the points (40, 28) and (20, 5).

$$d = \sqrt{(x_2 - x_1)^2 + (y_2 - y_1)^2}$$ Distance Formula

$$= \sqrt{(40 - 20)^2 + (28 - 5)^2}$$ Substitute for x_1, y_1, x_2, and y_2.

$$= \sqrt{20^2 + 23^2}$$ Simplify.

$$= \sqrt{400 + 529}$$ Simplify.

$$= \sqrt{929}$$ Simplify.

$$\approx 30$$ Use a calculator.

So, the pass is about 30 yards long.

✓ *Checkpoint*  *Audio-video solution in English & Spanish at LarsonPrecalculus.com.*

A football quarterback throws a pass from the 10-yard line, 10 yards from the sideline. A wide receiver catches the pass on the 32-yard line, 25 yards from the same sideline. How long is the pass?

In Example 6, the scale along the goal line does not normally appear on a football field. However, when you use coordinate geometry to solve real-life problems, you are free to place the coordinate system in any way that is convenient for the solution of the problem.

EXAMPLE 7   **Estimating Annual Sales**

Starbucks Corporation had annual sales of approximately $9.8 billion in 2009 and $11.7 billion in 2011. Without knowing any additional information, what would you estimate the 2010 sales to have been?   *(Source: Starbucks Corporation)*

Solution   One solution to the problem is to assume that sales followed a linear pattern. With this assumption, you can estimate the 2010 sales by finding the midpoint of the line segment connecting the points (2009, 9.8) and (2011, 11.7).

$$\text{Midpoint} = \left(\frac{x_1 + x_2}{2}, \frac{y_1 + y_2}{2} \right)$$ Midpoint Formula

$$= \left(\frac{2009 + 2011}{2}, \frac{9.8 + 11.7}{2} \right)$$ Substitute for x_1, x_2, y_1, and y_2.

$$= (2010, 10.75)$$ Simplify.

So, you would estimate the 2010 sales to have been about $10.75 billion, as shown in Figure 1.9. (The actual 2010 sales were about $10.71 billion.)

✓ *Checkpoint*  *Audio-video solution in English & Spanish at LarsonPrecalculus.com.*

Yahoo! Inc. had annual revenues of approximately $7.2 billon in 2008 and $6.3 billion in 2010. Without knowing any additional information, what would you estimate the 2009 revenue to have been?   *(Source: Yahoo! Inc.)*

Starbucks Corporation
Sales

(2011, 11.7)

(2010, 10.75) Midpoint

(2009, 9.8)

Year

Figure 1.9

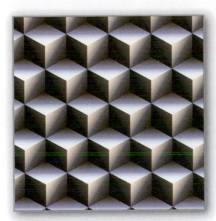

Much of computer graphics, including this computer-generated tessellation, consists of transformations of points in a coordinate plane. Example 8 illustrates one type of transformation called a translation. Other types include reflections, rotations, and stretches.

EXAMPLE 8 **Translating Points in the Plane**

The triangle in Figure 1.10 has vertices at the points $(-1, 2)$, $(1, -4)$, and $(2, 3)$. Shift the triangle three units to the right and two units up and find the vertices of the shifted triangle, as shown in Figure 1.11.

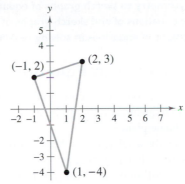

Figure 1.10

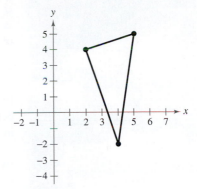

Figure 1.11

Solution To shift the vertices three units to the right, add 3 to each of the x-coordinates. To shift the vertices two units up, add 2 to each of the y-coordinates.

Original Point	Translated Point
$(-1, 2)$	$(-1 + 3, 2 + 2) = (2, 4)$
$(1, -4)$	$(1 + 3, -4 + 2) = (4, -2)$
$(2, 3)$	$(2 + 3, 3 + 2) = (5, 5)$

 Checkpoint *Audio-video solution in English & Spanish at LarsonPrecalculus.com.*

Find the vertices of the parallelogram shown after translating it two units to the left and four units down.

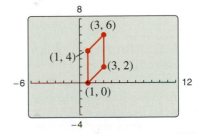

The figures in Example 8 were not really essential to the solution. Nevertheless, it is strongly recommended that you develop the habit of including sketches with your solutions—even when they are not required.

Summarize (Section 1.1)

1. Describe the Cartesian plane *(page 2)*. For an example of plotting points in the Cartesian plane, see Example 1.

2. State the Distance Formula *(page 4)*. For examples of using the Distance Formula to find the distance between two points, see Examples 3 and 4.

3. State the Midpoint Formula *(page 5)*. For an example of using the Midpoint Formula to find the midpoint of a line segment, see Example 5.

4. Describe examples of how to use a coordinate plane to model and solve real-life problems *(pages 6 and 7, Examples 6–8)*.

1.2 Graphs of Equations

The graph of an equation can help you see relationships between real-life quantities, such as the life expectancy of a child born in a given year.

- Sketch graphs of equations.
- Identify *x*- and *y*-intercepts of graphs of equations.
- Use symmetry to sketch graphs of equations.
- Write equations of and sketch graphs of circles.
- Use graphs of equations in solving real-life problems.

The Graph of an Equation

In Section 1.1, you used a coordinate system to graphically represent the relationship between two quantities. There, the graphical picture consisted of a collection of points in a coordinate plane.

Frequently, a relationship between two quantities is expressed as an **equation in two variables.** For instance, $y = 7 - 3x$ is an equation in x and y. An ordered pair (a, b) is a **solution** or **solution point** of an equation in x and y when the substitutions $x = a$ and $y = b$ result in a true statement. For instance, $(1, 4)$ is a solution of $y = 7 - 3x$ because $4 = 7 - 3(1)$ is a true statement.

In this section, you will review some basic procedures for sketching the graph of an equation in two variables. The **graph of an equation** is the set of all points that are solutions of the equation.

EXAMPLE 1 Determining Solution Points

Determine whether (a) $(2, 13)$ and (b) $(-1, -3)$ lie on the graph of $y = 10x - 7$.

Solution

a.

$$y = 10x - 7 \qquad \text{Write original equation.}$$
$$13 \overset{?}{=} 10(2) - 7 \qquad \text{Substitute 2 for } x \text{ and 13 for } y.$$
$$13 = 13 \qquad \text{(2, 13) is a solution. } \checkmark$$

The point $(2, 13)$ *does* lie on the graph of $y = 10x - 7$ because it is a solution point of the equation.

b.

$$y = 10x - 7 \qquad \text{Write original equation.}$$
$$-3 \overset{?}{=} 10(-1) - 7 \qquad \text{Substitute } -1 \text{ for } x \text{ and } -3 \text{ for } y.$$
$$-3 \neq -17 \qquad \text{(}-1, -3\text{) is not a solution.}$$

The point $(-1, -3)$ *does not* lie on the graph of $y = 10x - 7$ because it is *not* a solution point of the equation.

✓ *Checkpoint*  *Audio-video solution in English & Spanish at LarsonPrecalculus.com.*

Determine whether (a) $(3, -5)$ and (b) $(-2, 26)$ lie on the graph of $y = 14 - 6x$. ■

The basic technique used for sketching the graph of an equation is the **point-plotting method.**

> **The Point-Plotting Method of Graphing**
>
> **1.** When possible, isolate one of the variables.
>
> **2.** Construct a table of values showing several solution points.
>
> **3.** Plot these points in a rectangular coordinate system.
>
> **4.** Connect the points with a smooth curve or line.

▷ **ALGEBRA HELP** When evaluating an expression or an equation, remember to follow the Basic Rules of Algebra. To review these rules, see Appendix A.1.

It is important to use negative values, zero, and positive values for x when constructing a table.

EXAMPLE 2   Sketching the Graph of an Equation

Sketch the graph of

$$y = -3x + 7.$$

Solution

Because the equation is already solved for y, construct a table of values that consists of several solution points of the equation. For instance, when $x = -1$,

$$y = -3(-1) + 7$$
$$= 10$$

which implies that

$$(-1, 10)$$

is a solution point of the equation.

x	$y = -3x + 7$	(x, y)
-1	10	$(-1, 10)$
0	7	$(0, 7)$
1	4	$(1, 4)$
2	1	$(2, 1)$
3	-2	$(3, -2)$
4	-5	$(4, -5)$

From the table, it follows that

$$(-1, 10), (0, 7), (1, 4), (2, 1), (3, -2), \quad \text{and} \quad (4, -5)$$

are solution points of the equation. After plotting these points and connecting them, you can see that they appear to lie on a line, as shown below.

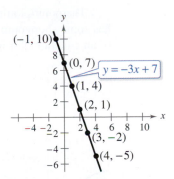

✓ **Checkpoint** *Audio-video solution in English & Spanish at LarsonPrecalculus.com.*

Sketch the graph of each equation.

a. $y = -3x + 2$

b. $y = 2x + 1$

EXAMPLE 3 **Sketching the Graph of an Equation**

Sketch the graph of

$$y = x^2 - 2.$$

Solution

Because the equation is already solved for y, begin by constructing a table of values.

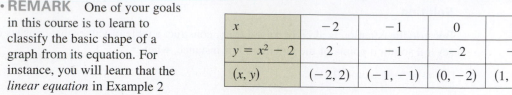

x	-2	-1	0	1	2	3
$y = x^2 - 2$	2	-1	-2	-1	2	7
(x, y)	$(-2, 2)$	$(-1, -1)$	$(0, -2)$	$(1, -1)$	$(2, 2)$	$(3, 7)$

Next, plot the points given in the table, as shown in Figure 1.12. Finally, connect the points with a smooth curve, as shown in Figure 1.13.

> •• **REMARK** One of your goals in this course is to learn to classify the basic shape of a graph from its equation. For instance, you will learn that the *linear equation* in Example 2 has the form
>
> $$y = mx + b$$
>
> and its graph is a line. Similarly, the *quadratic equation* in Example 3 has the form
>
> $$y = ax^2 + bx + c$$
>
> and its graph is a parabola.

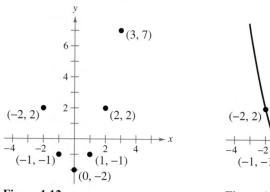

Figure 1.12   **Figure 1.13**

✓ **Checkpoint** *Audio-video solution in English & Spanish at LarsonPrecalculus.com.*

Sketch the graph of each equation.

a. $y = x^2 + 3$

b. $y = 1 - x^2$

The point-plotting method demonstrated in Examples 2 and 3 is easy to use, but it has some shortcomings. With too few solution points, you can misrepresent the graph of an equation. For instance, when you only plot the four points

$$(-2, 2), (-1, -1), (1, -1), \quad \text{and} \quad (2, 2)$$

in Figure 1.12, any one of the three graphs below is reasonable.

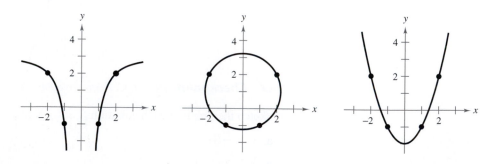

▷ **TECHNOLOGY** To graph an equation involving x and y on a graphing utility, use the following procedure.

1. Rewrite the equation so that y is isolated on the left side.

2. Enter the equation into the graphing utility.

3. Determine a *viewing window* that shows all important features of the graph.

4. Graph the equation.

Intercepts of a Graph

No x-intercepts; one y-intercept

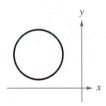

Three x-intercepts; one y-intercept

It is often easy to determine the solution points that have zero as either the x-coordinate or the y-coordinate. These points are called **intercepts** because they are the points at which the graph intersects or touches the x- or y-axis. It is possible for a graph to have no intercepts, one intercept, or several intercepts, as shown in Figure 1.14.

Note that an x-intercept can be written as the ordered pair $(a, 0)$ and a y-intercept can be written as the ordered pair $(0, b)$. Some texts denote the x-intercept as the x-coordinate of the point $(a, 0)$ [and the y-intercept as the y-coordinate of the point $(0, b)$] rather than the point itself. Unless it is necessary to make a distinction, the term *intercept* will refer to either the point or the coordinate.

Finding Intercepts

1. To find x-intercepts, let y be zero and solve the equation for x.

2. To find y-intercepts, let x be zero and solve the equation for y.

One x-intercept; two y-intercepts

No intercepts
Figure 1.14

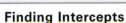 **EXAMPLE 4**   **Finding *x*- and *y*-Intercepts**

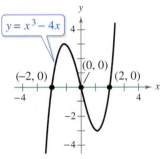

To find the x-intercepts of the graph of $y = x^3 - 4x$, let $y = 0$. Then $0 = x^3 - 4x = x(x^2 - 4)$ has solutions $x = 0$ and $x = \pm 2$.

x-intercepts: $(0, 0), (2, 0), (-2, 0)$ See figure.

To find the y-intercept of the graph of $y = x^3 - 4x$, let $x = 0$. Then $y = (0)^3 - 4(0)$ has one solution, $y = 0$.

y-intercept: $(0, 0)$ See figure.

✓ **Checkpoint** ◀))) *Audio-video solution in English & Spanish at LarsonPrecalculus.com.*

Find the x- and y-intercepts of the graph of $y = -x^2 - 5x$ shown in the figure below.

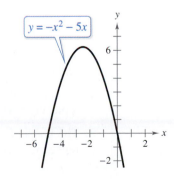

Symmetry

Graphs of equations can have **symmetry** with respect to one of the coordinate axes or with respect to the origin. Symmetry with respect to the x-axis means that when the Cartesian plane is folded along the x-axis, the portion of the graph above the x-axis coincides with the portion below the x-axis. Symmetry with respect to the y-axis or the origin can be described in a similar manner, as shown below.

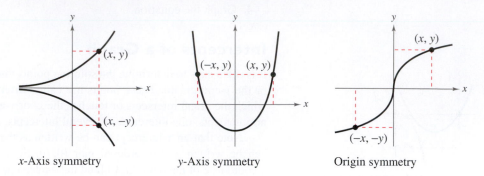

x-Axis symmetry y-Axis symmetry   Origin symmetry

Knowing the symmetry of a graph *before* attempting to sketch it is helpful, because then you need only half as many solution points to sketch the graph. There are three basic types of symmetry, described as follows.

Graphical Tests for Symmetry

1. A graph is **symmetric with respect to the x-axis** if, whenever (x, y) is on the graph, $(x, -y)$ is also on the graph.

2. A graph is **symmetric with respect to the y-axis** if, whenever (x, y) is on the graph, $(-x, y)$ is also on the graph.

3. A graph is **symmetric with respect to the origin** if, whenever (x, y) is on the graph, $(-x, -y)$ is also on the graph.

You can conclude that the graph of $y = x^2 - 2$ is symmetric with respect to the y-axis because the point $(-x, y)$ is also on the graph of $y = x^2 - 2$. (See the table below and Figure 1.15.)

x	-3	-2	-1	1	2	3
y	7	2	-1	-1	2	7
(x, y)	$(-3, 7)$	$(-2, 2)$	$(-1, -1)$	$(1, -1)$	$(2, 2)$	$(3, 7)$

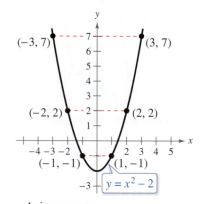

y-Axis symmetry
Figure 1.15

Algebraic Tests for Symmetry

1. The graph of an equation is symmetric with respect to the x-axis when replacing y with $-y$ yields an equivalent equation.

2. The graph of an equation is symmetric with respect to the y-axis when replacing x with $-x$ yields an equivalent equation.

3. The graph of an equation is symmetric with respect to the origin when replacing x with $-x$ and y with $-y$ yields an equivalent equation.

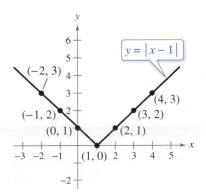

Figure 1.16

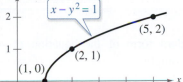

Figure 1.17

▷ **ALGEBRA HELP** In
Example 7, $|x - 1|$ is an
absolute value expression. You
can review the techniques for
evaluating an absolute value
expression in Appendix A.1.

Figure 1.18

EXAMPLE 5 **Testing for Symmetry**

Test $y = 2x^3$ for symmetry with respect to both axes and the origin.

Solution

x-Axis:	$y = 2x^3$	Write original equation.
	$-y = 2x^3$	Replace y with $-y$. Result is *not* an equivalent equation.
y-Axis:	$y = 2x^3$	Write original equation.
	$y = 2(-x)^3$	Replace x with $-x$.
	$y = -2x^3$	Simplify. Result is *not* an equivalent equation.
Origin:	$y = 2x^3$	Write original equation.
	$-y = 2(-x)^3$	Replace y with $-y$ and x with $-x$.
	$-y = -2x^3$	Simplify.
	$y = 2x^3$	Equivalent equation

Of the three tests for symmetry, the only one that is satisfied is the test for origin symmetry (see Figure 1.16).

✓ *Checkpoint* *Audio-video solution in English & Spanish at LarsonPrecalculus.com.*

Test $y^2 = 6 - x$ for symmetry with respect to both axes and the origin.

EXAMPLE 6 **Using Symmetry as a Sketching Aid**

Use symmetry to sketch the graph of $x - y^2 = 1$.

Solution Of the three tests for symmetry, the only one that is satisfied is the test for x-axis symmetry because $x - (-y)^2 = 1$ is equivalent to $x - y^2 = 1$. So, the graph is symmetric with respect to the x-axis. Using symmetry, you only need to find the solution points above the x-axis and then reflect them to obtain the graph, as shown in Figure 1.17.

✓ *Checkpoint* *Audio-video solution in English & Spanish at LarsonPrecalculus.com.*

Use symmetry to sketch the graph of $y = x^2 - 4$.

EXAMPLE 7 **Sketching the Graph of an Equation**

Sketch the graph of $y = |x - 1|$.

Solution This equation fails all three tests for symmetry, and consequently its graph is not symmetric with respect to either axis or to the origin. The absolute value bars indicate that y is always nonnegative. Construct a table of values. Then plot and connect the points, as shown in Figure 1.18. From the table, you can see that $x = 0$ when $y = 1$. So, the y-intercept is $(0, 1)$. Similarly, $y = 0$ when $x = 1$. So, the x-intercept is $(1, 0)$.

x	-2	-1	0	1	2	3	4
$y = |x - 1|$	3	2	1	0	1	2	3
(x, y)	$(-2, 3)$	$(-1, 2)$	$(0, 1)$	$(1, 0)$	$(2, 1)$	$(3, 2)$	$(4, 3)$

✓ *Checkpoint*  *Audio-video solution in English & Spanish at LarsonPrecalculus.com.*

Sketch the graph of $y = |x - 2|$.

Circles

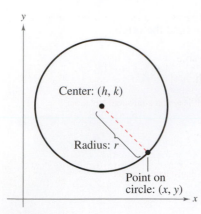

Figure 1.19

Throughout this course, you will learn to recognize several types of graphs from their equations. For instance, you will learn to recognize that the graph of a second-degree equation of the form

$$y = ax^2 + bx + c$$

is a parabola (see Example 3). The graph of a **circle** is also easy to recognize.

Consider the circle shown in Figure 1.19. A point (x, y) lies on the circle if and only if its distance from the center (h, k) is r. By the Distance Formula,

$$\sqrt{(x - h)^2 + (y - k)^2} = r.$$

By squaring each side of this equation, you obtain the **standard form of the equation of a circle.** For example, a circle with its center at $(h, k) = (1, 3)$ and radius $r = 4$ is given by

$$\sqrt{(x - 1)^2 + (y - 3)^2} = 4 \qquad \text{Substitute for } h, k, \text{ and } r.$$
$$(x - 1)^2 + (y - 3)^2 = 16. \qquad \text{Square each side.}$$

Standard Form of the Equation of a Circle

A point (x, y) lies on the circle of **radius** r and **center** (h, k) if and only if

$$(x - h)^2 + (y - k)^2 = r^2.$$

> **REMARK** Be careful when you are finding h and k from the standard form of the equation of a circle. For instance, to find h and k from the equation of the circle in Example 8, rewrite the quantities $(x + 1)^2$ and $(y - 2)^2$ using subtraction.
>
> $$(x + 1)^2 = [x - (-1)]^2,$$
> $$(y - 2)^2 = [y - (2)]^2$$
>
> So, $h = -1$ and $k = 2$.

From this result, you can see that the standard form of the equation of a circle *with its center at the origin*, $(h, k) = (0, 0)$, is simply

$$x^2 + y^2 = r^2. \qquad \text{Circle with center at origin}$$

EXAMPLE 8   **Writing the Equation of a Circle**

The point $(3, 4)$ lies on a circle whose center is at $(-1, 2)$, as shown in Figure 1.20. Write the standard form of the equation of this circle.

Solution

The radius of the circle is the distance between $(-1, 2)$ and $(3, 4)$.

$$r = \sqrt{(x - h)^2 + (y - k)^2} \qquad \text{Distance Formula}$$
$$= \sqrt{[3 - (-1)]^2 + (4 - 2)^2} \qquad \text{Substitute for } x, y, h, \text{ and } k.$$
$$= \sqrt{4^2 + 2^2} \qquad \text{Simplify.}$$
$$= \sqrt{16 + 4} \qquad \text{Simplify.}$$
$$= \sqrt{20} \qquad \text{Radius}$$

Using $(h, k) = (-1, 2)$ and $r = \sqrt{20}$, the equation of the circle is

$$(x - h)^2 + (y - k)^2 = r^2 \qquad \text{Equation of circle}$$
$$[x - (-1)]^2 + (y - 2)^2 = \left(\sqrt{20}\right)^2 \qquad \text{Substitute for } h, k, \text{ and } r.$$
$$(x + 1)^2 + (y - 2)^2 = 20. \qquad \text{Standard form}$$

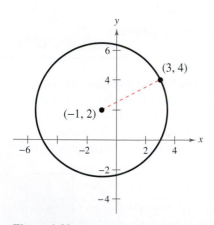

Figure 1.20

✓ **Checkpoint** 🔊 *Audio-video solution in English & Spanish at LarsonPrecalculus.com.*

The point $(1, -2)$ lies on a circle whose center is at $(-3, -5)$. Write the standard form of the equation of this circle.

Application

In this course, you will learn that there are many ways to approach a problem. Three common approaches are illustrated in Example 9.

A Numerical Approach: Construct and use a table.

A Graphical Approach: Draw and use a graph.

An Algebraic Approach: Use the rules of algebra.

EXAMPLE 9 Recommended Weight

The median recommended weights y (in pounds) for men of medium frame who are 25 to 59 years old can be approximated by the mathematical model

$$y = 0.073x^2 - 6.99x + 289.0, \quad 62 \le x \le 76$$

where x is a man's height (in inches). *(Source: Metropolitan Life Insurance Company)*

a. Construct a table of values that shows the median recommended weights for men with heights of 62, 64, 66, 68, 70, 72, 74, and 76 inches.

b. Use the table of values to sketch a graph of the model. Then use the graph to estimate *graphically* the median recommended weight for a man whose height is 71 inches.

c. Use the model to confirm *algebraically* the estimate you found in part (b).

Solution

a. You can use a calculator to construct the table, as shown on the left.

b. The table of values can be used to sketch the graph of the equation, as shown in Figure 1.21. From the graph, you can estimate that a height of 71 inches corresponds to a weight of about 161 pounds.

c. To confirm algebraically the estimate found in part (b), you can substitute 71 for x in the model.

$$y = 0.073(71)^2 - 6.99(71) + 289.0 \approx 160.70$$

So, the graphical estimate of 161 pounds is fairly good.

✓ *Checkpoint*  Audio-video solution in English & Spanish at LarsonPrecalculus.com.

Use Figure 1.21 to estimate *graphically* the median recommended weight for a man whose height is 75 inches. Then confirm the estimate *algebraically*. ◼

DATA	Height, x	Weight, y
	62	136.2
	64	140.6
	66	145.6
	68	151.2
	70	157.4
	72	164.2
	74	171.5
	76	179.4

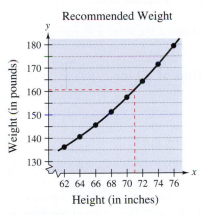

Figure 1.21

Summarize (Section 1.2)

1. Describe how to sketch the graph of an equation *(page 8)*. For examples of graphing equations, see Examples 1–3.

2. Describe how to identify the x- and y-intercepts of a graph *(page 11)*. For an example of identifying x- and y-intercepts, see Example 4.

3. Describe how to use symmetry to graph an equation *(page 12)*. For an example of using symmetry to graph an equation, see Example 6.

4. State the standard form of the equation of a circle *(page 14)* For an example of writing the standard form of the equation of a circle, see Example 8.

5. Describe how to use the graph of an equation to solve a real-life problem *(page 15, Example 9)*.

1.3 Linear Equations in Two Variables

- Use slope to graph linear equations in two variables.
- Find the slope of a line given two points on the line.
- Write linear equations in two variables.
- Use slope to identify parallel and perpendicular lines.
- Use slope and linear equations in two variables to model and solve real-life problems.

Using Slope

The simplest mathematical model for relating two variables is the **linear equation in two variables** $y = mx + b$. The equation is called *linear* because its graph is a line. (In mathematics, the term *line* means *straight line*.) By letting $x = 0$, you obtain

$$y = m(0) + b = b.$$

So, the line crosses the y-axis at $y = b$, as shown in the figures below. In other words, the y-intercept is $(0, b)$. The steepness or slope of the line is m.

$$y = mx + b$$

Slope ⟶   ⟵ y-Intercept

The **slope** of a nonvertical line is the number of units the line rises (or falls) vertically for each unit of horizontal change from left to right, as shown below.

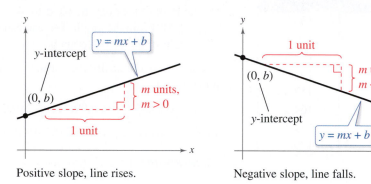

Positive slope, line rises.

Negative slope, line falls.

A linear equation written in **slope-intercept form** has the form $y = mx + b$.

> ### The Slope-Intercept Form of the Equation of a Line
> The graph of the equation
> $$y = mx + b$$
> is a line whose slope is m and whose y-intercept is $(0, b)$.

Once you have determined the slope and the y-intercept of a line, it is a relatively simple matter to sketch its graph. In the next example, note that none of the lines is vertical. A vertical line has an equation of the form

$$x = a. \qquad \text{Vertical line}$$

The equation of a vertical line cannot be written in the form $y = mx + b$ because the slope of a vertical line is undefined, as indicated in Figure 1.22.

Linear equations in two variables can help you model and solve real-life problems. For instance, a surveyor's measurements are used to find a linear equation that models a mountain road.

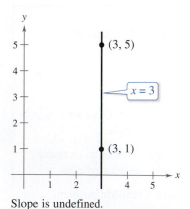

Slope is undefined.
Figure 1.22

EXAMPLE 1   **Graphing a Linear Equation**

Sketch the graph of each linear equation.

a. $y = 2x + 1$

b. $y = 2$

c. $x + y = 2$

Solution

a. Because $b = 1$, the y-intercept is $(0, 1)$. Moreover, because the slope is $m = 2$, the line *rises* two units for each unit the line moves to the right.

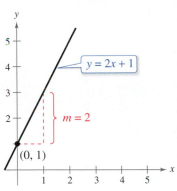

When m is positive, the line rises.

b. By writing this equation in the form $y = (0)x + 2$, you can see that the y-intercept is $(0, 2)$ and the slope is zero. A zero slope implies that the line is horizontal—that is, it does not rise *or* fall.

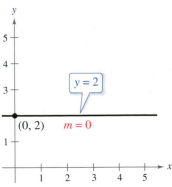

When m is 0, the line is horizontal.

c. By writing this equation in slope-intercept form

$$x + y = 2 \qquad \text{Write original equation.}$$
$$y = -x + 2 \qquad \text{Subtract } x \text{ from each side.}$$
$$y = (-1)x + 2 \qquad \text{Write in slope-intercept form.}$$

you can see that the y-intercept is $(0, 2)$. Moreover, because the slope is $m = -1$, the line *falls* one unit for each unit the line moves to the right.

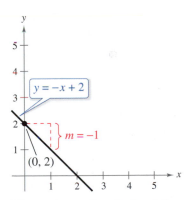

When m is negative, the line falls.

✓ *Checkpoint* ◀))) *Audio-video solution in English & Spanish at LarsonPrecalculus.com.*

Sketch the graph of each linear equation.

a. $y = -3x + 2$

b. $y = -3$

c. $4x + y = 5$

Finding the Slope of a Line

Given an equation of a line, you can find its slope by writing the equation in slope-intercept form. If you are not given an equation, then you can still find the slope of a line. For instance, suppose you want to find the slope of the line passing through the points (x_1, y_1) and (x_2, y_2), as shown below.

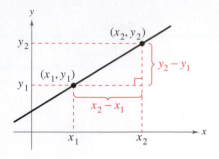

As you move from left to right along this line, a change of $(y_2 - y_1)$ units in the vertical direction corresponds to a change of $(x_2 - x_1)$ units in the horizontal direction.

$$y_2 - y_1 = \text{the change in } y = \text{rise}$$

and

$$x_2 - x_1 = \text{the change in } x = \text{run}$$

The ratio of $(y_2 - y_1)$ to $(x_2 - x_1)$ represents the slope of the line that passes through the points (x_1, y_1) and (x_2, y_2).

$$\text{Slope} = \frac{\text{change in } y}{\text{change in } x} = \frac{\text{rise}}{\text{run}} = \frac{y_2 - y_1}{x_2 - x_1}$$

The Slope of a Line Passing Through Two Points

The **slope** m of the nonvertical line through (x_1, y_1) and (x_2, y_2) is

$$m = \frac{y_2 - y_1}{x_2 - x_1}$$

where $x_1 \neq x_2$.

When using the formula for slope, the *order of subtraction* is important. Given two points on a line, you are free to label either one of them as (x_1, y_1) and the other as (x_2, y_2). However, once you have done this, you must form the numerator and denominator using the same order of subtraction.

$$m = \frac{y_2 - y_1}{x_2 - x_1} \qquad m = \frac{y_1 - y_2}{x_1 - x_2} \qquad m = \frac{y_2 - y_1}{x_1 - x_2}$$

$$\text{Correct} \qquad\qquad \text{Correct} \qquad\qquad \text{Incorrect}$$

For instance, the slope of the line passing through the points $(3, 4)$ and $(5, 7)$ can be calculated as

$$m = \frac{7 - 4}{5 - 3} = \frac{3}{2}$$

or, reversing the subtraction order in both the numerator and denominator, as

$$m = \frac{4 - 7}{3 - 5} = \frac{-3}{-2} = \frac{3}{2}.$$

EXAMPLE 2 **Finding the Slope of a Line Through Two Points**

Find the slope of the line passing through each pair of points.

a. $(-2, 0)$ and $(3, 1)$ **b.** $(-1, 2)$ and $(2, 2)$

c. $(0, 4)$ and $(1, -1)$ **d.** $(3, 4)$ and $(3, 1)$

Solution

a. Letting $(x_1, y_1) = (-2, 0)$ and $(x_2, y_2) = (3, 1)$, you obtain a slope of

$$m = \frac{y_2 - y_1}{x_2 - x_1} = \frac{1 - 0}{3 - (-2)} = \frac{1}{5}.$$ See Figure 1.23.

b. The slope of the line passing through $(-1, 2)$ and $(2, 2)$ is

$$m = \frac{2 - 2}{2 - (-1)} = \frac{0}{3} = 0.$$ See Figure 1.24.

c. The slope of the line passing through $(0, 4)$ and $(1, -1)$ is

$$m = \frac{-1 - 4}{1 - 0} = \frac{-5}{1} = -5.$$ See Figure 1.25.

d. The slope of the line passing through $(3, 4)$ and $(3, 1)$ is

$$m = \frac{1 - 4}{3 - 3} = \frac{-3}{0}.$$ See Figure 1.26.

Because division by 0 is undefined, the slope is undefined and the line is vertical.

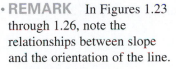

• • REMARK In Figures 1.23 through 1.26, note the relationships between slope and the orientation of the line.

a. Positive slope: line rises from left to right

b. Zero slope: line is horizontal

c. Negative slope: line falls from left to right

d. Undefined slope: line is vertical

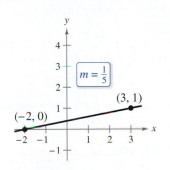

Figure 1.23

Figure 1.24

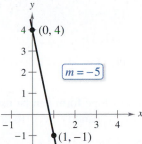

Figure 1.25

Figure 1.26

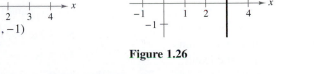 ✓ **Checkpoint** ◀))) Audio-video solution in English & Spanish at LarsonPrecalculus.com.

Find the slope of the line passing through each pair of points.

a. $(-5, -6)$ and $(2, 8)$ **b.** $(4, 2)$ and $(2, 5)$

c. $(0, 0)$ and $(0, -6)$ **d.** $(0, -1)$ and $(3, -1)$

Writing Linear Equations in Two Variables

If (x_1, y_1) is a point on a line of slope m and (x, y) is *any other* point on the line, then

$$\frac{y - y_1}{x - x_1} = m.$$

This equation involving the variables x and y, rewritten in the form

$$y - y_1 = m(x - x_1)$$

is the **point-slope form** of the equation of a line.

Point-Slope Form of the Equation of a Line

The equation of the line with slope m passing through the point (x_1, y_1) is

$$y - y_1 = m(x - x_1).$$

The point-slope form is most useful for *finding* the equation of a line. You should remember this form.

EXAMPLE 3   **Using the Point-Slope Form**

Find the slope-intercept form of the equation of the line that has a slope of 3 and passes through the point $(1, -2)$.

Solution   Use the point-slope form with $m = 3$ and $(x_1, y_1) = (1, -2)$.

$y - y_1 = m(x - x_1)$	Point-slope form
$y - (-2) = 3(x - 1)$	Substitute for m, x_1, and y_1.
$y + 2 = 3x - 3$	Simplify.
$y = 3x - 5$	Write in slope-intercept form.

The slope-intercept form of the equation of the line is $y = 3x - 5$. Figure 1.27 shows the graph of this equation.

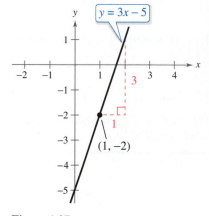

Figure 1.27

✓ *Checkpoint* ◀))) *Audio-video solution in English & Spanish at LarsonPrecalculus.com.*

Find the slope-intercept form of the equation of the line that has the given slope and passes through the given point.

a. $m = 2$, $\quad(3, -7)$

b. $m = -\frac{2}{3}$, $\quad(1, 1)$

c. $m = 0$, $\quad(1, 1)$

····**REMARK**   When you find an equation of the line that passes through two given points, you only need to substitute the coordinates of one of the points in the point-slope form. It does not matter which point you choose because both points will yield the same result. ·········▷

The point-slope form can be used to find an equation of the line passing through two points (x_1, y_1) and (x_2, y_2). To do this, first find the slope of the line

$$m = \frac{y_2 - y_1}{x_2 - x_1}, \quad x_1 \neq x_2$$

and then use the point-slope form to obtain the equation

$$y - y_1 = \frac{y_2 - y_1}{x_2 - x_1}(x - x_1). \qquad \text{Two-point form}$$

This is sometimes called the **two-point form** of the equation of a line.

Parallel and Perpendicular Lines

Slope can tell you whether two nonvertical lines in a plane are parallel, perpendicular, or neither.

Parallel and Perpendicular Lines

1. Two distinct nonvertical lines are **parallel** if and only if their slopes are equal. That is,

$$m_1 = m_2.$$

2. Two nonvertical lines are **perpendicular** if and only if their slopes are negative reciprocals of each other. That is,

$$m_1 = \frac{-1}{m_2}.$$

EXAMPLE 4   **Finding Parallel and Perpendicular Lines**

Find the slope-intercept form of the equations of the lines that pass through the point $(2, -1)$ and are (a) parallel to and (b) perpendicular to the line $2x - 3y = 5$.

Solution   By writing the equation of the given line in slope-intercept form

$2x - 3y = 5$	Write original equation.
$-3y = -2x + 5$	Subtract $2x$ from each side.
$y = \frac{2}{3}x - \frac{5}{3}$	Write in slope-intercept form.

you can see that it has a slope of $m = \frac{2}{3}$, as shown in Figure 1.28.

a. Any line parallel to the given line must also have a slope of $\frac{2}{3}$. So, the line through $(2, -1)$ that is parallel to the given line has the following equation.

$y - (-1) = \frac{2}{3}(x - 2)$	Write in point-slope form.
$3(y + 1) = 2(x - 2)$	Multiply each side by 3.
$3y + 3 = 2x - 4$	Distributive Property
$y = \frac{2}{3}x - \frac{7}{3}$	Write in slope-intercept form.

b. Any line perpendicular to the given line must have a slope of $-\frac{3}{2}$ $\left(\text{because } -\frac{3}{2} \text{ is the negative reciprocal of } \frac{2}{3}\right)$. So, the line through $(2, -1)$ that is perpendicular to the given line has the following equation.

$y - (-1) = -\frac{3}{2}(x - 2)$	Write in point-slope form.
$2(y + 1) = -3(x - 2)$	Multiply each side by 2.
$2y + 2 = -3x + 6$	Distributive Property
$y = -\frac{3}{2}x + 2$	Write in slope-intercept form.

✓ **Checkpoint** ◀))) *Audio-video solution in English & Spanish at LarsonPrecalculus.com.*

Find the slope-intercept form of the equations of the lines that pass through the point $(-4, 1)$ and are (a) parallel to and (b) perpendicular to the line $5x - 3y = 8$. ∎

Notice in Example 4 how the slope-intercept form is used to obtain information about the graph of a line, whereas the point-slope form is used to write the equation of a line.

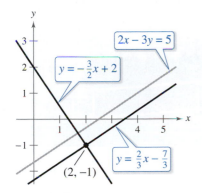

Figure 1.28

▷ **TECHNOLOGY** On a graphing utility, lines will not appear to have the correct slope unless you use a viewing window that has a square setting. For instance, try graphing the lines in Example 4 using the standard setting $-10 \leq x \leq 10$ and $-10 \leq y \leq 10$. Then reset the viewing window with the square setting $-9 \leq x \leq 9$ and $-6 \leq y \leq 6$. On which setting do the lines $y = \frac{2}{3}x - \frac{5}{3}$ and $y = -\frac{3}{2}x + 2$ appear to be perpendicular?

Applications

In real-life problems, the slope of a line can be interpreted as either a *ratio* or a *rate*. If the *x*-axis and *y*-axis have the same unit of measure, then the slope has no units and is a **ratio**. If the *x*-axis and *y*-axis have different units of measure, then the slope is a **rate** or **rate of change.**

EXAMPLE 5   Using Slope as a Ratio

The maximum recommended slope of a wheelchair ramp is $\frac{1}{12}$. A business is installing a wheelchair ramp that rises 22 inches over a horizontal length of 24 feet. Is the ramp steeper than recommended? *(Source: ADA Standards for Accessible Design)*

Solution   The horizontal length of the ramp is 24 feet or $12(24) = 288$ inches, as shown below. So, the slope of the ramp is

$$\text{Slope} = \frac{\text{vertical change}}{\text{horizontal change}} = \frac{22 \text{ in.}}{288 \text{ in.}} \approx 0.076.$$

Because $\frac{1}{12} \approx 0.083$, the slope of the ramp is not steeper than recommended.

22 in.

24 ft

✓ *Checkpoint* 🔊)) *Audio-video solution in English & Spanish at LarsonPrecalculus.com.*

The business in Example 5 installs a second ramp that rises 36 inches over a horizontal length of 32 feet. Is the ramp steeper than recommended?

EXAMPLE 6   Using Slope as a Rate of Change

A kitchen appliance manufacturing company determines that the total cost C (in dollars) of producing x units of a blender is

$$C = 25x + 3500. \qquad \text{Cost equation}$$

Describe the practical significance of the *y*-intercept and slope of this line.

Solution   The *y*-intercept $(0, 3500)$ tells you that the cost of producing zero units is \$3500. This is the *fixed cost* of production—it includes costs that must be paid regardless of the number of units produced. The slope of $m = 25$ tells you that the cost of producing each unit is \$25, as shown in Figure 1.29. Economists call the cost per unit the *marginal cost*. If the production increases by one unit, then the "margin," or extra amount of cost, is \$25. So, the cost increases at a rate of \$25 per unit.

✓ *Checkpoint* 🔊)) *Audio-video solution in English & Spanish at LarsonPrecalculus.com.*

An accounting firm determines that the value V (in dollars) of a copier t years after its purchase is

$$V = -300t + 1500.$$

Describe the practical significance of the *y*-intercept and slope of this line. ■

Manufacturing

Cost (in dollars)

$C = 25x + 3500$

Marginal cost: $m = \$25$

Fixed cost: \$3500

10,000
9,000
8,000
7,000
6,000
5,000
4,000
3,000
2,000
1,000

50   100   150

Number of units

Production cost
Figure 1.29

Businesses can deduct most of their expenses in the same year they occur. One exception is the cost of property that has a useful life of more than 1 year. Such costs must be *depreciated* (decreased in value) over the useful life of the property. Depreciating the *same amount* each year is called *linear* or *straight-line depreciation*. The *book value* is the difference between the original value and the total amount of depreciation accumulated to date.

EXAMPLE 7 Straight-Line Depreciation

A college purchased exercise equipment worth $12,000 for the new campus fitness center. The equipment has a useful life of 8 years. The salvage value at the end of 8 years is $2000. Write a linear equation that describes the book value of the equipment each year.

Solution Let V represent the value of the equipment at the end of year t. Represent the initial value of the equipment by the data point $(0, 12,000)$ and the salvage value of the equipment by the data point $(8, 2000)$. The slope of the line is

$$m = \frac{2000 - 12,000}{8 - 0}$$

$$= -\$1250$$

which represents the annual depreciation in *dollars per year*. Using the point-slope form, you can write the equation of the line as follows.

$$V - 12,000 = -1250(t - 0) \qquad \text{Write in point-slope form.}$$

$$V = -1250t + 12,000 \qquad \text{Write in slope-intercept form.}$$

The table shows the book value at the end of each year, and Figure 1.30 shows the graph of the equation.

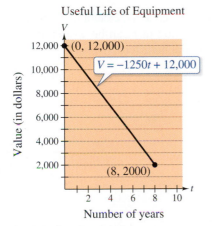

Useful Life of Equipment

$V = -1250t + 12,000$

Straight-line depreciation

Figure 1.30

Year, t	Value, V
0	12,000
1	10,750
2	9500
3	8250
4	7000
5	5750
6	4500
7	3250
8	2000

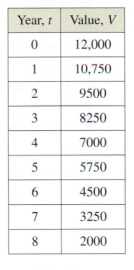

✓ *Checkpoint* ◀))) *Audio-video solution in English & Spanish at LarsonPrecalculus.com.*

A manufacturing firm purchased a machine worth $24,750. The machine has a useful life of 6 years. After 6 years, the machine will have to be discarded and replaced. That is, it will have no salvage value. Write a linear equation that describes the book value of the machine each year. ■

In many real-life applications, the two data points that determine the line are often given in a disguised form. Note how the data points are described in Example 7.

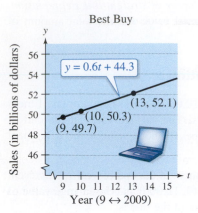

Figure 1.31

EXAMPLE 8 Predicting Sales

The sales for Best Buy were approximately \$49.7 billion in 2009 and \$50.3 billion in 2010. Using only this information, write a linear equation that gives the sales in terms of the year. Then predict the sales in 2013. *(Source: Best Buy Company, Inc.)*

Solution Let $t = 9$ represent 2009. Then the two given values are represented by the data points $(9, 49.7)$ and $(10, 50.3)$. The slope of the line through these points is

$$m = \frac{50.3 - 49.7}{10 - 9} = 0.6.$$

You can find the equation that relates the sales y and the year t to be

$$y - 49.7 = 0.6(t - 9) \qquad \text{Write in point-slope form.}$$
$$y = 0.6t + 44.3. \qquad \text{Write in slope-intercept form.}$$

According to this equation, the sales in 2013 will be

$$y = 0.6(13) + 44.3 = 7.8 + 44.3 = \$52.1 \text{ billion. (See Figure 1.31.)}$$

✓ **Checkpoint** ◀))) Audio-video solution in English & Spanish at LarsonPrecalculus.com.

The sales for Nokia were approximately \$58.6 billion in 2009 and \$56.6 billion in 2010. Repeat Example 8 using this information. *(Source: Nokia Corporation)* ■

The prediction method illustrated in Example 8 is called **linear extrapolation.** Note in Figure 1.32 that an extrapolated point does not lie between the given points. When the estimated point lies between two given points, as shown in Figure 1.33, the procedure is called **linear interpolation.**

Because the slope of a vertical line is not defined, its equation cannot be written in slope-intercept form. However, every line has an equation that can be written in the **general form** $Ax + By + C = 0$, where A and B are not both zero.

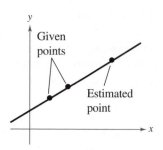

Linear extrapolation
Figure 1.32

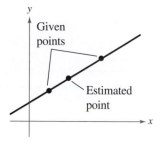

Linear interpolation
Figure 1.33

Summary of Equations of Lines

1. General form: $Ax + By + C = 0$
2. Vertical line: $x = a$
3. Horizontal line: $y = b$
4. Slope-intercept form: $y = mx + b$
5. Point-slope form: $y - y_1 = m(x - x_1)$
6. Two-point form: $y - y_1 = \dfrac{y_2 - y_1}{x_2 - x_1}(x - x_1)$

Summarize (Section 1.3)

1. Explain how to use slope to graph a linear equation in two variables *(page 16)* and how to find the slope of a line passing through two points *(page 18)*. For examples of using and finding slopes, see Examples 1 and 2.

2. State the point-slope form of the equation of a line *(page 20)*. For an example of using point-slope form, see Example 3.

3. Explain how to use slope to identify parallel and perpendicular lines *(page 21)*. For an example of finding parallel and perpendicular lines, see Example 4.

4. Describe examples of how to use slope and linear equations in two variables to model and solve real-life problems *(pages 22–24, Examples 5–8)*.

1.4 Functions

Functions can help you model and solve real-life problems, such as the force of water against the face of a dam.

- ▪ Determine whether relations between two variables are functions, and use function notation.
- ▪ Find the domains of functions.
- ▪ Use functions to model and solve real-life problems.
- ▪ Evaluate difference quotients.

Introduction to Functions and Function Notation

Many everyday phenomena involve two quantities that are related to each other by some rule of correspondence. The mathematical term for such a rule of correspondence is a **relation.** In mathematics, equations and formulas often represent relations. For instance, the simple interest I earned on $1000 for 1 year is related to the annual interest rate r by the formula $I = 1000r$.

The formula $I = 1000r$ represents a special kind of relation that matches each item from one set with *exactly one* item from a different set. Such a relation is called a **function.**

Definition of Function

A **function** f from a set A to a set B is a relation that assigns to each element x in the set A exactly one element y in the set B. The set A is the **domain** (or set of inputs) of the function f, and the set B contains the **range** (or set of outputs).

To help understand this definition, look at the function below, which relates the time of day to the temperature.

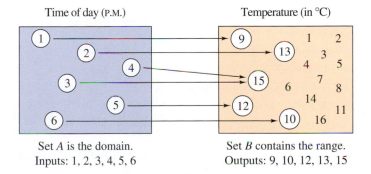

Set A is the domain.
Inputs: 1, 2, 3, 4, 5, 6

Set B contains the range.
Outputs: 9, 10, 12, 13, 15

The following ordered pairs can represent this function. The first coordinate (x-value) is the input and the second coordinate (y-value) is the output.

$$\{(1, 9), (2, 13), (3, 15), (4, 15), (5, 12), (6, 10)\}$$

Characteristics of a Function from Set A to Set B

1. Each element in A must be matched with an element in B.

2. Some elements in B may not be matched with any element in A.

3. Two or more elements in A may be matched with the same element in B.

4. An element in A (the domain) cannot be matched with two different elements in B.

Four common ways to represent functions are as follows.

Four Ways to Represent a Function

1. *Verbally* by a sentence that describes how the input variable is related to the output variable

2. *Numerically* by a table or a list of ordered pairs that matches input values with output values

3. *Graphically* by points on a graph in a coordinate plane in which the horizontal axis represents the input values and the vertical axis represents the output values

4. *Algebraically* by an equation in two variables

To determine whether a relation is a function, you must decide whether each input value is matched with exactly one output value. When any input value is matched with two or more output values, the relation is not a function.

EXAMPLE 1   **Testing for Functions**

Determine whether the relation represents *y* as a function of *x*.

a. The input value *x* is the number of representatives from a state, and the output value *y* is the number of senators.

b.

Input, *x*	Output, *y*
2	11
2	10
3	8
4	5
5	1

c.

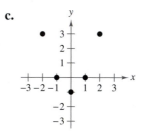

Solution

a. This verbal description *does* describe *y* as a function of *x*. Regardless of the value of *x*, the value of *y* is always 2. Such functions are called *constant functions*.

b. This table *does not* describe *y* as a function of *x*. The input value 2 is matched with two different *y*-values.

c. The graph *does* describe *y* as a function of *x*. Each input value is matched with exactly one output value.

✓ *Checkpoint*  Audio-video solution in English & Spanish at LarsonPrecalculus.com.

Determine whether the relation represents *y* as a function of *x*.

a. *Domain, x   Range, y*

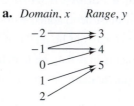

b.

Input, *x*	0	1	2	3	4
Output, *y*	−4	−2	0	2	4

Representing functions by sets of ordered pairs is common in *discrete mathematics*. In algebra, however, it is more common to represent functions by equations or formulas involving two variables. For instance, the equation

$$y = x^2 \qquad \text{\color{red} y is a function of x.}$$

represents the variable y as a function of the variable x. In this equation, x is the **independent variable** and y is the **dependent variable**. The domain of the function is the set of all values taken on by the independent variable x, and the range of the function is the set of all values taken on by the dependent variable y.

EXAMPLE 2 **Testing for Functions Represented Algebraically**

Which of the equations represent(s) y as a function of x?

a. $x^2 + y = 1$

b. $-x + y^2 = 1$

Solution To determine whether y is a function of x, try to solve for y in terms of x.

a. Solving for y yields

$$x^2 + y = 1 \qquad \text{\color{red} Write original equation.}$$
$$y = 1 - x^2. \qquad \text{\color{red} Solve for y.}$$

To each value of x there corresponds exactly one value of y. So, y is a function of x.

b. Solving for y yields

$$-x + y^2 = 1 \qquad \text{\color{red} Write original equation.}$$
$$y^2 = 1 + x \qquad \text{\color{red} Add x to each side.}$$
$$y = \pm\sqrt{1 + x}. \qquad \text{\color{red} Solve for y.}$$

The $\pm$ indicates that to a given value of x there correspond two values of y. So, y is not a function of x.

✓ **Checkpoint** *Audio-video solution in English & Spanish at LarsonPrecalculus.com.*

Which of the equations represent(s) y as a function of x?

a. $x^2 + y^2 = 8$ **b.** $y - 4x^2 = 36$

When using an equation to represent a function, it is convenient to name the function for easy reference. For example, you know that the equation $y = 1 - x^2$ describes y as a function of x. Suppose you give this function the name "f." Then you can use the following **function notation**.

Input	Output	Equation
x	$f(x)$	$f(x) = 1 - x^2$

The symbol $f(x)$ is read as *the value of f at x* or simply *f of x*. The symbol $f(x)$ corresponds to the y-value for a given x. So, you can write $y = f(x)$. Keep in mind that f is the *name* of the function, whereas $f(x)$ is the *value* of the function at x. For instance, the function $f(x) = 3 - 2x$ has *function values* denoted by $f(-1)$, $f(0)$, $f(2)$, and so on. To find these values, substitute the specified input values into the given equation.

For $x = -1$, $\qquad f(-1) = 3 - 2(-1) = 3 + 2 = 5.$

For $x = 0$, $\qquad f(0) = 3 - 2(0) = 3 - 0 = 3.$

For $x = 2$, $\qquad f(2) = 3 - 2(2) = 3 - 4 = -1.$

HISTORICAL NOTE

Many consider Leonhard Euler (1707–1783), a Swiss mathematician, the most prolific and productive mathematician in history. One of his greatest influences on mathematics was his use of symbols, or notation. Euler introduced the function notation $y = f(x)$.

Bettmann/Corbis

Although f is often used as a convenient function name and x is often used as the independent variable, you can use other letters. For instance,

$$f(x) = x^2 - 4x + 7, \quad f(t) = t^2 - 4t + 7, \quad \text{and} \quad g(s) = s^2 - 4s + 7$$

all define the same function. In fact, the role of the independent variable is that of a "placeholder." Consequently, the function could be described by

$$f(\blacksquare) = (\blacksquare)^2 - 4(\blacksquare) + 7.$$

EXAMPLE 3 Evaluating a Function

Let $g(x) = -x^2 + 4x + 1$. Find each function value.

a. $g(2)$ **b.** $g(t)$ **c.** $g(x + 2)$

Solution

a. Replacing x with 2 in $g(x) = -x^2 + 4x + 1$ yields the following.

$$g(2) = -(2)^2 + 4(2) + 1$$
$$= -4 + 8 + 1$$
$$= 5$$

b. Replacing x with t yields the following.

$$g(t) = -(t)^2 + 4(t) + 1$$
$$= -t^2 + 4t + 1$$

c. Replacing x with $x + 2$ yields the following.

$$g(x + 2) = -(x + 2)^2 + 4(x + 2) + 1$$
$$= -(x^2 + 4x + 4) + 4x + 8 + 1$$
$$= -x^2 - 4x - 4 + 4x + 8 + 1$$
$$= -x^2 + 5$$

REMARK In Example 3(c), note that $g(x + 2)$ is not equal to $g(x) + g(2)$. In general, $g(u + v) \neq g(u) + g(v)$.

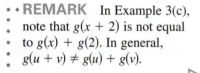 **Checkpoint** Audio-video solution in English & Spanish at LarsonPrecalculus.com.

Let $f(x) = 10 - 3x^2$. Find each function value.

a. $f(2)$ **b.** $f(-4)$ **c.** $f(x - 1)$

A function defined by two or more equations over a specified domain is called a **piecewise-defined function.**

EXAMPLE 4 A Piecewise-Defined Function

Evaluate the function when $x = -1, 0,$ and 1.

$$f(x) = \begin{cases} x^2 + 1, & x < 0 \\ x - 1, & x \geq 0 \end{cases}$$

Solution Because $x = -1$ is less than 0, use $f(x) = x^2 + 1$ to obtain $f(-1) = (-1)^2 + 1 = 2$. For $x = 0$, use $f(x) = x - 1$ to obtain $f(0) = (0) - 1 = -1$. For $x = 1$, use $f(x) = x - 1$ to obtain $f(1) = (1) - 1 = 0$.

Checkpoint Audio-video solution in English & Spanish at LarsonPrecalculus.com.

Evaluate the function given in Example 4 when $x = -2, 2,$ and 3.

EXAMPLE 5 **Finding Values for Which $f(x) = 0$**

Find all real values of x such that $f(x) = 0$.

a. $f(x) = -2x + 10$ **b.** $f(x) = x^2 - 5x + 6$

Solution   For each function, set $f(x) = 0$ and solve for x.

a. $-2x + 10 = 0$ Set $f(x)$ equal to 0.

$\qquad -2x = -10$ Subtract 10 from each side.

$\qquad\quad x = 5$ Divide each side by -2.

So, $f(x) = 0$ when $x = 5$.

b. $x^2 - 5x + 6 = 0$ Set $f(x)$ equal to 0.

$(x - 2)(x - 3) = 0$ Factor.

$\qquad x - 2 = 0 \implies x = 2$ Set 1st factor equal to 0.

$\qquad x - 3 = 0 \implies x = 3$ Set 2nd factor equal to 0.

So, $f(x) = 0$ when $x = 2$ or $x = 3$.

✓ **Checkpoint** *Audio-video solution in English & Spanish at LarsonPrecalculus.com.*

Find all real values of x such that $f(x) = 0$, where $f(x) = x^2 - 16$.

EXAMPLE 6 **Finding Values for Which $f(x) = g(x)$**

Find the values of x for which $f(x) = g(x)$.

a. $f(x) = x^2 + 1$ and $g(x) = 3x - x^2$

b. $f(x) = x^2 - 1$ and $g(x) = -x^2 + x + 2$

Solution

a. $\qquad x^2 + 1 = 3x - x^2$ Set $f(x)$ equal to $g(x)$.

$\quad 2x^2 - 3x + 1 = 0$ Write in general form.

$(2x - 1)(x - 1) = 0$ Factor.

$\qquad 2x - 1 = 0 \implies x = \frac{1}{2}$ Set 1st factor equal to 0.

$\qquad\; x - 1 = 0 \implies x = 1$ Set 2nd factor equal to 0.

So, $f(x) = g(x)$ when $x = \dfrac{1}{2}$ or $x = 1$.

b. $\qquad x^2 - 1 = -x^2 + x + 2$ Set $f(x)$ equal to $g(x)$.

$\quad 2x^2 - x - 3 = 0$ Write in general form.

$(2x - 3)(x + 1) = 0$ Factor.

$\qquad 2x - 3 = 0 \implies x = \frac{3}{2}$ Set 1st factor equal to 0.

$\qquad\; x + 1 = 0 \implies x = -1$ Set 2nd factor equal to 0.

So, $f(x) = g(x)$ when $x = \dfrac{3}{2}$ or $x = -1$.

✓ **Checkpoint** *Audio-video solution in English & Spanish at LarsonPrecalculus.com.*

Find the values of x for which $f(x) = g(x)$, where $f(x) = x^2 + 6x - 24$ and $g(x) = 4x - x^2$.

The Domain of a Function

The domain of a function can be described explicitly or it can be *implied* by the expression used to define the function. The **implied domain** is the set of all real numbers for which the expression is defined. For instance, the function

$$f(x) = \frac{1}{x^2 - 4} \qquad \text{Domain excludes } x\text{-values that result in division by zero.}$$

has an implied domain consisting of all real x other than $x = \pm 2$. These two values are excluded from the domain because division by zero is undefined. Another common type of implied domain is that used to avoid even roots of negative numbers. For example, the function

$$f(x) = \sqrt{x} \qquad \text{Domain excludes } x\text{-values that result in even roots of negative numbers.}$$

is defined only for $x \geq 0$. So, its implied domain is the interval $[0, \infty)$. In general, the domain of a function *excludes* values that would cause division by zero *or* that would result in the even root of a negative number.

EXAMPLE 7   **Finding the Domain of a Function**

Find the domain of each function.

a. f: $\{(-3, 0), (-1, 4), (0, 2), (2, 2), (4, -1)\}$ **b.** $g(x) = \dfrac{1}{x + 5}$

c. Volume of a sphere: $V = \frac{4}{3}\pi r^3$ **d.** $h(x) = \sqrt{4 - 3x}$

Solution

a. The domain of f consists of all first coordinates in the set of ordered pairs.

   Domain $= \{-3, -1, 0, 2, 4\}$

b. Excluding x-values that yield zero in the denominator, the domain of g is the set of all real numbers x except $x = -5$.

c. Because this function represents the volume of a sphere, the values of the radius r must be positive. So, the domain is the set of all real numbers r such that $r > 0$.

d. This function is defined only for x-values for which

   $4 - 3x \geq 0$.

By solving this inequality, you can conclude that $x \leq \frac{4}{3}$. So, the domain is the interval $\left(-\infty, \frac{4}{3}\right]$.

✓ **Checkpoint**  *Audio-video solution in English & Spanish at LarsonPrecalculus.com.*

Find the domain of each function.

a. f: $\{(-2, 2), (-1, 1), (0, 3), (1, 1), (2, 2)\}$ **b.** $g(x) = \dfrac{1}{3 - x}$

c. Circumference of a circle: $C = 2\pi r$ **d.** $h(x) = \sqrt{x - 16}$ ■

   In Example 7(c), note that the domain of a function may be implied by the physical context. For instance, from the equation

$$V = \frac{4}{3}\pi r^3$$

you would have no reason to restrict r to positive values, but the physical context implies that a sphere cannot have a negative or zero radius.

Applications

EXAMPLE 8 **The Dimensions of a Container**

You work in the marketing department of a soft-drink company and are experimenting with a new can for iced tea that is slightly narrower and taller than a standard can. For your experimental can, the ratio of the height to the radius is 4.

a. Write the volume of the can as a function of the radius r.

b. Write the volume of the can as a function of the height h.

Solution

a. $V(r) = \pi r^2 h = \pi r^2 (4r) = 4\pi r^3$ Write V as a function of r.

b. $V(h) = \pi r^2 h = \pi \left(\dfrac{h}{4}\right)^2 h = \dfrac{\pi h^3}{16}$ Write V as a function of h.

✓ **Checkpoint** *Audio-video solution in English & Spanish at LarsonPrecalculus.com.*

For the experimental can described in Example 8, write the *surface area* as a function of (a) the radius r and (b) the height h.

EXAMPLE 9 **The Path of a Baseball**

A batter hits a baseball at a point 3 feet above ground at a velocity of 100 feet per second and an angle of 45°. The path of the baseball is given by the function

$$f(x) = -0.0032x^2 + x + 3$$

where $f(x)$ is the height of the baseball (in feet) and x is the horizontal distance from home plate (in feet). Will the baseball clear a 10-foot fence located 300 feet from home plate?

Algebraic Solution

When $x = 300$, you can find the height of the baseball as follows.

$f(x) = -0.0032x^2 + x + 3$ Write original function.

$f(300) = -0.0032(300)^2 + 300 + 3$ Substitute 300 for x.

$= 15$ Simplify.

When $x = 300$, the height of the baseball is 15 feet. So, the baseball will clear a 10-foot fence.

Graphical Solution

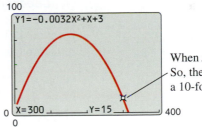

When $x = 300$, $y = 15$. So, the ball will clear a 10-foot fence.

✓ **Checkpoint** *Audio-video solution in English & Spanish at LarsonPrecalculus.com.*

A second baseman throws a baseball toward the first baseman 60 feet away. The path of the baseball is given by

$$f(x) = -0.004x^2 + 0.3x + 6$$

where $f(x)$ is the height of the baseball (in feet) and x is the horizontal distance from the second baseman (in feet). The first baseman can reach 8 feet high. Can the first baseman catch the baseball without jumping?

Alternative fuels for vehicles include electricity, ethanol, hydrogen, compressed natural gas, liquefied natural gas, and liquefied petroleum gas.

EXAMPLE 10 **Alternative-Fueled Vehicles**

The number V (in thousands) of alternative-fueled vehicles in the United States increased in a linear pattern from 2003 through 2005, and then increased in a different linear pattern from 2006 through 2009, as shown in the bar graph. These two patterns can be approximated by the function

$$V(t) = \begin{cases} 29.05t + 447.7, & 3 \le t \le 5 \\ 65.50t + 241.9, & 6 \le t \le 9 \end{cases}$$

where t represents the year, with $t = 3$ corresponding to 2003. Use this function to approximate the number of alternative-fueled vehicles for each year from 2003 through 2009. *(Source: U.S. Energy Information Administration)*

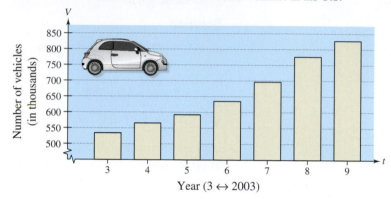

Number of Alternative-Fueled Vehicles in the U.S.

Solution From 2003 through 2005, use $V(t) = 29.05t + 447.7$.

534.9	563.9	593.0
2003	2004	2005

From 2006 to 2009, use $V(t) = 65.50t + 241.9$.

634.9	700.4	765.9	831.4
2006	2007	2008	2009

✓ *Checkpoint* ◀))) *Audio-video solution in English & Spanish at LarsonPrecalculus.com.*

The number V (in thousands) of 85%-ethanol-fueled vehicles in the United States from 2003 through 2009 can be approximated by the function

$$V(t) = \begin{cases} 33.65t + 77.8, & 3 \le t \le 5 \\ 70.75t - 126.6, & 6 \le t \le 9 \end{cases}$$

where t represents the year, with $t = 3$ corresponding to 2003. Use this function to approximate the number of 85%-ethanol-fueled vehicles for each year from 2003 through 2009. *(Source: U.S. Energy Information Administration)* ■

Difference Quotients

One of the basic definitions in calculus employs the ratio

$$\frac{f(x + h) - f(x)}{h}, \quad h \ne 0.$$

This ratio is called a **difference quotient,** as illustrated in Example 11.

EXAMPLE 11 Evaluating a Difference Quotient

For $f(x) = x^2 - 4x + 7$, find $\dfrac{f(x+h) - f(x)}{h}$.

Solution

$$\frac{f(x+h) - f(x)}{h} = \frac{[(x+h)^2 - 4(x+h) + 7] - (x^2 - 4x + 7)}{h}$$

$$= \frac{x^2 + 2xh + h^2 - 4x - 4h + 7 - x^2 + 4x - 7}{h}$$

$$= \frac{2xh + h^2 - 4h}{h} = \frac{h(2x + h - 4)}{h} = 2x + h - 4, \quad h \neq 0$$

> **REMARK** You may find it easier to calculate the difference quotient in Example 11 by first finding $f(x+h)$, and then substituting the resulting expression into the difference quotient
>
> $$\frac{f(x+h) - f(x)}{h}.$$

✓ *Checkpoint* Audio-video solution in English & Spanish at LarsonPrecalculus.com.

For $f(x) = x^2 + 2x - 3$, find $\dfrac{f(x+h) - f(x)}{h}$.

Summary of Function Terminology

Function: A **function** is a relationship between two variables such that to each value of the independent variable there corresponds exactly one value of the dependent variable.

Function Notation: $y = f(x)$

    f is the *name* of the function.

    y is the **dependent variable.**

    x is the **independent variable.**

    $f(x)$ is the *value of the function at x.*

Domain: The **domain** of a function is the set of all values (inputs) of the independent variable for which the function is defined. If *x* is in the domain of *f*, then *f* is said to be *defined* at *x*. If *x* is not in the domain of *f*, then *f* is said to be *undefined* at *x*.

Range: The **range** of a function is the set of all values (outputs) assumed by the dependent variable (that is, the set of all function values).

Implied Domain: If *f* is defined by an algebraic expression and the domain is not specified, then the **implied domain** consists of all real numbers for which the expression is defined.

Summarize (Section 1.4)

1. State the definition of a function and describe function notation *(pages 25–28)*. For examples of determining functions and using function notation, see Examples 1–6.

2. State the definition of the implied domain of a function *(page 30)*. For an example of finding the domains of functions, see Example 7.

3. Describe examples of how functions can model real-life problems *(pages 31 and 32, Examples 8–10)*.

4. State the definition of a difference quotient *(page 32)*. For an example of evaluating a difference quotient, see Example 11.

1.5   Analyzing Graphs of Functions

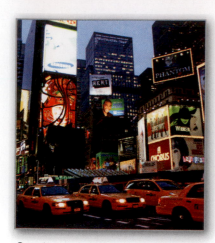

Graphs of functions can help you visualize relationships between variables in real life, such as the temperature of a city over a 24-hour period.

- ■ Use the Vertical Line Test for functions.
- ■ Find the zeros of functions.
- ■ Determine intervals on which functions are increasing or decreasing and determine relative maximum and relative minimum values of functions.
- ■ Determine the average rate of change of a function.
- ■ Identify even and odd functions.

The Graph of a Function

In Section 1.4, you studied functions from an algebraic point of view. In this section, you will study functions from a graphical perspective.

The **graph of a function** f is the collection of ordered pairs $(x, f(x))$ such that x is in the domain of f. As you study this section, remember that

x = the directed distance from the y-axis

$y = f(x)$ = the directed distance from the x-axis

as shown in the figure at the right.

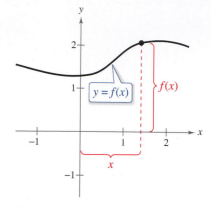

EXAMPLE 1   Finding the Domain and Range of a Function

Use the graph of the function f, shown in Figure 1.34, to find (a) the domain of f, (b) the function values $f(-1)$ and $f(2)$, and (c) the range of f.

Solution

a. The closed dot at $(-1, 1)$ indicates that $x = -1$ is in the domain of f, whereas the open dot at $(5, 2)$ indicates that $x = 5$ is not in the domain. So, the domain of f is all x in the interval $[-1, 5)$.

b. Because $(-1, 1)$ is a point on the graph of f, it follows that $f(-1) = 1$. Similarly, because $(2, -3)$ is a point on the graph of f, it follows that $f(2) = -3$.

c. Because the graph does not extend below $f(2) = -3$ or above $f(0) = 3$, the range of f is the interval $[-3, 3]$.

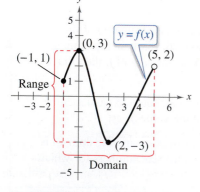

Figure 1.34

· · REMARK The use of dots (open or closed) at the extreme left and right points of a graph indicates that the graph does not extend beyond these points. If such dots are not on the graph, then assume that the graph extends beyond these points.

✓ **Checkpoint** ◀))) *Audio-video solution in English & Spanish at LarsonPrecalculus.com.*

Use the graph of the function f to find (a) the domain of f, (b) the function values $f(0)$ and $f(3)$, and (c) the range of f.

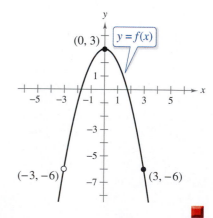

By the definition of a function, at most one *y*-value corresponds to a given *x*-value. This means that the graph of a function cannot have two or more different points with the same *x*-coordinate, and no two points on the graph of a function can be vertically above or below each other. It follows, then, that a vertical line can intersect the graph of a function at most once. This observation provides a convenient visual test called the **Vertical Line Test** for functions.

Vertical Line Test for Functions

A set of points in a coordinate plane is the graph of *y* as a function of *x* if and only if no *vertical* line intersects the graph at more than one point.

EXAMPLE 2   **Vertical Line Test for Functions**

Use the Vertical Line Test to decide whether each of the following graphs represents *y* as a function of *x*.

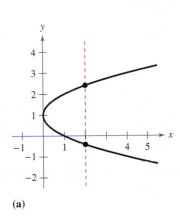

(a)

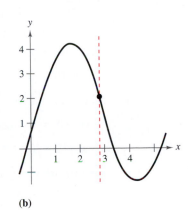

(b)

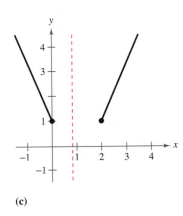

(c)

Solution

a. This *is not* a graph of *y* as a function of *x*, because there are vertical lines that intersect the graph twice. That is, for a particular input *x*, there is more than one output *y*.

b. This *is* a graph of *y* as a function of *x*, because every vertical line intersects the graph at most once. That is, for a particular input *x*, there is at most one output *y*.

c. This *is* a graph of *y* as a function of *x*. (Note that when a vertical line does not intersect the graph, it simply means that the function is undefined for that particular value of *x*.) That is, for a particular input *x*, there is at most one output *y*.

✓ **Checkpoint** 🔊) *Audio-video solution in English & Spanish at LarsonPrecalculus.com.*

Use the Vertical Line Test to decide whether the graph represents *y* as a function of *x*.

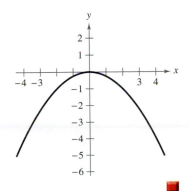

▷ **TECHNOLOGY** Most graphing utilities graph functions of *x* more easily than other types of equations. For instance, the graph shown in (a) above represents the equation $x - (y - 1)^2 = 0$. To use a graphing utility to duplicate this graph, you must first solve the equation for *y* to obtain $y = 1 \pm \sqrt{x}$, and then graph the two equations $y_1 = 1 + \sqrt{x}$ and $y_2 = 1 - \sqrt{x}$ in the same viewing window.

Zeros of a Function

▷ **ALGEBRA HELP** To do Example 3, you need to be able to solve equations. You can review the techniques for solving equations in Appendix A.5.

If the graph of a function of x has an x-intercept at $(a, 0)$, then a is a **zero** of the function.

> **Zeros of a Function**
>
> The **zeros of a function** f of x are the x-values for which $f(x) = 0$.

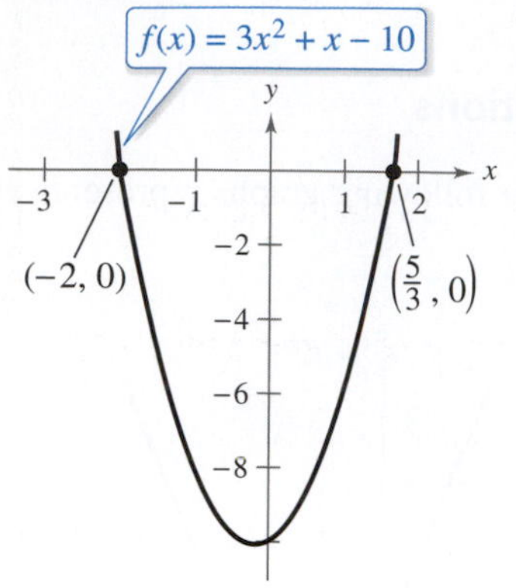

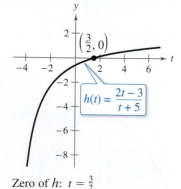

Zeros of f: $x = -2$, $x = \frac{5}{3}$
Figure 1.35

EXAMPLE 3 Finding the Zeros of a Function

Find the zeros of each function.

a. $f(x) = 3x^2 + x - 10$

b. $g(x) = \sqrt{10 - x^2}$

c. $h(t) = \dfrac{2t - 3}{t + 5}$

Solution To find the zeros of a function, set the function equal to zero and solve for the independent variable.

a. $\quad 3x^2 + x - 10 = 0$ Set $f(x)$ equal to 0.

$\quad (3x - 5)(x + 2) = 0$ Factor.

$\quad\quad\quad 3x - 5 = 0 \implies x = \frac{5}{3}$ Set 1st factor equal to 0.

$\quad\quad\quad\quad x + 2 = 0 \implies x = -2$ Set 2nd factor equal to 0.

The zeros of f are $x = \frac{5}{3}$ and $x = -2$. In Figure 1.35, note that the graph of f has $\left(\frac{5}{3}, 0\right)$ and $(-2, 0)$ as its x-intercepts.

b. $\quad \sqrt{10 - x^2} = 0$ Set $g(x)$ equal to 0.

$\quad\quad 10 - x^2 = 0$ Square each side.

$\quad\quad\quad\quad 10 = x^2$ Add x^2 to each side.

$\quad\quad\quad \pm\sqrt{10} = x$ Extract square roots.

The zeros of g are $x = -\sqrt{10}$ and $x = \sqrt{10}$. In Figure 1.36, note that the graph of g has $\left(-\sqrt{10}, 0\right)$ and $\left(\sqrt{10}, 0\right)$ as its x-intercepts.

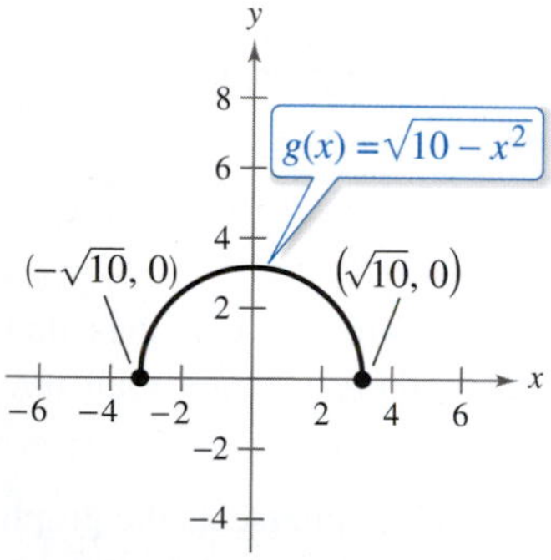

Zeros of g: $x = \pm\sqrt{10}$
Figure 1.36

c. $\quad \dfrac{2t - 3}{t + 5} = 0$ Set $h(t)$ equal to 0.

$\quad\quad 2t - 3 = 0$ Multiply each side by $t + 5$.

$\quad\quad\quad\quad 2t = 3$ Add 3 to each side.

$\quad\quad\quad\quad t = \dfrac{3}{2}$ Divide each side by 2.

The zero of h is $t = \frac{3}{2}$. In Figure 1.37, note that the graph of h has $\left(\frac{3}{2}, 0\right)$ as its t-intercept.

✓ **Checkpoint** ◀)) *Audio-video solution in English & Spanish at LarsonPrecalculus.com.*

Find the zeros of each function.

a. $f(x) = 2x^2 + 13x - 24$ **b.** $g(t) = \sqrt{t - 25}$ **c.** $h(x) = \dfrac{x^2 - 2}{x - 1}$ ■

Zero of h: $t = \frac{3}{2}$
Figure 1.37

Increasing and Decreasing Functions

The more you know about the graph of a function, the more you know about the function itself. Consider the graph shown in Figure 1.38. As you move from *left to right,* this graph falls from $x = -2$ to $x = 0$, is constant from $x = 0$ to $x = 2$, and rises from $x = 2$ to $x = 4$.

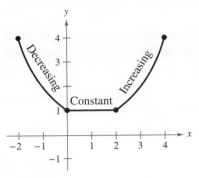

Figure 1.38

Increasing, Decreasing, and Constant Functions

A function f is **increasing** on an interval when, for any x_1 and x_2 in the interval,

$$x_1 < x_2 \quad \text{implies} \quad f(x_1) < f(x_2).$$

A function f is **decreasing** on an interval when, for any x_1 and x_2 in the interval,

$$x_1 < x_2 \quad \text{implies} \quad f(x_1) > f(x_2).$$

A function f is **constant** on an interval when, for any x_1 and x_2 in the interval,

$$f(x_1) = f(x_2).$$

EXAMPLE 4   **Describing Function Behavior**

Use the graphs to describe the increasing, decreasing, or constant behavior of each function.

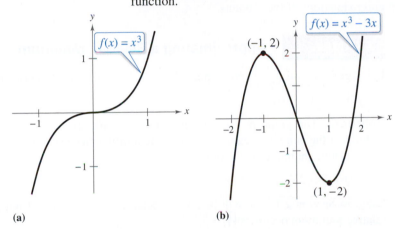

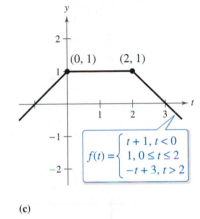

(a)   (b)   (c)

Solution

a. This function is increasing over the entire real line.

b. This function is increasing on the interval $(-\infty, -1)$, decreasing on the interval $(-1, 1)$, and increasing on the interval $(1, \infty)$.

c. This function is increasing on the interval $(-\infty, 0)$, constant on the interval $(0, 2)$, and decreasing on the interval $(2, \infty)$.

✓ *Checkpoint*   *Audio-video solution in English & Spanish at LarsonPrecalculus.com.*

Graph the function

$$f(x) = x^3 + 3x^2 - 1.$$

Then use the graph to describe the increasing and decreasing behavior of the function.

To help you decide whether a function is increasing, decreasing, or constant on an interval, you can evaluate the function for several values of x. However, you need calculus to determine, for certain, all intervals on which a function is increasing, decreasing, or constant.

The points at which a function changes its increasing, decreasing, or constant behavior are helpful in determining the **relative minimum** or **relative maximum** values of the function.

> **REMARK** A relative minimum or relative maximum is also referred to as a local minimum or local maximum.

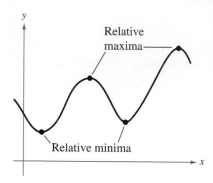

Figure 1.39

Definitions of Relative Minimum and Relative Maximum

A function value $f(a)$ is called a **relative minimum** of f when there exists an interval (x_1, x_2) that contains a such that

$$x_1 < x < x_2 \quad \text{implies} \quad f(a) \le f(x).$$

A function value $f(a)$ is called a **relative maximum** of f when there exists an interval (x_1, x_2) that contains a such that

$$x_1 < x < x_2 \quad \text{implies} \quad f(a) \ge f(x).$$

Figure 1.39 shows several different examples of relative minima and relative maxima. In Section 2.1, you will study a technique for finding the *exact point* at which a second-degree polynomial function has a relative minimum or relative maximum. For the time being, however, you can use a graphing utility to find reasonable approximations of these points.

EXAMPLE 5   **Approximating a Relative Minimum**

Use a graphing utility to approximate the relative minimum of the function

$$f(x) = 3x^2 - 4x - 2.$$

Solution   The graph of f is shown in Figure 1.40. By using the *zoom* and *trace* features or the *minimum* feature of a graphing utility, you can estimate that the function has a relative minimum at the point

$$(0.67, -3.33). \qquad \text{Relative minimum}$$

Later, in Section 2.1, you will be able to determine that the exact point at which the relative minimum occurs is $\left(\frac{2}{3}, -\frac{10}{3}\right)$.

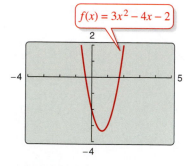

Figure 1.40

✓ *Checkpoint* 🔊))) *Audio-video solution in English & Spanish at LarsonPrecalculus.com.*

Use a graphing utility to approximate the relative maximum of the function

$$f(x) = -4x^2 - 7x + 3.$$ ∎

You can also use the *table* feature of a graphing utility to numerically approximate the relative minimum of the function in Example 5. Using a table that begins at 0.6 and increments the value of x by 0.01, you can approximate that the minimum of $f(x) = 3x^2 - 4x - 2$ occurs at the point $(0.67, -3.33)$.

▷ **TECHNOLOGY** When you use a graphing utility to estimate the x- and y-values of a relative minimum or relative maximum, the *zoom* feature will often produce graphs that are nearly flat. To overcome this problem, you can manually change the vertical setting of the viewing window. The graph will stretch vertically when the values of Ymin and Ymax are closer together.

Average Rate of Change

In Section 1.3, you learned that the slope of a line can be interpreted as a *rate of change*. For a nonlinear graph whose slope changes at each point, the **average rate of change** between any two points $(x_1, f(x_1))$ and $(x_2, f(x_2))$ is the slope of the line through the two points (see Figure 1.41). The line through the two points is called the **secant line**, and the slope of this line is denoted as m_{sec}.

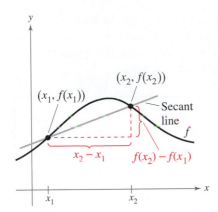

$$\text{Average rate of change of } f \text{ from } x_1 \text{ to } x_2 = \frac{f(x_2) - f(x_1)}{x_2 - x_1}$$

$$= \frac{\text{change in } y}{\text{change in } x}$$

$$= m_{\text{sec}}$$

Figure 1.41

EXAMPLE 6 **Average Rate of Change of a Function**

Find the average rates of change of $f(x) = x^3 - 3x$ (a) from $x_1 = -2$ to $x_2 = -1$ and (b) from $x_1 = 0$ to $x_2 = 1$ (see Figure 1.42).

Solution

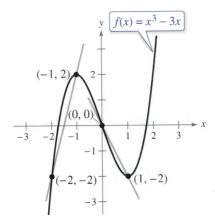

Figure 1.42

a. The average rate of change of f from $x_1 = -2$ to $x_2 = -1$ is

$$\frac{f(x_2) - f(x_1)}{x_2 - x_1} = \frac{f(-1) - f(-2)}{-1 - (-2)} = \frac{2 - (-2)}{1} = 4. \qquad \text{Secant line has positive slope.}$$

b. The average rate of change of f from $x_1 = 0$ to $x_2 = 1$ is

$$\frac{f(x_2) - f(x_1)}{x_2 - x_1} = \frac{f(1) - f(0)}{1 - 0} = \frac{-2 - 0}{1} = -2. \qquad \text{Secant line has negative slope.}$$

✓ ***Checkpoint***  *Audio-video solution in English & Spanish at LarsonPrecalculus.com.*

Find the average rates of change of $f(x) = x^2 + 2x$ (a) from $x_1 = -3$ to $x_2 = -2$ and (b) from $x_1 = -2$ to $x_2 = 0$.

EXAMPLE 7 **Finding Average Speed**

The distance s (in feet) a moving car is from a stoplight is given by the function

$$s(t) = 20t^{3/2}$$

where t is the time (in seconds). Find the average speed of the car (a) from $t_1 = 0$ to $t_2 = 4$ seconds and (b) from $t_1 = 4$ to $t_2 = 9$ seconds.

Solution

a. The average speed of the car from $t_1 = 0$ to $t_2 = 4$ seconds is

$$\frac{s(t_2) - s(t_1)}{t_2 - t_1} = \frac{s(4) - s(0)}{4 - 0} = \frac{160 - 0}{4} = 40 \text{ feet per second.}$$

b. The average speed of the car from $t_1 = 4$ to $t_2 = 9$ seconds is

$$\frac{s(t_2) - s(t_1)}{t_2 - t_1} = \frac{s(9) - s(4)}{9 - 4} = \frac{540 - 160}{5} = 76 \text{ feet per second.}$$

Average speed is an average rate of change.

✓ ***Checkpoint***  *Audio-video solution in English & Spanish at LarsonPrecalculus.com.*

In Example 7, find the average speed of the car (a) from $t_1 = 0$ to $t_2 = 1$ second and (b) from $t_1 = 1$ second to $t_2 = 4$ seconds.

sadwitch/Shutterstock.com

Even and Odd Functions

In Section 1.2, you studied different types of symmetry of a graph. In the terminology of functions, a function is said to be **even** when its graph is symmetric with respect to the y-axis and **odd** when its graph is symmetric with respect to the origin. The symmetry tests in Section 1.2 yield the following tests for even and odd functions.

> **Tests for Even and Odd Functions**
>
> A function $y = f(x)$ is **even** when, for each x in the domain of f, $f(-x) = f(x)$.
>
> A function $y = f(x)$ is **odd** when, for each x in the domain of f, $f(-x) = -f(x)$.

EXAMPLE 8   Even and Odd Functions

a. The function $g(x) = x^3 - x$ is odd because $g(-x) = -g(x)$, as follows.

$$g(-x) = (-x)^3 - (-x) \qquad \text{Substitute } -x \text{ for } x.$$

$$= -x^3 + x \qquad \text{Simplify.}$$

$$= -(x^3 - x) \qquad \text{Distributive Property}$$

$$= -g(x) \qquad \text{Test for odd function}$$

b. The function $h(x) = x^2 + 1$ is even because $h(-x) = h(x)$, as follows.

$$h(-x) = (-x)^2 + 1 \qquad \text{Substitute } -x \text{ for } x.$$

$$= x^2 + 1 \qquad \text{Simplify.}$$

$$= h(x) \qquad \text{Test for even function}$$

Figure 1.43 shows the graphs and symmetry of these two functions.

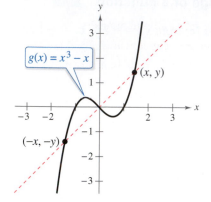

(a) Symmetric to origin: Odd Function

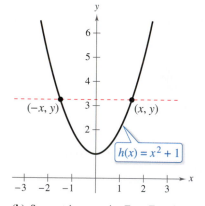

(b) Symmetric to y-axis: Even Function

Figure 1.43

✓ **Checkpoint** ◀))) *Audio-video solution in English & Spanish at LarsonPrecalculus.com.*

Determine whether the function is even, odd, or neither. Then describe the symmetry.

a. $f(x) = 5 - 3x$ **b.** $g(x) = x^4 - x^2 - 1$ **c.** $h(x) = 2x^3 + 3x$ ■

> **Summarize** (Section 1.5)
>
> 1. State the Vertical Line Test for functions (*page 35*). For an example of using the Vertical Line Test, see Example 2.
>
> 2. Explain how to find the zeros of a function (*page 36*). For an example of finding the zeros of functions, see Example 3.
>
> 3. Explain how to determine intervals on which functions are increasing or decreasing (*page 37*) and how to determine relative maximum and relative minimum values of functions (*page 38*). For an example of describing function behavior, see Example 4. For an example of approximating a relative minimum, see Example 5.
>
> 4. Explain how to determine the average rate of change of a function (*page 39*). For examples of determining average rates of change, see Examples 6 and 7.
>
> 5. State the definitions of an even function and an odd function (*page 40*). For an example of identifying even and odd functions, see Example 8.

1.6   A Library of Parent Functions

- Identify and graph linear and squaring functions.
- Identify and graph cubic, square root, and reciprocal functions.
- Identify and graph step and other piecewise-defined functions.
- Recognize graphs of parent functions.

Linear and Squaring Functions

One of the goals of this text is to enable you to recognize the basic shapes of the graphs of different types of functions. For instance, you know that the graph of the **linear function** $f(x) = ax + b$ is a line with slope $m = a$ and y-intercept at $(0, b)$. The graph of the linear function has the following characteristics.

- The domain of the function is the set of all real numbers.
- When $m \neq 0$, the range of the function is the set of all real numbers.
- The graph has an x-intercept at $(-b/m, 0)$ and a y-intercept at $(0, b)$.
- The graph is increasing when $m > 0$, decreasing when $m < 0$, and constant when $m = 0$.

Piecewise-defined functions can help you model real-life situations, such as the depth of snow during a snowstorm.

EXAMPLE 1   Writing a Linear Function

Write the linear function f for which $f(1) = 3$ and $f(4) = 0$.

Solution   To find the equation of the line that passes through $(x_1, y_1) = (1, 3)$ and $(x_2, y_2) = (4, 0)$, first find the slope of the line.

$$m = \frac{y_2 - y_1}{x_2 - x_1} = \frac{0 - 3}{4 - 1} = \frac{-3}{3} = -1$$

Next, use the point-slope form of the equation of a line.

$y - y_1 = m(x - x_1)$	Point-slope form
$y - 3 = -1(x - 1)$	Substitute for x_1, y_1, and m.
$y = -x + 4$	Simplify.
$f(x) = -x + 4$	Function notation

The figure below shows the graph of this function.

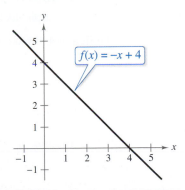

$f(x) = -x + 4$

 Checkpoint 🔊))) *Audio-video solution in English & Spanish at LarsonPrecalculus.com.*

Write the linear function f for which $f(-2) = 6$ and $f(4) = -9$.

There are two special types of linear functions, the **constant function** and the **identity function.** A constant function has the form

$$f(x) = c$$

and has the domain of all real numbers with a range consisting of a single real number c. The graph of a constant function is a horizontal line, as shown in Figure 1.44. The identity function has the form

$$f(x) = x.$$

Its domain and range are the set of all real numbers. The identity function has a slope of $m = 1$ and a y-intercept at $(0, 0)$. The graph of the identity function is a line for which each x-coordinate equals the corresponding y-coordinate. The graph is always increasing, as shown in Figure 1.45.

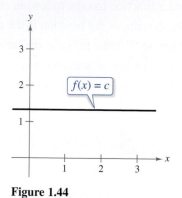

Figure 1.44   **Figure 1.45**

The graph of the **squaring function**

$$f(x) = x^2$$

is a U-shaped curve with the following characteristics.

- The domain of the function is the set of all real numbers.
- The range of the function is the set of all nonnegative real numbers.
- The function is even.
- The graph has an intercept at $(0, 0)$.
- The graph is decreasing on the interval $(-\infty, 0)$ and increasing on the interval $(0, \infty)$.
- The graph is symmetric with respect to the y-axis.
- The graph has a relative minimum at $(0, 0)$.

The figure below shows the graph of the squaring function.

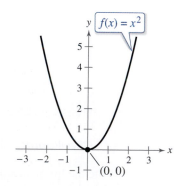

Cubic, Square Root, and Reciprocal Functions

The following summarizes the basic characteristics of the graphs of the **cubic, square root,** and **reciprocal functions.**

1. The graph of the *cubic* function

$$f(x) = x^3$$

has the following characteristics.

- The domain of the function is the set of all real numbers.
- The range of the function is the set of all real numbers.
- The function is odd.
- The graph has an intercept at $(0, 0)$.
- The graph is increasing on the interval $(-\infty, \infty)$.
- The graph is symmetric with respect to the origin.

The figure shows the graph of the cubic function.

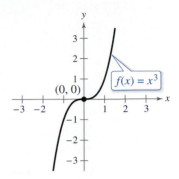

Cubic function

2. The graph of the *square root* function

$$f(x) = \sqrt{x}$$

has the following characteristics.

- The domain of the function is the set of all nonnegative real numbers.
- The range of the function is the set of all nonnegative real numbers.
- The graph has an intercept at $(0, 0)$.
- The graph is increasing on the interval $(0, \infty)$.

The figure shows the graph of the square root function.

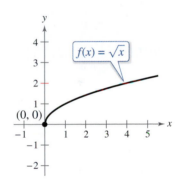

Square root function

3. The graph of the *reciprocal* function

$$f(x) = \frac{1}{x}$$

has the following characteristics.

- The domain of the function is $(-\infty, 0) \cup (0, \infty)$.
- The range of the function is $(-\infty, 0) \cup (0, \infty)$.
- The function is odd.
- The graph does not have any intercepts.
- The graph is decreasing on the intervals $(-\infty, 0)$ and $(0, \infty)$.
- The graph is symmetric with respect to the origin.

The figure shows the graph of the reciprocal function.

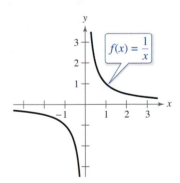

Reciprocal function

Step and Piecewise-Defined Functions

Functions whose graphs resemble sets of stairsteps are known as **step functions.** The most famous of the step functions is the **greatest integer function,** denoted by $[\![x]\!]$ and defined as

$$f(x) = [\![x]\!] = \textit{the greatest integer less than or equal to } x.$$

Some values of the greatest integer function are as follows.

$$[\![-1]\!] = (\text{greatest integer} \le -1) = -1$$

$$\left[\!\left[-\tfrac{1}{2}\right]\!\right] = \left(\text{greatest integer} \le -\tfrac{1}{2}\right) = -1$$

$$\left[\!\left[\tfrac{1}{10}\right]\!\right] = \left(\text{greatest integer} \le \tfrac{1}{10}\right) = 0$$

$$[\![1.5]\!] = (\text{greatest integer} \le 1.5) = 1$$

$$[\![1.9]\!] = (\text{greatest integer} \le 1.9) = 1$$

The graph of the greatest integer function

$$f(x) = [\![x]\!]$$

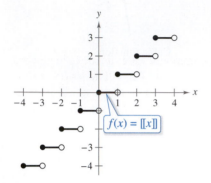

Figure 1.46

has the following characteristics, as shown in Figure 1.46.

- The domain of the function is the set of all real numbers.
- The range of the function is the set of all integers.
- The graph has a y-intercept at $(0, 0)$ and x-intercepts in the interval $[0, 1)$.
- The graph is constant between each pair of consecutive integer values of x.
- The graph jumps vertically one unit at each integer value of x.

▷ **TECHNOLOGY** When using your graphing utility to graph a step function, you should set your graphing utility to *dot* mode.

EXAMPLE 2 **Evaluating a Step Function**

Evaluate the function when $x = -1$, 2, and $\tfrac{3}{2}$.

$$f(x) = [\![x]\!] + 1$$

Solution For $x = -1$, the greatest integer ≤ -1 is -1, so

$$f(-1) = [\![-1]\!] + 1 = -1 + 1 = 0.$$

For $x = 2$, the greatest integer ≤ 2 is 2, so

$$f(2) = [\![2]\!] + 1 = 2 + 1 = 3.$$

For $x = \tfrac{3}{2}$, the greatest integer $\le \tfrac{3}{2}$ is 1, so

$$f\!\left(\tfrac{3}{2}\right) = \left[\!\left[\tfrac{3}{2}\right]\!\right] + 1 = 1 + 1 = 2.$$

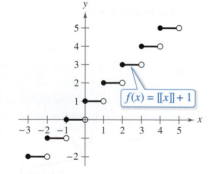

Figure 1.47

Verify your answers by examining the graph of $f(x) = [\![x]\!] + 1$ shown in Figure 1.47.

✓ **Checkpoint** ◀))) *Audio-video solution in English & Spanish at LarsonPrecalculus.com.*

Evaluate the function when $x = -\tfrac{3}{2}$, 1, and $-\tfrac{5}{2}$.

$$f(x) = [\![x + 2]\!]$$

Recall from Section 1.4 that a piecewise-defined function is defined by two or more equations over a specified domain. To graph a piecewise-defined function, graph each equation separately over the specified domain, as shown in Example 3.

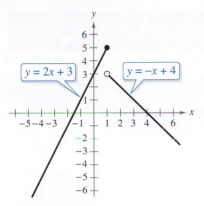

Figure 1.48

| EXAMPLE 3 | **Graphing a Piecewise-Defined Function** |

Sketch the graph of $f(x) = \begin{cases} 2x + 3, & x \le 1 \\ -x + 4, & x > 1 \end{cases}$.

Solution   This piecewise-defined function consists of two linear functions. At $x = 1$ and to the left of $x = 1$, the graph is the line $y = 2x + 3$, and to the right of $x = 1$ the graph is the line $y = -x + 4$, as shown in Figure 1.48. Notice that the point $(1, 5)$ is a solid dot and the point $(1, 3)$ is an open dot. This is because $f(1) = 2(1) + 3 = 5$.

✔ **Checkpoint** 🔊)) *Audio-video solution in English & Spanish at LarsonPrecalculus.com.*

Sketch the graph of $f(x) = \begin{cases} -\frac{1}{2}x - 6, & x \le -4 \\ x + 5, & x > -4 \end{cases}$.

Parent Functions

The eight graphs shown below represent the most commonly used functions in algebra. Familiarity with the basic characteristics of these simple graphs will help you analyze the shapes of more complicated graphs—in particular, graphs obtained from these graphs by the rigid and nonrigid transformations studied in the next section.

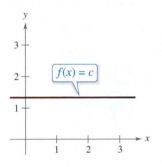

(a) Constant Function

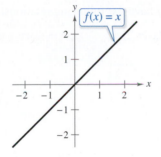

(b) Identity Function

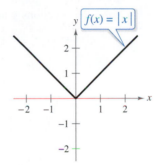

(c) Absolute Value Function

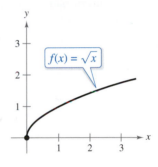

(d) Square Root Function

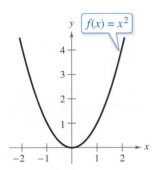

(e) Quadratic Function

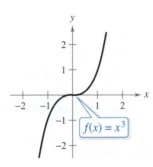

(f) Cubic Function

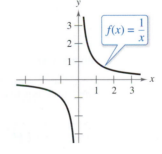

(g) Reciprocal Function

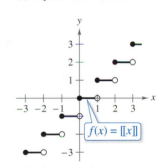

(h) Greatest Integer Function

Summarize   (Section 1.6)

1. Explain how to identify and graph linear and squaring functions *(pages 41 and 42)*. For an example involving a linear function, see Example 1.

2. Explain how to identify and graph cubic, square root, and reciprocal functions *(page 43)*.

3. Explain how to identify and graph step and other piecewise-defined functions *(page 44)*. For an example involving a step function, see Example 2. For an example of graphing a piecewise-defined function, see Example 3.

4. State and sketch the graphs of parent functions *(page 45)*.

1.7 Transformations of Functions

Transformations of functions can help you model real-life applications. For instance, a transformation of a function can model the number of horsepower required to overcome wind drag on an automobile.

- ■ Use vertical and horizontal shifts to sketch graphs of functions.
- ■ Use reflections to sketch graphs of functions.
- ■ Use nonrigid transformations to sketch graphs of functions.

Shifting Graphs

Many functions have graphs that are transformations of the parent graphs summarized in Section 1.6. For example, you can obtain the graph of

$$h(x) = x^2 + 2$$

by shifting the graph of $f(x) = x^2$ *up* two units, as shown in Figure 1.49. In function notation, h and f are related as follows.

$$h(x) = x^2 + 2 = f(x) + 2 \qquad \text{Upward shift of two units}$$

Similarly, you can obtain the graph of

$$g(x) = (x - 2)^2$$

by shifting the graph of $f(x) = x^2$ to the *right* two units, as shown in Figure 1.50. In this case, the functions g and f have the following relationship.

$$g(x) = (x - 2)^2 = f(x - 2) \qquad \text{Right shift of two units}$$

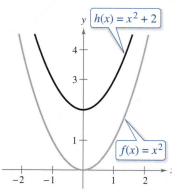

Figure 1.49

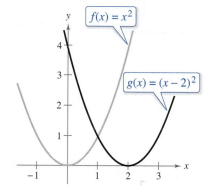

Figure 1.50

The following list summarizes this discussion about horizontal and vertical shifts.

• • REMARK In items 3 and 4, be sure you see that $h(x) = f(x - c)$ corresponds to a *right* shift and $h(x) = f(x + c)$ corresponds to a *left* shift for $c > 0$.

Vertical and Horizontal Shifts

Let c be a positive real number. **Vertical and horizontal shifts** in the graph of $y = f(x)$ are represented as follows.

1. Vertical shift c units *up:* $h(x) = f(x) + c$

2. Vertical shift c units *down:* $h(x) = f(x) - c$

3. Horizontal shift c units to the *right:* $h(x) = f(x - c)$

4. Horizontal shift c units to the *left:* $h(x) = f(x + c)$

Some graphs can be obtained from combinations of vertical and horizontal shifts, as demonstrated in Example 1(b). Vertical and horizontal shifts generate a *family of functions,* each with the same shape but at a different location in the plane.

EXAMPLE 1 **Shifts in the Graph of a Function**

Use the graph of $f(x) = x^3$ to sketch the graph of each function.

a. $g(x) = x^3 - 1$

b. $h(x) = (x + 2)^3 + 1$

Solution

a. Relative to the graph of $f(x) = x^3$, the graph of

$$g(x) = x^3 - 1$$

is a downward shift of one unit, as shown below.

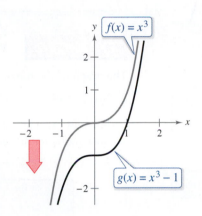

REMARK In Example 1(a), note that $g(x) = f(x) - 1$ and in Example 1(b), $h(x) = f(x + 2) + 1$.

b. Relative to the graph of $f(x) = x^3$, the graph of

$$h(x) = (x + 2)^3 + 1$$

involves a left shift of two units and an upward shift of one unit, as shown below.

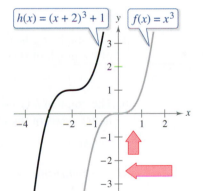

 Checkpoint Audio-video solution in English & Spanish at LarsonPrecalculus.com.

Use the graph of $f(x) = x^3$ to sketch the graph of each function.

a. $h(x) = x^3 + 5$

b. $g(x) = (x - 3)^3 + 2$

In Example 1(b), you obtain the same result when the vertical shift precedes the horizontal shift *or* when the horizontal shift precedes the vertical shift.

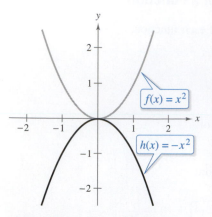

Figure 1.51

Reflecting Graphs

Another common type of transformation is a **reflection.** For instance, if you consider the x-axis to be a mirror, then the graph of $h(x) = -x^2$ is the mirror image (or reflection) of the graph of $f(x) = x^2$, as shown in Figure 1.51.

Reflections in the Coordinate Axes

Reflections in the coordinate axes of the graph of $y = f(x)$ are represented as follows.

1. Reflection in the x-axis: $h(x) = -f(x)$

2. Reflection in the y-axis: $h(x) = f(-x)$

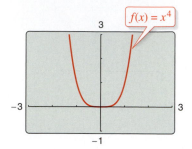

Figure 1.52

EXAMPLE 2   **Writing Equations from Graphs**

The graph of the function

$$f(x) = x^4$$

is shown in Figure 1.52. Each of the graphs below is a transformation of the graph of f. Write an equation for each of these functions.

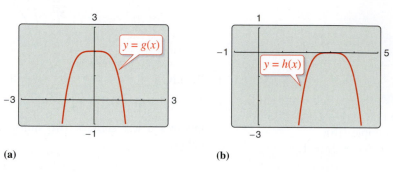

(a)   (b)

Solution

a. The graph of g is a reflection in the x-axis *followed by* an upward shift of two units of the graph of $f(x) = x^4$. So, the equation for g is

$$g(x) = -x^4 + 2.$$

b. The graph of h is a horizontal shift of three units to the right *followed by* a reflection in the x-axis of the graph of $f(x) = x^4$. So, the equation for h is

$$h(x) = -(x - 3)^4.$$

✓ **Checkpoint** ◄))) *Audio-video solution in English & Spanish at LarsonPrecalculus.com.*

The graph is a transformation of the graph of $f(x) = x^4$. Write an equation for the function.

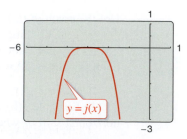

EXAMPLE 3   **Reflections and Shifts**

Compare the graph of each function with the graph of $f(x) = \sqrt{x}$.

a. $g(x) = -\sqrt{x}$ **b.** $h(x) = \sqrt{-x}$ **c.** $k(x) = -\sqrt{x + 2}$

Algebraic Solution

a. The graph of g is a reflection of the graph of f in the x-axis because

$$g(x) = -\sqrt{x}$$
$$= -f(x).$$

b. The graph of h is a reflection of the graph of f in the y-axis because

$$h(x) = \sqrt{-x}$$
$$= f(-x).$$

c. The graph of k is a left shift of two units followed by a reflection in the x-axis because

$$k(x) = -\sqrt{x + 2}$$
$$= -f(x + 2).$$

Graphical Solution

a. Graph f and g on the same set of coordinate axes. From the graph, you can see that the graph of g is a reflection of the graph of f in the x-axis.

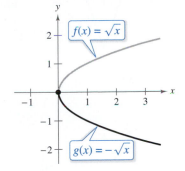

b. Graph f and h on the same set of coordinate axes. From the graph, you can see that the graph of h is a reflection of the graph of f in the y-axis.

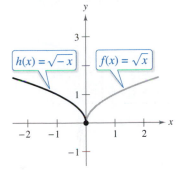

c. Graph f and k on the same set of coordinate axes. From the graph, you can see that the graph of k is a left shift of two units of the graph of f, followed by a reflection in the x-axis.

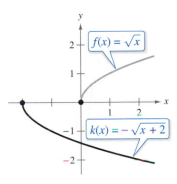

✓ *Checkpoint* 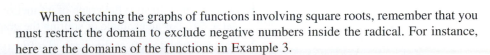 *Audio-video solution in English & Spanish at LarsonPrecalculus.com.*

Compare the graph of each function with the graph of

$$f(x) = \sqrt{x - 1}.$$

a. $g(x) = -\sqrt{x - 1}$ **b.** $h(x) = \sqrt{-x - 1}$ ■

When sketching the graphs of functions involving square roots, remember that you must restrict the domain to exclude negative numbers inside the radical. For instance, here are the domains of the functions in Example 3.

Domain of $g(x) = -\sqrt{x}$: $x \geq 0$

Domain of $h(x) = \sqrt{-x}$: $x \leq 0$

Domain of $k(x) = -\sqrt{x + 2}$: $x \geq -2$

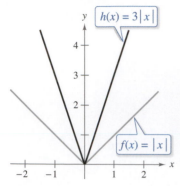

Figure 1.53

Nonrigid Transformations

Horizontal shifts, vertical shifts, and reflections are **rigid transformations** because the basic shape of the graph is unchanged. These transformations change only the *position* of the graph in the coordinate plane. **Nonrigid transformations** are those that cause a *distortion*—a change in the shape of the original graph. For instance, a nonrigid transformation of the graph of $y = f(x)$ is represented by $g(x) = cf(x)$, where the transformation is a **vertical stretch** when $c > 1$ and a **vertical shrink** when $0 < c < 1$. Another nonrigid transformation of the graph of $y = f(x)$ is represented by $h(x) = f(cx)$, where the transformation is a **horizontal shrink** when $c > 1$ and a **horizontal stretch** when $0 < c < 1$.

EXAMPLE 4   **Nonrigid Transformations**

Compare the graph of each function with the graph of $f(x) = |x|$.

a. $h(x) = 3|x|$ **b.** $g(x) = \frac{1}{3}|x|$

Solution

a. Relative to the graph of $f(x) = |x|$, the graph of $h(x) = 3|x| = 3f(x)$ is a vertical stretch (each y-value is multiplied by 3) of the graph of f. (See Figure 1.53.)

b. Similarly, the graph of $g(x) = \frac{1}{3}|x| = \frac{1}{3}f(x)$ is a vertical shrink $\left(\text{each } y\text{-value is multiplied by } \frac{1}{3}\right)$ of the graph of f. (See Figure 1.54.)

 ✓ *Checkpoint* 🔊))) *Audio-video solution in English & Spanish at LarsonPrecalculus.com.*

Compare the graph of each function with the graph of $f(x) = x^2$.

a. $g(x) = 4x^2$ **b.** $h(x) = \frac{1}{4}x^2$

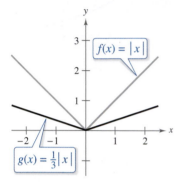

Figure 1.54

EXAMPLE 5   **Nonrigid Transformations**

Compare the graph of each function with the graph of $f(x) = 2 - x^3$.

a. $g(x) = f(2x)$ **b.** $h(x) = f\left(\frac{1}{2}x\right)$

Solution

a. Relative to the graph of $f(x) = 2 - x^3$, the graph of $g(x) = f(2x) = 2 - (2x)^3 = 2 - 8x^3$ is a horizontal shrink ($c > 1$) of the graph of f. (See Figure 1.55.)

b. Similarly, the graph of $h(x) = f\left(\frac{1}{2}x\right) = 2 - \left(\frac{1}{2}x\right)^3 = 2 - \frac{1}{8}x^3$ is a horizontal stretch ($0 < c < 1$) of the graph of f. (See Figure 1.56.)

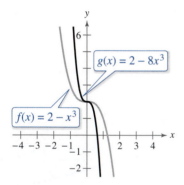

Figure 1.55

✓ *Checkpoint* 🔊))) *Audio-video solution in English & Spanish at LarsonPrecalculus.com.*

Compare the graph of each function with the graph of $f(x) = x^2 + 3$.

a. $g(x) = f(2x)$ **b.** $h(x) = f\left(\frac{1}{2}x\right)$ ◼

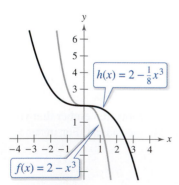

Figure 1.56

Summarize   (Section 1.7)

1. Describe how to shift the graph of a function vertically and horizontally *(page 46)*. For an example of shifting the graph of a function, see Example 1.

2. Describe how to reflect the graph of a function in the x-axis and in the y-axis *(page 48)*. For examples of reflecting graphs of functions, see Examples 2 and 3.

3. Describe nonrigid transformations of the graph of a function *(page 50)*. For examples of nonrigid transformations, see Examples 4 and 5.

1.8 Combinations of Functions: Composite Functions

■ **Add, subtract, multiply, and divide functions.**
■ **Find the composition of one function with another function.**
■ **Use combinations and compositions of functions to model and solve real-life problems.**

Arithmetic Combinations of Functions

Just as two real numbers can be combined by the operations of addition, subtraction, multiplication, and division to form other real numbers, two *functions* can be combined to create new functions. For example, the functions $f(x) = 2x - 3$ and $g(x) = x^2 - 1$ can be combined to form the sum, difference, product, and quotient of f and g.

$$f(x) + g(x) = (2x - 3) + (x^2 - 1) = x^2 + 2x - 4 \qquad \text{Sum}$$

$$f(x) - g(x) = (2x - 3) - (x^2 - 1) = -x^2 + 2x - 2 \qquad \text{Difference}$$

$$f(x)g(x) = (2x - 3)(x^2 - 1) = 2x^3 - 3x^2 - 2x + 3 \qquad \text{Product}$$

$$\frac{f(x)}{g(x)} = \frac{2x - 3}{x^2 - 1}, \quad x \neq \pm 1 \qquad \text{Quotient}$$

The domain of an **arithmetic combination** of functions f and g consists of all real numbers that are common to the domains of f and g. In the case of the quotient $f(x)/g(x)$, there is the further restriction that $g(x) \neq 0$.

Arithmetic combinations of functions can help you model and solve real-life problems, such as analyzing numbers of pets in the United States.

Sum, Difference, Product, and Quotient of Functions

Let f and g be two functions with overlapping domains. Then, for all x common to both domains, the *sum, difference, product,* and *quotient* of f and g are defined as follows.

1. Sum: $\quad (f + g)(x) = f(x) + g(x)$

2. Difference: $(f - g)(x) = f(x) - g(x)$

3. Product: $\quad (fg)(x) = f(x) \cdot g(x)$

4. Quotient: $\left(\dfrac{f}{g}\right)(x) = \dfrac{f(x)}{g(x)}, \quad g(x) \neq 0$

EXAMPLE 1 **Finding the Sum of Two Functions**

Given $f(x) = 2x + 1$ and $g(x) = x^2 + 2x - 1$, find $(f + g)(x)$. Then evaluate the sum when $x = 3$.

Solution The sum of f and g is

$$(f + g)(x) = f(x) + g(x) = (2x + 1) + (x^2 + 2x - 1) = x^2 + 4x.$$

When $x = 3$, the value of this sum is

$$(f + g)(3) = 3^2 + 4(3) = 21.$$

✓ *Checkpoint* ◀))) *Audio-video solution in English & Spanish at LarsonPrecalculus.com.*

Given $f(x) = x^2$ and $g(x) = 1 - x$, find $(f + g)(x)$. Then evaluate the sum when $x = 2$.

Michael Pettigrew/Shutterstock.com

EXAMPLE 2 Finding the Difference of Two Functions

Given $f(x) = 2x + 1$ and $g(x) = x^2 + 2x - 1$, find $(f - g)(x)$. Then evaluate the difference when $x = 2$.

Solution The difference of f and g is

$$(f - g)(x) = f(x) - g(x) = (2x + 1) - (x^2 + 2x - 1) = -x^2 + 2.$$

When $x = 2$, the value of this difference is

$$(f - g)(2) = -(2)^2 + 2 = -2.$$

✓ Checkpoint *Audio-video solution in English & Spanish at LarsonPrecalculus.com.*

Given $f(x) = x^2$ and $g(x) = 1 - x$, find $(f - g)(x)$. Then evaluate the difference when $x = 3$.

EXAMPLE 3 Finding the Product of Two Functions

Given $f(x) = x^2$ and $g(x) = x - 3$, find $(fg)(x)$. Then evaluate the product when $x = 4$.

Solution The product of f and g is

$$(fg)(x) = f(x)g(x) = (x^2)(x - 3) = x^3 - 3x^2.$$

When $x = 4$, the value of this product is

$$(fg)(4) = 4^3 - 3(4)^2 = 16.$$

✓ Checkpoint *Audio-video solution in English & Spanish at LarsonPrecalculus.com.*

Given $f(x) = x^2$ and $g(x) = 1 - x$, find $(fg)(x)$. Then evaluate the product when $x = 3$.

In Examples 1–3, both f and g have domains that consist of all real numbers. So, the domains of $f + g$, $f - g$, and fg are also the set of all real numbers. Remember to consider any restrictions on the domains of f and g when forming the sum, difference, product, or quotient of f and g.

EXAMPLE 4 Finding the Quotients of Two Functions

Find $(f/g)(x)$ and $(g/f)(x)$ for the functions $f(x) = \sqrt{x}$ and $g(x) = \sqrt{4 - x^2}$. Then find the domains of f/g and g/f.

Solution The quotient of f and g is

$$\left(\frac{f}{g}\right)(x) = \frac{f(x)}{g(x)} = \frac{\sqrt{x}}{\sqrt{4 - x^2}}$$

and the quotient of g and f is

$$\left(\frac{g}{f}\right)(x) = \frac{g(x)}{f(x)} = \frac{\sqrt{4 - x^2}}{\sqrt{x}}.$$

The domain of f is $[0, \infty)$ and the domain of g is $[-2, 2]$. The intersection of these domains is $[0, 2]$. So, the domains of f/g and g/f are as follows.

Domain of f/g: $[0, 2)$ Domain of g/f: $(0, 2]$

REMARK Note that the domain of f/g includes $x = 0$, but not $x = 2$, because $x = 2$ yields a zero in the denominator, whereas the domain of g/f includes $x = 2$, but not $x = 0$, because $x = 0$ yields a zero in the denominator.

✓ Checkpoint *Audio-video solution in English & Spanish at LarsonPrecalculus.com.*

Find $(f/g)(x)$ and $(g/f)(x)$ for the functions $f(x) = \sqrt{x - 3}$ and $g(x) = \sqrt{16 - x^2}$. Then find the domains of f/g and g/f.

Composition of Functions

Another way of combining two functions is to form the **composition** of one with the other. For instance, if $f(x) = x^2$ and $g(x) = x + 1,$ then the composition of f with g is

$$f(g(x)) = f(x + 1)$$
$$= (x + 1)^2.$$

This composition is denoted as $f \circ g$ and reads as "f composed with g."

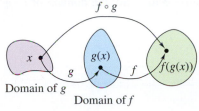

$f \circ g$

x $g(x)$ $f(g(x))$

g f

Domain of g

Domain of f

Figure 1.57

Definition of Composition of Two Functions

The **composition** of the function f with the function g is

$$(f \circ g)(x) = f(g(x)).$$

The domain of $f \circ g$ is the set of all x in the domain of g such that $g(x)$ is in the domain of f. (See Figure 1.57.)

EXAMPLE 5   **Composition of Functions**

Given $f(x) = x + 2$ and $g(x) = 4 - x^2,$ find the following.

a. $(f \circ g)(x)$

b. $(g \circ f)(x)$

c. $(g \circ f)(-2)$

Solution

a. The composition of f with g is as follows.

$$(f \circ g)(x) = f(g(x)) \qquad \text{Definition of } f \circ g$$
$$= f(4 - x^2) \qquad \text{Definition of } g(x)$$
$$= (4 - x^2) + 2 \qquad \text{Definition of } f(x)$$
$$= -x^2 + 6 \qquad \text{Simplify.}$$

b. The composition of g with f is as follows.

$$(g \circ f)(x) = g(f(x)) \qquad \text{Definition of } g \circ f$$
$$= g(x + 2) \qquad \text{Definition of } f(x)$$
$$= 4 - (x + 2)^2 \qquad \text{Definition of } g(x)$$
$$= 4 - (x^2 + 4x + 4) \qquad \text{Expand.}$$
$$= -x^2 - 4x \qquad \text{Simplify.}$$

Note that, in this case, $(f \circ g)(x) \neq (g \circ f)(x).$

c. Using the result of part (b), write the following.

$$(g \circ f)(-2) = -(-2)^2 - 4(-2) \qquad \text{Substitute.}$$
$$= -4 + 8 \qquad \text{Simplify.}$$
$$= 4 \qquad \text{Simplify.}$$

> ⋅⋅**REMARK**   The following tables of values help illustrate the composition $(f \circ g)(x)$ in Example 5(a).

x	0	1	2	3
$g(x)$	4	3	0	-5

$g(x)$	4	3	0	-5
$f(g(x))$	6	5	2	-3

x	0	1	2	3
$f(g(x))$	6	5	2	-3

Note that the first two tables can be combined (or "composed") to produce the values in the third table.

✓ **Checkpoint** ◀))) *Audio-video solution in English & Spanish at LarsonPrecalculus.com.*

Given $f(x) = 2x + 5$ and $g(x) = 4x^2 + 1,$ find the following.

a. $(f \circ g)(x)$ **b.** $(g \circ f)(x)$ **c.** $(f \circ g)\left(-\frac{1}{2}\right)$

EXAMPLE 6   **Finding the Domain of a Composite Function**

Find the domain of $(f \circ g)(x)$ for the functions

$$f(x) = x^2 - 9 \quad \text{and} \quad g(x) = \sqrt{9 - x^2}.$$

Algebraic Solution

The composition of the functions is as follows.

$$(f \circ g)(x) = f(g(x))$$

$$= f\left(\sqrt{9 - x^2}\right)$$

$$= \left(\sqrt{9 - x^2}\right)^2 - 9$$

$$= 9 - x^2 - 9$$

$$= -x^2$$

From this, it might appear that the domain of the composition is the set of all real numbers. This, however, is not true. Because the domain of f is the set of all real numbers and the domain of g is $[-3, 3]$, the domain of $f \circ g$ is $[-3, 3]$.

Graphical Solution

The x-coordinates of the points on the graph extend from -3 to 3. So, the domain of $f \circ g$ is $[-3, 3]$.

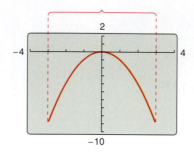

 Checkpoint *Audio-video solution in English & Spanish at LarsonPrecalculus.com.*

Find the domain of $(f \circ g)(x)$ for the functions $f(x) = \sqrt{x}$ and $g(x) = x^2 + 4$.   ■

In Examples 5 and 6, you formed the composition of two given functions. In calculus, it is also important to be able to identify two functions that make up a given composite function. For instance, the function $h(x) = (3x - 5)^3$ is the composition of $f(x) = x^3$ and $g(x) = 3x - 5$. That is,

$$h(x) = (3x - 5)^3 = [g(x)]^3 = f(g(x)).$$

Basically, to "decompose" a composite function, look for an "inner" function and an "outer" function. In the function h above, $g(x) = 3x - 5$ is the inner function and $f(x) = x^3$ is the outer function.

EXAMPLE 7   **Decomposing a Composite Function**

Write the function $h(x) = \dfrac{1}{(x - 2)^2}$ as a composition of two functions.

Solution   One way to write h as a composition of two functions is to take the inner function to be $g(x) = x - 2$ and the outer function to be

$$f(x) = \frac{1}{x^2} = x^{-2}.$$

Then write

$$h(x) = \frac{1}{(x - 2)^2}$$

$$= (x - 2)^{-2}$$

$$= f(x - 2)$$

$$= f(g(x)).$$

 Checkpoint *Audio-video solution in English & Spanish at LarsonPrecalculus.com.*

Write the function $h(x) = \dfrac{\sqrt[3]{8 - x}}{5}$ as a composition of two functions.   ■

Application

Refrigerated foods can have two types of bacteria: pathogenic bacteria, which can cause foodborne illness, and spoilage bacteria, which give foods an unpleasant look, smell, taste, or texture.

EXAMPLE 8 Bacteria Count

The number N of bacteria in a refrigerated food is given by

$$N(T) = 20T^2 - 80T + 500, \quad 2 \le T \le 14$$

where T is the temperature of the food in degrees Celsius. When the food is removed from refrigeration, the temperature of the food is given by

$$T(t) = 4t + 2, \quad 0 \le t \le 3$$

where t is the time in hours.

a. Find the composition $(N \circ T)(t)$ and interpret its meaning in context.

b. Find the time when the bacteria count reaches 2000.

Solution

a. $(N \circ T)(t) = N(T(t))$

$$= 20(4t + 2)^2 - 80(4t + 2) + 500$$

$$= 20(16t^2 + 16t + 4) - 320t - 160 + 500$$

$$= 320t^2 + 320t + 80 - 320t - 160 + 500$$

$$= 320t^2 + 420$$

The composite function $(N \circ T)(t)$ represents the number of bacteria in the food as a function of the amount of time the food has been out of refrigeration.

b. The bacteria count will reach 2000 when $320t^2 + 420 = 2000$. Solve this equation to find that the count will reach 2000 when $t \approx 2.2$ hours. Note that when you solve this equation, you reject the negative value because it is not in the domain of the composite function.

✓ **Checkpoint** *Audio-video solution in English & Spanish at LarsonPrecalculus.com.*

The number N of bacteria in a refrigerated food is given by

$$N(T) = 8T^2 - 14T + 200, \quad 2 \le T \le 12$$

where T is the temperature of the food in degrees Celsius. When the food is removed from refrigeration, the temperature of the food is given by

$$T(t) = 2t + 2, \quad 0 \le t \le 5$$

where t is the time in hours.

a. Find the composition $(N \circ T)(t)$.

b. Find the time when the bacteria count reaches 1000.

Summarize (Section 1.8)

1. Explain how to add, subtract, multiply, and divide functions *(page 51)*. For examples of finding arithmetic combinations of functions, see Examples 1–4.

2. Explain how to find the composition of one function with another function *(page 53)*. For examples that use compositions of functions, see Examples 5–7.

3. Describe a real-life example that uses a composition of functions *(page 55, Example 8)*.

1.9 Inverse Functions

■ Find inverse functions informally and verify that two functions are inverse functions of each other.
■ Use graphs of functions to determine whether functions have inverse functions.
■ Use the Horizontal Line Test to determine whether functions are one-to-one.
■ Find inverse functions algebraically.

Inverse Functions

Inverse functions can help you model and solve real-life problems. For instance, an inverse function can help you determine the percent load interval for a diesel engine.

Recall from Section 1.4 that a set of ordered pairs can represent a function. For instance, the function $f(x) = x + 4$ from the set $A = \{1, 2, 3, 4\}$ to the set $B = \{5, 6, 7, 8\}$ can be written as follows.

$$f(x) = x + 4: \ \{(1, 5), (2, 6), (3, 7), (4, 8)\}$$

In this case, by interchanging the first and second coordinates of each of these ordered pairs, you can form the **inverse function** of f, which is denoted by f^{-1}. It is a function from the set B to the set A, and can be written as follows.

$$f^{-1}(x) = x - 4: \ \{(5, 1), (6, 2), (7, 3), (8, 4)\}$$

Note that the domain of f is equal to the range of f^{-1}, and vice versa, as shown in the figure below. Also note that the functions f and f^{-1} have the effect of "undoing" each other. In other words, when you form the composition of f with f^{-1} or the composition of f^{-1} with f, you obtain the identity function.

$$f(f^{-1}(x)) = f(x - 4) = (x - 4) + 4 = x$$
$$f^{-1}(f(x)) = f^{-1}(x + 4) = (x + 4) - 4 = x$$

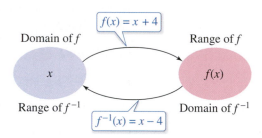

Domain of f $\boxed{f(x) = x + 4}$ Range of f

x $f(x)$

Range of f^{-1} $\boxed{f^{-1}(x) = x - 4}$ Domain of f^{-1}

EXAMPLE 1   **Finding an Inverse Function Informally**

Find the inverse function of $f(x) = 4x$. Then verify that both $f(f^{-1}(x))$ and $f^{-1}(f(x))$ are equal to the identity function.

Solution   The function f *multiplies* each input by 4. To "undo" this function, you need to *divide* each input by 4. So, the inverse function of $f(x) = 4x$ is

$$f^{-1}(x) = \frac{x}{4}.$$

Verify that $f(f^{-1}(x)) = x$ and $f^{-1}(f(x)) = x$ as follows.

$$f(f^{-1}(x)) = f\left(\frac{x}{4}\right) = 4\left(\frac{x}{4}\right) = x \qquad f^{-1}(f(x)) = f^{-1}(4x) = \frac{4x}{4} = x$$

✓ **Checkpoint** ◀ঠ))) *Audio-video solution in English & Spanish at LarsonPrecalculus.com.*

Find the inverse function of $f(x) = \frac{1}{5}x$. Then verify that both $f(f^{-1}(x))$ and $f^{-1}(f(x))$ are equal to the identity function. ∎

Definition of Inverse Function

Let f and g be two functions such that

$$f(g(x)) = x \qquad \text{for every } x \text{ in the domain of } g$$

and

$$g(f(x)) = x \qquad \text{for every } x \text{ in the domain of } f.$$

Under these conditions, the function g is the **inverse function** of the function f. The function g is denoted by f^{-1} (read "f-inverse"). So,

$$f(f^{-1}(x)) = x \quad \text{and} \quad f^{-1}(f(x)) = x.$$

The domain of f must be equal to the range of f^{-1}, and the range of f must be equal to the domain of f^{-1}.

Do not be confused by the use of -1 to denote the inverse function f^{-1}. In this text, whenever f^{-1} is written, it *always* refers to the inverse function of the function f and *not* to the reciprocal of $f(x)$.

If the function g is the inverse function of the function f, then it must also be true that the function f is the inverse function of the function g. For this reason, you can say that the functions f and g are *inverse functions of each other.*

EXAMPLE 2   **Verifying Inverse Functions**

Which of the functions is the inverse function of $f(x) = \dfrac{5}{x - 2}$?

$$g(x) = \dfrac{x - 2}{5} \qquad\qquad h(x) = \dfrac{5}{x} + 2$$

Solution   By forming the composition of f with g, you have

$$f(g(x)) = f\left(\dfrac{x - 2}{5}\right) = \dfrac{5}{\left(\dfrac{x - 2}{5}\right) - 2} = \dfrac{25}{x - 12} \neq x.$$

Because this composition is not equal to the identity function x, it follows that g is *not* the inverse function of f. By forming the composition of f with h, you have

$$f(h(x)) = f\left(\dfrac{5}{x} + 2\right) = \dfrac{5}{\left(\dfrac{5}{x} + 2\right) - 2} = \dfrac{5}{\left(\dfrac{5}{x}\right)} = x.$$

So, it appears that h *is* the inverse function of f. Confirm this by showing that the composition of h with f is also equal to the identity function, as follows.

$$h(f(x)) = h\left(\dfrac{5}{x - 2}\right) = \dfrac{5}{\left(\dfrac{5}{x - 2}\right)} + 2 = x - 2 + 2 = x$$

✓ *Checkpoint* ◆))) *Audio-video solution in English & Spanish at LarsonPrecalculus.com.*

Which of the functions is the inverse function of $f(x) = \dfrac{x - 4}{7}$?

$$g(x) = 7x + 4 \qquad\qquad h(x) = \dfrac{7}{x - 4}$$

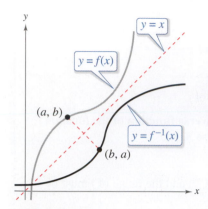

Figure 1.58

The Graph of an Inverse Function

The graphs of a function f and its inverse function f^{-1} are related to each other in the following way. If the point (a, b) lies on the graph of f, then the point (b, a) must lie on the graph of f^{-1}, and vice versa. This means that the graph of f^{-1} is a *reflection* of the graph of f in the line $y = x$, as shown in Figure 1.58.

EXAMPLE 3   **Verifying Inverse Functions Graphically**

Sketch the graphs of the inverse functions $f(x) = 2x - 3$ and $f^{-1}(x) = \frac{1}{2}(x + 3)$ on the same rectangular coordinate system and show that the graphs are reflections of each other in the line $y = x$.

Solution   The graphs of f and f^{-1} are shown in Figure 1.59. It appears that the graphs are reflections of each other in the line $y = x$. You can further verify this reflective property by testing a few points on each graph. Note in the following list that if the point (a, b) is on the graph of f, then the point (b, a) is on the graph of f^{-1}.

Graph of $f(x) = 2x - 3$	Graph of $f^{-1}(x) = \frac{1}{2}(x + 3)$
$(-1, -5)$	$(-5, -1)$
$(0, -3)$	$(-3, 0)$
$(1, -1)$	$(-1, 1)$
$(2, 1)$	$(1, 2)$
$(3, 3)$	$(3, 3)$

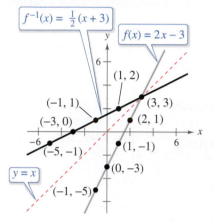

Figure 1.59

✓ *Checkpoint* 🔊)) *Audio-video solution in English & Spanish at LarsonPrecalculus.com.*

Sketch the graphs of the inverse functions $f(x) = 4x - 1$ and $f^{-1}(x) = \frac{1}{4}(x + 1)$ on the same rectangular coordinate system and show that the graphs are reflections of each other in the line $y = x$.

EXAMPLE 4   **Verifying Inverse Functions Graphically**

Sketch the graphs of the inverse functions $f(x) = x^2 \ (x \geq 0)$ and $f^{-1}(x) = \sqrt{x}$ on the same rectangular coordinate system and show that the graphs are reflections of each other in the line $y = x$.

Solution   The graphs of f and f^{-1} are shown in Figure 1.60. It appears that the graphs are reflections of each other in the line $y = x$. You can further verify this reflective property by testing a few points on each graph. Note in the following list that if the point (a, b) is on the graph of f, then the point (b, a) is on the graph of f^{-1}.

Graph of $f(x) = x^2, \quad x \geq 0$	Graph of $f^{-1}(x) = \sqrt{x}$
$(0, 0)$	$(0, 0)$
$(1, 1)$	$(1, 1)$
$(2, 4)$	$(4, 2)$
$(3, 9)$	$(9, 3)$

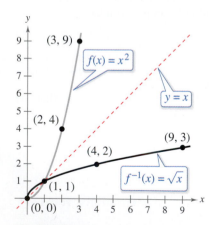

Figure 1.60

Try showing that $f(f^{-1}(x)) = x$ and $f^{-1}(f(x)) = x$.

✓ *Checkpoint* 🔊)) *Audio-video solution in English & Spanish at LarsonPrecalculus.com.*

Sketch the graphs of the inverse functions $f(x) = x^2 + 1 \ (x \geq 0)$ and $f^{-1}(x) = \sqrt{x - 1}$ on the same rectangular coordinate system and show that the graphs are reflections of each other in the line $y = x$.

One-to-One Functions

The reflective property of the graphs of inverse functions gives you a *geometric* test for determining whether a function has an inverse function. This test is called the **Horizontal Line Test** for inverse functions.

Horizontal Line Test for Inverse Functions

A function f has an inverse function if and only if no *horizontal* line intersects the graph of f at more than one point.

If no horizontal line intersects the graph of f at more than one point, then no y-value matches with more than one x-value. This is the essential characteristic of what are called **one-to-one functions.**

One-to-One Functions

A function f is **one-to-one** when each value of the dependent variable corresponds to exactly one value of the independent variable. A function f has an inverse function if and only if f is one-to-one.

Consider the function $f(x) = x^2$. The table on the left is a table of values for $f(x) = x^2$. The table on the right is the same as the table on the left but with the values in the columns interchanged. The table on the right does not represent a function because the input $x = 4$, for instance, matches with two different outputs: $y = -2$ and $y = 2$. So, $f(x) = x^2$ is not one-to-one and does not have an inverse function.

x	$f(x) = x^2$
-2	4
-1	1
0	0
1	1
2	4
3	9

x	y
4	-2
1	-1
0	0
1	1
4	2
9	3

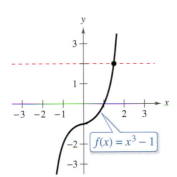

Figure 1.61

$f(x) = x^3 - 1$

Figure 1.62

$f(x) = x^2 - 1$

EXAMPLE 5   **Applying the Horizontal Line Test**

a. The graph of the function $f(x) = x^3 - 1$ is shown in Figure 1.61. Because no horizontal line intersects the graph of f at more than one point, f *is* a one-to-one function and *does* have an inverse function.

b. The graph of the function $f(x) = x^2 - 1$ is shown in Figure 1.62. Because it is possible to find a horizontal line that intersects the graph of f at more than one point, f *is not* a one-to-one function and *does not* have an inverse function.

✓ *Checkpoint* 🔊))) *Audio-video solution in English & Spanish at LarsonPrecalculus.com.*

Use the graph of f to determine whether the function has an inverse function.

a. $f(x) = \frac{1}{2}(3 - x)$

b. $f(x) = |x|$

Finding Inverse Functions Algebraically

For relatively simple functions (such as the one in Example 1), you can find inverse functions by inspection. For more complicated functions, however, it is best to use the following guidelines. The key step in these guidelines is Step 3—interchanging the roles of x and y. This step corresponds to the fact that inverse functions have ordered pairs with the coordinates reversed.

• **REMARK** Note what happens when you try to find the inverse function of a function that is not one-to-one.

$$f(x) = x^2 + 1 \qquad \text{Original function}$$

$$y = x^2 + 1 \qquad \text{Replace } f(x) \text{ with } y.$$

$$x = y^2 + 1 \qquad \text{Interchange } x \text{ and } y.$$

$$x - 1 = y^2 \qquad \text{Isolate } y\text{-term.}$$

$$y = \pm\sqrt{x - 1} \qquad \text{Solve for } y.$$

You obtain two y-values for each x.

Finding an Inverse Function

1. Use the Horizontal Line Test to decide whether f has an inverse function.

2. In the equation for $f(x)$, replace $f(x)$ with y.

3. Interchange the roles of x and y, and solve for y.

4. Replace y with $f^{-1}(x)$ in the new equation.

5. Verify that f and f^{-1} are inverse functions of each other by showing that the domain of f is equal to the range of f^{-1}, the range of f is equal to the domain of f^{-1}, and $f(f^{-1}(x)) = x$ and $f^{-1}(f(x)) = x$.

EXAMPLE 6 **Finding an Inverse Function Algebraically**

Find the inverse function of

$$f(x) = \frac{5 - x}{3x + 2}.$$

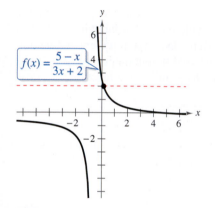

$f(x) = \dfrac{5 - x}{3x + 2}$

Figure 1.63

Solution The graph of f is shown in Figure 1.63. This graph passes the Horizontal Line Test. So, you know that f is one-to-one and has an inverse function.

$$f(x) = \frac{5 - x}{3x + 2} \qquad \text{Write original function.}$$

$$y = \frac{5 - x}{3x + 2} \qquad \text{Replace } f(x) \text{ with } y.$$

$$x = \frac{5 - y}{3y + 2} \qquad \text{Interchange } x \text{ and } y.$$

$$x(3y + 2) = 5 - y \qquad \text{Multiply each side by } 3y + 2.$$

$$3xy + 2x = 5 - y \qquad \text{Distributive Property}$$

$$3xy + y = 5 - 2x \qquad \text{Collect terms with } y.$$

$$y(3x + 1) = 5 - 2x \qquad \text{Factor.}$$

$$y = \frac{5 - 2x}{3x + 1} \qquad \text{Solve for } y.$$

$$f^{-1}(x) = \frac{5 - 2x}{3x + 1} \qquad \text{Replace } y \text{ with } f^{-1}(x).$$

Check that $f(f^{-1}(x)) = x$ and $f^{-1}(f(x)) = x$.

 Checkpoint  Audio-video solution in English & Spanish at LarsonPrecalculus.com.

Find the inverse function of

$$f(x) = \frac{5 - 3x}{x + 2}.$$

EXAMPLE 7 **Finding an Inverse Function Algebraically**

Find the inverse function of

$$f(x) = \sqrt{2x - 3}.$$

Solution The graph of f is a curve, as shown in the figure below. Because this graph passes the Horizontal Line Test, you know that f is one-to-one and has an inverse function.

$f(x) = \sqrt{2x - 3}$	Write original function.
$y = \sqrt{2x - 3}$	Replace $f(x)$ with y.
$x = \sqrt{2y - 3}$	Interchange x and y.
$x^2 = 2y - 3$	Square each side.
$2y = x^2 + 3$	Isolate y-term.
$y = \dfrac{x^2 + 3}{2}$	Solve for y.
$f^{-1}(x) = \dfrac{x^2 + 3}{2}, \quad x \geq 0$	Replace y with $f^{-1}(x)$.

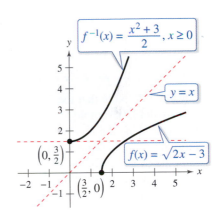

The graph of f^{-1} in the figure is the reflection of the graph of f in the line $y = x$. Note that the range of f is the interval $[0, \infty)$, which implies that the domain of f^{-1} is the interval $[0, \infty)$. Moreover, the domain of f is the interval $\left[\frac{3}{2}, \infty\right)$, which implies that the range of f^{-1} is the interval $\left[\frac{3}{2}, \infty\right)$. Verify that $f(f^{-1}(x)) = x$ and $f^{-1}(f(x)) = x$.

 Checkpoint *Audio-video solution in English & Spanish at LarsonPrecalculus.com.*

Find the inverse function of

$$f(x) = \sqrt[3]{10 + x}.$$

Summarize (Section 1.9)

1. State the definition of an inverse function *(page 57)*. For examples of finding inverse functions informally and verifying inverse functions, see Examples 1 and 2.

2. Explain how to use the graph of a function to determine whether the function has an inverse function *(page 58)*. For examples of verifying inverse functions graphically, see Examples 3 and 4.

3. Explain how to use the Horizontal Line Test to determine whether a function is one-to-one *(page 59)*. For an example of applying the Horizontal Line Test, see Example 5.

4. Explain how to find an inverse function algebraically *(page 60)*. For examples of finding inverse functions algebraically, see Examples 6 and 7.

1.10 Mathematical Modeling and Variation

- Use mathematical models to approximate sets of data points.
- Use the *regression* feature of a graphing utility to find the equation of a least squares regression line.
- Write mathematical models for direct variation, direct variation as an *n*th power, inverse variation, combined variation, and joint variation.

Introduction

In this section, you will study the following techniques for fitting models to data: *least squares regression* and *direct and inverse variation*.

You can use functions as models to represent a wide variety of real-life data sets. For instance, a variation model can be used to model the water temperatures of the ocean at various depths.

EXAMPLE 1 **A Mathematical Model**

The populations y (in millions) of the United States from 2003 through 2010 are shown in the table. *(Source: U.S. Census Bureau)*

Year, t	2003	2004	2005	2006	2007	2008	2009	2010
Population, y	290.1	292.8	295.5	298.4	301.2	304.1	306.8	308.7

Spreadsheet at LarsonPrecalculus.com

A linear model that approximates the data is

$$y = 2.72t + 282.0, \quad 3 \le t \le 10$$

where t represents the year, with $t = 3$ corresponding to 2003. Plot the actual data *and* the model on the same set of coordinate axes. How closely does the model represent the data?

Solution The actual data are plotted in Figure 1.64, along with the graph of the linear model. From the graph, it appears that the model is a "good fit" for the actual data. You can see how well the model fits by comparing the actual values of y with the values of y given by the model. The values given by the model are labeled $y*$ in the table below.

t	3	4	5	6	7	8	9	10
y	290.1	292.8	295.5	298.4	301.2	304.1	306.8	308.7
$y*$	290.2	292.9	295.6	298.3	301.0	303.8	306.5	309.2

✓ *Checkpoint* 🔊))) *Audio-video solution in English & Spanish at LarsonPrecalculus.com.*

The median sales prices y (in thousands of dollars) of new homes sold in a neighborhood from 2003 through 2010 are given by the following ordered pairs. *(Spreadsheet at LarsonPrecalculus.com)*

DATA
(2003, 179.4) (2005, 191.0) (2007, 202.6) (2009, 214.9)
(2004, 185.4) (2006, 196.7) (2008, 208.7) (2010, 221.4)

A linear model that approximates the data is

$$y = 5.96t + 161.3, \quad 3 \le t \le 10$$

where t represents the year, with $t = 3$ corresponding to 2003. Plot the actual data *and* the model on the same set of coordinate axes. How closely does the model represent the data?

• • • • • • • • • • • • • • • ▷

•• **REMARK** Note that the linear model in Example 1 was found using the *regression* feature of a graphing utility and is the line that *best* fits the data. This concept of a "best-fitting" line is discussed on the next page.

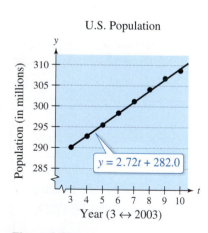

U.S. Population

$y = 2.72t + 282.0$

Figure 1.64

Least Squares Regression and Graphing Utilities

So far in this text, you have worked with many different types of mathematical models that approximate real-life data. In some instances the model was given (as in Example 1), whereas in other instances you were asked to find the model using simple algebraic techniques or a graphing utility.

To find a model that approximates a set of data most accurately, statisticians use a measure called the **sum of square differences,** which is the sum of the squares of the differences between actual data values and model values. The "best-fitting" linear model, called the **least squares regression line,** is the one with the least sum of square differences.

Recall that you can approximate this line visually by plotting the data points and drawing the line that appears to best fit the data—or you can enter the data points into a calculator or computer and use the *linear regression* feature of the calculator or computer.

When you use the *regression* feature of a graphing calculator or computer program, you will notice that the program may also output an "r-value." This r-value is the **correlation coefficient** of the data and gives a measure of how well the model fits the data. The closer the value of $|r|$ is to 1, the better the fit.

EXAMPLE 2 Finding a Least Squares Regression Line

The data in the table show the numbers E (in millions) of Medicare enrollees from 2003 through 2010. Construct a scatter plot that represents the data and find the least squares regression line for the data. *(Source: U.S. Centers for Medicare and Medicaid Services)*

Medicare Enrollees

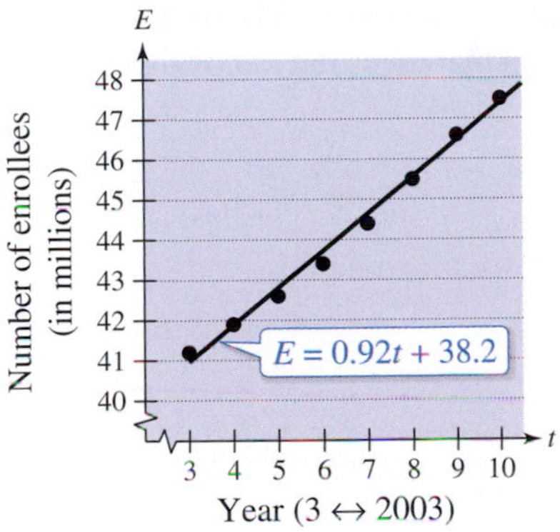

Figure 1.65

DATA

Spreadsheet at LarsonPrecalculus.com

Year	Medicare Enrollees, E
2003	41.2
2004	41.9
2005	42.6
2006	43.4
2007	44.4
2008	45.5
2009	46.6
2010	47.5

Solution Let $t = 3$ represent 2003. The scatter plot for the data is shown in Figure 1.65. Using the *regression* feature of a graphing utility, you can determine that the equation of the least squares regression line is

$$E = 0.92t + 38.2$$

To check this model, compare the actual E-values with the E-values given by the model, which are labeled $E*$ in the table at the left. The correlation coefficient for this model is $r \approx 0.996$, which implies that the model is a good fit.

t	E	$E*$
3	41.2	41.0
4	41.9	41.9
5	42.6	42.8
6	43.4	43.7
7	44.4	44.6
8	45.5	45.6
9	46.6	46.5
10	47.5	47.4

✓ **Checkpoint** *Audio-video solution in English & Spanish at LarsonPrecalculus.com.*

The total Medicare disbursements (in billions of dollars) are given by the following ordered pairs. *(Spreadsheet at LarsonPrecalculus.com)* Construct a scatter plot that represents the data and find the least squares regression line for the data. *(Source: U.S. Centers for Medicare and Medicaid Services)*

DATA

(2003, 277.8) (2005, 336.9) (2007, 434.8) (2009, 498.2)
(2004, 301.5) (2006, 380.4) (2008, 455.1) (2010, 521.1)

Direct Variation

There are two basic types of linear models. The more general model has a y-intercept that is nonzero.

$$y = mx + b, \quad b \neq 0$$

The simpler model

$$y = kx$$

has a y-intercept that is zero. In the simpler model, y is said to **vary directly** as x, or to be **directly proportional** to x.

Direct Variation

The following statements are equivalent.

1. y **varies directly** as x.
2. y is **directly proportional** to x.
3. $y = kx$ for some nonzero constant k.

k is the **constant of variation** or the **constant of proportionality**.

EXAMPLE 3 **Direct Variation**

In Pennsylvania, the state income tax is directly proportional to *gross income*. You are working in Pennsylvania and your state income tax deduction is $46.05 for a gross monthly income of $1500. Find a mathematical model that gives the Pennsylvania state income tax in terms of gross income.

Solution

Verbal Model: $\boxed{\text{State income tax}} = \boxed{k} \cdot \boxed{\text{Gross income}}$

Labels:
State income tax = y (dollars)
Gross income = x (dollars)
Income tax rate = k (percent in decimal form)

Equation: $y = kx$

To solve for k, substitute the given information into the equation $y = kx$, and then solve for k.

$y = kx$	Write direct variation model.
$46.05 = k(1500)$	Substitute 46.05 for y and 1500 for x.
$0.0307 = k$	Simplify.

So, the equation (or model) for state income tax in Pennsylvania is

$$y = 0.0307x.$$

In other words, Pennsylvania has a state income tax rate of 3.07% of gross income. The graph of this equation is shown in Figure 1.66.

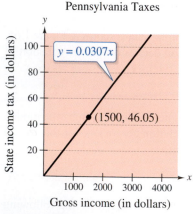

Pennsylvania Taxes

$y = 0.0307x$

(1500, 46.05)

State income tax (in dollars)

Gross income (in dollars)

Figure 1.66

✓ *Checkpoint*  *Audio-video solution in English & Spanish at LarsonPrecalculus.com.*

The simple interest on an investment is directly proportional to the amount of the investment. For instance, an investment of $2500 will earn $187.50 after 1 year. Find a mathematical model that gives the interest I after 1 year in terms of the amount invested P.

Direct Variation as an *n*th Power

Another type of direct variation relates one variable to a *power* of another variable. For example, in the formula for the area of a circle

$$A = \pi r^2$$

the area *A* is directly proportional to the square of the radius *r*. Note that for this formula, π is the constant of proportionality.

· · · · · · · · · · · · · · · ▷
· · REMARK   Note that the direct variation model $y = kx$ is a special case of $y = kx^n$ with $n = 1$.

Direct Variation as an *n*th Power

The following statements are equivalent.

1. *y* **varies directly as the *n*th power** of *x*.

2. *y* is **directly proportional to the *n*th power** of *x*.

3. $y = kx^n$ for some nonzero constant *k*.

EXAMPLE 4   **Direct Variation as an *n*th Power**

The distance a ball rolls down an inclined plane is directly proportional to the square of the time it rolls. During the first second, the ball rolls 8 feet. (See Figure 1.67.)

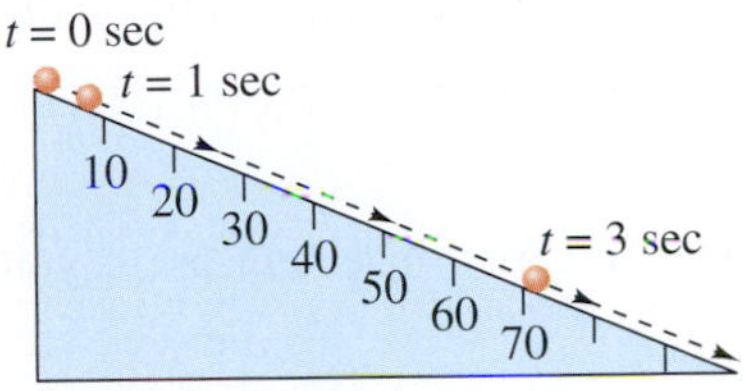

a. Write an equation relating the distance traveled to the time.

b. How far will the ball roll during the first 3 seconds?

Solution

a. Letting *d* be the distance (in feet) the ball rolls and letting *t* be the time (in seconds), you have

$$d = kt^2.$$

Now, because $d = 8$ when $t = 1$, you can see that $k = 8$, as follows.

$d = kt^2$	Write direct variation model.
$8 = k(1)^2$	Substitute 8 for *d* and 1 for *t*.
$8 = k$	Simplify.

So, the equation relating distance to time is

$$d = 8t^2.$$

b. When $t = 3$, the distance traveled is

$d = 8(3)^2$	Substitute 3 for *t*.
$= 8(9)$	Simplify.
$= 72$ feet.	Simplify.

So, the ball will roll 72 feet during the first 3 seconds.

✓ **Checkpoint** ◀))) *Audio-video solution in English & Spanish at LarsonPrecalculus.com.*

Neglecting air resistance, the distance *s* an object falls varies directly as the square of the duration *t* of the fall. An object falls a distance of 144 feet in 3 seconds. How far will it fall in 6 seconds? ■

t = 0 sec
t = 1 sec
10 20 30 40 50 60 70
t = 3 sec

Figure 1.67

In Examples 3 and 4, the direct variations are such that an *increase* in one variable corresponds to an *increase* in the other variable. This is also true in the model $d = \frac{1}{5}F$, $F > 0$, where an increase in *F* results in an increase in *d*. You should not, however, assume that this always occurs with direct variation. For example, in the model $y = -3x$, an increase in *x* results in a *decrease* in *y*, and yet *y* is said to vary directly as *x*.

Inverse Variation

> ### Inverse Variation
>
> The following statements are equivalent.
>
> 1. y **varies inversely** as x.
>
> 2. y is **inversely proportional** to x.
>
> 3. $y = \dfrac{k}{x}$ for some nonzero constant k.

If x and y are related by an equation of the form $y = k/x^n$, then y varies inversely as the nth power of x (or y is inversely proportional to the nth power of x).

EXAMPLE 5 **Inverse Variation**

A company has found that the demand for one of its products varies inversely as the price of the product. When the price is \$6.25, the demand is 400 units. Approximate the demand when the price is \$5.75.

Solution

Let p be the price and let x be the demand. Because x varies inversely as p, you have

$$x = \frac{k}{p}.$$

Now, because $x = 400$ when $p = 6.25$, you have

$$x = \frac{k}{p} \qquad \text{Write inverse variation model.}$$

$$400 = \frac{k}{6.25} \qquad \text{Substitute 400 for } x \text{ and 6.25 for } p.$$

$$(400)(6.25) = k \qquad \text{Multiply each side by 6.25.}$$

$$2500 = k. \qquad \text{Simplify.}$$

So, the equation relating price and demand is

$$x = \frac{2500}{p}.$$

When $p = 5.75$, the demand is

$$x = \frac{2500}{p} \qquad \text{Write inverse variation model.}$$

$$= \frac{2500}{5.75} \qquad \text{Substitute 5.75 for } p.$$

$$\approx 435 \text{ units.} \qquad \text{Simplify.}$$

So, the demand for the product is 435 units when the price of the product is \$5.75.

According to the National Association of Manufacturers, "manufacturing supports an estimated 17 million jobs in the U.S.—about one in six private sector jobs."

✓ **Checkpoint** *Audio-video solution in English & Spanish at LarsonPrecalculus.com.*

The company in Example 5 has found that the demand for another of its products varies inversely as the price of the product. When the price is \$2.75, the demand is 600 units. Approximate the demand when the price is \$3.25.

Combined Variation

Some applications of variation involve problems with *both* direct and inverse variations in the same model. These types of models are said to have **combined variation.**

EXAMPLE 6   **Combined Variation**

A gas law states that the volume of an enclosed gas varies directly as the temperature *and* inversely as the pressure, as shown in Figure 1.68. The pressure of a gas is 0.75 kilogram per square centimeter when the temperature is 294 K and the volume is 8000 cubic centimeters.

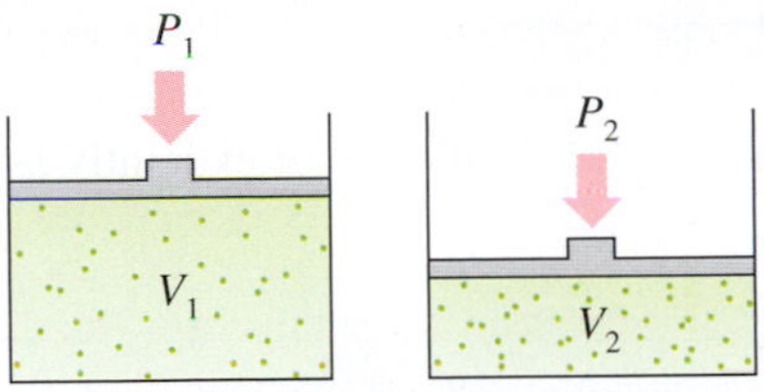

a. Write an equation relating pressure, temperature, and volume.

b. Find the pressure when the temperature is 300 K and the volume is 7000 cubic centimeters.

Solution

a. Let V be volume (in cubic centimeters), let P be pressure (in kilograms per square centimeter), and let T be temperature (in Kelvin). Because V varies directly as T and inversely as P, you have

$$V = \frac{kT}{P}.$$

Now, because $P = 0.75$ when $T = 294$ and $V = 8000$, you have

$$V = \frac{kT}{P} \qquad \text{Write combined variation model.}$$

$$8000 = \frac{k(294)}{0.75} \qquad \text{Substitute 8000 for } V, \text{ 294 for } T, \text{ and 0.75 for } P.$$

$$\frac{6000}{294} = k \qquad \text{Simplify.}$$

$$\frac{1000}{49} = k. \qquad \text{Simplify.}$$

So, the equation relating pressure, temperature, and volume is

$$V = \frac{1000}{49}\left(\frac{T}{P}\right).$$

b. Isolate P on one side of the equation by multiplying each side by P and dividing each side by V to obtain $P = \dfrac{1000}{49}\left(\dfrac{T}{V}\right)$. When $T = 300$ and $V = 7000$, the pressure is

$$P = \frac{1000}{49}\left(\frac{T}{V}\right) \qquad \text{Combined variation model solved for } P.$$

$$= \frac{1000}{49}\left(\frac{300}{7000}\right) \qquad \text{Substitute 300 for } T \text{ and 7000 for } V.$$

$$= \frac{300}{343} \qquad \text{Simplify.}$$

$$\approx 0.87 \text{ kilogram per square centimeter.} \qquad \text{Simplify.}$$

So, the pressure is about 0.87 kilogram per square centimeter when the temperature is 300 K and the volume is 7000 cubic centimeters.

✓ *Checkpoint*   *Audio-video solution in English & Spanish at LarsonPrecalculus.com.*

The resistance of a copper wire carrying an electrical current is directly proportional to its length and inversely proportional to its cross-sectional area. Given that #28 copper wire (which has a diameter of 0.0126 inch) has a resistance of 66.17 ohms per thousand feet, what length of #28 copper wire will produce a resistance of 33.5 ohms?

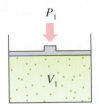

If $P_2 > P_1$, then $V_2 < V_1$.

If the temperature is held constant and pressure increases, then the volume decreases.

Figure 1.68

Joint Variation

> **Joint Variation**
>
> The following statements are equivalent.
>
> **1.** z **varies jointly** as x and y.   **2.** z is **jointly proportional** to x and y.
>
> **3.** $z = kxy$ for some nonzero constant k.

If x, y, and z are related by an equation of the form $z = kx^n y^m$, then z varies jointly as the nth power of x and the mth power of y.

EXAMPLE 7   **Joint Variation**

The *simple* interest for a certain savings account is jointly proportional to the time and the principal. After one quarter (3 months), the interest on a principal of $5000 is $43.75. (a) Write an equation relating the interest, principal, and time. (b) Find the interest after three quarters.

Solution

a. Let I = interest (in dollars), P = principal (in dollars), and t = time (in years). Because I is jointly proportional to P and t, you have

$$I = kPt.$$

For $I = 43.75$, $P = 5000$, and $t = \frac{1}{4}$, you have

$$43.75 = k(5000)\left(\frac{1}{4}\right)$$

which implies that $k = 4(43.75)/5000 = 0.035$. So, the equation relating interest, principal, and time is

$$I = 0.035Pt$$

which is the familiar equation for simple interest where the constant of proportionality, 0.035, represents an annual interest rate of 3.5%.

b. When $P = \$5000$ and $t = \frac{3}{4}$, the interest is

$$I = (0.035)(5000)\left(\frac{3}{4}\right) = \$131.25.$$

✓ ***Checkpoint*** ◀))) *Audio-video solution in English & Spanish at LarsonPrecalculus.com.*

The kinetic energy E of an object varies jointly with the object's mass m and the square of the object's velocity v. An object with a mass of 50 kilograms traveling at 16 meters per second has a kinetic energy of 6400 joules. What is the kinetic energy of an object with mass of 70 kilograms traveling at 20 meters per second?

Summarize   (Section 1.10)

1. Give an example of how you can use a mathematical model to approximate a set of data points *(page 62, Example 1)*.

2. Describe how you can use the *regression* feature of a graphing utility to find the equation of a least squares regression line *(page 63, Example 2)*.

3. Describe direct variation, direct variation as an nth power, inverse variation, combined variation, and joint variation *(pages 64–68, Examples 3–7)*.

Chapter Summary

	What Did You Learn?	Explanation/Examples	Review Exercises
Section 1.1	Plot points in the Cartesian plane (p. 2), use the Distance Formula (p. 4) and the Midpoint Formula (p. 5), and use a coordinate plane to model and solve real-life problems (p. 6).	For an ordered pair (x, y), the x-coordinate is the directed distance from the y-axis to the point, and the y-coordinate is the directed distance from the x-axis to the point. The coordinate plane can help you find the length of a football pass. (See Example 6.)	1–6
Section 1.2	Sketch graphs of equations (p. 8), identify intercepts (p. 11), and use symmetry to sketch graphs of equations (p. 12).	To find x-intercepts, let y be zero and solve for x. To find y-intercepts, let x be zero and solve for y. Graphs can have symmetry with respect to one of the coordinate axes or with respect to the origin.	7–22
	Write equations of and sketch graphs of circles (p. 14).	A point (x, y) lies on the circle of radius r and center (h, k) if and only if $(x - h)^2 + (y - k)^2 = r^2$.	23–27
	Use graphs of equations in solving real-life problems (p. 15).	The graph of an equation can help you estimate the recommended weight for a man. (See Example 9.)	28
Section 1.3	Use slope to graph linear equations in two variables (p. 16).	The graph of the equation $y = mx + b$ is a line whose slope is m and whose y-intercept is $(0, b)$.	29–32
	Find the slope of a line given two points on the line (p. 18).	The slope m of the nonvertical line through (x_1, y_1) and (x_2, y_2) is $m = (y_2 - y_1)/(x_2 - x_1)$, where $x_1 \neq x_2$.	33, 34
	Write linear equations in two variables (p. 20), and identify parallel and perpendicular lines (p. 21).	The equation of the line with slope m passing through the point (x_1, y_1) is $y - y_1 = m(x - x_1)$. **Parallel lines:** $m_1 = m_2$ **Perpendicular lines:** $m_1 = -1/m_2$	35–40
	Use slope and linear equations in two variables to model and solve real-life problems (p. 22).	A linear equation in two variables can help you describe the book value of exercise equipment in a given year. (See Example 7.)	41, 42
Section 1.4	Determine whether relations between two variables are functions and use function notation (p. 25), and find the domains of functions (p. 30).	A function f from a set A (domain) to a set B (range) is a relation that assigns to each element x in the set A exactly one element y in the set B. **Equation:** $f(x) = 5 - x^2$ **f(2):** $f(2) = 5 - 2^2 = 1$ **Domain of $f(x) = 5 - x^2$:** All real numbers	43–50
	Use functions to model and solve real-life problems (p. 31).	A function can model the number of alternative-fueled vehicles in the United States. (See Example 10.)	51, 52
	Evaluate difference quotients (p. 32).	Difference quotient: $[f(x + h) - f(x)]/h, \ h \neq 0$	53, 54
Section 1.5	Use the Vertical Line Test for functions (p. 35).	A set of points in a coordinate plane is the graph of y as a function of x if and only if no *vertical* line intersects the graph at more than one point.	55, 56
	Find the zeros of functions (p. 36).	**Zeros of $f(x)$:** x-values for which $f(x) = 0$	57–60

	What Did You Learn?	**Explanation/Examples**	**Review Exercises**
Section 1.5	Determine intervals on which functions are increasing or decreasing *(p. 37)*, relative minimum and maximum values of functions *(p. 38)*, and the average rates of change of functions *(p. 39)*.	To determine whether a function is increasing, decreasing, or constant on an interval, evaluate the function for several values of x. The points at which the behavior of a function changes can help determine a relative minimum or relative maximum. The average rate of change between any two points is the slope of the line (secant line) through the two points.	61–66
	Identify even and odd functions *(p. 40)*.	**Even:** For each x in the domain of f, $f(-x) = f(x)$. **Odd:** For each x in the domain of f, $f(-x) = -f(x)$.	67, 68
Section 1.6	Identify and graph different types of functions *(p. 41)*, and recognize graphs of parent functions *(p. 45)*.	**Linear:** $f(x) = ax + b$; **Squaring:** $f(x) = x^2$; **Cubic:** $f(x) = x^3$; **Square Root:** $f(x) = \sqrt{x}$; **Reciprocal:** $f(x) = 1/x$ Eight of the most commonly used functions in algebra are shown on page 45.	69–74
Section 1.7	Use vertical and horizontal shifts *(p. 46)*, reflections *(p. 48)*, and nonrigid transformations *(p. 50)* to sketch graphs of functions.	**Vertical shifts:** $h(x) = f(x) + c$ or $h(x) = f(x) - c$ **Horizontal shifts:** $h(x) = f(x - c)$ or $h(x) = f(x + c)$ **Reflection in x-axis:** $h(x) = -f(x)$ **Reflection in y-axis:** $h(x) = f(-x)$ **Nonrigid transformations:** $h(x) = cf(x)$ or $h(x) = f(cx)$	75–84
Section 1.8	Add, subtract, multiply, and divide functions *(p. 51)*, find compositions of functions *(p. 53)*, and use combinations and compositions of functions to model and solve real-life problems *(p. 55)*.	$(f + g)(x) = f(x) + g(x) \qquad (f - g)(x) = f(x) - g(x)$ $(fg)(x) = f(x) \cdot g(x) \qquad (f/g)(x) = f(x)/g(x),\ g(x) \neq 0$ The composition of the function f with the function g is $(f \circ g)(x) = f(g(x))$. A composite function can represent the number of bacteria in food as a function of the amount of time the food has been out of refrigeration. (See Example 8.)	85–90
Section 1.9	Find inverse functions informally and verify that two functions are inverse functions of each other *(p. 56)*.	Let f and g be two functions such that $f(g(x)) = x$ for every x in the domain of g and $g(f(x)) = x$ for every x in the domain of f. Under these conditions, the function g is the inverse function of the function f.	91, 92
	Use graphs to find inverse functions *(p. 58)*, use the Horizontal Line Test *(p. 59)*, and find inverse functions algebraically *(p. 60)*.	If the point (a, b) lies on the graph of f, then the point (b, a) must lie on the graph of f^{-1}, and vice versa. In short, the graph of f^{-1} is a reflection of the graph of f in the line $y = x$. To find an inverse function, replace $f(x)$ with y, interchange the roles of x and y, solve for y, and then replace y with $f^{-1}(x)$.	93–98
Section 1.10	Use mathematical models to approximate sets of data points *(p. 62)*, and use the *regression* feature of a graphing utility to find the equation of a least squares regression line *(p. 63)*.	To see how well a model fits a set of data, compare the actual values and model values of y. (See Example 1.) The sum of square differences is the sum of the squares of the differences between actual data values and model values. The least squares regression line is the linear model with the least sum of square differences.	99
	Write mathematical models for direct variation, direct variation as an *n*th power, inverse variation, combined variation, and joint variation *(pp. 64–68)*.	**Direct variation:** $y = kx$ for some nonzero constant k. **Direct variation as an *n*th power:** $y = kx^n$ for some nonzero constant k. **Inverse variation:** $y = k/x$ for some nonzero constant k. **Joint variation:** $z = kxy$ for some nonzero constant k.	100, 101

1.1 **Plotting Points in the Cartesian Plane** In Exercises 1 and 2, plot the points in the Cartesian plane.

1. $(5, 5), (-2, 0), (-3, 6), (-1, -7)$

2. $(0, 6), (8, 1), (5, -4), (-3, -3)$

Determining Quadrant(s) for a Point In Exercises 3 and 4, determine the quadrant(s) in which (x, y) is located so that the condition(s) is (are) satisfied.

3. $x > 0$ and $y = -2$ **4.** $xy = 4$

5. Plotting, Distance, and Midpoint Plot the points $(-2, 6)$ and $(4, -3)$. Then find the distance between the points and the midpoint of the line segment joining the points.

6. Sales Barnes & Noble had annual sales of $5.1 billion in 2008 and $7.0 billion in 2010. Use the Midpoint Formula to estimate the sales in 2009. Assume that the annual sales followed a linear pattern. (*Source: Barnes & Noble, Inc.*)

1.2 **Sketching the Graph of an Equation** In Exercises 7–10, construct a table of values. Use the resulting solution points to sketch the graph of the equation.

7. $y = 3x - 5$ **8.** $y = -\frac{1}{2}x + 2$

9. $y = x^2 - 3x$ **10.** $y = 2x^2 - x - 9$

Finding x- and y-Intercepts In Exercises 11–14, find the x- and y-intercepts of the graph of the equation.

11. $y = 2x + 7$ **12.** $y = |x + 1| - 3$

13. $y = (x - 3)^2 - 4$ **14.** $y = x\sqrt{4 - x^2}$

Intercepts, Symmetry, and Graphing In Exercises 15–22, identify any intercepts and test for symmetry. Then sketch the graph of the equation.

15. $y = -4x + 1$ **16.** $y = 5x - 6$

17. $y = 5 - x^2$ **18.** $y = x^2 - 10$

19. $y = x^3 + 3$ **20.** $y = -6 - x^3$

21. $y = \sqrt{x + 5}$ **22.** $y = |x| + 9$

Sketching the Graph of a Circle In Exercises 23–26, find the center and radius of the circle. Then sketch the graph of the circle.

23. $x^2 + y^2 = 9$ **24.** $x^2 + y^2 = 4$

25. $(x + 2)^2 + y^2 = 16$ **26.** $x^2 + (y - 8)^2 = 81$

27. Writing the Equation of a Circle Write the standard form of the equation of the circle for which the endpoints of a diameter are $(0, 0)$ and $(4, -6)$.

28. Physics The force F (in pounds) required to stretch a spring x inches from its natural length (see figure) is

$$F = \frac{5}{4}x, \ 0 \le x \le 20.$$

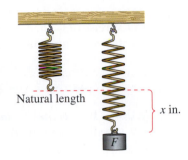

Natural length x in.

(a) Use the model to complete the table.

x	0	4	8	12	16	20
Force, F						

(b) Sketch a graph of the model.

(c) Use the graph to estimate the force necessary to stretch the spring 10 inches.

1.3 **Graphing a Linear Equation** In Exercises 29–32, find the slope and y-intercept (if possible) of the equation of the line. Sketch the line.

29. $y = 3x + 13$ **30.** $y = -10x + 9$

31. $y = 6$ **32.** $x = -3$

Finding the Slope of a Line Through Two Points In Exercises 33 and 34, plot the points and find the slope of the line passing through the pair of points.

33. $(6, 4), (-3, -4)$ **34.** $(-3, 2), (8, 2)$

Finding an Equation of a Line In Exercises 35 and 36, find an equation of the line that passes through the given point and has the indicated slope m. Sketch the line.

35. $(10, -3), \ m = -\frac{1}{2}$ **36.** $(-8, 5), \ m = 0$

Finding an Equation of a Line In Exercises 37 and 38, find an equation of the line passing through the points.

37. $(-1, 0), (6, 2)$ **38.** $(11, -2), (6, -1)$

Finding Parallel and Perpendicular Lines In Exercises 39 and 40, write the slope-intercept form of the equations of the lines through the given point (a) parallel to and (b) perpendicular to the given line.

39. $5x - 4y = 8, \ (3, -2)$ **40.** $2x + 3y = 5, \ (-8, 3)$

41. Sales A discount outlet is offering a 20% discount on all items. Write a linear equation giving the sale price S for an item with a list price L.

42. Hourly Wage A microchip manufacturer pays its assembly line workers $12.25 per hour. In addition, workers receive a piecework rate of $0.75 per unit produced. Write a linear equation for the hourly wage W in terms of the number of units x produced per hour.

1.4 **Testing for Functions Represented Algebraically** In Exercises 43–46, determine whether the equation represents y as a function of x.

43. $16x - y^4 = 0$ **44.** $2x - y - 3 = 0$

45. $y = \sqrt{1 - x}$ **46.** $|y| = x + 2$

Evaluating a Function In Exercises 47 and 48, evaluate the function at each specified value of the independent variable and simplify.

47. $f(x) = x^2 + 1$

   (a) $f(2)$ (b) $f(-4)$ (c) $f(t^2)$ (d) $f(t + 1)$

48. $h(x) = \begin{cases} 2x + 1, & x \le -1 \\ x^2 + 2, & x > -1 \end{cases}$

   (a) $h(-2)$ (b) $h(-1)$ (c) $h(0)$ (d) $h(2)$

Finding the Domain of a Function In Exercises 49 and 50, find the domain of the function. Verify your result with a graph.

49. $f(x) = \sqrt{25 - x^2}$ **50.** $h(x) = \dfrac{x}{x^2 - x - 6}$

Physics In Exercises 51 and 52, the velocity of a ball projected upward from ground level is given by $v(t) = -32t + 48$, where t is the time in seconds and v is the velocity in feet per second.

51. Find the velocity when $t = 1$.

52. Find the time when the ball reaches its maximum height. [*Hint:* Find the time when $v(t) = 0$.]

𝆑 Evaluating a Difference Quotient In Exercises 53 and 54, find the difference quotient and simplify your answer.

53. $f(x) = 2x^2 + 3x - 1$, $\dfrac{f(x + h) - f(x)}{h}$, $h \ne 0$

54. $f(x) = x^3 - 5x^2 + x$, $\dfrac{f(x + h) - f(x)}{h}$, $h \ne 0$

1.5 **Vertical Line Test for Functions** In Exercises 55 and 56, use the Vertical Line Test to determine whether y is a function of x. To print an enlarged copy of the graph, go to *MathGraphs.com*.

55. $y = (x - 3)^2$ **56.** $x = -|4 - y|$

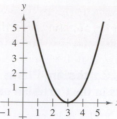

Figure for 55   Figure for 56

Finding the Zeros of a Function In Exercises 57–60, find the zeros of the function algebraically.

57. $f(x) = 3x^2 - 16x + 21$ **58.** $f(x) = 5x^2 + 4x - 1$

59. $f(x) = \dfrac{8x + 3}{11 - x}$ **60.** $f(x) = x^3 - x^2$

Describing Function Behavior In Exercises 61 and 62, use a graphing utility to graph the function and visually determine the intervals on which the function is increasing, decreasing, or constant.

61. $f(x) = |x| + |x + 1|$ **62.** $f(x) = (x^2 - 4)^2$

Approximating Relative Minima or Maxima In Exercises 63 and 64, use a graphing utility to graph the function and approximate (to two decimal places) any relative minima or maxima.

63. $f(x) = -x^2 + 2x + 1$ **64.** $f(x) = x^3 - 4x^2 - 1$

𝆑 Average Rate of Change of a Function In Exercises 65 and 66, find the average rate of change of the function from x_1 to x_2.

65. $f(x) = -x^2 + 8x - 4$, $x_1 = 0, x_2 = 4$

66. $f(x) = 2 - \sqrt{x + 1}$, $x_1 = 3, x_2 = 7$

Even, Odd, or Neither? In Exercises 67–68, determine whether the function is even, odd, or neither. Then describe the symmetry.

67. $f(x) = x^4 - 20x^2$ **68.** $f(x) = 2x\sqrt{x^2 + 3}$

1.6 **Writing a Linear Function** In Exercises 69 and 70, (a) write the linear function f such that it has the indicated function values, and (b) sketch the graph of the function.

69. $f(2) = -6$, $f(-1) = 3$

70. $f(0) = -5$, $f(4) = -8$

Graphing a Function In Exercises 71–74, sketch the graph of the function.

71. $f(x) = 3 - x^2$ **72.** $f(x) = \sqrt{x + 1}$

73. $g(x) = \dfrac{1}{x + 5}$

74. $f(x) = \begin{cases} 5x - 3, & x \ge -1 \\ -4x + 5, & x < -1 \end{cases}$

1.7 **Identifying a Parent Function** In Exercises 75–84, h is related to one of the parent functions described in this chapter. (a) Identify the parent function f. (b) Describe the sequence of transformations from f to h. (c) Sketch the graph of h. (d) Use function notation to write h in terms of f.

75. $h(x) = x^2 - 9$
76. $h(x) = (x - 2)^3 + 2$
77. $h(x) = -\sqrt{x} + 4$
78. $h(x) = |x + 3| - 5$
79. $h(x) = -(x + 2)^2 + 3$
80. $h(x) = \frac{1}{2}(x - 1)^2 - 2$
81. $h(x) = -[\![x]\!] + 6$
82. $h(x) = -\sqrt{x + 1} + 9$
83. $h(x) = 5[\![x - 9]\!]$
84. $h(x) = -\frac{1}{3}x^3$

1.8 **Finding Arithmetic Combinations of Functions** In Exercises 85 and 86, find (a) $(f + g)(x)$, (b) $(f - g)(x)$, (c) $(fg)(x)$ and (d) $(f/g)(x)$. What is the domain of f/g?

85. $f(x) = x^2 + 3$, $g(x) = 2x - 1$
86. $f(x) = x^2 - 4$, $g(x) = \sqrt{3 - x}$

Finding Domains of Functions and Composite Functions In Exercises 87 and 88, find (a) $f \circ g$ and (b) $g \circ f$. Find the domain of each function and each composite function.

87. $f(x) = \frac{1}{3}x - 3$, $g(x) = 3x + 1$
88. $f(x) = x^3 - 4$, $g(x) = \sqrt[3]{x + 7}$

Bacteria Count In Exercises 89 and 90, the number N of bacteria in a refrigerated food is given by $N(T) = 25T^2 - 50T + 300$, $1 \le T \le 19$, where T is the temperature of the food in degrees Celsius. When the food is removed from refrigeration, the temperature of the food is given by $T(t) = 2t + 1$, $0 \le t \le 9$, where t is the time in hours.

89. Find the composition $(N \circ T)(t)$ and interpret its meaning in context.

90. Find the time when the bacteria count reaches 750.

1.9 **Finding an Inverse Function Informally** In Exercises 91 and 92, find the inverse function of f informally. Verify that $f(f^{-1}(x)) = x$ and $f^{-1}(f(x)) = x$.

91. $f(x) = 3x + 8$

92. $f(x) = \dfrac{x - 4}{5}$

Applying the Horizontal Line Test In Exercises 93 and 94, use a graphing utility to graph the function, and use the Horizontal Line Test to determine whether the function has an inverse function.

93. $f(x) = (x - 1)^2$

94. $h(t) = \dfrac{2}{t - 3}$

Finding and Analyzing Inverse Functions In Exercises 95 and 96, (a) find the inverse function of f, (b) graph both f and f^{-1} on the same set of coordinate axes, (c) describe the relationship between the graphs of f and f^{-1}, and (d) state the domains and ranges of f and f^{-1}.

95. $f(x) = \frac{1}{2}x - 3$
96. $f(x) = \sqrt{x + 1}$

Restricting the Domain In Exercises 97 and 98, restrict the domain of the function f to an interval on which the function is increasing, and determine f^{-1} on that interval.

97. $f(x) = 2(x - 4)^2$
98. $f(x) = |x - 2|$

1.10

99. **Compact Discs** The values V (in billions of dollars) of shipments of compact discs in the United States from 2004 through 2010 are given by the following ordered pairs. (*Spreadsheet at LarsonPrecalculus.com*) (*Source: Recording Industry Association of America*)

DATA
(2004, 11.45) (2005, 10.52) (2006, 9.37)
(2007, 7.45)   (2008, 5.47)   (2009, 4.27)
(2010, 3.36)

(a) Use a graphing utility to create a scatter plot of the data. Let t represent the year, with $t = 4$ corresponding to 2004.

(b) Use the *regression* feature of the graphing utility to find the equation of the least squares regression line that fits the data. Then graph the model and the scatter plot you found in part (a) in the same viewing window. How closely does the model represent the data?

100. **Travel Time** The travel time between two cities is inversely proportional to the average speed. A train travels between the cities in 3 hours at an average speed of 65 miles per hour. How long would it take to travel between the cities at an average speed of 80 miles per hour?

101. **Cost** The cost of constructing a wooden box with a square base varies jointly as the height of the box and the square of the width of the box. A box of height 16 inches and width 6 inches costs $28.80. How much would a box of height 14 inches and width 8 inches cost?

Exploration

True or False? In Exercises 102 and 103, determine whether the statement is true or false. Justify your answer.

102. Relative to the graph of $f(x) = \sqrt{x}$, the function $h(x) = -\sqrt{x + 9} - 13$ is shifted 9 units to the left and 13 units down, then reflected in the x-axis.

103. If f and g are two inverse functions, then the domain of g is equal to the range of f.

Chapter Test

See **CalcChat.com** for tutorial help and worked-out solutions to odd-numbered exercises.

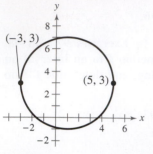

Figure for 6

Take this test as you would take a test in class. When you are finished, check your work against the answers given in the back of the book.

1. Plot the points $(-2, 5)$ and $(6, 0)$. Then find the distance between the points and the midpoint of the line segment joining the points.

2. A cylindrical can has a radius of 4 centimeters. Write the volume V of the can as a function of the height h.

In Exercises 3–5, use intercepts and symmetry to sketch the graph of the equation.

3. $y = 3 - 5x$

4. $y = 4 - |x|$

5. $y = x^2 - 1$

6. Write the standard form of the equation of the circle shown at the left.

In Exercises 7 and 8, find an equation of the line passing through the points.

7. $(2, -3), (-4, 9)$

8. $(3, 0.8), (7, -6)$

9. Write equations of the lines that pass through the point $(0, 4)$ and are (a) parallel to and (b) perpendicular to the line $5x + 2y = 3$.

10. Evaluate $f(x) = \dfrac{\sqrt{x + 9}}{x^2 - 81}$ at each value and simplify: (a) $f(7)$ (b) $f(-5)$ (c) $f(x - 9)$.

11. Find the domain of $f(x) = 10 - \sqrt{3 - x}$.

In Exercises 12–14, (a) find the zeros of the function, (b) use a graphing utility to graph the function, (c) approximate the intervals on which the function is increasing, decreasing, or constant, and (d) determine whether the function is even, odd, or neither.

12. $f(x) = 2x^6 + 5x^4 - x^2$

13. $f(x) = 4x\sqrt{3 - x}$

14. $f(x) = |x + 5|$

15. Sketch the graph of $f(x) = \begin{cases} 3x + 7, & x \le -3 \\ 4x^2 - 1, & x > -3 \end{cases}$.

In Exercises 16–18, identify the parent function in the transformation. Then sketch a graph of the function.

16. $h(x) = -[\![x]\!]$

17. $h(x) = -\sqrt{x + 5} + 8$

18. $h(x) = -2(x - 5)^3 + 3$

In Exercises 19 and 20, find (a) $(f + g)(x)$, (b) $(f - g)(x)$, (c) $(fg)(x)$, (d) $(f/g)(x)$, (e) $(f \circ g)(x)$, and (f) $(g \circ f)(x)$.

19. $f(x) = 3x^2 - 7$, $g(x) = -x^2 - 4x + 5$

20. $f(x) = 1/x$, $g(x) = 2\sqrt{x}$

In Exercises 21–23, determine whether the function has an inverse function. If so, find the inverse function.

21. $f(x) = x^3 + 8$

22. $f(x) = |x^2 - 3| + 6$

23. $f(x) = 3x\sqrt{x}$

In Exercises 24–26, find a mathematical model that represents the statement. (Determine the constant of proportionality.)

24. v varies directly as the square root of s. ($v = 24$ when $s = 16$.)

25. A varies jointly as x and y. ($A = 500$ when $x = 15$ and $y = 8$.)

26. b varies inversely as a. ($b = 32$ when $a = 1.5$.)

Proofs in Mathematics

What does the word *proof* mean to you? In mathematics, the word *proof* means a valid argument. When you are proving a statement or theorem, you must use facts, definitions, and accepted properties in a logical order. You can also use previously proved theorems in your proof. For instance, the proof of the Midpoint Formula below uses the Distance Formula. There are several different proof methods, which you will see in later chapters.

> **The Midpoint Formula**   *(p. 5)*
>
> The midpoint of the line segment joining the points (x_1, y_1) and (x_2, y_2) is given by the Midpoint Formula
>
> $$\text{Midpoint} = \left(\frac{x_1 + x_2}{2}, \frac{y_1 + y_2}{2}\right).$$

Proof

Using the figure, you must show that $d_1 = d_2$ and $d_1 + d_2 = d_3$.

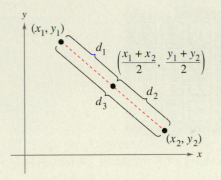

By the Distance Formula, you obtain

$$d_1 = \sqrt{\left(\frac{x_1 + x_2}{2} - x_1\right)^2 + \left(\frac{y_1 + y_2}{2} - y_1\right)^2}$$

$$= \frac{1}{2}\sqrt{(x_2 - x_1)^2 + (y_2 - y_1)^2}$$

$$d_2 = \sqrt{\left(x_2 - \frac{x_1 + x_2}{2}\right)^2 + \left(y_2 - \frac{y_1 + y_2}{2}\right)^2}$$

$$= \frac{1}{2}\sqrt{(x_2 - x_1)^2 + (y_2 - y_1)^2}$$

$$d_3 = \sqrt{(x_2 - x_1)^2 + (y_2 - y_1)^2}$$

So, it follows that $d_1 = d_2$ and $d_1 + d_2 = d_3$.

P.S. Problem Solving ■ ■ ■ ■ ■ ■ ■ ■ ■ ■ ■ ■ ■ ■

1. **Monthly Wages** As a salesperson, you receive a monthly salary of $2000, plus a commission of 7% of sales. You receive an offer for a new job at $2300 per month, plus a commission of 5% of sales.

   (a) Write a linear equation for your current monthly wage W_1 in terms of your monthly sales S.

   (b) Write a linear equation for the monthly wage W_2 of your new job offer in terms of the monthly sales S.

   (c) Use a graphing utility to graph both equations in the same viewing window. Find the point of intersection. What does it signify?

   (d) You think you can sell $20,000 per month. Should you change jobs? Explain.

2. **Telephone Keypad** For the numbers 2 through 9 on a telephone keypad (see figure), create two relations: one mapping numbers onto letters, and the other mapping letters onto numbers. Are both relations functions? Explain.

3. **Sums and Differences of Functions** What can be said about the sum and difference of each of the following?

   (a) Two even functions

   (b) Two odd functions

   (c) An odd function and an even function

4. **Inverse Functions** The two functions

   $$f(x) = x \quad \text{and} \quad g(x) = -x$$

   are their own inverse functions. Graph each function and explain why this is true. Graph other linear functions that are their own inverse functions. Find a general formula for a family of linear functions that are their own inverse functions.

5. **Proof** Prove that a function of the following form is even.

   $$y = a_{2n}x^{2n} + a_{2n-2}x^{2n-2} + \cdots + a_2x^2 + a_0$$

6. **Miniature Golf** A miniature golf professional is trying to make a hole-in-one on the miniature golf green shown. The golf ball is at the point $(2.5, 2)$ and the hole is at the point $(9.5, 2)$. The professional wants to bank the ball off the side wall of the green at the point (x, y). Find the coordinates of the point (x, y). Then write an equation for the path of the ball.

   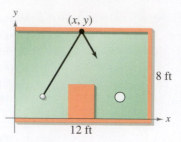

   Figure for 6

7. **Titanic** At 2:00 P.M. on April 11, 1912, the *Titanic* left Cobh, Ireland, on her voyage to New York City. At 11:40 P.M. on April 14, the *Titanic* struck an iceberg and sank, having covered only about 2100 miles of the approximately 3400-mile trip.

   (a) What was the total duration of the voyage in hours?

   (b) What was the average speed in miles per hour?

   (c) Write a function relating the distance of the *Titanic* from New York City and the number of hours traveled. Find the domain and range of the function.

   (d) Graph the function from part (c).

8. **Average Rate of Change** Consider the function $f(x) = -x^2 + 4x - 3$. Find the average rate of change of the function from x_1 to x_2.

   (a) $x_1 = 1, x_2 = 2$

   (b) $x_1 = 1, x_2 = 1.5$

   (c) $x_1 = 1, x_2 = 1.25$

   (d) $x_1 = 1, x_2 = 1.125$

   (e) $x_1 = 1, x_2 = 1.0625$

   (f) Does the average rate of change seem to be approaching one value? If so, then state the value.

   (g) Find the equations of the secant lines through the points $(x_1, f(x_1))$ and $(x_2, f(x_2))$ for parts (a)–(e).

   (h) Find the equation of the line through the point $(1, f(1))$ using your answer from part (f) as the slope of the line.

9. **Inverse of a Composition** Consider the functions $f(x) = 4x$ and $g(x) = x + 6$.

   (a) Find $(f \circ g)(x)$.

   (b) Find $(f \circ g)^{-1}(x)$.

   (c) Find $f^{-1}(x)$ and $g^{-1}(x)$.

   (d) Find $(g^{-1} \circ f^{-1})(x)$ and compare the result with that of part (b).

   (e) Repeat parts (a) through (d) for $f(x) = x^3 + 1$ and $g(x) = 2x$.

   (f) Write two one-to-one functions f and g, and repeat parts (a) through (d) for these functions.

   (g) Make a conjecture about $(f \circ g)^{-1}(x)$ and $(g^{-1} \circ f^{-1})(x)$.

76

10. Trip Time You are in a boat 2 miles from the nearest point on the coast. You are to travel to a point Q, 3 miles down the coast and 1 mile inland (see figure). You row at 2 miles per hour and walk at 4 miles per hour.

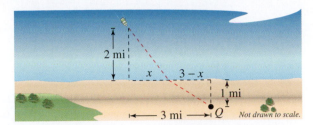

(a) Write the total time T of the trip as a function of x.

(b) Determine the domain of the function.

(c) Use a graphing utility to graph the function. Be sure to choose an appropriate viewing window.

(d) Find the value of x that minimizes T.

(e) Write a brief paragraph interpreting these values.

11. Heaviside Function The **Heaviside function** $H(x)$ is widely used in engineering applications. (See figure.) To print an enlarged copy of the graph, go to *MathGraphs.com*.

$$H(x) = \begin{cases} 1, & x \geq 0 \\ 0, & x < 0 \end{cases}$$

Sketch the graph of each function by hand.

(a) $H(x) - 2$

(b) $H(x - 2)$

(c) $-H(x)$

(d) $H(-x)$

(e) $\frac{1}{2}H(x)$

(f) $-H(x - 2) + 2$

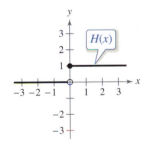

12. Repeated Composition Let $f(x) = \dfrac{1}{1-x}$.

(a) What are the domain and range of f?

(b) Find $f(f(x))$. What is the domain of this function?

(c) Find $f(f(f(x)))$. Is the graph a line? Why or why not?

13. Associative Property with Compositions Show that the Associative Property holds for compositions of functions—that is,

$$(f \circ (g \circ h))(x) = ((f \circ g) \circ h)(x).$$

14. Graphical Analysis Consider the graph of the function f shown in the figure. Use this graph to sketch the graph of each function. To print an enlarged copy of the graph, go to *MathGraphs.com*.

(a) $f(x + 1)$

(b) $f(x) + 1$

(c) $2f(x)$

(d) $f(-x)$

(e) $-f(x)$

(f) $|f(x)|$

(g) $f(|x|)$

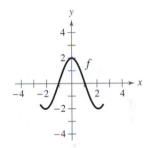

15. Graphical Analysis Use the graphs of f and f^{-1} to complete each table of function values.

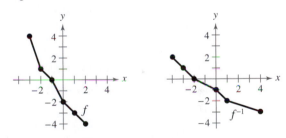

(a)

x	-4	-2	0	4
$(f(f^{-1}(x)))$				

(b)

x	-3	-2	0	1
$(f + f^{-1})(x)$				

(c)

x	-3	-2	0	1
$(f \cdot f^{-1})(x)$				

(d)

x	-4	-3	0	4
$|f^{-1}(x)|$				

2 Polynomial and Rational Functions

Candle Making

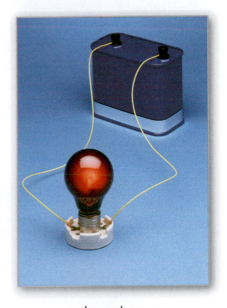

Impedance

Lyme Disease

Tree Growth

Path of a Diver

79

2.1 Quadratic Functions and Models

- Analyze graphs of quadratic functions.
- Write quadratic functions in standard form and use the results to sketch graphs of functions.
- Find minimum and maximum values of quadratic functions in real-life applications.

You can use quadratic functions to model the path of an object or person, such as the path of a diver.

The Graph of a Quadratic Function

In this and the next section, you will study the graphs of polynomial functions. In Section 1.6, you were introduced to the following basic functions.

$f(x) = ax + b$ Linear function

$f(x) = c$ Constant function

$f(x) = x^2$ Squaring function

These functions are examples of **polynomial functions.**

Definition of Polynomial Function

Let n be a nonnegative integer and let $a_n, a_{n-1}, \ldots, a_2, a_1, a_0$ be real numbers with $a_n \neq 0$. The function

$$f(x) = a_n x^n + a_{n-1} x^{n-1} + \cdots + a_2 x^2 + a_1 x + a_0$$

is called a **polynomial function of x with degree n.**

Polynomial functions are classified by degree. For instance, a constant function $f(x) = c$ with $c \neq 0$ has degree 0, and a linear function $f(x) = ax + b$ with $a \neq 0$ has degree 1. In this section, you will study second-degree polynomial functions, which are called **quadratic functions.**

For instance, each of the following functions is a quadratic function.

$f(x) = x^2 + 6x + 2$

$g(x) = 2(x + 1)^2 - 3$

$h(x) = 9 + \frac{1}{4}x^2$

$k(x) = -3x^2 + 4$

$m(x) = (x - 2)(x + 1)$

Note that the squaring function is a simple quadratic function that has degree 2.

Definition of Quadratic Function

Let a, b, and c be real numbers with $a \neq 0$. The function

$$f(x) = ax^2 + bx + c \qquad \text{Quadratic function}$$

is called a **quadratic function.**

The graph of a quadratic function is a special type of "U"-shaped curve called a **parabola.** Parabolas occur in many real-life applications—especially those involving reflective properties of satellite dishes and flashlight reflectors. You will study these properties in Section 10.2.

Photo credit   To come

All parabolas are symmetric with respect to a line called the **axis of symmetry,** or simply the **axis** of the parabola. The point where the axis intersects the parabola is the **vertex** of the parabola, as shown below. When the leading coefficient is positive, the graph of

$$f(x) = ax^2 + bx + c$$

is a parabola that opens upward. When the leading coefficient is negative, the graph of

$$f(x) = ax^2 + bx + c$$

is a parabola that opens downward.

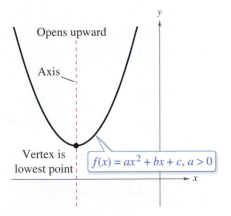

Leading coefficient is positive.

Leading coefficient is negative.

The simplest type of quadratic function is

$$f(x) = ax^2.$$

Its graph is a parabola whose vertex is $(0, 0)$. When $a > 0$, the vertex is the point with the *minimum* y-value on the graph, and when $a < 0$, the vertex is the point with the *maximum* y-value on the graph, as shown below.

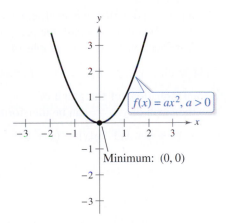

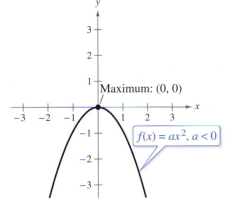

Leading coefficient is positive.

Leading coefficient is negative.

When sketching the graph of

$$f(x) = ax^2$$

it is helpful to use the graph of $y = x^2$ as a reference, as discussed in Section 1.7. There you learned that when $a > 1$, the graph of $y = af(x)$ is a vertical stretch of the graph of $y = f(x)$. When $0 < a < 1$, the graph of $y = af(x)$ is a vertical shrink of the graph of $y = f(x)$. This is demonstrated again in Example 1.

| EXAMPLE 1 | **Sketching Graphs of Quadratic Functions** |

Sketch the graph of each quadratic function and compare it with the graph of $y = x^2$.

a. $f(x) = \frac{1}{3}x^2$ **b.** $g(x) = 2x^2$

> ▷ **ALGEBRA HELP** You can review the techniques for shifting, reflecting, and stretching graphs in Section 1.7.

Solution

a. Compared with $y = x^2$, each output of $f(x) = \frac{1}{3}x^2$ "shrinks" by a factor of $\frac{1}{3}$, creating the broader parabola shown in Figure 2.1.

b. Compared with $y = x^2$, each output of $g(x) = 2x^2$ "stretches" by a factor of 2, creating the narrower parabola shown in Figure 2.2.

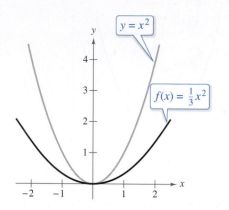

Figure 2.1   **Figure 2.2**

✓ *Checkpoint* *Audio-video solution in English & Spanish at LarsonPrecalculus.com.*

Sketch the graph of each quadratic function and compare it with the graph of $y = x^2$.

a. $f(x) = \frac{1}{4}x^2$ **b.** $g(x) = -\frac{1}{6}x^2$

c. $h(x) = \frac{5}{2}x^2$ **d.** $k(x) = -4x^2$ ■

In Example 1, note that the coefficient a determines how wide the parabola $f(x) = ax^2$ opens. When $|a|$ is small, the parabola opens wider than when $|a|$ is large.

Recall from Section 1.7 that the graphs of

$$y = f(x \pm c), \quad y = f(x) \pm c, \quad y = f(-x), \quad \text{and} \quad y = -f(x)$$

are rigid transformations of the graph of $y = f(x)$. For instance, in the figures below, notice how the graph of $y = x^2$ can be transformed to produce the graphs of

$$f(x) = -x^2 + 1 \quad \text{and} \quad g(x) = (x + 2)^2 - 3.$$

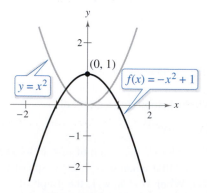

Reflection in x-axis followed by an upward shift of one unit

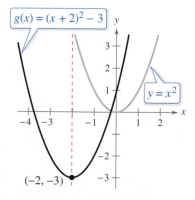

Left shift of two units followed by a downward shift of three units

The Standard Form of a Quadratic Function

The **standard form** of a quadratic function is $f(x) = a(x - h)^2 + k$. This form is especially convenient for sketching a parabola because it identifies the vertex of the parabola as (h, k).

> ## REMARK The standard form of a quadratic function identifies four basic transformations of the graph of $y = x^2$.
>
> **a.** The factor a produces a vertical stretch or shrink.
>
> **b.** When $a < 0$, the graph is reflected in the x-axis.
>
> **c.** The factor $(x - h)^2$ represents a horizontal shift of h units.
>
> **d.** The term k represents a vertical shift of k units.

> ### Standard Form of a Quadratic Function
>
> The quadratic function
>
> $$f(x) = a(x - h)^2 + k, \quad a \neq 0$$
>
> is in **standard form.** The graph of f is a parabola whose axis is the vertical line $x = h$ and whose vertex is the point (h, k). When $a > 0$, the parabola opens upward, and when $a < 0$, the parabola opens downward.

To graph a parabola, it is helpful to begin by writing the quadratic function in standard form using the process of completing the square, as illustrated in Example 2. In this example, notice that when completing the square, you *add and subtract* the square of half the coefficient of x within the parentheses instead of adding the value to each side of the equation as is done in Appendix A.5.

EXAMPLE 2   Using Standard Form to Graph a Parabola

Sketch the graph of

$$f(x) = 2x^2 + 8x + 7.$$

Identify the vertex and the axis of the parabola.

> ▷ **ALGEBRA HELP** You can review the techniques for completing the square in Appendix A.5.

Solution Begin by writing the quadratic function in standard form. Notice that the first step in completing the square is to factor out any coefficient of x^2 that is not 1.

$$f(x) = 2x^2 + 8x + 7 \qquad \text{\color{red}Write original function.}$$
$$= 2(x^2 + 4x) + 7 \qquad \text{\color{red}Factor 2 out of x-terms.}$$
$$= 2(x^2 + 4x + 4 - 4) + 7 \qquad \text{\color{red}Add and subtract 4 within parentheses.}$$

$$(4/2)^2$$

After adding and subtracting 4 within the parentheses, you must now regroup the terms to form a perfect square trinomial. The -4 can be removed from inside the parentheses; however, because of the 2 outside of the parentheses, you must multiply -4 by 2, as shown below.

$$f(x) = 2(x^2 + 4x + 4) - 2(4) + 7 \qquad \text{\color{red}Regroup terms.}$$
$$= 2(x^2 + 4x + 4) - 8 + 7 \qquad \text{\color{red}Simplify.}$$
$$= 2(x + 2)^2 - 1 \qquad \text{\color{red}Write in standard form.}$$

From this form, you can see that the graph of f is a parabola that opens upward and has its vertex at $(-2, -1)$. This corresponds to a left shift of two units and a downward shift of one unit relative to the graph of $y = 2x^2$, as shown in Figure 2.3. In the figure, you can see that the axis of the parabola is the vertical line through the vertex, $x = -2$.

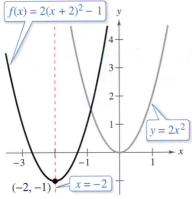

$$f(x) = 2(x + 2)^2 - 1$$
$$y = 2x^2$$
$$(-2, -1) \qquad x = -2$$

Figure 2.3

✓ *Checkpoint* 🔊)) *Audio-video solution in English & Spanish at LarsonPrecalculus.com.*

Sketch the graph of

$$f(x) = 3x^2 - 6x + 4.$$

Identify the vertex and the axis of the parabola.

▷ **ALGEBRA HELP** You can review the techniques for using the Quadratic Formula in Appendix A.5.

To find the x-intercepts of the graph of $f(x) = ax^2 + bx + c$, you must solve the equation $ax^2 + bx + c = 0$. When $ax^2 + bx + c$ does not factor, you can use the Quadratic Formula to find the x-intercepts. Remember, however, that a parabola may not have x-intercepts.

EXAMPLE 3 **Finding the Vertex and *x*-Intercepts of a Parabola**

Sketch the graph of $f(x) = -x^2 + 6x - 8$. Identify the vertex and x-intercepts.

Solution

$$f(x) = -x^2 + 6x - 8 \qquad \text{Write original function.}$$
$$= -(x^2 - 6x) - 8 \qquad \text{Factor } -1 \text{ out of } x\text{-terms.}$$
$$= -(x^2 - 6x + 9 - 9) - 8 \qquad \text{Add and subtract 9 within parentheses.}$$
$$\underset{(-6/2)^2}{\underline{\qquad}}$$
$$= -(x^2 - 6x + 9) - (-9) - 8 \qquad \text{Regroup terms.}$$
$$= -(x - 3)^2 + 1 \qquad \text{Write in standard form.}$$

From this form, you can see that f is a parabola that opens downward with vertex $(3, 1)$. The x-intercepts of the graph are determined as follows.

$$-(x^2 - 6x + 8) = 0 \qquad \text{Factor out } -1.$$
$$-(x - 2)(x - 4) = 0 \qquad \text{Factor.}$$
$$x - 2 = 0 \implies x = 2 \qquad \text{Set 1st factor equal to 0.}$$
$$x - 4 = 0 \implies x = 4 \qquad \text{Set 2nd factor equal to 0.}$$

So, the x-intercepts are $(2, 0)$ and $(4, 0)$, as shown in Figure 2.4.

Figure 2.4

✓ *Checkpoint* 🔊)) *Audio-video solution in English & Spanish at LarsonPrecalculus.com.*

Sketch the graph of $f(x) = x^2 - 4x + 3$. Identify the vertex and x-intercepts.

EXAMPLE 4 **Writing the Equation of a Parabola**

Write the standard form of the equation of the parabola whose vertex is $(1, 2)$ and that passes through the point $(3, -6)$.

Solution Because the vertex is $(h, k) = (1, 2)$, the equation has the form

$$f(x) = a(x - 1)^2 + 2. \qquad \text{Substitute for } h \text{ and } k \text{ in standard form.}$$

Because the parabola passes through the point $(3, -6)$, it follows that $f(3) = -6$. So,

$$f(x) = a(x - 1)^2 + 2 \qquad \text{Write in standard form.}$$
$$-6 = a(3 - 1)^2 + 2 \qquad \text{Substitute 3 for } x \text{ and } -6 \text{ for } f(x).$$
$$-6 = 4a + 2 \qquad \text{Simplify.}$$
$$-8 = 4a \qquad \text{Subtract 2 from each side.}$$
$$-2 = a. \qquad \text{Divide each side by 4.}$$

The equation in standard form is $f(x) = -2(x - 1)^2 + 2$. The graph of f is shown in Figure 2.5.

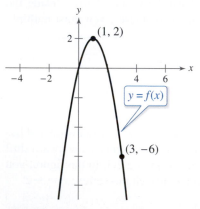
Figure 2.5

✓ *Checkpoint* 🔊)) *Audio-video solution in English & Spanish at LarsonPrecalculus.com.*

Write the standard form of the equation of the parabola whose vertex is $(-4, 11)$ and that passes through the point $(-6, 15)$.

Finding Minimum and Maximum Values

Many applications involve finding the maximum or minimum value of a quadratic function. By completing the square of the quadratic function $f(x) = ax^2 + bx + c$, you can rewrite the function in standard form.

$$f(x) = a\left(x + \frac{b}{2a}\right)^2 + \left(c - \frac{b^2}{4a}\right) \qquad \text{Standard form}$$

So, the vertex of the graph of f is $\left(-\dfrac{b}{2a}, f\left(-\dfrac{b}{2a}\right)\right)$, which implies the following.

Minimum and Maximum Values of Quadratic Functions

Consider the function $f(x) = ax^2 + bx + c$ with vertex $\left(-\dfrac{b}{2a}, f\left(-\dfrac{b}{2a}\right)\right)$.

1. When $a > 0$, f has a *minimum* at $x = -\dfrac{b}{2a}$. The minimum value is $f\left(-\dfrac{b}{2a}\right)$.

2. When $a < 0$, f has a *maximum* at $x = -\dfrac{b}{2a}$. The maximum value is $f\left(-\dfrac{b}{2a}\right)$.

EXAMPLE 5   The Maximum Height of a Baseball

The path of a baseball after being hit is given by the function $f(x) = -0.0032x^2 + x + 3$, where $f(x)$ is the height of the baseball (in feet) and x is the horizontal distance from home plate (in feet). What is the maximum height of the baseball?

Algebraic Solution

For this quadratic function, you have

$$f(x) = ax^2 + bx + c = -0.0032x^2 + x + 3$$

which implies that $a = -0.0032$ and $b = 1$. Because $a < 0$, the function has a maximum at $x = -b/(2a)$. So, the baseball reaches its maximum height when it is

$$x = -\frac{b}{2a} = -\frac{1}{2(-0.0032)} = 156.25 \text{ feet from home plate.}$$

At this distance, the maximum height is

$$f(156.25) = -0.0032(156.25)^2 + 156.25 + 3 = 81.125 \text{ feet.}$$

Graphical Solution

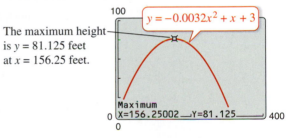

The maximum height is $y = 81.125$ feet at $x = 156.25$ feet.

$y = -0.0032x^2 + x + 3$

Maximum
X=156.25002   Y=81.125

✓ **Checkpoint** 🔊)) *Audio-video solution in English & Spanish at LarsonPrecalculus.com.*

Rework Example 5 when the path of the baseball is given by the function

$$f(x) = -0.007x^2 + x + 4.$$

Summarize   (Section 2.1)

1. State the definition of a quadratic function and describe its graph *(pages 80–82)*. For an example of sketching the graphs of quadratic functions, see Example 1.

2. State the standard form of a quadratic function *(page 83)*. For examples that use the standard form of a quadratic function, see Examples 2–4.

3. Describe how to find the minimum or maximum value of a quadratic function *(page 85)*. For a real-life example that involves finding the maximum height of a baseball, see Example 5.

2.2 Polynomial Functions of Higher Degree

You can use polynomial functions to analyze nature, such as the growth of a red oak tree.

■ Use transformations to sketch graphs of polynomial functions.
■ Use the Leading Coefficient Test to determine the end behaviors of graphs of polynomial functions.
■ Find and use the real zeros of polynomial functions as sketching aids.
■ Use the Intermediate Value Theorem to help locate the real zeros of polynomial functions.

Graphs of Polynomial Functions

In this section, you will study basic features of the graphs of polynomial functions. The first feature is that the graph of a polynomial function is **continuous.** Essentially, this means that the graph of a polynomial function has no breaks, holes, or gaps, as shown in Figure 2.6(a). The graph shown in Figure 2.6(b) is an example of a piecewise-defined function that is not continuous.

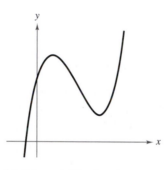

(a) Polynomial functions have continuous graphs.

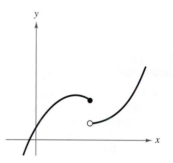

(b) Functions with graphs that are not continuous are not polynomial functions.

Figure 2.6

The second feature is that the graph of a polynomial function has only smooth, rounded turns, as shown in Figure 2.7. A polynomial function cannot have a sharp turn. For instance, the function $f(x) = |x|$, the graph of which has a sharp turn at the point $(0, 0)$, as shown in Figure 2.8, is not a polynomial function.

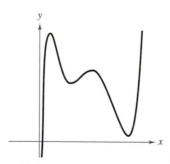

Polynomial functions have graphs with smooth, rounded turns.
Figure 2.7

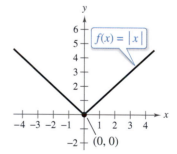

Graphs of polynomial functions cannot have sharp turns.
Figure 2.8

The graphs of polynomial functions of degree greater than 2 are more difficult to analyze than the graphs of polynomial functions of degree 0, 1, or 2. However, using the features presented in this section, coupled with your knowledge of point plotting, intercepts, and symmetry, you should be able to make reasonably accurate sketches *by hand.*

The polynomial functions that have the simplest graphs are monomial functions of the form $f(x) = x^n$, where n is an integer greater than zero. From Figure 2.9, you can see that when n is *even*, the graph is similar to the graph of $f(x) = x^2$, and when n is *odd*, the graph is similar to the graph of $f(x) = x^3$. Moreover, the greater the value of n, the flatter the graph near the origin.

▷ **REMARK** For functions given by $f(x) = x^n$, when n is even, the graph of the function is symmetric with respect to the y-axis, and when n is odd, the graph of the function is symmetric with respect to the origin.

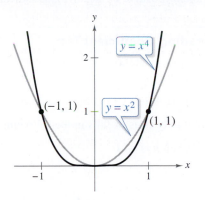

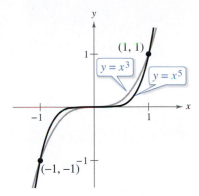

(a) When n is even, the graph of $y = x^n$ touches the axis at the x-intercept.

(b) When n is odd, the graph of $y = x^n$ crosses the axis at the x-intercept.

Figure 2.9

EXAMPLE 1   **Sketching Transformations of Monomial Functions**

Sketch the graph of each function.

a. $f(x) = -x^5$ **b.** $h(x) = (x + 1)^4$

Solution

a. Because the degree of $f(x) = -x^5$ is odd, its graph is similar to the graph of $y = x^3$. In Figure 2.10, note that the negative coefficient has the effect of reflecting the graph in the x-axis.

b. The graph of $h(x) = (x + 1)^4$, as shown in Figure 2.11, is a left shift by one unit of the graph of $y = x^4$.

▷ **ALGEBRA HELP** You can review the techniques for shifting, reflecting, and stretching graphs in Section 1.7.

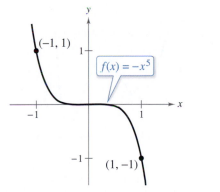

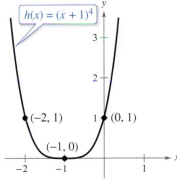

Figure 2.10   **Figure 2.11**

 *Checkpoint* ◀))) *Audio-video solution in English & Spanish at LarsonPrecalculus.com.*

Sketch the graph of each function.

a. $f(x) = (x + 5)^4$

b. $g(x) = x^4 - 7$

c. $h(x) = 7 - x^4$

d. $k(x) = \frac{1}{4}(x - 3)^4$

The Leading Coefficient Test

In Example 1, note that both graphs eventually rise or fall without bound as x moves to the left or to the right. Whether the graph of a polynomial function eventually rises or falls can be determined by the function's degree (even or odd) and by its leading coefficient (positive or negative), as indicated in the **Leading Coefficient Test.**

Leading Coefficient Test

As x moves without bound to the left or to the right, the graph of the polynomial function

$$f(x) = a_n x^n + \cdots + a_1 x + a_0$$

eventually rises or falls in the following manner.

1. When n is *odd:*

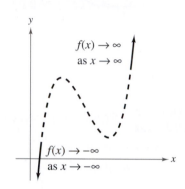

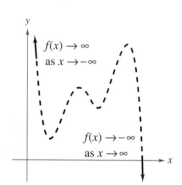

If the leading coefficient is positive $(a_n > 0)$, then the graph falls to the left and rises to the right.

If the leading coefficient is negative $(a_n < 0)$, then the graph rises to the left and falls to the right.

2. When n is *even:*

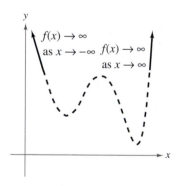

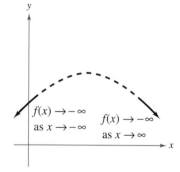

If the leading coefficient is positive $(a_n > 0)$, then the graph rises to the left and to the right.

If the leading coefficient is negative $(a_n < 0)$, then the graph falls to the left and to the right.

The dashed portions of the graphs indicate that the test determines *only* the right-hand and left-hand behavior of the graph.

> **• REMARK** The notation "$f(x) \to -\infty$ as $x \to -\infty$" indicates that the graph falls to the left. The notation "$f(x) \to \infty$ as $x \to \infty$" indicates that the graph rises to the right.

As you continue to study polynomial functions and their graphs, you will notice that the degree of a polynomial plays an important role in determining other characteristics of the polynomial function and its graph.

EXAMPLE 2 **Applying the Leading Coefficient Test**

Describe the right-hand and left-hand behavior of the graph of each function.

a. $f(x) = -x^3 + 4x$ **b.** $f(x) = x^4 - 5x^2 + 4$ **c.** $f(x) = x^5 - x$

Solution

a. Because the degree is odd and the leading coefficient is negative, the graph rises to the left and falls to the right, as shown below.

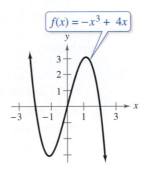

b. Because the degree is even and the leading coefficient is positive, the graph rises to the left and to the right, as shown below.

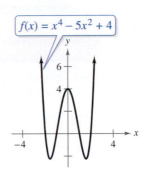

c. Because the degree is odd and the leading coefficient is positive, the graph falls to the left and rises to the right, as shown below.

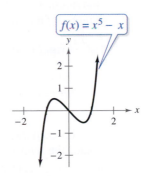

✓ *Checkpoint* ◀)) *Audio-video solution in English & Spanish at LarsonPrecalculus.com.*

Describe the right-hand and left-hand behavior of the graph of each function.

a. $f(x) = \frac{1}{4}x^3 - 2x$ **b.** $f(x) = -3.6x^5 + 5x^3 - 1$ ■

In Example 2, note that the Leading Coefficient Test tells you only whether the graph *eventually* rises or falls to the right or to the left. Other characteristics of the graph, such as intercepts and minimum and maximum points, must be determined by other tests.

Real Zeros of Polynomial Functions

It can be shown that for a polynomial function f of degree n, the following statements are true.

⋯⋯⋯⋯⋯⋯⋯⋯⋯▷

• • REMARK Remember that the *zeros* of a function of x are the x-values for which the function is zero.

1. The function f has, at most, n real zeros. (You will study this result in detail in the discussion of the Fundamental Theorem of Algebra in Section 2.5.)

2. The graph of f has, at most, $n - 1$ turning points. (Turning points, also called relative minima or relative maxima, are points at which the graph changes from increasing to decreasing or vice versa.)

Finding the zeros of a polynomial function is one of the most important problems in algebra. There is a strong interplay between graphical and algebraic approaches to this problem.

Real Zeros of Polynomial Functions

When f is a polynomial function and a is a real number, the following statements are equivalent.

▷ ALGEBRA HELP   To do Example 3 algebraically, you need to be able to completely factor polynomials. You can review the techniques for factoring in Appendix A.3.

1. $x = a$ is a *zero* of the function f.

2. $x = a$ is a *solution* of the polynomial equation $f(x) = 0$.

3. $(x - a)$ is a *factor* of the polynomial $f(x)$.

4. $(a, 0)$ is an *x-intercept* of the graph of f.

EXAMPLE 3   **Finding Real Zeros of a Polynomial Function**

Find all real zeros of $f(x) = -2x^4 + 2x^2$. Then determine the maximum possible number of turning points of the graph of the function.

Solution   To find the real zeros of the function, set $f(x)$ equal to zero and solve for x.

$$-2x^4 + 2x^2 = 0 \qquad \text{Set } f(x) \text{ equal to 0.}$$

$$-2x^2(x^2 - 1) = 0 \qquad \text{Remove common monomial factor.}$$

$$-2x^2(x - 1)(x + 1) = 0 \qquad \text{Factor completely.}$$

So, the real zeros are $x = 0$, $x = 1$, and $x = -1$. Because the function is a fourth-degree polynomial, the graph of f can have at most $4 - 1 = 3$ turning points. In this case, the graph of f has three turning points, as shown in Figure 2.12.

$f(x) = -2x^4 + 2x^2$

Turning point

Turning point

$(1, 0)$

$(-1, 0)$ $(0, 0)$

Turning point

Figure 2.12

✓ **Checkpoint**  *Audio-video solution in English & Spanish at LarsonPrecalculus.com.*

Find all real zeros of $f(x) = x^3 - 12x^2 + 36x$. Then determine the maximum possible number of turning points of the graph of the function. ∎

In Example 3, note that because the exponent is greater than 1, the factor $-2x^2$ yields the *repeated* zero $x = 0$.

Repeated Zeros

A factor $(x - a)^k$, $k > 1$, yields a **repeated zero** $x = a$ of **multiplicity** k.

1. When k is odd, the graph *crosses* the x-axis at $x = a$.

2. When k is even, the graph *touches* the x-axis (but does not cross the x-axis) at $x = a$.

To graph polynomial functions, use the fact that a polynomial function can change signs only at its zeros. Between two consecutive zeros, a polynomial must be entirely positive or entirely negative. (This follows from the Intermediate Value Theorem, which you will study later in this section.) This means that when you put the real zeros of a polynomial function in order, they divide the real number line into intervals in which the function has no sign changes. These resulting intervals are **test intervals** in which you choose a representative *x*-value to determine whether the value of the polynomial function is positive (the graph lies above the *x*-axis) or negative (the graph lies below the *x*-axis).

▷ **TECHNOLOGY** Example 4 uses an *algebraic approach* to describe the graph of the function. A graphing utility can complement this approach. Remember to find a viewing window that shows all significant features of the graph. For instance, viewing window (a) illustrates all of the significant features of the function in Example 4 while viewing window (b) does not.

(a)

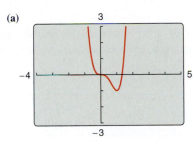

(b)

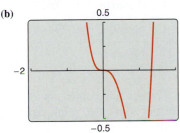

EXAMPLE 4   **Sketching the Graph of a Polynomial Function**

Sketch the graph of $f(x) = 3x^4 - 4x^3$.

Solution

1. *Apply the Leading Coefficient Test.* Because the leading coefficient is positive and the degree is even, you know that the graph eventually rises to the left and to the right (see Figure 2.13).

2. *Find the Real Zeros of the Polynomial.* By factoring $f(x) = 3x^4 - 4x^3$ as $f(x) = x^3(3x - 4)$, you can see that the real zeros of f are $x = 0$ and $x = \frac{4}{3}$ (both of odd multiplicity). So, the *x*-intercepts occur at $(0, 0)$ and $\left(\frac{4}{3}, 0\right)$. Add these points to your graph, as shown in Figure 2.13.

3. *Plot a Few Additional Points.* Use the zeros of the polynomial to find the test intervals. In each test interval, choose a representative *x*-value and evaluate the polynomial function, as shown in the table.

Test Interval	Representative *x*-Value	Value of f	Sign	Point on Graph
$(-\infty, 0)$	-1	$f(-1) = 7$	Positive	$(-1, 7)$
$\left(0, \frac{4}{3}\right)$	1	$f(1) = -1$	Negative	$(1, -1)$
$\left(\frac{4}{3}, \infty\right)$	1.5	$f(1.5) = 1.6875$	Positive	$(1.5, 1.6875)$

4. *Draw the Graph.* Draw a continuous curve through the points, as shown in Figure 2.14. Because both zeros are of odd multiplicity, you know that the graph should cross the *x*-axis at $x = 0$ and $x = \frac{4}{3}$.

REMARK If you are unsure of the shape of a portion of the graph of a polynomial function, then plot some additional points, such as the point $(0.5, -0.3125)$, as shown in Figure 2.14.

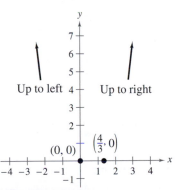

Figure 2.13

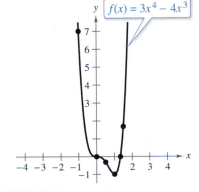

Figure 2.14

✓ **Checkpoint** ◀))) *Audio-video solution in English & Spanish at LarsonPrecalculus.com.*

Sketch the graph of $f(x) = 2x^3 - 6x^2$.

A polynomial function is in **standard form** when its terms are in descending order of exponents from left to right. Before applying the Leading Coefficient Test to a polynomial function, it is a good idea to make sure that the polynomial function is in standard form.

EXAMPLE 5   **Sketching the Graph of a Polynomial Function**

Sketch the graph of $f(x) = -2x^3 + 6x^2 - \frac{9}{2}x$.

Solution

1. *Apply the Leading Coefficient Test.*   Because the leading coefficient is negative and the degree is odd, you know that the graph eventually rises to the left and falls to the right (see Figure 2.15).

2. *Find the Real Zeros of the Polynomial.*   By factoring

$$f(x) = -2x^3 + 6x^2 - \tfrac{9}{2}x$$
$$= -\tfrac{1}{2}x(4x^2 - 12x + 9)$$
$$= -\tfrac{1}{2}x(2x - 3)^2$$

you can see that the real zeros of f are $x = 0$ (odd multiplicity) and $x = \frac{3}{2}$ (even multiplicity). So, the x-intercepts occur at $(0, 0)$ and $\left(\frac{3}{2}, 0\right)$. Add these points to your graph, as shown in Figure 2.15.

3. *Plot a Few Additional Points.*   Use the zeros of the polynomial to find the test intervals. In each test interval, choose a representative x-value and evaluate the polynomial function, as shown in the table.

Test Interval	Representative x-Value	Value of f	Sign	Point on Graph
$(-\infty, 0)$	-0.5	$f(-0.5) = 4$	Positive	$(-0.5, 4)$
$\left(0, \frac{3}{2}\right)$	0.5	$f(0.5) = -1$	Negative	$(0.5, -1)$
$\left(\frac{3}{2}, \infty\right)$	2	$f(2) = -1$	Negative	$(2, -1)$

• • • • • • • • • • • • • • ▷
•
• • **REMARK** Observe in Example 5 that the sign of $f(x)$ is positive to the left of and negative to the right of the zero $x = 0$. Similarly, the sign of $f(x)$ is negative to the left and to the right of the zero $x = \frac{3}{2}$. This suggests that if the zero of a polynomial function is of *odd* multiplicity, then the sign of $f(x)$ changes from one side to the other side of the zero. If the zero is of *even* multiplicity, then the sign of $f(x)$ does not change from one side to the other side of the zero.

4. *Draw the Graph.*   Draw a continuous curve through the points, as shown in Figure 2.16. As indicated by the multiplicities of the zeros, the graph crosses the x-axis at $(0, 0)$ but does not cross the x-axis at $\left(\frac{3}{2}, 0\right)$.

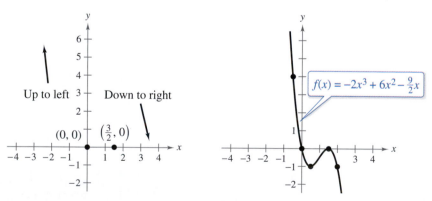

Figure 2.15   **Figure 2.16**

✓ *Checkpoint* ◀))) *Audio-video solution in English & Spanish at LarsonPrecalculus.com.*

Sketch the graph of $f(x) = -\frac{1}{4}x^4 + \frac{3}{2}x^3 - \frac{9}{4}x^2$.

The Intermediate Value Theorem

The **Intermediate Value Theorem** implies that if

$$(a, f(a)) \quad \text{and} \quad (b, f(b))$$

are two points on the graph of a polynomial function such that $f(a) \neq f(b)$, then for any number d between $f(a)$ and $f(b)$ there must be a number c between a and b such that $f(c) = d$. (See figure.)

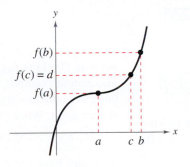

Intermediate Value Theorem

Let a and b be real numbers such that $a < b$. If f is a polynomial function such that $f(a) \neq f(b)$, then, in the interval $[a, b]$, f takes on every value between $f(a)$ and $f(b)$.

The Intermediate Value Theorem helps you locate the real zeros of a polynomial function in the following way. If you can find a value $x = a$ at which a polynomial function is positive, and another value $x = b$ at which it is negative, then you can conclude that the function has at least one real zero between these two values. For example, the function

$$f(x) = x^3 + x^2 + 1$$

is negative when $x = -2$ and positive when $x = -1$. So, it follows from the Intermediate Value Theorem that f must have a real zero somewhere between -2 and -1, as shown below.

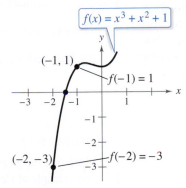

The function f must have a real zero somewhere between -2 and -1.

By continuing this line of reasoning, you can approximate any real zeros of a polynomial function to any desired accuracy. This concept is further demonstrated in Example 6.

▷**TECHNOLOGY** You can use the *table* feature of a graphing utility to approximate the real zero of the polynomial function in Example 6. Construct a table that shows function values for $-2 \le x \le 4$, as shown below.

X	Y1
-2	-11
-1	-1
0	1
1	1
2	5
3	19
4	49
X=0

Scroll through the table looking for consecutive values that differ in sign. From the table, you can see that $f(-1)$ and $f(0)$ differ in sign. So, you can conclude from the Intermediate Value Theorem that the function has a real zero between -1 and 0. You can adjust your table to show function values for $-1 \le x \le 0$ using increments of 0.1, as shown below.

X	Y1
-1	-1
-.9	-.539
-.8	-.152
-.7	.167
-.6	.424
-.5	.625
-.4	.776
X=-.7

By scrolling through the table, you can see that $f(-0.8)$ and $f(-0.7)$ differ in sign. So, the function has a real zero between -0.8 and -0.7. Repeating this process several times, you should obtain $x \approx -0.755$ as the real zero of the function. Use the *zero* or *root* feature of the graphing utility to confirm this result.

EXAMPLE 6   Using the Intermediate Value Theorem

Use the Intermediate Value Theorem to approximate the real zero of

$$f(x) = x^3 - x^2 + 1.$$

Solution   Begin by computing a few function values, as follows.

x	-2	-1	0	1
$f(x)$	-11	-1	1	1

Because $f(-1)$ is negative and $f(0)$ is positive, you can apply the Intermediate Value Theorem to conclude that the function has a real zero between -1 and 0. To pinpoint this zero more closely, divide the interval $[-1, 0]$ into tenths and evaluate the function at each point. When you do this, you will find that

$$f(-0.8) = -0.152 \quad \text{and} \quad f(-0.7) = 0.167.$$

So, f must have a real zero between -0.8 and -0.7, as shown below.

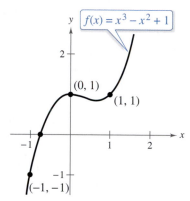

The function f has a real zero between -0.8 and -0.7.

For a more accurate approximation, compute function values between $f(-0.8)$ and $f(-0.7)$ and apply the Intermediate Value Theorem again. By continuing this process, you can approximate this zero to any desired accuracy.

✓ *Checkpoint*))) *Audio-video solution in English & Spanish at LarsonPrecalculus.com.*

Use the Intermediate Value Theorem to approximate the real zero of

$$f(x) = x^3 - 3x^2 - 2.$$ ■

Summarize   (Section 2.2)

1. Describe the graphs of monomial functions of the form $f(x) = x^n$ *(page 87)*. For an example of sketching transformations of monomial functions, see Example 1.

2. Describe the Leading Coefficient Test *(page 88)*. For an example of applying the Leading Coefficient Test, see Example 2.

3. Describe the real zeros of polynomial functions *(page 90)*. For examples involving the real zeros of polynomial functions, see Examples 3–5.

4. State the Intermediate Value Theorem *(page 93)*. For an example of using the Intermediate Value Theorem, see Example 6.

2.3 Polynomial and Synthetic Division

Synthetic division can help you evaluate polynomial functions, such as a model for the number of confirmed cases of Lyme disease in Maine.

■ Use long division to divide polynomials by other polynomials.
■ Use synthetic division to divide polynomials by binomials of the form $(x - k)$.
■ Use the Remainder Theorem and the Factor Theorem.

Long Division of Polynomials

Suppose you are given the graph of

$$f(x) = 6x^3 - 19x^2 + 16x - 4.$$

Notice that a zero of f occurs at $x = 2$, as shown at the right. Because $x = 2$ is a zero of f, you know that $(x - 2)$ is a factor of $f(x)$. This means that there exists a second-degree polynomial $q(x)$ such that

$$f(x) = (x - 2) \cdot q(x).$$

To find $q(x)$, you can use **long division,** as illustrated in Example 1.

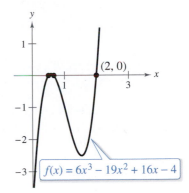

$f(x) = 6x^3 - 19x^2 + 16x - 4$

EXAMPLE 1   Long Division of Polynomials

Divide the polynomial $6x^3 - 19x^2 + 16x - 4$ by $x - 2$, and use the result to factor the polynomial completely.

Solution

Think $\dfrac{6x^3}{x} = 6x^2$.

Think $\dfrac{-7x^2}{x} = -7x$.

Think $\dfrac{2x}{x} = 2$.

$$
\begin{array}{r}
6x^2 - 7x + 2 \\
x - 2 \overline{)\, 6x^3 - 19x^2 + 16x - 4} \\
\underline{6x^3 - 12x^2} \qquad\qquad\qquad \\
-7x^2 + 16x \qquad\quad \\
\underline{-7x^2 + 14x} \qquad\quad \\
2x - 4 \\
\underline{2x - 4} \\
0
\end{array}
$$

Multiply: $6x^2(x - 2)$.
Subtract.
Multiply: $-7x(x - 2)$.
Subtract.
Multiply: $2(x - 2)$.
Subtract.

From this division, you can conclude that

$$6x^3 - 19x^2 + 16x - 4 = (x - 2)(6x^2 - 7x + 2)$$

and by factoring the quadratic $6x^2 - 7x + 2$, you have

$$6x^3 - 19x^2 + 16x - 4 = (x - 2)(2x - 1)(3x - 2).$$

✓ *Checkpoint* *Audio-video solution in English & Spanish at LarsonPrecalculus.com.*

Divide the polynomial $3x^2 + 19x + 28$ by $x + 4$, and use the result to factor the polynomial completely.

In Example 1, $x - 2$ is a factor of the polynomial

$$6x^3 - 19x^2 + 16x - 4$$

and the long division process produces a remainder of zero. Often, long division will produce a nonzero remainder. For instance, when you divide $x^2 + 3x + 5$ by $x + 1$, you obtain the following.

$$
\begin{array}{r}
x + 2 \quad \longleftarrow \text{ Quotient} \\
\text{Divisor} \longrightarrow x + 1 \overline{)\,x^2 + 3x + 5} \quad \longleftarrow \text{ Dividend} \\
\underline{x^2 + x} \\
2x + 5 \\
\underline{2x + 2} \\
3 \quad \longleftarrow \text{ Remainder}
\end{array}
$$

In fractional form, you can write this result as follows.

$$\frac{\overbrace{x^2 + 3x + 5}^{\text{Dividend}}}{\underbrace{x + 1}_{\text{Divisor}}} = \overbrace{x + 2}^{\text{Quotient}} + \frac{\overset{\text{Remainder}}{3}}{\underbrace{x + 1}_{\text{Divisor}}}$$

This implies that

$$x^2 + 3x + 5 = (x + 1)(x + 2) + 3 \qquad \text{Multiply each side by } (x + 1).$$

which illustrates the following theorem, called the **Division Algorithm.**

The Division Algorithm

If $f(x)$ and $d(x)$ are polynomials such that $d(x) \neq 0$, and the degree of $d(x)$ is less than or equal to the degree of $f(x)$, then there exist unique polynomials $q(x)$ and $r(x)$ such that

$$f(x) = d(x)q(x) + r(x)$$

$$\underset{\text{Dividend}}{\uparrow} \quad \underset{\underset{\text{Divisor}}{\uparrow}}{\uparrow} \quad \underset{\text{Quotient}}{} \quad \underset{\text{Remainder}}{\uparrow}$$

where $r(x) = 0$ or the degree of $r(x)$ is less than the degree of $d(x)$. If the remainder $r(x)$ is zero, $d(x)$ *divides evenly* into $f(x)$.

The Division Algorithm can also be written as

$$\frac{f(x)}{d(x)} = q(x) + \frac{r(x)}{d(x)}.$$

In the Division Algorithm, the rational expression $f(x)/d(x)$ is **improper** because the degree of $f(x)$ is greater than or equal to the degree of $d(x)$. On the other hand, the rational expression $r(x)/d(x)$ is **proper** because the degree of $r(x)$ is less than the degree of $d(x)$.

Before you apply the Division Algorithm, follow these steps.

1. Write the dividend and divisor in descending powers of the variable.

2. Insert placeholders with zero coefficients for missing powers of the variable.

Note how these steps are applied in the next example.

EXAMPLE 2 **Long Division of Polynomials**

Divide $x^3 - 1$ by $x - 1$.

Solution Because there is no x^2-term or x-term in the dividend $x^3 - 1$, you need to rewrite the dividend as $x^3 + 0x^2 + 0x - 1$ before you apply the Division Algorithm.

$$
\begin{array}{r}
x^2 + x + 1 \\
x - 1 \,\overline{)\, x^3 + 0x^2 + 0x - 1} \\
\underline{x^3 - x^2} \\
x^2 + 0x \\
\underline{x^2 - x} \\
x - 1 \\
\underline{x - 1} \\
0
\end{array}
$$

Multiply x^2 by $x - 1$.

Subtract and bring down $0x$.

Multiply x by $x - 1$.

Subtract and bring down -1.

Multiply 1 by $x - 1$.

Subtract.

So, $x - 1$ divides evenly into $x^3 - 1$, and you can write

$$\frac{x^3 - 1}{x - 1} = x^2 + x + 1, \quad x \neq 1.$$

You can check this result by multiplying.

$$(x - 1)(x^2 + x + 1) = x^3 + x^2 + x - x^2 - x - 1$$
$$= x^3 - 1$$

✓ **Checkpoint** *Audio-video solution in English & Spanish at LarsonPrecalculus.com.*

Divide $x^3 - 2x^2 - 9$ by $x - 3$.

Note that the condition $x \neq 1$ in Example 1 is equivalent to the condition in the Division Algorithm that $d(x) \neq 0$. Statement of this condition will be omitted for the remainder of this section, but you should be aware that the results of polynomial division are only valid when the divisor does not equal zero.

EXAMPLE 3 **Long Division of Polynomials**

Divide $-5x^2 - 2 + 3x + 2x^4 + 4x^3$ by $2x - 3 + x^2$.

Solution Begin by writing the dividend and divisor in descending powers of x.

$$
\begin{array}{r}
2x^2 \qquad + 1 \\
x^2 + 2x - 3 \,\overline{)\, 2x^4 + 4x^3 - 5x^2 + 3x - 2} \\
\underline{2x^4 + 4x^3 - 6x^2} \\
x^2 + 3x - 2 \\
\underline{x^2 + 2x - 3} \\
x + 1
\end{array}
$$

Multiply $2x^2$ by $x^2 + 2x - 3$.

Subtract and bring down $3x - 2$.

Multiply 1 by $x^2 + 2x - 3$.

Subtract.

Note that the first subtraction eliminated two terms from the dividend. When this happens, the quotient skips a term. You can write the result as

$$\frac{2x^4 + 4x^3 - 5x^2 + 3x - 2}{x^2 + 2x - 3} = 2x^2 + 1 + \frac{x + 1}{x^2 + 2x - 3}.$$

✓ **Checkpoint** *Audio-video solution in English & Spanish at LarsonPrecalculus.com.*

Divide $-x^3 + 9x + 6x^4 - x^2 - 3$ by $1 + 3x$.

Synthetic Division

There is a nice shortcut for long division of polynomials by divisors of the form $x - k$. This shortcut is called **synthetic division.** The pattern for synthetic division of a cubic polynomial is summarized as follows. (The pattern for higher-degree polynomials is similar.)

Synthetic Division (for a Cubic Polynomial)

To divide $ax^3 + bx^2 + cx + d$ by $x - k$, use the following pattern.

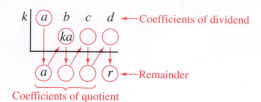

Vertical pattern: Add terms in columns.
Diagonal pattern: Multiply results by k.

This algorithm for synthetic division works only for divisors of the form $x - k$. Remember that $x + k = x - (-k)$.

EXAMPLE 4 **Using Synthetic Division**

Use synthetic division to divide $x^4 - 10x^2 - 2x + 4$ by $x + 3$.

Solution You should set up the array as follows. Note that a zero is included for the missing x^3-term in the dividend.

$$
\begin{array}{c|ccccc}
-3 & 1 & 0 & -10 & -2 & 4 \\
\end{array}
$$

Then, use the synthetic division pattern by adding terms in columns and multiplying the results by -3.

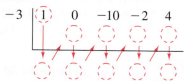

Divisor: $x + 3$ Dividend: $x^4 - 10x^2 - 2x + 4$

$$
\begin{array}{c|ccccc}
-3 & 1 & 0 & -10 & -2 & 4 \\
 & & -3 & 9 & 3 & -3 \\
\hline
 & 1 & -3 & -1 & 1 & 1 \\
\end{array}
$$

Remainder: 1

Quotient: $x^3 - 3x^2 - x + 1$

So, you have

$$
\frac{x^4 - 10x^2 - 2x + 4}{x + 3} = x^3 - 3x^2 - x + 1 + \frac{1}{x + 3}.
$$

✓ *Checkpoint* 🔊)) *Audio-video solution in English & Spanish at LarsonPrecalculus.com.*

Use synthetic division to divide $5x^3 + 8x^2 - x + 6$ by $x + 2$.

The Remainder and Factor Theorems

The remainder obtained in the synthetic division process has an important interpretation, as described in the **Remainder Theorem.**

The Remainder Theorem

If a polynomial $f(x)$ is divided by $x - k$, then the remainder is

$$r = f(k).$$

For a proof of the Remainder Theorem, see Proofs in Mathematics on page 137.

The Remainder Theorem tells you that synthetic division can be used to evaluate a polynomial function. That is, to evaluate a polynomial function $f(x)$ when $x = k$, divide $f(x)$ by $x - k$. The remainder will be $f(k)$, as illustrated in Example 5.

EXAMPLE 5   **Using the Remainder Theorem**

Use the Remainder Theorem to evaluate

$$f(x) = 3x^3 + 8x^2 + 5x - 7$$

when $x = -2$.

Solution   Using synthetic division, you obtain the following.

$$
\begin{array}{r|rrrr}
-2 & 3 & 8 & 5 & -7 \\
   &   & -6 & -4 & -2 \\
\hline
   & 3 & 2 & 1 & -9
\end{array}
$$

Because the remainder is $r = -9$, you can conclude that

$$f(-2) = -9. \qquad \color{red}{r = f(k)}$$

This means that $(-2, -9)$ is a point on the graph of f. You can check this by substituting $x = -2$ in the original function.

Check

$$
\begin{aligned}
f(-2) &= 3(-2)^3 + 8(-2)^2 + 5(-2) - 7 \\
      &= 3(-8) + 8(4) - 10 - 7 \\
      &= -24 + 32 - 10 - 7 \\
      &= -9
\end{aligned}
$$

✓ **Checkpoint** 🔊 *Audio-video solution in English & Spanish at LarsonPrecalculus.com.*

Use the Remainder Theorem to find each function value given

$$f(x) = 4x^3 + 10x^2 - 3x - 8.$$

a. $f(-1)$ **b.** $f(4)$

c. $f\left(\tfrac{1}{2}\right)$ **d.** $f(-3)$

▷ **TECHNOLOGY** You can evaluate a function with your graphing utility by entering the function in the equation editor and using the *table* feature in *ask* mode. When the table is set to *ask* mode, you can enter values in the X column and the corresponding function values will be displayed in the function column.

Another important theorem is the **Factor Theorem,** which is stated below.

The Factor Theorem

A polynomial $f(x)$ has a factor $(x - k)$ if and only if $f(k) = 0$.

For a proof of the Factor Theorem, see Proofs in Mathematics on page 137.

Using the Factor Theorem, you can test whether a polynomial has $(x - k)$ as a factor by evaluating the polynomial at $x = k$. If the result is 0, then $(x - k)$ is a factor.

EXAMPLE 6   Factoring a Polynomial: Repeated Division

Show that $(x - 2)$ and $(x + 3)$ are factors of $f(x) = 2x^4 + 7x^3 - 4x^2 - 27x - 18$. Then find the remaining factors of $f(x)$.

Algebraic Solution

Using synthetic division with the factor $(x - 2)$, you obtain the following.

$$
\begin{array}{r|rrrrr}
2 & 2 & 7 & -4 & -27 & -18 \\
  &   & 4 & 22 & 36 & 18 \\
\hline
  & 2 & 11 & 18 & 9 & 0
\end{array}
$$

0 remainder, so $f(2) = 0$ and $(x - 2)$ is a factor.

Take the result of this division and perform synthetic division again using the factor $(x + 3)$.

$$
\begin{array}{r|rrrr}
-3 & 2 & 11 & 18 & 9 \\
   &   & -6 & -15 & -9 \\
\hline
   & 2 & 5 & 3 & 0
\end{array}
$$

0 remainder, so $f(-3) = 0$ and $(x + 3)$ is a factor.

$2x^2 + 5x + 3$

Because the resulting quadratic expression factors as

$$2x^2 + 5x + 3 = (2x + 3)(x + 1)$$

the complete factorization of $f(x)$ is

$$f(x) = (x - 2)(x + 3)(2x + 3)(x + 1).$$

Graphical Solution

From the graph of $f(x) = 2x^4 + 7x^3 - 4x^2 - 27x - 18$, you can see that there are four x-intercepts (see figure below). These occur at $x = -3$, $x = -\frac{3}{2}$, $x = -1$, and $x = 2$. (Check this algebraically.) This implies that $(x + 3)$, $\left(x + \frac{3}{2}\right)$, $(x + 1)$, and $(x - 2)$ are factors of $f(x)$. [Note that $\left(x + \frac{3}{2}\right)$ and $(2x + 3)$ are equivalent factors because they both yield the same zero, $x = -\frac{3}{2}$.]

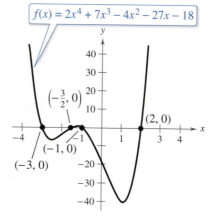

$f(x) = 2x^4 + 7x^3 - 4x^2 - 27x - 18$

✔ **Checkpoint** 🔊))) *Audio-video solution in English & Spanish at LarsonPrecalculus.com.*

Show that $(x + 3)$ is a factor of $f(x) = x^3 - 19x - 30$. Then find the remaining factors of $f(x)$.

Summarize   (Section 2.3)

1. Describe how to divide one polynomial by another polynomial using long division *(pages 95 and 96)*. For examples of long division of polynomials, see Examples 1–3.

2. Describe the algorithm for synthetic division *(page 98)*. For an example of using synthetic division, see Example 4.

3. State the Remainder Theorem and the Factor Theorem *(pages 99 and 100)*. For an example of using the Remainder Theorem, see Example 5. For an example of using the Factor Theorem, see Example 6.

2.4 Complex Numbers

You can use complex numbers to model and solve real-life problems in electronics, such as the impedance of an electrical circuit.

- ◼ Use the imaginary unit *i* to write complex numbers.
- ◼ Add, subtract, and multiply complex numbers.
- ◼ Use complex conjugates to write the quotient of two complex numbers in standard form.
- ◼ Find complex solutions of quadratic equations.

The Imaginary Unit *i*

You have learned that some quadratic equations have no real solutions. For instance, the quadratic equation

$$x^2 + 1 = 0$$

has no real solution because there is no real number x that can be squared to produce -1. To overcome this deficiency, mathematicians created an expanded system of numbers using the **imaginary unit *i*,** defined as

$$i = \sqrt{-1} \qquad \text{Imaginary unit}$$

where $i^2 = -1$. By adding real numbers to real multiples of this imaginary unit, the set of **complex numbers** is obtained. Each complex number can be written in the **standard form $a + bi$.** For instance, the standard form of the complex number $-5 + \sqrt{-9}$ is $-5 + 3i$ because

$$-5 + \sqrt{-9} = -5 + \sqrt{3^2(-1)} = -5 + 3\sqrt{-1} = -5 + 3i.$$

Definition of a Complex Number

Let a and b be real numbers. The number $a + bi$ is called a **complex number,** and it is said to be written in **standard form.** The real number a is called the **real part** and the real number b is called the **imaginary part** of the complex number.

When $b = 0$, the number $a + bi$ is a real number. When $b \neq 0$, the number $a + bi$ is called an **imaginary number.** A number of the form bi, where $b \neq 0$, is called a **pure imaginary number.**

The set of real numbers is a subset of the set of complex numbers, as shown below. This is true because every real number a can be written as a complex number using $b = 0$. That is, for every real number a, you can write $a = a + 0i$.

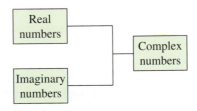

Equality of Complex Numbers

Two complex numbers $a + bi$ and $c + di$, written in standard form, are equal to each other

$$a + bi = c + di \qquad \text{Equality of two complex numbers}$$

if and only if $a = c$ and $b = d$.

Operations with Complex Numbers

To add (or subtract) two complex numbers, you add (or subtract) the real and imaginary parts of the numbers separately.

> ### Addition and Subtraction of Complex Numbers
>
> For two complex numbers $a + bi$ and $c + di$ written in standard form, the sum and difference are defined as follows.
>
> *Sum:* $(a + bi) + (c + di) = (a + c) + (b + d)i$
>
> *Difference:* $(a + bi) - (c + di) = (a - c) + (b - d)i$

The **additive identity** in the complex number system is zero (the same as in the real number system). Furthermore, the **additive inverse** of the complex number $a + bi$ is

$$-(a + bi) = -a - bi. \qquad \text{Additive inverse}$$

So, you have

$$(a + bi) + (-a - bi) = 0 + 0i = 0.$$

EXAMPLE 1   **Adding and Subtracting Complex Numbers**

a. $(4 + 7i) + (1 - 6i) = 4 + 7i + 1 - 6i$ Remove parentheses.

$\qquad\qquad\qquad\qquad = (4 + 1) + (7 - 6)i$ Group like terms.

$\qquad\qquad\qquad\qquad = 5 + i$ Write in standard form.

b. $(1 + 2i) + (3 - 2i) = 1 + 2i + 3 - 2i$ Remove parentheses.

$\qquad\qquad\qquad\qquad = (1 + 3) + (2 - 2)i$ Group like terms.

$\qquad\qquad\qquad\qquad = 4 + 0i$ Simplify.

$\qquad\qquad\qquad\qquad = 4$ Write in standard form.

REMARK Note that the sum of two complex numbers can be a real number.

c. $3i - (-2 + 3i) - (2 + 5i) = 3i + 2 - 3i - 2 - 5i$

$\qquad\qquad\qquad\qquad\qquad = (2 - 2) + (3 - 3 - 5)i$

$\qquad\qquad\qquad\qquad\qquad = 0 - 5i$

$\qquad\qquad\qquad\qquad\qquad = -5i$

d. $(3 + 2i) + (4 - i) - (7 + i) = 3 + 2i + 4 - i - 7 - i$

$\qquad\qquad\qquad\qquad\qquad\qquad = (3 + 4 - 7) + (2 - 1 - 1)i$

$\qquad\qquad\qquad\qquad\qquad\qquad = 0 + 0i$

$\qquad\qquad\qquad\qquad\qquad\qquad = 0$

✓ **Checkpoint**  Audio-video solution in English & Spanish at LarsonPrecalculus.com.

Perform each operation and write the result in standard form.

a. $(7 + 3i) + (5 - 4i)$

b. $(3 + 4i) - (5 - 3i)$

c. $2i + (-3 - 4i) - (-3 - 3i)$

d. $(5 - 3i) + (3 + 5i) - (8 + 2i)$

Many of the properties of real numbers are valid for complex numbers as well. Here are some examples.

Associative Properties of Addition and Multiplication

Commutative Properties of Addition and Multiplication

Distributive Property of Multiplication Over Addition

Notice below how these properties are used when two complex numbers are multiplied.

$(a + bi)(c + di) = a(c + di) + bi(c + di)$		Distributive Property
$\quad = ac + (ad)i + (bc)i + (bd)i^2$		Distributive Property
$\quad = ac + (ad)i + (bc)i + (bd)(-1)$		$i^2 = -1$
$\quad = ac - bd + (ad)i + (bc)i$		Commutative Property
$\quad = (ac - bd) + (ad + bc)i$		Associative Property

Rather than trying to memorize this multiplication rule, you should simply remember how to use the Distributive Property to multiply two complex numbers.

EXAMPLE 2   Multiplying Complex Numbers

a. $4(-2 + 3i) = 4(-2) + 4(3i)$	Distributive Property
$\qquad\qquad = -8 + 12i$	Simplify.
b. $(2 - i)(4 + 3i) = 2(4 + 3i) - i(4 + 3i)$	Distributive Property
$\qquad\qquad = 8 + 6i - 4i - 3i^2$	Distributive Property
$\qquad\qquad = 8 + 6i - 4i - 3(-1)$	$i^2 = -1$
$\qquad\qquad = (8 + 3) + (6 - 4)i$	Group like terms.
$\qquad\qquad = 11 + 2i$	Write in standard form.
c. $(3 + 2i)(3 - 2i) = 3(3 - 2i) + 2i(3 - 2i)$	Distributive Property
$\qquad\qquad = 9 - 6i + 6i - 4i^2$	Distributive Property
$\qquad\qquad = 9 - 6i + 6i - 4(-1)$	$i^2 = -1$
$\qquad\qquad = 9 + 4$	Simplify.
$\qquad\qquad = 13$	Write in standard form.
d. $(3 + 2i)^2 = (3 + 2i)(3 + 2i)$	Square of a binomial
$\qquad\qquad = 3(3 + 2i) + 2i(3 + 2i)$	Distributive Property
$\qquad\qquad = 9 + 6i + 6i + 4i^2$	Distributive Property
$\qquad\qquad = 9 + 6i + 6i + 4(-1)$	$i^2 = -1$
$\qquad\qquad = 9 + 12i - 4$	Simplify.
$\qquad\qquad = 5 + 12i$	Write in standard form.

REMARK The procedure described above is similar to multiplying two binomials and combining like terms, as in the FOIL Method shown in Appendix A.3. For instance, you can use the FOIL Method to multiply the two complex numbers from Example 2(b).

$$\overset{\text{F}\quad\text{O}\quad\text{I}\quad\text{L}}{(2 - i)(4 + 3i)} = 8 + 6i - 4i - 3i^2$$

✓ **Checkpoint** ◄))) *Audio-video solution in English & Spanish at LarsonPrecalculus.com.*

Perform each operation and write the result in standard form.

a. $(2 - 4i)(3 + 3i)$

b. $(4 + 5i)(4 - 5i)$

c. $(4 + 2i)^2$

Complex Conjugates

▷ **ALGEBRA HELP** You can compare complex conjugates with the method for rationalizing denominators in Appendix A.2.

Notice in Example 2(c) that the product of two complex numbers can be a real number. This occurs with pairs of complex numbers of the form $a + bi$ and $a - bi$, called **complex conjugates.**

$$(a + bi)(a - bi) = a^2 - abi + abi - b^2i^2$$
$$= a^2 - b^2(-1)$$
$$= a^2 + b^2$$

EXAMPLE 3 **Multiplying Conjugates**

Multiply each complex number by its complex conjugate.

a. $1 + i$ **b.** $4 - 3i$

Solution

a. The complex conjugate of $1 + i$ is $1 - i$.

$$(1 + i)(1 - i) = 1^2 - i^2 = 1 - (-1) = 2$$

b. The complex conjugate of $4 - 3i$ is $4 + 3i$.

$$(4 - 3i)(4 + 3i) = 4^2 - (3i)^2 = 16 - 9i^2 = 16 - 9(-1) = 25$$

✓ **Checkpoint** 🔊))) *Audio-video solution in English & Spanish at LarsonPrecalculus.com.*

Multiply each complex number by its complex conjugate.

a. $3 + 6i$

b. $2 - 5i$

To write the quotient of $a + bi$ and $c + di$ in standard form, where c and d are not both zero, multiply the numerator and denominator by the complex conjugate of the *denominator* to obtain

$$\frac{a + bi}{c + di} = \frac{a + bi}{c + di}\left(\frac{c - di}{c - di}\right) = \frac{(ac + bd) + (bc - ad)i}{c^2 + d^2}.$$

REMARK

Note that when you multiply the numerator and denominator of a quotient of complex numbers by

$$\frac{c - di}{c - di}$$

you are actually multiplying the quotient by a form of 1. You are not changing the original expression, you are only creating an expression that is equivalent to the original expression.

EXAMPLE 4 **Quotient of Complex Numbers in Standard Form**

$$\frac{2 + 3i}{4 - 2i} = \frac{2 + 3i}{4 - 2i}\left(\frac{4 + 2i}{4 + 2i}\right)$$ Multiply numerator and denominator by complex conjugate of denominator.

$$= \frac{8 + 4i + 12i + 6i^2}{16 - 4i^2}$$ Expand.

$$= \frac{8 - 6 + 16i}{16 + 4}$$ $i^2 = -1$

$$= \frac{2 + 16i}{20}$$ Simplify.

$$= \frac{1}{10} + \frac{4}{5}i$$ Write in standard form.

✓ **Checkpoint** 🔊))) *Audio-video solution in English & Spanish at LarsonPrecalculus.com.*

Write $\dfrac{2 + i}{2 - i}$ in standard form.

Complex Solutions of Quadratic Equations

You can write a number such as $\sqrt{-3}$ in standard form by factoring out $i = \sqrt{-1}$.

$$\sqrt{-3} = \sqrt{3(-1)} = \sqrt{3}\sqrt{-1} = \sqrt{3}i$$

The number $\sqrt{3}i$ is called the *principal square root* of -3.

▷ **REMARK**

The definition of principal square root uses the rule

$$\sqrt{ab} = \sqrt{a}\sqrt{b}$$

for $a > 0$ and $b < 0$. This rule is not valid if *both* a and b are negative. For example,

$$\sqrt{-5}\sqrt{-5} = \sqrt{5(-1)}\sqrt{5(-1)}$$
$$= \sqrt{5}i\sqrt{5}i$$
$$= \sqrt{25}i^2$$
$$= 5i^2$$
$$= -5$$

whereas

$$\sqrt{(-5)(-5)} = \sqrt{25} = 5.$$

To avoid problems with square roots of negative numbers, be sure to convert complex numbers to standard form *before* multiplying.

▷ **ALGEBRA HELP** You can review the techniques for using the Quadratic Formula in Appendix A.5.

Principal Square Root of a Negative Number

When a is a positive real number, the **principal square root** of $-a$ is defined as

$$\sqrt{-a} = \sqrt{a}i.$$

EXAMPLE 5   **Writing Complex Numbers in Standard Form**

a. $\sqrt{-3}\sqrt{-12} = \sqrt{3}i\sqrt{12}i = \sqrt{36}i^2 = 6(-1) = -6$

b. $\sqrt{-48} - \sqrt{-27} = \sqrt{48}i - \sqrt{27}i = 4\sqrt{3}i - 3\sqrt{3}i = \sqrt{3}i$

c. $\left(-1 + \sqrt{-3}\right)^2 = \left(-1 + \sqrt{3}i\right)^2 = (-1)^2 - 2\sqrt{3}i + \left(\sqrt{3}\right)^2(i^2)$
$$= 1 - 2\sqrt{3}i + 3(-1)$$
$$= -2 - 2\sqrt{3}i$$

✓ *Checkpoint*  Audio-video solution in English & Spanish at LarsonPrecalculus.com.

Write $\sqrt{-14}\sqrt{-2}$ in standard form.

EXAMPLE 6   **Complex Solutions of a Quadratic Equation**

Solve $3x^2 - 2x + 5 = 0$.

Solution

$$x = \frac{-(-2) \pm \sqrt{(-2)^2 - 4(3)(5)}}{2(3)} \qquad \text{Quadratic Formula}$$

$$= \frac{2 \pm \sqrt{-56}}{6} \qquad \text{Simplify.}$$

$$= \frac{2 \pm 2\sqrt{14}i}{6} \qquad \text{Write } \sqrt{-56} \text{ in standard form.}$$

$$= \frac{1}{3} \pm \frac{\sqrt{14}}{3}i \qquad \text{Write in standard form.}$$

✓ *Checkpoint* ◀)) Audio-video solution in English & Spanish at LarsonPrecalculus.com.

Solve $8x^2 + 14x + 9 = 0$. ◼

Summarize   (Section 2.4)

1. Describe how to write complex numbers using the imaginary unit i (*page 101*).

2. Describe how to add, subtract, and multiply complex numbers (*pages 102 and 103, Examples 1 and 2*).

3. Describe how to use complex conjugates to write the quotient of two complex numbers in standard form (*page 104, Example 4*).

4. Describe how to find complex solutions of a quadratic equation (*page 105, Example 6*).

2.5 Zeros of Polynomial Functions

- Use the Fundamental Theorem of Algebra to determine the number of zeros of polynomial functions.
- Find rational zeros of polynomial functions.
- Find conjugate pairs of complex zeros.
- Find zeros of polynomials by factoring.
- Use Descartes's Rule of Signs and the Upper and Lower Bound Rules to find zeros of polynomials.
- Find zeros of polynomials in real-life applications.

The Fundamental Theorem of Algebra

In the complex number system, every nth-degree polynomial function has *precisely n* zeros. This important result is derived from the **Fundamental Theorem of Algebra,** first proved by the German mathematician Carl Friedrich Gauss (1777–1855).

Finding zeros of polynomial functions is an important part of solving real-life problems. For instance, the zeros of a polynomial function can help you redesign a storage bin so that it can hold five times as much food.

> **The Fundamental Theorem of Algebra**
>
> If $f(x)$ is a polynomial of degree n, where $n > 0$, then f has at least one zero in the complex number system.

Using the Fundamental Theorem of Algebra and the equivalence of zeros and factors, you obtain the **Linear Factorization Theorem.**

> **Linear Factorization Theorem**
>
> If $f(x)$ is a polynomial of degree n, where $n > 0$, then $f(x)$ has precisely n linear factors
>
> $$f(x) = a_n(x - c_1)(x - c_2) \cdots (x - c_n)$$
>
> where $c_1, c_2, \ldots, c_n$ are complex numbers.

•• **REMARK** Recall that in order to find the zeros of a function f, set $f(x)$ equal to 0 and solve the resulting equation for x. For instance, the function in Example 1(a) has a zero at $x = 2$ because

$$x - 2 = 0$$

$$x = 2.$$

For a proof of the Linear Factorization Theorem, see Proofs in Mathematics on page 138.

Note that the Fundamental Theorem of Algebra and the Linear Factorization Theorem tell you only that the zeros or factors of a polynomial exist, not how to find them. Such theorems are called **existence theorems.**

EXAMPLE 1 **Zeros of Polynomial Functions**

a. The first-degree polynomial function $f(x) = x - 2$ has exactly *one* zero: $x = 2$.

b. Counting multiplicity, the second-degree polynomial function

$$f(x) = x^2 - 6x + 9 = (x - 3)(x - 3)$$

has exactly *two* zeros: $x = 3$ and $x = 3$. (This is called a *repeated zero.*)

c. The third-degree polynomial function

$$f(x) = x^3 + 4x = x(x^2 + 4) = x(x - 2i)(x + 2i)$$

has exactly *three* zeros: $x = 0$, $x = 2i$, and $x = -2i$.

▷ **ALGEBRA HELP** Examples 1(b) and 1(c) involve factoring polynomials. You can review the techniques for factoring polynomials in Appendix A.3.

✓ *Checkpoint* ◀)) *Audio-video solution in English & Spanish at LarsonPrecalculus.com.*

Determine the number of zeros of the polynomial function $f(x) = x^4 - 1$.

The Rational Zero Test

The **Rational Zero Test** relates the possible rational zeros of a polynomial (having integer coefficients) to the leading coefficient and to the constant term of the polynomial.

> **The Rational Zero Test**
>
> If the polynomial $f(x) = a_n x^n + a_{n-1} x^{n-1} + \cdots + a_2 x^2 + a_1 x + a_0$ has *integer* coefficients, then every rational zero of f has the form
>
> $$\text{Rational zero} = \frac{p}{q}$$
>
> where p and q have no common factors other than 1, and
>
> $p = $ a factor of the constant term a_0
>
> $q = $ a factor of the leading coefficient a_n.

To use the Rational Zero Test, you should first list all rational numbers whose numerators are factors of the constant term and whose denominators are factors of the leading coefficient.

$$\text{Possible rational zeros} = \frac{\text{factors of constant term}}{\text{factors of leading coefficient}}$$

Having formed this list of *possible rational zeros,* use a trial-and-error method to determine which, if any, are actual zeros of the polynomial. Note that when the leading coefficient is 1, the possible rational zeros are simply the factors of the constant term.

Although they were not contemporaries, Jean Le Rond d'Alembert (1717–1783) worked independently of Carl Gauss in trying to prove the Fundamental Theorem of Algebra. His efforts were such that, in France, the Fundamental Theorem of Algebra is frequently known as the Theorem of d'Alembert.

EXAMPLE 2   **Rational Zero Test with Leading Coefficient of 1**

Find the rational zeros of

$$f(x) = x^3 + x + 1.$$

Solution   Because the leading coefficient is 1, the possible rational zeros are the factors of the constant term.

Possible rational zeros: 1 and -1

By testing these possible zeros, you can see that neither works.

$$f(1) = (1)^3 + 1 + 1$$
$$= 3$$
$$f(-1) = (-1)^3 + (-1) + 1$$
$$= -1$$

So, you can conclude that the given polynomial has *no* rational zeros. Note from the graph of f in Figure 2.17 that f does have one real zero between -1 and 0. However, by the Rational Zero Test, you know that this real zero is *not* a rational number.

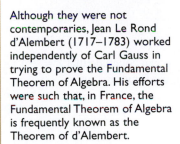

Figure 2.17

✓ *Checkpoint* 🔊 *Audio-video solution in English & Spanish at LarsonPrecalculus.com.*

Find the rational zeros of each function.

a. $f(x) = x^3 - 5x^2 + 2x + 8$

b. $f(x) = x^3 + 2x^2 + 6x - 4$

c. $f(x) = x^3 - 3x^2 + 2x - 6$

EXAMPLE 3   **Rational Zero Test with Leading Coefficient of 1**

Find the rational zeros of

$$f(x) = x^4 - x^3 + x^2 - 3x - 6.$$

Solution   Because the leading coefficient is 1, the possible rational zeros are the factors of the constant term.

Possible rational zeros: $\pm 1, \pm 2, \pm 3, \pm 6$

By applying synthetic division successively, you can determine that $x = -1$ and $x = 2$ are the only two rational zeros.

$$
\begin{array}{r|rrrrr}
-1 & 1 & -1 & 1 & -3 & -6 \\
   &   & -1 & 2 & -3 & 6 \\
\hline
   & 1 & -2 & 3 & -6 & 0
\end{array}
$$
    ⟶ 0 remainder, so $x = -1$ is a zero.

$$
\begin{array}{r|rrrr}
2 & 1 & -2 & 3 & -6 \\
  &   & 2 & 0 & 6 \\
\hline
  & 1 & 0 & 3 & 0
\end{array}
$$
    ⟶ 0 remainder, so $x = 2$ is a zero.

So, $f(x)$ factors as

$$f(x) = (x + 1)(x - 2)(x^2 + 3).$$

Because the factor $(x^2 + 3)$ produces no real zeros, you can conclude that $x = -1$ and $x = 2$ are the only *real* zeros of f, which is verified in the figure below.

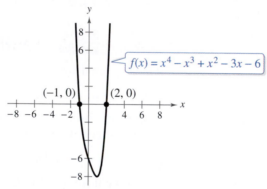

✔ **Checkpoint** 🔊)) *Audio-video solution in English & Spanish at LarsonPrecalculus.com.*

Find the rational zeros of

$$f(x) = x^3 - 15x^2 + 75x - 125.$$ ∎

When the leading coefficient of a polynomial is not 1, the list of possible rational zeros can increase dramatically. In such cases, the search can be shortened in several ways: (1) a programmable calculator can be used to speed up the calculations; (2) a graph, drawn either by hand or with a graphing utility, can give good estimates of the locations of the zeros; (3) the Intermediate Value Theorem along with a table generated by a graphing utility can give approximations of the zeros; and (4) synthetic division can be used to test the possible rational zeros.

Finding the first zero is often the most difficult part. After that, the search is simplified by working with the lower-degree polynomial obtained in synthetic division, as shown in Example 3.

........................▷

•• **REMARK**   When the list of possible rational zeros is small, as in Example 2, it may be quicker to test the zeros by evaluating the function. When the list of possible rational zeros is large, as in Example 3, it may be quicker to use a different approach to test the zeros, such as using synthetic division or sketching a graph.

Using the Rational Zero Test

Find the rational zeros of $f(x) = 2x^3 + 3x^2 - 8x + 3$.

Solution The leading coefficient is 2 and the constant term is 3.

$$\textit{Possible rational zeros:} \quad \frac{\text{Factors of 3}}{\text{Factors of 2}} = \frac{\pm 1, \pm 3}{\pm 1, \pm 2} = \pm 1, \pm 3, \pm \frac{1}{2}, \pm \frac{3}{2}$$

By synthetic division, you can determine that $x = 1$ is a rational zero.

$$
\begin{array}{r|rrrr}
1 & 2 & 3 & -8 & 3 \\
  &   & 2 & 5 & -3 \\
\hline
  & 2 & 5 & -3 & 0
\end{array}
$$

So, $f(x)$ factors as

$$f(x) = (x - 1)(2x^2 + 5x - 3)$$

$$= (x - 1)(2x - 1)(x + 3)$$

and you can conclude that the rational zeros of f are $x = 1$, $x = \frac{1}{2}$, and $x = -3$.

✓ Checkpoint ◀))) *Audio-video solution in English & Spanish at LarsonPrecalculus.com.*

Find the rational zeros of

$$f(x) = 2x^4 - 9x^3 - 18x^2 + 71x - 30.$$

> **REMARK** Remember that when you try to find the rational zeros of a polynomial function with many possible rational zeros, as in Example 4, you must use trial and error. There is no quick algebraic method to determine which of the possibilities is an actual zero; however, sketching a graph may be helpful.

Recall from Section 2.2 that if $x = a$ is a zero of the polynomial function f, then $x = a$ is a solution of the polynomial equation $f(x) = 0$.

Solving a Polynomial Equation

Find all real solutions of $-10x^3 + 15x^2 + 16x - 12 = 0$.

Solution The leading coefficient is -10 and the constant term is -12.

$$\textit{Possible rational solutions:} \quad \frac{\text{Factors of } -12}{\text{Factors of } -10} = \frac{\pm 1, \pm 2, \pm 3, \pm 4, \pm 6, \pm 12}{\pm 1, \pm 2, \pm 5, \pm 10}$$

With so many possibilities (32, in fact), it is worth your time to stop and sketch a graph. From Figure 2.18, it looks like three reasonable solutions would be $x = -\frac{6}{5}$, $x = \frac{1}{2}$, and $x = 2$. Testing these by synthetic division shows that $x = 2$ is the only rational solution. So, you have

$$(x - 2)(-10x^2 - 5x + 6) = 0.$$

Using the Quadratic Formula for the second factor, you find that the two additional solutions are irrational numbers.

$$x = \frac{-5 - \sqrt{265}}{20} \approx -1.0639$$

and

$$x = \frac{-5 + \sqrt{265}}{20} \approx 0.5639$$

✓ Checkpoint ◀))) *Audio-video solution in English & Spanish at LarsonPrecalculus.com.*

Find all real solutions of

$$x^3 + 4x^2 - 15x - 18 = 0.$$

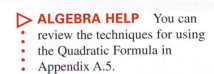

$f(x) = -10x^3 + 15x^2 + 16x - 12$

Figure 2.18

> ▷ **ALGEBRA HELP** You can review the techniques for using the Quadratic Formula in Appendix A.5.

Conjugate Pairs

In Example 1(c), note that the pair of complex zeros are conjugates. That is, they are of the form

$$a + bi$$

and

$$a - bi.$$

> ### Complex Zeros Occur in Conjugate Pairs
>
> Let f be a polynomial function that has *real coefficients*. If $a + bi$, where $b \neq 0$, is a zero of the function, then the conjugate $a - bi$ is also a zero of the function.

Be sure you see that this result is true only when the polynomial function has *real coefficients*. For instance, the result applies to the function

$$f(x) = x^2 + 1$$

but not to the function

$$g(x) = x - i.$$

EXAMPLE 6   Finding a Polynomial Function with Given Zeros

Find a fourth-degree polynomial function with real coefficients that has

$$-1, \quad -1, \quad \text{and} \quad 3i$$

as zeros.

Solution   Because $3i$ is a zero *and* the polynomial is stated to have real coefficients, you know that the conjugate $-3i$ must also be a zero. So, the four zeros are

$$-1, \quad -1, \quad 3i, \quad \text{and} \quad -3i.$$

Then, using the Linear Factorization Theorem, $f(x)$ can be written as

$$f(x) = a(x + 1)(x + 1)(x - 3i)(x + 3i).$$

For simplicity, let $a = 1$. Then multiply the factors with real coefficients to obtain

$$(x + 1)(x + 1) = x^2 + 2x + 1$$

and multiply the complex conjugates to obtain

$$(x - 3i)(x + 3i) = x^2 + 9.$$

So, you obtain the following fourth-degree polynomial function.

$$f(x) = (x^2 + 2x + 1)(x^2 + 9)$$
$$= x^4 + 2x^3 + 10x^2 + 18x + 9$$

✓ *Checkpoint*  *Audio-video solution in English & Spanish at LarsonPrecalculus.com.*

Find a fourth-degree polynomial function with real coefficients that has the given zeros.

a. 2, -2, and $-7i$

b. 1, 3, and $4 - i$

c. -1, 2, and $3 + i$

Factoring a Polynomial

The Linear Factorization Theorem shows that you can write any nth-degree polynomial as the product of n linear factors.

$$f(x) = a_n(x - c_1)(x - c_2)(x - c_3) \cdots (x - c_n)$$

However, this result includes the possibility that some of the values of c_i are complex. The following theorem says that even if you do not want to get involved with "complex factors," you can still write $f(x)$ as the product of linear and/or quadratic factors. For a proof of this theorem, see Proofs in Mathematics on page 138.

Factors of a Polynomial

Every polynomial of degree $n > 0$ with real coefficients can be written as the product of linear and quadratic factors with real coefficients, where the quadratic factors have no real zeros.

A quadratic factor with no real zeros is said to be *prime* or **irreducible over the reals**. Note that this is not the same as being *irreducible over the rationals*. For example, the quadratic $x^2 + 1 = (x - i)(x + i)$ is irreducible over the reals (and therefore over the rationals). On the other hand, the quadratic $x^2 - 2 = \left(x - \sqrt{2}\right)\left(x + \sqrt{2}\right)$ is irreducible over the rationals but *reducible* over the reals.

▷ TECHNOLOGY In Example 7, you can use a graphing utility to graph the function, as shown below.

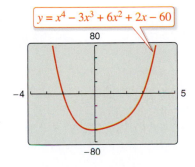

$y = x^4 - 3x^3 + 6x^2 + 2x - 60$

Then you can use the *zero* or *root* feature of the graphing utility to determine that $x = -2$ and $x = 3$ are zeros of the function.

▷ **ALGEBRA HELP** You can review the techniques for polynomial long division in Section 2.3.

EXAMPLE 7   **Finding the Zeros of a Polynomial Function**

Find all the zeros of $f(x) = x^4 - 3x^3 + 6x^2 + 2x - 60$ given that $1 + 3i$ is a zero of f.

Solution   Because complex zeros occur in conjugate pairs, you know that $1 - 3i$ is also a zero of f. This means that both $[x - (1 + 3i)]$ and $[x - (1 - 3i)]$ are factors of $f(x)$. Multiplying these two factors produces

$$[x - (1 + 3i)][x - (1 - 3i)] = [(x - 1) - 3i][(x - 1) + 3i]$$
$$= (x - 1)^2 - 9i^2$$
$$= x^2 - 2x + 10.$$

Using long division, you can divide $x^2 - 2x + 10$ into $f(x)$ to obtain the following.

$$
\begin{array}{r}
x^2 - x - 6 \\
x^2 - 2x + 10 \overline{\smash{\big)}\ x^4 - 3x^3 + 6x^2 + 2x - 60} \\
\underline{x^4 - 2x^3 + 10x^2} \\
-x^3 - 4x^2 + 2x \\
\underline{-x^3 + 2x^2 - 10x} \\
-6x^2 + 12x - 60 \\
\underline{-6x^2 + 12x - 60} \\
0
\end{array}
$$

So, you have

$$f(x) = (x^2 - 2x + 10)(x^2 - x - 6) = (x^2 - 2x + 10)(x - 3)(x + 2)$$

and you can conclude that the zeros of f are $x = 1 + 3i$, $x = 1 - 3i$, $x = 3$, and $x = -2$.

✓ **Checkpoint** ◀))) *Audio-video solution in English & Spanish at LarsonPrecalculus.com.*

Find all the zeros of $f(x) = 3x^3 - 2x^2 + 48x - 32$ given that $4i$ is a zero of f.

In Example 7, without being told that $1 + 3i$ is a zero of f, you could still find all the zeros of the function by using synthetic division to find the real zeros -2 and 3. Then you could factor the polynomial as

$$(x + 2)(x - 3)(x^2 - 2x + 10).$$

Finally, by using the Quadratic Formula, you could determine that the zeros are

$$x = -2, \quad x = 3, \quad x = 1 + 3i, \quad \text{and} \quad x = 1 - 3i.$$

Example 8 shows how to find all the zeros of a polynomial function, including complex zeros.

▷ TECHNOLOGY In Example 8, the fifth-degree polynomial function has three real zeros. In such cases, you can use the *zoom* and *trace* features or the *zero* or *root* feature of a graphing utility to approximate the real zeros. You can then use these real zeros to determine the complex zeros algebraically.

EXAMPLE 8   **Finding the Zeros of a Polynomial Function**

Write

$$f(x) = x^5 + x^3 + 2x^2 - 12x + 8$$

as the product of linear factors and list all the zeros of the function.

Solution   Because the leading coefficient is 1, the possible rational zeros are the factors of the constant term.

Possible rational zeros: $\pm 1, \pm 2, \pm 4,$ and ± 8

Synthetic division produces the following.

$$
\begin{array}{r|rrrrrr}
1 & 1 & 0 & 1 & 2 & -12 & 8 \\
  &   & 1 & 1 & 2 & 4 & -8 \\
\hline
  & 1 & 1 & 2 & 4 & -8 & 0 \\
\end{array}
\longrightarrow \text{1 is a zero.}
$$

$$
\begin{array}{r|rrrrr}
-2 & 1 & 1 & 2 & 4 & -8 \\
   &   & -2 & 2 & -8 & 8 \\
\hline
   & 1 & -1 & 4 & -4 & 0 \\
\end{array}
\longrightarrow \text{-2 is a zero.}
$$

So, you have

$$f(x) = x^5 + x^3 + 2x^2 - 12x + 8$$
$$= (x - 1)(x + 2)(x^3 - x^2 + 4x - 4).$$

You can factor $x^3 - x^2 + 4x - 4$ as $(x - 1)(x^2 + 4)$, and by factoring $x^2 + 4$ as

$$x^2 - (-4) = \left(x - \sqrt{-4}\right)\left(x + \sqrt{-4}\right)$$
$$= (x - 2i)(x + 2i)$$

you obtain

$$f(x) = (x - 1)(x - 1)(x + 2)(x - 2i)(x + 2i)$$

which gives the following five zeros of f.

$$x = 1, \quad x = 1, \quad x = -2, \quad x = 2i, \quad \text{and} \quad x = -2i$$

From the graph of f shown in Figure 2.19, you can see that the *real* zeros are the only ones that appear as x-intercepts. Note that $x = 1$ is a repeated zero.

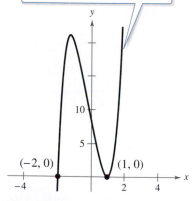

$f(x) = x^5 + x^3 + 2x^2 - 12x + 8$

$(-2, 0)$ $(1, 0)$

Figure 2.19

✓ **Checkpoint** ◀))) *Audio-video solution in English & Spanish at LarsonPrecalculus.com.*

Write each function as the product of linear factors and list all the zeros of the function.

a. $f(x) = x^4 + 8x^2 - 9$

b. $g(x) = x^3 - 3x^2 + 7x - 5$

c. $h(x) = x^3 - 11x^2 + 41x - 51$

Other Tests for Zeros of Polynomials

You know that an *n*th-degree polynomial function can have *at most n* real zeros. Of course, many *n*th-degree polynomials do not have that many real zeros. For instance, $f(x) = x^2 + 1$ has no real zeros, and $f(x) = x^3 + 1$ has only one real zero. The following theorem, called **Descartes's Rule of Signs,** sheds more light on the number of real zeros of a polynomial.

Descartes's Rule of Signs

Let $f(x) = a_n x^n + a_{n-1} x^{n-1} + \cdots + a_2 x^2 + a_1 x + a_0$ be a polynomial with real coefficients and $a_0 \neq 0$.

1. The number of *positive real zeros* of *f* is either equal to the number of variations in sign of $f(x)$ or less than that number by an even integer.

2. The number of *negative real zeros* of *f* is either equal to the number of variations in sign of $f(-x)$ or less than that number by an even integer.

A **variation in sign** means that two consecutive coefficients have opposite signs.

When using Descartes's Rule of Signs, a zero of multiplicity *k* should be counted as *k* zeros. For instance, the polynomial $x^3 - 3x + 2$ has two variations in sign, and so it has either two positive or no positive real zeros. Because

$$x^3 - 3x + 2 = (x - 1)(x - 1)(x + 2)$$

you can see that the two positive real zeros are $x = 1$ of multiplicity 2.

EXAMPLE 9   **Using Descartes's Rule of Signs**

Determine the possible numbers of positive and negative real zeros of

$$f(x) = 3x^3 - 5x^2 + 6x - 4.$$

Solution   The original polynomial has *three* variations in sign.

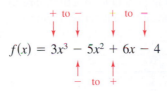

The polynomial

$$f(-x) = 3(-x)^3 - 5(-x)^2 + 6(-x) - 4$$
$$= -3x^3 - 5x^2 - 6x - 4$$

has no variations in sign. So, from Descartes's Rule of Signs, the polynomial

$$f(x) = 3x^3 - 5x^2 + 6x - 4$$

has either three positive real zeros or one positive real zero, and has no negative real zeros. From the graph in Figure 2.20, you can see that the function has only one real zero, $x = 1$.

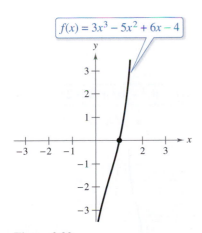

$f(x) = 3x^3 - 5x^2 + 6x - 4$

Figure 2.20

✓ **Checkpoint**))) *Audio-video solution in English & Spanish at LarsonPrecalculus.com.*

Determine the possible numbers of positive and negative real zeros of

$$f(x) = -2x^3 + 5x^2 - x + 8.$$

Another test for zeros of a polynomial function is related to the sign pattern in the last row of the synthetic division array. This test can give you an upper or lower bound for the real zeros of f. A real number c is an **upper bound** for the real zeros of f when no zeros are greater than c. Similarly, c is a **lower bound** when no real zeros of f are less than c.

Upper and Lower Bound Rules

Let $f(x)$ be a polynomial with real coefficients and a positive leading coefficient. Suppose $f(x)$ is divided by $x - c$, using synthetic division.

1. If $c > 0$ and each number in the last row is either positive or zero, then c is an **upper bound** for the real zeros of f.

2. If $c < 0$ and the numbers in the last row are alternately positive and negative (zero entries count as positive or negative), then c is a **lower bound** for the real zeros of f.

EXAMPLE 10   **Finding the Zeros of a Polynomial Function**

Find all real zeros of

$$f(x) = 6x^3 - 4x^2 + 3x - 2.$$

Solution   The possible real zeros are as follows.

$$\frac{\text{Factors of } -2}{\text{Factors of } 6} = \frac{\pm 1, \pm 2}{\pm 1, \pm 2, \pm 3, \pm 6} = \pm 1, \pm\frac{1}{2}, \pm\frac{1}{3}, \pm\frac{1}{6}, \pm\frac{2}{3}, \pm 2$$

The original polynomial $f(x)$ has three variations in sign. The polynomial

$$f(-x) = 6(-x)^3 - 4(-x)^2 + 3(-x) - 2$$
$$= -6x^3 - 4x^2 - 3x - 2$$

has no variations in sign. As a result of these two findings, you can apply Descartes's Rule of Signs to conclude that there are three positive real zeros or one positive real zero, and no negative real zeros. Trying $x = 1$ produces the following.

$$
\begin{array}{r|rrrr}
1 & 6 & -4 & 3 & -2 \\
  &   & 6 & 2 & 5 \\
\hline
  & 6 & 2 & 5 & 3
\end{array}
$$

So, $x = 1$ is not a zero, but because the last row has all positive entries, you know that $x = 1$ is an upper bound for the real zeros. So, you can restrict the search to zeros between 0 and 1. By trial and error, you can determine that $x = \frac{2}{3}$ is a zero. So,

$$f(x) = \left(x - \frac{2}{3}\right)(6x^2 + 3).$$

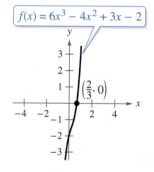

Because $6x^2 + 3$ has no real zeros, it follows that $x = \frac{2}{3}$ is the only real zero, as shown at the right.

✓ **Checkpoint** 🔊))) *Audio-video solution in English & Spanish at LarsonPrecalculus.com.*

Find all real zeros of

$$f(x) = 8x^3 - 4x^2 + 6x - 3.$$

Application

 EXAMPLE 11 Using a Polynomial Model

You are designing candle making kits. Each kit contains 25 cubic inches of candle wax and a mold for making a pyramid-shaped candle. You want the height of the candle to be 2 inches less than the length of each side of the candle's square base. What should the dimensions of your candle mold be?

Solution The volume of a pyramid is $V = \frac{1}{3}Bh$, where B is the area of the base and h is the height. The area of the base is x^2 and the height is $(x - 2)$. So, the volume of the pyramid is $V = \frac{1}{3}x^2(x - 2)$. Substituting 25 for the volume yields the following.

$$25 = \tfrac{1}{3}x^2(x - 2) \qquad \text{\color{red}Substitute 25 for } V.$$

$$75 = x^3 - 2x^2 \qquad \text{\color{red}Multiply each side by 3, and distribute } x^2.$$

$$0 = x^3 - 2x^2 - 75 \qquad \text{\color{red}Write in general form.}$$

The possible rational solutions are $x = \pm1, \pm3, \pm5, \pm15, \pm25, \pm75$. Use synthetic division to test some of the possible solutions. Note that in this case it makes sense to test only positive x-values. Using synthetic division, you can determine that $x = 5$ is a solution.

$$
\begin{array}{r|rrrr}
5 & 1 & -2 & 0 & -75 \\
  &   & 5 & 15 & 75 \\
\hline
  & 1 & 3 & 15 & 0
\end{array}
$$

The other two solutions that satisfy $x^2 + 3x + 15 = 0$ are imaginary and can be discarded. You can conclude that the base of the candle mold should be 5 inches by 5 inches and the height should be $5 - 2 = 3$ inches.

✓ *Checkpoint* *Audio-video solution in English & Spanish at LarsonPrecalculus.com.*

Rework Example 11 when each kit contains 147 cubic inches of candle wax and you want the height of the pyramid-shaped candle to be 2 inches more than the length of each side of the candle's square base. ■

Before concluding this section, here is an additional hint that can help you find the real zeros of a polynomial. When the terms of $f(x)$ have a common monomial factor, it should be factored out before applying the tests in this section. For instance, by writing

$$f(x) = x^4 - 5x^3 + 3x^2 + x = x(x^3 - 5x^2 + 3x + 1)$$

you can see that $x = 0$ is a zero of f and that the remaining zeros can be obtained by analyzing the cubic factor.

According to the National Candle Association, 7 out of 10 U.S. households use candles.

Summarize (Section 2.5)

1. State the Fundamental Theorem of Algebra and the Linear Factorization Theorem *(page 106, Example 1)*.

2. Describe the Rational Zero Test *(page 107, Examples 2–5)*.

3. State the definition of conjugates *(page 110, Example 6)*.

4. Describe what it means for a quadratic factor to be irreducible over the reals *(page 111, Examples 7 and 8)*.

5. State Descartes's Rule of Signs and the Upper and Lower Bound Rules *(pages 113 and 114, Examples 9 and 10)*.

6. Describe a real-life situation involving finding the zeros of a polynomial *(page 115, Example 11)*.

2.6 Rational Functions

Rational functions can help you model and solve real-life problems relating to environmental scenarios, such as determining the cost of supplying recycling bins to the population of a rural township.

- Find the domains of rational functions.
- Find the vertical and horizontal asymptotes of the graphs of rational functions.
- Sketch the graphs of rational functions.
- Sketch the graphs of rational functions that have slant asymptotes.
- Use rational functions to model and solve real-life problems.

Introduction

A **rational function** is a quotient of polynomial functions. It can be written in the form

$$f(x) = \frac{N(x)}{D(x)}$$

where $N(x)$ and $D(x)$ are polynomials and $D(x)$ is not the zero polynomial.

In general, the *domain* of a rational function of x includes all real numbers except x-values that make the denominator zero. Much of the discussion of rational functions will focus on their graphical behavior near the x-values excluded from the domain.

EXAMPLE 1 Finding the Domain of a Rational Function

Find the domain of $f(x) = \dfrac{1}{x}$ and discuss the behavior of f near any excluded x-values.

Solution Because the denominator is zero when $x = 0$, the domain of f is all real numbers except $x = 0$. To determine the behavior of f near this excluded value, evaluate $f(x)$ to the left and right of $x = 0$, as indicated in the following tables.

x	-1	-0.5	-0.1	-0.01	-0.001	$\longrightarrow 0$
$f(x)$	-1	-2	-10	-100	-1000	$\longrightarrow -\infty$

x	$0 \longleftarrow$	0.001	0.01	0.1	0.5	1
$f(x)$	$\infty \longleftarrow$	1000	100	10	2	1

Note that as x approaches 0 *from the left*, $f(x)$ decreases without bound. In contrast, as x approaches 0 *from the right*, $f(x)$ increases without bound. The graph of f is shown below.

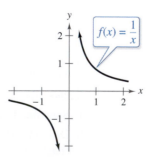

$f(x) = \dfrac{1}{x}$

✓ **Checkpoint** 🔊))) *Audio-video solution in English & Spanish at LarsonPrecalculus.com.*

Find the domain of the rational function $f(x) = \dfrac{3x}{x-1}$ and discuss the behavior of f near any excluded x-values.

ZQFotography/Shutterstock.com

Vertical and Horizontal Asymptotes

In Example 1, the behavior of f near $x = 0$ is denoted as follows.

$$f(x) \longrightarrow -\infty \text{ as } x \longrightarrow 0^- \qquad f(x) \longrightarrow \infty \text{ as } x \longrightarrow 0^+$$

$f(x)$ decreases without bound as x approaches 0 from the left.

$f(x)$ increases without bound as x approaches 0 from the right.

The line $x = 0$ is a **vertical asymptote** of the graph of f, as shown in Figure 2.21. From this figure, you can see that the graph of f also has a **horizontal asymptote**—the line $y = 0$. The behavior of f near $y = 0$ is denoted as follows.

$$f(x) \longrightarrow 0 \text{ as } x \longrightarrow -\infty \qquad f(x) \longrightarrow 0 \text{ as } x \longrightarrow \infty$$

$f(x)$ approaches 0 as x decreases without bound.

$f(x)$ approaches 0 as x increases without bound.

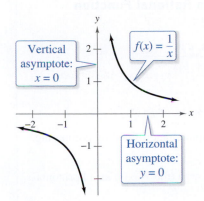

Figure 2.21

Definitions of Vertical and Horizontal Asymptotes

1. The line $x = a$ is a **vertical asymptote** of the graph of f when

$$f(x) \longrightarrow \infty \quad \text{or} \quad f(x) \longrightarrow -\infty$$

as $x \longrightarrow a$, either from the right or from the left.

2. The line $y = b$ is a **horizontal asymptote** of the graph of f when

$$f(x) \longrightarrow b$$

as $x \longrightarrow \infty$ or $x \longrightarrow -\infty$.

Eventually (as $x \longrightarrow \infty$ or $x \longrightarrow -\infty$), the distance between the horizontal asymptote and the points on the graph must approach zero. Figure 2.22 shows the vertical and horizontal asymptotes of the graphs of three rational functions.

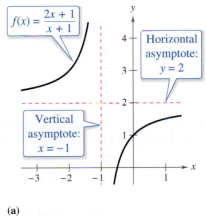

(a)

Figure 2.22

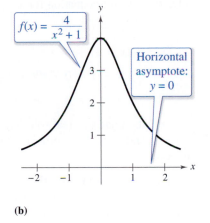

(b)

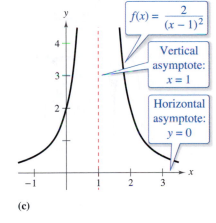

(c)

The graphs of

$$f(x) = \frac{1}{x}$$

in Figure 2.21 and

$$f(x) = \frac{2x + 1}{x + 1}$$

in Figure 2.22(a) are **hyperbolas.**

Vertical and Horizontal Asymptotes of a Rational Function

Let f be the rational function

$$f(x) = \frac{N(x)}{D(x)} = \frac{a_n x^n + a_{n-1} x^{n-1} + \cdots + a_1 x + a_0}{b_m x^m + b_{m-1} x^{m-1} + \cdots + b_1 x + b_0}$$

where $N(x)$ and $D(x)$ have no common factors.

1. The graph of f has *vertical* asymptotes at the zeros of $D(x)$.

2. The graph of f has one or no *horizontal* asymptote determined by comparing the degrees of $N(x)$ and $D(x)$.

    a. When $n < m$, the graph of f has the line $y = 0$ (the x-axis) as a horizontal asymptote.

    b. When $n = m$, the graph of f has the line $y = a_n / b_m$ (ratio of the leading coefficients) as a horizontal asymptote.

    c. When $n > m$, the graph of f has no horizontal asymptote.

EXAMPLE 2 **Finding Vertical and Horizontal Asymptotes**

Find all vertical and horizontal asymptotes of the graph of each rational function.

a. $f(x) = \dfrac{2x^2}{x^2 - 1}$ **b.** $f(x) = \dfrac{x^2 + x - 2}{x^2 - x - 6}$

Solution

a. For this rational function, the degree of the numerator is *equal* to the degree of the denominator. The leading coefficient of the numerator is 2 and the leading coefficient of the denominator is 1, so the graph has the line $y = 2$ as a horizontal asymptote. To find any vertical asymptotes, set the denominator equal to zero and solve the resulting equation for x.

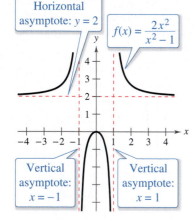

Horizontal asymptote: $y = 2$

$f(x) = \dfrac{2x^2}{x^2 - 1}$

Vertical asymptote: $x = -1$

Vertical asymptote: $x = 1$

Figure 2.23

$x^2 - 1 = 0$		Set denominator equal to zero.
$(x + 1)(x - 1) = 0$		Factor.
$x + 1 = 0$ $\Rightarrow$	$x = -1$	Set 1st factor equal to 0.
$x - 1 = 0$ $\Rightarrow$	$x = 1$	Set 2nd factor equal to 0.

This equation has two real solutions, $x = -1$ and $x = 1$, so the graph has the lines $x = -1$ and $x = 1$ as vertical asymptotes. Figure 2.23 shows the graph of this function.

b. For this rational function, the degree of the numerator is equal to the degree of the denominator. The leading coefficient of both the numerator and the denominator is 1, so the graph has the line $y = 1$ as a horizontal asymptote. To find any vertical asymptotes, first factor the numerator and denominator as follows.

$$f(x) = \frac{x^2 + x - 2}{x^2 - x - 6} = \frac{(x - 1)(x + 2)}{(x + 2)(x - 3)} = \frac{x - 1}{x - 3}, \quad x \neq -2$$

By setting the denominator $x - 3$ (of the simplified function) equal to zero, you can determine that the graph has the line $x = 3$ as a vertical asymptote.

> ▷ **ALGEBRA HELP** You can review the techniques for factoring in Appendix A.3.

✓ *Checkpoint* ◀))) *Audio-video solution in English & Spanish at LarsonPrecalculus.com.*

Find all vertical and horizontal asymptotes of the graph of $f(x) = \dfrac{3x^2 + 7x - 6}{x^2 + 4x + 3}$. ◼

Sketching the Graph of a Rational Function

To sketch the graph of a rational function, use the following guidelines.

•• **REMARK**   You may also
want to test for symmetry
when graphing rational
functions, especially for simple
rational functions. Recall from
Section 1.6 that the graph of
the reciprocal function

$$f(x) = \frac{1}{x}$$

is symmetric with respect to the
origin.

Guidelines for Graphing Rational Functions

Let $f(x) = \dfrac{N(x)}{D(x)}$, where $N(x)$ and $D(x)$ are polynomials and $D(x)$ is not the zero polynomial.

1. Simplify f, if possible.

2. Find and plot the y-intercept (if any) by evaluating $f(0)$.

3. Find the zeros of the numerator (if any) by solving the equation $N(x) = 0$. Then plot the corresponding x-intercepts.

4. Find the zeros of the denominator (if any) by solving the equation $D(x) = 0$. Then sketch the corresponding vertical asymptotes.

5. Find and sketch the horizontal asymptote (if any) by using the rule for finding the horizontal asymptote of a rational function.

6. Plot at least one point *between* and one point *beyond* each x-intercept and vertical asymptote.

7. Use smooth curves to complete the graph between and beyond the vertical asymptotes.

▷ **TECHNOLOGY** Some graphing utilities have difficulty graphing rational functions that have vertical asymptotes. Often, the graphing utility will connect portions of the graph that are not supposed to be connected. For instance, the screen on the left below shows the graph of $f(x) = 1/(x - 2)$. Notice that the graph should consist of two unconnected portions—one to the left of $x = 2$ and the other to the right of $x = 2$. To eliminate this problem, you can try changing the mode of the graphing utility to *dot* mode. The problem with this is that the graph is then represented as a collection of dots (as shown in the screen on the right) rather than as a smooth curve.

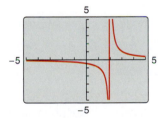 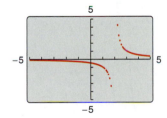

The concept of *test intervals* from Section 2.2 can be extended to graphing of rational functions. To do this, use the fact that a rational function can change signs only at its zeros and its undefined values (the x-values for which its denominator is zero). Between two consecutive zeros of the numerator and the denominator, a rational function must be entirely positive or entirely negative. This means that when put in order, the zeros of the numerator and the denominator of a rational function divide the real number line into test intervals in which the function has no sign changes. Choose a representative x-value to determine whether the value of the rational function is positive (the graph lies above the x-axis) or negative (the graph lies below the x-axis).

EXAMPLE 3   **Sketching the Graph of a Rational Function**

•• **REMARK**   You can use
transformations to help you
sketch graphs of rational
functions. For instance, the
graph of g in Example 3 is
a vertical stretch and a right
shift of the graph of $f(x) = 1/x$
because

$$g(x) = \frac{3}{x - 2}$$

$$= 3\left(\frac{1}{x - 2}\right)$$

$$= 3f(x - 2).$$

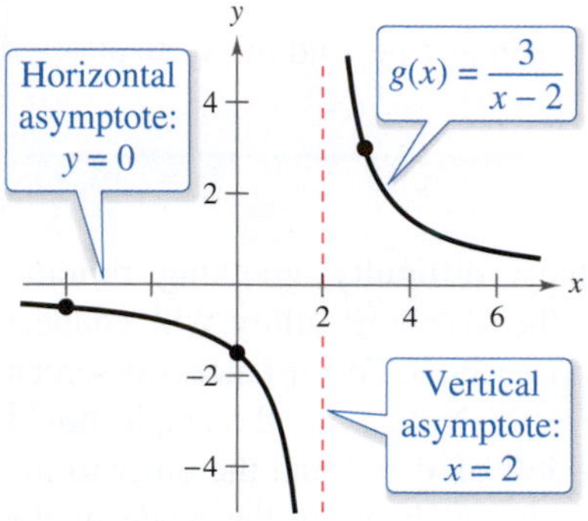

Sketch the graph of $g(x) = \dfrac{3}{x - 2}$ and state its domain.

Solution

y-intercept:	$\left(0, -\frac{3}{2}\right)$, because $g(0) = -\frac{3}{2}$
x-intercept:	none, because $3 \neq 0$
Vertical asymptote:	$x = 2$, zero of denominator
Horizontal asymptote:	$y = 0$, because degree of $N(x) <$ degree of $D(x)$

Additional points:

Test Interval	Representative x-Value	Value of g	Sign	Point on Graph
$(-\infty, 2)$	-4	$g(-4) = -0.5$	Negative	$(-4, -0.5)$
$(2, \infty)$	3	$g(3) = 3$	Positive	$(3, 3)$

By plotting the intercept, asymptotes, and a few additional points, you can obtain the
graph shown in Figure 2.24. The domain of g is all real numbers except $x = 2$.

✓ **Checkpoint**  *Audio-video solution in English & Spanish at LarsonPrecalculus.com.*

Sketch the graph of $f(x) = \dfrac{1}{x + 3}$ and state its domain.

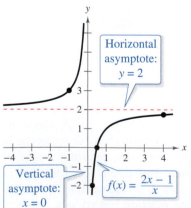

Figure 2.24

EXAMPLE 4   **Sketching the Graph of a Rational Function**

Sketch the graph of $f(x) = (2x - 1)/x$ and state its domain.

Solution

y-intercept:	none, because $x = 0$ is not in the domain
x-intercept:	$\left(\frac{1}{2}, 0\right)$, because $f\left(\frac{1}{2}\right) = 0$
Vertical asymptote:	$x = 0$, zero of denominator
Horizontal asymptote:	$y = 2$, because degree of $N(x) =$ degree of $D(x)$

Additional points:

Test Interval	Representative x-Value	Value of f	Sign	Point on Graph
$(-\infty, 0)$	-1	$f(-1) = 3$	Positive	$(-1, 3)$
$\left(0, \frac{1}{2}\right)$	$\frac{1}{4}$	$f\left(\frac{1}{4}\right) = -2$	Negative	$\left(\frac{1}{4}, -2\right)$
$\left(\frac{1}{2}, \infty\right)$	4	$f(4) = 1.75$	Positive	$(4, 1.75)$

By plotting the intercept, asymptotes, and a few additional points, you can obtain the
graph shown in Figure 2.25. The domain of f is all real numbers except $x = 0$.

Figure 2.25

✓ **Checkpoint** *Audio-video solution in English & Spanish at LarsonPrecalculus.com.*

Sketch the graph of $C(x) = (3 + 2x)/(1 + x)$ and state its domain.

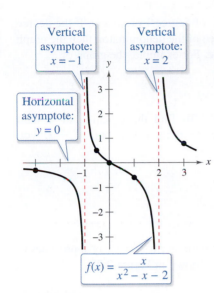

$$f(x) = \frac{x}{x^2 - x - 2}$$

Figure 2.26

REMARK If you are unsure of the shape of a portion of the graph of a rational function, then plot some additional points. Also note that when the numerator and the denominator of a rational function have a common factor, the graph of the function has a *hole* at the zero of the common factor (see Example 6).

EXAMPLE 5   Sketching the Graph of a Rational Function

Sketch the graph of $f(x) = x/(x^2 - x - 2)$.

Solution   Factoring the denominator, you have $f(x) = x/[(x + 1)(x - 2)]$.

Intercept: (0, 0), because $f(0) = 0$

Vertical asymptotes: $x = -1$, $x = 2$, zeros of denominator

Horizontal asymptote: $y = 0$, because degree of $N(x) <$ degree of $D(x)$

Additional points:

Test Interval	Representative x-Value	Value of f	Sign	Point on Graph
$(-\infty, -1)$	-3	$f(-3) = -0.3$	Negative	$(-3, -0.3)$
$(-1, 0)$	-0.5	$f(-0.5) = 0.4$	Positive	$(-0.5, 0.4)$
$(0, 2)$	1	$f(1) = -0.5$	Negative	$(1, -0.5)$
$(2, \infty)$	3	$f(3) = 0.75$	Positive	$(3, 0.75)$

Figure 2.26 shows the graph of this function.

✓ Checkpoint ◀))) Audio-video solution in English & Spanish at LarsonPrecalculus.com.

Sketch the graph of $f(x) = 3x/(x^2 + x - 2)$.

EXAMPLE 6   A Rational Function with Common Factors

Sketch the graph of $f(x) = (x^2 - 9)/(x^2 - 2x - 3)$.

Solution   By factoring the numerator and denominator, you have

$$f(x) = \frac{x^2 - 9}{x^2 - 2x - 3} = \frac{(x - 3)(x + 3)}{(x - 3)(x + 1)} = \frac{x + 3}{x + 1}, \quad x \neq 3.$$

y-intercept: (0, 3), because $f(0) = 3$

x-intercept: (-3, 0), because $f(-3) = 0$

Vertical asymptote: $x = -1$, zero of (simplified) denominator

Horizontal asymptote: $y = 1$, because degree of $N(x) =$ degree of $D(x)$

Additional points:

Test Interval	Representative x-Value	Value of f	Sign	Point on Graph
$(-\infty, -3)$	-4	$f(-4) = 0.33$	Positive	$(-4, 0.33)$
$(-3, -1)$	-2	$f(-2) = -1$	Negative	$(-2, -1)$
$(-1, \infty)$	2	$f(2) = 1.67$	Positive	$(2, 1.67)$

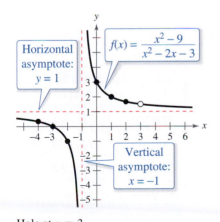

$$f(x) = \frac{x^2 - 9}{x^2 - 2x - 3}$$

Hole at $x = 3$

Figure 2.27

Figure 2.27 shows the graph of this function. Notice that there is a hole in the graph at $x = 3$, because the function is not defined when $x = 3$.

✓ Checkpoint ◀))) Audio-video solution in English & Spanish at LarsonPrecalculus.com.

Sketch the graph of $f(x) = (x^2 - 4)/(x^2 - x - 6)$.

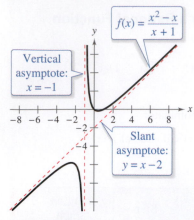

Figure 2.28

Slant Asymptotes

Consider a rational function whose denominator is of degree 1 or greater. If the degree of the numerator is exactly *one more* than the degree of the denominator, then the graph of the function has a **slant** (or **oblique**) **asymptote.** For example, the graph of

$$f(x) = \frac{x^2 - x}{x + 1}$$

has a slant asymptote, as shown in Figure 2.28. To find the equation of a slant asymptote, use long division. For instance, by dividing $x + 1$ into $x^2 - x$, you obtain

$$f(x) = \frac{x^2 - x}{x + 1} = \underbrace{x - 2} + \frac{2}{x + 1}.$$

Slant asymptote
$(y = x - 2)$

As x increases or decreases without bound, the remainder term $2/(x + 1)$ approaches 0, so the graph of f approaches the line $y = x - 2$, as shown in Figure 2.28.

EXAMPLE 7 **A Rational Function with a Slant Asymptote**

Sketch the graph of $f(x) = \dfrac{x^2 - x - 2}{x - 1}$.

Solution Factoring the numerator as $(x - 2)(x + 1)$ allows you to recognize the *x*-intercepts. Using long division

$$f(x) = \frac{x^2 - x - 2}{x - 1}$$

$$= x - \frac{2}{x - 1}$$

allows you to recognize that the line $y = x$ is a slant asymptote of the graph.

y-intercept: $(0, 2)$, because $f(0) = 2$

x-intercepts: $(-1, 0)$ and $(2, 0)$

Vertical asymptote: $x = 1$, zero of denominator

Slant asymptote: $y = x$

Additional points:

Test Interval	Representative x-Value	Value of f	Sign	Point on Graph
$(-\infty, -1)$	-2	$f(-2) = -1.33$	Negative	$(-2, -1.33)$
$(-1, 1)$	0.5	$f(0.5) = 4.5$	Positive	$(0.5, 4.5)$
$(1, 2)$	1.5	$f(1.5) = -2.5$	Negative	$(1.5, -2.5)$
$(2, \infty)$	3	$f(3) = 2$	Positive	$(3, 2)$

Figure 2.29 shows the graph of this function.

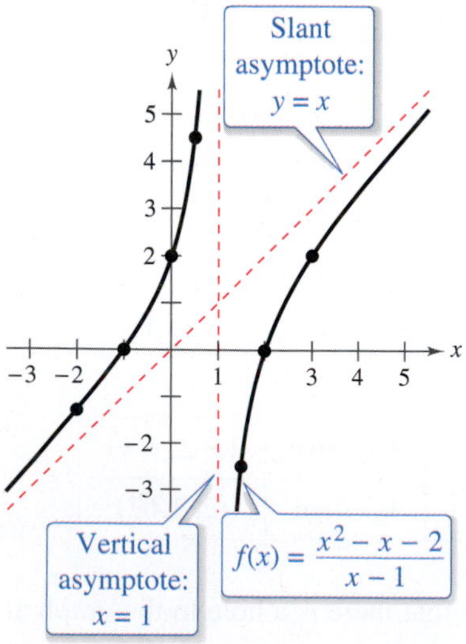

Figure 2.29

✓ *Checkpoint* ◀))) *Audio-video solution in English & Spanish at LarsonPrecalculus.com.*

Sketch the graph of $f(x) = \dfrac{3x^2 + 1}{x}$.

Applications

There are many examples of asymptotic behavior in real life. For instance, Example 8 shows how a vertical asymptote can help you to analyze the cost of removing pollutants from smokestack emissions.

EXAMPLE 8 **Cost-Benefit Model**

A utility company burns coal to generate electricity. The cost C (in dollars) of removing $p\%$ of the smokestack pollutants is given by

$$C = \frac{80,000p}{100 - p}, \quad 0 \le p < 100.$$

You are a member of a state legislature considering a law that would require utility companies to remove 90% of the pollutants from their smokestack emissions. The current law requires 85% removal. How much additional cost would the utility company incur as a result of the new law?

Algebraic Solution

Because the current law requires 85% removal, the current cost to the utility company is

$$C = \frac{80,000(85)}{100 - 85} \approx \$453,333. \qquad \text{Evaluate } C \text{ when } p = 85.$$

The cost to remove 90% of the pollutants would be

$$C = \frac{80,000(90)}{100 - 90} = \$720,000. \qquad \text{Evaluate } C \text{ when } p = 90.$$

So, the new law would require the utility company to spend an additional

$$720,000 - 453,333 = \$266,667. \qquad \begin{array}{l}\text{Subtract 85\% removal cost}\\ \text{from 90\% removal cost.}\end{array}$$

Graphical Solution

Use a graphing utility to graph the function

$$y_1 = \frac{80,000x}{100 - x}$$

as shown below. Note that the graph has a vertical asymptote at

$$x = 100.$$

Then approximate the values of y_1 when $x = 85$ and $x = 90$. You should obtain the following values.

When $x = 85$, $y_1 \approx 453,333$.

When $x = 90$, $y_1 = 720,000$.

So, the new law would require the utility company to spend an additional

$$720,000 - 453,333 = \$266,667.$$

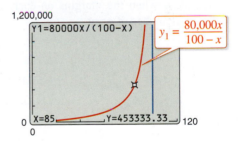

✓ *Checkpoint* 🔊))) *Audio-video solution in English & Spanish at LarsonPrecalculus.com.*

The cost C (in millions of dollars) of removing $p\%$ of the industrial and municipal pollutants discharged into a river is given by

$$C = \frac{255p}{100 - p}, \quad 0 \le p < 100.$$

a. Find the costs of removing 20%, 45%, and 80% of the pollutants.

b. According to the model, would it be possible to remove 100% of the pollutants? Explain.

EXAMPLE 9   **Finding a Minimum Area**

A rectangular page contains 48 square inches of print. The margins at the top and bottom of the page are each 1 inch deep. The margins on each side are $1\frac{1}{2}$ inches wide. What should the dimensions of the page be so that the least amount of paper is used?

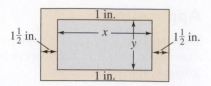

Figure 2.30

Graphical Solution

Let A be the area to be minimized. From Figure 2.30, you can write $A = (x + 3)(y + 2)$. The printed area inside the margins is modeled by $48 = xy$ or $y = 48/x$. To find the minimum area, rewrite the equation for A in terms of just one variable by substituting $48/x$ for y.

$$A = (x + 3)\left(\frac{48}{x} + 2\right) = \frac{(x + 3)(48 + 2x)}{x}, \quad x > 0$$

The graph of this rational function is shown below. Because x represents the width of the printed area, you need to consider only the portion of the graph for which x is positive. Using a graphing utility, you can approximate the minimum value of A to occur when $x \approx 8.5$ inches. The corresponding value of y is $48/8.5 \approx 5.6$ inches. So, the dimensions should be $x + 3 \approx 11.5$ inches by $y + 2 \approx 7.6$ inches.

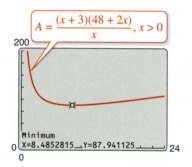

Numerical Solution

Let A be the area to be minimized. From Figure 2.30, you can write $A = (x + 3)(y + 2)$. The printed area inside the margins is modeled by $48 = xy$ or $y = 48/x$. To find the minimum area, rewrite the equation for A in terms of just one variable by substituting $48/x$ for y.

$$A = (x + 3)\left(\frac{48}{x} + 2\right) = \frac{(x + 3)(48 + 2x)}{x}, \quad x > 0$$

Use a graphing utility to create a table of values for the function $y_1 = [(x + 3)(48 + 2x)]/x$ beginning at $x = 1$. The minimum value of y_1 occurs when x is somewhere between 8 and 9, as shown in Figure 2.31. To approximate the minimum value of y_1 to one decimal place, change the table so that it starts at $x = 8$ and increases by 0.1. The minimum value of y_1 occurs when $x \approx 8.5$, as shown in Figure 2.32. The corresponding value of y is $48/8.5 \approx 5.6$ inches. So, the dimensions should be $x + 3 \approx 11.5$ inches by $y + 2 \approx 7.6$ inches.

X	Y₁		
6	90		
7	88.571		
8	88		
9	88		
10	88.4		
11	89.091		
12	90		
X=8			

Figure 2.31

X	Y₁		
8.2	87.961		
8.3	87.949		
8.4	87.943		
8.5	87.941		
8.6	87.944		
8.7	87.952		
8.8	87.964		
X=8.5			

Figure 2.32

✔ **Checkpoint** *Audio-video solution in English & Spanish at LarsonPrecalculus.com.*

Rework Example 9 when the margins on each side are 2 inches wide and the page contains 40 square inches of print.

• • • • • • • • • • • • • • • • • ▷

• •REMARK If you take a course in calculus, then you will learn an analytic technique for finding the exact value of x that produces a minimum area. In Example 9, that value is $x = 6\sqrt{2} \approx 8.485$.

Summarize   (Section 2.6)

1. State the definition and describe the domain of a rational function *(page 116)*. For an example of finding the domain of a rational function, see Example 1.

2. State the definitions of vertical and horizontal asymptotes *(page 117)*. For an example of finding vertical and horizontal asymptotes of the graphs of rational functions, see Example 2.

3. State the guidelines for graphing rational functions *(page 119)*. For examples of sketching the graphs of rational functions, see Examples 3–6.

4. Describe when a rational function has a slant asymptote *(page 122)*. For an example of sketching the graph of a rational function that has a slant asymptote, see Example 7.

5. Describe examples of how to use rational functions to model and solve real-life problems *(pages 123 and 124, Examples 8 and 9)*.

2.7 Nonlinear Inequalities

You can use inequalities to model and solve real-life problems, such as the height of a projectile.

■ Solve polynomial inequalities.
■ Solve rational inequalities.
■ Use inequalities to model and solve real-life problems.

Polynomial Inequalities

To solve a polynomial inequality such as $x^2 - 2x - 3 < 0$, you can use the fact that a polynomial can change signs only at its zeros (the x-values that make the polynomial equal to zero). Between two consecutive zeros, a polynomial must be entirely positive or entirely negative. This means that when the real zeros of a polynomial are put in order, they divide the real number line into intervals in which the polynomial has no sign changes. These zeros are the **key numbers** of the inequality, and the resulting intervals are the *test intervals* for the inequality. For instance, the polynomial above factors as

$$x^2 - 2x - 3 = (x + 1)(x - 3)$$

and has two zeros,

$$x = -1 \quad \text{and} \quad x = 3.$$

These zeros divide the real number line into three test intervals:

$$(-\infty, -1), \quad (-1, 3), \quad \text{and} \quad (3, \infty). \quad \text{(See figure.)}$$

So, to solve the inequality $x^2 - 2x - 3 < 0$, you need only test one value from each of these test intervals. When a value from a test interval satisfies the original inequality, you can conclude that the interval is a solution of the inequality.

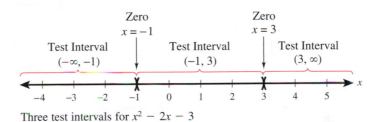

Three test intervals for $x^2 - 2x - 3$

You can use the same basic approach to determine the test intervals for any polynomial.

Finding Test Intervals for a Polynomial

To determine the intervals on which the values of a polynomial are entirely negative or entirely positive, use the following steps.

1. Find all real zeros of the polynomial, and arrange the zeros in increasing order (from least to greatest). These zeros are the key numbers of the polynomial.

2. Use the key numbers of the polynomial to determine its test intervals.

3. Choose one representative x-value in each test interval and evaluate the polynomial at that value. When the value of the polynomial is negative, the polynomial has negative values for every x-value in the interval. When the value of the polynomial is positive, the polynomial has positive values for every x-value in the interval.

EXAMPLE 1   **Solving a Polynomial Inequality**

Solve $x^2 - x - 6 < 0$. Then graph the solution set.

Solution   By factoring the polynomial as

$$x^2 - x - 6 = (x + 2)(x - 3)$$

you can see that the key numbers are $x = -2$ and $x = 3$. So, the polynomial's test intervals are

$$(-\infty, -2), \quad (-2, 3), \quad \text{and} \quad (3, \infty). \qquad \text{Test intervals}$$

In each test interval, choose a representative x-value and evaluate the polynomial.

Test Interval	x-Value	Polynomial Value	Conclusion
$(-\infty, -2)$	$x = -3$	$(-3)^2 - (-3) - 6 = 6$	Positive
$(-2, 3)$	$x = 0$	$(0)^2 - (0) - 6 = -6$	Negative
$(3, \infty)$	$x = 4$	$(4)^2 - (4) - 6 = 6$	Positive

From this, you can conclude that the inequality is satisfied for all x-values in $(-2, 3)$. This implies that the solution of the inequality

$$x^2 - x - 6 < 0$$

is the interval $(-2, 3)$, as shown below. Note that the original inequality contains a "less than" symbol. This means that the solution set does not contain the endpoints of the test interval $(-2, 3)$.

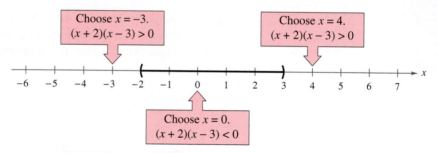

Choose $x = -3$.
$(x + 2)(x - 3) > 0$

Choose $x = 4$.
$(x + 2)(x - 3) > 0$

Choose $x = 0$.
$(x + 2)(x - 3) < 0$

✓ **Checkpoint** 🔊)) Audio-video solution in English & Spanish at LarsonPrecalculus.com.

Solve $x^2 - x - 20 < 0$. Then graph the solution set. ◼

As with linear inequalities, you can check the reasonableness of a solution by substituting x-values into the original inequality. For instance, to check the solution found in Example 1, try substituting several x-values from the interval $(-2, 3)$ into the inequality

$$x^2 - x - 6 < 0.$$

Regardless of which x-values you choose, the inequality should be satisfied.

You can also use a graph to check the result of Example 1. Sketch the graph of

$$y = x^2 - x - 6$$

as shown in Figure 2.33. Notice that the graph is below the x-axis on the interval $(-2, 3)$.

In Example 1, the polynomial inequality was given in general form (with the polynomial on one side and zero on the other). Whenever this is not the case, you should begin the solution process by writing the inequality in general form.

▷ **ALGEBRA HELP**   You can review the techniques for factoring polynomials in Appendix A.3.

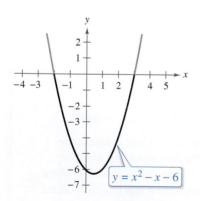

$y = x^2 - x - 6$

Figure 2.33

EXAMPLE 2 **Solving a Polynomial Inequality**

Solve $2x^3 - 3x^2 - 32x > -48$. Then graph the solution set.

Solution

$$2x^3 - 3x^2 - 32x + 48 > 0 \qquad \text{Write in general form.}$$

$$(x - 4)(x + 4)(2x - 3) > 0 \qquad \text{Factor.}$$

The key numbers are $x = -4$, $x = \frac{3}{2}$, and $x = 4$, and the test intervals are $(-\infty, -4)$, $\left(-4, \frac{3}{2}\right)$, $\left(\frac{3}{2}, 4\right)$, and $(4, \infty)$.

Test Interval	x-Value	Polynomial Value	Conclusion
$(-\infty, -4)$	$x = -5$	$2(-5)^3 - 3(-5)^2 - 32(-5) + 48 = -117$	Negative
$\left(-4, \frac{3}{2}\right)$	$x = 0$	$2(0)^3 - 3(0)^2 - 32(0) + 48 = 48$	Positive
$\left(\frac{3}{2}, 4\right)$	$x = 2$	$2(2)^3 - 3(2)^2 - 32(2) + 48 = -12$	Negative
$(4, \infty)$	$x = 5$	$2(5)^3 - 3(5)^2 - 32(5) + 48 = 63$	Positive

From this, you can conclude that the inequality is satisfied on the open intervals $\left(-4, \frac{3}{2}\right)$ and $(4, \infty)$. So, the solution set is $\left(-4, \frac{3}{2}\right) \cup (4, \infty)$, as shown below.

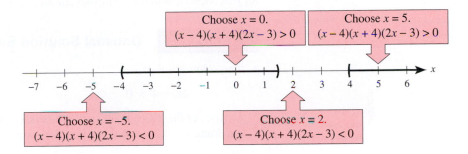

Choose $x = 0$.
$(x - 4)(x + 4)(2x - 3) > 0$

Choose $x = 5$.
$(x - 4)(x + 4)(2x - 3) > 0$

Choose $x = -5$.
$(x - 4)(x + 4)(2x - 3) < 0$

Choose $x = 2$.
$(x - 4)(x + 4)(2x - 3) < 0$

 Checkpoint Audio-video solution in English & Spanish at LarsonPrecalculus.com.

Solve $3x^3 - x^2 - 12x > -4$. Then graph the solution set.

EXAMPLE 3 **Solving a Polynomial Inequality**

Solve $4x^2 - 5x > 6$.

Algebraic Solution

$$4x^2 - 5x - 6 > 0 \qquad \text{Write in general form.}$$

$$(x - 2)(4x + 3) > 0 \qquad \text{Factor.}$$

Key Numbers: $x = -\frac{3}{4}$, $x = 2$

Test Intervals: $\left(-\infty, -\frac{3}{4}\right), \left(-\frac{3}{4}, 2\right), (2, \infty)$

Test: Is $(x - 2)(4x + 3) > 0$?

After testing these intervals, you can see that the polynomial $4x^2 - 5x - 6$ is positive on the open intervals $\left(-\infty, -\frac{3}{4}\right)$ and $(2, \infty)$. So, the solution set of the inequality is $\left(-\infty, -\frac{3}{4}\right) \cup (2, \infty)$.

Graphical Solution

First write the polynomial inequality $4x^2 - 5x > 6$ as $4x^2 - 5x - 6 > 0$. Then use a graphing utility to graph $y = 4x^2 - 5x - 6$. In the figure below, you can see that the graph is *above* the x-axis when x is less than $-\frac{3}{4}$ or when x is greater than 2. So, the solution set is $\left(-\infty, -\frac{3}{4}\right) \cup (2, \infty)$.

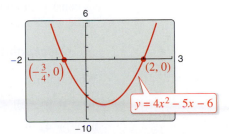

$\left(-\frac{3}{4}, 0\right)$ $(2, 0)$

$y = 4x^2 - 5x - 6$

 Checkpoint Audio-video solution in English & Spanish at LarsonPrecalculus.com.

Solve $2x^2 + 3x < 5$ (a) algebraically and (b) graphically.

You may find it easier to determine the sign of a polynomial from its *factored* form. For instance, in Example 3, when you substitute the test value $x = 1$ into the factored form

$$(x - 2)(4x + 3)$$

you can see that the sign pattern of the factors is

$$(-)(+)$$

which yields a negative result. Try using the factored forms of the polynomials to determine the signs of the polynomials in the test intervals of the other examples in this section.

When solving a polynomial inequality, be sure to account for the particular type of inequality symbol given in the inequality. For instance, in Example 3, note that the original inequality contained a "greater than" symbol and the solution consisted of two open intervals. If the original inequality had been

$$4x^2 - 5x \geq 6$$

then the solution set would have been $\left(-\infty, -\frac{3}{4}\right] \cup [2, \infty)$.

Each of the polynomial inequalities in Examples 1, 2, and 3 has a solution set that consists of a single interval or the union of two intervals. When solving the exercises for this section, watch for unusual solution sets, as illustrated in Example 4.

EXAMPLE 4   **Unusual Solution Sets**

a. The solution set of

$$x^2 + 2x + 4 > 0$$

consists of the entire set of real numbers, $(-\infty, \infty)$. In other words, the value of the quadratic $x^2 + 2x + 4$ is positive for every real value of x.

b. The solution set of

$$x^2 + 2x + 1 \leq 0$$

consists of the single real number $\{-1\}$, because the quadratic $x^2 + 2x + 1$ has only one key number, $x = -1$, and it is the only value that satisfies the inequality.

c. The solution set of

$$x^2 + 3x + 5 < 0$$

is empty. In other words, the quadratic $x^2 + 3x + 5$ is not less than zero for any value of x.

d. The solution set of

$$x^2 - 4x + 4 > 0$$

consists of all real numbers except $x = 2$. In interval notation, this solution set can be written as $(-\infty, 2) \cup (2, \infty)$.

✓ *Checkpoint* 🔊)) *Audio-video solution in English & Spanish at LarsonPrecalculus.com.*

What is unusual about the solution set of each inequality?

a. $x^2 + 6x + 9 < 0$

b. $x^2 + 4x + 4 \leq 0$

c. $x^2 - 6x + 9 > 0$

d. $x^2 - 2x + 1 \geq 0$

Rational Inequalities

The concepts of key numbers and test intervals can be extended to rational inequalities. Use the fact that the value of a rational expression can change sign only at its *zeros* (the *x*-values for which its numerator is zero) and its *undefined values* (the *x*-values for which its denominator is zero). These two types of numbers make up the *key numbers* of a rational inequality. When solving a rational inequality, begin by writing the inequality in general form with the rational expression on the left and zero on the right.

EXAMPLE 5   Solving a Rational Inequality

Solve $\dfrac{2x - 7}{x - 5} \le 3$. Then graph the solution set.

Solution

$$\frac{2x - 7}{x - 5} \le 3 \qquad \text{Write original inequality.}$$

$$\frac{2x - 7}{x - 5} - 3 \le 0 \qquad \text{Write in general form.}$$

$$\frac{2x - 7 - 3x + 15}{x - 5} \le 0 \qquad \text{Find the LCD and subtract fractions.}$$

$$\frac{-x + 8}{x - 5} \le 0 \qquad \text{Simplify.}$$

Key Numbers: $x = 5, x = 8$ Zeros and undefined values of rational expression

Test Intervals: $(-\infty, 5), (5, 8), (8, \infty)$

Test: Is $\dfrac{-x + 8}{x - 5} \le 0$?

After testing these intervals, as shown below, you can see that the inequality is satisfied on the open intervals $(-\infty, 5)$ and $(8, \infty)$. Moreover, because

$$\frac{-x + 8}{x - 5} = 0$$

when $x = 8$, you can conclude that the solution set consists of all real numbers in the intervals $(-\infty, 5) \cup [8, \infty)$. (Be sure to use a bracket to indicate that x can equal 8.)

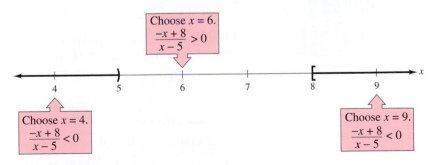

 Checkpoint ◀))) *Audio-video solution in English & Spanish at LarsonPrecalculus.com.*

Solve each inequality. Then graph the solution set.

a. $\dfrac{x - 2}{x - 3} \ge -3$

b. $\dfrac{4x - 1}{x - 6} > 3$

•• **REMARK**   In Example 5, write 3 as $\frac{3}{1}$. You should be able to see that the LCD (least common denominator) is $(x - 5)(1) = x - 5$. So, you can rewrite the general form as

$$\frac{2x - 7}{x - 5} - \frac{3(x - 5)}{x - 5} \le 0,$$

which simplifies as shown. ▷

Applications

One common application of inequalities comes from business and involves profit, revenue, and cost. The formula that relates these three quantities is

Profit	=	Revenue	−	Cost

$$P = R - C.$$

EXAMPLE 6   Increasing the Profit for a Product

The marketing department of a calculator manufacturer has determined that the demand for a new model of calculator is

$$p = 100 - 0.00001x, \quad 0 \le x \le 10{,}000{,}000 \qquad \text{Demand equation}$$

where p is the price per calculator (in dollars) and x represents the number of calculators sold. (According to this model, no one would be willing to pay \$100 for the calculator. At the other extreme, the company could not *give* away more than 10 million calculators.) The revenue for selling x calculators is

$$R = xp = x(100 - 0.00001x) \qquad \text{Revenue equation}$$

as shown in Figure 2.34. The total cost of producing x calculators is \$10 per calculator plus a one-time development cost of \$2,500,000. So, the total cost is

$$C = 10x + 2{,}500{,}000. \qquad \text{Cost equation}$$

What prices can the company charge per calculator to obtain a profit of at least \$190,000,000?

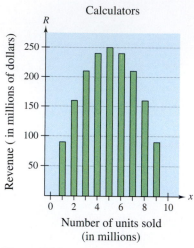

Calculators

Figure 2.34

Solution

Verbal Model:

Profit	=	Revenue	−	Cost

Equation: $P = R - C$

$$P = 100x - 0.00001x^2 - (10x + 2{,}500{,}000)$$

$$P = -0.00001x^2 + 90x - 2{,}500{,}000$$

To answer the question, solve the inequality

$$P \ge 190{,}000{,}000$$

$$-0.00001x^2 + 90x - 2{,}500{,}000 \ge 190{,}000{,}000.$$

When you write the inequality in general form, find the key numbers and the test intervals, and then test a value in each test interval, you can find the solution to be

$$3{,}500{,}000 \le x \le 5{,}500{,}000$$

as shown in Figure 2.35. Substituting the x-values in the original demand equation shows that prices of

$$\$45.00 \le p \le \$65.00$$

will yield a profit of at least \$190,000,000.

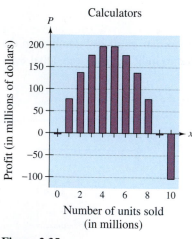

Calculators

Figure 2.35

✓ *Checkpoint*  *Audio-video solution in English & Spanish at LarsonPrecalculus.com.*

The revenue and cost equations for a product are

$$R = x(60 - 0.0001x) \quad \text{and} \quad C = 12x + 1{,}800{,}000$$

where R and C are measured in dollars and x represents the number of units sold. How many units must be sold to obtain a profit of at least \$3,600,000?

Another common application of inequalities is finding the domain of an expression that involves a square root, as shown in Example 7.

EXAMPLE 7   **Finding the Domain of an Expression**

Find the domain of $\sqrt{64 - 4x^2}$.

Algebraic Solution

Recall that the domain of an expression is the set of all x-values for which the expression is defined. Because $\sqrt{64 - 4x^2}$ is defined (has real values) only if $64 - 4x^2$ is nonnegative, the domain is given by $64 - 4x^2 \geq 0$.

$$64 - 4x^2 \geq 0 \qquad \text{Write in general form.}$$

$$16 - x^2 \geq 0 \qquad \text{Divide each side by 4.}$$

$$(4 - x)(4 + x) \geq 0 \qquad \text{Write in factored form.}$$

So, the inequality has two key numbers: $x = -4$ and $x = 4$. You can use these two numbers to test the inequality, as follows.

Key numbers: $x = -4$, $x = 4$

Test intervals: $(-\infty, -4)$, $(-4, 4)$, $(4, \infty)$

Test: Is $(4 - x)(4 + x) \geq 0$?

A test shows that the inequality is satisfied in the *closed interval* $[-4, 4]$. So, the domain of the expression $\sqrt{64 - 4x^2}$ is the interval $[-4, 4]$.

Graphical Solution

Begin by sketching the graph of the equation $y = \sqrt{64 - 4x^2}$, as shown below. From the graph, you can determine that the x-values extend from -4 to 4 (including -4 and 4). So, the domain of the expression $\sqrt{64 - 4x^2}$ is the interval $[-4, 4]$.

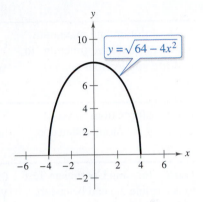

$y = \sqrt{64 - 4x^2}$

✓ **Checkpoint** 🔊)))   *Audio-video solution in English & Spanish at LarsonPrecalculus.com.*

Find the domain of $\sqrt{x^2 - 7x + 10}$.

To analyze a test interval, choose a representative x-value in the interval and evaluate the expression at that value. For instance, in Example 7, when you substitute any number from the interval $[-4, 4]$ into the expression $\sqrt{64 - 4x^2}$, you obtain a nonnegative number under the radical symbol that simplifies to a real number. When you substitute any number from the intervals $(-\infty, -4)$ and $(4, \infty)$, you obtain a complex number. It might be helpful to draw a visual representation of the intervals, as shown below.

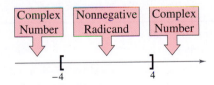

| Complex Number | Nonnegative Radicand | Complex Number |

$$[\underset{-4}{\quad} \qquad \underset{4}{\quad}]$$

Summarize   (Section 2.7)

1. Describe how to solve a polynomial inequality *(pages 125–128)*. For examples of solving polynomial inequalities, see Examples 1–4.

2. Describe how to solve a rational inequality *(page 129)*. For an example of solving a rational inequality, see Example 5.

3. Give an example of an application involving a polynomial inequality *(pages 130 and 131, Examples 6 and 7)*.

Chapter Summary

	What Did You Learn?	Explanation/Examples	Review Exercises
Section 2.1	Analyze graphs of quadratic functions (p. 80).	Let a, b, and c be real numbers with $a \neq 0$. The function $f(x) = ax^2 + bx + c$ is called a quadratic function. Its graph is a "U"-shaped curve called a parabola.	1, 2
	Write quadratic functions in standard form and use the results to sketch graphs of functions (p. 83).	The quadratic function $f(x) = a(x - h)^2 + k$, $a \neq 0$, is in standard form. The graph of f is a parabola whose axis is the vertical line $x = h$ and whose vertex is (h, k). When $a > 0$, the parabola opens upward. When $a < 0$, the parabola opens downward.	3–8
	Find minimum and maximum values of quadratic functions in real-life applications (p. 85).	Consider $f(x) = ax^2 + bx + c$ with vertex $\left(-\dfrac{b}{2a}, f\left(-\dfrac{b}{2a}\right)\right)$. When $a > 0$, f has a *minimum* at $x = -b/(2a)$. When $a < 0$, f has a *maximum* at $x = -b/(2a)$.	9, 10
Section 2.2	Use transformations to sketch graphs of polynomial functions (p. 86).	The graph of a polynomial function is continuous (no breaks, holes, or gaps) and has only smooth, rounded turns.	11, 12
	Use the Leading Coefficient Test to determine the end behaviors of graphs of polynomial functions (p. 88).	Consider the graph of $f(x) = a_n x^n + \cdots + a_1 x + a_0$. **When n is odd:** If $a_n > 0$, then the graph falls to the left and rises to the right. If $a_n < 0$, then the graph rises to the left and falls to the right. **When n is even:** If $a_n > 0$, then the graph rises to the left and to the right. If $a_n < 0$, then the graph falls to the left and to the right.	13–16
	Find and use the real zeros of polynomial functions as sketching aids (p. 90).	When f is a polynomial function and a is a real number, the following are equivalent: (1) $x = a$ is a *zero* of f, (2) $x = a$ is a *solution* of the equation $f(x) = 0$, (3) $(x - a)$ is a *factor* of $f(x)$, and (4) $(a, 0)$ is an *x-intercept* of the graph of f.	17–20
	Use the Intermediate Value Theorem to help locate the real zeros of polynomial functions (p. 93).	Let a and b be real numbers such that $a < b$. If f is a polynomial function such that $f(a) \neq f(b)$, then, in $[a, b]$, f takes on every value between $f(a)$ and $f(b)$.	21, 22
Section 2.3	Use long division to divide polynomials by other polynomials (p. 95).	Dividend ⟶ Quotient ⟶ Remainder $\dfrac{x^2 + 3x + 5}{x + 1} = x + 2 + \dfrac{3}{x + 1}$ ⟵ Divisor; Divisor ⟶	23, 24
	Use synthetic division to divide polynomials by binomials of the form $(x - k)$ (p. 98).	Divisor: $x + 3$ Dividend: $x^4 - 10x^2 - 2x + 4$ $$\begin{array}{r|rrrr} -3 & 1 & 0 & -10 & -2 & 4 \\ & & -3 & 9 & 3 & -3 \\ \hline & 1 & -3 & -1 & 1 & \boxed{1} \end{array}$$ ⟵ Remainder: 1 Quotient: $x^3 - 3x^2 - x + 1$	25, 26
	Use the Remainder Theorem and the Factor Theorem (p. 99).	**The Remainder Theorem:** If a polynomial $f(x)$ is divided by $x - k$, then the remainder is $r = f(k)$. **The Factor Theorem:** A polynomial $f(x)$ has a factor $(x - k)$ if and only if $f(k) = 0$.	27, 28

	What Did You Learn?	Explanation/Examples	Review Exercises
Section 2.4	Use the imaginary unit i to write complex numbers (p. 101).	When a and b are real numbers, $a + bi$ is a complex number. Two complex numbers $a + bi$ and $c + di$, written in standard form, are equal to each other if and only if $a = c$ and $b = d$.	29, 30
	Add, subtract, and multiply complex numbers (p. 102).	**Sum:** $(a + bi) + (c + di) = (a + c) + (b + d)i$ **Difference:** $(a + bi) - (c + di) = (a - c) + (b - d)i$	31–34
	Use complex conjugates to write the quotient of two complex numbers in standard form (p. 104).	The numbers $a + bi$ and $a - bi$ are complex conjugates. To write $(a + bi)/(c + di)$ in standard form, where c and d are not both zero, multiply the numerator and denominator by $c - di$.	35, 36
	Find complex solutions of quadratic equations (p. 105).	When a is a positive real number, the principal square root of $-a$ is defined as $\sqrt{-a} = \sqrt{a}i$.	37, 38
Section 2.5	Use the Fundamental Theorem of Algebra to determine the number of zeros of polynomial functions (p. 106).	**The Fundamental Theorem of Algebra** If $f(x)$ is a polynomial of degree n, where $n > 0$, then f has at least one zero in the complex number system.	39, 40
	Find rational zeros of polynomial functions (p. 107), and conjugate pairs of complex zeros (p. 110).	The Rational Zero Test relates the possible rational zeros of a polynomial to the leading coefficient and to the constant term of the polynomial. Let f be a polynomial function that has real coefficients. If $a + bi$ ($b \neq 0$) is a zero of the function, then the conjugate $a - bi$ is also a zero of the function.	41, 42
	Find zeros of polynomials by factoring (p. 111), using Descartes's Rule of Signs (p. 113), and using the Upper and Lower Bound Rules (p. 114).	Every polynomial of degree $n > 0$ with real coefficients can be written as the product of linear and quadratic factors with real coefficients, where the quadratic factors have no real zeros.	43–46
Section 2.6	Find the domains (p. 116), and vertical and horizontal asymptotes (p. 117), of the graphs of rational functions.	The domain of a rational function of x includes all real numbers except x-values that make the denominator zero. The line $x = a$ is a vertical asymptote of the graph of f when $f(x) \to \infty$ or $f(x) \to -\infty$ as $x \to a$, either from the right or from the left. The line $y = b$ is a horizontal asymptote of the graph of f when $f(x) \to b$ as $x \to \infty$ or $x \to -\infty$.	47, 48
	Sketch the graphs of rational functions (p. 119), including functions with slant asymptotes (p. 122).	Consider a rational function whose denominator is of degree 1 or greater. If the degree of the numerator is exactly *one more* than the degree of the denominator, then the graph of the function has a slant asymptote.	49–56
	Use rational functions to model and solve real-life problems (p. 123).	A rational function can help you model the cost of removing a given percent of the smokestack pollutants at a utility company that burns coal. (See Example 8.)	57, 58
Section 2.7	Solve polynomial (p. 125) and rational (p. 129) inequalities.	Use the concepts of key numbers and test intervals to solve both polynomial and rational inequalities.	59–62
	Use inequalities to model and solve real-life problems (p. 130).	A common application of inequalities involves profit P, revenue R, and cost C. (See Example 6.)	63

Review Exercises

See **CalcChat.com** for tutorial help and worked-out solutions to odd-numbered exercises.

2.1 Sketching Graphs of Quadratic Functions In Exercises 1 and 2, sketch the graph of each quadratic function and compare it with the graph of $y = x^2$.

1. (a) $g(x) = -2x^2$ **2.** (a) $h(x) = (x-3)^2$
 (b) $h(x) = x^2 + 2$ (b) $k(x) = \frac{1}{2}x^2 - 1$

Using Standard Form to Graph a Parabola In Exercises 3–8, write the quadratic function in standard form and sketch its graph. Identify the vertex, axis of symmetry, and x-intercept(s).

3. $g(x) = x^2 - 2x$ **4.** $f(x) = x^2 + 8x + 10$
5. $h(x) = 3 + 4x - x^2$ **6.** $f(t) = -2t^2 + 4t + 1$
7. $h(x) = 4x^2 + 4x + 13$ **8.** $f(x) = \frac{1}{3}(x^2 + 5x - 4)$

9. Geometry The perimeter of a rectangle is 1000 meters.

(a) Write the width y as a function of the length x. Use the result to write the area A as a function of x.

(b) Of all possible rectangles with perimeters of 1000 meters, find the dimensions of the one with the maximum area.

10. Minimum Cost A soft-drink manufacturer has daily production costs of $C = 70{,}000 - 120x + 0.055x^2$, where C is the total cost (in dollars) and x is the number of units produced. How many units should the manufacturer produce each day to yield a minimum cost?

2.2 Sketching a Transformation of a Monomial Function In Exercises 11 and 12, sketch the graphs of $y = x^n$ and the transformation.

11. $y = x^4$, $f(x) = 6 - x^4$ **12.** $y = x^5$, $f(x) = \frac{1}{2}x^5 + 3$

Applying the Leading Coefficient Test In Exercises 13–16, describe the right-hand and left-hand behavior of the graph of the polynomial function.

13. $f(x) = -2x^2 - 5x + 12$ **14.** $f(x) = \frac{1}{2}x^3 + 2x$
15. $g(x) = \frac{3}{4}(x^4 + 3x^2 + 2)$ **16.** $h(x) = -x^7 + 8x^2 - 8x$

Sketching the Graph of a Polynomial Function In Exercises 17–20, sketch the graph of the function by (a) applying the Leading Coefficient Test, (b) finding the real zeros of the polynomial, (c) plotting sufficient solution points, and (d) drawing a continuous curve through the points.

17. $g(x) = 2x^3 + 4x^2$
18. $h(x) = 3x^2 - x^4$
19. $f(x) = -x^3 + x^2 - 2$
20. $f(x) = x(x^3 + x^2 - 5x + 3)$

Using the Intermediate Value Theorem In Exercises 21 and 22, (a) use the Intermediate Value Theorem and the *table* feature of a graphing utility to find intervals one unit in length in which the polynomial function is guaranteed to have a zero. (b) Adjust the table to approximate the zeros of the function. Use the *zero* or *root* feature of the graphing utility to verify your results.

21. $f(x) = 3x^3 - x^2 + 3$ **22.** $f(x) = x^4 - 5x - 1$

2.3 Long Division of Polynomials In Exercises 23 and 24, use long division to divide.

23. $\dfrac{30x^2 - 3x + 8}{5x - 3}$ **24.** $\dfrac{5x^3 - 21x^2 - 25x - 4}{x^2 - 5x - 1}$

Using the Factor Theorem In Exercises 25 and 26, use synthetic division to determine whether the given values of x are zeros of the function.

25. $f(x) = 20x^4 + 9x^3 - 14x^2 - 3x$
 (a) $x = -1$ (b) $x = \frac{3}{4}$ (c) $x = 0$ (d) $x = 1$
26. $f(x) = 3x^3 - 8x^2 - 20x + 16$
 (a) $x = 4$ (b) $x = -4$ (c) $x = \frac{2}{3}$ (d) $x = -1$

Factoring a Polynomial In Exercises 27 and 28, (a) verify the given factor(s) of $f(x)$, (b) find the remaining factors of $f(x)$, (c) use your results to write the complete factorization of $f(x)$, (d) list all real zeros of f, and (e) confirm your results by using a graphing utility to graph the function.

Function	Factor(s)
27. $f(x) = 2x^3 + 11x^2 - 21x - 90$	$(x+6)$
28. $f(x) = x^4 - 4x^3 - 7x^2 + 22x + 24$	$(x+2), (x-3)$

2.4 Writing a Complex Number in Standard Form In Exercises 29 and 30, write the complex number in standard form.

29. $8 + \sqrt{-100}$ **30.** $-5i + i^2$

Performing Operations with Complex Numbers In Exercises 31–34, perform the operation and write the result in standard form.

31. $(7 + 5i) + (-4 + 2i)$
32. $(\sqrt{2} - \sqrt{2}i) - (\sqrt{2} + \sqrt{2}i)$
33. $7i(11 - 9i)$ **34.** $(1 + 6i)(5 - 2i)$

Performing Operations with Complex Numbers In Exercises 35 and 36, perform the operation and write the result in standard form.

35. $\dfrac{6+i}{4-i}$ **36.** $\dfrac{4}{2-3i} + \dfrac{2}{1+i}$

Complex Solutions of a Quadratic Equation In Exercises 37 and 38, use the Quadratic Formula to solve the quadratic equation.

37. $x^2 - 2x + 10 = 0$ 38. $6x^2 + 3x + 27 = 0$

2.5 **Zeros of Polynomial Functions** In Exercises 39 and 40, determine the number of zeros of the polynomial function.

39. $g(x) = x^2 - 2x - 8$ 40. $h(t) = t^2 - t^5$

Finding the Zeros of a Polynomial Function In Exercises 41 and 42, find all the zeros of the function.

41. $f(x) = x^3 + 3x^2 - 28x - 60$

42. $f(x) = 4x^4 - 11x^3 + 14x^2 - 6x$

Finding the Zeros of a Polynomial Function In Exercises 43 and 44, write the polynomial as the product of linear factors and list all the zeros of the function.

43. $g(x) = x^3 - 7x^2 + 36$

44. $f(x) = x^4 + 8x^3 + 8x^2 - 72x - 153$

45. **Using Descartes's Rule of Signs** Use Descartes's Rule of Signs to determine the possible numbers of positive and negative real zeros of $h(x) = -2x^5 + 4x^3 - 2x^2 + 5$.

46. **Verifying Upper and Lower Bounds** Use synthetic division to verify the upper and lower bounds of the real zeros of $f(x) = 4x^3 - 3x^2 + 4x - 3$.

 (a) Upper: $x = 1$ (b) Lower: $x = -\frac{1}{4}$

2.6 **Finding Domain and Asymptotes** In Exercises 47 and 48, find the domain and the vertical and horizontal asymptotes of the graph of the rational function.

47. $f(x) = \dfrac{3x}{x + 10}$

48. $f(x) = \dfrac{8}{x^2 - 10x + 24}$

Sketching the Graph of a Rational Function In Exercises 49–56, (a) state the domain of the function, (b) identify all intercepts, (c) find any asymptotes, and (d) plot additional solution points as needed to sketch the graph of the rational function.

49. $f(x) = \dfrac{4}{x}$ 50. $h(x) = \dfrac{x - 4}{x - 7}$

51. $f(x) = \dfrac{x}{x^2 + 1}$ 52. $f(x) = \dfrac{2x^2}{x^2 - 4}$

53. $f(x) = \dfrac{6x^2 - 11x + 3}{3x^2 - x}$ 54. $f(x) = \dfrac{6x^2 - 7x + 2}{4x^2 - 1}$

55. $f(x) = \dfrac{2x^3}{x^2 + 1}$ 56. $f(x) = \dfrac{x^2 + 1}{x + 1}$

57. **Seizure of Illegal Drugs** The cost C (in millions of dollars) for the federal government to seize $p\%$ of an illegal drug as it enters the country is given by

$$C = \frac{528p}{100 - p}, \quad 0 \le p < 100.$$

 (a) Use a graphing utility to graph the cost function.

 (b) Find the costs of seizing 25%, 50%, and 75% of the drug.

 (c) According to the model, would it be possible to seize 100% of the drug?

58. **Page Design** A page that is x inches wide and y inches high contains 30 square inches of print. The top and bottom margins are each 2 inches deep, and the margins on each side are 2 inches wide.

 (a) Write a function for the total area A of the page in terms of x.

 (b) Determine the domain of the function based on the physical constraints of the problem.

 (c) Use a graphing utility to graph the area function and approximate the page size for which the least amount of paper is used.

2.7 **Solving an Inequality** In Exercises 59–62, solve the inequality. Then graph the solution set.

59. $12x^2 + 5x < 2$

60. $x^3 - 16x \ge 0$

61. $\dfrac{2}{x + 1} \ge \dfrac{3}{x - 1}$

62. $\dfrac{x^2 - 9x + 20}{x} < 0$

63. **Population of a Species** A biologist introduces 200 ladybugs into a crop field. The population P of the ladybugs can be approximated by the model

$$P = \frac{1000(1 + 3t)}{5 + t}$$

where t is the time in days. Find the time required for the population to increase to at least 2000 ladybugs.

Exploration

64. **Writing** Describe what is meant by an asymptote of a graph.

True or False? In Exercises 65 and 66, determine whether the statement is true or false. Justify your answer.

65. A fourth-degree polynomial with real coefficients can have -5, $-8i$, $4i$, and 5 as its zeros.

66. The domain of a rational function can never be the set of all real numbers.

Chapter Test

See **CalcChat.com** for tutorial help and worked-out solutions to odd-numbered exercises.

Take this test as you would take a test in class. When you are finished, check your work against the answers given in the back of the book.

1. Sketch the graph of each quadratic function and compare it with the graph of $y = x^2$.

   (a) $g(x) = 2 - x^2$

   (b) $g(x) = \left(x - \frac{3}{2}\right)^2$

2. Write the standard form of the equation of the parabola shown in the figure.

3. The path of a ball is given by the function $f(x) = -\frac{1}{20}x^2 + 3x + 5$, where $f(x)$ is the height (in feet) of the ball and x is the horizontal distance (in feet) from where the ball was thrown.

   (a) What is the maximum height of the ball?

   (b) Which number determines the height at which the ball was thrown? Does changing this value change the coordinates of the maximum height of the ball? Explain.

4. Describe the right-hand and left-hand behavior of the graph of the function $h(t) = -\frac{3}{4}t^5 + 2t^2$. Then sketch its graph.

5. Divide using long division.

   $$\frac{3x^3 + 4x - 1}{x^2 + 1}$$

6. Divide using synthetic division.

   $$\frac{2x^4 - 5x^2 - 3}{x - 2}$$

7. Use synthetic division to show that $x = \frac{5}{2}$ is a zero of the function

   $$f(x) = 2x^3 - 5x^2 - 6x + 15.$$

   Use the result to factor the polynomial function completely and list all the zeros of the function.

8. Perform each operation and write the result in standard form.

   (a) $10i - \left(3 + \sqrt{-25}\right)$

   (b) $\left(2 + \sqrt{3}i\right)\left(2 - \sqrt{3}i\right)$

9. Write the quotient in standard form: $\dfrac{5}{2 + i}$.

In Exercises 10 and 11, find a polynomial function with real coefficients that has the given zeros. (There are many correct answers.)

10. $0, 3, 2 + i$

11. $1 - \sqrt{3}i, 2, 2$

In Exercises 12 and 13, find all the zeros of the function.

12. $f(x) = 3x^3 + 14x^2 - 7x - 10$

13. $f(x) = x^4 - 9x^2 - 22x - 24$

In Exercises 14–16, identify any intercepts and asymptotes of the graph of the function. Then sketch a graph of the function.

14. $h(x) = \dfrac{4}{x^2} - 1$

15. $f(x) = \dfrac{2x^2 - 5x - 12}{x^2 - 16}$

16. $g(x) = \dfrac{x^2 + 2}{x - 1}$

In Exercises 17 and 18, solve the inequality. Then graph the solution set.

17. $2x^2 + 5x > 12$

18. $\dfrac{2}{x} \leq \dfrac{1}{x + 6}$

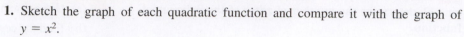

Figure for 2

Proofs in Mathematics ■ ■ ■ ■ ■ ■ ■ ■ ■ ■ ■ ■

These two pages contain proofs of four important theorems about polynomial functions. The first two theorems are from Section 2.3, and the second two theorems are from Section 2.5.

> ### The Remainder Theorem   *(p. 99)*
> If a polynomial $f(x)$ is divided by $x - k$, then the remainder is
> $$r = f(k).$$

Proof

From the Division Algorithm, you have

$$f(x) = (x - k)q(x) + r(x)$$

and because either $r(x) = 0$ or the degree of $r(x)$ is less than the degree of $x - k$, you know that $r(x)$ must be a constant. That is, $r(x) = r$. Now, by evaluating $f(x)$ at $x = k$, you have

$$f(k) = (k - k)q(k) + r$$
$$= (0)q(k) + r = r.$$

To be successful in algebra, it is important that you understand the connection among *factors* of a polynomial, *zeros* of a polynomial function, and *solutions* or *roots* of a polynomial equation. The Factor Theorem is the basis for this connection.

> ### The Factor Theorem   *(p. 100)*
> A polynomial $f(x)$ has a factor $(x - k)$ if and only if $f(k) = 0$.

Proof

Using the Division Algorithm with the factor $(x - k)$, you have

$$f(x) = (x - k)q(x) + r(x).$$

By the Remainder Theorem, $r(x) = r = f(k)$, and you have

$$f(x) = (x - k)q(x) + f(k)$$

where $q(x)$ is a polynomial of lesser degree than $f(x)$. If $f(k) = 0$, then

$$f(x) = (x - k)q(x)$$

and you see that $(x - k)$ is a factor of $f(x)$. Conversely, if $(x - k)$ is a factor of $f(x)$, then division of $f(x)$ by $(x - k)$ yields a remainder of 0. So, by the Remainder Theorem, you have $f(k) = 0$.

Linear Factorization Theorem (p. 106)

If $f(x)$ is a polynomial of degree n, where $n > 0$, then $f(x)$ has precisely n linear factors

$$f(x) = a_n(x - c_1)(x - c_2) \cdots (x - c_n)$$

where $c_1, c_2, \ldots, c_n$ are complex numbers.

THE FUNDAMENTAL THEOREM OF ALGEBRA

The Linear Factorization Theorem is closely related to the Fundamental Theorem of Algebra. The Fundamental Theorem of Algebra has a long and interesting history. In the early work with polynomial equations, the Fundamental Theorem of Algebra was thought to have been not true, because imaginary solutions were not considered. In fact, in the very early work by mathematicians such as Abu al-Khwarizmi (c. 800 A.D.), negative solutions were also not considered.

Once imaginary numbers were accepted, several mathematicians attempted to give a general proof of the Fundamental Theorem of Algebra. These included Gottfried von Leibniz (1702), Jean d'Alembert (1746), Leonhard Euler (1749), Joseph-Louis Lagrange (1772), and Pierre Simon Laplace (1795). The mathematician usually credited with the first correct proof of the Fundamental Theorem of Algebra is Carl Friedrich Gauss, who published the proof in his doctoral thesis in 1799.

Proof

Using the Fundamental Theorem of Algebra, you know that f must have at least one zero, c_1. Consequently, $(x - c_1)$ is a factor of $f(x)$, and you have

$$f(x) = (x - c_1)f_1(x).$$

If the degree of $f_1(x)$ is greater than zero, then you again apply the Fundamental Theorem of Algebra to conclude that f_1 must have a zero c_2, which implies that

$$f(x) = (x - c_1)(x - c_2)f_2(x).$$

It is clear that the degree of $f_1(x)$ is $n - 1$, that the degree of $f_2(x)$ is $n - 2$, and that you can repeatedly apply the Fundamental Theorem of Algebra n times until you obtain

$$f(x) = a_n(x - c_1)(x - c_2) \cdots (x - c_n)$$

where a_n is the leading coefficient of the polynomial $f(x)$. ■

Factors of a Polynomial (p. 111)

Every polynomial of degree $n > 0$ with real coefficients can be written as the product of linear and quadratic factors with real coefficients, where the quadratic factors have no real zeros.

Proof

To begin, you use the Linear Factorization Theorem to conclude that $f(x)$ can be *completely* factored in the form

$$f(x) = d(x - c_1)(x - c_2)(x - c_3) \cdots (x - c_n).$$

If each c_i is real, then there is nothing more to prove. If any c_i is complex ($c_i = a + bi$, $b \neq 0$), then, because the coefficients of $f(x)$ are real, you know that the conjugate $c_j = a - bi$ is also a zero. By multiplying the corresponding factors, you obtain

$$(x - c_i)(x - c_j) = [x - (a + bi)][x - (a - bi)]$$
$$= x^2 - 2ax + (a^2 + b^2)$$

where each coefficient is real. ■

P.S. Problem Solving ■ ■ ■ ■ ■ ■ ■ ■ ■ ■ ■ ■ ■ ■

1. **Verifying the Remainder Theorem** Show that if $f(x) = ax^3 + bx^2 + cx + d$, then $f(k) = r$, where $r = ak^3 + bk^2 + ck + d$, using long division. In other words, verify the Remainder Theorem for a third-degree polynomial function.

2. **Babylonian Mathematics** In 2000 B.C., the Babylonians solved polynomial equations by referring to tables of values. One such table gave the values of $y^3 + y^2$. To be able to use this table, the Babylonians sometimes had to manipulate the equation, as shown below.

$$ax^3 + bx^2 = c \qquad \text{Original equation}$$

$$\frac{a^3 x^3}{b^3} + \frac{a^2 x^2}{b^2} = \frac{a^2 c}{b^3} \qquad \text{Multiply each side by } \frac{a^2}{b^3}.$$

$$\left(\frac{ax}{b}\right)^3 + \left(\frac{ax}{b}\right)^2 = \frac{a^2 c}{b^3} \qquad \text{Rewrite.}$$

Then they would find $(a^2 c)/b^3$ in the $y^3 + y^2$ column of the table. Because they knew that the corresponding y-value was equal to $(ax)/b$, they could conclude that $x = (by)/a$.

(a) Calculate $y^3 + y^2$ for $y = 1, 2, 3, \ldots, 10$. Record the values in a table.

(b) Use the table from part (a) and the method above to solve each equation.

  (i) $x^3 + x^2 = 252$

  (ii) $x^3 + 2x^2 = 288$

  (iii) $3x^3 + x^2 = 90$

  (iv) $2x^3 + 5x^2 = 2500$

  (v) $7x^3 + 6x^2 = 1728$

  (vi) $10x^3 + 3x^2 = 297$

(c) Using the methods from this chapter, verify your solution of each equation.

3. **Finding Dimensions** At a glassware factory, molten cobalt glass is poured into molds to make paperweights. Each mold is a rectangular prism whose height is 3 inches greater than the length of each side of the square base. A machine pours 20 cubic inches of liquid glass into each mold. What are the dimensions of the mold?

4. **True or False?** Determine whether the statement is true or false. If false, provide one or more reasons why the statement is false and correct the statement. Let $f(x) = ax^3 + bx^2 + cx + d$, $a \neq 0$, and let $f(2) = -1$. Then

$$\frac{f(x)}{x + 1} = q(x) + \frac{2}{x + 1}$$

where $q(x)$ is a second-degree polynomial.

5. **Finding the Equation of a Parabola** The parabola shown in the figure has an equation of the form $y = ax^2 + bx + c$. Find the equation of this parabola using the following methods.

(a) Find the equation analytically.

(b) Use the regression feature of a graphing utility to find the equation.

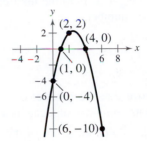

6. **Finding the Slope of a Tangent Line** One of the fundamental themes of calculus is to find the slope of the tangent line to a curve at a point. To see how this can be done, consider the point $(2, 4)$ on the graph of the quadratic function $f(x) = x^2$, as shown in the figure.

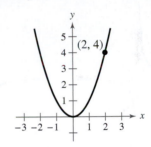

(a) Find the slope m_1 of the line joining $(2, 4)$ and $(3, 9)$. Is the slope of the tangent line at $(2, 4)$ greater than or less than the slope of the line through $(2, 4)$ and $(3, 9)$?

(b) Find the slope m_2 of the line joining $(2, 4)$ and $(1, 1)$. Is the slope of the tangent line at $(2, 4)$ greater than or less than the slope of the line through $(2, 4)$ and $(1, 1)$?

(c) Find the slope m_3 of the line joining $(2, 4)$ and $(2.1, 4.41)$. Is the slope of the tangent line at $(2, 4)$ greater than or less than the slope of the line through $(2, 4)$ and $(2.1, 4.41)$?

(d) Find the slope m_h of the line joining $(2, 4)$ and $(2 + h, f(2 + h))$ in terms of the nonzero number h.

(e) Evaluate the slope formula from part (d) for $h = -1$, 1, and 0.1. Compare these values with those in parts (a)–(c).

(f) What can you conclude the slope m_{tan} of the tangent line at $(2, 4)$ to be? Explain your answer.

7. **Writing Cubic Functions** Use the form $f(x) = (x - k)q(x) + r$ to write a cubic function that (a) passes through the point $(2, 5)$ and rises to the right and (b) passes through the point $(-3, 1)$ and falls to the right. (There are many correct answers.)

8. **Multiplicative Inverse of a Complex Number** The multiplicative inverse of z is a complex number z_m such that $z \cdot z_m = 1$. Find the multiplicative inverse of each complex number.

   (a) $z = 1 + i$

   (b) $z = 3 - i$

   (c) $z = -2 + 8i$

9. **Proof** Prove that the product of a complex number $a + bi$ and its complex conjugate is a real number.

10. **Matching** Match the graph of the rational function

    $$f(x) = \frac{ax + b}{cx + d}$$

    with the given conditions.

    (a)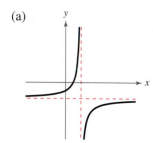

    (b)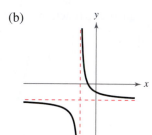

    (c)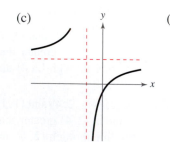

    (d)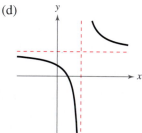

    (i) $a > 0$ (ii) $a > 0$ (iii) $a < 0$ (iv) $a > 0$
       $b < 0$ $b > 0$ $b > 0$ $b < 0$
       $c > 0$ $c < 0$ $c > 0$ $c > 0$
       $d < 0$ $d < 0$ $d < 0$ $d > 0$

11. **Effects of Values on a Graph** Consider the function

    $$f(x) = \frac{ax}{(x - b)^2}.$$

    (a) Determine the effect on the graph of f when $b \neq 0$ and a is varied. Consider cases in which a is positive and a is negative.

    (b) Determine the effect on the graph of f when $a \neq 0$ and b is varied.

12. **Distinct Vision** The endpoints of the interval over which distinct vision is possible are called the *near point* and *far point* of the eye (see figure). With increasing age, these points normally change. The table shows the approximate near points y (in inches) for various ages x (in years).

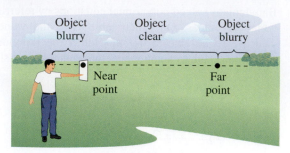

Age, x	Near Point, y
16	3.0
32	4.7
44	9.8
50	19.7
60	39.4

(a) Use the *regression* feature of a graphing utility to find a quadratic model for the data. Use the graphing utility to plot the data and graph the model in the same viewing window.

(b) Find a rational model for the data. Take the reciprocals of the near points to generate the points $(x, 1/y)$. Use the *regression* feature of the graphing utility to find a linear model for the data. The resulting line has the form

   $$\frac{1}{y} = ax + b.$$

   Solve for y. Use the graphing utility to plot the data and graph the model in the same viewing window.

(c) Use the *table* feature of the graphing utility to construct a table showing the predicted near point based on each model for each of the ages in the original table. How well do the models fit the original data?

(d) Use both models to estimate the near point for a person who is 25 years old. Which model is a better fit?

(e) Do you think either model can be used to predict the near point for a person who is 70 years old? Explain.

13. **Roots of a Cubic Equation** Can a cubic equation with real coefficients have two real zeros and one complex zero? Explain your reasoning.

3 Exponential and Logarithmic Functions

Trees per Acre

Earthquakes

Sound Intensity

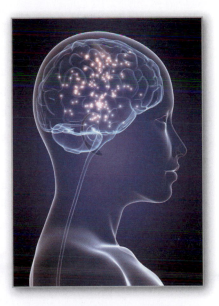

Human Memory Model

Nuclear Reactor Accident

141

3.1 Exponential Functions and Their Graphs

Exponential functions can help you model and solve real-life problems, such as the concentration of a drug in the bloodstream.

- ■ Recognize and evaluate exponential functions with base *a*.
- ■ Graph exponential functions and use the One-to-One Property.
- ■ Recognize, evaluate, and graph exponential functions with base e.
- ■ Use exponential functions to model and solve real-life problems.

Exponential Functions

So far, this text has dealt mainly with **algebraic functions,** which include polynomial functions and rational functions. In this chapter, you will study two types of nonalgebraic functions—*exponential functions* and *logarithmic functions*. These functions are examples of **transcendental functions.**

Definition of Exponential Function

The **exponential function** *f* with base *a* is denoted by

$$f(x) = a^x$$

where $a > 0$, $a \neq 1$, and x is any real number.

The base $a = 1$ is excluded because it yields $f(x) = 1^x = 1$. This is a constant function, not an exponential function.

You have evaluated a^x for integer and rational values of x. For example, you know that $4^3 = 64$ and $4^{1/2} = 2$. However, to evaluate 4^x for any real number x, you need to interpret forms with *irrational* exponents. For the purposes of this text, it is sufficient to think of $a^{\sqrt{2}}$ (where $\sqrt{2} \approx 1.41421356$) as the number that has the successively closer approximations

$$a^{1.4}, a^{1.41}, a^{1.414}, a^{1.4142}, a^{1.41421}, \ldots .$$

EXAMPLE 1   **Evaluating Exponential Functions**

Use a calculator to evaluate each function at the indicated value of *x*.

Function	Value
a. $f(x) = 2^x$	$x = -3.1$
b. $f(x) = 2^{-x}$	$x = \pi$
c. $f(x) = 0.6^x$	$x = \frac{3}{2}$

Solution

Function Value	Graphing Calculator Keystrokes	Display
a. $f(-3.1) = 2^{-3.1}$	2 ⌃ (−) 3.1 [ENTER]	0.1166291
b. $f(\pi) = 2^{-\pi}$	2 ⌃ (−) π [ENTER]	0.1133147
c. $f(\frac{3}{2}) = (0.6)^{3/2}$	.6 ⌃ (3 ÷ 2) [ENTER]	0.4647580

✓ *Checkpoint* 🔊)) *Audio-video solution in English & Spanish at LarsonPrecalculus.com.*

Use a calculator to evaluate $f(x) = 8^{-x}$ at $x = \sqrt{2}$.   ■

When evaluating exponential functions with a calculator, remember to enclose fractional exponents in parentheses. Because the calculator follows the order of operations, parentheses are crucial in order to obtain the correct result.

Graphs of Exponential Functions

The graphs of all exponential functions have similar characteristics, as shown in Examples 2, 3, and 5.

EXAMPLE 2 **Graphs of $y = a^x$**

In the same coordinate plane, sketch the graph of each function.

a. $f(x) = 2^x$

b. $g(x) = 4^x$

Solution The following table lists some values for each function, and Figure 3.1 shows the graphs of the two functions. Note that both graphs are increasing. Moreover, the graph of $g(x) = 4^x$ is increasing more rapidly than the graph of $f(x) = 2^x$.

> **ALGEBRA HELP** You can review the techniques for sketching the graph of an equation in Section 1.2.

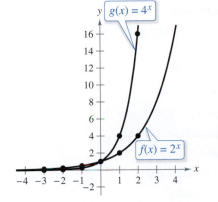

Figure 3.1

x	-3	-2	-1	0	1	2
2^x	$\frac{1}{8}$	$\frac{1}{4}$	$\frac{1}{2}$	1	2	4
4^x	$\frac{1}{64}$	$\frac{1}{16}$	$\frac{1}{4}$	1	4	16

✓ **Checkpoint**  Audio-video solution in English & Spanish at LarsonPrecalculus.com.

In the same coordinate plane, sketch the graph of each function.

a. $f(x) = 3^x$ **b.** $g(x) = 9^x$

The table in Example 2 was evaluated by hand. You could, of course, use a graphing utility to construct tables with even more values.

EXAMPLE 3 **Graphs of $y = a^{-x}$**

In the same coordinate plane, sketch the graph of each function.

a. $F(x) = 2^{-x}$

b. $G(x) = 4^{-x}$

Solution The following table lists some values for each function, and Figure 3.2 shows the graphs of the two functions. Note that both graphs are decreasing. Moreover, the graph of $G(x) = 4^{-x}$ is decreasing more rapidly than the graph of $F(x) = 2^{-x}$.

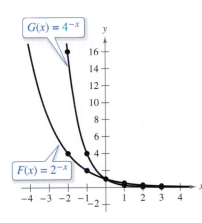

Figure 3.2

x	-2	-1	0	1	2	3
2^{-x}	4	2	1	$\frac{1}{2}$	$\frac{1}{4}$	$\frac{1}{8}$
4^{-x}	16	4	1	$\frac{1}{4}$	$\frac{1}{16}$	$\frac{1}{64}$

✓ **Checkpoint**  Audio-video solution in English & Spanish at LarsonPrecalculus.com.

In the same coordinate plane, sketch the graph of each function.

a. $f(x) = 3^{-x}$ **b.** $g(x) = 9^{-x}$

Note that it is possible to use one of the properties of exponents to rewrite the functions in Example 3 with positive exponents, as follows.

$$F(x) = 2^{-x} = \frac{1}{2^x} = \left(\frac{1}{2}\right)^x \quad \text{and} \quad G(x) = 4^{-x} = \frac{1}{4^x} = \left(\frac{1}{4}\right)^x$$

Comparing the functions in Examples 2 and 3, observe that

$$F(x) = 2^{-x} = f(-x) \quad \text{and} \quad G(x) = 4^{-x} = g(-x).$$

Consequently, the graph of F is a reflection (in the y-axis) of the graph of f. The graphs of G and g have the same relationship. The graphs in Figures 3.1 and 3.2 are typical of the exponential functions $y = a^x$ and $y = a^{-x}$. They have one y-intercept and one horizontal asymptote (the x-axis), and they are continuous. The following summarizes the basic characteristics of these exponential functions.

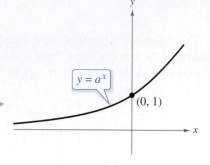

$y = a^x$

$(0, 1)$

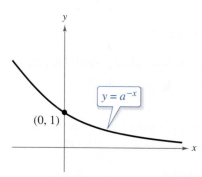

$y = a^{-x}$

$(0, 1)$

▷ •••••••••••••••••••••
• **REMARK**   Notice that the range of an exponential function is $(0, \infty)$, which means that $a^x > 0$ for all values of x.

Graph of $y = a^x$, $a > 1$
- Domain: $(-\infty, \infty)$
- Range: $(0, \infty)$
- y-intercept: $(0, 1)$
- Increasing
- x-axis is a horizontal asymptote
  ($a^x \to 0$ as $x \to -\infty$).
- Continuous

Graph of $y = a^{-x}$, $a > 1$
- Domain: $(-\infty, \infty)$
- Range: $(0, \infty)$
- y-intercept: $(0, 1)$
- Decreasing
- x-axis is a horizontal asymptote
  ($a^{-x} \to 0$ as $x \to \infty$).
- Continuous

Notice that the graph of an exponential function is always increasing or always decreasing. As a result, the graphs pass the Horizontal Line Test, and therefore the functions are one-to-one functions. You can use the following **One-to-One Property** to solve simple exponential equations.

For $a > 0$ and $a \neq 1$, $a^x = a^y$ if and only if $x = y$.   One-to-One Property

EXAMPLE 4   **Using the One-to-One Property**

a. $9 = 3^{x+1}$ Original equation

$3^2 = 3^{x+1}$ $9 = 3^2$

$2 = x + 1$ One-to-One Property

$1 = x$ Solve for x.

b. $\left(\frac{1}{2}\right)^x = 8$ Original equation

$2^{-x} = 2^3$ $\left(\frac{1}{2}\right)^x = 2^{-x}, 8 = 2^3$

$x = -3$ One-to-One Property

✓ **Checkpoint**   *Audio-video solution in English & Spanish at LarsonPrecalculus.com.*

Use the One-to-One Property to solve the equation for x.

a. $8 = 2^{2x-1}$ **b.** $\left(\frac{1}{3}\right)^{-x} = 27$

■

In the following example, notice how the graph of $y = a^x$ can be used to sketch the graphs of functions of the form $f(x) = b \pm a^{x+c}$.

▷ **ALGEBRA HELP** You can review the techniques for transforming the graph of a function in Section 1.7.

EXAMPLE 5 **Transformations of Graphs of Exponential Functions**

Each of the following graphs is a transformation of the graph of $f(x) = 3^x$.

a. Because $g(x) = 3^{x+1} = f(x + 1)$, the graph of g can be obtained by shifting the graph of f one unit to the *left*, as shown in Figure 3.3.

b. Because $h(x) = 3^x - 2 = f(x) - 2$, the graph of h can be obtained by shifting the graph of f *down* two units, as shown in Figure 3.4.

c. Because $k(x) = -3^x = -f(x)$, the graph of k can be obtained by *reflecting* the graph of f in the x-axis, as shown in Figure 3.5.

d. Because $j(x) = 3^{-x} = f(-x)$, the graph of j can be obtained by *reflecting* the graph of f in the y-axis, as shown in Figure 3.6.

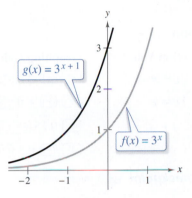

Figure 3.3 Horizontal shift

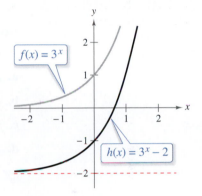

Figure 3.4 Vertical shift

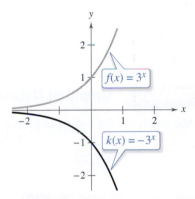

Figure 3.5 Reflection in x-axis

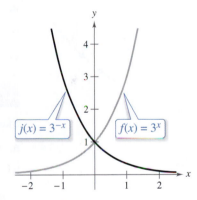

Figure 3.6 Reflection in y-axis

✓ *Checkpoint* ◀))) *Audio-video solution in English & Spanish at LarsonPrecalculus.com.*

Use the graph of $f(x) = 4^x$ to describe the transformation that yields the graph of each function.

a. $g(x) = 4^{x-2}$ **b.** $h(x) = 4^x + 3$ **c.** $k(x) = 4^{-x} - 3$ ∎

Notice that the transformations in Figures 3.3, 3.5, and 3.6 keep the x-axis as a horizontal asymptote, but the transformation in Figure 3.4 yields a new horizontal asymptote of $y = -2$. Also, be sure to note how each transformation affects the y-intercept.

The Natural Base e

In many applications, the most convenient choice for a base is the irrational number

$$e \approx 2.718281828 \ldots .$$

This number is called the **natural base**. The function $f(x) = e^x$ is called the **natural exponential function**. Figure 3.7 shows its graph. Be sure you see that for the exponential function $f(x) = e^x$, e is the constant $2.718281828 \ldots$, whereas x is the variable.

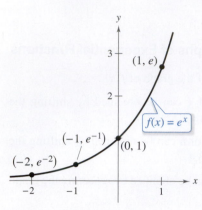

$f(x) = e^x$

Figure 3.7

EXAMPLE 6 **Evaluating the Natural Exponential Function**

Use a calculator to evaluate the function $f(x) = e^x$ at each value of x.

a. $x = -2$

b. $x = -1$

c. $x = 0.25$

d. $x = -0.3$

Solution

Function Value	Graphing Calculator Keystrokes	Display
a. $f(-2) = e^{-2}$	e^x $(-)$ 2 ENTER	0.1353353
b. $f(-1) = e^{-1}$	e^x $(-)$ 1 ENTER	0.3678794
c. $f(0.25) = e^{0.25}$	e^x 0.25 ENTER	1.2840254
d. $f(-0.3) = e^{-0.3}$	e^x $(-)$ 0.3 ENTER	0.7408182

✓ *Checkpoint* ◀))) *Audio-video solution in English & Spanish at LarsonPrecalculus.com.*

Use a calculator to evaluate the function $f(x) = e^x$ at each value of x.

a. $x = 0.3$

b. $x = -1.2$

c. $x = 6.2$

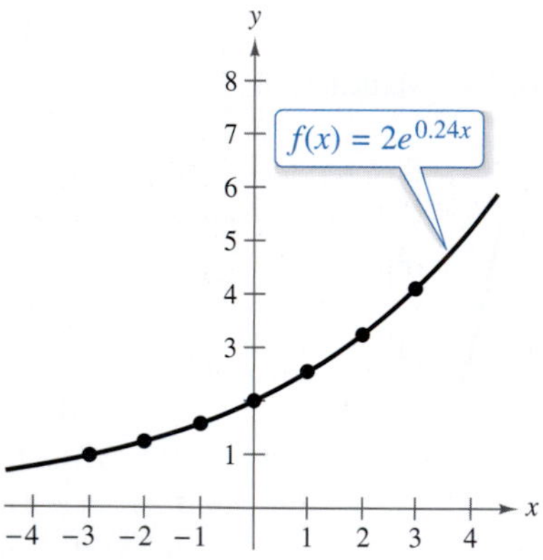

EXAMPLE 7 **Graphing Natural Exponential Functions**

Sketch the graph of each natural exponential function.

a. $f(x) = 2e^{0.24x}$

b. $g(x) = \frac{1}{2}e^{-0.58x}$

Solution To sketch these two graphs, use a graphing utility to construct a table of values, as follows. After constructing the table, plot the points and connect them with smooth curves, as shown in Figures 3.8 and 3.9. Note that the graph in Figure 3.8 is increasing, whereas the graph in Figure 3.9 is decreasing.

x	-3	-2	-1	0	1	2	3
$f(x)$	0.974	1.238	1.573	2.000	2.542	3.232	4.109
$g(x)$	2.849	1.595	0.893	0.500	0.280	0.157	0.088

✓ *Checkpoint* ◀))) *Audio-video solution in English & Spanish at LarsonPrecalculus.com.*

Sketch the graph of $f(x) = 5e^{0.17x}$.

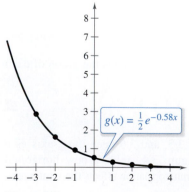

$f(x) = 2e^{0.24x}$

Figure 3.8

$g(x) = \frac{1}{2}e^{-0.58x}$

Figure 3.9

Applications

One of the most familiar examples of exponential growth is an investment earning *continuously compounded interest.* The formula for *interest compounded n times per year* is

$$A = P\left(1 + \frac{r}{n}\right)^{nt}.$$

In this formula, A is the balance in the account, P is the principal (or original deposit), r is the annual interest rate (in decimal form), n is the number of compoundings per year, and t is the time in years. Using exponential functions, you can *develop* this formula and show how it leads to continuous compounding.

Suppose you invest a principal P at an annual interest rate r, compounded once per year. If the interest is added to the principal at the end of the year, then the new balance P_1 is

$$P_1 = P + Pr$$
$$= P(1 + r).$$

This pattern of multiplying the previous principal by $1 + r$ repeats each successive year, as follows.

Year	Balance After Each Compounding
0	$P = P$
1	$P_1 = P(1 + r)$
2	$P_2 = P_1(1 + r) = P(1 + r)(1 + r) = P(1 + r)^2$
3	$P_3 = P_2(1 + r) = P(1 + r)^2(1 + r) = P(1 + r)^3$
$\vdots$	$\vdots$
t	$P_t = P(1 + r)^t$

To accommodate more frequent (quarterly, monthly, or daily) compounding of interest, let n be the number of compoundings per year and let t be the number of years. Then the rate per compounding is r/n and the account balance after t years is

$$A = P\left(1 + \frac{r}{n}\right)^{nt}. \qquad \text{\color{red}Amount (balance) with } n \text{ compoundings per year}$$

When you let the number of compoundings n increase without bound, the process approaches what is called **continuous compounding.** In the formula for n compoundings per year, let $m = n/r$. This produces

$$A = P\left(1 + \frac{r}{n}\right)^{nt} \qquad \text{\color{red}Amount with } n \text{ compoundings per year}$$

$$= P\left(1 + \frac{r}{mr}\right)^{mrt} \qquad \text{\color{red}Substitute } mr \text{ for } n.$$

$$= P\left(1 + \frac{1}{m}\right)^{mrt} \qquad \text{\color{red}Simplify.}$$

$$= P\left[\left(1 + \frac{1}{m}\right)^{m}\right]^{rt}. \qquad \text{\color{red}Property of exponents}$$

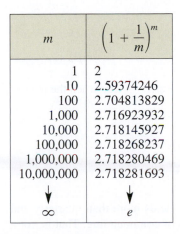

m	$\left(1 + \dfrac{1}{m}\right)^{m}$
1	2
10	2.59374246
100	2.704813829
1,000	2.716923932
10,000	2.718145927
100,000	2.718268237
1,000,000	2.718280469
10,000,000	2.718281693
$\downarrow$	$\downarrow$
∞	e

As m increases without bound (that is, as $m \to \infty$), the table at the left shows that $[1 + (1/m)]^m \to e$. From this, you can conclude that the formula for continuous compounding is

$$A = Pe^{rt}. \qquad \text{\color{red}Substitute } e \text{ for } (1 + 1/m)^m.$$

•• **REMARK** Be sure you see
that, when using the formulas
for compound interest, you must
write the annual interest rate in
decimal form. For instance, you
must write 6% as 0.06.

Formulas for Compound Interest

After t years, the balance A in an account with principal P and annual interest
rate r (in decimal form) is given by the following formulas.

1. For n compoundings per year: $A = P\left(1 + \dfrac{r}{n}\right)^{nt}$

2. For continuous compounding: $A = Pe^{rt}$

EXAMPLE 8 **Compound Interest**

You invest \$12,000 at an annual rate of 3%. Find the balance after 5 years when the
interest is compounded

a. quarterly.

b. monthly.

c. continuously.

Solution

a. For quarterly compounding, you have $n = 4$. So, in 5 years at 3%, the balance is

$$A = P\left(1 + \frac{r}{n}\right)^{nt} \qquad \text{Formula for compound interest}$$

$$= 12{,}000\left(1 + \frac{0.03}{4}\right)^{4(5)} \qquad \text{Substitute for } P, r, n, \text{ and } t.$$

$$\approx \$13{,}934.21. \qquad \text{Use a calculator.}$$

b. For monthly compounding, you have $n = 12$. So, in 5 years at 3%, the balance is

$$A = P\left(1 + \frac{r}{n}\right)^{nt} \qquad \text{Formula for compound interest}$$

$$= 12{,}000\left(1 + \frac{0.03}{12}\right)^{12(5)} \qquad \text{Substitute for } P, r, n, \text{ and } t.$$

$$\approx \$13{,}939.40. \qquad \text{Use a calculator.}$$

c. For continuous compounding, the balance is

$$A = Pe^{rt} \qquad \text{Formula for continuous compounding}$$

$$= 12{,}000e^{0.03(5)} \qquad \text{Substitute for } P, r, \text{ and } t.$$

$$\approx \$13{,}942.01. \qquad \text{Use a calculator.}$$

✓ *Checkpoint*  *Audio-video solution in English & Spanish at LarsonPrecalculus.com.*

You invest \$6000 at an annual rate of 4%. Find the balance after 7 years when the interest
is compounded

a. quarterly.

b. monthly.

c. continuously.

In Example 8, note that continuous compounding yields more than quarterly and
monthly compounding. This is typical of the two types of compounding. That is, for a
given principal, interest rate, and time, continuous compounding will always yield a
larger balance than compounding n times per year.

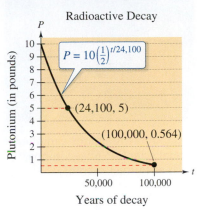

The International Atomic Energy Authority ranks nuclear incidents and accidents by severity using a scale from 1 to 7 called the International Nuclear and Radiological Event Scale (INES). A level 7 ranking is the most severe. To date, the Chernobyl accident is the only nuclear accident in history to be given an INES level 7 ranking.

EXAMPLE 9   **Radioactive Decay**

In 1986, a nuclear reactor accident occurred in Chernobyl in what was then the Soviet Union. The explosion spread highly toxic radioactive chemicals, such as plutonium (^{239}Pu), over hundreds of square miles, and the government evacuated the city and the surrounding area. To see why the city is now uninhabited, consider the model

$$P = 10\left(\frac{1}{2}\right)^{t/24,100}$$

which represents the amount of plutonium P that remains (from an initial amount of 10 pounds) after t years. Sketch the graph of this function over the interval from $t = 0$ to $t = 100,000$, where $t = 0$ represents 1986. How much of the 10 pounds will remain in the year 2017? How much of the 10 pounds will remain after 100,000 years?

Solution   The graph of this function is shown in the figure at the right. Note from this graph that plutonium has a *half-life* of about 24,100 years. That is, after 24,100 years, *half* of the original amount will remain. After another 24,100 years, one-quarter of the original amount will remain, and so on. In the year 2017 ($t = 31$), there will still be

$$P = 10\left(\frac{1}{2}\right)^{31/24,100}$$

$$\approx 10\left(\frac{1}{2}\right)^{0.0012863}$$

$$\approx 9.991 \text{ pounds}$$

of plutonium remaining. After 100,000 years, there will still be

$$P = 10\left(\frac{1}{2}\right)^{100,000/24,100}$$

$$\approx 0.564 \text{ pound}$$

of plutonium remaining.

Radioactive Decay

[Graph: vertical axis labeled "Plutonium (in pounds)" with P, marked from 1 to 10; curve labeled $P = 10\left(\frac{1}{2}\right)^{t/24,100}$; points $(24,100, 5)$ and $(100,000, 0.564)$; horizontal axis labeled "Years of decay" with t, marked 50,000 and 100,000.]

✓ *Checkpoint* *Audio-video solution in English & Spanish at LarsonPrecalculus.com.*

In Example 9, how much of the 10 pounds will remain in the year 2089? How much of the 10 pounds will remain after 125,000 years?

Summarize   (Section 3.1)

1. State the definition of an exponential function f with base a *(page 142)*. For an example of evaluating exponential functions, see Example 1.

2. Describe the basic characteristics of the exponential functions $y = a^x$ and $y = a^{-x}$, $a > 1$ *(page 144)*. For examples of graphing exponential functions, see Examples 2, 3, and 5.

3. State the definitions of the natural base and the natural exponential function *(page 146)*. For examples of evaluating and graphing natural exponential functions, see Examples 6 and 7.

4. Describe examples of how to use exponential functions to model and solve real-life problems *(pages 148 and 149, Examples 8 and 9)*.

3.2 Logarithmic Functions and Their Graphs

- Recognize and evaluate logarithmic functions with base a.
- Graph logarithmic functions.
- Recognize, evaluate, and graph natural logarithmic functions.
- Use logarithmic functions to model and solve real-life problems.

Logarithmic Functions

In Section 1.9, you studied the concept of an inverse function. There, you learned that when a function is one-to-one—that is, when the function has the property that no horizontal line intersects the graph of the function more than once—the function must have an inverse function. By looking back at the graphs of the exponential functions introduced in Section 3.1, you will see that every function of the form $f(x) = a^x$ passes the Horizontal Line Test and therefore must have an inverse function. This inverse function is called the **logarithmic function with base a.**

Logarithmic functions can often model scientific observations, such as human memory.

Definition of Logarithmic Function with Base a

For $x > 0$, $a > 0$, and $a \neq 1$,

$\quad y = \log_a x$ if and only if $x = a^y$.

The function

$\quad f(x) = \log_a x \qquad$ Read as "log base a of x."

is called the **logarithmic function with base a.**

The equations $y = \log_a x$ and $x = a^y$ are equivalent. The first equation is in logarithmic form and the second is in exponential form. For example, $2 = \log_3 9$ is equivalent to $9 = 3^2$, and $5^3 = 125$ is equivalent to $\log_5 125 = 3$.

When evaluating logarithms, remember that *a logarithm is an exponent.* This means that $\log_a x$ is the exponent to which a must be raised to obtain x. For instance, $\log_2 8 = 3$ because 2 raised to the third power is 8.

EXAMPLE 1 **Evaluating Logarithms**

Evaluate each logarithm at the indicated value of x.

a. $f(x) = \log_2 x, \quad x = 32$ 　　　　**b.** $f(x) = \log_3 x, \quad x = 1$

c. $f(x) = \log_4 x, \quad x = 2$ 　　　　**d.** $f(x) = \log_{10} x, \quad x = \frac{1}{100}$

Solution

a. $f(32) = \log_2 32 = 5 \qquad$ because $\quad 2^5 = 32$.

b. $f(1) = \log_3 1 = 0 \qquad$ because $\quad 3^0 = 1$.

c. $f(2) = \log_4 2 = \frac{1}{2} \qquad$ because $\quad 4^{1/2} = \sqrt{4} = 2$.

d. $f\left(\frac{1}{100}\right) = \log_{10} \frac{1}{100} = -2 \quad$ because $\quad 10^{-2} = \frac{1}{10^2} = \frac{1}{100}$.

✓ ***Checkpoint*** 🔊 *Audio-video solution in English & Spanish at LarsonPrecalculus.com.*

Evaluate each logarithm at the indicated value of x.

a. $f(x) = \log_6 x, \quad x = 1$ 　　**b.** $f(x) = \log_5 x, \quad x = \frac{1}{125}$ 　　**c.** $f(x) = \log_{10} x, \quad x = 10{,}000$

The logarithmic function with base 10 is called the **common logarithmic function.** It is denoted by $\log_{10}$ or simply by log. On most calculators, this function is denoted by $\boxed{\text{LOG}}$. Example 2 shows how to use a calculator to evaluate common logarithmic functions. You will learn how to use a calculator to calculate logarithms to any base in the next section.

EXAMPLE 2   **Evaluating Common Logarithms on a Calculator**

Use a calculator to evaluate the function $f(x) = \log x$ at each value of x.

a. $x = 10$ **b.** $x = \frac{1}{3}$ **c.** $x = 2.5$ **d.** $x = -2$

Solution

Function Value	Graphing Calculator Keystrokes	Display
a. $f(10) = \log 10$	$\boxed{\text{LOG}}$ 10 $\boxed{\text{ENTER}}$	1
b. $f\left(\frac{1}{3}\right) = \log\frac{1}{3}$	$\boxed{\text{LOG}}$ $\boxed{(}$ 1 $\boxed{\div}$ 3 $\boxed{)}$ $\boxed{\text{ENTER}}$	-0.4771213
c. $f(2.5) = \log 2.5$	$\boxed{\text{LOG}}$ 2.5 $\boxed{\text{ENTER}}$	0.3979400
d. $f(-2) = \log(-2)$	$\boxed{\text{LOG}}$ $\boxed{(-)}$ 2 $\boxed{\text{ENTER}}$	ERROR

Note that the calculator displays an error message (or a complex number) when you try to evaluate $\log(-2)$. The reason for this is that there is no real number power to which 10 can be raised to obtain -2.

✓ **Checkpoint** *Audio-video solution in English & Spanish at LarsonPrecalculus.com.*

Use a calculator to evaluate the function $f(x) = \log x$ at each value of x.

a. $x = 275$ **b.** $x = 0.275$ **c.** $x = -\frac{1}{2}$ **d.** $x = \frac{1}{2}$ ■

The following properties follow directly from the definition of the logarithmic function with base a.

Properties of Logarithms

1. $\log_a 1 = 0$ because $a^0 = 1$.

2. $\log_a a = 1$ because $a^1 = a$.

3. $\log_a a^x = x$ and $a^{\log_a x} = x$ Inverse Properties

4. If $\log_a x = \log_a y$, then $x = y$.   One-to-One Property

---

EXAMPLE 3   **Using Properties of Logarithms**

a. Simplify $\log_4 1$.   **b.** Simplify $\log_{\sqrt{7}} \sqrt{7}$.   **c.** Simplify $6^{\log_6 20}$.

Solution

a. Using Property 1, $\log_4 1 = 0$.

b. Using Property 2, $\log_{\sqrt{7}} \sqrt{7} = 1$.

c. Using the Inverse Property (Property 3), $6^{\log_6 20} = 20$.

✓ **Checkpoint**  *Audio-video solution in English & Spanish at LarsonPrecalculus.com.*

a. Simplify $\log_9 9$.   **b.** Simplify $20^{\log_{20} 3}$.   **c.** Simplify $\log_{\sqrt{3}} 1$.   ■

You can use the One-to-One Property (Property 4) to solve simple logarithmic equations, as shown in Example 4.

EXAMPLE 4 **Using the One-to-One Property**

a. $\log_3 x = \log_3 12$ Original equation

        $x = 12$ One-to-One Property

b. $\log(2x + 1) = \log 3x \implies 2x + 1 = 3x \implies 1 = x$

c. $\log_4(x^2 - 6) = \log_4 10 \implies x^2 - 6 = 10 \implies x^2 = 16 \implies x = \pm 4$

✓ **Checkpoint** 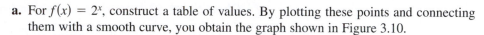 *Audio-video solution in English & Spanish at LarsonPrecalculus.com.*

Solve $\log_5(x^2 + 3) = \log_5 12$ for x.

Graphs of Logarithmic Functions

To sketch the graph of $y = \log_a x$, use the fact that the graphs of inverse functions are reflections of each other in the line $y = x$.

EXAMPLE 5 **Graphs of Exponential and Logarithmic Functions**

In the same coordinate plane, sketch the graph of each function.

a. $f(x) = 2^x$ **b.** $g(x) = \log_2 x$

Solution

a. For $f(x) = 2^x$, construct a table of values. By plotting these points and connecting them with a smooth curve, you obtain the graph shown in Figure 3.10.

x	-2	-1	0	1	2	3
$f(x) = 2^x$	$\frac{1}{4}$	$\frac{1}{2}$	1	2	4	8

b. Because $g(x) = \log_2 x$ is the inverse function of $f(x) = 2^x$, the graph of g is obtained by plotting the points $(f(x), x)$ and connecting them with a smooth curve. The graph of g is a reflection of the graph of f in the line $y = x$, as shown in Figure 3.10.

✓ **Checkpoint**  *Audio-video solution in English & Spanish at LarsonPrecalculus.com.*

In the same coordinate plane, sketch the graphs of (a) $f(x) = 8^x$ and (b) $g(x) = \log_8 x$.

EXAMPLE 6 **Sketching the Graph of a Logarithmic Function**

Sketch the graph of $f(x) = \log x$. Identify the vertical asymptote.

Solution Begin by constructing a table of values. Note that some of the values can be obtained without a calculator by using the properties of logarithms. Others require a calculator. Next, plot the points and connect them with a smooth curve, as shown in Figure 3.11. The vertical asymptote is $x = 0$ (y-axis).

	Without calculator				With calculator		
x	$\frac{1}{100}$	$\frac{1}{10}$	1	10	2	5	8
$f(x) = \log x$	-2	-1	0	1	0.301	0.699	0.903

✓ **Checkpoint**  *Audio-video solution in English & Spanish at LarsonPrecalculus.com.*

Sketch the graph of $f(x) = \log_9 x$. Identify the vertical asymptote.

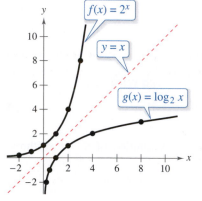

Figure 3.10

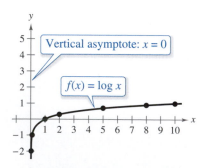

Figure 3.11

The nature of the graph in Figure 3.11 is typical of functions of the form $f(x) = \log_a x$, $a > 1$. They have one x-intercept and one vertical asymptote. Notice how slowly the graph rises for $x > 1$. The following summarizes the basic characteristics of logarithmic graphs.

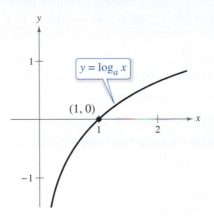

Graph of $y = \log_a x$, $a > 1$

- Domain: $(0, \infty)$
- Range: $(-\infty, \infty)$
- x-intercept: $(1, 0)$
- Increasing
- One-to-one, therefore has an inverse function
- y-axis is a vertical asymptote
  ($\log_a x \to -\infty$ as $x \to 0^+$).
- Continuous
- Reflection of graph of $y = a^x$ in the line $y = x$

The basic characteristics of the graph of $f(x) = a^x$ are shown below to illustrate the inverse relation between $f(x) = a^x$ and $g(x) = \log_a x$.

- Domain: $(-\infty, \infty)$
- y-intercept: $(0, 1)$
- Range: $(0, \infty)$
- x-axis is a horizontal asymptote ($a^x \to 0$ as $x \to -\infty$).

The next example uses the graph of $y = \log_a x$ to sketch the graphs of functions of the form $f(x) = b \pm \log_a(x + c)$. Notice how a horizontal shift of the graph results in a horizontal shift of the vertical asymptote.

EXAMPLE 7 **Shifting Graphs of Logarithmic Functions**

The graph of each of the functions is similar to the graph of $f(x) = \log x$.

a. Because $g(x) = \log(x - 1) = f(x - 1)$, the graph of g can be obtained by shifting the graph of f one unit to the right, as shown in Figure 3.12.

b. Because $h(x) = 2 + \log x = 2 + f(x)$, the graph of h can be obtained by shifting the graph of f two units up, as shown in Figure 3.13.

REMARK Use your understanding of transformations to identify vertical asymptotes of logarithmic functions. For instance, in Example 7(a), the graph of $g(x) = f(x - 1)$ shifts the graph of $f(x)$ one unit to the right. So, the vertical asymptote of the graph of $g(x)$ is $x = 1$, one unit to the right of the vertical asymptote of the graph of $f(x)$.

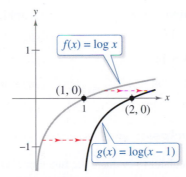

Figure 3.12

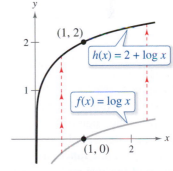

Figure 3.13

✓ *Checkpoint* 🔊 *Audio-video solution in English & Spanish at LarsonPrecalculus.com.*

▷ **ALGEBRA HELP** You can review the techniques for shifting, reflecting, and stretching graphs in Section 1.7.

Use the graph of $f(x) = \log_3 x$ to sketch the graph of each function.

a. $g(x) = -1 + \log_3 x$

b. $h(x) = \log_3(x + 3)$

The Natural Logarithmic Function

By looking back at the graph of the natural exponential function introduced on page 146 in Section 3.1, you will see that $f(x) = e^x$ is one-to-one and so has an inverse function. This inverse function is called the **natural logarithmic function** and is denoted by the special symbol ln x, read as "the natural log of x" or "el en of x." Note that the natural logarithm is written without a base. The base is understood to be e.

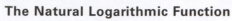

> **The Natural Logarithmic Function**
>
> The function defined by
>
> $$f(x) = \log_e x = \ln x, \quad x > 0$$
>
> is called the **natural logarithmic function.**

The above definition implies that the natural logarithmic function and the natural exponential function are inverse functions of each other. So, every logarithmic equation can be written in an equivalent exponential form, and every exponential equation can be written in an equivalent logarithmic form. That is, $y = \ln x$ and $x = e^y$ are equivalent equations.

Because the functions $f(x) = e^x$ and $g(x) = \ln x$ are inverse functions of each other, their graphs are reflections of each other in the line $y = x$. Figure 3.14 illustrates this reflective property.

On most calculators, ⌈LN⌉ denotes the natural logarithm, as illustrated in Example 8.

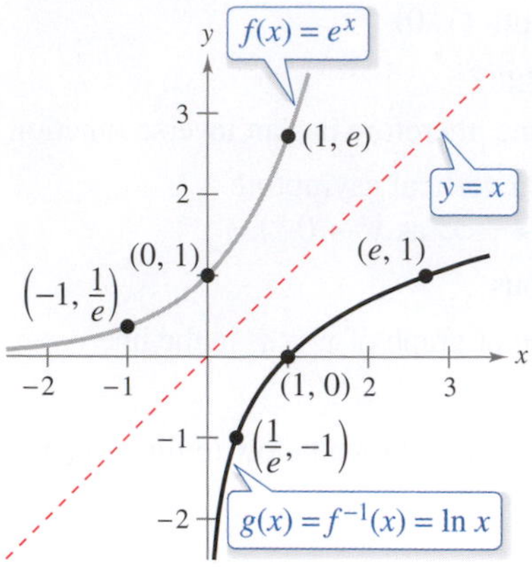

Reflection of graph of $f(x) = e^x$ in the line $y = x$.
Figure 3.14

EXAMPLE 8   **Evaluating the Natural Logarithmic Function**

Use a calculator to evaluate the function $f(x) = \ln x$ at each value of x.

a. $x = 2$

b. $x = 0.3$

c. $x = -1$

d. $x = 1 + \sqrt{2}$

Solution

	Function Value	Graphing Calculator Keystrokes	Display
a.	$f(2) = \ln 2$	⌈LN⌉ 2 ⌈ENTER⌉	0.6931472
b.	$f(0.3) = \ln 0.3$	⌈LN⌉ .3 ⌈ENTER⌉	−1.2039728
c.	$f(-1) = \ln(-1)$	⌈LN⌉ ⌈(−)⌉ 1 ⌈ENTER⌉	ERROR
d.	$f(1 + \sqrt{2}) = \ln(1 + \sqrt{2})$	⌈LN⌉ ⌈(⌉ 1 ⌈+⌉ ⌈√⌉ 2 ⌈)⌉ ⌈ENTER⌉	0.8813736

✓ *Checkpoint* *Audio-video solution in English & Spanish at LarsonPrecalculus.com.*

Use a calculator to evaluate the function $f(x) = \ln x$ at each value of x.

a. $x = 0.01$ **b.** $x = 4$ **c.** $x = \sqrt{3} + 2$ **d.** $x = \sqrt{3} - 2$ ■

In Example 8, be sure you see that $\ln(-1)$ gives an error message on most calculators. (Some calculators may display a complex number.) This occurs because the domain of $\ln x$ is the set of positive real numbers (see Figure 3.14). So, $\ln(-1)$ is undefined.

The four properties of logarithms listed on page 151 are also valid for natural logarithms.

⋯ REMARK Notice that as with every other logarithmic function, the domain of the natural logarithmic function is the set of *positive real numbers*—be sure you see that ln x is not defined for zero or for negative numbers.

> **Properties of Natural Logarithms**
>
> 1. $\ln 1 = 0$ because $e^0 = 1$.
> 2. $\ln e = 1$ because $e^1 = e$.
> 3. $\ln e^x = x$ and $e^{\ln x} = x$ Inverse Properties
> 4. If $\ln x = \ln y$, then $x = y$.   One-to-One Property

EXAMPLE 9   **Using Properties of Natural Logarithms**

Use the properties of natural logarithms to simplify each expression.

a. $\ln \dfrac{1}{e}$ **b.** $e^{\ln 5}$ **c.** $\dfrac{\ln 1}{3}$ **d.** $2 \ln e$

Solution

a. $\ln \dfrac{1}{e} = \ln e^{-1} = -1$ Inverse Property   **b.** $e^{\ln 5} = 5$ Inverse Property

c. $\dfrac{\ln 1}{3} = \dfrac{0}{3} = 0$ Property 1   **d.** $2 \ln e = 2(1) = 2$ Property 2

✓ **Checkpoint** 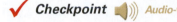 Audio-video solution in English & Spanish at LarsonPrecalculus.com.

Use the properties of natural logarithms to simplify each expression.

a. $\ln e^{1/3}$ **b.** $5 \ln 1$ **c.** $\tfrac{3}{4} \ln e$ **d.** $e^{\ln 7}$

EXAMPLE 10   **Finding the Domains of Logarithmic Functions**

Find the domain of each function.

a. $f(x) = \ln(x - 2)$ **b.** $g(x) = \ln(2 - x)$ **c.** $h(x) = \ln x^2$

Solution

a. Because $\ln(x - 2)$ is defined only when $x - 2 > 0$, it follows that the domain of f is $(2, \infty)$. Figure 3.15 shows the graph of f.

b. Because $\ln(2 - x)$ is defined only when $2 - x > 0$, it follows that the domain of g is $(-\infty, 2)$. Figure 3.16 shows the graph of g.

c. Because $\ln x^2$ is defined only when $x^2 > 0$, it follows that the domain of h is all real numbers except $x = 0$. Figure 3.17 shows the graph of h.

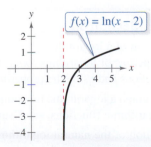

Figure 3.15

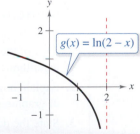

Figure 3.16

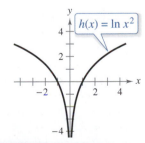

Figure 3.17

✓ **Checkpoint**   Audio-video solution in English & Spanish at LarsonPrecalculus.com.

Find the domain of $f(x) = \ln(x + 3)$.

Application

EXAMPLE 11 Human Memory Model

Students participating in a psychology experiment attended several lectures on a subject and took an exam. Every month for a year after the exam, the students took a retest to see how much of the material they remembered. The average scores for the group are given by the *human memory model* $f(t) = 75 - 6 \ln(t + 1)$, $0 \le t \le 12$, where t is the time in months.

a. What was the average score on the original exam ($t = 0$)?

b. What was the average score at the end of $t = 2$ months?

c. What was the average score at the end of $t = 6$ months?

Algebraic Solution

a. The original average score was

$$f(0) = 75 - 6 \ln(0 + 1)$$ Substitute 0 for t.

$$= 75 - 6 \ln 1$$ Simplify.

$$= 75 - 6(0)$$ Property of natural logarithms

$$= 75.$$ Solution

b. After 2 months, the average score was

$$f(2) = 75 - 6 \ln(2 + 1)$$ Substitute 2 for t.

$$= 75 - 6 \ln 3$$ Simplify.

$$\approx 75 - 6(1.0986)$$ Use a calculator.

$$\approx 68.41.$$ Solution

c. After 6 months, the average score was

$$f(6) = 75 - 6 \ln(6 + 1)$$ Substitute 6 for t.

$$= 75 - 6 \ln 7$$ Simplify.

$$\approx 75 - 6(1.9459)$$ Use a calculator.

$$\approx 63.32.$$ Solution

Graphical Solution

a.

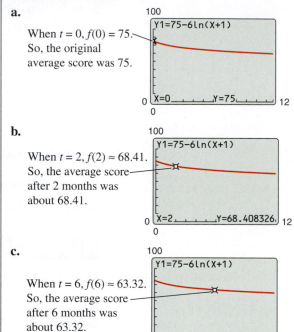

When $t = 0$, $f(0) = 75$. So, the original average score was 75.

b.

When $t = 2$, $f(2) \approx 68.41$. So, the average score after 2 months was about 68.41.

c.

When $t = 6$, $f(6) \approx 63.32$. So, the average score after 6 months was about 63.32.

✓ *Checkpoint*  Audio-video solution in English & Spanish at LarsonPrecalculus.com.

In Example 11, find the average score at the end of (a) $t = 1$ month, (b) $t = 9$ months, and (c) $t = 12$ months. ◼

Summarize (Section 3.2)

1. State the definition of a logarithmic function with base a *(page 150)* and make a list of the properties of logarithms *(page 151)*. For examples of evaluating logarithmic functions and using the properties of logarithms, see Examples 1–4.

2. Explain how to graph a logarithmic function *(pages 152 and 153)*. For examples of graphing logarithmic functions, see Examples 5–7.

3. State the definition of the natural logarithmic function *(page 154)* and make a list of the properties of natural logarithms *(page 155)*. For examples of evaluating natural logarithmic functions and using the properties of natural logarithms, see Examples 8 and 9.

4. Describe an example of how to use a logarithmic function to model and solve a real-life problem *(page 156, Example 11)*.

3.3 Properties of Logarithms

- Use the change-of-base formula to rewrite and evaluate logarithmic expressions.
- Use properties of logarithms to evaluate or rewrite logarithmic expressions.
- Use properties of logarithms to expand or condense logarithmic expressions.
- Use logarithmic functions to model and solve real-life problems.

Logarithmic functions can help you model and solve real-life problems, such as the relationship between the number of decibels and the intensity of a sound.

Change of Base

Most calculators have only two types of log keys, one for common logarithms (base 10) and one for natural logarithms (base e). Although common logarithms and natural logarithms are the most frequently used, you may occasionally need to evaluate logarithms with other bases. To do this, use the following **change-of-base formula.**

Change-of-Base Formula

Let a, b, and x be positive real numbers such that $a \neq 1$ and $b \neq 1$. Then $\log_a x$ can be converted to a different base as follows.

Base b	**Base 10**	**Base e**
$\log_a x = \dfrac{\log_b x}{\log_b a}$	$\log_a x = \dfrac{\log x}{\log a}$	$\log_a x = \dfrac{\ln x}{\ln a}$

One way to look at the change-of-base formula is that logarithms with base a are *constant multiples* of logarithms with base b. The constant multiplier is

$$\frac{1}{\log_b a}.$$

EXAMPLE 1   Changing Bases Using Common Logarithms

$$\log_4 25 = \frac{\log 25}{\log 4} \qquad \log_a x = \frac{\log x}{\log a}$$

$$\approx \frac{1.39794}{0.60206} \qquad \text{Use a calculator.}$$

$$\approx 2.3219 \qquad \text{Simplify.}$$

✓ **Checkpoint**  *Audio-video solution in English & Spanish at LarsonPrecalculus.com.*

Evaluate $\log_2 12$ using the change-of-base formula and common logarithms.

EXAMPLE 2   Changing Bases Using Natural Logarithms

$$\log_4 25 = \frac{\ln 25}{\ln 4} \qquad \log_a x = \frac{\ln x}{\ln a}$$

$$\approx \frac{3.21888}{1.38629} \qquad \text{Use a calculator.}$$

$$\approx 2.3219 \qquad \text{Simplify.}$$

✓ **Checkpoint** *Audio-video solution in English & Spanish at LarsonPrecalculus.com.*

Evaluate $\log_2 12$ using the change-of-base formula and natural logarithms.

Properties of Logarithms

You know from the preceding section that the logarithmic function with base a is the *inverse function* of the exponential function with base a. So, it makes sense that the properties of exponents should have corresponding properties involving logarithms. For instance, the exponential property $a^u a^v = a^{u+v}$ has the corresponding logarithmic property $\log_a(uv) = \log_a u + \log_a v$.

• • REMARK There is no property that can be used to rewrite $\log_a(u \pm v)$. Specifically, $\log_a(u + v)$ is *not* equal to $\log_a u + \log_a v$.

Properties of Logarithms

Let a be a positive number such that $a \neq 1$, and let n be a real number. If u and v are positive real numbers, then the following properties are true.

	Logarithm with Base a	**Natural Logarithm**
1. Product Property:	$\log_a(uv) = \log_a u + \log_a v$	$\ln(uv) = \ln u + \ln v$
2. Quotient Property:	$\log_a \dfrac{u}{v} = \log_a u - \log_a v$	$\ln \dfrac{u}{v} = \ln u - \ln v$
3. Power Property:	$\log_a u^n = n \log_a u$	$\ln u^n = n \ln u$

For proofs of the properties listed above, see Proofs in Mathematics on page 183.

EXAMPLE 3 **Using Properties of Logarithms**

Write each logarithm in terms of $\ln 2$ and $\ln 3$.

a. $\ln 6$ **b.** $\ln \dfrac{2}{27}$

Solution

a. $\ln 6 = \ln(2 \cdot 3)$ Rewrite 6 as 2 · 3.

      $= \ln 2 + \ln 3$ Product Property

b. $\ln \dfrac{2}{27} = \ln 2 - \ln 27$ Quotient Property

      $= \ln 2 - \ln 3^3$ Rewrite 27 as 3^3.

      $= \ln 2 - 3 \ln 3$ Power Property

✓ **Checkpoint** 🔊))) *Audio-video solution in English & Spanish at LarsonPrecalculus.com.*

Write each logarithm in terms of $\log 3$ and $\log 5$.

a. $\log 75$ **b.** $\log \dfrac{9}{125}$

EXAMPLE 4 **Using Properties of Logarithms**

Find the exact value of $\log_5 \sqrt[3]{5}$ without using a calculator.

Solution

$\log_5 \sqrt[3]{5} = \log_5 5^{1/3} = \frac{1}{3} \log_5 5 = \frac{1}{3}(1) = \frac{1}{3}$

✓ **Checkpoint** 🔊))) *Audio-video solution in English & Spanish at LarsonPrecalculus.com.*

Find the exact value of $\ln e^6 - \ln e^2$ without using a calculator.

HISTORICAL NOTE

John Napier, a Scottish mathematician, developed logarithms as a way to simplify tedious calculations. Beginning in 1594, Napier worked about 20 years on the development of logarithms. Napier only partially succeeded in his quest to simplify tedious calculations. Nonetheless, the development of logarithms was a step forward and received immediate recognition.

Rewriting Logarithmic Expressions

The properties of logarithms are useful for rewriting logarithmic expressions in forms that simplify the operations of algebra. This is true because these properties convert complicated products, quotients, and exponential forms into simpler sums, differences, and products, respectively.

EXAMPLE 5 **Expanding Logarithmic Expressions**

Expand each logarithmic expression.

a. $\log_4 5x^3y$ **b.** $\ln \dfrac{\sqrt{3x - 5}}{7}$

Solution

a. $\log_4 5x^3y = \log_4 5 + \log_4 x^3 + \log_4 y$ Product Property

$\qquad\qquad = \log_4 5 + 3 \log_4 x + \log_4 y$ Power Property

▷ **ALGEBRA HELP** You can review rewriting radicals and rational exponents in Appendix A.2.

b. $\ln \dfrac{\sqrt{3x - 5}}{7} = \ln \dfrac{(3x - 5)^{1/2}}{7}$ Rewrite using rational exponent.

$\qquad\qquad = \ln(3x - 5)^{1/2} - \ln 7$ Quotient Property

$\qquad\qquad = \dfrac{1}{2} \ln(3x - 5) - \ln 7$ Power Property

✓ **Checkpoint**  *Audio-video solution in English & Spanish at LarsonPrecalculus.com.*

Expand the expression $\log_3 \dfrac{4x^2}{\sqrt{y}}$.

Example 5 uses the properties of logarithms to *expand* logarithmic expressions. Example 6 reverses this procedure and uses the properties of logarithms to *condense* logarithmic expressions.

EXAMPLE 6 **Condensing Logarithmic Expressions**

Condense each logarithmic expression.

a. $\dfrac{1}{2} \log x + 3 \log(x + 1)$ **b.** $2 \ln(x + 2) - \ln x$ **c.** $\dfrac{1}{3}[\log_2 x + \log_2(x + 1)]$

Solution

a. $\dfrac{1}{2} \log x + 3 \log(x + 1) = \log x^{1/2} + \log(x + 1)^3$ Power Property

$\qquad\qquad = \log\left[\sqrt{x}(x + 1)^3\right]$ Product Property

b. $2 \ln(x + 2) - \ln x = \ln(x + 2)^2 - \ln x$ Power Property

$\qquad\qquad = \ln \dfrac{(x + 2)^2}{x}$ Quotient Property

c. $\dfrac{1}{3}[\log_2 x + \log_2(x + 1)] = \dfrac{1}{3} \log_2[x(x + 1)]$ Product Property

$\qquad\qquad = \log_2[x(x + 1)]^{1/3}$ Power Property

$\qquad\qquad = \log_2 \sqrt[3]{x(x + 1)}$ Rewrite with a radical.

✓ **Checkpoint** 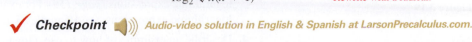 *Audio-video solution in English & Spanish at LarsonPrecalculus.com.*

Condense the expression $2[\log(x + 3) - 2 \log(x - 2)]$.

Application

One method of determining how the x- and y-values for a set of nonlinear data are related is to take the natural logarithm of each of the x- and y-values. If the points, when graphed, fall on a line, then you can determine that the x- and y-values are related by the equation $\ln y = m \ln x$ where m is the slope of the line.

EXAMPLE 7   Finding a Mathematical Model

The table shows the mean distance from the sun x and the period y (the time it takes a planet to orbit the sun) for each of the six planets that are closest to the sun. In the table, the mean distance is given in terms of astronomical units (where Earth's mean distance is defined as 1.0), and the period is given in years. Find an equation that relates y and x.

DATA	Planet	Mean Distance, x	Period, y
	Mercury	0.387	0.241
	Venus	0.723	0.615
	Earth	1.000	1.000
	Mars	1.524	1.881
	Jupiter	5.203	11.860
	Saturn	9.555	29.420

Spreadsheet at LarsonPrecalculus.com

Solution   Figure 3.18 shows the plots of the points given by the above table. From this figure, it is not clear how to find an equation that relates y and x. To solve this problem, take the natural logarithm of each of the x- and y-values, as shown in the table at the left. Now, by plotting the points in the table at the left, you can see that all six of the points appear to lie in a line (see Figure 3.19). Choose any two points to determine the slope of the line. Using the points $(0.421, 0.632)$ and $(0, 0)$, the slope of the line is

$$m = \frac{0.632 - 0}{0.421 - 0} \approx 1.5 = \frac{3}{2}.$$

By the point-slope form, the equation of the line is $Y = \frac{3}{2}X$, where $Y = \ln y$ and $X = \ln x$. So, $\ln y = \frac{3}{2} \ln x$.

✓ **Checkpoint**) Audio-video solution in English & Spanish at LarsonPrecalculus.com.

Find a logarithmic equation that relates y and x for the following ordered pairs.

$(0.37, 0.51), (1.00, 1.00), (2.72, 1.95) (7.39, 3.79), (20.09, 7.39)$ ◼

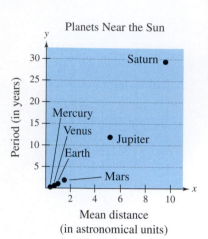

Planets Near the Sun

Mean distance (in astronomical units)

Figure 3.18

Planet	$\ln x$	$\ln y$
Mercury	-0.949	-1.423
Venus	-0.324	-0.486
Earth	0.000	0.000
Mars	0.421	0.632
Jupiter	1.649	2.473
Saturn	2.257	3.382

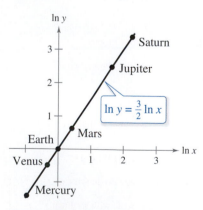

$\ln y = \frac{3}{2} \ln x$

Figure 3.19

Summarize   (Section 3.3)

1. State the change-of-base formula (*page 157*). For examples that use the change-of-base formula to rewrite and evaluate logarithmic expressions, see Examples 1 and 2.

2. Make a list of the properties of logarithms (*page 158*). For examples that use the properties of logarithms to evaluate or rewrite logarithmic expressions, see Examples 3 and 4.

3. Explain how to use the properties of logarithms to expand or condense logarithmic expressions (*page 159*). For examples of expanding and condensing logarithmic expressions, see Examples 5 and 6.

4. Describe an example of how to use a logarithmic function to model and solve a real-life problem (*page 160, Example 7*).

3.4 Exponential and Logarithmic Equations

- ■ Solve simple exponential and logarithmic equations.
- ■ Solve more complicated exponential equations.
- ■ Solve more complicated logarithmic equations.
- ■ Use exponential and logarithmic equations to model and solve real-life problems.

Introduction

So far in this chapter, you have studied the definitions, graphs, and properties of exponential and logarithmic functions. In this section, you will study procedures for *solving equations* involving these exponential and logarithmic functions.

There are two basic strategies for solving exponential or logarithmic equations. The first is based on the One-to-One Properties and was used to solve simple exponential and logarithmic equations in Sections 3.1 and 3.2. The second is based on the Inverse Properties. For $a > 0$ and $a \neq 1$, the following properties are true for all x and y for which $\log_a x$ and $\log_a y$ are defined.

Exponential and logarithmic equations can help you model and solve life science applications. For instance, an exponential function can be used to model the number of trees per acre given the average diameter of the trees.

One-to-One Properties

$a^x = a^y$ if and only if $x = y$.

$\log_a x = \log_a y$ if and only if $x = y$.

Inverse Properties

$a^{\log_a x} = x$

$\log_a a^x = x$

| EXAMPLE 1 | **Solving Simple Equations** |

Original Equation	Rewritten Equation	Solution	Property
a. $2^x = 32$	$2^x = 2^5$	$x = 5$	One-to-One
b. $\ln x - \ln 3 = 0$	$\ln x = \ln 3$	$x = 3$	One-to-One
c. $\left(\frac{1}{3}\right)^x = 9$	$3^{-x} = 3^2$	$x = -2$	One-to-One
d. $e^x = 7$	$\ln e^x = \ln 7$	$x = \ln 7$	Inverse
e. $\ln x = -3$	$e^{\ln x} = e^{-3}$	$x = e^{-3}$	Inverse
f. $\log x = -1$	$10^{\log x} = 10^{-1}$	$x = 10^{-1} = \frac{1}{10}$	Inverse
g. $\log_3 x = 4$	$3^{\log_3 x} = 3^4$	$x = 81$	Inverse

✓ *Checkpoint* ◀))) *Audio-video solution in English & Spanish at LarsonPrecalculus.com.*

Solve each equation for x.

a. $2^x = 512$ **b.** $\log_6 x = 3$ **c.** $5 - e^x = 0$ **d.** $9^x = \frac{1}{3}$

The strategies used in Example 1 are summarized as follows.

Strategies for Solving Exponential and Logarithmic Equations

1. Rewrite the original equation in a form that allows the use of the One-to-One Properties of exponential or logarithmic functions.

2. Rewrite an *exponential* equation in logarithmic form and apply the Inverse Property of logarithmic functions.

3. Rewrite a *logarithmic* equation in exponential form and apply the Inverse Property of exponential functions.

Solving Exponential Equations

Solving Exponential Equations

Solve each equation and approximate the result to three decimal places, if necessary.

a. $e^{-x^2} = e^{-3x-4}$ **b.** $3(2^x) = 42$

Solution

a.

$e^{-x^2} = e^{-3x-4}$	Write original equation.
$-x^2 = -3x - 4$	One-to-One Property
$x^2 - 3x - 4 = 0$	Write in general form.
$(x + 1)(x - 4) = 0$	Factor.
$(x + 1) = 0 \implies x = -1$	Set 1st factor equal to 0.
$(x - 4) = 0 \implies x = 4$	Set 2nd factor equal to 0.

The solutions are $x = -1$ and $x = 4$. Check these in the original equation.

b.

$3(2^x) = 42$	Write original equation.
$2^x = 14$	Divide each side by 3.
$\log_2 2^x = \log_2 14$	Take log (base 2) of each side.
$x = \log_2 14$	Inverse Property
$x = \dfrac{\ln 14}{\ln 2} \approx 3.807$	Change-of-base formula

The solution is $x = \log_2 14 \approx 3.807$. Check this in the original equation.

> **REMARK**
> Another way to solve Example 2(b) is by taking the natural log of each side and then applying the Power Property, as follows.
>
> $$3(2^x) = 42$$
> $$2^x = 14$$
> $$\ln 2^x = \ln 14$$
> $$x \ln 2 = \ln 14$$
> $$x = \frac{\ln 14}{\ln 2} \approx 3.807$$
>
> Notice that you obtain the same result as in Example 2(b).

✓ Checkpoint  Audio-video solution in English & Spanish at LarsonPrecalculus.com.

Solve each equation and approximate the result to three decimal places, if necessary.

a. $e^{2x} = e^{x^2-8}$ **b.** $2(5^x) = 32$

In Example 2(b), the exact solution is $x = \log_2 14$, and the approximate solution is $x \approx 3.807$. An exact answer is preferred when the solution is an intermediate step in a larger problem. For a final answer, an approximate solution is easier to comprehend.

Solving an Exponential Equation

Solve $e^x + 5 = 60$ and approximate the result to three decimal places.

Solution

$e^x + 5 = 60$	Write original equation.
$e^x = 55$	Subtract 5 from each side.
$\ln e^x = \ln 55$	Take natural log of each side.
$x = \ln 55 \approx 4.007$	Inverse Property

> **REMARK** Remember that the natural logarithmic function has a base of e.

The solution is $x = \ln 55 \approx 4.007$. Check this in the original equation.

✓ Checkpoint Audio-video solution in English & Spanish at LarsonPrecalculus.com.

Solve $e^x - 7 = 23$ and approximate the result to three decimal places.

EXAMPLE 4 **Solving an Exponential Equation**

Solve $2(3^{2t-5}) - 4 = 11$ and approximate the result to three decimal places.

Solution

$2(3^{2t-5}) - 4 = 11$	Write original equation.
$2(3^{2t-5}) = 15$	Add 4 to each side.
$3^{2t-5} = \dfrac{15}{2}$	Divide each side by 2.
$\log_3 3^{2t-5} = \log_3 \dfrac{15}{2}$	Take log (base 3) of each side.
$2t - 5 = \log_3 \dfrac{15}{2}$	Inverse Property
$2t = 5 + \log_3 7.5$	Add 5 to each side.
$t = \dfrac{5}{2} + \dfrac{1}{2}\log_3 7.5$	Divide each side by 2.
$t \approx 3.417$	Use a calculator.

> **REMARK** Remember that to evaluate a logarithm such as $\log_3 7.5$, you need to use the change-of-base formula.
>
> $$\log_3 7.5 = \frac{\ln 7.5}{\ln 3} \approx 1.834$$

The solution is $t = \frac{5}{2} + \frac{1}{2}\log_3 7.5 \approx 3.417$. Check this in the original equation.

✓ **Checkpoint** *Audio-video solution in English & Spanish at LarsonPrecalculus.com.*

Solve $6(2^{t+5}) + 4 = 11$ and approximate the result to three decimal places.

When an equation involves two or more exponential expressions, you can still use a procedure similar to that demonstrated in Examples 2, 3, and 4. However, the algebra is a bit more complicated.

EXAMPLE 5 **Solving an Exponential Equation of Quadratic Type**

Solve $e^{2x} - 3e^x + 2 = 0$.

Algebraic Solution

$e^{2x} - 3e^x + 2 = 0$	Write original equation.
$(e^x)^2 - 3e^x + 2 = 0$	Write in quadratic form.
$(e^x - 2)(e^x - 1) = 0$	Factor.
$e^x - 2 = 0$	Set 1st factor equal to 0.
$x = \ln 2$	Solution
$e^x - 1 = 0$	Set 2nd factor equal to 0.
$x = 0$	Solution

The solutions are $x = \ln 2 \approx 0.693$ and $x = 0$. Check these in the original equation.

Graphical Solution

Use a graphing utility to graph $y = e^{2x} - 3e^x + 2$ and then find the zeros.

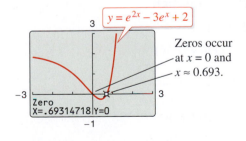

Zeros occur at $x = 0$ and $x \approx 0.693$.

So, you can conclude that the solutions are $x = 0$ and $x \approx 0.693$.

✓ **Checkpoint** *Audio-video solution in English & Spanish at LarsonPrecalculus.com.*

Solve $e^{2x} - 7e^x + 12 = 0$.

Solving Logarithmic Equations

To solve a logarithmic equation, you can write it in exponential form.

$$\ln x = 3 \qquad\qquad \text{Logarithmic form}$$

$$e^{\ln x} = e^3 \qquad\qquad \text{Exponentiate each side.}$$

$$x = e^3 \qquad\qquad \text{Exponential form}$$

This procedure is called *exponentiating* each side of an equation.

EXAMPLE 6 **Solving Logarithmic Equations**

a. $\ln x = 2$ Original equation

$\quad e^{\ln x} = e^2$ Exponentiate each side.

$\quad\quad x = e^2$ Inverse Property

b. $\log_3(5x - 1) = \log_3(x + 7)$ Original equation

$\quad\quad\quad 5x - 1 = x + 7$ One-to-One Property

$\quad\quad\quad\quad x = 2$ Solution

c. $\log_6(3x + 14) - \log_6 5 = \log_6 2x$ Original equation

$\quad \log_6\left(\dfrac{3x + 14}{5}\right) = \log_6 2x$ Quotient Property of Logarithms

$\quad\quad\quad \dfrac{3x + 14}{5} = 2x$ One-to-One Property

$\quad\quad\quad 3x + 14 = 10x$ Multiply each side by 5.

$\quad\quad\quad\quad x = 2$ Solution

✓ **Checkpoint** *Audio-video solution in English & Spanish at LarsonPrecalculus.com.*

Solve each equation.

a. $\ln x = \frac{2}{3}$ **b.** $\log_2(2x - 3) = \log_2(x + 4)$ **c.** $\log 4x - \log(12 + x) = \log 2$

EXAMPLE 7 **Solving a Logarithmic Equation**

Solve $5 + 2 \ln x = 4$ and approximate the result to three decimal places.

Algebraic Solution

$5 + 2 \ln x = 4$ Write original equation.

$2 \ln x = -1$ Subtract 5 from each side.

$\ln x = -\dfrac{1}{2}$ Divide each side by 2.

$e^{\ln x} = e^{-1/2}$ Exponentiate each side.

$x = e^{-1/2}$ Inverse Property

$x \approx 0.607$ Use a calculator.

Graphical Solution

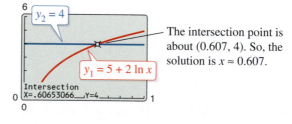

The intersection point is about $(0.607, 4)$. So, the solution is $x \approx 0.607$.

✓ **Checkpoint** *Audio-video solution in English & Spanish at LarsonPrecalculus.com.*

Solve $7 + 3 \ln x = 5$ and approximate the result to three decimal places.

EXAMPLE 8 **Solving a Logarithmic Equation**

Solve $2 \log_5 3x = 4$.

Solution

$2 \log_5 3x = 4$	Write original equation.
$\log_5 3x = 2$	Divide each side by 2.
$5^{\log_5 3x} = 5^2$	Exponentiate each side (base 5).
$3x = 25$	Inverse Property
$x = \dfrac{25}{3}$	Divide each side by 3.

The solution is $x = \frac{25}{3}$. Check this in the original equation.

✓ *Checkpoint* *Audio-video solution in English & Spanish at LarsonPrecalculus.com.*

Solve $3 \log_4 6x = 9$.

Because the domain of a logarithmic function generally does not include all real numbers, you should be sure to check for extraneous solutions of logarithmic equations.

EXAMPLE 9 **Checking for Extraneous Solutions**

Solve $\log 5x + \log(x - 1) = 2$.

Algebraic Solution

$\log 5x + \log(x - 1) = 2$	Write original equation.
$\log[5x(x - 1)] = 2$	Product Property of Logarithms
$10^{\log(5x^2 - 5x)} = 10^2$	Exponentiate each side (base 10).
$5x^2 - 5x = 100$	Inverse Property
$x^2 - x - 20 = 0$	Write in general form.
$(x - 5)(x + 4) = 0$	Factor.
$x - 5 = 0$	Set 1st factor equal to 0.
$x = 5$	Solution
$x + 4 = 0$	Set 2nd factor equal to 0.
$x = -4$	Solution

The solutions appear to be $x = 5$ and $x = -4$. However, when you check these in the original equation, you can see that $x = 5$ is the only solution.

Graphical Solution

First, rewrite the original equation as

$$\log 5x + \log(x - 1) - 2 = 0.$$

Then use a graphing utility to graph the equation

$$y = \log 5x + \log(x - 1) - 2$$

and find the zero(s).

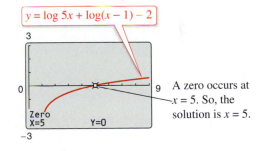

A zero occurs at $x = 5$. So, the solution is $x = 5$.

✓ *Checkpoint* *Audio-video solution in English & Spanish at LarsonPrecalculus.com.*

Solve $\log x + \log(x - 9) = 1$.

▷ •• **REMARK** Notice in Example 9 that the logarithmic part of the equation is condensed into a single logarithm before exponentiating each side of the equation.

In Example 9, the domain of $\log 5x$ is $x > 0$ and the domain of $\log(x - 1)$ is $x > 1$, so the domain of the original equation is

$$x > 1.$$

Because the domain is all real numbers greater than 1, the solution $x = -4$ is extraneous. The graphical solution verifies this conclusion.

Applications

EXAMPLE 10 **Doubling an Investment**

You invest $500 at an annual interest rate of 6.75%, compounded continuously. How long will it take your money to double?

Solution Using the formula for continuous compounding, the balance is

$$A = Pe^{rt}$$

$$A = 500e^{0.0675t}.$$

To find the time required for the balance to double, let $A = 1000$ and solve the resulting equation for t.

$500e^{0.0675t} = 1000$	Let $A = 1000$.
$e^{0.0675t} = 2$	Divide each side by 500.
$\ln e^{0.0675t} = \ln 2$	Take natural log of each side.
$0.0675t = \ln 2$	Inverse Property
$t = \dfrac{\ln 2}{0.0675}$	Divide each side by 0.0675.
$t \approx 10.27$	Use a calculator.

The balance in the account will double after approximately 10.27 years. This result is demonstrated graphically below.

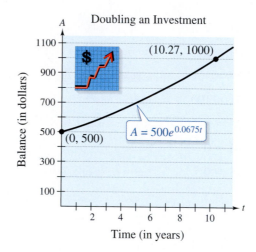

Doubling an Investment

$A = 500e^{0.0675t}$

(10.27, 1000)

(0, 500)

Balance (in dollars)

Time (in years)

✓ **Checkpoint** *Audio-video solution in English & Spanish at LarsonPrecalculus.com.*

You invest $500 at an annual interest rate of 5.25%, compounded continuously. How long will it take your money to double? Compare your result with that of Example 10.

In Example 10, an approximate answer of 10.27 years is given. Within the context of the problem, the exact solution

$$t = \frac{\ln 2}{0.0675}$$

does not make sense as an answer.

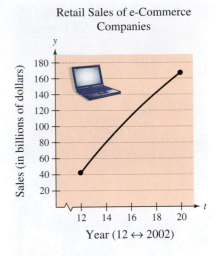

EXAMPLE 11 Retail Sales

The retail sales y (in billions of dollars) of e-commerce companies in the United States from 2002 through 2010 can be modeled by

$$y = -566 + 244.7 \ln t, \quad 12 \le t \le 20$$

where t represents the year, with $t = 12$ corresponding to 2002 (see figure). During which year did the sales reach \$141 billion? (*Source: U.S. Census Bureau*)

Retail Sales of e-Commerce Companies

Solution

$-566 + 244.7 \ln t = y$	Write original equation.
$-566 + 244.7 \ln t = 141$	Substitute 141 for y.
$244.7 \ln t = 707$	Add 566 to each side.
$\ln t = \dfrac{707}{244.7}$	Divide each side by 244.7.
$e^{\ln t} = e^{707/244.7}$	Exponentiate each side.
$t = e^{707/244.7}$	Inverse Property
$t \approx 18$	Use a calculator.

The solution is $t \approx 18$. Because $t = 12$ represents 2002, it follows that the sales reached \$141 billion in 2008.

 Checkpoint *Audio-video solution in English & Spanish at LarsonPrecalculus.com.*

In Example 11, during which year did the sales reach \$80 billion?

Summarize (Section 3.4)

1. State the One-to-One Properties and the Inverse Properties that can help you solve simple exponential and logarithmic equations (*page 161*). For an example of solving simple exponential and logarithmic equations, see Example 1.

2. Describe strategies for solving exponential equations (*pages 162 and 163*). For examples of solving exponential equations, see Examples 2–5.

3. Describe strategies for solving logarithmic equations (*pages 164 and 165*). For examples of solving logarithmic equations, see Examples 6–9.

4. Describe examples of how to use exponential and logarithmic equations to model and solve real-life problems (*pages 166 and 167, Examples 10 and 11*).

3.5 Exponential and Logarithmic Models

Exponential growth and decay models can often represent the populations of countries.

■ Recognize the five most common types of models involving exponential and logarithmic functions.
■ Use exponential growth and decay functions to model and solve real-life problems.
■ Use Gaussian functions to model and solve real-life problems.
■ Use logistic growth functions to model and solve real-life problems.
■ Use logarithmic functions to model and solve real-life problems.

Introduction

The five most common types of mathematical models involving exponential functions and logarithmic functions are as follows.

1. **Exponential growth model:** $y = ae^{bx}, \quad b > 0$

2. **Exponential decay model:** $y = ae^{-bx}, \quad b > 0$

3. **Gaussian model:** $y = ae^{-(x-b)^2/c}$

4. **Logistic growth model:** $y = \dfrac{a}{1 + be^{-rx}}$

5. **Logarithmic models:** $y = a + b \ln x, \quad y = a + b \log x$

The basic shapes of the graphs of these functions are as follows.

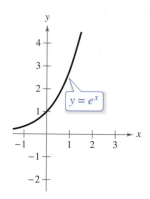

Exponential growth model

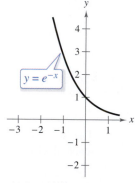

Exponential decay model

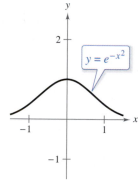

Gaussian model

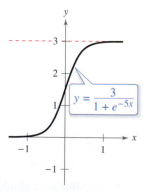

Logistic growth model

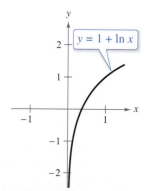

Natural logarithmic model

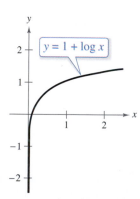

Common logarithmic model

You often gain insight into a situation modeled by an exponential or logarithmic function by identifying and interpreting the asymptotes of the graph of the function. Identify the asymptote(s) of the graph of each function shown above.

Alan Becker/Stone/Getty Images

Exponential Growth and Decay

EXAMPLE 1 **Online Advertising**

Estimates of the amounts (in billions of dollars) of U.S. online advertising spending from 2011 through 2015 are shown in the table. A scatter plot of the data is shown at the right. (*Source: eMarketer*)

Year	2011	2012	2013	2014	2015
Advertising Spending	31.3	36.8	41.2	45.5	49.5

An exponential growth model that approximates the data is given by

$$S = 9.30e^{0.1129t}, \quad 11 \le t \le 15$$

where S is the amount of spending (in billions of dollars) and $t = 11$ represents 2011. Compare the values given by the model with the estimates shown in the table. According to this model, when will the amount of U.S. online advertising spending reach $80 billion?

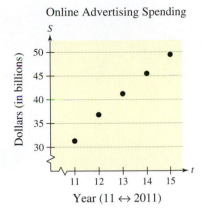

Online Advertising Spending

Year ($11 \leftrightarrow 2011$)

Algebraic Solution

The following table compares the two sets of advertising spending amounts.

Year	2011	2012	2013	2014	2015
Advertising Spending	31.3	36.8	41.2	45.5	49.5
Model	32.2	36.0	40.4	45.2	50.6

To find when the amount of U.S. online advertising spending will reach $80 billion, let $S = 80$ in the model and solve for t.

$9.30e^{0.1129t} = S$	Write original model.
$9.30e^{0.1129t} = 80$	Substitute 80 for S.
$e^{0.1129t} \approx 8.6022$	Divide each side by 9.30.
$\ln e^{0.1129t} \approx \ln 8.6022$	Take natural log of each side.
$0.1129t \approx 2.1520$	Inverse Property
$t \approx 19.1$	Divide each side by 0.1129.

According to the model, the amount of U.S. online advertising spending will reach $80 billion in 2019.

Graphical Solution

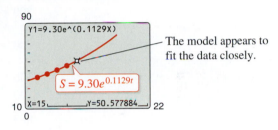

The model appears to fit the data closely.

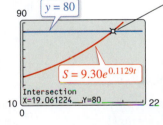

The intersection point of the model and the line $y = 80$ is about (19.1, 80). So, according to the model, the amount of U.S. online advertising spending will reach $80 billion in 2019.

✓ **Checkpoint** 🔊))) *Audio-video solution in English & Spanish at LarsonPrecalculus.com.*

In Example 1, when will the amount of U.S. online advertising spending reach $100 billion?

▷ **TECHNOLOGY** Some graphing utilities have an *exponential regression* feature that can help you find exponential models to represent data. If you have such a graphing utility, try using it to find an exponential model for the data given in Example 1. How does your model compare with the model given in Example 1?

In Example 1, the exponential growth model is given. But if this model were not given, then how would you find such a model? Example 2 demonstrates one technique.

EXAMPLE 2 **Modeling Population Growth**

In a research experiment, a population of fruit flies is increasing according to the law of exponential growth. After 2 days there are 100 flies, and after 4 days there are 300 flies. How many flies will there be after 5 days?

Solution Let y be the number of flies at time t. From the given information, you know that $y = 100$ when $t = 2$ and $y = 300$ when $t = 4$. Substituting this information into the model $y = ae^{bt}$ produces

$$100 = ae^{2b} \quad \text{and} \quad 300 = ae^{4b}.$$

To solve for b, solve for a in the first equation.

$$100 = ae^{2b} \qquad \text{Write first equation.}$$

$$\frac{100}{e^{2b}} = a \qquad \text{Solve for } a.$$

Then substitute the result into the second equation.

$$300 = ae^{4b} \qquad \text{Write second equation.}$$

$$300 = \left(\frac{100}{e^{2b}}\right)e^{4b} \qquad \text{Substitute } \frac{100}{e^{2b}} \text{ for } a.$$

$$\frac{300}{100} = e^{2b} \qquad \text{Simplify, and divide each side by 100.}$$

$$\ln 3 = 2b \qquad \text{Take natural log of each side.}$$

$$\frac{1}{2}\ln 3 = b \qquad \text{Solve for } b.$$

Using $b = \frac{1}{2}\ln 3$ and the equation you found for a,

$$a = \frac{100}{e^{2[(1/2)\ln 3]}} \qquad \text{Substitute } \frac{1}{2}\ln 3 \text{ for } b.$$

$$= \frac{100}{e^{\ln 3}} \qquad \text{Simplify.}$$

$$= \frac{100}{3} \qquad \text{Inverse Property}$$

$$\approx 33.33. \qquad \text{Divide.}$$

So, with $a \approx 33.33$ and $b = \frac{1}{2}\ln 3 \approx 0.5493$, the exponential growth model is

$$y = 33.33e^{0.5493t}$$

as shown in Figure 3.20. After 5 days, the population will be

$$y = 33.33e^{0.5493(5)}$$

$$\approx 520 \text{ flies.}$$

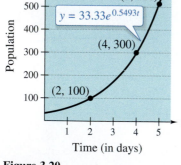

Fruit Flies

$y = 33.33e^{0.5493t}$

(5, 520)

(4, 300)

(2, 100)

Time (in days)

Figure 3.20

✓ *Checkpoint* *Audio-video solution in English & Spanish at LarsonPrecalculus.com.*

The number of bacteria in a culture is increasing according to the law of exponential growth. After 1 hour there are 100 bacteria, and after 2 hours there are 200 bacteria. How many bacteria will there be after 3 hours?

In living organic material, the ratio of the number of radioactive carbon isotopes (carbon 14) to the number of nonradioactive carbon isotopes (carbon 12) is about 1 to 10^{12}. When organic material dies, its carbon 12 content remains fixed, whereas its radioactive carbon 14 begins to decay with a half-life of about 5700 years. To estimate the age of dead organic material, scientists use the following formula, which denotes the ratio of carbon 14 to carbon 12 present at any time t (in years).

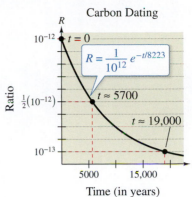

Carbon Dating

$$R = \frac{1}{10^{12}}e^{-t/8223} \qquad \text{\color{red}Carbon dating model}$$

The graph of R is shown at the right. Note that R decreases as t increases.

EXAMPLE 3 **Carbon Dating**

Estimate the age of a newly discovered fossil for which the ratio of carbon 14 to carbon 12 is $R = 1/10^{13}$.

Algebraic Solution

In the carbon dating model, substitute the given value of R to obtain the following.

$\dfrac{1}{10^{12}}e^{-t/8223} = R$	Write original model.
$\dfrac{e^{-t/8223}}{10^{12}} = \dfrac{1}{10^{13}}$	Substitute $\dfrac{1}{10^{13}}$ for R.
$e^{-t/8223} = \dfrac{1}{10}$	Multiply each side by 10^{12}.
$\ln e^{-t/8223} = \ln \dfrac{1}{10}$	Take natural log of each side.
$-\dfrac{t}{8223} \approx -2.3026$	Inverse Property
$t \approx 18{,}934$	Multiply each side by -8223.

So, to the nearest thousand years, the age of the fossil is about 19,000 years.

Graphical Solution

Use a graphing utility to graph the formula for the ratio of carbon 14 to carbon 12 at any time t as

$$y_1 = \frac{1}{10^{12}}e^{-x/8223}.$$

In the same viewing window, graph $y_2 = 1/10^{13}$.

Use the *intersect* feature to estimate that $x \approx 18{,}934$ when $y = 1/10^{13}$.

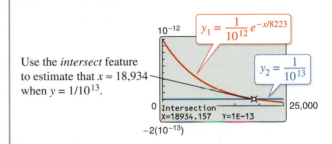

So, to the nearest thousand years, the age of the fossil is about 19,000 years.

✓ **Checkpoint** *Audio-video solution in English & Spanish at LarsonPrecalculus.com.*

Estimate the age of a newly discovered fossil for which the ratio of carbon 14 to carbon 12 is $R = 1/10^{14}$.

The value of b in the exponential decay model $y = ae^{-bt}$ determines the *decay* of radioactive isotopes. For instance, to find how much of an initial 10 grams of ^{226}Ra isotope with a half-life of 1599 years is left after 500 years, substitute this information into the model $y = ae^{-bt}$.

$$\frac{1}{2}(10) = 10e^{-b(1599)} \implies \ln \frac{1}{2} = -1599b \implies b = -\frac{\ln \frac{1}{2}}{1599}$$

Using the value of b found above and $a = 10$, the amount left is

$$y = 10e^{-[-\ln(1/2)/1599](500)} \approx 8.05 \text{ grams.}$$

Gaussian Models

As mentioned at the beginning of this section, Gaussian models are of the form

$$y = ae^{-(x-b)^2/c}.$$

In probability and statistics, this type of model commonly represents populations that are **normally distributed.** The graph of a Gaussian model is called a **bell-shaped curve.** Try graphing the normal distribution with a graphing utility. Can you see why it is called a bell-shaped curve?

For *standard* normal distributions, the model takes the form

$$y = \frac{1}{\sqrt{2\pi}}e^{-x^2/2}.$$

The **average value** of a population can be found from the bell-shaped curve by observing where the maximum y-value of the function occurs. The x-value corresponding to the maximum y-value of the function represents the average value of the independent variable—in this case, x.

EXAMPLE 4 SAT Scores

In 2011, the SAT mathematics scores for high school graduates in the United States roughly followed the normal distribution given by

$$y = 0.0034e^{-(x-514)^2/27,378}, \quad 200 \le x \le 800$$

where x is the SAT score for mathematics. Sketch the graph of this function. From the graph, estimate the average SAT mathematics score. *(Source: The College Board)*

Solution The graph of the function is shown below. On this bell-shaped curve, the maximum value of the curve represents the average score. From the graph, you can estimate that the average mathematics score for high school graduates in 2011 was 514.

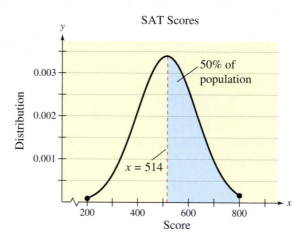

Checkpoint  *Audio-video solution in English & Spanish at LarsonPrecalculus.com.*

In 2011, the SAT critical reading scores for high school graduates in the United States roughly followed the normal distribution given by

$$y = 0.0035e^{-(x-497)^2/25,992}, \quad 200 \le x \le 800$$

where x is the SAT score for critical reading. Sketch the graph of this function. From the graph, estimate the average SAT critical reading score. *(Source: The College Board)*

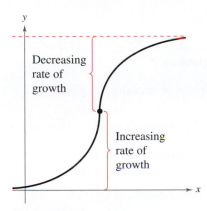

Figure 3.21

Logistic Growth Models

Some populations initially have rapid growth, followed by a declining rate of growth, as indicated by the graph in Figure 3.21. One model for describing this type of growth pattern is the **logistic curve** given by the function

$$y = \frac{a}{1 + be^{-rx}}$$

where y is the population size and x is the time. An example is a bacteria culture that is initially allowed to grow under ideal conditions and then under less favorable conditions that inhibit growth. A logistic growth curve is also called a **sigmoidal curve.**

EXAMPLE 5   **Spread of a Virus**

On a college campus of 5000 students, one student returns from vacation with a contagious and long-lasting flu virus. The spread of the virus is modeled by

$$y = \frac{5000}{1 + 4999e^{-0.8t}}, \quad t \geq 0$$

where y is the total number of students infected after t days. The college will cancel classes when 40% or more of the students are infected.

a. How many students are infected after 5 days?

b. After how many days will the college cancel classes?

Algebraic Solution

a. After 5 days, the number of students infected is

$$y = \frac{5000}{1 + 4999e^{-0.8(5)}} = \frac{5000}{1 + 4999e^{-4}} \approx 54.$$

b. The college will cancel classes when the number of infected students is $(0.40)(5000) = 2000$.

$$2000 = \frac{5000}{1 + 4999e^{-0.8t}}$$

$$1 + 4999e^{-0.8t} = 2.5$$

$$e^{-0.8t} = \frac{1.5}{4999}$$

$$-0.8t = \ln\frac{1.5}{4999}$$

$$t = -\frac{1}{0.8}\ln\frac{1.5}{4999}$$

$$t \approx 10.14$$

So, after about 10 days, at least 40% of the students will be infected, and the college will cancel classes.

Graphical Solution

a.

Use the *value* feature to estimate that $y \approx 54$ when $x = 5$. So, after 5 days, about 54 students are infected.

> Y1=5000/(1+4999e^(−.8X))
> 6000
> $y = \dfrac{5000}{1 + 4999e^{-0.8x}}$
> 0   20
> X=5   Y=54.019085
> −1000

b. The college will cancel classes when the number of infected students is $(0.40)(5000) = 2000$. Use a graphing utility to graph

$$y_1 = \frac{5000}{1 + 4999e^{-0.8x}} \quad \text{and} \quad y_2 = 2000$$

in the same viewing window. Use the *intersect* feature of the graphing utility to find the point of intersection of the graphs.

The point of intersection occurs near $x \approx 10.14$. So, after about 10 days, at least 40% of the students will be infected, and the college will cancel classes.

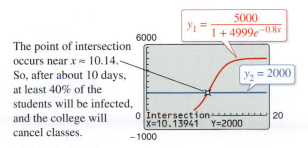

> $y_1 = \dfrac{5000}{1 + 4999e^{-0.8x}}$
> 6000
> $y_2 = 2000$
> 0   Intersection   20
> X=10.13941   Y=2000
> −1000

 ✔ *Checkpoint* *Audio-video solution in English & Spanish at LarsonPrecalculus.com.*

In Example 5, after how many days are 250 students infected?

Logarithmic Models

On February 22, 2011, an earthquake of magnitude 6.3 struck Christchurch, New Zealand. The total economic loss was estimated at 15.5 billion U.S. dollars.

EXAMPLE 6 **Magnitudes of Earthquakes**

On the Richter scale, the magnitude R of an earthquake of intensity I is given by

$$R = \log \frac{I}{I_0}$$

where $I_0 = 1$ is the minimum intensity used for comparison. Find the intensity of each earthquake. (Intensity is a measure of the wave energy of an earthquake.)

a. Alaska in 2012: $R = 4.0$ **b.** Christchurch, New Zealand, in 2011: $R = 6.3$

Solution

a. Because $I_0 = 1$ and $R = 4.0$, you have

$$4.0 = \log \frac{I}{1}$$ Substitute 1 for I_0 and 4.0 for R.

$$10^{4.0} = 10^{\log I}$$ Exponentiate each side.

$$10^{4.0} = I$$ Inverse Property

$$10{,}000 = I.$$ Simplify.

b. For $R = 6.3$, you have

$$6.3 = \log \frac{I}{1}$$ Substitute 1 for I_0 and 6.3 for R.

$$10^{6.3} = 10^{\log I}$$ Exponentiate each side.

$$10^{6.3} = I$$ Inverse Property

$$2{,}000{,}000 \approx I.$$ Use a calculator.

Note that an increase of 2.3 units on the Richter scale (from 4.0 to 6.3) represents an increase in intensity by a factor of $2{,}000{,}000/10{,}000 = 200$. In other words, the intensity of the earthquake in Christchurch was about 200 times as great as that of the earthquake in Alaska.

✓ *Checkpoint* *Audio-video solution in English & Spanish at LarsonPrecalculus.com.*

Find the intensities of earthquakes whose magnitudes are (a) $R = 6.0$ and (b) $R = 7.9$.

Summarize (Section 3.5)

1. State the five most common types of models involving exponential and logarithmic functions *(page 168)*.

2. Describe examples of how to use exponential growth and decay functions to model and solve real-life problems *(pages 169–171, Examples 1–3)*.

3. Describe an example of how to use a Gaussian function to model and solve a real-life problem *(page 172, Example 4)*.

4. Describe an example of how to use a logistic growth function to model and solve a real-life problem *(page 173, Example 5)*.

5. Describe an example of how to use a logarithmic function to model and solve a real-life problem *(page 174, Example 6)*.

Chapter Summary

What Did You Learn?	Explanation/Examples	Review Exercises
Section 3.1		
Recognize and evaluate exponential functions with base a *(p. 142)*.	The exponential function f with base a is denoted by $f(x) = a^x$, where $a > 0$, $a \neq 1$, and x is any real number.	1–6
Graph exponential functions and use the One-to-One Property *(p. 143)*.	**One-to-One Property:** For $a > 0$ and $a \neq 1$, $a^x = a^y$ if and only if $x = y$.	7–20
Recognize, evaluate, and graph exponential functions with base e *(p. 146)*.	The function $f(x) = e^x$ is called the natural exponential function.	21–28
Use exponential functions to model and solve real-life problems *(p. 147)*.	Exponential functions are used in compound interest formulas (see Example 8) and in radioactive decay models (see Example 9).	29–32
Section 3.2		
Recognize and evaluate logarithmic functions with base a *(p. 150)*.	For $x > 0$, $a > 0$, and $a \neq 1$, $y = \log_a x$ if and only if $x = a^y$. The function $f(x) = \log_a x$ is called the logarithmic function with base a. The logarithmic function with base 10 is the common logarithmic function. It is denoted by $\log_{10}$ or $\log$.	33–44
Graph logarithmic functions *(p. 152)*, and recognize, evaluate, and graph natural logarithmic functions *(p. 154)*.	The graph of $y = \log_a x$ is a reflection of the graph of $y = a^x$ in the line $y = x$. The function $f(x) = \ln x$, $x > 0$, is the natural logarithmic function. Its graph is a reflection of the graph of $f(x) = e^x$ in the line $y = x$.	45–56
Use logarithmic functions to model and solve real-life problems *(p. 156)*.	A logarithmic function can model human memory. (See Example 11.)	57, 58

	What Did You Learn?	**Explanation/Examples**	**Review Exercises**
Section 3.3	Use the change-of-base formula to rewrite and evaluate logarithmic expressions *(p. 157)*.	Let a, b, and x be positive real numbers such that $a \neq 1$ and $b \neq 1$. Then $\log_a x$ can be converted to a different base as follows. **Base b** $\log_a x = \dfrac{\log_b x}{\log_b a}$ **Base 10** $\log_a x = \dfrac{\log x}{\log a}$ **Base e** $\log_a x = \dfrac{\ln x}{\ln a}$	59–62
	Use properties of logarithms to evaluate, rewrite, expand, or condense logarithmic expressions *(p. 158)*.	Let a be a positive number such that $a \neq 1$, let n be a real number, and let u and v be positive real numbers. **1. Product Property:** $\log_a(uv) = \log_a u + \log_a v$ $\ln(uv) = \ln u + \ln v$ **2. Quotient Property:** $\log_a(u/v) = \log_a u - \log_a v$ $\ln(u/v) = \ln u - \ln v$ **3. Power Property:** $\log_a u^n = n \log_a u$, $\ln u^n = n \ln u$	63–78
	Use logarithmic functions to model and solve real-life problems *(p. 160)*.	Logarithmic functions can help you find an equation that relates the periods of several planets and their distances from the sun. (See Example 7.)	79, 80
Section 3.4	Solve simple exponential and logarithmic equations *(p. 161)*.	One-to-One Properties and Inverse Properties of exponential or logarithmic functions can help you solve exponential or logarithmic equations.	81–86
	Solve more complicated exponential equations *(p. 162)* and logarithmic equations *(p. 164)*.	To solve more complicated equations, rewrite the equations to allow the use of the One-to-One Properties or Inverse Properties of exponential or logarithmic functions. (See Examples 2–9.)	87–104
	Use exponential and logarithmic equations to model and solve real-life problems *(p. 166)*.	Exponential and logarithmic equations can help you find how long it will take to double an investment (see Example 10) and find the year in which companies reached a given amount of sales (see Example 11).	105, 106
Section 3.5	Recognize the five most common types of models involving exponential and logarithmic functions *(p. 168)*.	**1. Exponential growth model:** $y = ae^{bx}$, $b > 0$ **2. Exponential decay model:** $y = ae^{-bx}$, $b > 0$ **3. Gaussian model:** $y = ae^{-(x-b)^2/c}$ **4. Logistic growth model:** $y = \dfrac{a}{1 + be^{-rx}}$ **5. Logarithmic models:** $y = a + b \ln x$, $y = a + b \log x$	107–112
	Use exponential growth and decay functions to model and solve real-life problems *(p. 169)*.	An exponential growth function can help you model a population of fruit flies (see Example 2), and an exponential decay function can help you estimate the age of a fossil (see Example 3).	113, 114
	Use Gaussian functions *(p. 172)*, logistic growth functions *(p. 173)*, and logarithmic functions *(p. 174)* to model and solve real-life problems.	A Gaussian function can help you model SAT mathematics scores for high school graduates. (See Example 4.) A logistic growth function can help you model the spread of a flu virus. (See Example 5.) A logarithmic function can help you find the intensity of an earthquake given its magnitude. (See Example 6.)	115–117

Review Exercises   See **CalcChat.com** for tutorial help and worked-out solutions to odd-numbered exercises.

3.1 **Evaluating an Exponential Function** In Exercises 1–6, evaluate the function at the indicated value of x. Round your result to three decimal places.

1. $f(x) = 0.3^x$, $x = 1.5$ **2.** $f(x) = 30^x$, $x = \sqrt{3}$

3. $f(x) = 2^{-0.5x}$, $x = \pi$ **4.** $f(x) = 1278^{x/5}$, $x = 1$

5. $f(x) = 7(0.2^x)$, $x = -\sqrt{11}$

6. $f(x) = -14(5^x)$, $x = -0.8$

Transforming the Graph of an Exponential Function In Exercises 7–10, use the graph of f to describe the transformation that yields the graph of g.

7. $f(x) = 5^x$, $g(x) = 5^x + 1$

8. $f(x) = 6^x$, $g(x) = 6^{x+1}$

9. $f(x) = 3^x$, $g(x) = 1 - 3^x$

10. $f(x) = \left(\frac{1}{2}\right)^x$, $g(x) = -\left(\frac{1}{2}\right)^{x+2}$

Graphing an Exponential Function In Exercises 11–16, use a graphing utility to construct a table of values for the function. Then sketch the graph of the function.

11. $f(x) = 4^{-x} + 4$ **12.** $f(x) = 2.65^{x-1}$

13. $f(x) = 5^{x-2} + 4$ **14.** $f(x) = 2^{x-6} - 5$

15. $f(x) = \left(\frac{1}{2}\right)^{-x} + 3$ **16.** $f(x) = \left(\frac{1}{8}\right)^{x+2} - 5$

Using the One-to-One Property In Exercises 17–20, use the One-to-One Property to solve the equation for x.

17. $\left(\frac{1}{3}\right)^{x-3} = 9$ **18.** $3^{x+3} = \frac{1}{81}$

19. $e^{3x-5} = e^7$ **20.** $e^{8-2x} = e^{-3}$

Evaluating the Natural Exponential Function In Exercises 21–24, evaluate $f(x) = e^x$ at the indicated value of x. Round your result to three decimal places.

21. $x = 8$ **22.** $x = \frac{5}{8}$

23. $x = -1.7$ **24.** $x = 0.278$

Graphing a Natural Exponential Function In Exercises 25–28, use a graphing utility to construct a table of values for the function. Then sketch the graph of the function.

25. $h(x) = e^{-x/2}$ **26.** $h(x) = 2 - e^{-x/2}$

27. $f(x) = e^{x+2}$ **28.** $s(t) = 4e^{-2/t}$, $t > 0$

29. Waiting Times The average time between incoming calls at a switchboard is 3 minutes. The probability F of waiting less than t minutes until the next incoming call is approximated by the model $F(t) = 1 - e^{-t/3}$. The switchboard has just received a call. Find the probability that the next call will be within

(a) $\frac{1}{2}$ minute.   (b) 2 minutes.   (c) 5 minutes.

30. Depreciation After t years, the value V of a car that originally cost $23,970 is given by $V(t) = 23{,}970\left(\frac{3}{4}\right)^t$.

(a) Use a graphing utility to graph the function.

(b) Find the value of the car 2 years after it was purchased.

(c) According to the model, when does the car depreciate most rapidly? Is this realistic? Explain.

(d) According to the model, when will the car have no value?

Compound Interest In Exercises 31 and 32, complete the table to determine the balance A for P dollars invested at rate r for t years and compounded n times per year.

n	1	2	4	12	365	Continuous
A						

31. $P = \$5000$, $r = 3\%$, $t = 10$ years

32. $P = \$4500$, $r = 2.5\%$, $t = 30$ years

3.2 **Writing a Logarithmic Equation** In Exercises 33–36, write the exponential equation in logarithmic form. For example, the logarithmic form of $2^3 = 8$ is $\log_2 8 = 3$.

33. $3^3 = 27$ **34.** $25^{3/2} = 125$

35. $e^{0.8} = 2.2255\ldots$ **36.** $e^0 = 1$

Evaluating a Logarithmic Function In Exercises 37–40, evaluate the function at the indicated value of x without using a calculator.

37. $f(x) = \log x$, $x = 1000$ **38.** $g(x) = \log_9 x$, $x = 3$

39. $g(x) = \log_2 x$, $x = \frac{1}{4}$ **40.** $f(x) = \log_3 x$, $x = \frac{1}{81}$

Using the One-to-One Property In Exercises 41–44, use the One-to-One Property to solve the equation for x.

41. $\log_4(x + 7) = \log_4 14$ **42.** $\log_8(3x - 10) = \log_8 5$

43. $\ln(x + 9) = \ln 4$ **44.** $\ln(2x - 1) = \ln 11$

Sketching the Graph of a Logarithmic Function In Exercises 45–48, find the domain, x-intercept, and vertical asymptote of the logarithmic function and sketch its graph.

45. $g(x) = \log_7 x$

46. $f(x) = \log\left(\frac{x}{3}\right)$

47. $f(x) = 4 - \log(x + 5)$

48. $f(x) = \log(x - 3) + 1$

Evaluating a Logarithmic Function on a Calculator In Exercises 49–52, use a calculator to evaluate the function at the indicated value of x. Round your result to three decimal places, if necessary.

49. $f(x) = \ln x, \quad x = 22.6$ **50.** $f(x) = \ln x, \quad x = e^{-12}$

51. $f(x) = \frac{1}{2} \ln x, \quad x = \sqrt{e}$

52. $f(x) = 5 \ln x, \quad x = 0.98$

Graphing a Natural Logarithmic Function In Exercises 53–56, find the domain, x-intercept, and vertical asymptote of the logarithmic function and sketch its graph.

53. $f(x) = \ln x + 3$ **54.** $f(x) = \ln(x - 3)$

55. $h(x) = \ln(x^2)$ **56.** $f(x) = \frac{1}{4} \ln x$

57. Antler Spread The antler spread a (in inches) and shoulder height h (in inches) of an adult male American elk are related by the model

$$h = 116 \log(a + 40) - 176.$$

Approximate the shoulder height of a male American elk with an antler spread of 55 inches.

58. Snow Removal The number of miles s of roads cleared of snow is approximated by the model

$$s = 25 - \frac{13 \ln(h/12)}{\ln 3}, \quad 2 \le h \le 15$$

where h is the depth of the snow in inches. Use this model to find s when $h = 10$ inches.

3.3 **Using the Change-of-Base Formula** In Exercises 59–62, evaluate the logarithm using the change-of-base formula (a) with common logarithms and (b) with natural logarithms. Round your results to three decimal places.

59. $\log_2 6$ **60.** $\log_{12} 200$

61. $\log_{1/2} 5$ **62.** $\log_3 0.28$

Using Properties of Logarithms In Exercises 63–66, use the properties of logarithms to rewrite and simplify the logarithmic expression.

63. $\log 18$ **64.** $\log_2\left(\frac{1}{12}\right)$

65. $\ln 20$ **66.** $\ln(3e^{-4})$

Expanding a Logarithmic Expression In Exercises 67–72, use the properties of logarithms to expand the expression as a sum, difference, and/or constant multiple of logarithms. (Assume all variables are positive.)

67. $\log_5 5x^2$ **68.** $\log 7x^4$

69. $\log_3 \dfrac{9}{\sqrt{x}}$ **70.** $\log_7 \dfrac{\sqrt[3]{x}}{14}$

71. $\ln x^2 y^2 z$ **72.** $\ln\left(\dfrac{y - 1}{4}\right)^2, \quad y > 1$

Condensing a Logarithmic Expression In Exercises 73–78, condense the expression to the logarithm of a single quantity.

73. $\log_2 5 + \log_2 x$

74. $\log_6 y - 2 \log_6 z$

75. $\ln x - \frac{1}{4} \ln y$

76. $3 \ln x + 2 \ln(x + 1)$

77. $\frac{1}{2} \log_3 x - 2 \log_3(y + 8)$

78. $5 \ln(x - 2) - \ln(x + 2) - 3 \ln x$

79. Climb Rate The time t (in minutes) for a small plane to climb to an altitude of h feet is modeled by $t = 50 \log[18{,}000/(18{,}000 - h)]$, where 18,000 feet is the plane's absolute ceiling.

(a) Determine the domain of the function in the context of the problem.

(b) Use a graphing utility to graph the function and identify any asymptotes.

(c) As the plane approaches its absolute ceiling, what can be said about the time required to increase its altitude?

(d) Find the time for the plane to climb to an altitude of 4000 feet.

80. Human Memory Model Students in a learning theory study took an exam and then retested monthly for 6 months with an equivalent exam. The data obtained in the study are given by the ordered pairs (t, s), where t is the time in months after the initial exam and s is the average score for the class. Use the data to find a logarithmic equation that relates t and s.

$(1, 84.2), (2, 78.4), (3, 72.1),$
$(4, 68.5), (5, 67.1), (6, 65.3)$

3.4 **Solving a Simple Equation** In Exercises 81–86, solve for x.

81. $5^x = 125$ **82.** $6^x = \frac{1}{216}$

83. $e^x = 3$ **84.** $\log_6 x = -1$

85. $\ln x = 4$ **86.** $\ln x = -1.6$

Solving an Exponential Equation In Exercises 87–90, solve the exponential equation algebraically. Approximate the result to three decimal places.

87. $e^{4x} = e^{x^2 + 3}$ **88.** $e^{3x} = 25$

89. $2^x - 3 = 29$ **90.** $e^{2x} - 6e^x + 8 = 0$

Graphing and Solving an Exponential Equation In Exercises 91 and 92, use a graphing utility to graph and solve the equation. Approximate the result to three decimal places. Verify your result algebraically.

91. $25e^{-0.3x} = 12$

92. $2^x = 3 + x - e^x$

Solving a Logarithmic Equation In Exercises 93–100, solve the logarithmic equation algebraically. Approximate the result to three decimal places.

93. $\ln 3x = 8.2$

94. $4 \ln 3x = 15$

95. $\ln x - \ln 3 = 2$

96. $\ln \sqrt{x + 8} = 3$

97. $\log_8(x - 1) = \log_8(x - 2) - \log_8(x + 2)$

98. $\log_6(x + 2) - \log_6 x = \log_6(x + 5)$

99. $\log(1 - x) = -1$

100. $\log(-x - 4) = 2$

Graphing and Solving a Logarithmic Equation In Exercises 101–104, use a graphing utility to graph and solve the equation. Approximate the result to three decimal places. Verify your result algebraically.

101. $2 \ln(x + 3) - 3 = 0$

102. $x - 2 \log(x + 4) = 0$

103. $6 \log(x^2 + 1) - x = 0$

104. $3 \ln x + 2 \log x = e^x - 25$

105. Compound Interest You deposit $8500 in an account that pays 1.5% interest, compounded continuously. How long will it take for the money to triple?

106. Meteorology The speed of the wind S (in miles per hour) near the center of a tornado and the distance d (in miles) the tornado travels are related by the model $S = 93 \log d + 65$. On March 18, 1925, a large tornado struck portions of Missouri, Illinois, and Indiana with a wind speed at the center of about 283 miles per hour. Approximate the distance traveled by this tornado.

3.5 **Matching a Function with Its Graph** In Exercises 107–112, match the function with its graph. [The graphs are labeled (a), (b), (c), (d), (e), and (f).]

(a)

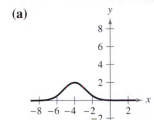

(b)

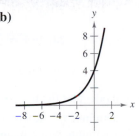

(c)

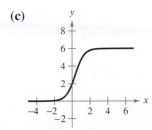

(d)

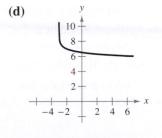

(e)

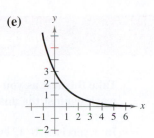

(f)

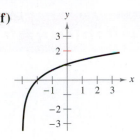

107. $y = 3e^{-2x/3}$

108. $y = 4e^{2x/3}$

109. $y = \ln(x + 3)$

110. $y = 7 - \log(x + 3)$

111. $y = 2e^{-(x+4)^2/3}$

112. $y = \dfrac{6}{1 + 2e^{-2x}}$

113. Finding an Exponential Model Find the exponential model $y = ae^{bx}$ that fits the points $(0, 2)$ and $(4, 3)$.

114. Wildlife Population A species of bat is in danger of becoming extinct. Five years ago, the total population of the species was 2000. Two years ago, the total population of the species was 1400. What was the total population of the species one year ago?

115. Test Scores The test scores for a biology test follow a normal distribution modeled by

$$y = 0.0499e^{-(x-71)^2/128}, \quad 40 \leq x \leq 100$$

where x is the test score. Use a graphing utility to graph the equation and estimate the average test score.

116. Typing Speed In a typing class, the average number N of words per minute typed after t weeks of lessons is

$$N = 157/(1 + 5.4e^{-0.12t}).$$

Find the time necessary to type (a) 50 words per minute and (b) 75 words per minute.

117. Sound Intensity The relationship between the number of decibels β and the intensity of a sound I in watts per square meter is

$$\beta = 10 \log(I/10^{-12}).$$

Find I for each decibel level β.

(a) $\beta = 60$ (b) $\beta = 135$ (c) $\beta = 1$

Exploration

118. Graph of an Exponential Function Consider the graph of $y = e^{kt}$. Describe the characteristics of the graph when k is positive and when k is negative.

True or False? In Exercises 119 and 120, determine whether the equation is true or false. Justify your answer.

119. $\log_b b^{2x} = 2x$

120. $\ln(x + y) = \ln x + \ln y$

Chapter Test

See **CalcChat.com** for tutorial help and worked-out solutions to odd-numbered exercises.

Take this test as you would take a test in class. When you are finished, check your work against the answers given in the back of the book.

In Exercises 1–4, evaluate the expression. Round your result to three decimal places.

1. $4.2^{0.6}$ **2.** $4^{3\pi/2}$ **3.** $e^{-7/10}$ **4.** $e^{3.1}$

In Exercises 5–7, construct a table of values for the function. Then sketch the graph of the function.

5. $f(x) = 10^{-x}$ **6.** $f(x) = -6^{x-2}$ **7.** $f(x) = 1 - e^{2x}$

8. Evaluate (a) $\log_7 7^{-0.89}$ and (b) $4.6 \ln e^2$.

In Exercises 9–11, find the domain, x-intercept, and vertical asymptote of the logarithmic function and sketch its graph.

9. $f(x) = -\log x - 6$ **10.** $f(x) = \ln(x - 4)$ **11.** $f(x) = 1 + \ln(x + 6)$

In Exercises 12–14, evaluate the logarithm using the change-of-base formula. Round your result to three decimal places.

12. $\log_7 44$ **13.** $\log_{16} 0.63$ **14.** $\log_{3/4} 24$

In Exercises 15–17, use the properties of logarithms to expand the expression as a sum, difference, and/or constant multiple of logarithms. (Assume all variables are positive.)

15. $\log_2 3a^4$ **16.** $\ln \dfrac{5\sqrt{x}}{6}$ **17.** $\log \dfrac{(x - 1)^3}{y^2 z}$

In Exercises 18–20, condense the expression to the logarithm of a single quantity.

18. $\log_3 13 + \log_3 y$ **19.** $4 \ln x - 4 \ln y$

20. $3 \ln x - \ln(x + 3) + 2 \ln y$

In Exercises 21–26, solve the equation algebraically. Approximate the result to three decimal places, if necessary.

21. $5^x = \dfrac{1}{25}$ **22.** $3e^{-5x} = 132$

23. $\dfrac{1025}{8 + e^{4x}} = 5$ **24.** $\ln x = \dfrac{1}{2}$

25. $18 + 4 \ln x = 7$ **26.** $\log x + \log(x - 15) = 2$

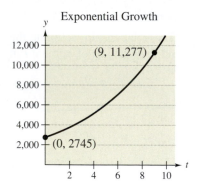

Exponential Growth

(9, 11,277)

(0, 2745)

Figure for 27

27. Find the exponential growth model that fits the points shown in the graph.

28. The half-life of radioactive actinium (^{227}Ac) is 21.77 years. What percent of a present amount of radioactive actinium will remain after 19 years?

29. A model that can predict the height H (in centimeters) of a child based on his or her age is $H = 70.228 + 5.104x + 9.222 \ln x$, $\frac{1}{4} \le x \le 6$, where x is the age of the child in years.   *(Source: Snapshots of Applications in Mathematics)*

   (a) Construct a table of values for the model. Then sketch the graph of the model.

   (b) Use the graph from part (a) to predict the height of a child when he or she is four years old. Then confirm your prediction algebraically.

Cumulative Test for Chapters 1–3

See **CalcChat.com** for tutorial help and worked-out solutions to odd-numbered exercises.

Take this test as you would take a test in class. When you are finished, check your work against the answers given in the back of the book.

1. Plot the points $(-2, 5)$ and $(3, -1)$. Find the coordinates of the midpoint of the line segment joining the points and the distance between the points.

In Exercises 2–4, sketch the graph of the equation.

2. $x - 3y + 12 = 0$ 3. $y = x^2 - 9$ 4. $y = \sqrt{4 - x}$

5. Find an equation of the line passing through $\left(-\frac{1}{2}, 1\right)$ and $(3, 8)$.

6. Explain why the graph at the left does not represent y as a function of x.

7. Evaluate (if possible) the function $f(x) = \dfrac{x}{x - 2}$ at each specified value of the independent variable.

   (a) $f(6)$ (b) $f(2)$ (c) $f(s + 2)$

8. Compare the graph of each function with the graph of $y = \sqrt[3]{x}$. (*Note:* It is not necessary to sketch the graphs.)

   (a) $r(x) = \frac{1}{2}\sqrt[3]{x}$ (b) $h(x) = \sqrt[3]{x} + 2$ (c) $g(x) = \sqrt[3]{x + 2}$

In Exercises 9 and 10, find (a) $(f + g)(x)$, (b) $(f - g)(x)$, (c) $(fg)(x)$, and (d) $(f/g)(x)$. What is the domain of f/g?

9. $f(x) = x - 3$, $g(x) = 4x + 1$

10. $f(x) = \sqrt{x - 1}$, $g(x) = x^2 + 1$

In Exercises 11 and 12, find (a) $f \circ g$ and (b) $g \circ f$. Find the domain of each composite function.

11. $f(x) = 2x^2$, $g(x) = \sqrt{x + 6}$

12. $f(x) = x - 2$, $g(x) = |x|$

13. Determine whether $h(x) = -5x + 3$ has an inverse function. If so, find the inverse function.

14. The power P produced by a wind turbine is proportional to the cube of the wind speed S. A wind speed of 27 miles per hour produces a power output of 750 kilowatts. Find the output for a wind speed of 40 miles per hour.

15. Find the quadratic function whose graph has a vertex at $(-8, 5)$ and passes through the point $(-4, -7)$.

In Exercises 16–18, sketch the graph of the function.

16. $h(x) = -(x^2 + 4x)$

17. $f(t) = \frac{1}{4}t(t - 2)^2$

18. $g(s) = s^2 + 4s + 10$

In Exercises 19–21, find all the zeros of the function and write the function as a product of linear factors.

19. $f(x) = x^3 + 2x^2 + 4x + 8$

20. $f(x) = x^4 + 4x^3 - 21x^2$

21. $f(x) = 2x^4 - 11x^3 + 30x^2 - 62x - 40$

Figure for 6

22. Use long division to divide $6x^3 - 4x^2$ by $2x^2 + 1$.

23. Use synthetic division to divide $3x^4 + 2x^2 - 5x + 3$ by $x - 2$.

24. Use the Intermediate Value Theorem and the *table* feature of a graphing utility to find intervals one unit in length in which the function $g(x) = x^3 + 3x^2 - 6$ is guaranteed to have a zero. Adjust the table to approximate the real zeros of the function.

In Exercises 25–27, sketch the graph of the rational function. Identify all intercepts and asymptotes.

25. $f(x) = \dfrac{2x}{x^2 + 2x - 3}$

26. $f(x) = \dfrac{x^2 - 4}{x^2 + x - 2}$

27. $f(x) = \dfrac{x^3 - 2x^2 - 9x + 18}{x^2 + 4x + 3}$

In Exercises 28 and 29, solve the inequality. Then graph the solution set.

28. $2x^3 - 18x \le 0$

29. $\dfrac{1}{x + 1} \ge \dfrac{1}{x + 5}$

In Exercises 30 and 31, use the graph of f to describe the transformation that yields the graph of g.

30. $f(x) = \left(\frac{2}{5}\right)^x$, $\quad g(x) = -\left(\frac{2}{5}\right)^{-x+3}$

31. $f(x) = 2.2^x$, $\quad g(x) = -2.2^x + 4$

In Exercises 32–35, use a calculator to evaluate the expression. Round your result to three decimal places.

32. $\log 98$

33. $\log \frac{6}{7}$

34. $\ln \sqrt{31}$

35. $\ln\left(\sqrt{40} - 5\right)$

36. Use the properties of logarithms to expand $\ln\left(\dfrac{x^2 - 16}{x^4}\right)$, where $x > 4$.

37. Condense $2 \ln x - \frac{1}{2} \ln(x + 5)$ to the logarithm of a single quantity.

In Exercises 38–40, solve the equation algebraically. Approximate the result to three decimal places.

38. $6e^{2x} = 72$

39. $e^{2x} - 13e^x + 42 = 0$

40. $\ln\sqrt{x + 2} = 3$

41. The sales S (in billions of dollars) of lottery tickets in the United States from 2000 through 2010 are shown in the table. *(Source: TLF Publications, Inc.)*

(a) Use a graphing utility to create a scatter plot of the data. Let t represent the year, with $t = 0$ corresponding to 2000.

(b) Use the *regression* feature of the graphing utility to find a cubic model for the data.

(c) Use the graphing utility to graph the model in the same viewing window used for the scatter plot. How well does the model fit the data?

(d) Use the model to predict the sales of lottery tickets in 2018. Does your answer seem reasonable? Explain.

42. The number N of bacteria in a culture is given by the model $N = 175e^{kt}$, where t is the time in hours. If $N = 420$ when $t = 8$, then estimate the time required for the population to double in size.

Year	Sales, S
2000	37.2
2001	38.4
2002	42.0
2003	43.5
2004	47.7
2005	47.4
2006	51.6
2007	52.4
2008	53.4
2009	53.1
2010	54.2

Spreadsheet at LarsonPrecalculus.com

Table for 41

Proofs in Mathematics ▪ ▪ ▪ ▪ ▪ ▪ ▪ ▪ ▪ ▪ ▪ ▪ ▪ ▪

Each of the following three properties of logarithms can be proved by using properties of exponential functions.

Properties of Logarithms (p. 158)

Let a be a positive number such that $a \neq 1$, and let n be a real number. If u and v are positive real numbers, then the following properties are true.

	Logarithm with Base a	Natural Logarithm
1. **Product Property:**	$\log_a(uv) = \log_a u + \log_a v$	$\ln(uv) = \ln u + \ln v$
2. **Quotient Property:**	$\log_a \dfrac{u}{v} = \log_a u - \log_a v$	$\ln \dfrac{u}{v} = \ln u - \ln v$
3. **Power Property:**	$\log_a u^n = n \log_a u$	$\ln u^n = n \ln u$

Proof

Let

$$x = \log_a u \quad \text{and} \quad y = \log_a v.$$

The corresponding exponential forms of these two equations are

$$a^x = u \quad \text{and} \quad a^y = v.$$

To prove the Product Property, multiply u and v to obtain

$$uv = a^x a^y$$
$$= a^{x+y}.$$

The corresponding logarithmic form of $uv = a^{x+y}$ is $\log_a(uv) = x + y$. So,

$$\log_a(uv) = \log_a u + \log_a v.$$

To prove the Quotient Property, divide u by v to obtain

$$\frac{u}{v} = \frac{a^x}{a^y}$$
$$= a^{x-y}.$$

The corresponding logarithmic form of $\dfrac{u}{v} = a^{x-y}$ is $\log_a \dfrac{u}{v} = x - y$. So,

$$\log_a \frac{u}{v} = \log_a u - \log_a v.$$

To prove the Power Property, substitute a^x for u in the expression $\log_a u^n$, as follows.

$$\log_a u^n = \log_a (a^x)^n \qquad \text{Substitute } a^x \text{ for } u.$$
$$= \log_a a^{nx} \qquad \text{Property of Exponents}$$
$$= nx \qquad \text{Inverse Property of Logarithms}$$
$$= n \log_a u \qquad \text{Substitute } \log_a u \text{ for } x.$$

So, $\log_a u^n = n \log_a u$. ▪

P.S. Problem Solving ■ ■ ■ ■ ■ ■ ■ ■ ■ ■ ■ ■

1. **Graphical Analysis** Graph the exponential function $y = a^x$ for $a = 0.5$, 1.2, and 2.0. Which of these curves intersects the line $y = x$? Determine all positive numbers a for which the curve $y = a^x$ intersects the line $y = x$.

2. **Graphical Analysis** Use a graphing utility to graph $y_1 = e^x$ and each of the functions $y_2 = x^2$, $y_3 = x^3$, $y_4 = \sqrt{x}$, and $y_5 = |x|$. Which function increases at the greatest rate as x approaches $+\infty$?

3. **Conjecture** Use the result of Exercise 2 to make a conjecture about the rate of growth of $y_1 = e^x$ and $y = x^n$, where n is a natural number and x approaches $+\infty$.

4. **Implication of "Growing Exponentially"** Use the results of Exercises 2 and 3 to describe what is implied when it is stated that a quantity is growing exponentially.

5. **Exponential Function** Given the exponential function

   $$f(x) = a^x$$

   show that

   (a) $f(u + v) = f(u) \cdot f(v)$. (b) $f(2x) = [f(x)]^2$.

6. **Hyperbolic Functions** Given that

   $$f(x) = \frac{e^x + e^{-x}}{2} \quad \text{and} \quad g(x) = \frac{e^x - e^{-x}}{2}$$

   show that

   $$[f(x)]^2 - [g(x)]^2 = 1.$$

7. **Graphical Analysis** Use a graphing utility to compare the graph of the function $y = e^x$ with the graph of each given function. [$n!$ (read "n factorial") is defined as $n! = 1 \cdot 2 \cdot 3 \cdots (n - 1) \cdot n$.]

   (a) $y_1 = 1 + \dfrac{x}{1!}$

   (b) $y_2 = 1 + \dfrac{x}{1!} + \dfrac{x^2}{2!}$

   (c) $y_3 = 1 + \dfrac{x}{1!} + \dfrac{x^2}{2!} + \dfrac{x^3}{3!}$

8. **Identifying a Pattern** Identify the pattern of successive polynomials given in Exercise 7. Extend the pattern one more term and compare the graph of the resulting polynomial function with the graph of $y = e^x$. What do you think this pattern implies?

9. **Finding an Inverse Function** Graph the function

   $$f(x) = e^x - e^{-x}.$$

   From the graph, the function appears to be one-to-one. Assuming that the function has an inverse function, find $f^{-1}(x)$.

10. **Finding a Pattern for an Inverse Function** Find a pattern for $f^{-1}(x)$ when

    $$f(x) = \frac{a^x + 1}{a^x - 1}$$

    where $a > 0$, $a \neq 1$.

11. **Determining the Equation of a Graph** By observation, determine whether equation (a), (b), or (c) corresponds to the graph. Explain your reasoning.

    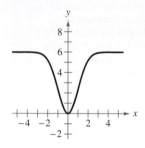

    (a) $y = 6e^{-x^2/2}$

    (b) $y = \dfrac{6}{1 + e^{-x/2}}$

    (c) $y = 6(1 - e^{-x^2/2})$

12. **Simple and Compound Interest** You have two options for investing $500. The first earns 7% compounded annually, and the second earns 7% simple interest. The figure shows the growth of each investment over a 30-year period.

    (a) Identify which graph represents each type of investment. Explain your reasoning.

    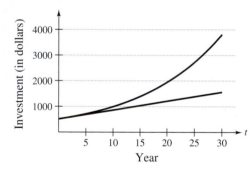

    (b) Verify your answer in part (a) by finding the equations that model the investment growth and by graphing the models.

    (c) Which option would you choose? Explain your reasoning.

13. **Radioactive Decay** Two different samples of radioactive isotopes are decaying. The isotopes have initial amounts of c_1 and c_2, as well as half-lives of k_1 and k_2, respectively. Find the time t required for the samples to decay to equal amounts.

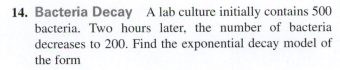

14. Bacteria Decay A lab culture initially contains 500 bacteria. Two hours later, the number of bacteria decreases to 200. Find the exponential decay model of the form

$$B = B_0 a^{kt}$$

that approximates the number of bacteria after t hours.

15. Colonial Population The table shows the colonial population estimates of the American colonies from 1700 through 1780. (*Source: U.S. Census Bureau*)

DATA	Year	Population
	1700	250,900
	1710	331,700
	1720	466,200
	1730	629,400
	1740	905,600
	1750	1,170,800
	1760	1,593,600
	1770	2,148,400
	1780	2,780,400

Spreadsheet at LarsonPrecalculus.com

In each of the following, let y represent the population in the year t, with $t = 0$ corresponding to 1700.

(a) Use the *regression* feature of a graphing utility to find an exponential model for the data.

(b) Use the *regression* feature of the graphing utility to find a quadratic model for the data.

(c) Use the graphing utility to plot the data and the models from parts (a) and (b) in the same viewing window.

(d) Which model is a better fit for the data? Would you use this model to predict the population of the United States in 2018? Explain your reasoning.

16. Ratio of Logarithms Show that

$$\frac{\log_a x}{\log_{a/b} x} = 1 + \log_a \frac{1}{b}.$$

17. Solving a Logarithmic Equation Solve

$$(\ln x)^2 = \ln x^2.$$

18. Graphical Analysis Use a graphing utility to compare the graph of the function $y = \ln x$ with the graph of each given function.

(a) $y_1 = x - 1$

(b) $y_2 = (x - 1) - \frac{1}{2}(x - 1)^2$

(c) $y_3 = (x - 1) - \frac{1}{2}(x - 1)^2 + \frac{1}{3}(x - 1)^3$

19. Identifying a Pattern Identify the pattern of successive polynomials given in Exercise 18. Extend the pattern one more term and compare the graph of the resulting polynomial function with the graph of $y = \ln x$. What do you think the pattern implies?

20. Finding Slope and y-Intercept Using

$$y = ab^x \quad \text{and} \quad y = ax^b$$

take the natural logarithm of each side of each equation. What are the slope and y-intercept of the line relating x and $\ln y$ for $y = ab^x$? What are the slope and y-intercept of the line relating $\ln x$ and $\ln y$ for $y = ax^b$?

Ventilation Rate In Exercises 21 and 22, use the model

$$y = 80.4 - 11 \ln x, \quad 100 \leq x \leq 1500$$

which approximates the minimum required ventilation rate in terms of the air space per child in a public school classroom. In the model, x is the air space per child in cubic feet and y is the ventilation rate per child in cubic feet per minute.

21. Use a graphing utility to graph the model and approximate the required ventilation rate when there are 300 cubic feet of air space per child.

22. In a classroom designed for 30 students, the air conditioning system can move 450 cubic feet of air per minute.

(a) Determine the ventilation rate per child in a full classroom.

(b) Estimate the air space required per child.

(c) Determine the minimum number of square feet of floor space required for the room when the ceiling height is 30 feet.

Data Analysis In Exercises 23–26, (a) use a graphing utility to create a scatter plot of the data, (b) decide whether the data could best be modeled by a linear model, an exponential model, or a logarithmic model, (c) explain why you chose the model you did in part (b), (d) use the *regression* feature of the graphing utility to find the model you chose in part (b) for the data and graph the model with the scatter plot, and (e) determine how well the model you chose fits the data.

23. (1, 2.0), (1.5, 3.5), (2, 4.0), (4, 5.8), (6, 7.0), (8, 7.8)

24. (1, 4.4), (1.5, 4.7), (2, 5.5), (4, 9.9), (6, 18.1), (8, 33.0)

25. (1, 7.5), (1.5, 7.0), (2, 6.8), (4, 5.0), (6, 3.5), (8, 2.0)

26. (1, 5.0), (1.5, 6.0), (2, 6.4), (4, 7.8), (6, 8.6), (8, 9.0)

4 Trigonometry

Waterslide Design

Television Coverage

Respiratory Cycle

Meteorology

Skateboarding

4.1 Radian and Degree Measure

Angles can help you model and solve real-life problems, such as finding the speed of a bicycle.

■ Describe angles.
■ Use radian measure.
■ Use degree measure.
■ Use angles to model and solve real-life problems.

Angles

As derived from the Greek language, the word **trigonometry** means "measurement of triangles." Initially, trigonometry dealt with relationships among the sides and angles of triangles and was used in the development of astronomy, navigation, and surveying. With the development of calculus and the physical sciences in the 17th century, a different perspective arose—one that viewed the classic trigonometric relationships as *functions* with the set of real numbers as their domains. Consequently, the applications of trigonometry expanded to include a vast number of physical phenomena, such as sound waves, planetary orbits, vibrating strings, pendulums, and orbits of atomic particles.

This text incorporates *both* perspectives, starting with angles and their measures.

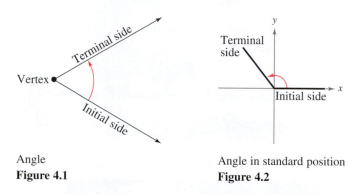

Angle
Figure 4.1

Angle in standard position
Figure 4.2

An **angle** is determined by rotating a ray (half-line) about its endpoint. The starting position of the ray is the **initial side** of the angle, and the position after rotation is the **terminal side,** as shown in Figure 4.1. The endpoint of the ray is the **vertex** of the angle. This perception of an angle fits a coordinate system in which the origin is the vertex and the initial side coincides with the positive *x*-axis. Such an angle is in **standard position,** as shown in Figure 4.2. Counterclockwise rotation generates **positive angles** and clockwise rotation generates **negative angles,** as shown in Figure 4.3. Angles are labeled with Greek letters such as

α (alpha), β (beta),   and θ (theta)

as well as uppercase letters such as

A, B,   and C.

In Figure 4.4, note that angles α and β have the same initial and terminal sides. Such angles are **coterminal.**

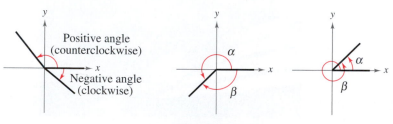

Figure 4.3

Coterminal angles
Figure 4.4

Radian Measure

You determine the **measure of an angle** by the amount of rotation from the initial side to the terminal side. One way to measure angles is in *radians*. This type of measure is especially useful in calculus. To define a radian, you can use a **central angle** of a circle, one whose vertex is the center of the circle, as shown in Figure 4.5.

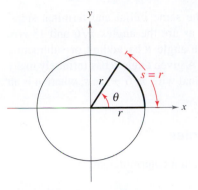

Arc length = radius when $\theta = 1$ radian.
Figure 4.5

> ### Definition of Radian
>
> One **radian** is the measure of a central angle θ that intercepts an arc s equal in length to the radius r of the circle. See Figure 4.5. Algebraically, this means that
>
> $$\theta = \frac{s}{r}$$
>
> where θ is measured in radians. (Note that $\theta = 1$ when $s = r$.)

Because the circumference of a circle is $2\pi r$ units, it follows that a central angle of one full revolution (counterclockwise) corresponds to an arc length of $s = 2\pi r$. Moreover, because $2\pi \approx 6.28$, there are just over six radius lengths in a full circle, as shown in Figure 4.6. Because the units of measure for s and r are the same, the ratio s/r has no units—it is a real number.

Because the measure of an angle of one full revolution is $s/r = 2\pi r/r = 2\pi$ radians, you can obtain the following.

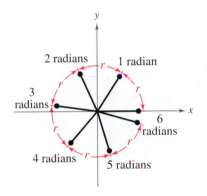

Figure 4.6

$$\frac{1}{2}\text{ revolution} = \frac{2\pi}{2} = \pi \text{ radians} \qquad \frac{1}{4}\text{ revolution} = \frac{2\pi}{4} = \frac{\pi}{2}\text{ radians}$$

$$\frac{1}{6}\text{ revolution} = \frac{2\pi}{6} = \frac{\pi}{3}\text{ radians}$$

These and other common angles are shown below.

• • **REMARK** The phrase "the terminal side of θ lies in a quadrant" is often abbreviated by the phrase "θ lies in a quadrant." The terminal sides of the "quadrant angles" 0, $\pi/2$, π, and $3\pi/2$ do not lie within quadrants.

Recall that the four quadrants in a coordinate system are numbered I, II, III, and IV. The figure below shows which angles between 0 and 2π lie in each of the four quadrants. Note that angles between 0 and $\pi/2$ are **acute** angles and angles between $\pi/2$ and π are **obtuse** angles.

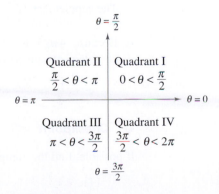

Two angles are coterminal when they have the same initial and terminal sides. For instance, the angles 0 and 2π are coterminal, as are the angles $\pi/6$ and $13\pi/6$. You can find an angle that is coterminal to a given angle θ by adding or subtracting 2π (one revolution), as demonstrated in Example 1. A given angle θ has infinitely many coterminal angles. For instance, $\theta = \pi/6$ is coterminal with $\pi/6 + 2n\pi$, where n is an integer.

EXAMPLE 1   Finding Coterminal Angles

▷ **ALGEBRA HELP**   You can review operations involving fractions in Appendix A.1.

a. For the positive angle $13\pi/6$, subtract 2π to obtain a coterminal angle

$$\frac{13\pi}{6} - 2\pi = \frac{\pi}{6}.$$ See Figure 4.7.

b. For the negative angle $-2\pi/3$, add 2π to obtain a coterminal angle

$$-\frac{2\pi}{3} + 2\pi = \frac{4\pi}{3}.$$ See Figure 4.8.

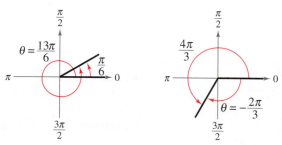

Figure 4.7   **Figure 4.8**

✔ **Checkpoint** 🔊 *Audio-video solution in English & Spanish at LarsonPrecalculus.com.*

Determine two coterminal angles (one positive and one negative) for each angle.

a. $\theta = \dfrac{9\pi}{4}$ **b.** $\theta = -\dfrac{\pi}{3}$

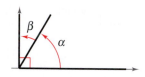

Complementary angles

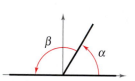

Supplementary angles
Figure 4.9

Two positive angles α and β are **complementary** (complements of each other) when their sum is $\pi/2$. Two positive angles are **supplementary** (supplements of each other) when their sum is π. See Figure 4.9.

EXAMPLE 2   Complementary and Supplementary Angles

a. The complement of $\dfrac{2\pi}{5}$ is $\dfrac{\pi}{2} - \dfrac{2\pi}{5} = \dfrac{5\pi}{10} - \dfrac{4\pi}{10} = \dfrac{\pi}{10}.$

The supplement of $\dfrac{2\pi}{5}$ is $\pi - \dfrac{2\pi}{5} = \dfrac{5\pi}{5} - \dfrac{2\pi}{5} = \dfrac{3\pi}{5}.$

b. Because $4\pi/5$ is greater than $\pi/2$, it has no complement. (Remember that complements are *positive* angles.) The supplement of $4\pi/5$ is

$$\pi - \frac{4\pi}{5} = \frac{5\pi}{5} - \frac{4\pi}{5} = \frac{\pi}{5}.$$

✔ **Checkpoint** 🔊 *Audio-video solution in English & Spanish at LarsonPrecalculus.com.*

If possible, find the complement and the supplement of (a) $\pi/6$ and (b) $5\pi/6$.

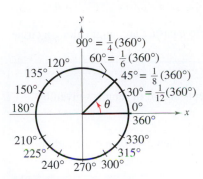

Figure 4.10

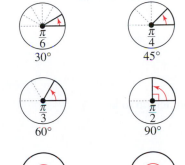

Figure 4.11

▷ **TECHNOLOGY**

With calculators, it is convenient to use *decimal* degrees to denote fractional parts of degrees. Historically, however, fractional parts of degrees were expressed in *minutes* and *seconds*, using the prime (′) and double prime (″) notations, respectively. That is,

$$1' = \text{one minute} = \tfrac{1}{60}(1°)$$

$$1'' = \text{one second} = \tfrac{1}{3600}(1°).$$

Consequently, an angle of 64 degrees, 32 minutes, and 47 seconds is represented by $\theta = 64° 32' 47''$. Many calculators have special keys for converting an angle in degrees, minutes, and seconds (D° M′ S″) to decimal degree form, and vice versa.

Degree Measure

A second way to measure angles is in **degrees,** denoted by the symbol °. A measure of one degree (1°) is equivalent to a rotation of $\frac{1}{360}$ of a complete revolution about the vertex. To measure angles, it is convenient to mark degrees on the circumference of a circle, as shown in Figure 4.10. So, a full revolution (counterclockwise) corresponds to 360°, a half revolution to 180°, a quarter revolution to 90°, and so on.

Because 2π radians corresponds to one complete revolution, degrees and radians are related by the equations

$$360° = 2\pi \text{ rad} \quad \text{and} \quad 180° = \pi \text{ rad}.$$

From the latter equation, you obtain

$$1° = \frac{\pi}{180} \text{ rad} \quad \text{and} \quad 1 \text{ rad} = \left(\frac{180}{\pi}\right)°$$

which lead to the following conversion rules.

Conversions Between Degrees and Radians

1. To convert degrees to radians, multiply degrees by $\dfrac{\pi \text{ rad}}{180°}$.

2. To convert radians to degrees, multiply radians by $\dfrac{180°}{\pi \text{ rad}}$.

To apply these two conversion rules, use the basic relationship $\pi \text{ rad} = 180°$. (See Figure 4.11.)

When no units of angle measure are specified, *radian measure is implied.* For instance, $\theta = 2$ implies that $\theta = 2$ radians.

EXAMPLE 3 **Converting from Degrees to Radians**

a. $135° = (135 \text{ deg})\left(\dfrac{\pi \text{ rad}}{180 \text{ deg}}\right) = \dfrac{3\pi}{4} \text{ radians}$ Multiply by $\frac{\pi \text{ rad}}{180°}$.

b. $540° = (540 \text{ deg})\left(\dfrac{\pi \text{ rad}}{180 \text{ deg}}\right) = 3\pi \text{ radians}$ Multiply by $\frac{\pi \text{ rad}}{180°}$.

✓ **Checkpoint** ◀))) *Audio-video solution in English & Spanish at LarsonPrecalculus.com.*

Rewrite each angle in radian measure as a multiple of π. (Do not use a calculator.)

a. $\theta = 60°$ **b.** $\theta = 320°$

EXAMPLE 4 **Converting from Radians to Degrees**

a. $-\dfrac{\pi}{2} \text{ rad} = \left(-\dfrac{\pi}{2} \text{ rad}\right)\left(\dfrac{180 \text{ deg}}{\pi \text{ rad}}\right) = -90°$ Multiply by $\frac{180°}{\pi \text{ rad}}$.

b. $2 \text{ rad} = (2 \text{ rad})\left(\dfrac{180 \text{ deg}}{\pi \text{ rad}}\right) = \dfrac{360°}{\pi} \approx 114.59°$ Multiply by $\frac{180°}{\pi \text{ rad}}$.

✓ **Checkpoint** ◀))) *Audio-video solution in English & Spanish at LarsonPrecalculus.com.*

Rewrite each angle in degree measure. (Do not use a calculator.)

a. $\pi/6$ **b.** $5\pi/3$

Applications

The *radian measure* formula, $\theta = s/r$, can be used to measure arc length along a circle.

Arc Length

For a circle of radius r, a central angle θ intercepts an arc of length s given by

$$s = r\theta \qquad \text{Length of circular arc}$$

where θ is measured in radians. Note that if $r = 1$, then $s = \theta$, and the radian measure of θ equals the arc length.

EXAMPLE 5   **Finding Arc Length**

A circle has a radius of 4 inches. Find the length of the arc intercepted by a central angle of 240°, as shown in Figure 4.12.

Solution   To use the formula $s = r\theta$, first convert 240° to radian measure.

$$240° = (240 \, \text{deg})\left(\frac{\pi \, \text{rad}}{180 \, \text{deg}}\right)$$

$$= \frac{4\pi}{3} \text{ radians}$$

Then, using a radius of $r = 4$ inches, you can find the arc length to be

$$s = r\theta$$

$$= 4\left(\frac{4\pi}{3}\right)$$

$$\approx 16.76 \text{ inches.}$$

Note that the units for r determine the units for $r\theta$ because θ is given in radian measure, which has no units.

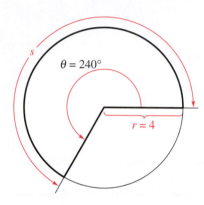

$\theta = 240°$

$r = 4$

Figure 4.12

✓ **Checkpoint**   *Audio-video solution in English & Spanish at LarsonPrecalculus.com.*

A circle has a radius of 27 inches. Find the length of the arc intercepted by a central angle of 160°.

The formula for the length of a circular arc can help you analyze the motion of a particle moving at a *constant speed* along a circular path.

• • **REMARK**   Linear speed measures how fast the particle moves, and angular speed measures how fast the angle changes. By dividing each side of the formula for arc length by t, you can establish a relationship between linear speed v and angular speed ω, as shown.

$$s = r\theta$$

$$\frac{s}{t} = \frac{r\theta}{t}$$

$$v = r\omega$$

Linear and Angular Speeds

Consider a particle moving at a constant speed along a circular arc of radius r. If s is the length of the arc traveled in time t, then the **linear speed** v of the particle is

$$\text{Linear speed } v = \frac{\text{arc length}}{\text{time}} = \frac{s}{t}.$$

Moreover, if θ is the angle (in radian measure) corresponding to the arc length s, then the **angular speed** ω (the lowercase Greek letter omega) of the particle is

$$\text{Angular speed } \omega = \frac{\text{central angle}}{\text{time}} = \frac{\theta}{t}.$$

Figure 4.13

EXAMPLE 6 **Finding Linear Speed**

The second hand of a clock is 10.2 centimeters long, as shown in Figure 4.13. Find the linear speed of the tip of this second hand as it passes around the clock face.

Solution In one revolution, the arc length traveled is

$$s = 2\pi r$$
$$= 2\pi(10.2) \qquad \text{Substitute for } r.$$
$$= 20.4\pi \text{ centimeters.}$$

The time required for the second hand to travel this distance is

$$t = 1 \text{ minute}$$
$$= 60 \text{ seconds.}$$

So, the linear speed of the tip of the second hand is

$$\text{Linear speed} = \frac{s}{t}$$
$$= \frac{20.4\pi \text{ centimeters}}{60 \text{ seconds}}$$
$$\approx 1.068 \text{ centimeters per second.}$$

✓ *Checkpoint* *Audio-video solution in English & Spanish at LarsonPrecalculus.com.*

The second hand of a clock is 8 centimeters long. Find the linear speed of the tip of this second hand as it passes around the clock face.

EXAMPLE 7 **Finding Angular and Linear Speeds**

The blades of a wind turbine are 116 feet long (see Figure 4.14). The propeller rotates at 15 revolutions per minute.

a. Find the angular speed of the propeller in radians per minute.

b. Find the linear speed of the tips of the blades.

Solution

a. Because each revolution generates 2π radians, it follows that the propeller turns

$$(15)(2\pi) = 30\pi \text{ radians per minute.}$$

In other words, the angular speed is

$$\text{Angular speed} = \frac{\theta}{t} = \frac{30\pi \text{ radians}}{1 \text{ minute}} = 30\pi \text{ radians per minute.}$$

b. The linear speed is

$$\text{Linear speed} = \frac{s}{t} = \frac{r\theta}{t} = \frac{(116)(30\pi) \text{ feet}}{1 \text{ minute}} \approx 10{,}933 \text{ feet per minute.}$$

Figure 4.14

✓ *Checkpoint* *Audio-video solution in English & Spanish at LarsonPrecalculus.com.*

The circular blade on a saw rotates at 2400 revolutions per minute.

a. Find the angular speed of the blade in radians per minute.

b. The blade has a radius of 4 inches. Find the linear speed of a blade tip.

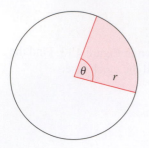

Figure 4.15

A **sector** of a circle is the region bounded by two radii of the circle and their intercepted arc (see Figure 4.15).

> ### Area of a Sector of a Circle
>
> For a circle of radius r, the area A of a sector of the circle with central angle θ is
>
> $$A = \frac{1}{2}r^2\theta$$
>
> where θ is measured in radians.

EXAMPLE 8 Area of a Sector of a Circle

A sprinkler on a golf course fairway sprays water over a distance of 70 feet and rotates through an angle of 120° (see Figure 4.16). Find the area of the fairway watered by the sprinkler.

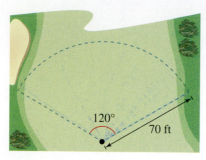

Figure 4.16

Solution

First convert 120° to radian measure as follows.

$$\theta = 120°$$

$$= (120 \text{ deg})\left(\frac{\pi \text{ rad}}{180 \text{ deg}}\right) \qquad \text{Multiply by } \frac{\pi \text{ rad}}{180°}.$$

$$= \frac{2\pi}{3} \text{ radians}$$

Then, using $\theta = 2\pi/3$ and $r = 70$, the area is

$$A = \frac{1}{2}r^2\theta \qquad\qquad \text{Formula for the area of a sector of a circle}$$

$$= \frac{1}{2}(70)^2\left(\frac{2\pi}{3}\right) \qquad \text{Substitute for } r \text{ and } \theta.$$

$$= \frac{4900\pi}{3} \qquad\qquad \text{Multiply.}$$

$$\approx 5131 \text{ square feet.} \qquad \text{Simplify.}$$

✓ *Checkpoint* 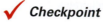 *Audio-video solution in English & Spanish at LarsonPrecalculus.com.*

A sprinkler sprays water over a distance of 40 feet and rotates through an angle of 80°. Find the area watered by the sprinkler.

> ### Summarize (Section 4.1)
>
> **1.** Describe an angle *(page 188).*
>
> **2.** Describe how to determine the measure of an angle using radians *(page 189).* For examples involving radian measure, see Examples 1 and 2.
>
> **3.** Describe how to determine the measure of an angle using degrees *(page 191).* For examples involving degree measure, see Examples 3 and 4.
>
> **4.** Describe examples of how to use angles to model and solve real-life problems *(pages 192–194, Examples 5–8).*

4.2 Trigonometric Functions: The Unit Circle

- Identify a unit circle and describe its relationship to real numbers.
- Evaluate trigonometric functions using the unit circle.
- Use domain and period to evaluate sine and cosine functions, and use a calculator to evaluate trigonometric functions.

The Unit Circle

The two historical perspectives of trigonometry incorporate different methods for introducing the trigonometric functions. One such perspective follows and is based on the unit circle.

Consider the **unit circle** given by

$$x^2 + y^2 = 1 \qquad \text{Unit circle}$$

as shown below.

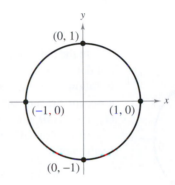

Imagine wrapping the real number line around this circle, with positive numbers corresponding to a counterclockwise wrapping and negative numbers corresponding to a clockwise wrapping, as shown below.

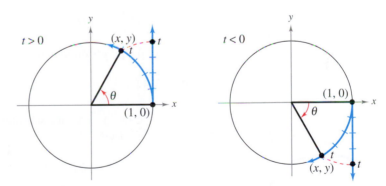

As the real number line wraps around the unit circle, each real number t corresponds to a point (x, y) on the circle. For example, the real number 0 corresponds to the point $(1, 0)$. Moreover, because the unit circle has a circumference of 2π, the real number 2π also corresponds to the point $(1, 0)$.

In general, each real number t also corresponds to a central angle θ (in standard position) whose radian measure is t. With this interpretation of t, the arc length formula

$$s = r\theta \quad \text{(with } r = 1\text{)}$$

indicates that the real number t is the (directional) length of the arc intercepted by the angle θ, given in radians.

Trigonometric functions can help you analyze the displacement of an oscillating weight suspended by a spring.

Richard Megna/Fundamental Photographs

The Trigonometric Functions

From the preceding discussion, it follows that the coordinates x and y are two functions of the real variable t. You can use these coordinates to define the six trigonometric functions of t.

| sine | cosecant | cosine | secant | tangent | cotangent |

These six functions are normally abbreviated sin, csc, cos, sec, tan, and cot, respectively.

Definitions of Trigonometric Functions

Let t be a real number and let (x, y) be the point on the unit circle corresponding to t.

$$\sin t = y \qquad \cos t = x \qquad \tan t = \frac{y}{x}, \quad x \neq 0$$

$$\csc t = \frac{1}{y}, \quad y \neq 0 \qquad \sec t = \frac{1}{x}, \quad x \neq 0 \qquad \cot t = \frac{x}{y}, \quad y \neq 0$$

•• **REMARK** Note that the functions in the second row are the *reciprocals* of the corresponding functions in the first row.

In the definitions of the trigonometric functions, note that the tangent and secant are not defined when $x = 0$. For instance, because $t = \pi/2$ corresponds to $(x, y) = (0, 1)$, it follows that $\tan(\pi/2)$ and $\sec(\pi/2)$ are *undefined*. Similarly, the cotangent and cosecant are not defined when $y = 0$. For instance, because $t = 0$ corresponds to $(x, y) = (1, 0)$, cot 0 and csc 0 are *undefined*.

In Figure 4.17, the unit circle is divided into eight equal arcs, corresponding to t-values of

$$0, \frac{\pi}{4}, \frac{\pi}{2}, \frac{3\pi}{4}, \pi, \frac{5\pi}{4}, \frac{3\pi}{2}, \frac{7\pi}{4}, \text{ and } 2\pi.$$

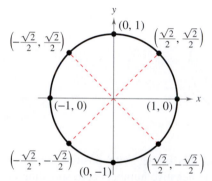

Figure 4.17

Similarly, in Figure 4.18, the unit circle is divided into 12 equal arcs, corresponding to t-values of

$$0, \frac{\pi}{6}, \frac{\pi}{3}, \frac{\pi}{2}, \frac{2\pi}{3}, \frac{5\pi}{6}, \pi, \frac{7\pi}{6}, \frac{4\pi}{3}, \frac{3\pi}{2}, \frac{5\pi}{3}, \frac{11\pi}{6}, \text{ and } 2\pi.$$

To verify the points on the unit circle in Figure 4.17, note that

$$\left(\frac{\sqrt{2}}{2}, \frac{\sqrt{2}}{2} \right)$$

also lies on the line $y = x$. So, substituting x for y in the equation of the unit circle produces the following.

$$x^2 + x^2 = 1 \implies 2x^2 = 1 \implies x^2 = \frac{1}{2} \implies x = \pm\frac{\sqrt{2}}{2}$$

Because the point is in the first quadrant and $y = x$, you have

$$x = \frac{\sqrt{2}}{2} \quad \text{and} \quad y = \frac{\sqrt{2}}{2}.$$

You can use similar reasoning to verify the rest of the points in Figure 4.17 and the points in Figure 4.18.

Using the (x, y) coordinates in Figures 4.17 and 4.18, you can evaluate the trigonometric functions for common t-values. Examples 1 and 2 demonstrate this procedure. You should study and learn these exact function values for common t-values because they will help you in later sections to perform calculations.

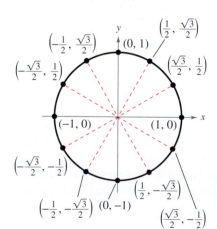

Figure 4.18

EXAMPLE 1   **Evaluating Trigonometric Functions**

▷ **ALGEBRA HELP**   You can review dividing fractions and rationalizing denominators in Appendix A.1 and Appendix A.2, respectively.

Evaluate the six trigonometric functions at each real number.

a. $t = \dfrac{\pi}{6}$ **b.** $t = \dfrac{5\pi}{4}$ **c.** $t = \pi$ **d.** $t = -\dfrac{\pi}{3}$

Solution   For each t-value, begin by finding the corresponding point (x, y) on the unit circle. Then use the definitions of trigonometric functions listed on page 196.

a. $t = \pi/6$ corresponds to the point $(x, y) = \left(\sqrt{3}/2,\ 1/2\right)$.

$$\sin\frac{\pi}{6} = y = \frac{1}{2} \qquad\qquad \csc\frac{\pi}{6} = \frac{1}{y} = \frac{1}{1/2} = 2$$

$$\cos\frac{\pi}{6} = x = \frac{\sqrt{3}}{2} \qquad\qquad \sec\frac{\pi}{6} = \frac{1}{x} = \frac{2}{\sqrt{3}} = \frac{2\sqrt{3}}{3}$$

$$\tan\frac{\pi}{6} = \frac{y}{x} = \frac{1/2}{\sqrt{3}/2} = \frac{1}{\sqrt{3}} = \frac{\sqrt{3}}{3} \qquad\qquad \cot\frac{\pi}{6} = \frac{x}{y} = \frac{\sqrt{3}/2}{1/2} = \sqrt{3}$$

b. $t = 5\pi/4$ corresponds to the point $(x, y) = \left(-\sqrt{2}/2,\ -\sqrt{2}/2\right)$.

$$\sin\frac{5\pi}{4} = y = -\frac{\sqrt{2}}{2} \qquad\qquad \csc\frac{5\pi}{4} = \frac{1}{y} = -\frac{2}{\sqrt{2}} = -\sqrt{2}$$

$$\cos\frac{5\pi}{4} = x = -\frac{\sqrt{2}}{2} \qquad\qquad \sec\frac{5\pi}{4} = \frac{1}{x} = -\frac{2}{\sqrt{2}} = -\sqrt{2}$$

$$\tan\frac{5\pi}{4} = \frac{y}{x} = \frac{-\sqrt{2}/2}{-\sqrt{2}/2} = 1 \qquad\qquad \cot\frac{5\pi}{4} = \frac{x}{y} = \frac{-\sqrt{2}/2}{-\sqrt{2}/2} = 1$$

c. $t = \pi$ corresponds to the point $(x, y) = (-1, 0)$.

$$\sin\pi = y = 0 \qquad\qquad \csc\pi = \frac{1}{y} \text{ is undefined.}$$

$$\cos\pi = x = -1 \qquad\qquad \sec\pi = \frac{1}{x} = \frac{1}{-1} = -1$$

$$\tan\pi = \frac{y}{x} = \frac{0}{-1} = 0 \qquad\qquad \cot\pi = \frac{x}{y} \text{ is undefined.}$$

d. Moving *clockwise* around the unit circle, it follows that $t = -\pi/3$ corresponds to the point $(x, y) = \left(1/2,\ -\sqrt{3}/2\right)$.

$$\sin\left(-\frac{\pi}{3}\right) = y = -\frac{\sqrt{3}}{2} \qquad\qquad \csc\left(-\frac{\pi}{3}\right) = \frac{1}{y} = -\frac{2}{\sqrt{3}} = -\frac{2\sqrt{3}}{3}$$

$$\cos\left(-\frac{\pi}{3}\right) = x = \frac{1}{2} \qquad\qquad \sec\left(-\frac{\pi}{3}\right) = \frac{1}{x} = \frac{1}{1/2} = 2$$

$$\tan\left(-\frac{\pi}{3}\right) = \frac{y}{x} = \frac{-\sqrt{3}/2}{1/2} = -\sqrt{3}$$

$$\cot\left(-\frac{\pi}{3}\right) = \frac{x}{y} = \frac{1/2}{-\sqrt{3}/2} = -\frac{1}{\sqrt{3}} = -\frac{\sqrt{3}}{3}$$

✓ *Checkpoint*   *Audio-video solution in English & Spanish at LarsonPrecalculus.com.*

Evaluate the six trigonometric functions at each real number.

a. $t = \pi/2$ **b.** $t = 0$ **c.** $t = -5\pi/6$ **d.** $t = -3\pi/4$

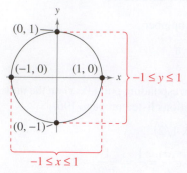

Figure 4.19

Domain and Period of Sine and Cosine

The *domain* of the sine and cosine functions is the set of all real numbers. To determine the *range* of these two functions, consider the unit circle shown in Figure 4.19. By definition, $\sin t = y$ and $\cos t = x$. Because (x, y) is on the unit circle, you know that $-1 \le y \le 1$ and $-1 \le x \le 1$. So, the values of sine and cosine also range between -1 and 1.

$$-1 \le \ y \ \le 1 \qquad\qquad -1 \le \ x \ \le 1$$
$$\text{and}$$
$$-1 \le \sin t \le 1 \qquad -1 \le \cos t \le 1$$

Adding 2π to each value of t in the interval $[0, 2\pi]$ results in a revolution around the unit circle, as shown below.

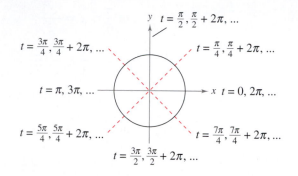

The values of $\sin(t + 2\pi)$ and $\cos(t + 2\pi)$ correspond to those of $\sin t$ and $\cos t$. Similar results can be obtained for repeated revolutions (positive or negative) on the unit circle. This leads to the general result

$$\sin(t + 2\pi n) = \sin t \quad \text{and} \quad \cos(t + 2\pi n) = \cos t$$

for any integer n and real number t. Functions that behave in such a repetitive (or cyclic) manner are called **periodic.**

•• **REMARK** From this definition, it follows that the sine and cosine functions are periodic and have a period of 2π. The other four trigonometric functions are also periodic and will be discussed further in Section 4.6.

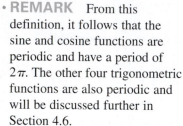

Definition of Periodic Function

A function f is **periodic** when there exists a positive real number c such that

$$f(t + c) = f(t)$$

for all t in the domain of f. The smallest number c for which f is periodic is called the **period** of f.

Recall from Section 1.5 that a function f is *even* when $f(-t) = f(t)$ and is *odd* when $f(-t) = -f(t)$.

Even and Odd Trigonometric Functions

The cosine and secant functions are *even.*

$$\cos(-t) = \cos t \qquad \sec(-t) = \sec t$$

The sine, cosecant, tangent, and cotangent functions are *odd.*

$$\sin(-t) = -\sin t \qquad \csc(-t) = -\csc t$$

$$\tan(-t) = -\tan t \qquad \cot(-t) = -\cot t$$

EXAMPLE 2 **Evaluating Sine and Cosine**

a. Because $\dfrac{13\pi}{6} = 2\pi + \dfrac{\pi}{6}$, you have $\sin\dfrac{13\pi}{6} = \sin\left(2\pi + \dfrac{\pi}{6}\right) = \sin\dfrac{\pi}{6} = \dfrac{1}{2}$.

b. Because $-\dfrac{7\pi}{2} = -4\pi + \dfrac{\pi}{2}$, you have

$$\cos\left(-\dfrac{7\pi}{2}\right) = \cos\left(-4\pi + \dfrac{\pi}{2}\right) = \cos\dfrac{\pi}{2} = 0.$$

c. For $\sin t = \dfrac{4}{5}$, $\sin(-t) = -\dfrac{4}{5}$ because the sine function is odd.

✓ *Checkpoint*  *Audio-video solution in English & Spanish at LarsonPrecalculus.com.*

a. Use the period of the cosine function to evaluate $\cos(9\pi/2)$.

b. Use the period of the sine function to evaluate $\sin(-7\pi/3)$.

c. Evaluate $\cos t$ given that $\cos(-t) = 0.3$.

▷ **TECHNOLOGY** When evaluating trigonometric functions with a calculator, remember to enclose all fractional angle measures in parentheses. For instance, to evaluate $\sin t$ for $t = \pi/6$, you should enter

⟨SIN⟩ ⟨(⟩ ⟨π⟩ ⟨÷⟩ 6 ⟨)⟩ ⟨ENTER⟩.

These keystrokes yield the correct value of 0.5. Note that some calculators automatically place a left parenthesis after trigonometric functions.

When evaluating a trigonometric function with a calculator, you need to set the calculator to the desired *mode* of measurement (*degree* or *radian*). Most calculators do not have keys for the cosecant, secant, and cotangent functions. To evaluate these functions, you can use the ⟨x⁻¹⟩ key with their respective reciprocal functions: sine, cosine, and tangent. For instance, to evaluate $\csc(\pi/8)$, use the fact that

$$\csc\dfrac{\pi}{8} = \dfrac{1}{\sin(\pi/8)}$$

and enter the following keystroke sequence in *radian* mode.

⟨(⟩ ⟨SIN⟩ ⟨(⟩ ⟨π⟩ ⟨÷⟩ 8 ⟨)⟩ ⟨)⟩ ⟨x⁻¹⟩ ⟨ENTER⟩   Display 2.6131259

EXAMPLE 3 **Using a Calculator**

Function	Mode	Calculator Keystrokes	Display
a. $\sin\dfrac{2\pi}{3}$	Radian	⟨SIN⟩ ⟨(⟩ 2 ⟨π⟩ ⟨÷⟩ 3 ⟨)⟩ ⟨ENTER⟩	0.8660254
b. $\cot 1.5$	Radian	⟨(⟩ ⟨TAN⟩ ⟨(⟩ 1.5 ⟨)⟩ ⟨)⟩ ⟨x⁻¹⟩ ⟨ENTER⟩	0.0709148

✓ *Checkpoint*  *Audio-video solution in English & Spanish at LarsonPrecalculus.com.*

Use a calculator to evaluate (a) $\sin(5\pi/7)$ and (b) $\csc 2.0$.

Summarize **(Section 4.2)**

1. Explain how to identify a unit circle and describe its relationship to real numbers *(page 195)*.

2. State the unit circle definitions of the trigonometric functions *(page 196)*. For an example of evaluating trigonometric functions using the unit circle, see Example 1.

3. Explain how to use domain and period to evaluate sine and cosine functions *(page 198)* and describe how to use a calculator to evaluate trigonometric functions *(page 199)*. For an example of using period and an odd trigonometric function to evaluate sine and cosine functions, see Example 2. For an example of using a calculator to evaluate trigonometric functions, see Example 3.

4.3 Right Triangle Trigonometry

Trigonometric functions can help you analyze real-life situations, such as finding the height of a helium-filled balloon.

■ Evaluate trigonometric functions of acute angles, and use a calculator to evaluate trigonometric functions.
■ Use the fundamental trigonometric identities.
■ Use trigonometric functions to model and solve real-life problems.

The Six Trigonometric Functions

This section introduces the trigonometric functions from a *right triangle* perspective. Consider a right triangle with one acute angle labeled θ, as shown below. Relative to the angle θ, the three sides of the triangle are the **hypotenuse,** the **opposite side** (the side opposite the angle θ), and the **adjacent side** (the side adjacent to the angle θ).

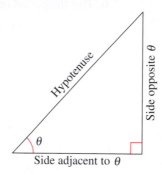

Side adjacent to θ

Using the lengths of these three sides, you can form six ratios that define the six trigonometric functions of the acute angle θ.

sine   cosecant   cosine   secant   tangent   cotangent

In the following definitions, it is important to see that

$$0° < \theta < 90°$$

(θ lies in the first quadrant) and that for such angles the value of each trigonometric function is *positive.*

Right Triangle Definitions of Trigonometric Functions

Let θ be an *acute* angle of a right triangle. The six trigonometric functions of the angle θ are defined as follows. (Note that the functions in the second row are the *reciprocals* of the corresponding functions in the first row.)

$$\sin \theta = \frac{\text{opp}}{\text{hyp}} \qquad \cos \theta = \frac{\text{adj}}{\text{hyp}} \qquad \tan \theta = \frac{\text{opp}}{\text{adj}}$$

$$\csc \theta = \frac{\text{hyp}}{\text{opp}} \qquad \sec \theta = \frac{\text{hyp}}{\text{adj}} \qquad \cot \theta = \frac{\text{adj}}{\text{opp}}$$

The abbreviations

opp,   adj,   and   hyp

represent the lengths of the three sides of a right triangle.

opp = the length of the side *opposite* θ

adj = the length of the side *adjacent to* θ

hyp = the length of the *hypotenuse*

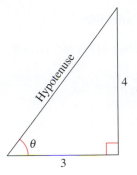

Figure 4.20

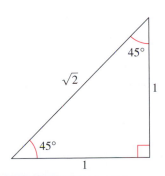

Figure 4.21

EXAMPLE 1   **Evaluating Trigonometric Functions**

Use the triangle in Figure 4.20 to find the values of the six trigonometric functions of θ.

Solution   By the Pythagorean Theorem,

$$(\text{hyp})^2 = (\text{opp})^2 + (\text{adj})^2$$

it follows that

$$\text{hyp} = \sqrt{4^2 + 3^2}$$
$$= \sqrt{25}$$
$$= 5.$$

So, the six trigonometric functions of θ are

$$\sin\theta = \frac{\text{opp}}{\text{hyp}} = \frac{4}{5} \qquad \csc\theta = \frac{\text{hyp}}{\text{opp}} = \frac{5}{4}$$

$$\cos\theta = \frac{\text{adj}}{\text{hyp}} = \frac{3}{5} \qquad \sec\theta = \frac{\text{hyp}}{\text{adj}} = \frac{5}{3}$$

$$\tan\theta = \frac{\text{opp}}{\text{adj}} = \frac{4}{3} \qquad \cot\theta = \frac{\text{adj}}{\text{opp}} = \frac{3}{4}.$$

✓ *Checkpoint* ◀))) *Audio-video solution in English & Spanish at LarsonPrecalculus.com.*

Use the triangle at the right to find
the values of the six trigonometric
functions of θ.

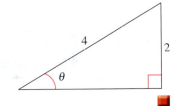

In Example 1, you were given the lengths of two sides of the right triangle, but not
the angle θ. Often, you will be asked to find the trigonometric functions of a *given* acute
angle θ. To do this, construct a right triangle having θ as one of its angles.

EXAMPLE 2   **Evaluating Trigonometric Functions of 45°**

Find the values of $\sin 45°$, $\cos 45°$, and $\tan 45°$.

Solution   Construct a right triangle having 45° as one of its acute angles, as shown
in Figure 4.21. Choose 1 as the length of the adjacent side. From geometry, you know
that the other acute angle is also 45°. So, the triangle is isosceles and the length of the
opposite side is also 1. Using the Pythagorean Theorem, you find the length of the
hypotenuse to be $\sqrt{2}$.

$$\sin 45° = \frac{\text{opp}}{\text{hyp}} = \frac{1}{\sqrt{2}} = \frac{\sqrt{2}}{2}$$

$$\cos 45° = \frac{\text{adj}}{\text{hyp}} = \frac{1}{\sqrt{2}} = \frac{\sqrt{2}}{2}$$

$$\tan 45° = \frac{\text{opp}}{\text{adj}} = \frac{1}{1} = 1$$

✓ *Checkpoint* ◀))) *Audio-video solution in English & Spanish at LarsonPrecalculus.com.*

Find the value of $\sec 45°$.

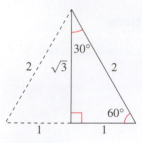

Figure 4.22

•••••••••••••••••••▷
•• **REMARK** Because the angles 30°, 45°, and 60° ($\pi/6$, $\pi/4$, and $\pi/3$, respectively) occur frequently in trigonometry, you should learn to construct the triangles shown in Figures 4.21 and 4.22.

•••••••••••••••••••▷
•• **REMARK** Throughout this text, angles are assumed to be measured in radians unless noted otherwise. For example, sin 1 means the sine of 1 radian and sin 1° means the sine of 1 degree.

EXAMPLE 3 **Trigonometric Functions of 30° and 60°**

Use the equilateral triangle shown in Figure 4.22 to find the values of sin 60°, cos 60°, sin 30°, and cos 30°.

Solution For $\theta = 60°$, you have adj = 1, opp = $\sqrt{3}$, and hyp = 2. So,

$$\sin 60° = \frac{\text{opp}}{\text{hyp}} = \frac{\sqrt{3}}{2} \quad \text{and} \quad \cos 60° = \frac{\text{adj}}{\text{hyp}} = \frac{1}{2}.$$

For $\theta = 30°$, adj = $\sqrt{3}$, opp = 1, and hyp = 2. So,

$$\sin 30° = \frac{\text{opp}}{\text{hyp}} = \frac{1}{2} \quad \text{and} \quad \cos 30° = \frac{\text{adj}}{\text{hyp}} = \frac{\sqrt{3}}{2}.$$

✓ **Checkpoint** 🔊)) *Audio-video solution in English & Spanish at LarsonPrecalculus.com.*

Use the equilateral triangle shown in Figure 4.22 to find the values of tan 60° and tan 30°.

Sines, Cosines, and Tangents of Special Angles

$$\sin 30° = \sin \frac{\pi}{6} = \frac{1}{2} \qquad \cos 30° = \cos \frac{\pi}{6} = \frac{\sqrt{3}}{2} \qquad \tan 30° = \tan \frac{\pi}{6} = \frac{\sqrt{3}}{3}$$

$$\sin 45° = \sin \frac{\pi}{4} = \frac{\sqrt{2}}{2} \qquad \cos 45° = \cos \frac{\pi}{4} = \frac{\sqrt{2}}{2} \qquad \tan 45° = \tan \frac{\pi}{4} = 1$$

$$\sin 60° = \sin \frac{\pi}{3} = \frac{\sqrt{3}}{2} \qquad \cos 60° = \cos \frac{\pi}{3} = \frac{1}{2} \qquad \tan 60° = \tan \frac{\pi}{3} = \sqrt{3}$$

In the box, note that $\sin 30° = \frac{1}{2} = \cos 60°$. This occurs because 30° and 60° are complementary angles. In general, it can be shown from the right triangle definitions that *cofunctions of complementary angles are equal.* That is, if θ is an acute angle, then the following relationships are true.

$$\sin(90° - \theta) = \cos \theta \qquad \cos(90° - \theta) = \sin \theta \qquad \tan(90° - \theta) = \cot \theta$$

$$\cot(90° - \theta) = \tan \theta \qquad \sec(90° - \theta) = \csc \theta \qquad \csc(90° - \theta) = \sec \theta$$

To use a calculator to evaluate trigonometric functions of angles measured in degrees, first set the calculator to *degree* mode and then proceed as demonstrated in Section 4.2.

EXAMPLE 4 **Using a Calculator**

Use a calculator to evaluate sec 5° 40′ 12″.

Solution Begin by converting to decimal degree form. [Recall that $1' = \frac{1}{60}(1°)$ and $1'' = \frac{1}{3600}(1°)$.]

$$5° \, 40' \, 12'' = 5° + \left(\frac{40}{60}\right)° + \left(\frac{12}{3600}\right)° = 5.67°$$

Then, use a calculator to evaluate sec 5.67°.

Function	Calculator Keystrokes	Display
sec 5° 40′ 12″ = sec 5.67°	(COS (5.67)) x⁻¹ ENTER	1.0049166

✓ **Checkpoint** 🔊)) *Audio-video solution in English & Spanish at LarsonPrecalculus.com.*

Use a calculator to evaluate csc 34° 30′ 36″.

Trigonometric Identities

In trigonometry, a great deal of time is spent studying relationships between trigonometric functions (identities).

Fundamental Trigonometric Identities

Reciprocal Identities

$$\sin \theta = \frac{1}{\csc \theta} \qquad \cos \theta = \frac{1}{\sec \theta} \qquad \tan \theta = \frac{1}{\cot \theta}$$

$$\csc \theta = \frac{1}{\sin \theta} \qquad \sec \theta = \frac{1}{\cos \theta} \qquad \cot \theta = \frac{1}{\tan \theta}$$

Quotient Identities

$$\tan \theta = \frac{\sin \theta}{\cos \theta} \qquad \cot \theta = \frac{\cos \theta}{\sin \theta}$$

Pythagorean Identities

$$\sin^2 \theta + \cos^2 \theta = 1$$

$$1 + \tan^2 \theta = \sec^2 \theta$$

$$1 + \cot^2 \theta = \csc^2 \theta$$

Note that $\sin^2 \theta$ represents $(\sin \theta)^2$, $\cos^2 \theta$ represents $(\cos \theta)^2$, and so on.

EXAMPLE 5 **Applying Trigonometric Identities**

Let θ be an acute angle such that $\sin \theta = 0.6$. Find the values of (a) $\cos \theta$ and (b) $\tan \theta$ using trigonometric identities.

Solution

a. To find the value of $\cos \theta$, use the Pythagorean identity

$$\sin^2 \theta + \cos^2 \theta = 1.$$

So, you have

$$(0.6)^2 + \cos^2 \theta = 1 \qquad \text{Substitute 0.6 for } \sin \theta.$$

$$\cos^2 \theta = 1 - (0.6)^2 \qquad \text{Subtract } (0.6)^2 \text{ from each side.}$$

$$\cos^2 \theta = 0.64 \qquad \text{Simplify.}$$

$$\cos \theta = \sqrt{0.64} \qquad \text{Extract positive square root.}$$

$$\cos \theta = 0.8. \qquad \text{Simplify.}$$

b. Now, knowing the sine and cosine of θ, you can find the tangent of θ to be

$$\tan \theta = \frac{\sin \theta}{\cos \theta} = \frac{0.6}{0.8} = 0.75.$$

Use the definitions of $\cos \theta$ and $\tan \theta$ and the triangle shown in Figure 4.23 to check these results.

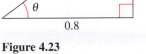

Figure 4.23

✓ **Checkpoint** 🔊))) *Audio-video solution in English & Spanish at LarsonPrecalculus.com.*

Let θ be an acute angle such that $\cos \theta = 0.25$. Find the values of (a) $\sin \theta$ and (b) $\tan \theta$ using trigonometric identities.

EXAMPLE 6   Applying Trigonometric Identities

Let θ be an acute angle such that $\tan \theta = \frac{1}{3}$. Find the values of (a) $\cot \theta$ and (b) $\sec \theta$ using trigonometric identities.

Solution

a. $\cot \theta = \dfrac{1}{\tan \theta}$ Reciprocal identity

$= \dfrac{1}{1/3}$

$= 3$

b. $\sec^2 \theta = 1 + \tan^2 \theta$ Pythagorean identity

$= 1 + (1/3)^2$

$= 10/9$

$\sec \theta = \sqrt{10}/3$

Use the definitions of $\cot \theta$ and $\sec \theta$ and the triangle shown below to check these results.

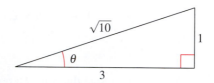

 Checkpoint ◀))) Audio-video solution in English & Spanish at *LarsonPrecalculus.com*.

Let β be an acute angle such that $\tan \beta = 2$. Find the values of (a) $\cot \beta$ and (b) $\sec \beta$ using trigonometric identities.

EXAMPLE 7   Using Trigonometric Identities

Use trigonometric identities to transform the left side of the equation into the right side $(0 < \theta < \pi/2)$.

a. $\sin \theta \csc \theta = 1$ **b.** $(\csc \theta + \cot \theta)(\csc \theta - \cot \theta) = 1$

Solution

a. $\sin \theta \csc \theta = \left(\dfrac{1}{\csc \theta} \right) \csc \theta = 1$ Use a reciprocal identity and simplify.

b. $(\csc \theta + \cot \theta)(\csc \theta - \cot \theta)$

$= \csc^2 \theta - \csc \theta \cot \theta + \csc \theta \cot \theta - \cot^2 \theta$ FOIL Method

$= \csc^2 \theta - \cot^2 \theta$ Simplify.

$= 1$ Pythagorean identity

 **Checkpoint** ◀))) Audio-video solution in English & Spanish at *LarsonPrecalculus.com*.

Use trigonometric identities to transform the left side of the equation into the right side $(0 < \theta < \pi/2)$.

a. $\tan \theta \csc \theta = \sec \theta$

b. $(\csc \theta + 1)(\csc \theta - 1) = \cot^2 \theta$

Applications Involving Right Triangles

Many applications of trigonometry involve a process called **solving right triangles.** In this type of application, you are usually given one side of a right triangle and one of the acute angles and are asked to find one of the other sides, *or* you are given two sides and are asked to find one of the acute angles.

In Example 8, the angle you are given is the **angle of elevation,** which represents the angle from the horizontal upward to an object. In other applications you may be given the **angle of depression,** which represents the angle from the horizontal downward to an object. (See Figure 4.24.)

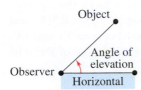
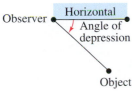

Figure 4.24

EXAMPLE 8   **Using Trigonometry to Solve a Right Triangle**

A surveyor is standing 115 feet from the base of the Washington Monument, as shown in the figure at the right. The surveyor measures the angle of elevation to the top of the monument as 78.3°. How tall is the Washington Monument?

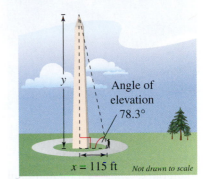

Solution   From the figure, you can see that

$$\tan 78.3° = \frac{\text{opp}}{\text{adj}} = \frac{y}{x}$$

where $x = 115$ and y is the height of the monument. So, the height of the Washington Monument is

$$y = x \tan 78.3°$$
$$\approx 115(4.82882)$$
$$\approx 555 \text{ feet.}$$

✓ **Checkpoint** 🔊))) *Audio-video solution in English & Spanish at LarsonPrecalculus.com.*

How tall is the flagpole in Figure 4.25?

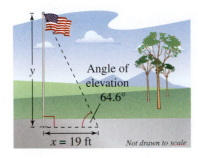

Figure 4.25

EXAMPLE 9   **Using Trigonometry to Solve a Right Triangle**

A historic lighthouse is 200 yards from a bike path along the edge of a lake. A walkway to the lighthouse is 400 yards long. Find the acute angle θ between the bike path and the walkway, as illustrated in the figure at the right.

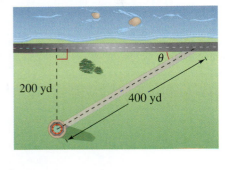

Solution   From the figure, you can see that the sine of the angle θ is

$$\sin \theta = \frac{\text{opp}}{\text{hyp}} = \frac{200}{400} = \frac{1}{2}.$$

Now you should recognize that $\theta = 30°$.

✓ **Checkpoint** 🔊))) *Audio-video solution in English & Spanish at LarsonPrecalculus.com.*

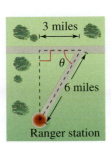

Figure 4.26

Find the acute angle θ between the two paths, as illustrated in Figure 4.26.   ■

In Example 9, you were able to recognize that $\theta = 30°$ is the acute angle that satisfies the equation $\sin \theta = \frac{1}{2}$. Suppose, however, that you were given the equation $\sin \theta = 0.6$ and were asked to find the acute angle θ. Because $\sin 30° = \frac{1}{2} = 0.5000$ and $\sin 45° = 1/\sqrt{2} \approx 0.7071$, you might guess that θ lies somewhere between $30°$ and $45°$. In a later section, you will study a method by which a more precise value of θ can be determined.

EXAMPLE 10   **Solving a Right Triangle**

Find the length c of the skateboard ramp shown in the figure below. Find the horizontal length a of the ramp.

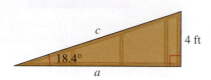

Skateboarders can go to a skatepark, which is a recreational environment built with many different types of ramps and rails.

Solution   From the figure, you can see that

$$\sin 18.4° = \frac{\text{opp}}{\text{hyp}} = \frac{4}{c}.$$

So, the length of the skateboard ramp is

$$c = \frac{4}{\sin 18.4°} \approx \frac{4}{0.3156} \approx 12.7 \text{ feet.}$$

Also from the figure, you can see that

$$\tan 18.4° = \frac{\text{opp}}{\text{adj}} = \frac{4}{a}.$$

So, the horizontal length is

$$a = \frac{4}{\tan 18.4°} \approx 12.0 \text{ feet.}$$

✓ **Checkpoint** 🔊))) *Audio-video solution in English & Spanish at LarsonPrecalculus.com.*

Find the length c of the loading ramp shown in the figure below. Find the horizontal length a of the ramp.

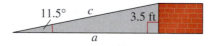

Summarize   (Section 4.3)

1. State the right triangle definitions of the trigonometric functions *(page 200)* and describe how to use a calculator to evaluate trigonometric functions *(page 202)*. For examples of evaluating trigonometric functions of acute angles, see Examples 1–3. For an example of using a calculator to evaluate a trigonometric function, see Example 4.

2. List the fundamental trigonometric identities *(page 203)*. For examples of using the fundamental trigonometric identities, see Examples 5–7.

3. Describe examples of how to use trigonometric functions to model and solve real-life problems *(pages 205 and 206, Examples 8–10)*.

4.4 Trigonometric Functions of Any Angle

■ Evaluate trigonometric functions of any angle.
■ Find reference angles.
■ Evaluate trigonometric functions of real numbers.

Introduction

In Section 4.3, the definitions of trigonometric functions were restricted to acute angles. In this section, the definitions are extended to cover *any* angle. When θ is an *acute* angle, the definitions here coincide with those given in the preceding section.

Trigonometric functions can help you model and solve real-life problems, such as the monthly normal temperatures in New York City and Fairbanks, Alaska.

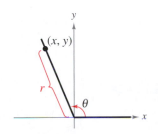

Definitions of Trigonometric Functions of Any Angle

Let θ be an angle in standard position with (x, y) a point on the terminal side of θ and $r = \sqrt{x^2 + y^2} \neq 0$.

$$\sin \theta = \frac{y}{r} \qquad \cos \theta = \frac{x}{r}$$

$$\tan \theta = \frac{y}{x}, \quad x \neq 0 \qquad \cot \theta = \frac{x}{y}, \quad y \neq 0$$

$$\sec \theta = \frac{r}{x}, \quad x \neq 0 \qquad \csc \theta = \frac{r}{y}, \quad y \neq 0$$

Because $r = \sqrt{x^2 + y^2}$ *cannot* be zero, it follows that the sine and cosine functions are defined for any real value of θ. However, when $x = 0$, the tangent and secant of θ are undefined. For example, the tangent of $90°$ is undefined. Similarly, when $y = 0$, the cotangent and cosecant of θ are undefined.

EXAMPLE 1 Evaluating Trigonometric Functions

Let $(-3, 4)$ be a point on the terminal side of θ. Find the sine, cosine, and tangent of θ.

Solution Referring to Figure 4.27, you can see that $x = -3$, $y = 4$, and

$$r = \sqrt{x^2 + y^2}$$
$$= \sqrt{(-3)^2 + 4^2}$$
$$= \sqrt{25}$$
$$= 5.$$

So, you have the following.

$$\sin \theta = \frac{y}{r} = \frac{4}{5}$$

$$\cos \theta = \frac{x}{r} = -\frac{3}{5}$$

$$\tan \theta = \frac{y}{x} = -\frac{4}{3}$$

Figure 4.27

▷ **ALGEBRA HELP** The formula $r = \sqrt{x^2 + y^2}$ is a result of the Distance Formula. You can review the Distance Formula in Section 1.1.

✓ *Checkpoint* ◀))) *Audio-video solution in English & Spanish at LarsonPrecalculus.com.*

Let $(-2, 3)$ be a point on the terminal side of θ. Find the sine, cosine, and tangent of θ.

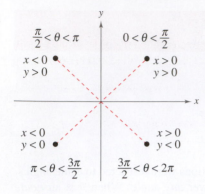

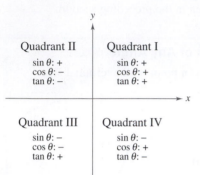

Figure 4.28

The *signs* of the trigonometric functions in the four quadrants can be determined from the definitions of the functions. For instance, because $\cos \theta = x/r$, it follows that $\cos \theta$ is positive wherever $x > 0$, which is in Quadrants I and IV. (Remember, r is always positive.) In a similar manner, you can verify the results shown in Figure 4.28.

EXAMPLE 2   Evaluating Trigonometric Functions

Given $\tan \theta = -\frac{5}{4}$ and $\cos \theta > 0$, find $\sin \theta$ and $\sec \theta$.

Solution   Note that θ lies in Quadrant IV because that is the only quadrant in which the tangent is negative and the cosine is positive. Moreover, using

$$\tan \theta = \frac{y}{x} = -\frac{5}{4}$$

and the fact that y is negative in Quadrant IV, you can let $y = -5$ and $x = 4$. So, $r = \sqrt{16 + 25} = \sqrt{41}$ and you have the following.

$$\sin \theta = \frac{y}{r}$$

$$= \frac{-5}{\sqrt{41}} \qquad \text{Exact value}$$

$$\approx -0.7809 \qquad \text{Approximate value}$$

$$\sec \theta = \frac{r}{x}$$

$$= \frac{\sqrt{41}}{4} \qquad \text{Exact value}$$

$$\approx 1.6008 \qquad \text{Approximate value}$$

✓ *Checkpoint* *Audio-video solution in English & Spanish at LarsonPrecalculus.com.*

Given $\sin \theta = \frac{4}{5}$ and $\tan \theta < 0$, find $\cos \theta$.

EXAMPLE 3   Trigonometric Functions of Quadrant Angles

Evaluate the cosine and tangent functions at the four quadrant angles 0, $\frac{\pi}{2}$, π, and $\frac{3\pi}{2}$.

Solution   To begin, choose a point on the terminal side of each angle, as shown in Figure 4.29. For each of the four points, $r = 1$ and you have the following.

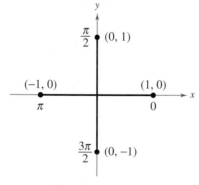

Figure 4.29

$$\cos 0 = \frac{x}{r} = \frac{1}{1} = 1 \qquad \tan 0 = \frac{y}{x} = \frac{0}{1} = 0 \qquad (x, y) = (1, 0)$$

$$\cos \frac{\pi}{2} = \frac{x}{r} = \frac{0}{1} = 0 \qquad \tan \frac{\pi}{2} = \frac{y}{x} = \frac{1}{0} \implies \text{undefined} \qquad (x, y) = (0, 1)$$

$$\cos \pi = \frac{x}{r} = \frac{-1}{1} = -1 \qquad \tan \pi = \frac{y}{x} = \frac{0}{-1} = 0 \qquad (x, y) = (-1, 0)$$

$$\cos \frac{3\pi}{2} = \frac{x}{r} = \frac{0}{1} = 0 \qquad \tan \frac{3\pi}{2} = \frac{y}{x} = \frac{-1}{0} \implies \text{undefined} \qquad (x, y) = (0, -1)$$

✓ *Checkpoint* *Audio-video solution in English & Spanish at LarsonPrecalculus.com.*

Evaluate the sine and cotangent functions at the quadrant angle $\frac{3\pi}{2}$.

Reference Angles

The values of the trigonometric functions of angles greater than 90° (or less than 0°) can be determined from their values at corresponding acute angles called **reference angles.**

> ### Definition of Reference Angle
>
> Let θ be an angle in standard position. Its **reference angle** is the acute angle θ' formed by the terminal side of θ and the horizontal axis.

The reference angles for θ in Quadrants II, III, and IV are shown below.

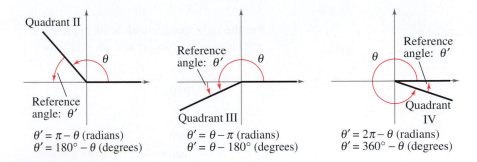

$$\theta' = \pi - \theta \text{ (radians)}$$
$$\theta' = 180° - \theta \text{ (degrees)}$$

$$\theta' = \theta - \pi \text{ (radians)}$$
$$\theta' = \theta - 180° \text{ (degrees)}$$

$$\theta' = 2\pi - \theta \text{ (radians)}$$
$$\theta' = 360° - \theta \text{ (degrees)}$$

EXAMPLE 4   **Finding Reference Angles**

Find the reference angle θ'.

a. $\theta = 300°$ **b.** $\theta = 2.3$ **c.** $\theta = -135°$

Solution

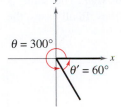

Figure 4.30

a. Because 300° lies in Quadrant IV, the angle it makes with the x-axis is

$$\theta' = 360° - 300°$$
$$= 60°. \qquad \text{Degrees}$$

Figure 4.30 shows the angle $\theta = 300°$ and its reference angle $\theta' = 60°$.

b. Because 2.3 lies between $\pi/2 \approx 1.5708$ and $\pi \approx 3.1416$, it follows that it is in Quadrant II and its reference angle is

$$\theta' = \pi - 2.3$$
$$\approx 0.8416. \qquad \text{Radians}$$

Figure 4.31 shows the angle $\theta = 2.3$ and its reference angle $\theta' = \pi - 2.3$.

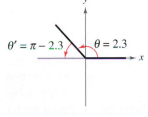

Figure 4.31

c. First, determine that $-135°$ is coterminal with 225°, which lies in Quadrant III. So, the reference angle is

$$\theta' = 225° - 180°$$
$$= 45°. \qquad \text{Degrees}$$

Figure 4.32 shows the angle $\theta = -135°$ and its reference angle $\theta' = 45°$.

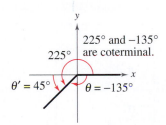

Figure 4.32

✓ **Checkpoint**))) *Audio-video solution in English & Spanish at LarsonPrecalculus.com.*

Find the reference angle θ'.

a. 213°   **b.** $\dfrac{14\pi}{9}$ **c.** $\dfrac{4\pi}{5}$

Trigonometric Functions of Real Numbers

To see how a reference angle is used to evaluate a trigonometric function, consider the point (x, y) on the terminal side of θ, as shown at the right. By definition, you know that

$$\sin \theta = \frac{y}{r}$$

and

$$\tan \theta = \frac{y}{x}.$$

For the right triangle with acute angle θ' and sides of lengths $|x|$ and $|y|$, you have

$$\sin \theta' = \frac{\text{opp}}{\text{hyp}} = \frac{|y|}{r}$$

and

$$\tan \theta' = \frac{\text{opp}}{\text{adj}} = \frac{|y|}{|x|}.$$

$$\text{opp} = |y|, \text{adj} = |x|$$

So, it follows that $\sin \theta$ and $\sin \theta'$ are equal, *except possibly in sign*. The same is true for $\tan \theta$ and $\tan \theta'$ *and* for the other four trigonometric functions. In all cases, the quadrant in which θ lies determines the sign of the function value.

Evaluating Trigonometric Functions of Any Angle

To find the value of a trigonometric function of any angle θ:

1. Determine the function value of the associated reference angle θ'.

2. Depending on the quadrant in which θ lies, affix the appropriate sign to the function value.

> • • **REMARK** Learning the table of values at the right is worth the effort because doing so will increase both your efficiency and your confidence. Here is a pattern for the sine function that may help you remember the values.
>
> | θ | 0° | 30° | 45° | 60° | 90° |
> |---|---|---|---|---|---|
> | $\sin \theta$ | $\frac{\sqrt{0}}{2}$ | $\frac{\sqrt{1}}{2}$ | $\frac{\sqrt{2}}{2}$ | $\frac{\sqrt{3}}{2}$ | $\frac{\sqrt{4}}{2}$ |
>
> Reverse the order to get cosine values of the same angles.

By using reference angles and the special angles discussed in the preceding section, you can greatly extend the scope of *exact* trigonometric values. For instance, knowing the function values of 30° means that you know the function values of all angles for which 30° is a reference angle. For convenience, the table below shows the exact values of the sine, cosine, and tangent functions of special angles and quadrant angles.

Trigonometric Values of Common Angles

θ (degrees)	0°	30°	45°	60°	90°	180°	270°
θ (radians)	0	$\frac{\pi}{6}$	$\frac{\pi}{4}$	$\frac{\pi}{3}$	$\frac{\pi}{2}$	π	$\frac{3\pi}{2}$
$\sin \theta$	0	$\frac{1}{2}$	$\frac{\sqrt{2}}{2}$	$\frac{\sqrt{3}}{2}$	1	0	-1
$\cos \theta$	1	$\frac{\sqrt{3}}{2}$	$\frac{\sqrt{2}}{2}$	$\frac{1}{2}$	0	-1	0
$\tan \theta$	0	$\frac{\sqrt{3}}{3}$	1	$\sqrt{3}$	Undef.	0	Undef.

EXAMPLE 5   **Using Reference Angles**

Evaluate each trigonometric function.

a. $\cos \dfrac{4\pi}{3}$ **b.** $\tan(-210°)$ **c.** $\csc \dfrac{11\pi}{4}$

Solution

a. Because $\theta = 4\pi/3$ lies in Quadrant III, the reference angle is

$$\theta' = \frac{4\pi}{3} - \pi = \frac{\pi}{3}$$

as shown in Figure 4.33. Moreover, the cosine is negative in Quadrant III, so

$$\cos \frac{4\pi}{3} = (-) \cos \frac{\pi}{3}$$

$$= -\frac{1}{2}.$$

b. Because $-210° + 360° = 150°$, it follows that $-210°$ is coterminal with the second-quadrant angle $150°$. So, the reference angle is $\theta' = 180° - 150° = 30°$, as shown in Figure 4.34. Finally, because the tangent is negative in Quadrant II, you have

$$\tan(-210°) = (-) \tan 30°$$

$$= -\frac{\sqrt{3}}{3}.$$

c. Because $(11\pi/4) - 2\pi = 3\pi/4$, it follows that $11\pi/4$ is coterminal with the second-quadrant angle $3\pi/4$. So, the reference angle is $\theta' = \pi - (3\pi/4) = \pi/4$, as shown in Figure 4.35. Because the cosecant is positive in Quadrant II, you have

$$\csc \frac{11\pi}{4} = (+) \csc \frac{\pi}{4}$$

$$= \frac{1}{\sin(\pi/4)}$$

$$= \sqrt{2}.$$

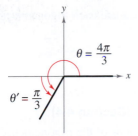

Figure 4.33

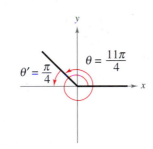

Figure 4.34

Figure 4.35

✓ **Checkpoint** ◀))) *Audio-video solution in English & Spanish at LarsonPrecalculus.com.*

Evaluate each trigonometric function.

a. $\sin \dfrac{7\pi}{4}$ **b.** $\cos(-120°)$ **c.** $\tan \dfrac{11\pi}{6}$

EXAMPLE 6 **Using Trigonometric Identities**

Let θ be an angle in Quadrant II such that $\sin \theta = \frac{1}{3}$. Find (a) $\cos \theta$ and (b) $\tan \theta$ by using trigonometric identities.

Solution

a. Using the Pythagorean identity $\sin^2 \theta + \cos^2 \theta = 1$, you obtain

$$\left(\tfrac{1}{3}\right)^2 + \cos^2 \theta = 1 \quad \Longrightarrow \quad \cos^2 \theta = 1 - \tfrac{1}{9} = \tfrac{8}{9}.$$

Because $\cos \theta < 0$ in Quadrant II, use the negative root to obtain

$$\cos \theta = -\frac{\sqrt{8}}{\sqrt{9}} = -\frac{2\sqrt{2}}{3}.$$

b. Using the trigonometric identity $\tan \theta = \dfrac{\sin \theta}{\cos \theta}$, you obtain

$$\tan \theta = \frac{1/3}{-2\sqrt{2}/3} = -\frac{1}{2\sqrt{2}} = -\frac{\sqrt{2}}{4}.$$

✓ ***Checkpoint***  *Audio-video solution in English & Spanish at LarsonPrecalculus.com.*

Let θ be an angle in Quadrant III such that $\sin \theta = -\frac{4}{5}$. Find (a) $\cos \theta$ and (b) $\tan \theta$ by using trigonometric identities.

EXAMPLE 7 **Using a Calculator**

Use a calculator to evaluate each trigonometric function.

a. $\cot 410°$ **b.** $\sin(-7)$ **c.** $\sec \dfrac{\pi}{9}$

Solution

Function	Mode	Calculator Keystrokes	Display
a. $\cot 410°$	Degree	(TAN (410)) x^{-1} ENTER	0.8390996
b. $\sin(-7)$	Radian	SIN ((−) 7) ENTER	−0.6569866
c. $\sec(\pi/9)$	Radian	(COS (π ÷ 9)) x^{-1} ENTER	1.0641778

✓ ***Checkpoint***  *Audio-video solution in English & Spanish at LarsonPrecalculus.com.*

Use a calculator to evaluate each trigonometric function.

a. $\tan 119°$ **b.** $\csc 5$ **c.** $\cos \dfrac{\pi}{5}$

Summarize **(Section 4.4)**

1. State the definitions of the trigonometric functions of any angle *(page 207)*. For examples of evaluating trigonometric functions, see Examples 1–3.

2. Explain how to find a reference angle *(page 209)*. For an example of finding reference angles, see Example 4.

3. Explain how to evaluate a trigonometric function of a real number *(page 210)*. For examples of evaluating trigonometric functions of real numbers, see Examples 5–7.

4.5   Graphs of Sine and Cosine Functions

You can use sine and cosine functions in scientific calculations. For instance, a trigonometric function can be used to model the airflow of your respiratory cycle.

■ Sketch the graphs of basic sine and cosine functions.
■ Use amplitude and period to help sketch the graphs of sine and cosine functions.
■ Sketch translations of the graphs of sine and cosine functions.
■ Use sine and cosine functions to model real-life data.

Basic Sine and Cosine Curves

In this section, you will study techniques for sketching the graphs of the sine and cosine functions. The graph of the sine function is a **sine curve.** In Figure 4.36, the black portion of the graph represents one period of the function and is called **one cycle** of the sine curve. The gray portion of the graph indicates that the basic sine curve repeats indefinitely to the left and right. The graph of the cosine function is shown in Figure 4.37.

Recall from Section 4.2 that the domain of the sine and cosine functions is the set of all real numbers. Moreover, the range of each function is the interval $[-1, 1]$, and each function has a period of 2π. Do you see how this information is consistent with the basic graphs shown in Figures 4.36 and 4.37?

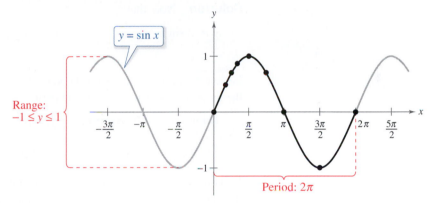

Figure 4.36

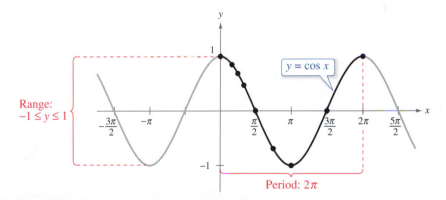

Figure 4.37

Note in Figures 4.36 and 4.37 that the sine curve is symmetric with respect to the *origin*, whereas the cosine curve is symmetric with respect to the *y-axis*. These properties of symmetry follow from the fact that the sine function is odd and the cosine function is even.

To sketch the graphs of the basic sine and cosine functions by hand, it helps to note five **key points** in one period of each graph: the *intercepts, maximum points,* and *minimum points* (see below).

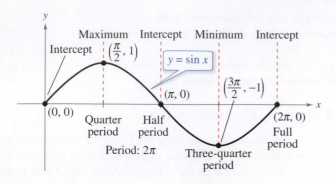
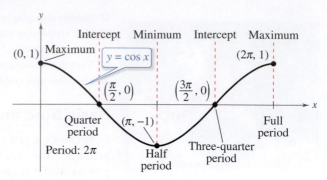

EXAMPLE 1   **Using Key Points to Sketch a Sine Curve**

Sketch the graph of

$$y = 2 \sin x$$

on the interval $[-\pi, 4\pi]$.

Solution   Note that

$$y = 2 \sin x$$
$$= 2(\sin x)$$

indicates that the y-values for the key points will have twice the magnitude of those on the graph of $y = \sin x$. Divide the period 2π into four equal parts to get the key points

Intercept	**Maximum**	**Intercept**	**Minimum**	**Intercept**
$(0, 0),$	$\left(\dfrac{\pi}{2}, 2\right),$	$(\pi, 0),$	$\left(\dfrac{3\pi}{2}, -2\right),$ and	$(2\pi, 0).$

By connecting these key points with a smooth curve and extending the curve in both directions over the interval $[-\pi, 4\pi]$, you obtain the graph shown below.

▷ **TECHNOLOGY** When using a graphing utility to graph trigonometric functions, pay special attention to the viewing window you use. For instance, try graphing $y = [\sin(10x)]/10$ in the standard viewing window in *radian* mode. What do you observe? Use the *zoom* feature to find a viewing window that displays a good view of the graph.

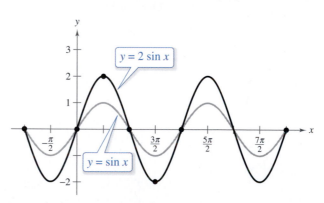

✓ *Checkpoint* ◀)) *Audio-video solution in English & Spanish at LarsonPrecalculus.com.*

Sketch the graph of

$$y = 2 \cos x$$

on the interval $\left[-\dfrac{\pi}{2}, \dfrac{9\pi}{2}\right]$.

Amplitude and Period

In the rest of this section, you will study the graphic effect of each of the constants a, b, c, and d in equations of the forms

$$y = d + a \sin(bx - c)$$

and

$$y = d + a \cos(bx - c).$$

A quick review of the transformations you studied in Section 1.7 should help in this investigation.

The constant factor a in $y = a \sin x$ acts as a *scaling factor*—a *vertical stretch* or *vertical shrink* of the basic sine curve. When $|a| > 1$, the basic sine curve is stretched, and when $|a| < 1$, the basic sine curve is shrunk. The result is that the graph of $y = a \sin x$ ranges between $-a$ and a instead of between -1 and 1. The absolute value of a is the **amplitude** of the function $y = a \sin x$. The range of the function $y = a \sin x$ for $a > 0$ is $-a \le y \le a$.

Definition of Amplitude of Sine and Cosine Curves

The **amplitude** of $y = a \sin x$ and $y = a \cos x$ represents half the distance between the maximum and minimum values of the function and is given by

$$\text{Amplitude} = |a|.$$

EXAMPLE 2   **Scaling: Vertical Shrinking and Stretching**

In the same coordinate plane, sketch the graph of each function.

a. $y = \frac{1}{2} \cos x$

b. $y = 3 \cos x$

Solution

a. Because the amplitude of $y = \frac{1}{2} \cos x$ is $\frac{1}{2}$, the maximum value is $\frac{1}{2}$ and the minimum value is $-\frac{1}{2}$. Divide one cycle, $0 \le x \le 2\pi$, into four equal parts to get the key points

Maximum	Intercept	Minimum	Intercept	Maximum
$\left(0, \frac{1}{2}\right),$	$\left(\frac{\pi}{2}, 0\right),$	$\left(\pi, -\frac{1}{2}\right),$	$\left(\frac{3\pi}{2}, 0\right),$ and	$\left(2\pi, \frac{1}{2}\right).$

b. A similar analysis shows that the amplitude of $y = 3 \cos x$ is 3, and the key points are

Maximum	Intercept	Minimum	Intercept	Maximum
$(0, 3),$	$\left(\frac{\pi}{2}, 0\right),$	$(\pi, -3),$	$\left(\frac{3\pi}{2}, 0\right),$ and	$(2\pi, 3).$

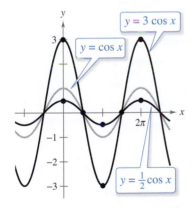

Figure 4.38

The graphs of these two functions are shown in Figure 4.38. Notice that the graph of $y = \frac{1}{2} \cos x$ is a vertical *shrink* of the graph of $y = \cos x$ and the graph of $y = 3 \cos x$ is a vertical *stretch* of the graph of $y = \cos x$.

✓ **Checkpoint** 🔊))) *Audio-video solution in English & Spanish at LarsonPrecalculus.com.*

In the same coordinate plane, sketch the graph of each function.

a. $y = \frac{1}{3} \sin x$

b. $y = 3 \sin x$

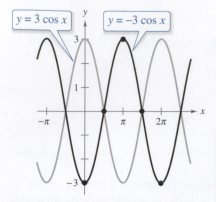

Figure 4.39

You know from Section 1.7 that the graph of $y = -f(x)$ is a **reflection** in the x-axis of the graph of $y = f(x)$. For instance, the graph of $y = -3 \cos x$ is a reflection of the graph of $y = 3 \cos x$, as shown in Figure 4.39.

Because $y = a \sin x$ completes one cycle from $x = 0$ to $x = 2\pi$, it follows that $y = a \sin bx$ completes one cycle from $x = 0$ to $x = 2\pi/b$, where b is a positive real number.

Period of Sine and Cosine Functions

Let b be a positive real number. The **period** of $y = a \sin bx$ and $y = a \cos bx$ is given by

$$\text{Period} = \frac{2\pi}{b}.$$

Note that when $0 < b < 1$, the period of $y = a \sin bx$ is greater than 2π and represents a *horizontal stretching* of the graph of $y = a \sin x$. Similarly, when $b > 1$, the period of $y = a \sin bx$ is less than 2π and represents a *horizontal shrinking* of the graph of $y = a \sin x$. When b is negative, the identities $\sin(-x) = -\sin x$ and $\cos(-x) = \cos x$ are used to rewrite the function.

EXAMPLE 3 **Scaling: Horizontal Stretching**

Sketch the graph of

$$y = \sin \frac{x}{2}.$$

Solution The amplitude is 1. Moreover, because $b = \frac{1}{2}$, the period is

$$\frac{2\pi}{b} = \frac{2\pi}{\frac{1}{2}} = 4\pi. \qquad \text{\textcolor{red}{Substitute for } } b.$$

> **REMARK** In general, to divide a period-interval into four equal parts, successively add "period/4," starting with the left endpoint of the interval. For instance, for the period-interval $[-\pi/6, \pi/2]$ of length $2\pi/3$, you would successively add
>
> $$\frac{2\pi/3}{4} = \frac{\pi}{6}$$
>
> to get $-\pi/6$, 0, $\pi/6$, $\pi/3$, and $\pi/2$ as the x-values for the key points on the graph.

Now, divide the period-interval $[0, 4\pi]$ into four equal parts using the values π, 2π, and 3π to obtain the key points

Intercept	**Maximum**	**Intercept**	**Minimum**	**Intercept**
$(0, 0)$,	$(\pi, 1)$,	$(2\pi, 0)$,	$(3\pi, -1)$, and	$(4\pi, 0)$.

The graph is shown below.

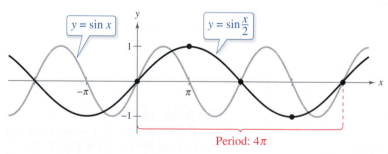

✓ **Checkpoint** ◀))) *Audio-video solution in English & Spanish at LarsonPrecalculus.com.*

Sketch the graph of

$$y = \cos \frac{x}{3}.$$

Translations of Sine and Cosine Curves

▷ **ALGEBRA HELP**   You
can review the techniques
for shifting, reflecting,
and stretching graphs in
Section 1.7.

The constant c in the general equations

$$y = a \sin(bx - c) \quad \text{and} \quad y = a \cos(bx - c)$$

creates *horizontal translations* (shifts) of the basic sine and cosine curves. Comparing $y = a \sin bx$ with $y = a \sin(bx - c)$, you find that the graph of $y = a \sin(bx - c)$ completes one cycle from $bx - c = 0$ to $bx - c = 2\pi$. By solving for x, you can find the interval for one cycle to be

Left endpoint   Right endpoint

$$\frac{c}{b} \le x \le \frac{c}{b} + \frac{2\pi}{b}.$$

Period

This implies that the period of $y = a \sin(bx - c)$ is $2\pi/b$, and the graph of $y = a \sin bx$ is shifted by an amount c/b. The number c/b is the **phase shift.**

Graphs of Sine and Cosine Functions

The graphs of $y = a \sin(bx - c)$ and $y = a \cos(bx - c)$ have the following characteristics. (Assume $b > 0$.)

$$\text{Amplitude} = |a| \qquad \text{Period} = \frac{2\pi}{b}$$

The left and right endpoints of a one-cycle interval can be determined by solving the equations $bx - c = 0$ and $bx - c = 2\pi$.

EXAMPLE 4   **Horizontal Translation**

Analyze the graph of $y = \dfrac{1}{2} \sin\left(x - \dfrac{\pi}{3}\right)$.

Algebraic Solution

The amplitude is $\frac{1}{2}$ and the period is 2π. By solving the equations

$$x - \frac{\pi}{3} = 0 \quad \Longrightarrow \quad x = \frac{\pi}{3}$$

and

$$x - \frac{\pi}{3} = 2\pi \quad \Longrightarrow \quad x = \frac{7\pi}{3}$$

you see that the interval $[\pi/3, 7\pi/3]$ corresponds to one cycle of the graph. Dividing this interval into four equal parts produces the key points

Intercept	Maximum	Intercept	Minimum		Intercept
$\left(\dfrac{\pi}{3}, 0\right),$	$\left(\dfrac{5\pi}{6}, \dfrac{1}{2}\right),$	$\left(\dfrac{4\pi}{3}, 0\right),$	$\left(\dfrac{11\pi}{6}, -\dfrac{1}{2}\right),$	and	$\left(\dfrac{7\pi}{3}, 0\right).$

Graphical Solution

Use a graphing utility set in *radian* mode to graph $y = (1/2)\sin(x - \pi/3)$, as shown below. Use the *minimum, maximum,* and *zero* or *root* features of the graphing utility to approximate the key points $(1.05, 0)$, $(2.62, 0.5)$, $(4.19, 0)$, $(5.76, -0.5)$, and $(7.33, 0)$.

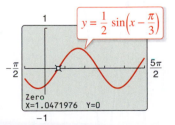

$y = \dfrac{1}{2}\sin\left(x - \dfrac{\pi}{3}\right)$

Zero
X=1.0471976   Y=0

✓ *Checkpoint* ◄))) *Audio-video solution in English & Spanish at LarsonPrecalculus.com.*

Analyze the graph of $y = 2 \cos\left(x - \dfrac{\pi}{2}\right)$.

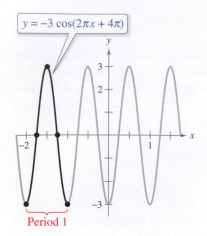

$y = -3\cos(2\pi x + 4\pi)$

Period 1

Figure 4.40

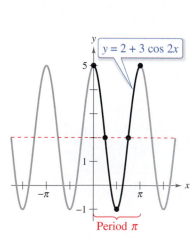

$y = 2 + 3\cos 2x$

Period π

Figure 4.41

EXAMPLE 5 **Horizontal Translation**

Sketch the graph of

$$y = -3\cos(2\pi x + 4\pi).$$

Solution The amplitude is 3 and the period is $2\pi/2\pi = 1$. By solving the equations

$$2\pi x + 4\pi = 0$$
$$2\pi x = -4\pi$$
$$x = -2$$

and

$$2\pi x + 4\pi = 2\pi$$
$$2\pi x = -2\pi$$
$$x = -1$$

you see that the interval $[-2, -1]$ corresponds to one cycle of the graph. Dividing this interval into four equal parts produces the key points

Minimum	Intercept	Maximum	Intercept	Minimum
$(-2, -3)$,	$\left(-\frac{7}{4}, 0\right)$,	$\left(-\frac{3}{2}, 3\right)$,	$\left(-\frac{5}{4}, 0\right)$, and	$(-1, -3)$.

The graph is shown in Figure 4.40.

✓ *Checkpoint* 🔊))) *Audio-video solution in English & Spanish at LarsonPrecalculus.com.*

Sketch the graph of

$$y = -\frac{1}{2}\sin(\pi x + \pi).$$

The final type of transformation is the *vertical translation* caused by the constant d in the equations

$$y = d + a\sin(bx - c) \quad \text{and} \quad y = d + a\cos(bx - c).$$

The shift is d units up for $d > 0$ and d units down for $d < 0$. In other words, the graph oscillates about the horizontal line $y = d$ instead of about the x-axis.

EXAMPLE 6 **Vertical Translation**

Sketch the graph of

$$y = 2 + 3\cos 2x.$$

Solution The amplitude is 3 and the period is π. The key points over the interval $[0, \pi]$ are

$$(0, 5), \quad \left(\frac{\pi}{4}, 2\right), \quad \left(\frac{\pi}{2}, -1\right), \quad \left(\frac{3\pi}{4}, 2\right), \quad \text{and} \quad (\pi, 5).$$

The graph is shown in Figure 4.41. Compared with the graph of $f(x) = 3\cos 2x$, the graph of $y = 2 + 3\cos 2x$ is shifted up two units.

✓ *Checkpoint* 🔊))) *Audio-video solution in English & Spanish at LarsonPrecalculus.com.*

Sketch the graph of

$$y = 2\cos x - 5.$$

Mathematical Modeling

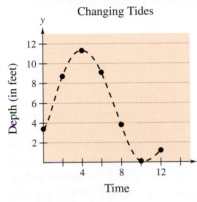

DATA	Time, t	Depth, y
	0	3.4
	2	8.7
	4	11.3
	6	9.1
	8	3.8
	10	0.1
	12	1.2

Spreadsheet at
LarsonPrecalculus.com

EXAMPLE 7   Finding a Trigonometric Model

The table shows the depths (in feet) of the water at the end of a dock at various times during the morning, where $t = 0$ corresponds to midnight.

a. Use a trigonometric function to model the data.

b. Find the depths at 9 A.M. and 3 P.M.

c. A boat needs at least 10 feet of water to moor at the dock. During what times in the afternoon can it safely dock?

Solution

a. Begin by graphing the data, as shown in Figure 4.42. You can use either a sine or cosine model. Suppose you use a cosine model of the form $y = a \cos(bt - c) + d$. The difference between the maximum value and minimum value is twice the amplitude of the function. So, the amplitude is

$$a = \tfrac{1}{2}[(\text{maximum depth}) - (\text{minimum depth})] = \tfrac{1}{2}(11.3 - 0.1) = 5.6.$$

The cosine function completes one half of a cycle between the times at which the maximum and minimum depths occur. So, the period p is

$$p = 2[(\text{time of min. depth}) - (\text{time of max. depth})] = 2(10 - 4) = 12$$

which implies that $b = 2\pi/p \approx 0.524$. Because high tide occurs 4 hours after midnight, consider the left endpoint to be $c/b = 4$, so $c \approx 2.094$. Moreover, because the average depth is $\tfrac{1}{2}(11.3 + 0.1) = 5.7$, it follows that $d = 5.7$. So, you can model the depth with the function $y = 5.6 \cos(0.524t - 2.094) + 5.7$.

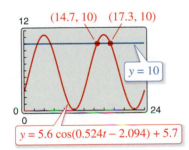

Changing Tides

Figure 4.42

b. The depths at 9 A.M. and 3 P.M. are as follows.

$$y = 5.6 \cos(0.524 \cdot 9 - 2.094) + 5.7 \approx 0.84 \text{ foot} \qquad \text{9 A.M.}$$

$$y = 5.6 \cos(0.524 \cdot 15 - 2.094) + 5.7 \approx 10.57 \text{ feet} \qquad \text{3 P.M.}$$

c. Using a graphing utility, graph the model with the line $y = 10$. Using the *intersect* feature, you can determine that the depth is at least 10 feet between 2:42 P.M. ($t \approx 14.7$) and 5:18 P.M. ($t \approx 17.3$), as shown in Figure 4.43.

(14.7, 10)   (17.3, 10)

$y = 10$

$y = 5.6 \cos(0.524t - 2.094) + 5.7$

Figure 4.43

✓ *Checkpoint*  *Audio-video solution in English & Spanish at LarsonPrecalculus.com.*

Find a sine model for the data in Example 7.

Summarize   *(Section 4.5)*

1. Describe how to sketch the graphs of basic sine and cosine functions *(pages 213 and 214)*. For an example of sketching the graph of a sine function, see Example 1.

2. Describe how you can use amplitude and period to help sketch the graphs of sine and cosine functions *(pages 215 and 216)*. For examples of using amplitude and period to sketch graphs of sine and cosine functions, see Examples 2 and 3.

3. Describe how to sketch translations of the graphs of sine and cosine functions *(pages 217 and 218)*. For examples of translating the graphs of sine and cosine functions, see Examples 4–6.

4. Give an example of how to use sine and cosine functions to model real-life data *(page 219, Example 7)*.

4.6 Graphs of Other Trigonometric Functions

You can use graphs of trigonometric functions to model real-life situations, such as the distance from a television camera to a unit in a parade.

■ Sketch the graphs of tangent functions.
■ Sketch the graphs of cotangent functions.
■ Sketch the graphs of secant and cosecant functions.
■ Sketch the graphs of damped trigonometric functions.

Graph of the Tangent Function

Recall that the tangent function is odd. That is, $\tan(-x) = -\tan x$. Consequently, the graph of $y = \tan x$ is symmetric with respect to the origin. You also know from the identity $\tan x = \sin x / \cos x$ that the tangent is undefined for values at which $\cos x = 0$. Two such values are $x = \pm \pi/2 \approx \pm 1.5708$.

x	$-\dfrac{\pi}{2}$	-1.57	-1.5	$-\dfrac{\pi}{4}$	0	$\dfrac{\pi}{4}$	1.5	1.57	$\dfrac{\pi}{2}$
$\tan x$	Undef.	-1255.8	-14.1	-1	0	1	14.1	1255.8	Undef.

As indicated in the table, $\tan x$ increases without bound as x approaches $\pi/2$ from the left and decreases without bound as x approaches $-\pi/2$ from the right. So, the graph of $y = \tan x$ has *vertical asymptotes* at $x = \pi/2$ and $x = -\pi/2$, as shown below. Moreover, because the period of the tangent function is π, vertical asymptotes also occur at $x = \pi/2 + n\pi$, where n is an integer. The domain of the tangent function is the set of all real numbers other than $x = \pi/2 + n\pi$, and the range is the set of all real numbers.

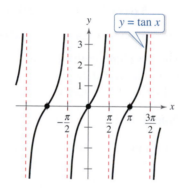

Period: π

Domain: all $x \neq \dfrac{\pi}{2} + n\pi$

Range: $(-\infty, \infty)$

Vertical asymptotes: $x = \dfrac{\pi}{2} + n\pi$

Symmetry: origin

▷ **ALGEBRA HELP**
• You can review odd and even functions in Section 1.5.
• You can review symmetry of a graph in Section 1.2.
• You can review trigonometric identities in Section 4.3.
• You can review asymptotes in Section 2.6.
• You can review domain and range of a function in Section 1.4.
• You can review intercepts of a graph in Section 1.2.

Sketching the graph of $y = a \tan(bx - c)$ is similar to sketching the graph of $y = a \sin(bx - c)$ in that you locate key points that identify the intercepts and asymptotes. Two consecutive vertical asymptotes can be found by solving the equations

$$bx - c = -\frac{\pi}{2} \quad \text{and} \quad bx - c = \frac{\pi}{2}.$$

The midpoint between two consecutive vertical asymptotes is an x-intercept of the graph. The period of the function $y = a \tan(bx - c)$ is the distance between two consecutive vertical asymptotes. The amplitude of a tangent function is not defined. After plotting the asymptotes and the x-intercept, plot a few additional points between the two asymptotes and sketch one cycle. Finally, sketch one or two additional cycles to the left and right.

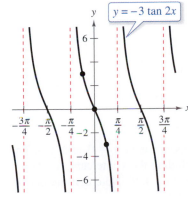

$y = \tan \dfrac{x}{2}$

Figure 4.44

Sketching the Graph of a Tangent Function

Sketch the graph of $y = \tan \dfrac{x}{2}$.

Solution

By solving the equations

$$\frac{x}{2} = -\frac{\pi}{2} \quad \text{and} \quad \frac{x}{2} = \frac{\pi}{2}$$

$$x = -\pi \qquad\qquad x = \pi$$

you can see that two consecutive vertical asymptotes occur at $x = -\pi$ and $x = \pi$. Between these two asymptotes, plot a few points, including the x-intercept, as shown in the table. Three cycles of the graph are shown in Figure 4.44.

x	$-\pi$	$-\dfrac{\pi}{2}$	0	$\dfrac{\pi}{2}$	π
$\tan \dfrac{x}{2}$	Undef.	-1	0	1	Undef.

✓ **Checkpoint** *Audio-video solution in English & Spanish at LarsonPrecalculus.com.*

Sketch the graph of $y = \tan \dfrac{x}{4}$.

Sketching the Graph of a Tangent Function

Sketch the graph of $y = -3 \tan 2x$.

Solution

By solving the equations

$$2x = -\frac{\pi}{2} \quad \text{and} \quad 2x = \frac{\pi}{2}$$

$$x = -\frac{\pi}{4} \qquad\qquad x = \frac{\pi}{4}$$

you can see that two consecutive vertical asymptotes occur at $x = -\pi/4$ and $x = \pi/4$. Between these two asymptotes, plot a few points, including the x-intercept, as shown in the table. Three cycles of the graph are shown in Figure 4.45.

$y = -3 \tan 2x$

Figure 4.45

x	$-\dfrac{\pi}{4}$	$-\dfrac{\pi}{8}$	0	$\dfrac{\pi}{8}$	$\dfrac{\pi}{4}$
$-3 \tan 2x$	Undef.	3	0	-3	Undef.

By comparing the graphs in Examples 1 and 2, you can see that the graph of $y = a \tan(bx - c)$ increases between consecutive vertical asymptotes when $a > 0$ and decreases between consecutive vertical asymptotes when $a < 0$. In other words, the graph for $a < 0$ is a reflection in the x-axis of the graph for $a > 0$.

✓ **Checkpoint** *Audio-video solution in English & Spanish at LarsonPrecalculus.com.*

Sketch the graph of $y = \tan 2x$.

Graph of the Cotangent Function

The graph of the cotangent function is similar to the graph of the tangent function. It also has a period of π. However, from the identity

$$y = \cot x = \frac{\cos x}{\sin x}$$

▷ **TECHNOLOGY** Some graphing utilities have difficulty graphing trigonometric functions that have vertical asymptotes. Your graphing utility may connect parts of the graphs of tangent, cotangent, secant, and cosecant functions that are not supposed to be connected. To eliminate this problem, change the mode of the graphing utility to *dot* mode.

you can see that the cotangent function has vertical asymptotes when $\sin x$ is zero, which occurs at $x = n\pi$, where n is an integer. The graph of the cotangent function is shown below. Note that two consecutive vertical asymptotes of the graph of $y = a \cot(bx - c)$ can be found by solving the equations $bx - c = 0$ and $bx - c = \pi$.

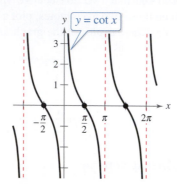

Period: π
Domain: all $x \neq n\pi$
Range: $(-\infty, \infty)$
Vertical asymptotes: $x = n\pi$
Symmetry: origin

EXAMPLE 3   **Sketching the Graph of a Cotangent Function**

Sketch the graph of

$$y = 2 \cot \frac{x}{3}.$$

Solution

By solving the equations

$$\frac{x}{3} = 0 \quad \text{and} \quad \frac{x}{3} = \pi$$

$$x = 0 \qquad\qquad x = 3\pi$$

you can see that two consecutive vertical asymptotes occur at $x = 0$ and $x = 3\pi$. Between these two asymptotes, plot a few points, including the x-intercept, as shown in the table. Three cycles of the graph are shown in Figure 4.46. Note that the period is 3π, the distance between consecutive asymptotes.

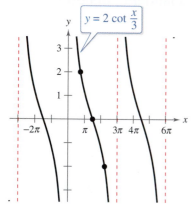

Figure 4.46

x	0	$\dfrac{3\pi}{4}$	$\dfrac{3\pi}{2}$	$\dfrac{9\pi}{4}$	3π
$2 \cot \dfrac{x}{3}$	Undef.	2	0	-2	Undef.

 **Checkpoint**  *Audio-video solution in English & Spanish at LarsonPrecalculus.com.*

Sketch the graph of

$$y = \cot \frac{x}{4}.$$

Graphs of the Reciprocal Functions

You can obtain the graphs of the two remaining trigonometric functions from the graphs of the sine and cosine functions using the reciprocal identities

$$\csc x = \frac{1}{\sin x} \quad \text{and} \quad \sec x = \frac{1}{\cos x}.$$

For instance, at a given value of x, the y-coordinate of $\sec x$ is the reciprocal of the y-coordinate of $\cos x$. Of course, when $\cos x = 0$, the reciprocal does not exist. Near such values of x, the behavior of the secant function is similar to that of the tangent function. In other words, the graphs of

$$\tan x = \frac{\sin x}{\cos x} \quad \text{and} \quad \sec x = \frac{1}{\cos x}$$

have vertical asymptotes where $\cos x = 0$—that is, at $x = \pi/2 + n\pi$, where n is an integer. Similarly,

$$\cot x = \frac{\cos x}{\sin x} \quad \text{and} \quad \csc x = \frac{1}{\sin x}$$

have vertical asymptotes where $\sin x = 0$—that is, at $x = n\pi$, where n is an integer.

To sketch the graph of a secant or cosecant function, you should first make a sketch of its reciprocal function. For instance, to sketch the graph of $y = \csc x$, first sketch the graph of $y = \sin x$. Then take reciprocals of the y-coordinates to obtain points on the graph of $y = \csc x$. You can use this procedure to obtain the graphs shown below.

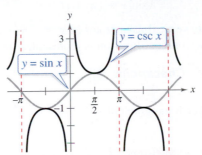

Period: 2π
Domain: all $x \neq n\pi$
Range: $(-\infty, -1] \cup [1, \infty)$
Vertical asymptotes: $x = n\pi$
Symmetry: origin

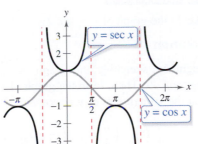

Period: 2π
Domain: all $x \neq \dfrac{\pi}{2} + n\pi$
Range: $(-\infty, -1] \cup [1, \infty)$
Vertical asymptotes: $x = \dfrac{\pi}{2} + n\pi$
Symmetry: y-axis

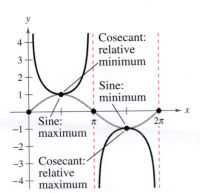

Figure 4.47

In comparing the graphs of the cosecant and secant functions with those of the sine and cosine functions, respectively, note that the "hills" and "valleys" are interchanged. For instance, a hill (or maximum point) on the sine curve corresponds to a valley (a relative minimum) on the cosecant curve, and a valley (or minimum point) on the sine curve corresponds to a hill (a relative maximum) on the cosecant curve, as shown in Figure 4.47. Additionally, x-intercepts of the sine and cosine functions become vertical asymptotes of the cosecant and secant functions, respectively (see Figure 4.47).

EXAMPLE 4   **Sketching the Graph of a Cosecant Function**

Sketch the graph of $y = 2 \csc\left(x + \dfrac{\pi}{4}\right)$.

Solution

Begin by sketching the graph of

$$y = 2 \sin\left(x + \frac{\pi}{4}\right).$$

For this function, the amplitude is 2 and the period is 2π. By solving the equations

$$x + \frac{\pi}{4} = 0 \qquad \text{and} \qquad x + \frac{\pi}{4} = 2\pi$$

$$x = -\frac{\pi}{4} \qquad\qquad x = \frac{7\pi}{4}$$

you can see that one cycle of the sine function corresponds to the interval from $x = -\pi/4$ to $x = 7\pi/4$. The graph of this sine function is represented by the gray curve in Figure 4.48. Because the sine function is zero at the midpoint and endpoints of this interval, the corresponding cosecant function

$$y = 2 \csc\left(x + \frac{\pi}{4}\right)$$

$$= 2\left(\frac{1}{\sin[x + (\pi/4)]}\right)$$

has vertical asymptotes at $x = -\pi/4$, $x = 3\pi/4$, $x = 7\pi/4$, and so on. The graph of the cosecant function is represented by the black curve in Figure 4.48.

 Checkpoint *Audio-video solution in English & Spanish at LarsonPrecalculus.com.*

Sketch the graph of $y = 2 \csc\left(x + \dfrac{\pi}{2}\right)$.

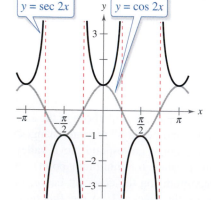

Figure 4.48

EXAMPLE 5   **Sketching the Graph of a Secant Function**

Sketch the graph of $y = \sec 2x$.

Solution

Begin by sketching the graph of $y = \cos 2x$, as indicated by the gray curve in Figure 4.49. Then, form the graph of $y = \sec 2x$ as the black curve in the figure. Note that the x-intercepts of $y = \cos 2x$

$$\left(-\frac{\pi}{4}, 0\right), \quad \left(\frac{\pi}{4}, 0\right), \quad \left(\frac{3\pi}{4}, 0\right), \ldots$$

correspond to the vertical asymptotes

$$x = -\frac{\pi}{4}, \quad x = \frac{\pi}{4}, \quad x = \frac{3\pi}{4}, \ldots$$

of the graph of $y = \sec 2x$. Moreover, notice that the period of $y = \cos 2x$ and $y = \sec 2x$ is π.

Figure 4.49

Checkpoint *Audio-video solution in English & Spanish at LarsonPrecalculus.com.*

Sketch the graph of $y = \sec\dfrac{x}{2}$.

Damped Trigonometric Graphs

You can graph a *product* of two functions using properties of the individual functions. For instance, consider the function

$$f(x) = x \sin x$$

as the product of the functions $y = x$ and $y = \sin x$. Using properties of absolute value and the fact that $|\sin x| \le 1$, you have

$$0 \le |x||\sin x| \le |x|.$$

Consequently,

$$-|x| \le x \sin x \le |x|$$

which means that the graph of $f(x) = x \sin x$ lies between the lines $y = -x$ and $y = x$. Furthermore, because

$$f(x) = x \sin x = \pm x \quad \text{at} \quad x = \frac{\pi}{2} + n\pi$$

and

$$f(x) = x \sin x = 0 \quad \text{at} \quad x = n\pi$$

where n is an integer, the graph of f touches the line $y = -x$ or the line $y = x$ at $x = \pi/2 + n\pi$ and has x-intercepts at $x = n\pi$. A sketch of f is shown at the right. In the function $f(x) = x \sin x$, the factor x is called the **damping factor.**

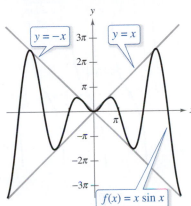

•• **REMARK** Do you see why the graph of $f(x) = x \sin x$ touches the lines $y = \pm x$ at $x = \pi/2 + n\pi$ and why the graph has x-intercepts at $x = n\pi$? Recall that the sine function is equal to 1 at $\dots, -3\pi/2, \pi/2, 5\pi/2, \dots$ $(x = \pi/2 + 2n\pi)$ and -1 at $\dots, -\pi/2, 3\pi/2, 7\pi/2, \dots$ $(x = -\pi/2 + 2n\pi)$ and is equal to 0 at $\dots, -\pi, 0, \pi, 2\pi, 3\pi, \dots$ $(x = n\pi)$.

EXAMPLE 6   Damped Sine Wave

Sketch the graph of $f(x) = e^{-x} \sin 3x$.

Solution

Consider $f(x)$ as the product of the two functions

$$y = e^{-x} \quad \text{and} \quad y = \sin 3x$$

each of which has the set of real numbers as its domain. For any real number x, you know that $e^{-x} > 0$ and $|\sin 3x| \le 1$. So,

$$e^{-x} |\sin 3x| \le e^{-x}$$

which means that

$$-e^{-x} \le e^{-x} \sin 3x \le e^{-x}.$$

Furthermore, because

$$f(x) = e^{-x} \sin 3x = \pm e^{-x} \quad \text{at} \quad x = \frac{\pi}{6} + \frac{n\pi}{3}$$

and

$$f(x) = e^{-x} \sin 3x = 0 \quad \text{at} \quad x = \frac{n\pi}{3}$$

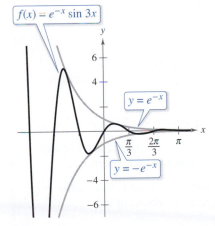

Figure 4.50

the graph of f touches the curve $y = -e^{-x}$ or the curve $y = e^{-x}$ at $x = \pi/6 + n\pi/3$ and has intercepts at $x = n\pi/3$. A sketch of f is shown in Figure 4.50.

✓ **Checkpoint** 🔊 *Audio-video solution in English & Spanish at LarsonPrecalculus.com.*

Sketch the graph of $f(x) = e^x \sin 4x$.

Below is a summary of the characteristics of the six basic trigonometric functions.

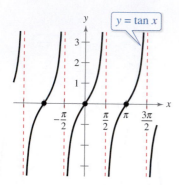

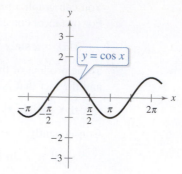

Domain: $(-\infty, \infty)$
Range: $[-1, 1]$
Period: 2π

Domain: $(-\infty, \infty)$
Range: $[-1, 1]$
Period: 2π

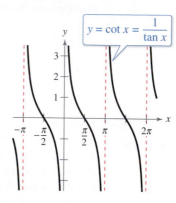

Domain: all $x \neq \dfrac{\pi}{2} + n\pi$
Range: $(-\infty, \infty)$
Period: π

Domain: all $x \neq n\pi$
Range: $(-\infty, \infty)$
Period: π

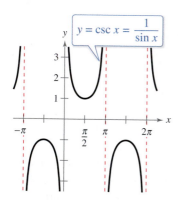

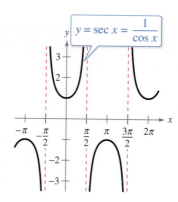

Domain: all $x \neq n\pi$
Range:
$(-\infty, -1] \cup [1, \infty)$
Period: 2π

Domain: all $x \neq \dfrac{\pi}{2} + n\pi$
Range:
$(-\infty, -1] \cup [1, \infty)$
Period: 2π

Summarize (Section 4.6)

1. Describe how to sketch the graph of $y = a \tan(bx - c)$ *(page 220)*. For examples of sketching the graphs of tangent functions, see Examples 1 and 2.

2. Describe how to sketch the graph of $y = a \cot(bx - c)$ *(page 222)*. For an example of sketching the graph of a cotangent function, see Example 3.

3. Describe how to sketch the graphs of $y = a \csc(bx - c)$ and $y = a \sec(bx - c)$ *(page 223)*. For examples of sketching the graphs of cosecant and secant functions, see Examples 4 and 5.

4. Describe how to sketch the graph of a damped trigonometric function *(page 225)*. For an example of sketching the graph of a damped trigonometric function, see Example 6.

4.7 Inverse Trigonometric Functions

■ Evaluate and graph the inverse sine function.
■ Evaluate and graph the other inverse trigonometric functions.
■ Evaluate the compositions of trigonometric functions.

Inverse Sine Function

Recall from Section 1.9 that for a function to have an inverse function, it must be one-to-one—that is, it must pass the Horizontal Line Test. From Figure 4.51, you can see that $y = \sin x$ does not pass the test because different values of x yield the same y-value.

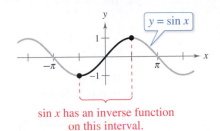

sin x has an inverse function
on this interval.

Figure 4.51

However, when you restrict the domain to the interval $-\pi/2 \leq x \leq \pi/2$ (corresponding to the black portion of the graph in Figure 4.51), the following properties hold.

1. On the interval $[-\pi/2, \pi/2]$, the function $y = \sin x$ is increasing.

2. On the interval $[-\pi/2, \pi/2]$, $y = \sin x$ takes on its full range of values, $-1 \leq \sin x \leq 1$.

3. On the interval $[-\pi/2, \pi/2]$, $y = \sin x$ is one-to-one.

So, on the restricted domain $-\pi/2 \leq x \leq \pi/2$, $y = \sin x$ has a unique inverse function called the **inverse sine function.** It is denoted by

$$y = \arcsin x \quad \text{or} \quad y = \sin^{-1} x.$$

The notation $\sin^{-1} x$ is consistent with the inverse function notation $f^{-1}(x)$. The arcsin x notation (read as "the arcsine of x") comes from the association of a central angle with its intercepted *arc length* on a unit circle. So, arcsin x means the angle (or arc) whose sine is x. Both notations, arcsin x and $\sin^{-1} x$, are commonly used in mathematics, so remember that $\sin^{-1} x$ denotes the *inverse* sine function rather than $1/\sin x$. The values of arcsin x lie in the interval

$$-\frac{\pi}{2} \leq \arcsin x \leq \frac{\pi}{2}.$$

The graph of $y = \arcsin x$ is shown in Example 2.

You can use inverse trigonometric functions to model and solve real-life problems, such as the angle of elevation from a television camera to a space shuttle launch.

••**REMARK** When evaluating the inverse sine function, it helps to remember the phrase "the arcsine of x is the angle (or number) whose sine is x."

Definition of Inverse Sine Function

The **inverse sine function** is defined by

$$y = \arcsin x \quad \text{if and only if} \quad \sin y = x$$

where $-1 \leq x \leq 1$ and $-\pi/2 \leq y \leq \pi/2$. The domain of $y = \arcsin x$ is $[-1, 1]$, and the range is $[-\pi/2, \pi/2]$.

▷

REMARK As with the trigonometric functions, much of the work with the inverse trigonometric functions can be done by *exact* calculations rather than by calculator approximations. Exact calculations help to increase your understanding of the inverse functions by relating them to the right triangle definitions of the trigonometric functions.

EXAMPLE 1 Evaluating the Inverse Sine Function

If possible, find the exact value.

a. $\arcsin\left(-\dfrac{1}{2}\right)$ **b.** $\sin^{-1}\dfrac{\sqrt{3}}{2}$ **c.** $\sin^{-1}2$

Solution

a. Because $\sin\left(-\dfrac{\pi}{6}\right) = -\dfrac{1}{2}$ and $-\dfrac{\pi}{6}$ lies in $\left[-\dfrac{\pi}{2}, \dfrac{\pi}{2}\right]$, it follows that

$$\arcsin\left(-\frac{1}{2}\right) = -\frac{\pi}{6}.$$ Angle whose sine is $-\frac{1}{2}$

b. Because $\sin\dfrac{\pi}{3} = \dfrac{\sqrt{3}}{2}$ and $\dfrac{\pi}{3}$ lies in $\left[-\dfrac{\pi}{2}, \dfrac{\pi}{2}\right]$, it follows that

$$\sin^{-1}\frac{\sqrt{3}}{2} = \frac{\pi}{3}.$$ Angle whose sine is $\sqrt{3}/2$

c. It is not possible to evaluate $y = \sin^{-1}x$ when $x = 2$ because there is no angle whose sine is 2. Remember that the domain of the inverse sine function is $[-1, 1]$.

✔ *Checkpoint* ◀))) Audio-video solution in English & Spanish at LarsonPrecalculus.com.

If possible, find the exact value.

a. $\arcsin 1$ **b.** $\sin^{-1}(-2)$

EXAMPLE 2 Graphing the Arcsine Function

Sketch a graph of

$$y = \arcsin x.$$

Solution

By definition, the equations $y = \arcsin x$ and $\sin y = x$ are equivalent for $-\pi/2 \le y \le \pi/2$. So, their graphs are the same. From the interval $[-\pi/2, \pi/2]$, you can assign values to y in the equation $\sin y = x$ to make a table of values. Then plot the points and connect them with a smooth curve.

y	$-\dfrac{\pi}{2}$	$-\dfrac{\pi}{4}$	$-\dfrac{\pi}{6}$	0	$\dfrac{\pi}{6}$	$\dfrac{\pi}{4}$	$\dfrac{\pi}{2}$
$x = \sin y$	-1	$-\dfrac{\sqrt{2}}{2}$	$-\dfrac{1}{2}$	0	$\dfrac{1}{2}$	$\dfrac{\sqrt{2}}{2}$	1

The resulting graph of $y = \arcsin x$ is shown in Figure 4.52. Note that it is the reflection (in the line $y = x$) of the black portion of the graph in Figure 4.51. Be sure you see that Figure 4.52 shows the *entire* graph of the inverse sine function. Remember that the domain of $y = \arcsin x$ is the closed interval $[-1, 1]$ and the range is the closed interval $[-\pi/2, \pi/2]$.

✔ *Checkpoint* ◀))) Audio-video solution in English & Spanish at LarsonPrecalculus.com.

Use a graphing utility to graph $f(x) = \sin x$, $g(x) = \arcsin x$, and $y = x$ in the same viewing window to verify geometrically that g is the inverse function of f. (Be sure to restrict the domain of f properly.) ■

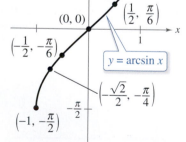

Figure 4.52

Other Inverse Trigonometric Functions

The cosine function is decreasing and one-to-one on the interval $0 \le x \le \pi$, as shown below.

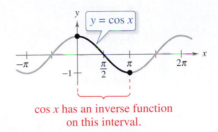

cos x has an inverse function
on this interval.

Consequently, on this interval the cosine function has an inverse function—the **inverse cosine function**—denoted by

$$y = \arccos x \quad \text{or} \quad y = \cos^{-1} x.$$

Similarly, you can define an **inverse tangent function** by restricting the domain of $y = \tan x$ to the interval $(-\pi/2, \pi/2)$. The following list summarizes the definitions of the three most common inverse trigonometric functions. The remaining three are defined in the exercises.

Definitions of the Inverse Trigonometric Functions

Function	Domain	Range
$y = \arcsin x$ if and only if $\sin y = x$	$-1 \le x \le 1$	$-\dfrac{\pi}{2} \le y \le \dfrac{\pi}{2}$
$y = \arccos x$ if and only if $\cos y = x$	$-1 \le x \le 1$	$0 \le y \le \pi$
$y = \arctan x$ if and only if $\tan y = x$	$-\infty < x < \infty$	$-\dfrac{\pi}{2} < y < \dfrac{\pi}{2}$

The graphs of these three inverse trigonometric functions are shown below.

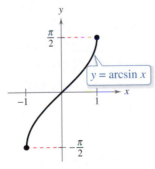

Domain: $[-1, 1]$

Range: $\left[-\dfrac{\pi}{2}, \dfrac{\pi}{2}\right]$

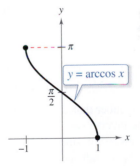

Domain: $[-1, 1]$
Range: $[0, \pi]$

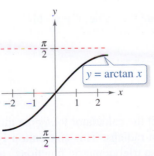

Domain: $(-\infty, \infty)$

Range: $\left(-\dfrac{\pi}{2}, \dfrac{\pi}{2}\right)$

EXAMPLE 3 **Evaluating Inverse Trigonometric Functions**

Find the exact value.

a. $\arccos \dfrac{\sqrt{2}}{2}$

b. $\arctan 0$

c. $\tan^{-1}(-1)$

Solution

a. Because $\cos(\pi/4) = \sqrt{2}/2$ and $\pi/4$ lies in $[0, \pi]$, it follows that

$$\arccos \frac{\sqrt{2}}{2} = \frac{\pi}{4}. \qquad \color{red}{\text{Angle whose cosine is } \sqrt{2}/2}$$

b. Because $\tan 0 = 0$ and 0 lies in $(-\pi/2, \pi/2)$, it follows that

$$\arctan 0 = 0. \qquad \color{red}{\text{Angle whose tangent is } 0}$$

c. Because $\tan(-\pi/4) = -1$ and $-\pi/4$ lies in $(-\pi/2, \pi/2)$, it follows that

$$\tan^{-1}(-1) = -\frac{\pi}{4}. \qquad \color{red}{\text{Angle whose tangent is } -1}$$

✓ **Checkpoint**))) *Audio-video solution in English & Spanish at LarsonPrecalculus.com.*

Find the exact value of $\cos^{-1}(-1)$.

EXAMPLE 4 **Calculators and Inverse Trigonometric Functions**

Use a calculator to approximate the value, if possible.

a. $\arctan(-8.45)$

b. $\sin^{-1} 0.2447$

c. $\arccos 2$

Solution

Function	Mode	Calculator Keystrokes
a. $\arctan(-8.45)$	Radian	(TAN⁻¹) (⃞ (−) 8.45 ⃞) (ENTER)

From the display, it follows that $\arctan(-8.45) \approx -1.453001$.

b. $\sin^{-1} 0.2447$	Radian	(SIN⁻¹) (⃞ 0.2447 ⃞) (ENTER)

From the display, it follows that $\sin^{-1} 0.2447 \approx 0.2472103$.

c. $\arccos 2$	Radian	(COS⁻¹) (⃞ 2 ⃞) (ENTER)

> ▷ **REMARK** Remember that the domain of the inverse sine function and the inverse cosine function is $[-1, 1]$, as indicated in Example 4(c).

In *radian* mode, the calculator should display an *error message* because the domain of the inverse cosine function is $[-1, 1]$.

✓ **Checkpoint**))) *Audio-video solution in English & Spanish at LarsonPrecalculus.com.*

Use a calculator to approximate the value, if possible.

a. $\arctan 4.84$

b. $\arcsin(-1.1)$

c. $\arccos(-0.349)$

In Example 4, had you set the calculator to *degree* mode, the displays would have been in degrees rather than in radians. This convention is peculiar to calculators. By definition, the values of inverse trigonometric functions are *always in radians*.

Compositions of Functions

▷ **ALGEBRA HELP** You can review the composition of functions in Section 1.8.

Recall from Section 1.9 that for all x in the domains of f and f^{-1}, inverse functions have the properties

$$f(f^{-1}(x)) = x \quad \text{and} \quad f^{-1}(f(x)) = x.$$

Inverse Properties of Trigonometric Functions

If $-1 \le x \le 1$ and $-\pi/2 \le y \le \pi/2$, then

$$\sin(\arcsin x) = x \quad \text{and} \quad \arcsin(\sin y) = y.$$

If $-1 \le x \le 1$ and $0 \le y \le \pi$, then

$$\cos(\arccos x) = x \quad \text{and} \quad \arccos(\cos y) = y.$$

If x is a real number and $-\pi/2 < y < \pi/2$, then

$$\tan(\arctan x) = x \quad \text{and} \quad \arctan(\tan y) = y.$$

Keep in mind that these inverse properties do not apply for arbitrary values of x and y. For instance,

$$\arcsin\left(\sin\frac{3\pi}{2}\right) = \arcsin(-1) = -\frac{\pi}{2} \ne \frac{3\pi}{2}.$$

In other words, the property $\arcsin(\sin y) = y$ is not valid for values of y outside the interval $[-\pi/2, \pi/2]$.

EXAMPLE 5   **Using Inverse Properties**

If possible, find the exact value.

a. $\tan[\arctan(-5)]$ **b.** $\arcsin\left(\sin\dfrac{5\pi}{3}\right)$ **c.** $\cos(\cos^{-1}\pi)$

Solution

a. Because -5 lies in the domain of the arctangent function, the inverse property applies, and you have

$$\tan[\arctan(-5)] = -5.$$

b. In this case, $5\pi/3$ does not lie in the range of the arcsine function, $-\pi/2 \le y \le \pi/2$. However, $5\pi/3$ is coterminal with

$$\frac{5\pi}{3} - 2\pi = -\frac{\pi}{3}$$

which does lie in the range of the arcsine function, and you have

$$\arcsin\left(\sin\frac{5\pi}{3}\right) = \arcsin\left[\sin\left(-\frac{\pi}{3}\right)\right] = -\frac{\pi}{3}.$$

c. The expression $\cos(\cos^{-1}\pi)$ is not defined because $\cos^{-1}\pi$ is not defined. Remember that the domain of the inverse cosine function is $[-1, 1]$.

✓ **Checkpoint**  *Audio-video solution in English & Spanish at LarsonPrecalculus.com.*

If possible, find the exact value.

a. $\tan[\tan^{-1}(-14)]$ **b.** $\sin^{-1}\left(\sin\dfrac{7\pi}{4}\right)$ **c.** $\cos(\arccos 0.54)$

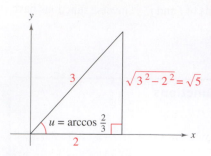

Angle whose cosine is $\frac{2}{3}$
Figure 4.53

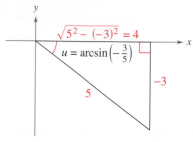

Angle whose sine is $-\frac{3}{5}$
Figure 4.54

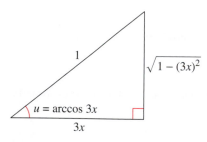

Angle whose cosine is $3x$
Figure 4.55

EXAMPLE 6 **Evaluating Compositions of Functions**

Find the exact value.

a. $\tan\left(\arccos \frac{2}{3}\right)$ **b.** $\cos\left[\arcsin\left(-\frac{3}{5}\right)\right]$

Solution

a. If you let $u = \arccos \frac{2}{3}$, then $\cos u = \frac{2}{3}$. Because the range of the inverse cosine function is the first and second quadrants and $\cos u$ is positive, u is a *first*-quadrant angle. You can sketch and label angle u, as shown in Figure 4.53. Consequently,

$$\tan\left(\arccos \frac{2}{3}\right) = \tan u = \frac{\text{opp}}{\text{adj}} = \frac{\sqrt{5}}{2}.$$

b. If you let $u = \arcsin\left(-\frac{3}{5}\right)$, then $\sin u = -\frac{3}{5}$. Because the range of the inverse sine function is the first and fourth quadrants and $\sin u$ is negative, u is a *fourth*-quadrant angle. You can sketch and label angle u, as shown in Figure 4.54. Consequently,

$$\cos\left[\arcsin\left(-\frac{3}{5}\right)\right] = \cos u = \frac{\text{adj}}{\text{hyp}} = \frac{4}{5}.$$

✓ **Checkpoint** Audio-video solution in English & Spanish at *LarsonPrecalculus.com.*

Find the exact value of $\cos\left[\arctan\left(-\frac{3}{4}\right)\right]$.

EXAMPLE 7 **Some Problems from Calculus**

Write each of the following as an algebraic expression in x.

a. $\sin(\arccos 3x), \quad 0 \leq x \leq \frac{1}{3}$ **b.** $\cot(\arccos 3x), \quad 0 \leq x < \frac{1}{3}$

Solution

If you let $u = \arccos 3x$, then $\cos u = 3x$, where $-1 \leq 3x \leq 1$. Because

$$\cos u = \frac{\text{adj}}{\text{hyp}} = \frac{3x}{1}$$

you can sketch a right triangle with acute angle u, as shown in Figure 4.55. From this triangle, you can easily convert each expression to algebraic form.

a. $\sin(\arccos 3x) = \sin u = \frac{\text{opp}}{\text{hyp}} = \sqrt{1 - 9x^2}, \quad 0 \leq x \leq \frac{1}{3}$

b. $\cot(\arccos 3x) = \cot u = \frac{\text{adj}}{\text{opp}} = \frac{3x}{\sqrt{1 - 9x^2}}, \quad 0 \leq x < \frac{1}{3}$

✓ **Checkpoint** Audio-video solution in English & Spanish at *LarsonPrecalculus.com.*

Write $\sec(\arctan x)$ as an algebraic expression in x.

Summarize (Section 4.7)

1. State the definition of the inverse sine function (*page 227*). For examples of evaluating and graphing the inverse sine function, see Examples 1 and 2.

2. State the definitions of the inverse cosine and inverse tangent functions (*page 229*). For examples of evaluating and graphing inverse trigonometric functions, see Examples 3 and 4.

3. State the inverse properties of trigonometric functions (*page 231*). For examples involving the compositions of trigonometric functions, see Examples 5–7.

4.8   Applications and Models

- Solve real-life problems involving right triangles.
- Solve real-life problems involving directional bearings.
- Solve real-life problems involving harmonic motion.

Applications Involving Right Triangles

In this section, the three angles of a right triangle are denoted by the letters A, B, and C (where C is the right angle), and the lengths of the sides opposite these angles by the letters a, b, and c, respectively (where c is the hypotenuse).

Right triangles often occur in real-life situations. For instance, right triangles can be used to analyze the design of a new slide at a water park.

EXAMPLE 1   Solving a Right Triangle

Solve the right triangle shown at the right for all unknown sides and angles.

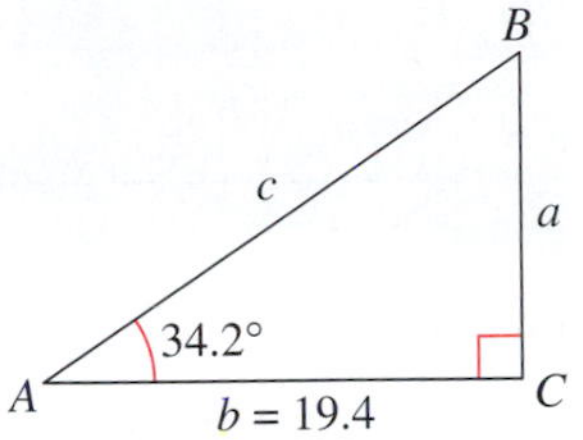

Solution   Because $C = 90°$, it follows that

$$A + B = 90° \quad \text{and} \quad B = 90° - 34.2° = 55.8°.$$

To solve for a, use the fact that

$$\tan A = \frac{\text{opp}}{\text{adj}} = \frac{a}{b} \quad \Longrightarrow \quad a = b \tan A.$$

So, $a = 19.4 \tan 34.2° \approx 13.18$. Similarly, to solve for c, use the fact that

$$\cos A = \frac{\text{adj}}{\text{hyp}} = \frac{b}{c} \quad \Longrightarrow \quad c = \frac{b}{\cos A}.$$

So, $c = \dfrac{19.4}{\cos 34.2°} \approx 23.46$.

✓ **Checkpoint** *Audio-video solution in English & Spanish at LarsonPrecalculus.com.*

Solve the right triangle shown in Figure 4.56 for all unknown sides and angles.

Figure 4.56

EXAMPLE 2   Finding a Side of a Right Triangle

A safety regulation states that the maximum angle of elevation for a rescue ladder is 72°. A fire department's longest ladder is 110 feet. What is the maximum safe rescue height?

Solution   A sketch is shown in Figure 4.57. From the equation $\sin A = a/c$, it follows that

$$a = c \sin A$$
$$= 110 \sin 72°$$
$$\approx 104.6.$$

So, the maximum safe rescue height is about 104.6 feet above the height of the fire truck.

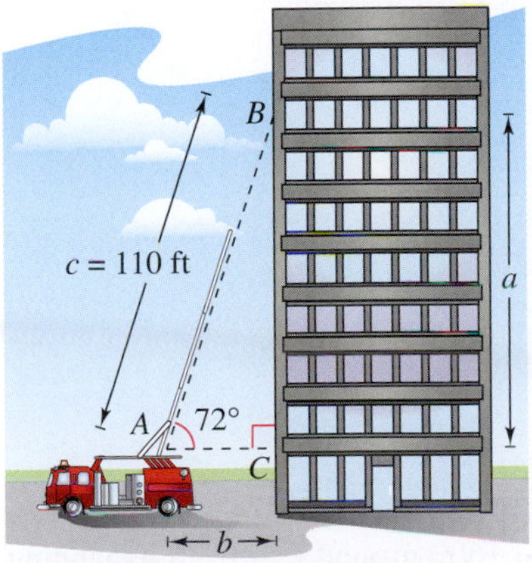

✓ **Checkpoint** *Audio-video solution in English & Spanish at LarsonPrecalculus.com.*

A ladder that is 16 feet long leans against the side of a house. The angle of elevation of the ladder is 80°. Find the height from the top of the ladder to the ground.

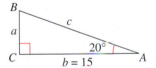

Figure 4.57

Figure 4.58

EXAMPLE 3 **Finding a Side of a Right Triangle**

At a point 200 feet from the base of a building, the angle of elevation to the *bottom* of a smokestack is 35°, whereas the angle of elevation to the *top* is 53°, as shown in Figure 4.58. Find the height *s* of the smokestack alone.

Solution

Note from Figure 4.58 that this problem involves two right triangles. For the smaller right triangle, use the fact that

$$\tan 35° = \frac{a}{200}$$

to conclude that the height of the building is

$$a = 200 \tan 35°.$$

For the larger right triangle, use the equation

$$\tan 53° = \frac{a + s}{200}$$

to conclude that $a + s = 200 \tan 53°$. So, the height of the smokestack is

$$s = 200 \tan 53° - a$$

$$= 200 \tan 53° - 200 \tan 35°$$

$$\approx 125.4 \text{ feet.}$$

✓ *Checkpoint*)) *Audio-video solution in English & Spanish at LarsonPrecalculus.com.*

At a point 65 feet from the base of a church, the angles of elevation to the bottom of the steeple and the top of the steeple are 35° and 43°, respectively. Find the height of the steeple.

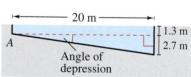

Figure 4.59

EXAMPLE 4 **Finding an Acute Angle of a Right Triangle**

A swimming pool is 20 meters long and 12 meters wide. The bottom of the pool is slanted so that the water depth is 1.3 meters at the shallow end and 4 meters at the deep end, as shown in Figure 4.59. Find the angle of depression (in degrees) of the bottom of the pool.

Solution Using the tangent function, you can see that

$$\tan A = \frac{\text{opp}}{\text{adj}}$$

$$= \frac{2.7}{20}$$

$$= 0.135.$$

So, the angle of depression is

$$A = \arctan 0.135$$

$$\approx 0.13419 \text{ radian}$$

$$\approx 7.69°.$$

✓ *Checkpoint*)) *Audio-video solution in English & Spanish at LarsonPrecalculus.com.*

From the time a small airplane is 100 feet high and 1600 ground feet from its landing runway, the plane descends in a straight line to the runway. Determine the angle of descent (in degrees) of the plane.

Trigonometry and Bearings

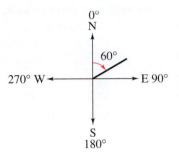

In surveying and navigation, directions can be given in terms of **bearings.** A bearing measures the acute angle that a path or line of sight makes with a fixed north-south line. For instance, the bearing S 35° E, shown below, means 35 degrees east of south.

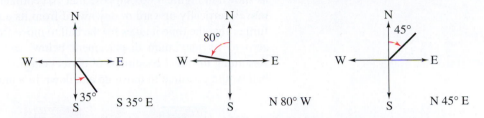

EXAMPLE 5   Finding Directions in Terms of Bearings

A ship leaves port at noon and heads due west at 20 knots, or 20 nautical miles (nm) per hour. At 2 P.M. the ship changes course to N 54° W, as shown below. Find the ship's bearing and distance from the port of departure at 3 P.M.

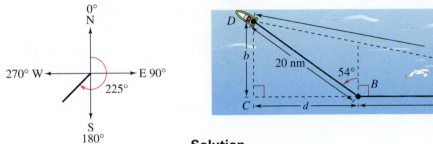

Solution

For triangle *BCD*, you have

$$B = 90° - 54° = 36°.$$

The two sides of this triangle can be determined to be

$$b = 20 \sin 36° \quad \text{and} \quad d = 20 \cos 36°.$$

For triangle *ACD*, you can find angle *A* as follows.

$$\tan A = \frac{b}{d + 40} = \frac{20 \sin 36°}{20 \cos 36° + 40} \approx 0.2092494$$

$$A \approx \arctan 0.2092494 \approx 0.2062732 \text{ radian} \approx 11.82°$$

The angle with the north-south line is $90° - 11.82° = 78.18°$. So, the bearing of the ship is N 78.18° W. Finally, from triangle *ACD*, you have $\sin A = b/c$, which yields

$$c = \frac{b}{\sin A}$$

$$= \frac{20 \sin 36°}{\sin 11.82°}$$

$$\approx 57.4 \text{ nautical miles.} \qquad \text{Distance from port}$$

✓ **Checkpoint**  *Audio-video solution in English & Spanish at LarsonPrecalculus.com.*

A sailboat leaves a pier heading due west at 8 knots. After 15 minutes, the sailboat tacks, changing course to N 16° W at 10 knots. Find the sailboat's distance and bearing from the pier after 12 minutes on this course.

Harmonic Motion

The periodic nature of the trigonometric functions is useful for describing the motion of a point on an object that vibrates, oscillates, rotates, or is moved by wave motion.

For example, consider a ball that is bobbing up and down on the end of a spring, as shown in Figure 4.60. Suppose that 10 centimeters is the maximum distance the ball moves vertically upward or downward from its equilibrium (at rest) position. Suppose further that the time it takes for the ball to move from its maximum displacement above zero to its maximum displacement below zero and back again is $t = 4$ seconds. Assuming the ideal conditions of perfect elasticity and no friction or air resistance, the ball would continue to move up and down in a uniform and regular manner.

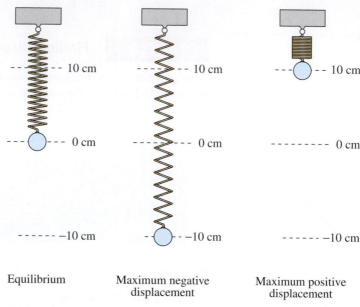

| Equilibrium | Maximum negative displacement | Maximum positive displacement |

Figure 4.60

From this spring you can conclude that the period (time for one complete cycle) of the motion is

Period = 4 seconds

its amplitude (maximum displacement from equilibrium) is

Amplitude = 10 centimeters

and its **frequency** (number of cycles per second) is

Frequency = $\dfrac{1}{4}$ cycle per second.

Motion of this nature can be described by a sine or cosine function and is called **simple harmonic motion.**

Definition of Simple Harmonic Motion

A point that moves on a coordinate line is in **simple harmonic motion** when its distance d from the origin at time t is given by either

$$d = a \sin \omega t \quad \text{or} \quad d = a \cos \omega t$$

where a and ω are real numbers such that $\omega > 0$. The motion has amplitude $|a|$, period $\dfrac{2\pi}{\omega}$, and frequency $\dfrac{\omega}{2\pi}$.

EXAMPLE 6   **Simple Harmonic Motion**

Write an equation for the simple harmonic motion of the ball described in Figure 4.60, where the period is 4 seconds. What is the frequency of this harmonic motion?

Solution

Because the spring is at equilibrium $(d = 0)$ when $t = 0$, use the equation

$$d = a \sin \omega t.$$

Moreover, because the maximum displacement from zero is 10 and the period is 4, you have the following.

$$\text{Amplitude} = |a|$$

$$= 10$$

$$\text{Period} = \frac{2\pi}{\omega} = 4 \quad \Longrightarrow \quad \omega = \frac{\pi}{2}$$

Consequently, an equation of motion is

$$d = 10 \sin \frac{\pi}{2} t.$$

Note that the choice of

$$a = 10 \quad \text{or} \quad a = -10$$

depends on whether the ball initially moves up or down. The frequency is

$$\text{Frequency} = \frac{\omega}{2\pi}$$

$$= \frac{\pi/2}{2\pi}$$

$$= \frac{1}{4} \text{ cycle per second.}$$

✓ *Checkpoint* ◀))) *Audio-video solution in English & Spanish at LarsonPrecalculus.com.*

Find a model for simple harmonic motion that satisfies the following conditions: $d = 0$ when $t = 0$, the amplitude is 6 centimeters, and the period is 3 seconds. Then find the frequency.

One illustration of the relationship between sine waves and harmonic motion is in the wave motion that results when a stone is dropped into a calm pool of water. The waves move outward in roughly the shape of sine (or cosine) waves, as shown in Figure 4.61. As an example, suppose you are fishing and your fishing bobber is attached so that it does not move horizontally. As the waves move outward from the dropped stone, your fishing bobber will move up and down in simple harmonic motion, as shown in Figure 4.62.

Figure 4.61

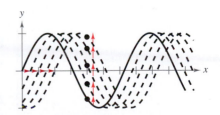

Figure 4.62

EXAMPLE 7 **Simple Harmonic Motion**

Given the equation for simple harmonic motion $d = 6 \cos \frac{3\pi}{4}t$, find (a) the maximum

displacement, (b) the frequency, (c) the value of d when $t = 4$, and (d) the least positive value of t for which $d = 0$.

Algebraic Solution

The given equation has the form $d = a \cos \omega t$, with $a = 6$ and $\omega = 3\pi/4$.

a. The maximum displacement (from the point of equilibrium) is given by the amplitude. So, the maximum displacement is 6.

b. Frequency $= \dfrac{\omega}{2\pi}$

$= \dfrac{3\pi/4}{2\pi}$

$= \dfrac{3}{8}$ cycle per unit of time

c. $d = 6 \cos\left[\dfrac{3\pi}{4}(4)\right] = 6 \cos 3\pi = 6(-1) = -6$

d. To find the least positive value of t for which $d = 0$, solve the equation

$$6 \cos \frac{3\pi}{4}t = 0.$$

First divide each side by 6 to obtain

$$\cos \frac{3\pi}{4}t = 0.$$

This equation is satisfied when

$$\frac{3\pi}{4}t = \frac{\pi}{2}, \frac{3\pi}{2}, \frac{5\pi}{2}, \ldots \ldots$$

Multiply these values by $4/(3\pi)$ to obtain

$$t = \frac{2}{3}, 2, \frac{10}{3}, \ldots \ldots$$

So, the least positive value of t is $t = \frac{2}{3}$.

Graphical Solution

Use a graphing utility set in *radian* mode.

a.

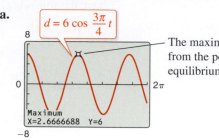

The maximum displacement from the point of equilibrium ($d = 0$) is 6.

b.

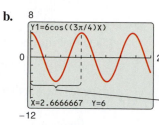

The period is the time for the graph to complete one cycle, which is $t \approx 2.67$. So, the frequency is about $1/2.67 \approx 0.37$ per unit of time.

c.

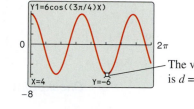

The value of d when $t = 4$ is $d = -6$.

d. The least positive value of t for which $d = 0$ is $t \approx 0.67$.

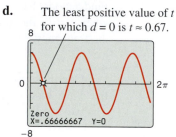

✓ *Checkpoint*))) *Audio-video solution in English & Spanish at LarsonPrecalculus.com.*

Rework Example 7 for the equation $d = 4 \cos 6\pi t$.

Summarize (Section 4.8)

1. Describe real-life problems that can be solved using right triangles (*pages 233 and 234, Examples 1–4*).

2. State the definition of a bearing (*page 235, Example 5*).

3. State the definition of simple harmonic motion (*page 236, Examples 6 and 7*).

Chapter Summary

What Did You Learn?	Explanation/Examples	Review Exercises

Section 4.1

Describe angles *(p. 188)*.		1–4
Convert between degrees and radians *(p. 191)*.	To convert degrees to radians, multiply degrees by $(\pi\,\text{rad})/180°$. To convert radians to degrees, multiply radians by $180°/(\pi\,\text{rad})$.	5–14
Use angles to model and solve real-life problems *(p. 192)*.	Angles can help you find the length of a circular arc and the area of a sector of a circle. (See Examples 5 and 8.)	15–18

Section 4.2

Identify a unit circle and describe its relationship to real numbers *(p. 195)*.		19–22
Evaluate trigonometric functions using the unit circle *(p. 196)*.	$t = \dfrac{2\pi}{3}$ corresponds to $(x, y) = \left(-\dfrac{1}{2}, \dfrac{\sqrt{3}}{2}\right)$. So $\cos\dfrac{2\pi}{3} = -\dfrac{1}{2}$, $\sin\dfrac{2\pi}{3} = \dfrac{\sqrt{3}}{2}$, and $\tan\dfrac{2\pi}{3} = -\sqrt{3}$.	23, 24
Use domain and period to evaluate sine and cosine functions *(p. 198)*, and use a calculator to evaluate trigonometric functions *(p. 199)*.	Because $\dfrac{13\pi}{6} = 2\pi + \dfrac{\pi}{6}$, $\sin\dfrac{13\pi}{6} = \sin\dfrac{\pi}{6} = \dfrac{1}{2}$. $\sin\dfrac{3\pi}{8} \approx 0.9239$, $\cot(-1.2) \approx -0.3888$	25–32

Section 4.3

Evaluate trigonometric functions of acute angles *(p. 200)*, and use a calculator to evaluate trigonometric functions *(p. 202)*.	$\sin\theta = \dfrac{\text{opp}}{\text{hyp}}$, $\cos\theta = \dfrac{\text{adj}}{\text{hyp}}$, $\tan\theta = \dfrac{\text{opp}}{\text{adj}}$ $\csc\theta = \dfrac{\text{hyp}}{\text{opp}}$, $\sec\theta = \dfrac{\text{hyp}}{\text{adj}}$, $\cot\theta = \dfrac{\text{adj}}{\text{opp}}$ $\tan 34.7° \approx 0.6924$, $\csc 29°\,15' \approx 2.0466$	33–38
Use the fundamental trigonometric identities *(p. 203)*.	$\sin\theta = \dfrac{1}{\csc\theta}$, $\tan\theta = \dfrac{\sin\theta}{\cos\theta}$, $\sin^2\theta + \cos^2\theta = 1$	39, 40
Use trigonometric functions to model and solve real-life problems *(p. 205)*.	Trigonometric functions can help you find the height of a monument, the angle between two paths, and the length of a ramp. (See Examples 8–10.)	41, 42

What Did You Learn?	Explanation/Examples	Review Exercises

Section 4.4

Evaluate trigonometric functions of any angle *(p. 207)*.	Let $(3, 4)$ be a point on the terminal side of θ. Then $\sin \theta = \frac{4}{5}$, $\cos \theta = \frac{3}{5}$, and $\tan \theta = \frac{4}{3}$.	43–50
Find reference angles *(p. 209)*.	Let θ be an angle in standard position. Its reference angle is the acute angle θ' formed by the terminal side of θ and the horizontal axis.	51–54
Evaluate trigonometric functions of real numbers *(p. 210)*.	$\cos \dfrac{7\pi}{3} = \dfrac{1}{2}$ because $\theta' = \dfrac{7\pi}{3} - 2\pi = \dfrac{\pi}{3}$ and $\cos \dfrac{\pi}{3} = \dfrac{1}{2}$.	55–62

Section 4.5

Sketch the graphs of sine and cosine functions using amplitude and period *(p. 213)*.		63, 64
Sketch translations of the graphs of sine and cosine functions *(p. 217)*.	For $y = d + a \sin(bx - c)$ and $y = d + a \cos(bx - c)$, the constant c creates a horizontal translation. The constant d creates a vertical translation. (See Examples 4–6.)	65–68
Use sine and cosine functions to model real-life data *(p. 219)*.	A cosine function can help you model the depth of the water at the end of a dock at various times. (See Example 7.)	69, 70

Section 4.6

| Sketch the graphs of tangent *(p. 220)*, cotangent *(p. 222)*, secant *(p. 223)*, and cosecant *(p. 223)* functions. | | 71–74 |
| Sketch the graphs of damped trigonometric functions *(p. 225)*. | In $f(x) = x \cos 2x$, the factor x is called the damping factor. | 75, 76 |

Section 4.7

| Evaluate and graph inverse trigonometric functions *(p. 227)*. | $\sin^{-1} \dfrac{\sqrt{3}}{2} = \dfrac{\pi}{3}$, $\cos^{-1}\left(-\dfrac{\sqrt{2}}{2}\right) = \dfrac{3\pi}{4}$, $\tan^{-1}(-1) = -\dfrac{\pi}{4}$ | 77–86 |
| Evaluate the compositions of trigonometric functions *(p. 231)*. | $\cos\left[\arctan\left(\frac{5}{12}\right)\right] = \frac{12}{13}$, $\sin(\sin^{-1} 0.4) = 0.4$ | 87–92 |

Section 4.8

Solve real-life problems involving right triangles *(p. 233)*.	A trigonometric function can help you find the height of a smokestack on top of a building. (See Example 3.)	93, 94
Solve real-life problems involving directional bearings *(p. 235)*.	Trigonometric functions can help you find a ship's bearing and distance from a port at a given time. (See Example 5.)	95
Solve real-life problems involving harmonic motion *(p. 236)*.	Sine or cosine functions can help you describe the motion of an object that vibrates, oscillates, rotates, or is moved by wave motion. (See Examples 6 and 7.)	96

Review Exercises   See **CalcChat.com** for tutorial help and worked-out solutions to odd-numbered exercises.

4.1 **Using Radian or Degree Measure** In Exercises 1–4, (a) sketch the angle in standard position, (b) determine the quadrant in which the angle lies, and (c) determine one positive and one negative coterminal angle.

1. $\dfrac{15\pi}{4}$

2. $-\dfrac{4\pi}{3}$

3. $-110°$

4. $280°$

Converting from Degrees to Radians In Exercises 5–8, convert the angle measure from degrees to radians. Round to three decimal places.

5. $450°$

6. $-112.5°$

7. $-33°\,45'$

8. $197°\,17'$

Converting from Radians to Degrees In Exercises 9–12, convert the angle measure from radians to degrees. Round to three decimal places.

9. $\dfrac{3\pi}{10}$

10. $-\dfrac{11\pi}{6}$

11. -3.5

12. 5.7

Converting to D° M′ S″ Form In Exercises 13 and 14, convert the angle measure to degrees, minutes, and seconds without using a calculator. Then check your answer using a calculator.

13. $198.4°$

14. $-5.96°$

15. **Arc Length** Find the length of the arc on a circle of radius 20 inches intercepted by a central angle of $138°$.

16. **Phonograph** Phonograph records are vinyl discs that rotate on a turntable. A typical record album is 12 inches in diameter and plays at $33\frac{1}{3}$ revolutions per minute.
    (a) What is the angular speed of a record album?
    (b) What is the linear speed of the outer edge of a record album?

17. **Circular Sector** Find the area of the sector of a circle of radius 18 inches and central angle $\theta = 120°$.

18. **Circular Sector** Find the area of the sector of a circle of radius 6.5 millimeters and central angle $\theta = 5\pi/6$.

4.2 **Finding a Point on the Unit Circle** In Exercises 19–22, find the point (x, y) on the unit circle that corresponds to the real number t.

19. $t = \dfrac{2\pi}{3}$

20. $t = \dfrac{7\pi}{4}$

21. $t = \dfrac{7\pi}{6}$

22. $t = -\dfrac{4\pi}{3}$

Evaluating Trigonometric Functions In Exercises 23 and 24, evaluate (if possible) the six trigonometric functions at the real number.

23. $t = \dfrac{3\pi}{4}$

24. $t = -\dfrac{2\pi}{3}$

Using Period to Evaluate Sine and Cosine In Exercises 25–28, evaluate the trigonometric function using its period as an aid.

25. $\sin\dfrac{11\pi}{4}$

26. $\cos 4\pi$

27. $\sin\!\left(-\dfrac{17\pi}{6}\right)$

28. $\cos\!\left(-\dfrac{13\pi}{3}\right)$

Using a Calculator In Exercises 29–32, use a calculator to evaluate the trigonometric function. Round your answer to four decimal places. (Be sure the calculator is in the correct mode.)

29. $\tan 33$

30. $\csc 10.5$

31. $\sec\dfrac{12\pi}{5}$

32. $\sin\!\left(-\dfrac{\pi}{9}\right)$

4.3 **Evaluating Trigonometric Functions** In Exercises 33 and 34, find the exact values of the six trigonometric functions of the angle θ shown in the figure. (Use the Pythagorean Theorem to find the third side of the triangle.)

33.
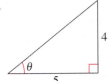

34.
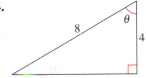

Using a Calculator In Exercises 35–38, use a calculator to evaluate the trigonometric function. Round your answer to four decimal places. (Be sure the calculator is in the correct mode.)

35. $\tan 33°$

36. $\sec 79.3°$

37. $\cot 15°\,14'$

38. $\cos 78°\,11'\,58''$

Applying Trigonometric Identities In Exercises 39 and 40, use the given function value and the trigonometric identities to find the indicated trigonometric functions.

39. $\sin\theta = \frac{1}{3}$
    (a) $\csc\theta$ (b) $\cos\theta$
    (c) $\sec\theta$ (d) $\tan\theta$

40. $\csc\theta = 5$
    (a) $\sin\theta$ (b) $\cot\theta$
    (c) $\tan\theta$ (d) $\sec(90° - \theta)$

41. Railroad Grade A train travels 3.5 kilometers on a straight track with a grade of 1° 10′ (see figure). What is the vertical rise of the train in that distance?

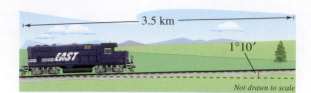

Not drawn to scale

42. Guy Wire A guy wire runs from the ground to the top of a 25-foot telephone pole. The angle formed between the wire and the ground is 52°. How far from the base of the pole is the wire attached to the ground? Assume the pole is perpendicular to the ground.

4.4 **Evaluating Trigonometric Functions** In Exercises 43–46, the point is on the terminal side of an angle in standard position. Determine the exact values of the six trigonometric functions of the angle.

43. $(12, 16)$ **44.** $(3, -4)$

45. $(0.3, 0.4)$ **46.** $\left(-\frac{10}{3}, -\frac{2}{3}\right)$

Evaluating Trigonometric Functions In Exercises 47–50, find the values of the remaining five trigonometric functions of θ with the given constraint.

Function Value	Constraint
47. $\sec \theta = \frac{6}{5}$	$\tan \theta < 0$
48. $\csc \theta = \frac{3}{2}$	$\cos \theta < 0$
49. $\cos \theta = -\frac{2}{5}$	$\sin \theta > 0$
50. $\sin \theta = -\frac{1}{2}$	$\cos \theta > 0$

Finding a Reference Angle In Exercises 51–54, find the reference angle θ' and sketch θ and θ' in standard position.

51. $\theta = 264°$ **52.** $\theta = 635°$

53. $\theta = -6\pi/5$ **54.** $\theta = 17\pi/3$

Using a Reference Angle In Exercises 55–58, evaluate the sine, cosine, and tangent of the angle without using a calculator.

55. $\pi/3$ **56.** $-5\pi/4$

57. $-150°$ **58.** $495°$

Using a Calculator In Exercises 59–62, use a calculator to evaluate the trigonometric function. Round your answer to four decimal places. (Be sure the calculator is in the correct mode.)

59. $\sin 4$

60. $\cot(-4.8)$

61. $\sin(12\pi/5)$

62. $\tan(-25\pi/7)$

4.5 **Sketching the Graph of a Sine or Cosine Function** In Exercises 63–68, sketch the graph of the function. (Include two full periods.)

63. $y = \sin 6x$

64. $f(x) = 5 \sin(2x/5)$

65. $y = 5 + \sin x$

66. $y = -4 - \cos \pi x$

67. $g(t) = \frac{5}{2} \sin(t - \pi)$

68. $g(t) = 3 \cos(t + \pi)$

69. Sound Waves Sine functions of the form $y = a \sin bx$, where x is measured in seconds, can model sine waves.

(a) Write an equation of a sound wave whose amplitude is 2 and whose period is $\frac{1}{264}$ second.

(b) What is the frequency of the sound wave described in part (a)?

70. Data Analysis: Meteorology The times S of sunset (Greenwich Mean Time) at 40° north latitude on the 15th of each month are: 1(16:59), 2(17:35), 3(18:06), 4(18:38), 5(19:08), 6(19:30), 7(19:28), 8(18:57), 9(18:09), 10(17:21), 11(16:44), 12(16:36). The month is represented by t, with $t = 1$ corresponding to January. A model (in which minutes have been converted to the decimal parts of an hour) for the data is $S(t) = 18.09 + 1.41 \sin[(\pi t/6) + 4.60]$.

(a) Use a graphing utility to graph the data points and the model in the same viewing window.

(b) What is the period of the model? Is it what you expected? Explain.

(c) What is the amplitude of the model? What does it represent in the model? Explain.

4.6 **Sketching the Graph of a Trigonometric Function** In Exercises 71–74, sketch the graph of the function. (Include two full periods.)

71. $f(t) = \tan\left(t + \dfrac{\pi}{2}\right)$

72. $f(x) = \dfrac{1}{2} \cot x$

73. $f(x) = \dfrac{1}{2} \csc \dfrac{x}{2}$

74. $h(t) = \sec\left(t - \dfrac{\pi}{4}\right)$

Analyzing a Damped Trigonometric Graph In Exercises 75 and 76, use a graphing utility to graph the function and the damping factor of the function in the same viewing window. Describe the behavior of the function as x increases without bound.

75. $f(x) = x \cos x$ **76.** $g(x) = x^4 \cos x$

4.7 Evaluating an Inverse Trigonometric Function   In Exercises 77–80, evaluate the expression without using a calculator.

77. $\arcsin(-1)$

78. $\cos^{-1} 1$

79. $\operatorname{arccot} \sqrt{3}$

80. $\operatorname{arcsec}(-\sqrt{2})$

Calculators and Inverse Trigonometric Functions   In Exercises 81–84, use a calculator to evaluate the expression. Round your result to two decimal places.

81. $\tan^{-1}(-1.5)$

82. $\arccos 0.324$

83. $\operatorname{arccot} 10.5$

84. $\operatorname{arccsc}(-2.01)$

Graphing an Inverse Trigonometric Function   In Exercises 85 and 86, use a graphing utility to graph the function.

85. $f(x) = \arctan(x/2)$

86. $f(x) = -\arcsin 2x$

Evaluating a Composition of Functions   In Exercises 87–90, find the exact value of the expression. (*Hint:* Sketch a right triangle.)

87. $\cos\left(\arctan \frac{3}{4}\right)$

88. $\sec\left(\tan^{-1} \frac{12}{5}\right)$

89. $\sec\left[\sin^{-1}\left(-\frac{1}{4}\right)\right]$

90. $\cot\left[\arcsin\left(-\frac{12}{13}\right)\right]$

Writing an Expression   In Exercises 91 and 92, write an algebraic expression that is equivalent to the given expression. (*Hint:* Sketch a right triangle.)

91. $\tan[\arccos(x/2)]$

92. $\sec(\arcsin x)$

4.8

93. **Angle of Elevation** The height of a radio transmission tower is 70 meters, and it casts a shadow of length 30 meters. Draw a right triangle that gives a visual representation of the problem. Label the known and unknown quantities. Then find the angle of elevation of the sun.

94. **Height** Your football has landed at the edge of the roof of your school building. When you are 25 feet from the base of the building, the angle of elevation to your football is 21°. How high off the ground is your football?

95. **Distance** From city A to city B, a plane flies 650 miles at a bearing of 48°. From city B to city C, the plane flies 810 miles at a bearing of 115°. Find the distance from city A to city C and the bearing from city A to city C.

96. **Wave Motion** Your fishing bobber oscillates in simple harmonic motion from the waves in the lake where you fish. Your bobber moves a total of 1.5 inches from its high point to its low point and returns to its high point every 3 seconds. Write an equation modeling the motion of your bobber, where the high point corresponds to the time $t = 0$.

Exploration

True or False?   In Exercises 97 and 98, determine whether the statement is true or false. Justify your answer.

97. $y = \sin \theta$ is not a function because $\sin 30° = \sin 150°$.

98. Because $\tan 3\pi/4 = -1$, $\arctan(-1) = 3\pi/4$.

99. **Writing** Describe the behavior of $f(\theta) = \sec \theta$ at the zeros of $g(\theta) = \cos \theta$. Explain your reasoning.

100. **Conjecture**
(a) Use a graphing utility to complete the table.

θ	0.1	0.4	0.7	1.0	1.3
$\tan\left(\theta - \dfrac{\pi}{2}\right)$					
$-\cot \theta$					

(b) Make a conjecture about the relationship between $\tan[\theta - (\pi/2)]$ and $-\cot \theta$.

101. **Writing** When graphing the sine and cosine functions, determining the amplitude is part of the analysis. Explain why this is not true for the other four trigonometric functions.

102. **Oscillation of a Spring** A weight is suspended from a ceiling by a steel spring. The weight is lifted (positive direction) from the equilibrium position and released. The resulting motion of the weight is modeled by $y = Ae^{-kt} \cos bt = \frac{1}{5}e^{-t/10} \cos 6t$, where y is the distance (in feet) from equilibrium and t is the time (in seconds). The figure shows the graph of the function. For each of the following, describe the change in the system without graphing the resulting function.

(a) A is changed from $\frac{1}{5}$ to $\frac{1}{3}$.

(b) k is changed from $\frac{1}{10}$ to $\frac{1}{3}$.

(c) b is changed from 6 to 9.

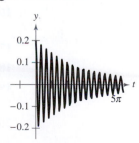

Chapter Test

See **CalcChat.com** for tutorial help and worked-out solutions to odd-numbered exercises.

Take this test as you would take a test in class. When you are finished, check your work against the answers given in the back of the book.

1. Consider an angle that measures $\dfrac{5\pi}{4}$ radians.

   (a) Sketch the angle in standard position.

   (b) Determine two coterminal angles (one positive and one negative).

   (c) Convert the angle to degree measure.

2. A truck is moving at a rate of 105 kilometers per hour, and the diameter of each of its wheels is 1 meter. Find the angular speed of the wheels in radians per minute.

3. A water sprinkler sprays water on a lawn over a distance of 25 feet and rotates through an angle of 130°. Find the area of the lawn watered by the sprinkler.

4. Find the exact values of the six trigonometric functions of the angle θ shown in the figure.

5. Given that $\tan \theta = \frac{3}{2}$, find the other five trigonometric functions of θ.

6. Determine the reference angle θ' of the angle $\theta = 205°$ and sketch θ and θ' in standard position.

7. Determine the quadrant in which θ lies when $\sec \theta < 0$ and $\tan \theta > 0$.

8. Find two exact values of θ in degrees ($0 \le \theta < 360°$) for which $\cos \theta = -\sqrt{3}/2$. (Do not use a calculator.)

9. Use a calculator to approximate two values of θ in radians ($0 \le \theta < 2\pi$) for which $\csc \theta = 1.030$. Round the results to two decimal places.

In Exercises 10 and 11, find the values of the remaining five trigonometric functions of θ with the given constraint.

10. $\cos \theta = \frac{3}{5}, \quad \tan \theta < 0$

11. $\sec \theta = -\frac{29}{20}, \quad \sin \theta > 0$

In Exercises 12 and 13, sketch the graph of the function. (Include two full periods.)

12. $g(x) = -2 \sin\left(x - \dfrac{\pi}{4}\right)$

13. $f(\alpha) = \dfrac{1}{2} \tan 2\alpha$

In Exercises 14 and 15, use a graphing utility to graph the function. If the function is periodic, then find its period.

14. $y = \sin 2\pi x + 2 \cos \pi x$

15. $y = 6e^{-0.12t} \cos(0.25t), \quad 0 \le t \le 32$

16. Find a, b, and c for the function $f(x) = a \sin(bx + c)$ such that the graph of f matches the figure.

17. Find the exact value of $\cot\left(\arcsin \frac{3}{8}\right)$ without using a calculator.

18. Graph the function $f(x) = 2 \arcsin\left(\frac{1}{2}x\right)$.

19. A plane is 90 miles south and 110 miles east of London Heathrow Airport. What bearing should be taken to fly directly to the airport?

20. Write the equation for the simple harmonic motion of a ball on a spring that starts at its lowest point of 6 inches below equilibrium, bounces to its maximum height of 6 inches above equilibrium, and returns to its lowest point in a total of 2 seconds.

(−2, 6)

θ

Figure for 4

f

Figure for 16

Proofs in Mathematics ■ ■ ■ ■ ■ ■ ■ ■ ■ ■ ■ ■ ■ ■

The Pythagorean Theorem

The Pythagorean Theorem is one of the most famous theorems in mathematics. More than 100 different proofs now exist. James A. Garfield, the twentieth president of the United States, developed a proof of the Pythagorean Theorem in 1876. His proof, shown below, involves the fact that a trapezoid can be formed from two congruent right triangles and an isosceles right triangle.

> **The Pythagorean Theorem**
>
> In a right triangle, the sum of the squares of the lengths of the legs is equal to the square of the length of the hypotenuse, where a and b are the legs and c is the hypotenuse.
>
> $a^2 + b^2 = c^2$
>
>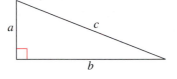

Proof

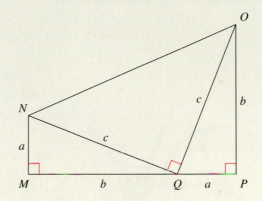

$$\text{Area of trapezoid } MNOP = \text{Area of } \triangle MNQ + \text{Area of } \triangle PQO + \text{Area of } \triangle NOQ$$

$$\frac{1}{2}(a + b)(a + b) = \frac{1}{2}ab + \frac{1}{2}ab + \frac{1}{2}c^2$$

$$\frac{1}{2}(a + b)(a + b) = ab + \frac{1}{2}c^2$$

$$(a + b)(a + b) = 2ab + c^2$$

$$a^2 + 2ab + b^2 = 2ab + c^2$$

$$a^2 + b^2 = c^2$$

P.S. Problem Solving ▪ ▪ ▪ ▪ ▪ ▪ ▪ ▪ ▪ ▪ ▪ ▪ ▪ ▪ ▪ ▪ ▪

1. **Angle of Rotation** The restaurant at the top of the Space Needle in Seattle, Washington, is circular and has a radius of 47.25 feet. The dining part of the restaurant revolves, making about one complete revolution every 48 minutes. A dinner party, seated at the edge of the revolving restaurant at 6:45 P.M., finishes at 8:57 P.M.

   (a) Find the angle through which the dinner party rotated.

   (b) Find the distance the party traveled during dinner.

2. **Bicycle Gears** A bicycle's gear ratio is the number of times the freewheel turns for every one turn of the chainwheel (see figure). The table shows the numbers of teeth in the freewheel and chainwheel for the first five gears of an 18-speed touring bicycle. The chainwheel completes one rotation for each gear. Find the angle through which the freewheel turns for each gear. Give your answers in both degrees and radians.

Gear Number	Number of Teeth in Freewheel	Number of Teeth in Chainwheel
1	32	24
2	26	24
3	22	24
4	32	40
5	19	24

DATA

Spreadsheet at LarsonPrecalculus.com

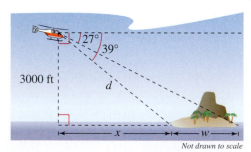

Freewheel

Chainwheel

3. **Surveying** A surveyor in a helicopter is trying to determine the width of an island, as shown in the figure.

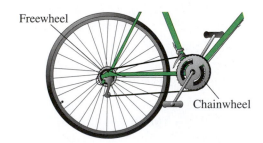

27° 39°

3000 ft

d

x   w

Not drawn to scale

   (a) What is the shortest distance d the helicopter would have to travel to land on the island?

   (b) What is the horizontal distance x the helicopter would have to travel before it would be directly over the nearer end of the island?

   (c) Find the width w of the island. Explain how you found your answer.

4. **Similar Triangles and Trigonometric Functions** Use the figure below.

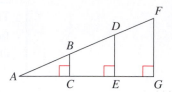

   (a) Explain why $\triangle ABC$, $\triangle ADE$, and $\triangle AFG$ are similar triangles.

   (b) What does similarity imply about the ratios

   $$\frac{BC}{AB}, \quad \frac{DE}{AD}, \quad \text{and} \quad \frac{FG}{AF}?$$

   (c) Does the value of sin A depend on which triangle from part (a) you use to calculate it? Would using a different right triangle similar to the three given triangles change the value of sin A?

   (d) Do your conclusions from part (c) apply to the other five trigonometric functions? Explain.

5. **Graphical Analysis** Use a graphing utility to graph h, and use the graph to decide whether h is even, odd, or neither.

   (a) $h(x) = \cos^2 x$

   (b) $h(x) = \sin^2 x$

6. **Squares of Even and Odd Functions** Given that f is an even function and g is an odd function, use the results of Exercise 5 to make a conjecture about h, where

   (a) $h(x) = [f(x)]^2$

   (b) $h(x) = [g(x)]^2$.

7. **Height of a Ferris Wheel Car** The model for the height h (in feet) of a Ferris wheel car is

   $$h = 50 + 50 \sin 8\pi t$$

   where t is the time (in minutes). (The Ferris wheel has a radius of 50 feet.) This model yields a height of 50 feet when $t = 0$. Alter the model so that the height of the car is 1 foot when $t = 0$.

8. **Periodic Function** The function f is periodic, with period c. So, $f(t + c) = f(t)$. Are the following statements true? Explain.

   (a) $f(t - 2c) = f(t)$

   (b) $f(t + \frac{1}{2}c) = f(\frac{1}{2}t)$

   (c) $f(\frac{1}{2}(t + c)) = f(\frac{1}{2}t)$

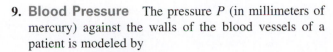

9. Blood Pressure The pressure P (in millimeters of mercury) against the walls of the blood vessels of a patient is modeled by

$$P = 100 - 20 \cos\left(\frac{8\pi}{3}t\right)$$

where t is time (in seconds).

(a) Use a graphing utility to graph the model.

(b) What is the period of the model? What does the period tell you about this situation?

(c) What is the amplitude of the model? What does it tell you about this situation?

(d) If one cycle of this model is equivalent to one heartbeat, what is the pulse of this patient?

(e) A physician wants this patient's pulse rate to be 64 beats per minute or less. What should the period be? What should the coefficient of t be?

10. Biorhythms A popular theory that attempts to explain the ups and downs of everyday life states that each of us has three cycles, called biorhythms, which begin at birth. These three cycles can be modeled by sine waves.

Physical (23 days): $P = \sin\dfrac{2\pi t}{23}, \quad t \geq 0$

Emotional (28 days): $E = \sin\dfrac{2\pi t}{28}, \quad t \geq 0$

Intellectual (33 days): $I = \sin\dfrac{2\pi t}{33}, \quad t \geq 0$

where t is the number of days since birth. Consider a person who was born on July 20, 1990.

(a) Use a graphing utility to graph the three models in the same viewing window for $7300 \leq t \leq 7380$.

(b) Describe the person's biorhythms during the month of September 2010.

(c) Calculate the person's three energy levels on September 22, 2010.

11. (a) **Graphical Reasoning** Use a graphing utility to graph the functions

$f(x) = 2 \cos 2x + 3 \sin 3x$ and

$g(x) = 2 \cos 2x + 3 \sin 4x$.

(b) Use the graphs from part (a) to find the period of each function.

(c) Is the function $h(x) = A \cos \alpha x + B \sin \beta x$, where α and β are positive integers, periodic? Explain your reasoning.

12. Analyzing Trigonometric Functions Two trigonometric functions f and g have periods of 2, and their graphs intersect at $x = 5.35$.

(a) Give one positive value of x less than 5.35 and one value of x greater than 5.35 at which the functions have the same value.

(b) Determine one negative value of x at which the graphs intersect.

(c) Is it true that $f(13.35) = g(-4.65)$? Explain your reasoning.

13. Refraction When you stand in shallow water and look at an object below the surface of the water, the object will look farther away from you than it really is. This is because when light rays pass between air and water, the water refracts, or bends, the light rays. The index of refraction for water is 1.333. This is the ratio of the sine of θ_1 and the sine of θ_2 (see figure).

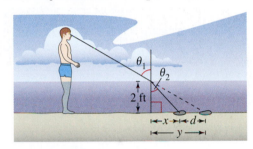

(a) While standing in water that is 2 feet deep, you look at a rock at angle $\theta_1 = 60°$ (measured from a line perpendicular to the surface of the water). Find θ_2.

(b) Find the distances x and y.

(c) Find the distance d between where the rock is and where it appears to be.

(d) What happens to d as you move closer to the rock? Explain your reasoning.

14. Polynomial Approximation In calculus, it can be shown that the arctangent function can be approximated by the polynomial

$$\arctan x \approx x - \frac{x^3}{3} + \frac{x^5}{5} - \frac{x^7}{7}$$

where x is in radians.

(a) Use a graphing utility to graph the arctangent function and its polynomial approximation in the same viewing window. How do the graphs compare?

(b) Study the pattern in the polynomial approximation of the arctangent function and guess the next term. Then repeat part (a). How does the accuracy of the approximation change when you add additional terms?

247

5 Analytic Trigonometry

Standing Waves

Projectile Motion

Honeycomb Cell

Shadow Length

Friction

5.1 Using Fundamental Identities

■ Recognize and write the fundamental trigonometric identities.
■ Use the fundamental trigonometric identities to evaluate trigonometric functions, simplify trigonometric expressions, and rewrite trigonometric expressions.

Introduction

In Chapter 4, you studied the basic definitions, properties, graphs, and applications of the individual trigonometric functions. In this chapter, you will learn how to use the fundamental identities to do the following.

1. Evaluate trigonometric functions.

2. Simplify trigonometric expressions.

3. Develop additional trigonometric identities.

4. Solve trigonometric equations.

Fundamental trigonometric identities can help you simplify a trigonometric expression for the coefficient of friction.

Fundamental Trigonometric Identities

Reciprocal Identities

$$\sin u = \frac{1}{\csc u} \qquad \cos u = \frac{1}{\sec u} \qquad \tan u = \frac{1}{\cot u}$$

$$\csc u = \frac{1}{\sin u} \qquad \sec u = \frac{1}{\cos u} \qquad \cot u = \frac{1}{\tan u}$$

Quotient Identities

$$\tan u = \frac{\sin u}{\cos u} \qquad \cot u = \frac{\cos u}{\sin u}$$

Pythagorean Identities

$$\sin^2 u + \cos^2 u = 1 \qquad 1 + \tan^2 u = \sec^2 u \qquad 1 + \cot^2 u = \csc^2 u$$

Cofunction Identities

$$\sin\left(\frac{\pi}{2} - u\right) = \cos u \qquad \cos\left(\frac{\pi}{2} - u\right) = \sin u$$

$$\tan\left(\frac{\pi}{2} - u\right) = \cot u \qquad \cot\left(\frac{\pi}{2} - u\right) = \tan u$$

$$\sec\left(\frac{\pi}{2} - u\right) = \csc u \qquad \csc\left(\frac{\pi}{2} - u\right) = \sec u$$

Even/Odd Identities

$$\sin(-u) = -\sin u \qquad \cos(-u) = \cos u \qquad \tan(-u) = -\tan u$$

$$\csc(-u) = -\csc u \qquad \sec(-u) = \sec u \qquad \cot(-u) = -\cot u$$

•• **REMARK** You should learn the fundamental trigonometric identities well, because you will use them frequently in trigonometry and they will also appear in calculus. Note that u can be an angle, a real number, or a variable.

Pythagorean identities are sometimes used in radical form such as

$$\sin u = \pm\sqrt{1 - \cos^2 u}$$

or

$$\tan u = \pm\sqrt{\sec^2 u - 1}$$

where the sign depends on the choice of u.

Using the Fundamental Identities

One common application of trigonometric identities is to use given values of trigonometric functions to evaluate other trigonometric functions.

EXAMPLE 1 **Using Identities to Evaluate a Function**

Use the values $\sec u = -\frac{3}{2}$ and $\tan u > 0$ to find the values of all six trigonometric functions.

Solution Using a reciprocal identity, you have

$$\cos u = \frac{1}{\sec u} = \frac{1}{-3/2} = -\frac{2}{3}.$$

Using a Pythagorean identity, you have

$$\sin^2 u = 1 - \cos^2 u \qquad \text{Pythagorean identity}$$
$$= 1 - \left(-\tfrac{2}{3}\right)^2 \qquad \text{Substitute } -\tfrac{2}{3} \text{ for } \cos u.$$
$$= \tfrac{5}{9}. \qquad \text{Simplify.}$$

Because $\sec u < 0$ and $\tan u > 0$, it follows that u lies in Quadrant III. Moreover, because $\sin u$ is negative when u is in Quadrant III, choose the negative root and obtain $\sin u = -\sqrt{5}/3$. Knowing the values of the sine and cosine enables you to find the values of all six trigonometric functions.

$$\sin u = -\frac{\sqrt{5}}{3} \qquad\qquad \csc u = \frac{1}{\sin u} = -\frac{3}{\sqrt{5}} = -\frac{3\sqrt{5}}{5}$$

$$\cos u = -\frac{2}{3} \qquad\qquad \sec u = \frac{1}{\cos u} = -\frac{3}{2}$$

$$\tan u = \frac{\sin u}{\cos u} = \frac{-\sqrt{5}/3}{-2/3} = \frac{\sqrt{5}}{2} \qquad \cot u = \frac{1}{\tan u} = \frac{2}{\sqrt{5}} = \frac{2\sqrt{5}}{5}$$

✓ **Checkpoint** 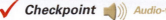 *Audio-video solution in English & Spanish at LarsonPrecalculus.com.*

Use the values $\tan x = \frac{1}{3}$ and $\cos x < 0$ to find the values of all six trigonometric functions.

> **TECHNOLOGY** To use a graphing utility to check the result of Example 2, graph
>
> $$y_1 = \sin x \cos^2 x - \sin x$$
>
> and
>
> $$y_2 = -\sin^3 x$$
>
> in the same viewing window, as shown below. Because Example 2 shows the equivalence algebraically and the two graphs appear to coincide, you can conclude that the expressions are equivalent.
>
>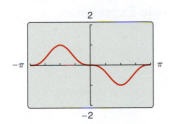

EXAMPLE 2 **Simplifying a Trigonometric Expression**

Simplify

$$\sin x \cos^2 x - \sin x.$$

Solution First factor out a common monomial factor and then use a fundamental identity.

$$\sin x \cos^2 x - \sin x = \sin x(\cos^2 x - 1) \qquad \text{Factor out common monomial factor.}$$
$$= -\sin x(1 - \cos^2 x) \qquad \text{Factor out } -1.$$
$$= -\sin x(\sin^2 x) \qquad \text{Pythagorean identity}$$
$$= -\sin^3 x \qquad \text{Multiply.}$$

✓ **Checkpoint** 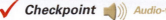 *Audio-video solution in English & Spanish at LarsonPrecalculus.com.*

Simplify

$$\cos^2 x \csc x - \csc x.$$

When factoring trigonometric expressions, it is helpful to find a special polynomial factoring form that fits the expression, as shown in Example 3.

EXAMPLE 3   **Factoring Trigonometric Expressions**

Factor each expression.

a. $\sec^2 \theta - 1$ **b.** $4 \tan^2 \theta + \tan \theta - 3$

Solution

a. This expression has the form $u^2 - v^2$, which is the difference of two squares. It factors as

$$\sec^2 \theta - 1 = (\sec \theta + 1)(\sec \theta - 1).$$

b. This expression has the polynomial form $ax^2 + bx + c$, and it factors as

$$4 \tan^2 \theta + \tan \theta - 3 = (4 \tan \theta - 3)(\tan \theta + 1).$$

✓ **Checkpoint**  Audio-video solution in English & Spanish at LarsonPrecalculus.com.

Factor each expression.

a. $1 - \cos^2 \theta$ **b.** $2 \csc^2 \theta - 7 \csc \theta + 6$ ■

On occasion, factoring or simplifying can best be done by first rewriting the expression in terms of just *one* trigonometric function or in terms of *sine and cosine only*. Examples 4 and 5, respectively, show these strategies.

EXAMPLE 4   **Factoring a Trigonometric Expression**

Factor $\csc^2 x - \cot x - 3$.

Solution   Use the identity $\csc^2 x = 1 + \cot^2 x$ to rewrite the expression.

$$\csc^2 x - \cot x - 3 = (1 + \cot^2 x) - \cot x - 3 \qquad \text{Pythagorean identity}$$

$$= \cot^2 x - \cot x - 2 \qquad \text{Combine like terms.}$$

$$= (\cot x - 2)(\cot x + 1) \qquad \text{Factor.}$$

✓ **Checkpoint**  Audio-video solution in English & Spanish at LarsonPrecalculus.com.

Factor $\sec^2 x + 3 \tan x + 1$.

EXAMPLE 5   **Simplifying a Trigonometric Expression**

$$\sin t + \cot t \cos t = \sin t + \left(\frac{\cos t}{\sin t} \right) \cos t \qquad \text{Quotient identity}$$

$$= \frac{\sin^2 t + \cos^2 t}{\sin t} \qquad \text{Add fractions.}$$

$$= \frac{1}{\sin t} \qquad \text{Pythagorean identity}$$

$$= \csc t \qquad \text{Reciprocal identity}$$

✓ **Checkpoint**  Audio-video solution in English & Spanish at LarsonPrecalculus.com.

Simplify $\csc x - \cos x \cot x$.   ■

▷ **ALGEBRA HELP**   In Example 3, you need to be able to factor the difference of two squares and factor a trinomial. You can review the techniques for factoring in Appendix A.3.

•• **REMARK**   Remember that when adding rational expressions, you must first find the least common denominator (LCD). In Example 5, the LCD is $\sin t$.

EXAMPLE 6   **Adding Trigonometric Expressions**

Perform the addition $\dfrac{\sin \theta}{1 + \cos \theta} + \dfrac{\cos \theta}{\sin \theta}$ and simplify.

Solution

$$\dfrac{\sin \theta}{1 + \cos \theta} + \dfrac{\cos \theta}{\sin \theta} = \dfrac{(\sin \theta)(\sin \theta) + (\cos \theta)(1 + \cos \theta)}{(1 + \cos \theta)(\sin \theta)}$$

$$= \dfrac{\sin^2 \theta + \cos^2 \theta + \cos \theta}{(1 + \cos \theta)(\sin \theta)} \qquad \text{Multiply.}$$

$$= \dfrac{1 + \cos \theta}{(1 + \cos \theta)(\sin \theta)} \qquad \begin{array}{l}\text{Pythagorean identity:}\\ \sin^2 \theta + \cos^2 \theta = 1\end{array}$$

$$= \dfrac{1}{\sin \theta} \qquad \text{Divide out common factor.}$$

$$= \csc \theta \qquad \text{Reciprocal identity}$$

✓ **Checkpoint**   *Audio-video solution in English & Spanish at LarsonPrecalculus.com.*

Perform the addition $\dfrac{1}{1 + \sin \theta} + \dfrac{1}{1 - \sin \theta}$ and simplify.   ■

The next two examples involve techniques for rewriting expressions in forms that are used in calculus.

EXAMPLE 7   **Rewriting a Trigonometric Expression**

Rewrite $\dfrac{1}{1 + \sin x}$ so that it is *not* in fractional form.

Solution   From the Pythagorean identity

$$\cos^2 x = 1 - \sin^2 x = (1 - \sin x)(1 + \sin x)$$

multiplying both the numerator and the denominator by $(1 - \sin x)$ will produce a monomial denominator.

$$\dfrac{1}{1 + \sin x} = \dfrac{1}{1 + \sin x} \cdot \dfrac{1 - \sin x}{1 - \sin x} \qquad \begin{array}{l}\text{Multiply numerator and}\\ \text{denominator by } (1 - \sin x).\end{array}$$

$$= \dfrac{1 - \sin x}{1 - \sin^2 x} \qquad \text{Multiply.}$$

$$= \dfrac{1 - \sin x}{\cos^2 x} \qquad \text{Pythagorean identity}$$

$$= \dfrac{1}{\cos^2 x} - \dfrac{\sin x}{\cos^2 x} \qquad \text{Write as separate fractions.}$$

$$= \dfrac{1}{\cos^2 x} - \dfrac{\sin x}{\cos x} \cdot \dfrac{1}{\cos x} \qquad \text{Product of fractions}$$

$$= \sec^2 x - \tan x \sec x \qquad \text{Reciprocal and quotient identities}$$

✓ **Checkpoint**   *Audio-video solution in English & Spanish at LarsonPrecalculus.com.*

Rewrite $\dfrac{\cos^2 \theta}{1 - \sin \theta}$ so that it is *not* in fractional form.   ■

EXAMPLE 8   **Trigonometric Substitution**

Use the substitution $x = 2 \tan \theta$, $0 < \theta < \pi/2$, to write $\sqrt{4 + x^2}$ as a trigonometric function of θ.

Solution   Begin by letting $x = 2 \tan \theta$. Then, you obtain

$$\sqrt{4 + x^2} = \sqrt{4 + (2 \tan \theta)^2} \qquad \text{Substitute } 2 \tan \theta \text{ for } x.$$

$$= \sqrt{4 + 4 \tan^2 \theta} \qquad \text{Rule of exponents}$$

$$= \sqrt{4(1 + \tan^2 \theta)} \qquad \text{Factor.}$$

$$= \sqrt{4 \sec^2 \theta} \qquad \text{Pythagorean identity}$$

$$= 2 \sec \theta. \qquad \sec \theta > 0 \text{ for } 0 < \theta < \pi/2$$

✓ *Checkpoint*))) *Audio-video solution in English & Spanish at LarsonPrecalculus.com.*

Use the substitution $x = 3 \sin \theta$, $0 < \theta < \pi/2$, to write $\sqrt{9 - x^2}$ as a trigonometric function of θ.

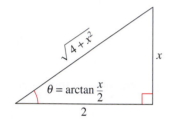

$\theta = \arctan \dfrac{x}{2}$

Angle whose tangent is $x/2$
Figure 5.1

Figure 5.1 shows the right triangle illustration of the trigonometric substitution $x = 2 \tan \theta$ in Example 8. Use this triangle to check the solution of Example 8, as follows. For $0 < \theta < \pi/2$, you have

$$\text{opp} = x, \quad \text{adj} = 2, \quad \text{and} \quad \text{hyp} = \sqrt{4 + x^2}.$$

With these expressions, you can write

$$\sec \theta = \frac{\text{hyp}}{\text{adj}} = \frac{\sqrt{4 + x^2}}{2}.$$

So, $2 \sec \theta = \sqrt{4 + x^2}$, and the solution checks.

EXAMPLE 9   **Rewriting a Logarithmic Expression**

Rewrite $\ln|\csc \theta| + \ln|\tan \theta|$ as a single logarithm and simplify the result.

Solution

$$\ln|\csc \theta| + \ln|\tan \theta| = \ln|\csc \theta \tan \theta| \qquad \text{Product Property of Logarithms}$$

$$= \ln \left| \frac{1}{\sin \theta} \cdot \frac{\sin \theta}{\cos \theta} \right| \qquad \text{Reciprocal and quotient identities}$$

$$= \ln \left| \frac{1}{\cos \theta} \right| \qquad \text{Simplify.}$$

$$= \ln|\sec \theta| \qquad \text{Reciprocal identity}$$

▷ **ALGEBRA HELP**   Recall that for positive real numbers u and v,

$$\ln u + \ln v = \ln(uv).$$

You can review the properties of logarithms in Section 3.3.

✓ *Checkpoint*))) *Audio-video solution in English & Spanish at LarsonPrecalculus.com.*

Rewrite $\ln|\sec x| + \ln|\sin x|$ as a single logarithm and simplify the result.

Summarize   **(Section 5.1)**

1. State the fundamental trigonometric identities *(page 250)*.

2. Explain how to use the fundamental trigonometric identities to evaluate trigonometric functions, simplify trigonometric expressions, and rewrite trigonometric expressions *(pages 251–254)*. For examples of these concepts, see Examples 1–9.

5.2 Verifying Trigonometric Identities

■ Verify trigonometric identities.

Introduction

In this section, you will study techniques for verifying trigonometric identities. In the next section, you will study techniques for solving trigonometric equations. The key to verifying identities *and* solving equations is the ability to use the fundamental identities and the rules of algebra to rewrite trigonometric expressions.

Remember that a *conditional equation* is an equation that is true for only some of the values in its domain. For example, the conditional equation

$$\sin x = 0 \qquad \text{Conditional equation}$$

is true only for

$$x = n\pi$$

where n is an integer. When you find these values, you are *solving* the equation.

On the other hand, an equation that is true for all real values in the domain of the variable is an *identity*. For example, the familiar equation

$$\sin^2 x = 1 - \cos^2 x \qquad \text{Identity}$$

is true for all real numbers x. So, it is an identity.

Trigonometric identities enable you to rewrite trigonometric equations that model real-life situations, such as the equation for the length of a shadow cast by a gnomon (a device used to tell time).

Verifying Trigonometric Identities

Although there are similarities, verifying that a trigonometric equation is an identity is quite different from solving an equation. There is no well-defined set of rules to follow in verifying trigonometric identities, and it is best to learn the process by practicing.

Guidelines for Verifying Trigonometric Identities

1. Work with one side of the equation at a time. It is often better to work with the more complicated side first.

2. Look for opportunities to factor an expression, add fractions, square a binomial, or create a monomial denominator.

3. Look for opportunities to use the fundamental identities. Note which functions are in the final expression you want. Sines and cosines pair up well, as do secants and tangents, and cosecants and cotangents.

4. If the preceding guidelines do not help, then try converting all terms to sines and cosines.

5. Always try *something*. Even making an attempt that leads to a dead end can provide insight.

Verifying trigonometric identities is a useful process when you need to convert a trigonometric expression into a form that is more useful algebraically. When you verify an identity, you cannot *assume* that the two sides of the equation are equal because you are trying to verify that they *are* equal. As a result, when verifying identities, you cannot use operations such as adding the same quantity to each side of the equation or cross multiplication.

EXAMPLE 1   **Verifying a Trigonometric Identity**

Verify the identity $\dfrac{\sec^2 \theta - 1}{\sec^2 \theta} = \sin^2 \theta$.

Solution   Start with the left side because it is more complicated.

$$\frac{\sec^2 \theta - 1}{\sec^2 \theta} = \frac{(\tan^2 \theta + 1) - 1}{\sec^2 \theta} \qquad \text{Pythagorean identity}$$

$$= \frac{\tan^2 \theta}{\sec^2 \theta} \qquad \text{Simplify.}$$

$$= \tan^2 \theta (\cos^2 \theta) \qquad \text{Reciprocal identity}$$

$$= \frac{\sin^2 \theta}{(\cos^2 \theta)} (\cos^2 \theta) \qquad \text{Quotient identity}$$

$$= \sin^2 \theta \qquad \text{Simplify.}$$

Notice that you verify the identity by starting with the left side of the equation (the more complicated side) and using the fundamental trigonometric identities to simplify it until you obtain the right side.

✓ **Checkpoint** *Audio-video solution in English & Spanish at LarsonPrecalculus.com.*

Verify the identity $\dfrac{\sin^2 \theta + \cos^2 \theta}{\cos^2 \theta \sec^2 \theta} = 1$.

There can be more than one way to verify an identity. Here is another way to verify the identity in Example 1.

$$\frac{\sec^2 \theta - 1}{\sec^2 \theta} = \frac{\sec^2 \theta}{\sec^2 \theta} - \frac{1}{\sec^2 \theta} \qquad \text{Write as separate fractions.}$$

$$= 1 - \cos^2 \theta \qquad \text{Reciprocal identity}$$

$$= \sin^2 \theta \qquad \text{Pythagorean identity}$$

EXAMPLE 2   **Verifying a Trigonometric Identity**

Verify the identity $2 \sec^2 \alpha = \dfrac{1}{1 - \sin \alpha} + \dfrac{1}{1 + \sin \alpha}$.

Algebraic Solution

Start with the right side because it is more complicated.

$$\frac{1}{1 - \sin \alpha} + \frac{1}{1 + \sin \alpha} = \frac{1 + \sin \alpha + 1 - \sin \alpha}{(1 - \sin \alpha)(1 + \sin \alpha)} \qquad \text{Add fractions.}$$

$$= \frac{2}{1 - \sin^2 \alpha} \qquad \text{Simplify.}$$

$$= \frac{2}{\cos^2 \alpha} \qquad \text{Pythagorean identity}$$

$$= 2 \sec^2 \alpha \qquad \text{Reciprocal identity}$$

Numerical Solution

Use a graphing utility to create a table that shows the values of $y_1 = 2/\cos^2 x$ and $y_2 = 1/(1 - \sin x) + 1/(1 + \sin x)$ for different values of x.

X	Y1	Y2
-.5	2.5969	2.5969
-.25	2.1304	2.1304
0	2	2
.25	2.1304	2.1304
.5	2.5969	2.5969
.75	3.7357	3.7357
1	6.851	6.851
X=-.5		

The values for y_1 and y_2 appear to be identical, so the equation appears to be an identity.

✓ **Checkpoint**  *Audio-video solution in English & Spanish at LarsonPrecalculus.com.*

Verify the identity $2 \csc^2 \beta = \dfrac{1}{1 - \cos \beta} + \dfrac{1}{1 + \cos \beta}$.

In Example 2, you needed to write the Pythagorean identity $\sin^2 u + \cos^2 u = 1$ in the equivalent form $\cos^2 u = 1 - \sin^2 u$. When verifying identities, you may find it useful to write the Pythagorean identities in one of these equivalent forms.

Pythagorean Identities   **Equivalent Forms**

$$\sin^2 u + \cos^2 u = 1$$
$$\sin^2 u = 1 - \cos^2 u$$
$$\cos^2 u = 1 - \sin^2 u$$

$$1 + \tan^2 u = \sec^2 u$$
$$1 = \sec^2 u - \tan^2 u$$
$$\tan^2 u = \sec^2 u - 1$$

$$1 + \cot^2 u = \csc^2 u$$
$$1 = \csc^2 u - \cot^2 u$$
$$\cot^2 u = \csc^2 u - 1$$

EXAMPLE 3   Verifying a Trigonometric Identity

Verify the identity $(\tan^2 x + 1)(\cos^2 x - 1) = -\tan^2 x$.

Algebraic Solution

By applying identities before multiplying, you obtain the following.

$$(\tan^2 x + 1)(\cos^2 x - 1) = (\sec^2 x)(-\sin^2 x) \qquad \text{Pythagorean identities}$$

$$= -\frac{\sin^2 x}{\cos^2 x} \qquad \text{Reciprocal identity}$$

$$= -\left(\frac{\sin x}{\cos x}\right)^2 \qquad \text{Property of exponents}$$

$$= -\tan^2 x \qquad \text{Quotient identity}$$

Graphical Solution

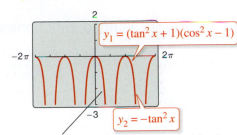

Because the graphs appear to coincide, the given equation appears to be an identity.

✓ **Checkpoint** ◀)) *Audio-video solution in English & Spanish at LarsonPrecalculus.com.*

Verify the identity $(\sec^2 x - 1)(\sin^2 x - 1) = -\sin^2 x$.

• • • • • • • • • • • • • • • • • ▷

REMARK Although a graphing utility can be useful in helping to verify an identity, you **must** use algebraic techniques to produce a *valid* proof.

EXAMPLE 4   Converting to Sines and Cosines

Verify the identity $\tan x + \cot x = \sec x \csc x$.

Solution Convert the left side into sines and cosines.

$$\tan x + \cot x = \frac{\sin x}{\cos x} + \frac{\cos x}{\sin x} \qquad \text{Quotient identities}$$

$$= \frac{\sin^2 x + \cos^2 x}{\cos x \sin x} \qquad \text{Add fractions.}$$

$$= \frac{1}{\cos x \sin x} \qquad \text{Pythagorean identity}$$

$$= \frac{1}{\cos x} \cdot \frac{1}{\sin x} \qquad \text{Product of fractions}$$

$$= \sec x \csc x \qquad \text{Reciprocal identities}$$

✓ **Checkpoint** ◀)) *Audio-video solution in English & Spanish at LarsonPrecalculus.com.*

Verify the identity $\csc x - \sin x = \cos x \cot x$.

Recall from algebra that *rationalizing the denominator* using conjugates is, on occasion, a powerful simplification technique. A related form of this technique works for simplifying trigonometric expressions as well. For instance, to simplify

$$\frac{1}{1 - \cos x}$$

multiply the numerator and the denominator by $1 + \cos x$.

$$\frac{1}{1 - \cos x} = \frac{1}{1 - \cos x}\left(\frac{1 + \cos x}{1 + \cos x}\right)$$

$$= \frac{1 + \cos x}{1 - \cos^2 x}$$

$$= \frac{1 + \cos x}{\sin^2 x}$$

$$= \csc^2 x(1 + \cos x)$$

The expression $\csc^2 x(1 + \cos x)$ is considered a simplified form of

$$\frac{1}{1 - \cos x}$$

because $\csc^2 x(1 + \cos x)$ does not contain fractions.

EXAMPLE 5   Verifying a Trigonometric Identity

Verify the identity $\sec x + \tan x = \dfrac{\cos x}{1 - \sin x}$.

Algebraic Solution

Begin with the *right* side and create a monomial denominator by multiplying the numerator and the denominator by $1 + \sin x$.

$$\frac{\cos x}{1 - \sin x} = \frac{\cos x}{1 - \sin x}\left(\frac{1 + \sin x}{1 + \sin x}\right) \qquad \text{Multiply numerator and denominator by } 1 + \sin x.$$

$$= \frac{\cos x + \cos x \sin x}{1 - \sin^2 x} \qquad \text{Multiply.}$$

$$= \frac{\cos x + \cos x \sin x}{\cos^2 x} \qquad \text{Pythagorean identity}$$

$$= \frac{\cos x}{\cos^2 x} + \frac{\cos x \sin x}{\cos^2 x} \qquad \text{Write as separate fractions.}$$

$$= \frac{1}{\cos x} + \frac{\sin x}{\cos x} \qquad \text{Simplify.}$$

$$= \sec x + \tan x \qquad \text{Identities}$$

Graphical Solution

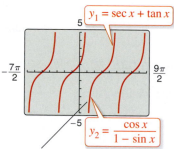

Because the graphs appear to coincide, the given equation appears to be an identity.

✓ *Checkpoint*  Audio-video solution in English & Spanish at LarsonPrecalculus.com.

Verify the identity $\csc x + \cot x = \dfrac{\sin x}{1 - \cos x}$.

◼

In Examples 1 through 5, you have been verifying trigonometric identities by working with one side of the equation and converting to the form given on the other side. On occasion, it is practical to work with each side *separately*, to obtain one common form that is equivalent to both sides. This is illustrated in Example 6.

EXAMPLE 6   **Working with Each Side Separately**

Verify the identity $\dfrac{\cot^2 \theta}{1 + \csc \theta} = \dfrac{1 - \sin \theta}{\sin \theta}$.

Algebraic Solution

Working with the left side, you have

$$\frac{\cot^2 \theta}{1 + \csc \theta} = \frac{\csc^2 \theta - 1}{1 + \csc \theta} \qquad \text{Pythagorean identity}$$

$$= \frac{(\csc \theta - 1)(\csc \theta + 1)}{1 + \csc \theta} \qquad \text{Factor.}$$

$$= \csc \theta - 1. \qquad \text{Simplify.}$$

Now, simplifying the right side, you have

$$\frac{1 - \sin \theta}{\sin \theta} = \frac{1}{\sin \theta} - \frac{\sin \theta}{\sin \theta} = \csc \theta - 1.$$

This verifies the identity because both sides are equal to $\csc \theta - 1$.

Numerical Solution

Use a graphing utility to create a table that shows the values of

$$y_1 = \frac{\cot^2 x}{1 + \csc x} \quad \text{and} \quad y_2 = \frac{1 - \sin x}{\sin x}$$

for different values of x.

X	Y1	Y2
-.5	-3.086	-3.086
-.25	-5.042	-5.042
0	ERROR	ERROR
.25	3.042	3.042
.5	1.0858	1.0858
.75	.46705	.46705
1	.1884	.1884

X=1

The values for y_1 and y_2 appear to be identical, so the equation appears to be an identity.

✓ **Checkpoint** *Audio-video solution in English & Spanish at LarsonPrecalculus.com.*

Verify the identity $\dfrac{\tan^2 \theta}{1 + \sec \theta} = \dfrac{1 - \cos \theta}{\cos \theta}$.

Example 7 shows powers of trigonometric functions rewritten as more complicated sums of products of trigonometric functions. This is a common procedure used in calculus.

EXAMPLE 7   **Two Examples from Calculus** $\int$

Verify each identity.

a. $\tan^4 x = \tan^2 x \sec^2 x - \tan^2 x$ **b.** $\csc^4 x \cot x = \csc^2 x (\cot x + \cot^3 x)$

Solution

a. $\tan^4 x = (\tan^2 x)(\tan^2 x)$ Write as separate factors.

$\qquad = \tan^2 x (\sec^2 x - 1)$ Pythagorean identity

$\qquad = \tan^2 x \sec^2 x - \tan^2 x$ Multiply.

b. $\csc^4 x \cot x = \csc^2 x \csc^2 x \cot x$ Write as separate factors.

$\qquad = \csc^2 x (1 + \cot^2 x) \cot x$ Pythagorean identity

$\qquad = \csc^2 x (\cot x + \cot^3 x)$ Multiply.

✓ **Checkpoint** *Audio-video solution in English & Spanish at LarsonPrecalculus.com.*

Verify each identity.

a. $\tan^3 x = \tan x \sec^2 x - \tan x$ **b.** $\sin^3 x \cos^4 x = (\cos^4 x - \cos^6 x)\sin x$

Summarize   (Section 5.2)

1. State the guidelines for verifying trigonometric identities *(page 255)*. For examples of verifying trigonometric identities, see Examples 1–7.

5.3 Solving Trigonometric Equations

- Use standard algebraic techniques to solve trigonometric equations.
- Solve trigonometric equations of quadratic type.
- Solve trigonometric equations involving multiple angles.
- Use inverse trigonometric functions to solve trigonometric equations.

Introduction

To solve a trigonometric equation, use standard algebraic techniques (when possible) such as collecting like terms and factoring. Your preliminary goal in solving a trigonometric equation is to *isolate* the trigonometric function on one side of the equation. For example, to solve the equation $2 \sin x = 1$, divide each side by 2 to obtain

$$\sin x = \frac{1}{2}.$$

To solve for x, note in the figure below that the equation $\sin x = \frac{1}{2}$ has solutions $x = \pi/6$ and $x = 5\pi/6$ in the interval $[0, 2\pi)$. Moreover, because $\sin x$ has a period of 2π, there are infinitely many other solutions, which can be written as

$$x = \frac{\pi}{6} + 2n\pi \quad \text{and} \quad x = \frac{5\pi}{6} + 2n\pi \qquad \text{General solution}$$

Trigonometric equations can help you solve a variety of real-life problems, such as determining the height above ground of a seat on a Ferris wheel.

where n is an integer, as shown below.

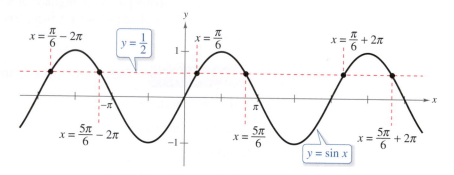

The figure below illustrates another way to show that the equation $\sin x = \frac{1}{2}$ has infinitely many solutions. Any angles that are coterminal with $\pi/6$ or $5\pi/6$ will also be solutions of the equation.

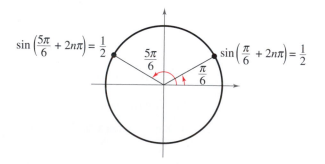

When solving trigonometric equations, you should write your answer(s) using exact values, when possible, rather than decimal approximations.

EXAMPLE 1 Collecting Like Terms

Solve

$$\sin x + \sqrt{2} = -\sin x.$$

Solution Begin by isolating $\sin x$ on one side of the equation.

$\sin x + \sqrt{2} = -\sin x$	Write original equation.
$\sin x + \sin x + \sqrt{2} = 0$	Add $\sin x$ to each side.
$\sin x + \sin x = -\sqrt{2}$	Subtract $\sqrt{2}$ from each side.
$2 \sin x = -\sqrt{2}$	Combine like terms.
$\sin x = -\dfrac{\sqrt{2}}{2}$	Divide each side by 2.

Because $\sin x$ has a period of 2π, first find all solutions in the interval $[0, 2\pi)$. These solutions are $x = 5\pi/4$ and $x = 7\pi/4$. Finally, add multiples of 2π to each of these solutions to obtain the general form

$$x = \frac{5\pi}{4} + 2n\pi \quad \text{and} \quad x = \frac{7\pi}{4} + 2n\pi \qquad \text{General solution}$$

where n is an integer.

✓ *Checkpoint* 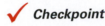 *Audio-video solution in English & Spanish at LarsonPrecalculus.com.*

Solve $\sin x - \sqrt{2} = -\sin x$.

EXAMPLE 2 Extracting Square Roots

Solve

$$3 \tan^2 x - 1 = 0.$$

Solution Begin by isolating $\tan x$ on one side of the equation.

> • **REMARK** When you extract
> square roots, make sure you
> account for both the positive
> and negative solutions.

$3 \tan^2 x - 1 = 0$	Write original equation.
$3 \tan^2 x = 1$	Add 1 to each side.
$\tan^2 x = \dfrac{1}{3}$	Divide each side by 3.
$\tan x = \pm\dfrac{1}{\sqrt{3}}$	Extract square roots.
$\tan x = \pm\dfrac{\sqrt{3}}{3}$	Rationalize the denominator.

Because $\tan x$ has a period of π, first find all solutions in the interval $[0, \pi)$. These solutions are $x = \pi/6$ and $x = 5\pi/6$. Finally, add multiples of π to each of these solutions to obtain the general form

$$x = \frac{\pi}{6} + n\pi \quad \text{and} \quad x = \frac{5\pi}{6} + n\pi \qquad \text{General solution}$$

where n is an integer.

✓ *Checkpoint*  *Audio-video solution in English & Spanish at LarsonPrecalculus.com.*

Solve $4 \sin^2 x - 3 = 0$.

The equations in Examples 1 and 2 involved only one trigonometric function. When two or more functions occur in the same equation, collect all terms on one side and try to separate the functions by factoring or by using appropriate identities. This may produce factors that yield no solutions, as illustrated in Example 3.

> **EXAMPLE 3**   Factoring

Solve $\cot x \cos^2 x = 2 \cot x$.

Solution   Begin by collecting all terms on one side of the equation and factoring.

$$\cot x \cos^2 x = 2 \cot x \qquad \text{Write original equation.}$$

$$\cot x \cos^2 x - 2 \cot x = 0 \qquad \text{Subtract } 2 \cot x \text{ from each side.}$$

$$\cot x (\cos^2 x - 2) = 0 \qquad \text{Factor.}$$

By setting each of these factors equal to zero, you obtain

$$\cot x = 0 \quad \text{and} \quad \cos^2 x - 2 = 0$$

$$\cos^2 x = 2$$

$$\cos x = \pm \sqrt{2}.$$

In the interval $(0, \pi)$, the equation $\cot x = 0$ has the solution

$$x = \frac{\pi}{2}.$$

No solution exists for $\cos x = \pm \sqrt{2}$ because $\pm \sqrt{2}$ are outside the range of the cosine function. Because $\cot x$ has a period of π, you obtain the general form of the solution by adding multiples of π to $x = \pi/2$ to get

$$x = \frac{\pi}{2} + n\pi \qquad \text{General solution}$$

where n is an integer. Confirm this graphically by sketching the graph of $y = \cot x \cos^2 x - 2 \cot x$, as shown below.

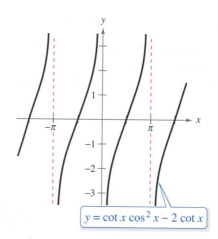

$$y = \cot x \cos^2 x - 2 \cot x$$

Notice that the x-intercepts occur at

$$-\frac{3\pi}{2}, \quad -\frac{\pi}{2}, \quad \frac{\pi}{2}, \quad \frac{3\pi}{2}$$

and so on. These x-intercepts correspond to the solutions of $\cot x \cos^2 x - 2 \cot x = 0$.

✓ **Checkpoint**   *Audio-video solution in English & Spanish at LarsonPrecalculus.com.*

Solve $\sin^2 x = 2 \sin x$.

Equations of Quadratic Type

Many trigonometric equations are of quadratic type $ax^2 + bx + c = 0$, as shown below. To solve equations of this type, factor the quadratic or, when this is not possible, use the Quadratic Formula.

Quadratic in sin x	**Quadratic in sec x**
$2 \sin^2 x - \sin x - 1 = 0$	$\sec^2 x - 3 \sec x - 2 = 0$
$2(\sin x)^2 - \sin x - 1 = 0$	$(\sec x)^2 - 3(\sec x) - 2 = 0$

EXAMPLE 4 Factoring an Equation of Quadratic Type

Find all solutions of $2 \sin^2 x - \sin x - 1 = 0$ in the interval $[0, 2\pi)$.

Algebraic Solution

Treat the equation as a quadratic in $\sin x$ and factor.

$2 \sin^2 x - \sin x - 1 = 0$ Write original equation.

$(2 \sin x + 1)(\sin x - 1) = 0$ Factor.

Setting each factor equal to zero, you obtain the following solutions in the interval $[0, 2\pi)$.

$2 \sin x + 1 = 0$ and $\sin x - 1 = 0$

$\sin x = -\dfrac{1}{2}$ $\sin x = 1$

$x = \dfrac{7\pi}{6}, \dfrac{11\pi}{6}$ $x = \dfrac{\pi}{2}$

Graphical Solution

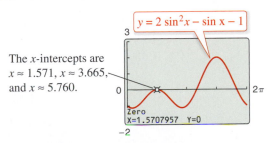

The x-intercepts are $x \approx 1.571$, $x \approx 3.665$, and $x \approx 5.760$.

From the above figure, you can conclude that the approximate solutions of $2 \sin^2 x - \sin x - 1 = 0$ in the interval $[0, 2\pi)$ are

$x \approx 1.571 \approx \dfrac{\pi}{2}$, $x \approx 3.665 \approx \dfrac{7\pi}{6}$, and $x \approx 5.760 \approx \dfrac{11\pi}{6}$.

✓ *Checkpoint* 🔊 *Audio-video solution in English & Spanish at LarsonPrecalculus.com.*

Find all solutions of $2 \sin^2 x - 3 \sin x + 1 = 0$ in the interval $[0, 2\pi)$.

EXAMPLE 5 Rewriting with a Single Trigonometric Function

Solve $2 \sin^2 x + 3 \cos x - 3 = 0$.

Solution This equation contains both sine and cosine functions. You can rewrite the equation so that it has only cosine functions by using the identity $\sin^2 x = 1 - \cos^2 x$.

$2 \sin^2 x + 3 \cos x - 3 = 0$ Write original equation.

$2(1 - \cos^2 x) + 3 \cos x - 3 = 0$ Pythagorean identity

$2 \cos^2 x - 3 \cos x + 1 = 0$ Multiply each side by -1.

$(2 \cos x - 1)(\cos x - 1) = 0$ Factor.

By setting each factor equal to zero, you can find the solutions in the interval $[0, 2\pi)$ to be $x = 0$, $x = \pi/3$, and $x = 5\pi/3$. Because $\cos x$ has a period of 2π, the general solution is

$x = 2n\pi$, $x = \dfrac{\pi}{3} + 2n\pi$, $x = \dfrac{5\pi}{3} + 2n\pi$ General solution

where n is an integer.

✓ *Checkpoint* 🔊 *Audio-video solution in English & Spanish at LarsonPrecalculus.com.*

Solve $3 \sec^2 x - 2 \tan^2 x - 4 = 0$.

Sometimes you must square each side of an equation to obtain a quadratic, as demonstrated in the next example. Because this procedure can introduce extraneous solutions, you should check any solutions in the original equation to see whether they are valid or extraneous.

REMARK You square each side of the equation in Example 6 because the squares of the sine and cosine functions are related by a Pythagorean identity. The same is true for the squares of the secant and tangent functions and for the squares of the cosecant and cotangent functions.

▷

EXAMPLE 6   Squaring and Converting to Quadratic Type

Find all solutions of $\cos x + 1 = \sin x$ in the interval $[0, 2\pi)$.

Solution It is not clear how to rewrite this equation in terms of a single trigonometric function. Notice what happens when you square each side of the equation.

$\cos x + 1 = \sin x$	Write original equation.
$\cos^2 x + 2 \cos x + 1 = \sin^2 x$	Square each side.
$\cos^2 x + 2 \cos x + 1 = 1 - \cos^2 x$	Pythagorean identity
$\cos^2 x + \cos^2 x + 2 \cos x + 1 - 1 = 0$	Rewrite equation.
$2 \cos^2 x + 2 \cos x = 0$	Combine like terms.
$2 \cos x(\cos x + 1) = 0$	Factor.

Setting each factor equal to zero produces

$$2 \cos x = 0 \qquad \text{and} \qquad \cos x + 1 = 0$$

$$\cos x = 0 \qquad\qquad\qquad \cos x = -1$$

$$x = \frac{\pi}{2}, \frac{3\pi}{2} \qquad\qquad\qquad x = \pi.$$

Because you squared the original equation, check for extraneous solutions.

Check $x = \dfrac{\pi}{2}$

$$\cos \frac{\pi}{2} + 1 \stackrel{?}{=} \sin \frac{\pi}{2} \qquad\qquad \text{Substitute } \frac{\pi}{2} \text{ for } x.$$

$$0 + 1 = 1 \qquad\qquad \text{Solution checks. } ✔$$

Check $x = \dfrac{3\pi}{2}$

$$\cos \frac{3\pi}{2} + 1 \stackrel{?}{=} \sin \frac{3\pi}{2} \qquad\qquad \text{Substitute } \frac{3\pi}{2} \text{ for } x.$$

$$0 + 1 \neq -1 \qquad\qquad \text{Solution does not check.}$$

Check $x = \pi$

$$\cos \pi + 1 \stackrel{?}{=} \sin \pi \qquad\qquad \text{Substitute } \pi \text{ for } x.$$

$$-1 + 1 = 0 \qquad\qquad \text{Solution checks. } ✔$$

Of the three possible solutions, $x = 3\pi/2$ is extraneous. So, in the interval $[0, 2\pi)$, the only two solutions are

$$x = \frac{\pi}{2} \quad \text{and} \quad x = \pi.$$

✓ *Checkpoint* ◀))) *Audio-video solution in English & Spanish at LarsonPrecalculus.com.*

Find all solutions of $\sin x + 1 = \cos x$ in the interval $[0, 2\pi)$. ■

Functions Involving Multiple Angles

The next two examples involve trigonometric functions of multiple angles of the forms $\cos ku$ and $\tan ku$. To solve equations of these forms, first solve the equation for ku, and then divide your result by k.

EXAMPLE 7   Solving a Multiple-Angle Equation

Solve $2 \cos 3t - 1 = 0$.

Solution

$$2 \cos 3t - 1 = 0 \qquad \text{Write original equation.}$$

$$2 \cos 3t = 1 \qquad \text{Add 1 to each side.}$$

$$\cos 3t = \frac{1}{2} \qquad \text{Divide each side by 2.}$$

In the interval $[0, 2\pi)$, you know that $3t = \pi/3$ and $3t = 5\pi/3$ are the only solutions, so, in general, you have

$$3t = \frac{\pi}{3} + 2n\pi \quad \text{and} \quad 3t = \frac{5\pi}{3} + 2n\pi.$$

Dividing these results by 3, you obtain the general solution

$$t = \frac{\pi}{9} + \frac{2n\pi}{3} \quad \text{and} \quad t = \frac{5\pi}{9} + \frac{2n\pi}{3} \qquad \text{General solution}$$

where n is an integer.

✓ **Checkpoint**  *Audio-video solution in English & Spanish at LarsonPrecalculus.com.*

Solve $2 \sin 2t - \sqrt{3} = 0$.

EXAMPLE 8   Solving a Multiple-Angle Equation

$$3 \tan \frac{x}{2} + 3 = 0 \qquad \text{Original equation}$$

$$3 \tan \frac{x}{2} = -3 \qquad \text{Subtract 3 from each side.}$$

$$\tan \frac{x}{2} = -1 \qquad \text{Divide each side by 3.}$$

In the interval $[0, \pi)$, you know that $x/2 = 3\pi/4$ is the only solution, so, in general, you have

$$\frac{x}{2} = \frac{3\pi}{4} + n\pi.$$

Multiplying this result by 2, you obtain the general solution

$$x = \frac{3\pi}{2} + 2n\pi \qquad \text{General solution}$$

where n is an integer.

✓ **Checkpoint**  *Audio-video solution in English & Spanish at LarsonPrecalculus.com.*

Solve $2 \tan \frac{x}{2} - 2 = 0$.

Using Inverse Functions

It is possible to find the minimum surface area of a honeycomb cell using a graphing utility or using calculus and the arccosine function.

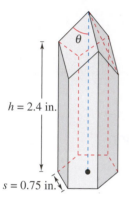

$h = 2.4$ in.

$s = 0.75$ in.

Figure 5.2

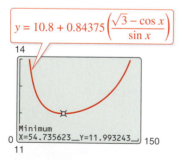

$y = 10.8 + 0.84375\left(\dfrac{\sqrt{3} - \cos x}{\sin x}\right)$

Minimum
X=54.735623 Y=11.993243

Figure 5.3

EXAMPLE 9 **Using Inverse Functions**

$$\sec^2 x - 2 \tan x = 4 \qquad \text{Original equation}$$
$$1 + \tan^2 x - 2 \tan x - 4 = 0 \qquad \text{Pythagorean identity}$$
$$\tan^2 x - 2 \tan x - 3 = 0 \qquad \text{Combine like terms.}$$
$$(\tan x - 3)(\tan x + 1) = 0 \qquad \text{Factor.}$$

Setting each factor equal to zero, you obtain two solutions in the interval $(-\pi/2, \pi/2)$. [Recall that the range of the inverse tangent function is $(-\pi/2, \pi/2)$.]

$$x = \arctan 3 \quad \text{and} \quad x = \arctan(-1) = -\pi/4$$

Finally, because $\tan x$ has a period of π, you add multiples of π to obtain

$$x = \arctan 3 + n\pi \quad \text{and} \quad x = (-\pi/4) + n\pi \qquad \text{General solution}$$

where n is an integer. You can use a calculator to approximate the value of $\arctan 3$.

✓ **Checkpoint**  *Audio-video solution in English & Spanish at LarsonPrecalculus.com.*

Solve $4 \tan^2 x + 5 \tan x - 6 = 0$.

EXAMPLE 10 **Surface Area of a Honeycomb Cell**

The surface area S (in square inches) of a honeycomb cell is given by

$$S = 6hs + 1.5s^2\left[\left(\sqrt{3} - \cos \theta\right)/\sin \theta\right], \quad 0 < \theta \le 90°$$

where $h = 2.4$ inches, $s = 0.75$ inch, and θ is the angle shown in Figure 5.2. What value of θ gives the minimum surface area?

Solution Letting $h = 2.4$ and $s = 0.75$, you obtain

$$S = 10.8 + 0.84375\left[\left(\sqrt{3} - \cos \theta\right)/\sin \theta\right].$$

Graph this function using a graphing utility. The minimum point on the graph, which occurs at $\theta \approx 54.7°$, is shown in Figure 5.3. By using calculus, the exact minimum point on the graph can be shown to occur at $\theta = \arccos\left(1/\sqrt{3}\right) \approx 0.9553 \approx 54.7356°$.

✓ **Checkpoint**  *Audio-video solution in English & Spanish at LarsonPrecalculus.com.*

In Example 10, for what value(s) of θ is the surface area 12 square inches? ◼

Summarize **(Section 5.3)**

1. Describe how to use standard algebraic techniques to solve trigonometric equations *(page 260)*. For examples of using standard algebraic techniques to solve trigonometric equations, see Examples 1–3.

2. Explain how to solve a trigonometric equation of quadratic type *(page 263)*. For examples of solving trigonometric equations of quadratic type, see Examples 4–6.

3. Explain how to solve a trigonometric equation involving multiple angles *(page 265)*. For examples of solving trigonometric equations involving multiple angles, see Examples 7 and 8.

4. Explain how to use inverse trigonometric functions to solve trigonometric equations *(page 266)*. For examples of using inverse trigonometric functions to solve trigonometric equations, see Examples 9 and 10.

5.4 Sum and Difference Formulas

Trigonometric identities enable you to rewrite trigonometric expressions. For instance, an identity can be used to rewrite a trigonometric expression in a form that helps you analyze a harmonic motion equation.

■ **Use sum and difference formulas to evaluate trigonometric functions, verify identities, and solve trigonometric equations.**

Using Sum and Difference Formulas

In this and the following section, you will study the uses of several trigonometric identities and formulas.

Sum and Difference Formulas

$$\sin(u + v) = \sin u \cos v + \cos u \sin v$$

$$\sin(u - v) = \sin u \cos v - \cos u \sin v$$

$$\cos(u + v) = \cos u \cos v - \sin u \sin v$$

$$\cos(u - v) = \cos u \cos v + \sin u \sin v$$

$$\tan(u + v) = \frac{\tan u + \tan v}{1 - \tan u \tan v} \qquad \tan(u - v) = \frac{\tan u - \tan v}{1 + \tan u \tan v}$$

For a proof of the sum and difference formulas for $\cos(u \pm v)$ and $\tan(u \pm v)$, see Proofs in Mathematics on page 283.

Examples 1 and 2 show how **sum and difference formulas** can enable you to find exact values of trigonometric functions involving sums or differences of special angles.

EXAMPLE 1   **Evaluating a Trigonometric Function**

Find the exact value of $\sin \dfrac{\pi}{12}$.

Solution   To find the *exact* value of $\sin \pi/12$, use the fact that

$$\frac{\pi}{12} = \frac{\pi}{3} - \frac{\pi}{4}.$$

Consequently, the formula for $\sin(u - v)$ yields

$$\sin \frac{\pi}{12} = \sin\left(\frac{\pi}{3} - \frac{\pi}{4}\right)$$

$$= \sin \frac{\pi}{3} \cos \frac{\pi}{4} - \cos \frac{\pi}{3} \sin \frac{\pi}{4}$$

$$= \frac{\sqrt{3}}{2}\left(\frac{\sqrt{2}}{2}\right) - \frac{1}{2}\left(\frac{\sqrt{2}}{2}\right)$$

$$= \frac{\sqrt{6} - \sqrt{2}}{4}.$$

Try checking this result on your calculator. You will find that $\sin \pi/12 \approx 0.259$.

✓ *Checkpoint*  *Audio-video solution in English & Spanish at LarsonPrecalculus.com.*

Find the exact value of $\cos \dfrac{\pi}{12}$.

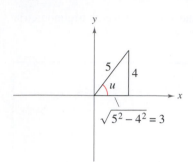

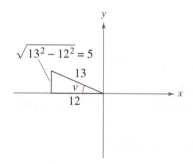

Figure 5.4

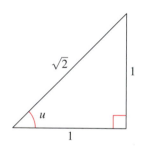

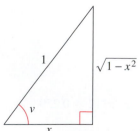

Figure 5.5

Figure 5.6

> **EXAMPLE 2**   **Evaluating a Trigonometric Function**

Find the exact value of $\cos 75°$.

Solution Using the fact that $75° = 30° + 45°$, together with the formula for $\cos(u + v)$, you obtain

$$\cos 75° = \cos(30° + 45°)$$

$$= \cos 30° \cos 45° - \sin 30° \sin 45°$$

$$= \frac{\sqrt{3}}{2}\left(\frac{\sqrt{2}}{2}\right) - \frac{1}{2}\left(\frac{\sqrt{2}}{2}\right)$$

$$= \frac{\sqrt{6} - \sqrt{2}}{4}.$$

✓ **Checkpoint** 🔊)) Audio-video solution in English & Spanish at LarsonPrecalculus.com.

Find the exact value of $\sin 75°$.

> **EXAMPLE 3**   **Evaluating a Trigonometric Expression**

Find the exact value of $\sin(u + v)$ given $\sin u = 4/5$, where $0 < u < \pi/2$, and $\cos v = -12/13$, where $\pi/2 < v < \pi$.

Solution Because $\sin u = 4/5$ and u is in Quadrant I, $\cos u = 3/5$, as shown in Figure 5.4. Because $\cos v = -12/13$ and v is in Quadrant II, $\sin v = 5/13$, as shown in Figure 5.5. You can find $\sin(u + v)$ as follows.

$$\sin(u + v) = \sin u \cos v + \cos u \sin v$$

$$= \frac{4}{5}\left(-\frac{12}{13}\right) + \frac{3}{5}\left(\frac{5}{13}\right)$$

$$= -\frac{33}{65}$$

✓ **Checkpoint** 🔊)) Audio-video solution in English & Spanish at LarsonPrecalculus.com.

Find the exact value of $\cos(u + v)$ given $\sin u = 12/13$, where $0 < u < \pi/2$, and $\cos v = -3/5$, where $\pi/2 < v < \pi$.

> **EXAMPLE 4**   **An Application of a Sum Formula**

Write $\cos(\arctan 1 + \arccos x)$ as an algebraic expression.

Solution This expression fits the formula for $\cos(u + v)$. Figure 5.6 shows angles $u = \arctan 1$ and $v = \arccos x$. So,

$$\cos(u + v) = \cos(\arctan 1)\cos(\arccos x) - \sin(\arctan 1)\sin(\arccos x)$$

$$= \frac{1}{\sqrt{2}} \cdot x - \frac{1}{\sqrt{2}} \cdot \sqrt{1 - x^2}$$

$$= \frac{x - \sqrt{1 - x^2}}{\sqrt{2}}.$$

✓ **Checkpoint** 🔊)) Audio-video solution in English & Spanish at LarsonPrecalculus.com.

Write $\sin(\arctan 1 + \arccos x)$ as an algebraic expression.

Hipparchus, considered the most eminent of Greek astronomers, was born about 190 B.C. in Nicaea. He is credited with the invention of trigonometry. He also derived the sum and difference formulas for $\sin(A \pm B)$ and $\cos(A \pm B)$.

 EXAMPLE 5   **Proving a Cofunction Identity**

Use a difference formula to prove the cofunction identity $\cos\left(\dfrac{\pi}{2} - x\right) = \sin x$.

Solution   Using the formula for $\cos(u - v)$, you have

$$\cos\left(\frac{\pi}{2} - x\right) = \cos\frac{\pi}{2}\cos x + \sin\frac{\pi}{2}\sin x$$

$$= (0)(\cos x) + (1)(\sin x)$$

$$= \sin x.$$

✓ **Checkpoint** ◀))) *Audio-video solution in English & Spanish at LarsonPrecalculus.com.*

Use a difference formula to prove the cofunction identity $\sin\left(x - \dfrac{\pi}{2}\right) = -\cos x.$

Sum and difference formulas can be used to rewrite expressions such as

$$\sin\left(\theta + \frac{n\pi}{2}\right) \quad \text{and} \quad \cos\left(\theta + \frac{n\pi}{2}\right), \quad \text{where } n \text{ is an integer}$$

as expressions involving only $\sin\theta$ or $\cos\theta$. The resulting formulas are called **reduction formulas.**

EXAMPLE 6   **Deriving Reduction Formulas**

Simplify each expression.

a. $\cos\left(\theta - \dfrac{3\pi}{2}\right)$

b. $\tan(\theta + 3\pi)$

Solution

a. Using the formula for $\cos(u - v)$, you have

$$\cos\left(\theta - \frac{3\pi}{2}\right) = \cos\theta\cos\frac{3\pi}{2} + \sin\theta\sin\frac{3\pi}{2}$$

$$= (\cos\theta)(0) + (\sin\theta)(-1)$$

$$= -\sin\theta.$$

b. Using the formula for $\tan(u + v)$, you have

$$\tan(\theta + 3\pi) = \frac{\tan\theta + \tan 3\pi}{1 - \tan\theta\tan 3\pi}$$

$$= \frac{\tan\theta + 0}{1 - (\tan\theta)(0)}$$

$$= \tan\theta.$$

✓ **Checkpoint** ◀))) *Audio-video solution in English & Spanish at LarsonPrecalculus.com.*

Simplify each expression.

a. $\sin\left(\dfrac{3\pi}{2} - \theta\right)$ **b.** $\tan\left(\theta - \dfrac{\pi}{4}\right)$

EXAMPLE 7 **Solving a Trigonometric Equation**

Find all solutions of $\sin[x + (\pi/4)] + \sin[x - (\pi/4)] = -1$ in the interval $[0, 2\pi)$.

Algebraic Solution

Using sum and difference formulas, rewrite the equation as

$$\sin x \cos \frac{\pi}{4} + \cos x \sin \frac{\pi}{4} + \sin x \cos \frac{\pi}{4} - \cos x \sin \frac{\pi}{4} = -1$$

$$2 \sin x \cos \frac{\pi}{4} = -1$$

$$2(\sin x)\left(\frac{\sqrt{2}}{2}\right) = -1$$

$$\sin x = -\frac{1}{\sqrt{2}}$$

$$\sin x = -\frac{\sqrt{2}}{2}.$$

So, the only solutions in the interval $[0, 2\pi)$ are $x = 5\pi/4$ and $x = 7\pi/4$.

Graphical Solution

$$y = \sin\left(x + \frac{\pi}{4}\right) + \sin\left(x - \frac{\pi}{4}\right) + 1$$

The x-intercepts are $x \approx 3.927$ and $x \approx 5.498$.

Zero
X=3.9269908 Y=0

From the above figure, you can conclude that the approximate solutions in the interval $[0, 2\pi)$ are

$$x \approx 3.927 \approx \frac{5\pi}{4} \quad \text{and} \quad x \approx 5.498 \approx \frac{7\pi}{4}.$$

✓ **Checkpoint** ◀))) *Audio-video solution in English & Spanish at LarsonPrecalculus.com.*

Find all solutions of $\sin[x + (\pi/2)] + \sin[x - (3\pi/2)] = 1$ in the interval $[0, 2\pi)$.

The next example is an application from calculus.

EXAMPLE 8 **An Application from Calculus** $\int$

Verify that $\dfrac{\sin(x + h) - \sin x}{h} = (\cos x)\left(\dfrac{\sin h}{h}\right) - (\sin x)\left(\dfrac{1 - \cos h}{h}\right)$, where $h \neq 0$.

Solution Using the formula for $\sin(u + v)$, you have

$$\frac{\sin(x + h) - \sin x}{h} = \frac{\sin x \cos h + \cos x \sin h - \sin x}{h}$$

$$= \frac{\cos x \sin h - \sin x(1 - \cos h)}{h}$$

$$= (\cos x)\left(\frac{\sin h}{h}\right) - (\sin x)\left(\frac{1 - \cos h}{h}\right).$$

One application of the sum and difference formulas is in the analysis of standing waves, such as those that can be produced when plucking a guitar string.

✓ **Checkpoint** ◀))) *Audio-video solution in English & Spanish at LarsonPrecalculus.com.*

Verify that $\dfrac{\cos(x + h) - \cos x}{h} = (\cos x)\left(\dfrac{\cos h - 1}{h}\right) - (\sin x)\left(\dfrac{\sin h}{h}\right)$, where $h \neq 0$.

Summarize (Section 5.4)

1. State the sum and difference formulas for sine, cosine, and tangent (*page 267*). For examples of using the sum and difference formulas to evaluate trigonometric functions, verify identities, and solve trigonometric equations, see Examples 1–8.

5.5   Multiple-Angle and Product-to-Sum Formulas

A variety of trigonometric formulas enable you to rewrite trigonometric functions in more convenient forms. For instance, a half-angle formula can be used to relate the Mach number of a supersonic airplane to the apex angle of the cone formed by the sound waves behind the airplane.

- Use multiple-angle formulas to rewrite and evaluate trigonometric functions.
- Use power-reducing formulas to rewrite and evaluate trigonometric functions.
- Use half-angle formulas to rewrite and evaluate trigonometric functions.
- Use product-to-sum and sum-to-product formulas to rewrite and evaluate trigonometric functions.
- Use trigonometric formulas to rewrite real-life models.

Multiple-Angle Formulas

In this section, you will study four other categories of trigonometric identities.

1. The first category involves *functions of multiple angles* such as $\sin ku$ and $\cos ku$.

2. The second category involves *squares of trigonometric functions* such as $\sin^2 u$.

3. The third category involves *functions of half-angles* such as $\sin(u/2)$.

4. The fourth category involves *products of trigonometric functions* such as $\sin u \cos v$.

You should learn the **double-angle formulas** because they are used often in trigonometry and calculus. For proofs of these formulas, see Proofs in Mathematics on page 284.

Double-Angle Formulas

$$\sin 2u = 2 \sin u \cos u \qquad\qquad \cos 2u = \cos^2 u - \sin^2 u$$

$$\tan 2u = \frac{2 \tan u}{1 - \tan^2 u} \qquad\qquad\qquad = 2 \cos^2 u - 1$$

$$\qquad\qquad\qquad\qquad\qquad\qquad = 1 - 2 \sin^2 u$$

EXAMPLE 1   **Solving a Multiple-Angle Equation**

Solve $2 \cos x + \sin 2x = 0$.

Solution   Begin by rewriting the equation so that it involves functions of x (rather than $2x$). Then factor and solve.

$2 \cos x + \sin 2x = 0$	Write original equation.
$2 \cos x + 2 \sin x \cos x = 0$	Double-angle formula
$2 \cos x(1 + \sin x) = 0$	Factor.
$2 \cos x = 0 \quad \text{and} \quad 1 + \sin x = 0$	Set factors equal to zero.
$x = \dfrac{\pi}{2}, \dfrac{3\pi}{2} \qquad\qquad x = \dfrac{3\pi}{2}$	Solutions in $[0, 2\pi)$

So, the general solution is

$$x = \frac{\pi}{2} + 2n\pi \quad \text{and} \quad x = \frac{3\pi}{2} + 2n\pi$$

where n is an integer. Try verifying these solutions graphically.

✓ *Checkpoint*  *Audio-video solution in English & Spanish at LarsonPrecalculus.com.*

Solve $\cos 2x + \cos x = 0$.

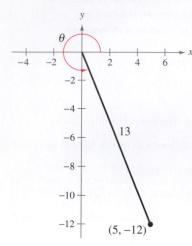

Figure 5.7

EXAMPLE 2   **Evaluating Functions Involving Double Angles**

Use the following to find $\sin 2\theta$, $\cos 2\theta$, and $\tan 2\theta$.

$$\cos \theta = \frac{5}{13}, \quad \frac{3\pi}{2} < \theta < 2\pi$$

Solution   From Figure 5.7,

$$\sin \theta = \frac{y}{r} = -\frac{12}{13} \quad \text{and} \quad \tan \theta = \frac{y}{x} = -\frac{12}{5}.$$

Consequently, using each of the double-angle formulas, you can write

$$\sin 2\theta = 2 \sin \theta \cos \theta = 2\left(-\frac{12}{13}\right)\left(\frac{5}{13}\right) = -\frac{120}{169}$$

$$\cos 2\theta = 2 \cos^2 \theta - 1 = 2\left(\frac{25}{169}\right) - 1 = -\frac{119}{169}$$

$$\tan 2\theta = \frac{2 \tan \theta}{1 - \tan^2 \theta} = \frac{2\left(-\dfrac{12}{5}\right)}{1 - \left(-\dfrac{12}{5}\right)^2} = \frac{120}{119}.$$

✓ *Checkpoint*  Audio-video solution in English & Spanish at *LarsonPrecalculus.com.*

Use the following to find $\sin 2\theta$, $\cos 2\theta$, and $\tan 2\theta$.

$$\sin \theta = \frac{3}{5}, \quad 0 < \theta < \frac{\pi}{2}$$

The double-angle formulas are not restricted to angles 2θ and θ. Other *double* combinations, such as 4θ and 2θ or 6θ and 3θ, are also valid. Here are two examples.

$$\sin 4\theta = 2 \sin 2\theta \cos 2\theta \quad \text{and} \quad \cos 6\theta = \cos^2 3\theta - \sin^2 3\theta$$

By using double-angle formulas together with the sum formulas given in the preceding section, you can form other multiple-angle formulas.

EXAMPLE 3   **Deriving a Triple-Angle Formula**

Rewrite $\sin 3x$ in terms of $\sin x$.

Solution

$\sin 3x = \sin(2x + x)$	Rewrite as a sum.
$= \sin 2x \cos x + \cos 2x \sin x$	Sum formula
$= 2 \sin x \cos x \cos x + (1 - 2 \sin^2 x) \sin x$	Double-angle formulas
$= 2 \sin x \cos^2 x + \sin x - 2 \sin^3 x$	Distributive Property
$= 2 \sin x(1 - \sin^2 x) + \sin x - 2 \sin^3 x$	Pythagorean identity
$= 2 \sin x - 2 \sin^3 x + \sin x - 2 \sin^3 x$	Distributive Property
$= 3 \sin x - 4 \sin^3 x$	Simplify.

✓ *Checkpoint*  Audio-video solution in English & Spanish at *LarsonPrecalculus.com.*

Rewrite $\cos 3x$ in terms of $\cos x$.

Power-Reducing Formulas

The double-angle formulas can be used to obtain the following **power-reducing formulas.**

Power-Reducing Formulas

$$\sin^2 u = \frac{1 - \cos 2u}{2}$$

$$\cos^2 u = \frac{1 + \cos 2u}{2}$$

$$\tan^2 u = \frac{1 - \cos 2u}{1 + \cos 2u}$$

For a proof of the power-reducing formulas, see Proofs in Mathematics on page 281. Example 4 shows a typical power reduction used in calculus.

EXAMPLE 4   **Reducing a Power**

Rewrite $\sin^4 x$ in terms of first powers of the cosines of multiple angles.

Solution   Note the repeated use of power-reducing formulas.

$\sin^4 x = (\sin^2 x)^2$	Property of exponents
$= \left(\dfrac{1 - \cos 2x}{2}\right)^2$	Power-reducing formula
$= \dfrac{1}{4}(1 - 2\cos 2x + \cos^2 2x)$	Expand.
$= \dfrac{1}{4}\left(1 - 2\cos 2x + \dfrac{1 + \cos 4x}{2}\right)$	Power-reducing formula
$= \dfrac{1}{4} - \dfrac{1}{2}\cos 2x + \dfrac{1}{8} + \dfrac{1}{8}\cos 4x$	Distributive Property
$= \dfrac{3}{8} - \dfrac{1}{2}\cos 2x + \dfrac{1}{8}\cos 4x$	Simplify.
$= \dfrac{1}{8}(3 - 4\cos 2x + \cos 4x)$	Factor out common factor.

You can use a graphing utility to check this result, as shown below. Notice that the graphs coincide.

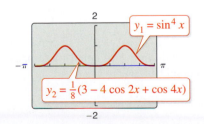

$$y_1 = \sin^4 x$$
$$y_2 = \frac{1}{8}(3 - 4\cos 2x + \cos 4x)$$

 ✓ *Checkpoint*   *Audio-video solution in English & Spanish at LarsonPrecalculus.com.*

Rewrite $\tan^4 x$ in terms of first powers of the cosines of multiple angles.   ◼

Half-Angle Formulas

You can derive some useful alternative forms of the power-reducing formulas by replacing u with $u/2$. The results are called **half-angle formulas.**

> **REMARK** To find the exact value of a trigonometric function with an angle measure in D°M′S″ form using a half-angle formula, first convert the angle measure to decimal degree form. Then multiply the resulting angle measure by 2.

Half-Angle Formulas

$$\sin \frac{u}{2} = \pm\sqrt{\frac{1 - \cos u}{2}} \qquad \cos \frac{u}{2} = \pm\sqrt{\frac{1 + \cos u}{2}}$$

$$\tan \frac{u}{2} = \frac{1 - \cos u}{\sin u} = \frac{\sin u}{1 + \cos u}$$

The signs of $\sin \dfrac{u}{2}$ and $\cos \dfrac{u}{2}$ depend on the quadrant in which $\dfrac{u}{2}$ lies.

> **REMARK** Use your calculator to verify the result obtained in Example 5. That is, evaluate $\sin 105°$ and $\left(\sqrt{2 + \sqrt{3}}\right)/2$. Note that both values are approximately 0.9659258.

EXAMPLE 5   **Using a Half-Angle Formula**

Find the exact value of $\sin 105°$.

Solution   Begin by noting that $105°$ is half of $210°$. Then, using the half-angle formula for $\sin(u/2)$ and the fact that $105°$ lies in Quadrant II, you have

$$\sin 105° = \sqrt{\frac{1 - \cos 210°}{2}} = \sqrt{\frac{1 + \left(\sqrt{3}/2\right)}{2}} = \frac{\sqrt{2 + \sqrt{3}}}{2}.$$

The positive square root is chosen because $\sin \theta$ is positive in Quadrant II.

✓ *Checkpoint* ))) *Audio-video solution in English & Spanish at LarsonPrecalculus.com.*

Find the exact value of $\cos 105°$.

EXAMPLE 6   **Solving a Trigonometric Equation**

Find all solutions of $1 + \cos^2 x = 2\cos^2 \dfrac{x}{2}$ in the interval $[0, 2\pi)$.

Algebraic Solution

$$1 + \cos^2 x = 2\cos^2 \frac{x}{2} \qquad \text{Write original equation.}$$

$$1 + \cos^2 x = 2\left(\pm\sqrt{\frac{1 + \cos x}{2}}\right)^2 \qquad \text{Half-angle formula}$$

$$1 + \cos^2 x = 1 + \cos x \qquad \text{Simplify.}$$

$$\cos^2 x - \cos x = 0 \qquad \text{Simplify.}$$

$$\cos x(\cos x - 1) = 0 \qquad \text{Factor.}$$

By setting the factors $\cos x$ and $\cos x - 1$ equal to zero, you find that the solutions in the interval $[0, 2\pi)$ are

$$x = \frac{\pi}{2}, \quad x = \frac{3\pi}{2}, \quad \text{and} \quad x = 0.$$

Graphical Solution

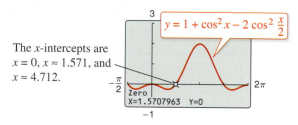

The x-intercepts are $x = 0$, $x \approx 1.571$, and $x \approx 4.712$.

From the above figure, you can conclude that the approximate solutions of $1 + \cos^2 x = 2\cos^2 x/2$ in the interval $[0, 2\pi)$ are

$$x = 0, \quad x \approx 1.571 \approx \frac{\pi}{2}, \quad \text{and} \quad x \approx 4.712 \approx \frac{3\pi}{2}.$$

✓ *Checkpoint* ))) *Audio-video solution in English & Spanish at LarsonPrecalculus.com.*

Find all solutions of $\cos^2 x = \sin^2 \dfrac{x}{2}$ in the interval $[0, 2\pi)$.

Product-to-Sum Formulas

Each of the following **product-to-sum formulas** can be verified using the sum and difference formulas discussed in the preceding section.

Product-to-Sum Formulas

$$\sin u \sin v = \frac{1}{2}[\cos(u - v) - \cos(u + v)]$$

$$\cos u \cos v = \frac{1}{2}[\cos(u - v) + \cos(u + v)]$$

$$\sin u \cos v = \frac{1}{2}[\sin(u + v) + \sin(u - v)]$$

$$\cos u \sin v = \frac{1}{2}[\sin(u + v) - \sin(u - v)]$$

Product-to-sum formulas are used in calculus to solve problems involving the products of sines and cosines of two different angles.

EXAMPLE 7   **Writing Products as Sums**

Rewrite the product $\cos 5x \sin 4x$ as a sum or difference.

Solution   Using the appropriate product-to-sum formula, you obtain

$$\cos 5x \sin 4x = \frac{1}{2}[\sin(5x + 4x) - \sin(5x - 4x)]$$

$$= \frac{1}{2}\sin 9x - \frac{1}{2}\sin x.$$

✓ *Checkpoint*   *Audio-video solution in English & Spanish at LarsonPrecalculus.com.*

Rewrite the product $\sin 5x \cos 3x$ as a sum or difference.   ■

Occasionally, it is useful to reverse the procedure and write a sum of trigonometric functions as a product. This can be accomplished with the following **sum-to-product formulas.**

Sum-to-Product Formulas

$$\sin u + \sin v = 2 \sin\left(\frac{u + v}{2}\right) \cos\left(\frac{u - v}{2}\right)$$

$$\sin u - \sin v = 2 \cos\left(\frac{u + v}{2}\right) \sin\left(\frac{u - v}{2}\right)$$

$$\cos u + \cos v = 2 \cos\left(\frac{u + v}{2}\right) \cos\left(\frac{u - v}{2}\right)$$

$$\cos u - \cos v = -2 \sin\left(\frac{u + v}{2}\right) \sin\left(\frac{u - v}{2}\right)$$

For a proof of the sum-to-product formulas, see Proofs in Mathematics on page 282.

EXAMPLE 8 Using a Sum-to-Product Formula

Find the exact value of $\cos 195° + \cos 105°$.

Solution Using the appropriate sum-to-product formula, you obtain

$$\cos 195° + \cos 105° = 2 \cos\left(\frac{195° + 105°}{2}\right) \cos\left(\frac{195° - 105°}{2}\right)$$

$$= 2 \cos 150° \cos 45°$$

$$= 2\left(-\frac{\sqrt{3}}{2}\right)\left(\frac{\sqrt{2}}{2}\right)$$

$$= -\frac{\sqrt{6}}{2}.$$

✓ **Checkpoint** 🔊 *Audio-video solution in English & Spanish at LarsonPrecalculus.com.*

Find the exact value of $\sin 195° + \sin 105°$.

EXAMPLE 9 Solving a Trigonometric Equation

Solve $\sin 5x + \sin 3x = 0$.

Solution

$$\sin 5x + \sin 3x = 0 \qquad \text{Write original equation.}$$

$$2 \sin\left(\frac{5x + 3x}{2}\right) \cos\left(\frac{5x - 3x}{2}\right) = 0 \qquad \text{Sum-to-product formula}$$

$$2 \sin 4x \cos x = 0 \qquad \text{Simplify.}$$

By setting the factor $2 \sin 4x$ equal to zero, you can find that the solutions in the interval $[0, 2\pi)$ are

$$x = 0, \frac{\pi}{4}, \frac{\pi}{2}, \frac{3\pi}{4}, \pi, \frac{5\pi}{4}, \frac{3\pi}{2}, \frac{7\pi}{4}.$$

The equation $\cos x = 0$ yields no additional solutions, so you can conclude that the solutions are of the form $x = n\pi/4$ where n is an integer. To confirm this graphically, sketch the graph of $y = \sin 5x + \sin 3x$, as shown below.

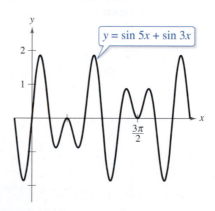

Notice from the graph that the x-intercepts occur at multiples of $\pi/4$.

✓ **Checkpoint** 🔊 *Audio-video solution in English & Spanish at LarsonPrecalculus.com.*

Solve $\sin 4x - \sin 2x = 0$.

Application

EXAMPLE 10   **Projectile Motion**

Ignoring air resistance, the range of a projectile fired at an angle θ with the horizontal and with an initial velocity of v_0 feet per second is given by

$$r = \frac{1}{16}v_0^2 \sin \theta \cos \theta$$

where r is the horizontal distance (in feet) that the projectile travels. A football player can kick a football from ground level with an initial velocity of 80 feet per second.

a. Write the projectile motion model in a simpler form.

b. At what angle must the player kick the football so that the football travels 200 feet?

Solution

a. You can use a double-angle formula to rewrite the projectile motion model as

$$r = \frac{1}{32}v_0^2(2 \sin \theta \cos \theta) \qquad \text{Rewrite original projectile motion model.}$$

$$= \frac{1}{32}v_0^2 \sin 2\theta. \qquad \text{Rewrite model using a double-angle formula.}$$

b. $r = \frac{1}{32}v_0^2 \sin 2\theta \qquad \text{Write projectile motion model.}$

$200 = \frac{1}{32}(80)^2 \sin 2\theta \qquad \text{Substitute 200 for } r \text{ and 80 for } v_0.$

$200 = 200 \sin 2\theta \qquad \text{Simplify.}$

$1 = \sin 2\theta \qquad \text{Divide each side by 200.}$

You know that $2\theta = \pi/2$, so dividing this result by 2 produces $\theta = \pi/4$. Because $\pi/4 = 45°$, the player must kick the football at an angle of 45° so that the football travels 200 feet.

Kicking a football with an initial velocity of 80 feet per second at an angle of 45° with the horizontal results in a distance traveled of 200 feet.

✓ *Checkpoint* *Audio-video solution in English & Spanish at LarsonPrecalculus.com.*

In Example 10, for what angle is the horizontal distance the football travels a maximum? ■

Summarize   (Section 5.5)

1. State the double-angle formulas *(page 271)*. For examples of using multiple-angle formulas to rewrite and evaluate trigonometric functions, see Examples 1–3.

2. State the power-reducing formulas *(page 273)*. For an example of using power-reducing formulas to rewrite a trigonometric function, see Example 4.

3. State the half-angle formulas *(page 274)*. For examples of using half-angle formulas to rewrite and evaluate trigonometric functions, see Examples 5 and 6.

4. State the product-to-sum and sum-to-product formulas *(page 275)*. For an example of using a product-to-sum formula to rewrite a trigonometric function, see Example 7. For examples of using sum-to-product formulas to rewrite and evaluate trigonometric functions, see Examples 8 and 9.

5. Describe an example of how to use a trigonometric formula to rewrite a real-life model *(page 277, Example 10)*.

Chapter Summary

	What Did You Learn?	**Explanation/Examples**	**Review Exercises**
Section 5.1	Recognize and write the fundamental trigonometric identities *(p. 250)*.	**Reciprocal Identities** $\sin u = 1/\csc u \qquad \cos u = 1/\sec u \qquad \tan u = 1/\cot u$ $\csc u = 1/\sin u \qquad \sec u = 1/\cos u \qquad \cot u = 1/\tan u$ **Quotient Identities:** $\tan u = \dfrac{\sin u}{\cos u}, \quad \cot u = \dfrac{\cos u}{\sin u}$ **Pythagorean Identities:** $\sin^2 u + \cos^2 u = 1$, $1 + \tan^2 u = \sec^2 u, \quad 1 + \cot^2 u = \csc^2 u$ **Cofunction Identities** $\sin[(\pi/2) - u] = \cos u \qquad \cos[(\pi/2) - u] = \sin u$ $\tan[(\pi/2) - u] = \cot u \qquad \cot[(\pi/2) - u] = \tan u$ $\sec[(\pi/2) - u] = \csc u \qquad \csc[(\pi/2) - u] = \sec u$ **Even/Odd Identities** $\sin(-u) = -\sin u \qquad \cos(-u) = \cos u \qquad \tan(-u) = -\tan u$ $\csc(-u) = -\csc u \qquad \sec(-u) = \sec u \qquad \cot(-u) = -\cot u$	1–4
	Use the fundamental trigonometric identities to evaluate trigonometric functions, simplify trigonometric expressions, and rewrite trigonometric expressions *(p. 251)*.	In some cases, when factoring or simplifying trigonometric expressions, it is helpful to rewrite the expression in terms of just *one* trigonometric function or in terms of *sine and cosine only*.	5–18
Section 5.2	Verify trigonometric identities *(p. 255)*.	**Guidelines for Verifying Trigonometric Identities** **1.** Work with one side of the equation at a time. **2.** Look to factor an expression, add fractions, square a binomial, or create a monomial denominator. **3.** Look to use the fundamental identities. Note which functions are in the final expression you want. Sines and cosines pair up well, as do secants and tangents, and cosecants and cotangents. **4.** If the preceding guidelines do not help, then try converting all terms to sines and cosines. **5.** Always try *something*.	19–26
Section 5.3	Use standard algebraic techniques to solve trigonometric equations *(p. 260)*.	Use standard algebraic techniques (when possible) such as collecting like terms, extracting square roots, and factoring to solve trigonometric equations.	27–32
	Solve trigonometric equations of quadratic type *(p. 263)*.	To solve trigonometric equations of quadratic type $ax^2 + bx + c = 0$, factor the quadratic or, when this is not possible, use the Quadratic Formula.	33–36
	Solve trigonometric equations involving multiple angles *(p. 265)*.	To solve equations that contain forms such as $\sin ku$ or $\cos ku$, first solve the equation for ku, and then divide your result by k.	37–42
	Use inverse trigonometric functions to solve trigonometric equations *(p. 266)*.	After factoring an equation, you may get an equation such as $(\tan x - 3)(\tan x + 1) = 0$. In such cases, use inverse trigonometric functions to solve. (See Example 9.)	43–46

What Did You Learn?	Explanation/Examples	Review Exercises
Section 5.4 Use sum and difference formulas to evaluate trigonometric functions, verify identities, and solve trigonometric equations (p. 267).	**Sum and Difference Formulas** $$\sin(u + v) = \sin u \cos v + \cos u \sin v$$ $$\sin(u - v) = \sin u \cos v - \cos u \sin v$$ $$\cos(u + v) = \cos u \cos v - \sin u \sin v$$ $$\cos(u - v) = \cos u \cos v + \sin u \sin v$$ $$\tan(u + v) = \frac{\tan u + \tan v}{1 - \tan u \tan v}$$ $$\tan(u - v) = \frac{\tan u - \tan v}{1 + \tan u \tan v}$$	47–62
Section 5.5 Use multiple-angle formulas to rewrite and evaluate trigonometric functions (p. 271).	**Double-Angle Formulas** $$\sin 2u = 2 \sin u \cos u \qquad \cos 2u = \cos^2 u - \sin^2 u$$ $$\tan 2u = \frac{2 \tan u}{1 - \tan^2 u} \qquad\quad = 2 \cos^2 u - 1$$ $$= 1 - 2 \sin^2 u$$	63–66
Use power-reducing formulas to rewrite and evaluate trigonometric functions (p. 273).	**Power-Reducing Formulas** $$\sin^2 u = \frac{1 - \cos 2u}{2}, \quad \cos^2 u = \frac{1 + \cos 2u}{2}$$ $$\tan^2 u = \frac{1 - \cos 2u}{1 + \cos 2u}$$	67, 68
Use half-angle formulas to rewrite and evaluate trigonometric functions (p. 274).	**Half-Angle Formulas** $$\sin \frac{u}{2} = \pm \sqrt{\frac{1 - \cos u}{2}}, \quad \cos \frac{u}{2} = \pm \sqrt{\frac{1 + \cos u}{2}}$$ $$\tan \frac{u}{2} = \frac{1 - \cos u}{\sin u} = \frac{\sin u}{1 + \cos u}$$ The signs of $\sin(u/2)$ and $\cos(u/2)$ depend on the quadrant in which $u/2$ lies.	69–74
Use product-to-sum and sum-to-product formulas to rewrite and evaluate trigonometric functions (p. 275).	**Product-to-Sum Formulas** $$\sin u \sin v = (1/2)[\cos(u - v) - \cos(u + v)]$$ $$\cos u \cos v = (1/2)[\cos(u - v) + \cos(u + v)]$$ $$\sin u \cos v = (1/2)[\sin(u + v) + \sin(u - v)]$$ $$\cos u \sin v = (1/2)[\sin(u + v) - \sin(u - v)]$$ **Sum-to-Product Formulas** $$\sin u + \sin v = 2 \sin\left(\frac{u + v}{2}\right) \cos\left(\frac{u - v}{2}\right)$$ $$\sin u - \sin v = 2 \cos\left(\frac{u + v}{2}\right) \sin\left(\frac{u - v}{2}\right)$$ $$\cos u + \cos v = 2 \cos\left(\frac{u + v}{2}\right) \cos\left(\frac{u - v}{2}\right)$$ $$\cos u - \cos v = -2 \sin\left(\frac{u + v}{2}\right) \sin\left(\frac{u - v}{2}\right)$$	75–78
Use trigonometric formulas to rewrite real-life models (p. 277).	A trigonometric formula can be used to rewrite the projectile motion model $r = (1/16)v_0^2 \sin \theta \cos \theta$. (See Example 10.)	79, 80

Review Exercises See **CalcChat.com** for tutorial help and worked-out solutions to odd-numbered exercises.

5.1 **Recognizing a Fundamental Identity** In Exercises 1–4, name the trigonometric function that is equivalent to the expression.

1. $\dfrac{\sin x}{\cos x}$

2. $\dfrac{1}{\sin x}$

3. $\dfrac{1}{\tan x}$

4. $\sqrt{\cot^2 x + 1}$

Using Identities to Evaluate a Function In Exercises 5 and 6, use the given values and fundamental trigonometric identities to find the values (if possible) of all six trigonometric functions.

5. $\tan \theta = \dfrac{2}{3}, \quad \sec \theta = \dfrac{\sqrt{13}}{3}$

6. $\sin\left(\dfrac{\pi}{2} - x\right) = \dfrac{\sqrt{2}}{2}, \quad \sin x = -\dfrac{\sqrt{2}}{2}$

Simplifying a Trigonometric Expression In Exercises 7–16, use the fundamental trigonometric identities to simplify the expression. There is more than one correct form of each answer.

7. $\dfrac{1}{\cot^2 x + 1}$

8. $\dfrac{\tan \theta}{1 - \cos^2 \theta}$

9. $\tan^2 x(\csc^2 x - 1)$

10. $\cot^2 x(\sin^2 x)$

11. $\dfrac{\cot\left(\dfrac{\pi}{2} - u\right)}{\cos u}$

12. $\dfrac{\sec^2(-\theta)}{\csc^2 \theta}$

13. $\cos^2 x + \cos^2 x \cot^2 x$

14. $(\tan x + 1)^2 \cos x$

15. $\dfrac{1}{\csc \theta + 1} - \dfrac{1}{\csc \theta - 1}$

16. $\dfrac{\tan^2 x}{1 + \sec x}$

Trigonometric Substitution In Exercises 17 and 18, use the trigonometric substitution to write the algebraic expression as a trigonometric function of θ, where $0 < \theta < \pi/2$.

17. $\sqrt{25 - x^2}, \, x = 5 \sin \theta$

18. $\sqrt{x^2 - 16}, \, x = 4 \sec \theta$

5.2 **Verifying a Trigonometric Identity** In Exercises 19–26, verify the identity.

19. $\cos x(\tan^2 x + 1) = \sec x$

20. $\sec^2 x \cot x - \cot x = \tan x$

21. $\sec\left(\dfrac{\pi}{2} - \theta\right) = \csc \theta$

22. $\cot\left(\dfrac{\pi}{2} - x\right) = \tan x$

23. $\dfrac{1}{\tan \theta \csc \theta} = \cos \theta$

24. $\dfrac{1}{\tan x \csc x \sin x} = \cot x$

25. $\sin^5 x \cos^2 x = (\cos^2 x - 2 \cos^4 x + \cos^6 x) \sin x$

26. $\cos^3 x \sin^2 x = (\sin^2 x - \sin^4 x) \cos x$

5.3 **Solving a Trigonometric Equation** In Exercises 27–32, solve the equation.

27. $\sin x = \sqrt{3} - \sin x$

28. $4 \cos \theta = 1 + 2 \cos \theta$

29. $3\sqrt{3} \tan u = 3$

30. $\frac{1}{2} \sec x - 1 = 0$

31. $3 \csc^2 x = 4$

32. $4 \tan^2 u - 1 = \tan^2 u$

Solving a Trigonometric Equation In Exercises 33–42, find all solutions of the equation in the interval $[0, 2\pi)$.

33. $2 \cos^2 x - \cos x = 1$

34. $2 \cos^2 x + 3 \cos x = 0$

35. $\cos^2 x + \sin x = 1$

36. $\sin^2 x + 2 \cos x = 2$

37. $2 \sin 2x - \sqrt{2} = 0$

38. $2 \cos \dfrac{x}{2} + 1 = 0$

39. $3 \tan^2\left(\dfrac{x}{3}\right) - 1 = 0$

40. $\sqrt{3} \tan 3x = 0$

41. $\cos 4x(\cos x - 1) = 0$

42. $3 \csc^2 5x = -4$

Using Inverse Functions In Exercises 43–46, use inverse functions where needed to find all solutions of the equation in the interval $[0, 2\pi)$.

43. $\tan^2 x - 2 \tan x = 0$

44. $2 \tan^2 x - 3 \tan x = -1$

45. $\tan^2 \theta + \tan \theta - 6 = 0$

46. $\sec^2 x + 6 \tan x + 4 = 0$

5.4 **Evaluating Trigonometric Functions** In Exercises 47–50, find the exact values of the sine, cosine, and tangent of the angle.

47. $285° = 315° - 30°$

48. $345° = 300° + 45°$

49. $\dfrac{25\pi}{12} = \dfrac{11\pi}{6} + \dfrac{\pi}{4}$

50. $\dfrac{19\pi}{12} = \dfrac{11\pi}{6} - \dfrac{\pi}{4}$

Rewriting a Trigonometric Expression In Exercises 51 and 52, write the expression as the sine, cosine, or tangent of an angle.

51. $\sin 60° \cos 45° - \cos 60° \sin 45°$

52. $\dfrac{\tan 68° - \tan 115°}{1 + \tan 68° \tan 115°}$

Evaluating a Trigonometric Expression In Exercises 53–56, find the exact value of the trigonometric expression given that $\tan u = \frac{3}{4}$ and $\cos v = -\frac{4}{5}$. (u is in Quadrant I and v is in Quadrant III.)

53. $\sin(u + v)$

54. $\tan(u + v)$

55. $\cos(u - v)$

56. $\sin(u - v)$

Proving a Trigonometric Identity In Exercises 57–60, prove the identity.

57. $\cos\left(x + \dfrac{\pi}{2}\right) = -\sin x$ **58.** $\tan\left(x - \dfrac{\pi}{2}\right) = -\cot x$

59. $\tan(\pi - x) = -\tan x$

60. $\cos 3x = 4\cos^3 x - 3\cos x$

Solving a Trigonometric Equation In Exercises 61 and 62, find all solutions of the equation in the interval $[0, 2\pi)$.

61. $\sin\left(x + \dfrac{\pi}{4}\right) - \sin\left(x - \dfrac{\pi}{4}\right) = 1$

62. $\cos\left(x + \dfrac{\pi}{6}\right) - \cos\left(x - \dfrac{\pi}{6}\right) = 1$

5.5 Evaluating Functions Involving Double Angles In Exercises 63 and 64, find the exact values of $\sin 2u$, $\cos 2u$, and $\tan 2u$ using the double-angle formulas.

63. $\sin u = -\dfrac{4}{5}, \quad \pi < u < 3\pi/2$

64. $\cos u = -2/\sqrt{5}, \quad \pi/2 < u < \pi$

Verifying a Trigonometric Identity In Exercises 65 and 66, use the double-angle formulas to verify the identity algebraically and use a graphing utility to confirm your result graphically.

65. $\sin 4x = 8\cos^3 x \sin x - 4\cos x \sin x$

66. $\tan^2 x = \dfrac{1 - \cos 2x}{1 + \cos 2x}$

Reducing Powers In Exercises 67 and 68, use the power-reducing formulas to rewrite the expression in terms of the first power of the cosine.

67. $\tan^2 2x$ **68.** $\sin^2 x \tan^2 x$

Using Half-Angle Formulas In Exercises 69 and 70, use the half-angle formulas to determine the exact values of the sine, cosine, and tangent of the angle.

69. $-75°$ **70.** $\dfrac{19\pi}{12}$

Using Half-Angle Formulas In Exercises 71 and 72, (a) determine the quadrant in which $u/2$ lies, and (b) find the exact values of $\sin(u/2)$, $\cos(u/2)$, and $\tan(u/2)$ using the half-angle formulas.

71. $\tan u = \dfrac{4}{3}, \quad \pi < u < 3\pi/2$

72. $\cos u = -\dfrac{2}{7}, \quad \pi/2 < u < \pi$

Using Half-Angle Formulas In Exercises 73 and 74, use the half-angle formulas to simplify the expression.

73. $-\sqrt{\dfrac{1 + \cos 10x}{2}}$ **74.** $\dfrac{\sin 6x}{1 + \cos 6x}$

Using Product-to-Sum Formulas In Exercises 75 and 76, use the product-to-sum formulas to rewrite the product as a sum or difference.

75. $\cos 4\theta \sin 6\theta$ **76.** $2 \sin 7\theta \cos 3\theta$

Using Sum-to-Product Formulas In Exercises 77 and 78, use the sum-to-product formulas to rewrite the sum or difference as a product.

77. $\cos 6\theta + \cos 5\theta$

78. $\sin\left(x + \dfrac{\pi}{4}\right) - \sin\left(x - \dfrac{\pi}{4}\right)$

79. Projectile Motion A baseball leaves the hand of a player at first base at an angle of θ with the horizontal and at an initial velocity of $v_0 = 80$ feet per second. A player at second base 100 feet away catches the ball. Find θ when the range r of a projectile is

$$r = \dfrac{1}{32}v_0^2 \sin 2\theta.$$

80. Geometry A trough for feeding cattle is 4 meters long and its cross sections are isosceles triangles with the two equal sides being $\frac{1}{2}$ meter (see figure). The angle between the two sides is θ.

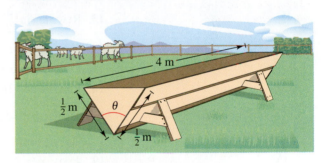

(a) Write the trough's volume as a function of $\theta/2$.

(b) Write the volume of the trough as a function of θ and determine the value of θ such that the volume is maximum.

Exploration

True or False? In Exercises 81–84, determine whether the statement is true or false. Justify your answer.

81. If $\dfrac{\pi}{2} < \theta < \pi$, then $\cos\dfrac{\theta}{2} < 0$.

82. $\sin(x + y) = \sin x + \sin y$

83. $4 \sin(-x) \cos(-x) = -2 \sin 2x$

84. $4 \sin 45° \cos 15° = 1 + \sqrt{3}$

85. Think About It When a trigonometric equation has an infinite number of solutions, is it true that the equation is an identity? Explain.

Chapter Test

See **CalcChat.com** for tutorial help and worked-out solutions to odd-numbered exercises.

Take this test as you would take a test in class. When you are finished, check your work against the answers given in the back of the book.

1. When $\tan \theta = \frac{6}{5}$ and $\cos \theta < 0$, evaluate (if possible) all six trigonometric functions of θ.

2. Use the fundamental identities to simplify $\csc^2 \beta (1 - \cos^2 \beta)$.

3. Factor and simplify $\dfrac{\sec^4 x - \tan^4 x}{\sec^2 x + \tan^2 x}$.

4. Add and simplify $\dfrac{\cos \theta}{\sin \theta} + \dfrac{\sin \theta}{\cos \theta}$.

5. Determine the values of θ, $0 \le \theta < 2\pi$, for which $\tan \theta = -\sqrt{\sec^2 \theta - 1}$.

6. Use a graphing utility to graph the functions $y_1 = \cos x + \sin x \tan x$ and $y_2 = \sec x$ in the same viewing window. Make a conjecture about y_1 and y_2. Verify the result algebraically.

In Exercises 7–12, verify the identity.

7. $\sin \theta \sec \theta = \tan \theta$

8. $\sec^2 x \tan^2 x + \sec^2 x = \sec^4 x$

9. $\dfrac{\csc \alpha + \sec \alpha}{\sin \alpha + \cos \alpha} = \cot \alpha + \tan \alpha$

10. $\tan\left(x + \dfrac{\pi}{2}\right) = -\cot x$

11. $\sin(n\pi + \theta) = (-1)^n \sin \theta$, n is an integer.

12. $(\sin x + \cos x)^2 = 1 + \sin 2x$

13. Rewrite $\sin^4 \dfrac{x}{2}$ in terms of the first power of the cosine.

14. Use a half-angle formula to simplify the expression $(\sin 4\theta)/(1 + \cos 4\theta)$.

15. Rewrite $4 \sin 3\theta \cos 2\theta$ as a sum or difference.

16. Rewrite $\cos 3\theta - \cos \theta$ as a product.

In Exercises 17–20, find all solutions of the equation in the interval $[0, 2\pi)$.

17. $\tan^2 x + \tan x = 0$

18. $\sin 2\alpha - \cos \alpha = 0$

19. $4 \cos^2 x - 3 = 0$

20. $\csc^2 x - \csc x - 2 = 0$

21. Use a graphing utility to approximate the solutions (to three decimal places) of $5 \sin x - x = 0$ in the interval $[0, 2\pi)$.

22. Find the exact value of $\cos 105°$ using the fact that $105° = 135° - 30°$.

23. Use the figure to find the exact values of $\sin 2u$, $\cos 2u$, and $\tan 2u$.

24. Cheyenne, Wyoming, has a latitude of 41°N. At this latitude, the position of the sun at sunrise can be modeled by

$$D = 31 \sin\left(\frac{2\pi}{365}t - 1.4\right)$$

where t is the time (in days), with $t = 1$ representing January 1. In this model, D represents the number of degrees north or south of due east that the sun rises. Use a graphing utility to determine the days on which the sun is more than 20° north of due east at sunrise.

25. The heights above ground h_1 and h_2 (in feet) of two people in different seats on a Ferris wheel can be modeled by

$$h_1 = 28 \cos 10t + 38 \quad \text{and} \quad h_2 = 28 \cos\left[10\left(t - \frac{\pi}{6}\right)\right] + 38, \; 0 \le t \le 2$$

where t is the time (in minutes). When are the two people at the same height?

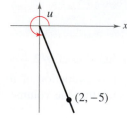

(2, −5)

Figure for 23

Proofs in Mathematics ■ ■ ■ ■ ■ ■ ■ ■ ■ ■ ■ ■ ■

Sum and Difference Formulas   *(p. 267)*

$$\sin(u + v) = \sin u \cos v + \cos u \sin v \qquad \tan(u + v) = \frac{\tan u + \tan v}{1 - \tan u \tan v}$$

$$\sin(u - v) = \sin u \cos v - \cos u \sin v$$

$$\cos(u + v) = \cos u \cos v - \sin u \sin v \qquad \tan(u - v) = \frac{\tan u - \tan v}{1 + \tan u \tan v}$$

$$\cos(u - v) = \cos u \cos v + \sin u \sin v$$

Proof

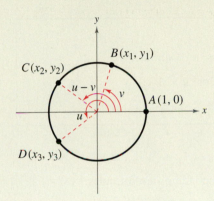

Use the figures at the left for the proofs of the formulas for $\cos(u \pm v)$. In the top figure, let A be the point $(1, 0)$ and then use u and v to locate the points $B(x_1, y_1)$, $C(x_2, y_2)$, and $D(x_3, y_3)$ on the unit circle. So, $x_i^2 + y_i^2 = 1$ for $i = 1, 2,$ and 3. For convenience, assume that $0 < v < u < 2\pi$. In the bottom figure, note that arcs AC and BD have the same length. So, line segments AC and BD are also equal in length, which implies that

$$\sqrt{(x_2 - 1)^2 + (y_2 - 0)^2} = \sqrt{(x_3 - x_1)^2 + (y_3 - y_1)^2}$$

$$x_2^2 - 2x_2 + 1 + y_2^2 = x_3^2 - 2x_1x_3 + x_1^2 + y_3^2 - 2y_1y_3 + y_1^2$$

$$(x_2^2 + y_2^2) + 1 - 2x_2 = (x_3^2 + y_3^2) + (x_1^2 + y_1^2) - 2x_1x_3 - 2y_1y_3$$

$$1 + 1 - 2x_2 = 1 + 1 - 2x_1x_3 - 2y_1y_3$$

$$x_2 = x_3x_1 + y_3y_1.$$

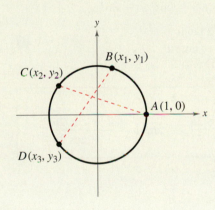

Finally, by substituting the values $x_2 = \cos(u - v)$, $x_3 = \cos u$, $x_1 = \cos v$, $y_3 = \sin u$, and $y_1 = \sin v$, you obtain $\cos(u - v) = \cos u \cos v + \sin u \sin v$. To establish the formula for $\cos(u + v)$, consider $u + v = u - (-v)$ and use the formula just derived to obtain

$$\cos(u + v) = \cos[u - (-v)]$$

$$= \cos u \cos(-v) + \sin u \sin(-v)$$

$$= \cos u \cos v - \sin u \sin v.$$

You can use the sum and difference formulas for sine and cosine to prove the formulas for $\tan(u \pm v)$.

$$\tan(u \pm v) = \frac{\sin(u \pm v)}{\cos(u \pm v)} \qquad \text{Quotient identity}$$

$$= \frac{\sin u \cos v \pm \cos u \sin v}{\cos u \cos v \mp \sin u \sin v} \qquad \text{Sum and difference formulas}$$

$$= \frac{\dfrac{\sin u \cos v \pm \cos u \sin v}{\cos u \cos v}}{\dfrac{\cos u \cos v \mp \sin u \sin v}{\cos u \cos v}} \qquad \text{Divide numerator and denominator by } \cos u \cos v.$$

$$= \dfrac{\dfrac{\sin u \cos v}{\cos u \cos v} \pm \dfrac{\cos u \sin v}{\cos u \cos v}}{\dfrac{\cos u \cos v}{\cos u \cos v} \mp \dfrac{\sin u \sin v}{\cos u \cos v}}$$ Write as separate fractions.

$$= \dfrac{\dfrac{\sin u}{\cos u} \pm \dfrac{\sin v}{\cos v}}{1 \mp \dfrac{\sin u}{\cos u} \cdot \dfrac{\sin v}{\cos v}}$$ Simplify.

$$= \dfrac{\tan u \pm \tan v}{1 \mp \tan u \tan v}$$ Quotient identity

Double-Angle Formulas   *(p. 271)*

$$\sin 2u = 2 \sin u \cos u \qquad \cos 2u = \cos^2 u - \sin^2 u$$

$$\tan 2u = \dfrac{2 \tan u}{1 - \tan^2 u} \qquad\qquad = 2 \cos^2 u - 1$$

$$= 1 - 2 \sin^2 u$$

TRIGONOMETRY AND ASTRONOMY

Early astronomers used trigonometry to calculate measurements in the universe. For instance, they used trigonometry to calculate the circumference of Earth and the distance from Earth to the moon. Another major accomplishment in astronomy using trigonometry was computing distances to stars.

Proof

To prove all three formulas, let $v = u$ in the corresponding sum formulas.

$$\sin 2u = \sin(u + u) = \sin u \cos u + \cos u \sin u = 2 \sin u \cos u$$

$$\cos 2u = \cos(u + u) = \cos u \cos u - \sin u \sin u = \cos^2 u - \sin^2 u$$

$$\tan 2u = \tan(u + u) = \dfrac{\tan u + \tan u}{1 - \tan u \tan u} = \dfrac{2 \tan u}{1 - \tan^2 u}$$

Power-Reducing Formulas   *(p. 273)*

$$\sin^2 u = \dfrac{1 - \cos 2u}{2} \qquad \cos^2 u = \dfrac{1 + \cos 2u}{2} \qquad \tan^2 u = \dfrac{1 - \cos 2u}{1 + \cos 2u}$$

Proof

To prove the first formula, solve for $\sin^2 u$ in the double-angle formula $\cos 2u = 1 - 2 \sin^2 u$, as follows.

$$\cos 2u = 1 - 2 \sin^2 u$$ Write double-angle formula.

$$2 \sin^2 u = 1 - \cos 2u$$ Subtract cos 2u from, and add 2 sin² u to, each side.

$$\sin^2 u = \dfrac{1 - \cos 2u}{2}$$ Divide each side by 2.

In a similar way, you can prove the second formula by solving for $\cos^2 u$ in the double-angle formula

$$\cos 2u = 2\cos^2 u - 1.$$

To prove the third formula, use a quotient identity, as follows.

$$\tan^2 u = \frac{\sin^2 u}{\cos^2 u}$$

$$= \frac{\dfrac{1 - \cos 2u}{2}}{\dfrac{1 + \cos 2u}{2}}$$

$$= \frac{1 - \cos 2u}{1 + \cos 2u}$$

Sum-to-Product Formulas *(p. 275)*

$$\sin u + \sin v = 2 \sin\left(\frac{u + v}{2}\right) \cos\left(\frac{u - v}{2}\right)$$

$$\sin u - \sin v = 2 \cos\left(\frac{u + v}{2}\right) \sin\left(\frac{u - v}{2}\right)$$

$$\cos u + \cos v = 2 \cos\left(\frac{u + v}{2}\right) \cos\left(\frac{u - v}{2}\right)$$

$$\cos u - \cos v = -2 \sin\left(\frac{u + v}{2}\right) \sin\left(\frac{u - v}{2}\right)$$

Proof

To prove the first formula, let $x = u + v$ and $y = u - v$. Then substitute $u = (x + y)/2$ and $v = (x - y)/2$ in the product-to-sum formula.

$$\sin u \cos v = \frac{1}{2}[\sin(u + v) + \sin(u - v)]$$

$$\sin\left(\frac{x + y}{2}\right) \cos\left(\frac{x - y}{2}\right) = \frac{1}{2}(\sin x + \sin y)$$

$$2 \sin\left(\frac{x + y}{2}\right) \cos\left(\frac{x - y}{2}\right) = \sin x + \sin y$$

The other sum-to-product formulas can be proved in a similar manner.

P.S. Problem Solving ▪ ▪ ▪ ▪ ▪ ▪ ▪ ▪ ▪ ▪ ▪ ▪ ▪ ▪

1. **Writing Trigonometric Functions in Terms of Cosine** Write each of the other trigonometric functions of θ in terms of $\cos \theta$.

2. **Verifying a Trigonometric Identity** Verify that for all integers n,

$$\cos\left[\frac{(2n+1)\pi}{2}\right] = 0.$$

3. **Verifying a Trigonometric Identity** Verify that for all integers n,

$$\sin\left[\frac{(12n+1)\pi}{6}\right] = \frac{1}{2}.$$

4. **Sound Wave** A sound wave is modeled by

$$p(t) = \frac{1}{4\pi}\left[p_1(t) + 30p_2(t) + p_3(t) + p_5(t) + 30p_6(t)\right]$$

where $p_n(t) = \dfrac{1}{n}\sin(524n\pi t)$, and t is the time (in seconds).

(a) Find the sine components $p_n(t)$ and use a graphing utility to graph the components. Then verify the graph of p shown below.

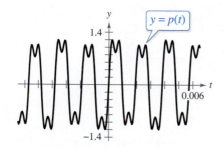

(b) Find the period of each sine component of p. Is p periodic? If so, then what is its period?

(c) Use the graphing utility to find the t-intercepts of the graph of p over one cycle.

(d) Use the graphing utility to approximate the absolute maximum and absolute minimum values of p over one cycle.

5. **Geometry** Three squares of side s are placed side by side (see figure). Make a conjecture about the relationship between the sum $u + v$ and w. Prove your conjecture by using the identity for the tangent of the sum of two angles.

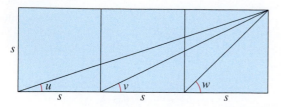

6. **Projectile Motion** The path traveled by an object (neglecting air resistance) that is projected at an initial height of h_0 feet, an initial velocity of v_0 feet per second, and an initial angle θ is given by

$$y = -\frac{16}{v_0^2 \cos^2 \theta}x^2 + (\tan \theta)x + h_0$$

where x and y are measured in feet. Find a formula for the maximum height of an object projected from ground level at velocity v_0 and angle θ. To do this, find half of the horizontal distance

$$\frac{1}{32}v_0^2 \sin 2\theta$$

and then substitute it for x in the general model for the path of a projectile (where $h_0 = 0$).

7. **Geometry** The length of each of the two equal sides of an isosceles triangle is 10 meters (see figure). The angle between the two sides is θ.

(a) Write the area of the triangle as a function of $\theta/2$.

(b) Write the area of the triangle as a function of θ. Determine the value of θ such that the area is a maximum.

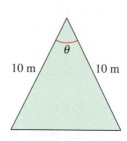

Figure for 7   Figure for 8

8. **Geometry** Use the figure to derive the formulas for

$$\sin\frac{\theta}{2}, \cos\frac{\theta}{2}, \text{ and } \tan\frac{\theta}{2}$$

where θ is an acute angle.

9. **Force** The force F (in pounds) on a person's back when he or she bends over at an angle θ is modeled by

$$F = \frac{0.6W \sin(\theta + 90°)}{\sin 12°}$$

where W is the person's weight (in pounds).

(a) Simplify the model.

(b) Use a graphing utility to graph the model, where $W = 185$ and $0° < \theta < 90°$.

(c) At what angle is the force a maximum? At what angle is the force a minimum?

10. **Hours of Daylight** The number of hours of daylight that occur at any location on Earth depends on the time of year and the latitude of the location. The following equations model the numbers of hours of daylight in Seward, Alaska (60° latitude), and New Orleans, Louisiana (30° latitude).

$$D = 12.2 - 6.4 \cos\left[\frac{\pi(t + 0.2)}{182.6}\right] \quad \text{Seward}$$

$$D = 12.2 - 1.9 \cos\left[\frac{\pi(t + 0.2)}{182.6}\right] \quad \text{New Orleans}$$

In these models, D represents the number of hours of daylight and t represents the day, with $t = 0$ corresponding to January 1.

(a) Use a graphing utility to graph both models in the same viewing window. Use a viewing window of $0 \le t \le 365$.

(b) Find the days of the year on which both cities receive the same amount of daylight.

(c) Which city has the greater variation in the number of daylight hours? Which constant in each model would you use to determine the difference between the greatest and least numbers of hours of daylight?

(d) Determine the period of each model.

11. **Ocean Tide** The tide, or depth of the ocean near the shore, changes throughout the day. The water depth d (in feet) of a bay can be modeled by

$$d = 35 - 28 \cos\frac{\pi}{6.2}t$$

where t is the time in hours, with $t = 0$ corresponding to 12:00 A.M.

(a) Algebraically find the times at which the high and low tides occur.

(b) If possible, algebraically find the time(s) at which the water depth is 3.5 feet.

(c) Use a graphing utility to verify your results from parts (a) and (b).

12. **Piston Heights** The heights h (in inches) of pistons 1 and 2 in an automobile engine can be modeled by

$$h_1 = 3.75 \sin 733t + 7.5$$

and

$$h_2 = 3.75 \sin 733(t + 4\pi/3) + 7.5$$

respectively, where t is measured in seconds.

(a) Use a graphing utility to graph the heights of these pistons in the same viewing window for $0 \le t \le 1$.

(b) How often are the pistons at the same height?

13. **Index of Refraction** The index of refraction n of a transparent material is the ratio of the speed of light in a vacuum to the speed of light in the material. Some common materials and their indices of refraction are air (1.00), water (1.33), and glass (1.50). Triangular prisms are often used to measure the index of refraction based on the formula

$$n = \frac{\sin\left(\dfrac{\theta}{2} + \dfrac{\alpha}{2}\right)}{\sin\dfrac{\theta}{2}}.$$

For the prism shown in the figure, $\alpha = 60°$.

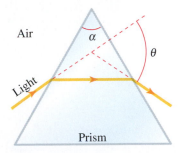

(a) Write the index of refraction as a function of $\cot(\theta/2)$.

(b) Find θ for a prism made of glass.

14. **Sum Formulas**

(a) Write a sum formula for $\sin(u + v + w)$.

(b) Write a sum formula for $\tan(u + v + w)$.

15. **Solving Trigonometric Inequalities** Find the solution of each inequality in the interval $[0, 2\pi)$.

(a) $\sin x \ge 0.5$ (b) $\cos x \le -0.5$

(c) $\tan x < \sin x$ (d) $\cos x \ge \sin x$

16. **Sum of Fourth Powers** Consider the function $f(x) = \sin^4 x + \cos^4 x$.

(a) Use the power-reducing formulas to write the function in terms of cosine to the first power.

(b) Determine another way of rewriting the function. Use a graphing utility to rule out incorrectly rewritten functions.

(c) Add a trigonometric term to the function so that it becomes a perfect square trinomial. Rewrite the function as a perfect square trinomial minus the term that you added. Use the graphing utility to rule out incorrectly rewritten functions.

(d) Rewrite the result of part (c) in terms of the sine of a double angle. Use the graphing utility to rule out incorrectly rewritten functions.

(e) When you rewrite a trigonometric expression, the result may not be the same as a friend's. Does this mean that one of you is wrong? Explain.

6 Additional Topics in Trigonometry

Electrical Engineering

Work

Navigation

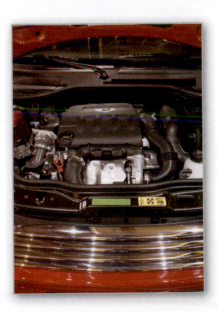

Engine Design

Surveying

6.1 Law of Sines

- ■ **Use the Law of Sines to solve oblique triangles (AAS or ASA).**
- ■ **Use the Law of Sines to solve oblique triangles (SSA).**
- ■ **Find the areas of oblique triangles.**
- ■ **Use the Law of Sines to model and solve real-life problems.**

Introduction

In Chapter 4, you studied techniques for solving right triangles. In this section and the next, you will solve **oblique triangles**—triangles that have no right angles. As standard notation, the angles of a triangle are labeled A, B, and C, and their opposite sides are labeled a, b, and c, as shown below.

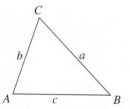

You can use the Law of Sines to solve real-life problems involving oblique triangles, such as determining the distance from a boat to the shoreline.

To solve an oblique triangle, you need to know the measure of at least one side and any two other measures of the triangle—either two sides, two angles, or one angle and one side. This breaks down into the following four cases.

1. Two angles and any side (AAS or ASA)
2. Two sides and an angle opposite one of them (SSA)
3. Three sides (SSS)
4. Two sides and their included angle (SAS)

The first two cases can be solved using the **Law of Sines,** whereas the last two cases require the Law of Cosines (see Section 6.2).

Law of Sines

If ABC is a triangle with sides a, b, and c, then

$$\frac{a}{\sin A} = \frac{b}{\sin B} = \frac{c}{\sin C}.$$

A is acute.

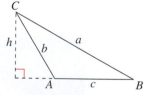

A is obtuse.

The Law of Sines can also be written in the reciprocal form

$$\frac{\sin A}{a} = \frac{\sin B}{b} = \frac{\sin C}{c}.$$

For a proof of the Law of Sines, see Proofs in Mathematics on page 332.

Figure 6.1

b = 28 ft

EXAMPLE 1 Given Two Angles and One Side—AAS

For the triangle in Figure 6.1, $C = 102°$, $B = 29°$, and $b = 28$ feet. Find the remaining angle and sides.

Solution The third angle of the triangle is

$$A = 180° - B - C$$
$$= 180° - 29° - 102°$$
$$= 49°.$$

By the Law of Sines, you have

$$\frac{a}{\sin A} = \frac{b}{\sin B} = \frac{c}{\sin C}.$$

Using $b = 28$ produces

$$a = \frac{b}{\sin B}(\sin A) = \frac{28}{\sin 29°}(\sin 49°) \approx 43.59 \text{ feet}$$

and

$$c = \frac{b}{\sin B}(\sin C) = \frac{28}{\sin 29°}(\sin 102°) \approx 56.49 \text{ feet}.$$

✓ **Checkpoint** 🔊 *Audio-video solution in English & Spanish at LarsonPrecalculus.com.*

For the triangle in Figure 6.2, $A = 30°$, $B = 45°$, and $a = 32$. Find the remaining angle and sides.

Figure 6.2

In the 1850s, surveyors used the Law of Sines to calculate the height of Mount Everest. Their calculation was within 30 feet of the currently accepted value.

EXAMPLE 2 Given Two Angles and One Side—ASA

A pole tilts *toward* the sun at an 8° angle from the vertical, and it casts a 22-foot shadow. The angle of elevation from the tip of the shadow to the top of the pole is 43°. How tall is the pole?

Solution From the figure at the right, note that $A = 43°$ and

$$B = 90° + 8° = 98°.$$

So, the third angle is

$$C = 180° - A - B$$
$$= 180° - 43° - 98°$$
$$= 39°.$$

By the Law of Sines, you have

$$\frac{a}{\sin A} = \frac{c}{\sin C}.$$

Because $c = 22$ feet, the height of the pole is

$$a = \frac{c}{\sin C}(\sin A) = \frac{22}{\sin 39°}(\sin 43°) \approx 23.84 \text{ feet}.$$

✓ **Checkpoint** 🔊 *Audio-video solution in English & Spanish at LarsonPrecalculus.com.*

Find the height of the tree shown in Figure 6.3.

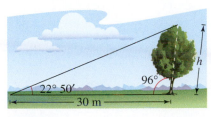

Figure 6.3

The Ambiguous Case (SSA)

In Examples 1 and 2, you saw that two angles and one side determine a unique triangle. However, if two sides and one opposite angle are given, then three possible situations can occur: (1) no such triangle exists, (2) one such triangle exists, or (3) two distinct triangles may satisfy the conditions.

The Ambiguous Case (SSA)

Consider a triangle in which you are given a, b, and A. ($h = b \sin A$)

	A is acute.	A is acute.	A is acute.	A is acute.	A is obtuse.	A is obtuse.
Sketch						
Necessary condition	$a < h$	$a = h$	$a \geq b$	$h < a < b$	$a \leq b$	$a > b$
Triangles possible	None	One	One	Two	None	One

EXAMPLE 3   Single-Solution Case—SSA

For the triangle in Figure 6.4, $a = 22$ inches, $b = 12$ inches, and $A = 42°$. Find the remaining side and angles.

Solution   By the Law of Sines, you have

$$\frac{\sin B}{b} = \frac{\sin A}{a} \qquad \text{\color{red}Reciprocal form}$$

$$\sin B = b\left(\frac{\sin A}{a}\right) \qquad \text{\color{red}Multiply each side by } b.$$

$$\sin B = 12\left(\frac{\sin 42°}{22}\right) \qquad \text{\color{red}Substitute for } A, a, \text{ and } b.$$

$$B \approx 21.41°.$$

Now, you can determine that

$$C \approx 180° - 42° - 21.41° = 116.59°.$$

Then, the remaining side is

$$\frac{c}{\sin C} = \frac{a}{\sin A}$$

$$c = \frac{a}{\sin A}(\sin C)$$

$$= \frac{22}{\sin 42°}(\sin 116.59°)$$

$$\approx 29.40 \text{ inches.}$$

b = 12 in.   a = 22 in.   42°

One solution: $a \geq b$
Figure 6.4

 Checkpoint ◄))) *Audio-video solution in English & Spanish at LarsonPrecalculus.com.*

Given $A = 31°$, $a = 12$, and $b = 5$, find the remaining side and angles of the triangle.

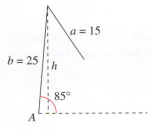

No solution: $a < h$

Figure 6.5

EXAMPLE 4 No-Solution Case—SSA

Show that there is no triangle for which $a = 15$, $b = 25$, and $A = 85°$.

Solution Begin by making the sketch shown in Figure 6.5. From this figure, it appears that no triangle is formed. You can verify this using the Law of Sines.

$$\frac{\sin B}{b} = \frac{\sin A}{a} \qquad \text{\color{red}Reciprocal form}$$

$$\sin B = b\left(\frac{\sin A}{a}\right) \qquad \text{\color{red}Multiply each side by } b.$$

$$\sin B = 25\left(\frac{\sin 85°}{15}\right) \approx 1.6603 > 1$$

This contradicts the fact that

$$|\sin B| \leq 1.$$

So, no triangle can be formed having sides $a = 15$ and $b = 25$ and angle $A = 85°$.

✓ **Checkpoint** *Audio-video solution in English & Spanish at LarsonPrecalculus.com.*

Show that there is no triangle for which $a = 4$, $b = 14$, and $A = 60°$.

EXAMPLE 5 Two-Solution Case—SSA

Find two triangles for which $a = 12$ meters, $b = 31$ meters, and $A = 20.5°$.

Solution By the Law of Sines, you have

$$\frac{\sin B}{b} = \frac{\sin A}{a} \qquad \text{\color{red}Reciprocal form}$$

$$\sin B = b\left(\frac{\sin A}{a}\right) = 31\left(\frac{\sin 20.5°}{12}\right) \approx 0.9047.$$

There are two angles, $B_1 \approx 64.8°$ and $B_2 \approx 180° - 64.8° = 115.2°$, between $0°$ and $180°$ whose sine is 0.9047. For $B_1 \approx 64.8°$, you obtain

$$C \approx 180° - 20.5° - 64.8° = 94.7°$$

$$c = \frac{a}{\sin A}(\sin C) = \frac{12}{\sin 20.5°}(\sin 94.7°) \approx 34.15 \text{ meters.}$$

For $B_2 \approx 115.2°$, you obtain

$$C \approx 180° - 20.5° - 115.2° = 44.3°$$

$$c = \frac{a}{\sin A}(\sin C) = \frac{12}{\sin 20.5°}(\sin 44.3°) \approx 23.93 \text{ meters.}$$

The resulting triangles are shown below.

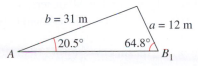

Two solutions: $h < a < b$

✓ **Checkpoint** *Audio-video solution in English & Spanish at LarsonPrecalculus.com.*

Find two triangles for which $a = 4.5$ feet, $b = 5$ feet, and $A = 58°$.

Area of an Oblique Triangle

The procedure used to prove the Law of Sines leads to a simple formula for the area of an oblique triangle. Referring to the triangles below, note that each triangle has a height of $h = b \sin A$. Consequently, the area of each triangle is

$$\text{Area} = \frac{1}{2}(\text{base})(\text{height})$$

$$= \frac{1}{2}(c)(b \sin A)$$

$$= \frac{1}{2}bc \sin A.$$

By similar arguments, you can develop the formulas

$$\text{Area} = \frac{1}{2}ab \sin C = \frac{1}{2}ac \sin B.$$

REMARK To see how to obtain the height of the obtuse triangle, notice the use of the reference angle $180° - A$ and the difference formula for sine, as follows.

$$h = b \sin(180° - A)$$

$$= b(\sin 180° \cos A$$

$$- \cos 180° \sin A)$$

$$= b[0 \cdot \cos A - (-1) \cdot \sin A]$$

$$= b \sin A$$

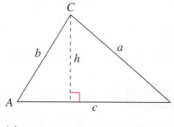

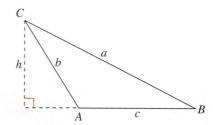

A is acute. A is obtuse.

Area of an Oblique Triangle

The area of any triangle is one-half the product of the lengths of two sides times the sine of their included angle. That is,

$$\text{Area} = \frac{1}{2}bc \sin A = \frac{1}{2}ab \sin C = \frac{1}{2}ac \sin B.$$

Note that when angle A is $90°$, the formula gives the area of a right triangle:

$$\text{Area} = \frac{1}{2}bc(\sin 90°) = \frac{1}{2}bc = \frac{1}{2}(\text{base})(\text{height}). \qquad \textcolor{red}{\sin 90° = 1}$$

Similar results are obtained for angles C and B equal to $90°$.

EXAMPLE 6 **Finding the Area of a Triangular Lot**

Find the area of a triangular lot having two sides of lengths 90 meters and 52 meters and an included angle of $102°$.

Solution Consider $a = 90$ meters, $b = 52$ meters, and angle $C = 102°$, as shown in Figure 6.6. Then, the area of the triangle is

$$\text{Area} = \frac{1}{2}ab \sin C = \frac{1}{2}(90)(52)(\sin 102°) \approx 2289 \text{ square meters.}$$

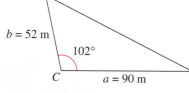

Figure 6.6

✓ **Checkpoint** ◀))) *Audio-video solution in English & Spanish at LarsonPrecalculus.com.*

Find the area of a triangular lot having two sides of lengths 24 inches and 18 inches and an included angle of $80°$.

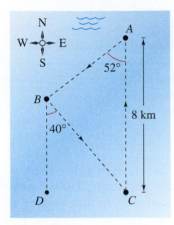

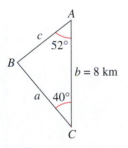

Figure 6.7

Figure 6.8

Application

EXAMPLE 7   An Application of the Law of Sines

The course for a boat race starts at point A and proceeds in the direction S 52° W to point B, then in the direction S 40° E to point C, and finally back to point A, as shown in Figure 6.7. Point C lies 8 kilometers directly south of point A. Approximate the total distance of the race course.

Solution   Because lines BD and AC are parallel, it follows that $\angle BCA \cong \angle CBD$. Consequently, triangle ABC has the measures shown in Figure 6.8. The measure of angle B is $180° - 52° - 40° = 88°$. Using the Law of Sines,

$$\frac{a}{\sin 52°} = \frac{b}{\sin 88°} = \frac{c}{\sin 40°}.$$

Because $b = 8$,

$$a = \frac{8}{\sin 88°}(\sin 52°) \approx 6.31 \quad \text{and} \quad c = \frac{8}{\sin 88°}(\sin 40°) \approx 5.15.$$

The total distance of the course is approximately

Length $\approx 8 + 6.31 + 5.15 = 19.46$ kilometers.

✓ **Checkpoint** *Audio-video solution in English & Spanish at LarsonPrecalculus.com.*

On a small lake, you swim from point A to point B at a bearing of N 28° E, then to point C at a bearing of N 58° W, and finally back to point A, as shown in the figure below. Point C lies 800 meters directly north of point A. Approximate the total distance that you swim.

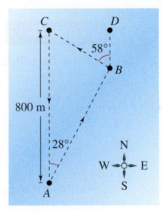

Summarize   (Section 6.1)

1. State the Law of Sines *(page 390)*. For examples of using the Law of Sines to solve oblique triangles (AAS or ASA), see Examples 1 and 2.

2. List the necessary conditions and the numbers of possible triangles for the ambiguous case (SSA) *(page 392)*. For examples of using the Law of Sines to solve oblique triangles (SSA), see Examples 3–5.

3. State the formula for the area of an oblique triangle *(page 394)*. For an example of finding the area of an oblique triangle, see Example 6.

4. Describe how you can use the Law of Sines to model and solve a real-life problem *(page 395, Example 7)*.

6.2 Law of Cosines

You can use the Law of Cosines to solve real-life problems involving oblique triangles, such as determining the total distance a piston moves in an engine.

■ Use the Law of Cosines to solve oblique triangles (SSS or SAS).
■ Use the Law of Cosines to model and solve real-life problems.
■ Use Heron's Area Formula to find the area of a triangle.

Introduction

Two cases remain in the list of conditions needed to solve an oblique triangle—SSS and SAS. When you are given three sides (SSS), or two sides and their included angle (SAS), none of the ratios in the Law of Sines would be complete. In such cases, you can use the **Law of Cosines.**

Law of Cosines

Standard Form	**Alternative Form**
$a^2 = b^2 + c^2 - 2bc \cos A$	$\cos A = \dfrac{b^2 + c^2 - a^2}{2bc}$
$b^2 = a^2 + c^2 - 2ac \cos B$	$\cos B = \dfrac{a^2 + c^2 - b^2}{2ac}$
$c^2 = a^2 + b^2 - 2ab \cos C$	$\cos C = \dfrac{a^2 + b^2 - c^2}{2ab}$

For a proof of the Law of Cosines, see Proofs in Mathematics on page 333.

EXAMPLE 1   **Three Sides of a Triangle—SSS**

Find the three angles of the triangle shown below.

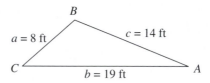

Solution   It is a good idea first to find the angle opposite the longest side—side b in this case. Using the alternative form of the Law of Cosines, you find that

$$\cos B = \frac{a^2 + c^2 - b^2}{2ac} = \frac{8^2 + 14^2 - 19^2}{2(8)(14)} \approx -0.45089.$$

Because $\cos B$ is negative, B is an *obtuse* angle given by $B \approx 116.80°$. At this point, it is simpler to use the Law of Sines to determine A.

$$\sin A = a\left(\frac{\sin B}{b}\right) \approx 8\left(\frac{\sin 116.80°}{19}\right) \approx 0.37583$$

Because B is obtuse and a triangle can have at most one obtuse angle, you know that A must be acute. So, $A \approx 22.08°$ and $C \approx 180° - 22.08° - 116.80° = 41.12°$.

✓ *Checkpoint* ◀))) *Audio-video solution in English & Spanish at LarsonPrecalculus.com.*

Find the three angles of the triangle whose sides have lengths $a = 6$, $b = 8$, and $c = 12$.

■

Smart-foto/Shutterstock.com

Do you see why it was wise to find the largest angle *first* in Example 1? Knowing the cosine of an angle, you can determine whether the angle is acute or obtuse. That is,

$$\cos \theta > 0 \quad \text{for} \quad 0° < \theta < 90° \qquad \text{Acute}$$

$$\cos \theta < 0 \quad \text{for} \quad 90° < \theta < 180°. \qquad \text{Obtuse}$$

So, in Example 1, once you found that angle *B* was obtuse, you knew that angles *A* and *C* were both acute. Furthermore, if the largest angle is acute, then the remaining two angles are also acute.

• • REMARK   When solving an oblique triangle given three sides, you use the alternative form of the Law of Cosines to solve for an angle. When solving an oblique triangle given two sides and their included angle, you use the standard form of the Law of Cosines to solve for an unknown.

EXAMPLE 2   **Two Sides and the Included Angle—SAS**

Find the remaining angles and side of the triangle shown below.

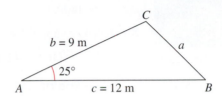

Solution   Use the Law of Cosines to find the unknown side *a* in the figure.

$$a^2 = b^2 + c^2 - 2bc \cos A$$

$$a^2 = 9^2 + 12^2 - 2(9)(12) \cos 25°$$

$$a^2 \approx 29.2375$$

$$a \approx 5.4072$$

Because $a \approx 5.4072$ meters, you now know the ratio $(\sin A)/a$, and you can use the reciprocal form of the Law of Sines to solve for *B*.

$$\frac{\sin B}{b} = \frac{\sin A}{a} \qquad \text{Reciprocal form}$$

$$\sin B = b\left(\frac{\sin A}{a}\right) \qquad \text{Multiply each side by } b.$$

$$\sin B \approx 9\left(\frac{\sin 25°}{5.4072}\right) \qquad \text{Substitute for } A, a, \text{ and } b.$$

$$\sin B \approx 0.7034 \qquad \text{Use a calculator.}$$

There are two angles between 0° and 180° whose sine is 0.7034, $B_1 \approx 44.7°$ and $B_2 \approx 180° - 44.7° = 135.3°$.

For $B_1 \approx 44.7°$,

$$C_1 \approx 180° - 25° - 44.7° = 110.3°.$$

For $B_2 \approx 135.3°$,

$$C_2 \approx 180° - 25° - 135.3° = 19.7°.$$

Because side *c* is the longest side of the triangle, *C* must be the largest angle of the triangle. So, $B \approx 44.7°$ and $C \approx 110.3°$.

✓ *Checkpoint* ◀))) *Audio-video solution in English & Spanish at LarsonPrecalculus.com.*

Given $A = 80°$, $b = 16$, and $c = 12$, find the remaining angles and side of the triangle.

Applications

EXAMPLE 3 **An Application of the Law of Cosines**

The pitcher's mound on a women's softball field is 43 feet from home plate and the distance between the bases is 60 feet, as shown in Figure 6.9. (The pitcher's mound is *not* halfway between home plate and second base.) How far is the pitcher's mound from first base?

Solution In triangle *HPF*, $H = 45°$ (line *HP* bisects the right angle at *H*), $f = 43$, and $p = 60$. Using the Law of Cosines for this SAS case, you have

$$h^2 = f^2 + p^2 - 2fp \cos H$$
$$= 43^2 + 60^2 - 2(43)(60) \cos 45°$$
$$\approx 1800.3.$$

So, the approximate distance from the pitcher's mound to first base is

$$h \approx \sqrt{1800.3} \approx 42.43 \text{ feet.}$$

✓ *Checkpoint* 🔊))) *Audio-video solution in English & Spanish at LarsonPrecalculus.com.*

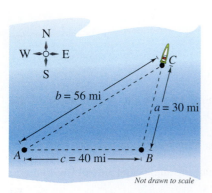

Figure 6.9

In a softball game, a batter hits a ball to dead center field, a distance of 240 feet from home plate. The center fielder then throws the ball to third base and gets a runner out. The distance between the bases is 60 feet. How far is the center fielder from third base?

EXAMPLE 4 **An Application of the Law of Cosines**

A ship travels 60 miles due east and then adjusts its course, as shown below. After traveling 80 miles in this new direction, the ship is 139 miles from its point of departure. Describe the bearing from point *B* to point *C*.

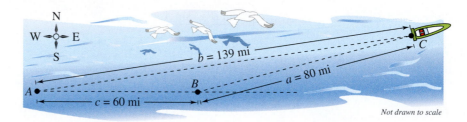

Solution You have $a = 80$, $b = 139$, and $c = 60$. So, using the alternative form of the Law of Cosines, you have

$$\cos B = \frac{a^2 + c^2 - b^2}{2ac}$$
$$= \frac{80^2 + 60^2 - 139^2}{2(80)(60)}$$
$$\approx -0.97094.$$

So, $B \approx 166.15°$, and thus the bearing measured from due north from point *B* to point *C* is $166.15° - 90° = 76.15°$, or N 76.15° E.

✓ *Checkpoint* 🔊))) *Audio-video solution in English & Spanish at LarsonPrecalculus.com.*

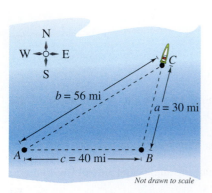

Figure 6.10

A ship travels 40 miles due east and then changes direction, as shown in Figure 6.10. After traveling 30 miles in this new direction, the ship is 56 miles from its point of departure. Describe the bearing from point *B* to point *C*.

HISTORICAL NOTE

Heron of Alexandria (ca. 100 B.C.) was a Greek geometer and inventor. His works describe how to find the areas of triangles, quadrilaterals, regular polygons having 3 to 12 sides, and circles, as well as the surface areas and volumes of three-dimensional objects.

Heron's Area Formula

The Law of Cosines can be used to establish the following formula for the area of a triangle. This formula is called **Heron's Area Formula** after the Greek mathematician Heron (ca. 100 B.C.).

Heron's Area Formula

Given any triangle with sides of lengths a, b, and c, the area of the triangle is

$$\text{Area} = \sqrt{s(s-a)(s-b)(s-c)}$$

where

$$s = \frac{a+b+c}{2}.$$

For a proof of Heron's Area Formula, see Proofs in Mathematics on page 334.

EXAMPLE 5   **Using Heron's Area Formula**

Find the area of a triangle having sides of lengths $a = 43$ meters, $b = 53$ meters, and $c = 72$ meters.

Solution   Because $s = (a+b+c)/2 = 168/2 = 84$, Heron's Area Formula yields

$$\begin{aligned}
\text{Area} &= \sqrt{s(s-a)(s-b)(s-c)} \\
&= \sqrt{84(84-43)(84-53)(84-72)} \\
&= \sqrt{84(41)(31)(12)} \\
&\approx 1131.89 \text{ square meters.}
\end{aligned}$$

✓ **Checkpoint** *Audio-video solution in English & Spanish at LarsonPrecalculus.com.*

Given $a = 5$, $b = 9$, and $c = 8$, use Heron's Area Formula to find the area of the triangle.

You have now studied three different formulas for the area of a triangle.

Standard Formula: $\text{Area} = \dfrac{1}{2}bh$

Oblique Triangle: $\text{Area} = \dfrac{1}{2}bc \sin A = \dfrac{1}{2}ab \sin C = \dfrac{1}{2}ac \sin B$

Heron's Area Formula: $\text{Area} = \sqrt{s(s-a)(s-b)(s-c)}$

Summarize   (Section 6.2)

1. State the Law of Cosines *(page 296)*. For examples of using the Law of Cosines to solve oblique triangles (SSS or SAS), see Examples 1 and 2.

2. Describe real-life problems that can be modeled and solved using the Law of Cosines *(page 298, Examples 3 and 4)*.

3. State Heron's Area Formula *(page 299)*. For an example of using Heron's Area Formula to find the area of a triangle, see Example 5.

6.3 Vectors in the Plane

You can use vectors to model and solve real-life problems involving magnitude and direction, such as determining the true direction of a commercial jet.

■ Represent vectors as directed line segments.
■ Write the component forms of vectors.
■ Perform basic vector operations and represent them graphically.
■ Write vectors as linear combinations of unit vectors.
■ Find the direction angles of vectors.
■ Use vectors to model and solve real-life problems.

Introduction

Quantities such as force and velocity involve both *magnitude* and *direction* and cannot be completely characterized by a single real number. To represent such a quantity, you can use a **directed line segment,** as shown in Figure 6.11. The directed line segment $\overrightarrow{PQ}$ has **initial point** P and **terminal point** Q. Its **magnitude** (or length) is denoted by $\|\overrightarrow{PQ}\|$ and can be found using the Distance Formula.

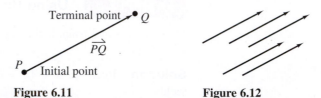

Figure 6.11   **Figure 6.12**

Two directed line segments that have the same magnitude and direction are equivalent. For example, the directed line segments in Figure 6.12 are all equivalent. The set of all directed line segments that are equivalent to the directed line segment $\overrightarrow{PQ}$ is a **vector v in the plane,** written $\mathbf{v} = \overrightarrow{PQ}$. Vectors are denoted by lowercase, boldface letters such as $\mathbf{u}$, $\mathbf{v}$, and $\mathbf{w}$.

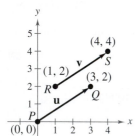

Figure 6.13

EXAMPLE 1   **Showing That Two Vectors Are Equivalent**

Show that $\mathbf{u}$ and $\mathbf{v}$ in Figure 6.13 are equivalent.

Solution   From the Distance Formula, it follows that $\overrightarrow{PQ}$ and $\overrightarrow{RS}$ have the *same magnitude*.

$$\|\overrightarrow{PQ}\| = \sqrt{(3-0)^2 + (2-0)^2} = \sqrt{13} \qquad \|\overrightarrow{RS}\| = \sqrt{(4-1)^2 + (4-2)^2} = \sqrt{13}$$

Moreover, both line segments have the *same direction* because they are both directed toward the upper right on lines having a slope of

$$\frac{4-2}{4-1} = \frac{2-0}{3-0} = \frac{2}{3}.$$

Because $\overrightarrow{PQ}$ and $\overrightarrow{RS}$ have the same magnitude and direction, $\mathbf{u}$ and $\mathbf{v}$ are equivalent.

✓ *Checkpoint*  *Audio-video solution in English & Spanish at LarsonPrecalculus.com.*

Show that $\mathbf{u}$ and $\mathbf{v}$ in the figure at the right are equivalent.

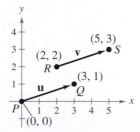

Component Form of a Vector

The directed line segment whose initial point is the origin is often the most convenient representative of a set of equivalent directed line segments. This representative of the vector **v** is in **standard position.**

A vector whose initial point is the origin $(0, 0)$ can be uniquely represented by the coordinates of its terminal point (v_1, v_2). This is the **component form of a vector v,** written as $\mathbf{v} = \langle v_1, v_2 \rangle$. The coordinates v_1 and v_2 are the *components* of **v**. If both the initial point and the terminal point lie at the origin, then **v** is the **zero vector** and is denoted by $\mathbf{0} = \langle 0, 0 \rangle$.

▷ **TECHNOLOGY** You can graph vectors with a graphing utility by graphing directed line segments. Consult the user's guide for your graphing utility for specific instructions.

Component Form of a Vector

The component form of the vector with initial point $P(p_1, p_2)$ and terminal point $Q(q_1, q_2)$ is given by

$$\overrightarrow{PQ} = \langle q_1 - p_1, q_2 - p_2 \rangle = \langle v_1, v_2 \rangle = \mathbf{v}.$$

The **magnitude** (or length) of **v** is given by

$$\|\mathbf{v}\| = \sqrt{(q_1 - p_1)^2 + (q_2 - p_2)^2} = \sqrt{v_1^2 + v_2^2}.$$

If $\|\mathbf{v}\| = 1$, then **v** is a **unit vector.** Moreover, $\|\mathbf{v}\| = 0$ if and only if **v** is the zero vector **0**.

Two vectors $\mathbf{u} = \langle u_1, u_2 \rangle$ and $\mathbf{v} = \langle v_1, v_2 \rangle$ are *equal* if and only if $u_1 = v_1$ and $u_2 = v_2$. For instance, in Example 1, the vector **u** from $P(0, 0)$ to $Q(3, 2)$ is $\mathbf{u} = \overrightarrow{PQ} = \langle 3 - 0, 2 - 0 \rangle = \langle 3, 2 \rangle$, and the vector **v** from $R(1, 2)$ to $S(4, 4)$ is $\mathbf{v} = \overrightarrow{RS} = \langle 4 - 1, 4 - 2 \rangle = \langle 3, 2 \rangle$.

EXAMPLE 2 Finding the Component Form of a Vector

Find the component form and magnitude of the vector **v** that has initial point $(4, -7)$ and terminal point $(-1, 5)$.

Algebraic Solution

Let

$$P(4, -7) = (p_1, p_2)$$

and

$$Q(-1, 5) = (q_1, q_2).$$

Then, the components of $\mathbf{v} = \langle v_1, v_2 \rangle$ are

$$v_1 = q_1 - p_1 = -1 - 4 = -5$$

$$v_2 = q_2 - p_2 = 5 - (-7) = 12.$$

So, $\mathbf{v} = \langle -5, 12 \rangle$ and the magnitude of **v** is

$$\|\mathbf{v}\| = \sqrt{(-5)^2 + 12^2}$$

$$= \sqrt{169}$$

$$= 13.$$

Graphical Solution

Use centimeter graph paper to plot the points $P(4, -7)$ and $Q(-1, 5)$. Carefully sketch the vector **v**. Use the sketch to find the components of $\mathbf{v} = \langle v_1, v_2 \rangle$. Then use a centimeter ruler to find the magnitude of **v**. The figure at the right shows that the components of **v** are $v_1 = -5$ and $v_2 = 12$, so $\mathbf{v} = \langle -5, 12 \rangle$. The figure also shows that the magnitude of **v** is $\|\mathbf{v}\| = 13$.

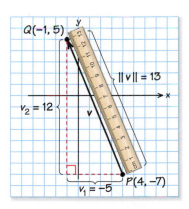

 **Checkpoint** *Audio-video solution in English & Spanish at LarsonPrecalculus.com.*

Find the component form and magnitude of the vector **v** that has initial point $(-2, 3)$ and terminal point $(-7, 9)$.

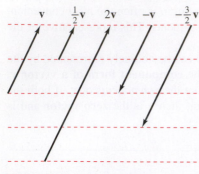

Figure 6.14

Vector Operations

The two basic vector operations are **scalar multiplication** and **vector addition.** In operations with vectors, numbers are usually referred to as **scalars.** In this text, scalars will always be real numbers. Geometrically, the product of a vector **v** and a scalar k is the vector that is $|k|$ times as long as **v**. When k is positive, $k\mathbf{v}$ has the same direction as **v**, and when k is negative, $k\mathbf{v}$ has the direction opposite that of **v**, as shown in Figure 6.14.

To add two vectors **u** and **v** geometrically, first position them (without changing their lengths or directions) so that the initial point of the second vector **v** coincides with the terminal point of the first vector **u**. The sum

$$\mathbf{u} + \mathbf{v}$$

is the vector formed by joining the initial point of the first vector **u** with the terminal point of the second vector **v**, as shown below. This technique is called the **parallelogram law** for vector addition because the vector **u** + **v**, often called the **resultant** of vector addition, is the diagonal of a parallelogram having adjacent sides **u** and **v**.

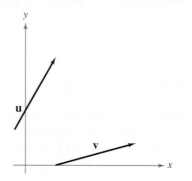
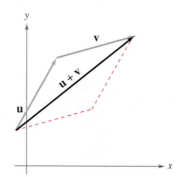

Definitions of Vector Addition and Scalar Multiplication

Let $\mathbf{u} = \langle u_1, u_2 \rangle$ and $\mathbf{v} = \langle v_1, v_2 \rangle$ be vectors and let k be a scalar (a real number). Then the **sum** of **u** and **v** is the vector

$$\mathbf{u} + \mathbf{v} = \langle u_1 + v_1, u_2 + v_2 \rangle \qquad \text{Sum}$$

and the **scalar multiple** of k times **u** is the vector

$$k\mathbf{u} = k\langle u_1, u_2 \rangle = \langle ku_1, ku_2 \rangle. \qquad \text{Scalar multiple}$$

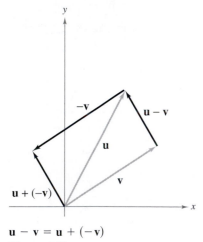

$$\mathbf{u} - \mathbf{v} = \mathbf{u} + (-\mathbf{v})$$

Figure 6.15

The **negative** of $\mathbf{v} = \langle v_1, v_2 \rangle$ is

$$-\mathbf{v} = (-1)\mathbf{v}$$

$$= \langle -v_1, -v_2 \rangle \qquad \text{Negative}$$

and the **difference** of **u** and **v** is

$$\mathbf{u} - \mathbf{v} = \mathbf{u} + (-\mathbf{v}) \qquad \text{Add } (-\mathbf{v}). \text{ See Figure 6.15.}$$

$$= \langle u_1 - v_1, u_2 - v_2 \rangle. \qquad \text{Difference}$$

To represent **u** − **v** geometrically, you can use directed line segments with the *same* initial point. The difference **u** − **v** is the vector from the terminal point of **v** to the terminal point of **u**, which is equal to

$$\mathbf{u} + (-\mathbf{v})$$

as shown in Figure 6.15.

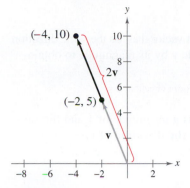

Figure 6.16

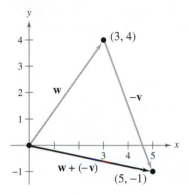

Figure 6.17

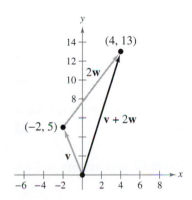

Figure 6.18

The component definitions of vector addition and scalar multiplication are illustrated in Example 3. In this example, notice that each of the vector operations can be interpreted geometrically.

EXAMPLE 3 **Vector Operations**

Let $\mathbf{v} = \langle -2, 5 \rangle$ and $\mathbf{w} = \langle 3, 4 \rangle$. Find each of the following vectors.

a. $2\mathbf{v}$ **b.** $\mathbf{w} - \mathbf{v}$ **c.** $\mathbf{v} + 2\mathbf{w}$

Solution

a. Because $\mathbf{v} = \langle -2, 5 \rangle$, you have

$$2\mathbf{v} = 2\langle -2, 5 \rangle = \langle 2(-2), 2(5) \rangle = \langle -4, 10 \rangle.$$

A sketch of $2\mathbf{v}$ is shown in Figure 6.16.

b. The difference of $\mathbf{w}$ and $\mathbf{v}$ is

$$\mathbf{w} - \mathbf{v} = \langle 3, 4 \rangle - \langle -2, 5 \rangle$$
$$= \langle 3 - (-2), 4 - 5 \rangle$$
$$= \langle 5, -1 \rangle.$$

A sketch of $\mathbf{w} - \mathbf{v}$ is shown in Figure 6.17. Note that the figure shows the vector difference $\mathbf{w} - \mathbf{v}$ as the sum $\mathbf{w} + (-\mathbf{v})$.

c. The sum of $\mathbf{v}$ and $2\mathbf{w}$ is

$$\mathbf{v} + 2\mathbf{w} = \langle -2, 5 \rangle + 2\langle 3, 4 \rangle$$
$$= \langle -2, 5 \rangle + \langle 2(3), 2(4) \rangle$$
$$= \langle -2, 5 \rangle + \langle 6, 8 \rangle$$
$$= \langle -2 + 6, 5 + 8 \rangle$$
$$= \langle 4, 13 \rangle.$$

A sketch of $\mathbf{v} + 2\mathbf{w}$ is shown in Figure 6.18.

✓ *Checkpoint* ◀))) *Audio-video solution in English & Spanish at LarsonPrecalculus.com.*

Let $\mathbf{u} = \langle 1, 4 \rangle$ and $\mathbf{v} = \langle 3, 2 \rangle$. Find each of the following vectors.

a. $\mathbf{u} + \mathbf{v}$ **b.** $\mathbf{u} - \mathbf{v}$ **c.** $2\mathbf{u} - 3\mathbf{v}$ ◼

Vector addition and scalar multiplication share many of the properties of ordinary arithmetic.

Properties of Vector Addition and Scalar Multiplication

Let $\mathbf{u}$, $\mathbf{v}$, and $\mathbf{w}$ be vectors and let c and d be scalars. Then the following properties are true.

1. $\mathbf{u} + \mathbf{v} = \mathbf{v} + \mathbf{u}$ **2.** $(\mathbf{u} + \mathbf{v}) + \mathbf{w} = \mathbf{u} + (\mathbf{v} + \mathbf{w})$

3. $\mathbf{u} + \mathbf{0} = \mathbf{u}$ **4.** $\mathbf{u} + (-\mathbf{u}) = \mathbf{0}$

5. $c(d\mathbf{u}) = (cd)\mathbf{u}$ **6.** $(c + d)\mathbf{u} = c\mathbf{u} + d\mathbf{u}$

7. $c(\mathbf{u} + \mathbf{v}) = c\mathbf{u} + c\mathbf{v}$ **8.** $1(\mathbf{u}) = \mathbf{u}, \quad 0(\mathbf{u}) = \mathbf{0}$

9. $\|c\mathbf{v}\| = |c|\,\|\mathbf{v}\|$

•• **REMARK** Property 9 can be stated as follows: The magnitude of the vector $c\mathbf{v}$ is the absolute value of c times the magnitude of $\mathbf{v}$.

William Rowan Hamilton (1805–1865), an Irish mathematician, did some of the earliest work with vectors. Hamilton spent many years developing a system of vector-like quantities called quaternions. Although Hamilton was convinced of the benefits of quaternions, the operations he defined did not produce good models for physical phenomena. It was not until the latter half of the nineteenth century that the Scottish physicist James Maxwell (1831–1879) restructured Hamilton's quaternions in a form useful for representing physical quantities such as force, velocity, and acceleration.

Unit Vectors

In many applications of vectors, it is useful to find a unit vector that has the same direction as a given nonzero vector $\mathbf{v}$. To do this, you can divide $\mathbf{v}$ by its magnitude to obtain

$$\mathbf{u} = \text{unit vector} = \frac{\mathbf{v}}{\|\mathbf{v}\|} = \left(\frac{1}{\|\mathbf{v}\|}\right)\mathbf{v}. \qquad \textcolor{red}{\text{Unit vector in direction of } \mathbf{v}}$$

Note that $\mathbf{u}$ is a scalar multiple of $\mathbf{v}$. The vector $\mathbf{u}$ has a magnitude of 1 and the same direction as $\mathbf{v}$. The vector $\mathbf{u}$ is called a **unit vector in the direction of v.**

EXAMPLE 4   **Finding a Unit Vector**

Find a unit vector in the direction of $\mathbf{v} = \langle -2, 5 \rangle$ and verify that the result has a magnitude of 1.

Solution   The unit vector in the direction of $\mathbf{v}$ is

$$\frac{\mathbf{v}}{\|\mathbf{v}\|} = \frac{\langle -2, 5 \rangle}{\sqrt{(-2)^2 + 5^2}}$$

$$= \frac{1}{\sqrt{29}}\langle -2, 5 \rangle$$

$$= \left\langle \frac{-2}{\sqrt{29}}, \frac{5}{\sqrt{29}} \right\rangle.$$

This vector has a magnitude of 1 because

$$\sqrt{\left(\frac{-2}{\sqrt{29}}\right)^2 + \left(\frac{5}{\sqrt{29}}\right)^2} = \sqrt{\frac{4}{29} + \frac{25}{29}}$$

$$= \sqrt{\frac{29}{29}}$$

$$= 1.$$

✓ **Checkpoint** *Audio-video solution in English & Spanish at LarsonPrecalculus.com.*

Find a unit vector $\mathbf{u}$ in the direction of $\mathbf{v} = \langle 6, -1 \rangle$ and verify that the result has a magnitude of 1.

The unit vectors $\langle 1, 0 \rangle$ and $\langle 0, 1 \rangle$ are called the **standard unit vectors** and are denoted by

$$\mathbf{i} = \langle 1, 0 \rangle \quad \text{and} \quad \mathbf{j} = \langle 0, 1 \rangle$$

as shown in Figure 6.19. (Note that the lowercase letter $\mathbf{i}$ is written in boldface to distinguish it from the imaginary unit $i = \sqrt{-1}$.) These vectors can be used to represent any vector $\mathbf{v} = \langle v_1, v_2 \rangle$, as follows.

$$\mathbf{v} = \langle v_1, v_2 \rangle$$

$$= v_1 \langle 1, 0 \rangle + v_2 \langle 0, 1 \rangle$$

$$= v_1 \mathbf{i} + v_2 \mathbf{j}$$

The scalars v_1 and v_2 are called the **horizontal** and **vertical components of v,** respectively. The vector sum

$$v_1 \mathbf{i} + v_2 \mathbf{j}$$

is called a **linear combination** of the vectors $\mathbf{i}$ and $\mathbf{j}$. Any vector in the plane can be written as a linear combination of the standard unit vectors $\mathbf{i}$ and $\mathbf{j}$.

Figure 6.19

EXAMPLE 5 **Writing a Linear Combination of Unit Vectors**

Let **u** be the vector with initial point $(2, -5)$ and terminal point $(-1, 3)$. Write **u** as a linear combination of the standard unit vectors **i** and **j**.

Solution Begin by writing the component form of the vector **u**.

$$\mathbf{u} = \langle -1 - 2, 3 - (-5) \rangle$$
$$= \langle -3, 8 \rangle$$
$$= -3\mathbf{i} + 8\mathbf{j}$$

This result is shown graphically below.

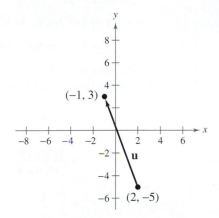

 Checkpoint *Audio-video solution in English & Spanish at LarsonPrecalculus.com.*

Let **u** be the vector with initial point $(-2, 6)$ and terminal point $(-8, 3)$. Write **u** as a linear combination of the standard unit vectors **i** and **j**.

EXAMPLE 6 **Vector Operations**

Let

$$\mathbf{u} = -3\mathbf{i} + 8\mathbf{j} \quad \text{and} \quad \mathbf{v} = 2\mathbf{i} - \mathbf{j}.$$

Find

$$2\mathbf{u} - 3\mathbf{v}.$$

Solution You could solve this problem by converting **u** and **v** to component form. This, however, is not necessary. It is just as easy to perform the operations in unit vector form.

$$2\mathbf{u} - 3\mathbf{v} = 2(-3\mathbf{i} + 8\mathbf{j}) - 3(2\mathbf{i} - \mathbf{j})$$
$$= -6\mathbf{i} + 16\mathbf{j} - 6\mathbf{i} + 3\mathbf{j}$$
$$= -12\mathbf{i} + 19\mathbf{j}$$

 Checkpoint *Audio-video solution in English & Spanish at LarsonPrecalculus.com.*

Let

$$\mathbf{u} = \mathbf{i} - 2\mathbf{j} \quad \text{and} \quad \mathbf{v} = -3\mathbf{i} + 2\mathbf{j}.$$

Find

$$5\mathbf{u} - 2\mathbf{v}.$$

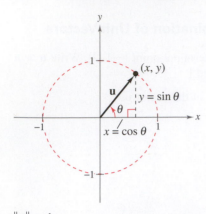

$\|\mathbf{u}\| = 1$
Figure 6.20

Direction Angles

If **u** is a *unit vector* such that θ is the angle (measured counterclockwise) from the positive x-axis to **u**, then the terminal point of **u** lies on the unit circle and you have

$$\mathbf{u} = \langle x, y \rangle$$
$$= \langle \cos \theta, \sin \theta \rangle$$
$$= (\cos \theta)\mathbf{i} + (\sin \theta)\mathbf{j}$$

as shown in Figure 6.20. The angle θ is the **direction angle** of the vector **u**.

Suppose that **u** is a unit vector with direction angle θ. If $\mathbf{v} = a\mathbf{i} + b\mathbf{j}$ is any vector that makes an angle θ with the positive x-axis, then it has the same direction as **u** and you can write

$$\mathbf{v} = \|\mathbf{v}\|\langle \cos \theta, \sin \theta \rangle$$
$$= \|\mathbf{v}\|(\cos \theta)\mathbf{i} + \|\mathbf{v}\|(\sin \theta)\mathbf{j}.$$

Because $\mathbf{v} = a\mathbf{i} + b\mathbf{j} = \|\mathbf{v}\|(\cos \theta)\mathbf{i} + \|\mathbf{v}\|(\sin \theta)\mathbf{j}$, it follows that the direction angle θ for **v** is determined from

$$\tan \theta = \frac{\sin \theta}{\cos \theta} \qquad \text{Quotient identity}$$

$$= \frac{\|\mathbf{v}\| \sin \theta}{\|\mathbf{v}\| \cos \theta} \qquad \text{Multiply numerator and denominator by } \|\mathbf{v}\|.$$

$$= \frac{b}{a}. \qquad \text{Simplify.}$$

EXAMPLE 7 **Finding Direction Angles of Vectors**

Find the direction angle of each vector.

a. $\mathbf{u} = 3\mathbf{i} + 3\mathbf{j}$ **b.** $\mathbf{v} = 3\mathbf{i} - 4\mathbf{j}$

Solution

a. The direction angle is determined from

$$\tan \theta = \frac{b}{a} = \frac{3}{3} = 1.$$

So, $\theta = 45°$, as shown in Figure 6.21.

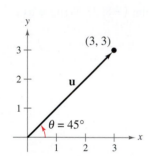

Figure 6.21

b. The direction angle is determined from

$$\tan \theta = \frac{b}{a} = \frac{-4}{3}.$$

Moreover, because $\mathbf{v} = 3\mathbf{i} - 4\mathbf{j}$ lies in Quadrant IV, θ lies in Quadrant IV, and its reference angle is

$$\theta' = \left| \arctan\left(-\frac{4}{3}\right) \right| \approx \left| -0.9273 \text{ radian} \right| \approx \left| -53.13° \right| = 53.13°.$$

So, it follows that $\theta \approx 360° - 53.13° = 306.87°$, as shown in Figure 6.22.

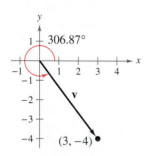

Figure 6.22

✓ **Checkpoint** ◀))) *Audio-video solution in English & Spanish at LarsonPrecalculus.com.*

Find the direction angle of each vector.

a. $\mathbf{v} = -6\mathbf{i} + 6\mathbf{j}$ **b.** $\mathbf{v} = -7\mathbf{i} - 4\mathbf{j}$

Applications

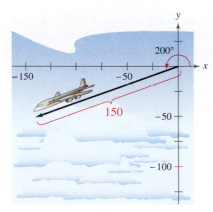

Figure 6.23

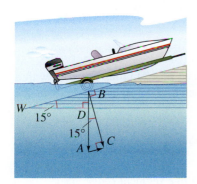

Figure 6.24

EXAMPLE 8   **Finding the Component Form of a Vector**

Find the component form of the vector that represents the velocity of an airplane descending at a speed of 150 miles per hour at an angle 20° below the horizontal, as shown in Figure 6.23.

Solution   The velocity vector **v** has a magnitude of 150 and a direction angle of $\theta = 200°$.

$$\mathbf{v} = \|\mathbf{v}\|(\cos\theta)\mathbf{i} + \|\mathbf{v}\|(\sin\theta)\mathbf{j}$$
$$= 150(\cos 200°)\mathbf{i} + 150(\sin 200°)\mathbf{j}$$
$$\approx 150(-0.9397)\mathbf{i} + 150(-0.3420)\mathbf{j}$$
$$\approx -140.96\mathbf{i} - 51.30\mathbf{j}$$
$$= \langle -140.96, -51.30 \rangle$$

You can check that **v** has a magnitude of 150, as follows.

$$\|\mathbf{v}\| \approx \sqrt{(-140.96)^2 + (-51.30)^2} \approx \sqrt{22,501.41} \approx 150$$

✓ *Checkpoint* ◀))) *Audio-video solution in English & Spanish at LarsonPrecalculus.com.*

Find the component form of the vector that represents the velocity of an airplane descending at a speed of 100 miles per hour at an angle 45° below the horizontal ($\theta = 225°$).

EXAMPLE 9   **Using Vectors to Determine Weight**

A force of 600 pounds is required to pull a boat and trailer up a ramp inclined at 15° from the horizontal. Find the combined weight of the boat and trailer.

Solution   Based on Figure 6.24, you can make the following observations.

$\|\overrightarrow{BA}\|$ = force of gravity = combined weight of boat and trailer

$\|\overrightarrow{BC}\|$ = force against ramp

$\|\overrightarrow{AC}\|$ = force required to move boat up ramp = 600 pounds

By construction, triangles BWD and ABC are similar. So, angle ABC is 15°. In triangle ABC, you have

$$\sin 15° = \frac{\|\overrightarrow{AC}\|}{\|\overrightarrow{BA}\|}$$

$$\sin 15° = \frac{600}{\|\overrightarrow{BA}\|}$$

$$\|\overrightarrow{BA}\| = \frac{600}{\sin 15°}$$

$$\|\overrightarrow{BA}\| \approx 2318.$$

So, the combined weight is approximately 2318 pounds. (In Figure 6.24, note that $\overrightarrow{AC}$ is parallel to the ramp.)

✓ *Checkpoint* ◀))) *Audio-video solution in English & Spanish at LarsonPrecalculus.com.*

A force of 500 pounds is required to pull a boat and trailer up a ramp inclined at 12° from the horizontal. Find the combined weight of the boat and trailer. ◼

• • **REMARK** Recall from
Section 4.8 that in air navigation,
bearings can be measured in
degrees clockwise from north.

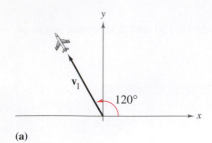

(a)

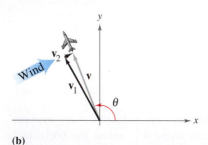

(b)
Figure 6.25

Airplanes can take advantage of
fast-moving air currents called jet
streams to decrease travel time.

EXAMPLE 10 **Using Vectors to Find Speed and Direction**

An airplane is traveling at a speed of 500 miles per hour with a bearing of 330° at a fixed
altitude with a negligible wind velocity, as shown in Figure 6.25(a). When the airplane
reaches a certain point, it encounters a wind with a velocity of 70 miles per hour in
the direction N 45° E, as shown in Figure 6.25(b). What are the resultant speed and
direction of the airplane?

Solution Using Figure 6.25, the velocity of the airplane (alone) is

$$\mathbf{v}_1 = 500\langle \cos 120°, \sin 120° \rangle = \langle -250, 250\sqrt{3} \rangle$$

and the velocity of the wind is

$$\mathbf{v}_2 = 70\langle \cos 45°, \sin 45° \rangle = \langle 35\sqrt{2}, 35\sqrt{2} \rangle.$$

So, the velocity of the airplane (in the wind) is

$$\mathbf{v} = \mathbf{v}_1 + \mathbf{v}_2$$
$$= \langle -250 + 35\sqrt{2}, 250\sqrt{3} + 35\sqrt{2} \rangle$$
$$\approx \langle -200.5, 482.5 \rangle$$

and the resultant speed of the airplane is

$$\|\mathbf{v}\| \approx \sqrt{(-200.5)^2 + (482.5)^2} \approx 522.5 \text{ miles per hour.}$$

Finally, given that θ is the direction angle of the flight path, you have

$$\tan \theta \approx \frac{482.5}{-200.5} \approx -2.4065$$

which implies that

$$\theta \approx 180° - 67.4° = 112.6°.$$

So, the true direction of the airplane is approximately

$$270° + (180° - 112.6°) = 337.4°.$$

✓ *Checkpoint*)) *Audio-video solution in English & Spanish at LarsonPrecalculus.com.*

Repeat Example 10 for an airplane traveling at a speed of 450 miles per hour with a
bearing of 300° that encounters a wind with a velocity of 40 miles per hour in the
direction N 30° E. ∎

Summarize (Section 6.3)

1. Describe how to represent a vector as a directed line segment (*page 300*).
   For an example involving vectors represented as directed line segments, see
   Example 1.

2. Describe how to write a vector in component form (*page 301*). For an
   example of finding the component form of a vector, see Example 2.

3. State the definitions of vector addition and scalar multiplication (*page 302*).
   For an example of performing vector operations, see Example 3.

4. Describe how to write a vector as a linear combination of unit vectors
   (*page 304*). For examples involving unit vectors, see Examples 4–6.

5. Describe how to find the direction angle of a vector (*page 306*). For an
   example of finding the direction angles of vectors, see Example 7.

6. Describe real-life situations that can be modeled and solved using vectors
   (*pages 307 and 308, Examples 8–10*).

6.4 Vectors and Dot Products

- Find the dot product of two vectors and use the properties of the dot product.
- Find the angle between two vectors and determine whether two vectors are orthogonal.
- Write a vector as the sum of two vector components.
- Use vectors to find the work done by a force.

The Dot Product of Two Vectors

So far you have studied two vector operations—vector addition and multiplication by a scalar—each of which yields another vector. In this section, you will study a third vector operation, the **dot product.** This operation yields a scalar, rather than a vector.

You can use the dot product of two vectors to solve real-life problems involving two vector quantities. For instance, the dot product can be used to find the force necessary to keep a sport utility vehicle from rolling down a hill.

> ### Definition of the Dot Product
>
> The **dot product** of $\mathbf{u} = \langle u_1, u_2 \rangle$ and $\mathbf{v} = \langle v_1, v_2 \rangle$ is
>
> $$\mathbf{u} \cdot \mathbf{v} = u_1 v_1 + u_2 v_2.$$

> ### Properties of the Dot Product
>
> Let $\mathbf{u}$, $\mathbf{v}$, and $\mathbf{w}$ be vectors in the plane or in space and let c be a scalar.
>
> 1. $\mathbf{u} \cdot \mathbf{v} = \mathbf{v} \cdot \mathbf{u}$
> 2. $\mathbf{0} \cdot \mathbf{v} = 0$
> 3. $\mathbf{u} \cdot (\mathbf{v} + \mathbf{w}) = \mathbf{u} \cdot \mathbf{v} + \mathbf{u} \cdot \mathbf{w}$
> 4. $\mathbf{v} \cdot \mathbf{v} = \|\mathbf{v}\|^2$
> 5. $c(\mathbf{u} \cdot \mathbf{v}) = c\mathbf{u} \cdot \mathbf{v} = \mathbf{u} \cdot c\mathbf{v}$

For proofs of the properties of the dot product, see Proofs in Mathematics on page 335.

· · REMARK In Example 1, be sure you see that the dot product of two vectors is a scalar (a real number), not a vector. Moreover, notice that the dot product can be positive, zero, or negative.

EXAMPLE 1 Finding Dot Products

a. $\langle 4, 5 \rangle \cdot \langle 2, 3 \rangle = 4(2) + 5(3)$

$\qquad\qquad\qquad = 8 + 15$

$\qquad\qquad\qquad = 23$

b. $\langle 2, -1 \rangle \cdot \langle 1, 2 \rangle = 2(1) + (-1)(2)$

$\qquad\qquad\qquad = 2 - 2$

$\qquad\qquad\qquad = 0$

c. $\langle 0, 3 \rangle \cdot \langle 4, -2 \rangle = 0(4) + 3(-2)$

$\qquad\qquad\qquad = 0 - 6$

$\qquad\qquad\qquad = -6$

✓ *Checkpoint* ◀)) *Audio-video solution in English & Spanish at LarsonPrecalculus.com.*

Find the dot product of $\mathbf{u} = \langle 3, 4 \rangle$ and $\mathbf{v} = \langle 2, -3 \rangle$.

Anthony Berenyi/Shutterstock.com

EXAMPLE 2   **Using Properties of Dot Products**

Let $\mathbf{u} = \langle -1, 3 \rangle$, $\mathbf{v} = \langle 2, -4 \rangle$, and $\mathbf{w} = \langle 1, -2 \rangle$. Use the vectors and the properties of the dot product to find each quantity.

a. $(\mathbf{u} \cdot \mathbf{v})\mathbf{w}$ **b.** $\mathbf{u} \cdot 2\mathbf{v}$ **c.** $\|\mathbf{u}\|$

Solution   Begin by finding the dot product of $\mathbf{u}$ and $\mathbf{v}$ and the dot product of $\mathbf{u}$ and $\mathbf{u}$.

$$\mathbf{u} \cdot \mathbf{v} = \langle -1, 3 \rangle \cdot \langle 2, -4 \rangle = -1(2) + 3(-4) = -14$$

$$\mathbf{u} \cdot \mathbf{u} = \langle -1, 3 \rangle \cdot \langle -1, 3 \rangle = -1(-1) + 3(3) = 10$$

a. $(\mathbf{u} \cdot \mathbf{v})\mathbf{w} = -14\langle 1, -2 \rangle = \langle -14, 28 \rangle$

b. $\mathbf{u} \cdot 2\mathbf{v} = 2(\mathbf{u} \cdot \mathbf{v}) = 2(-14) = -28$

c. Because $\|\mathbf{u}\|^2 = \mathbf{u} \cdot \mathbf{u} = 10$, it follows that $\|\mathbf{u}\| = \sqrt{\mathbf{u} \cdot \mathbf{u}} = \sqrt{10}$.

In Example 2, notice that the product in part (a) is a vector, whereas the product in part (b) is a scalar. Can you see why?

✓ **Checkpoint**  *Audio-video solution in English & Spanish at LarsonPrecalculus.com.*

Let $\mathbf{u} = \langle 3, 4 \rangle$ and $\mathbf{v} = \langle -2, 6 \rangle$. Use the vectors and the properties of the dot product to find each quantity.

a. $(\mathbf{u} \cdot \mathbf{v})\mathbf{v}$ **b.** $\mathbf{u} \cdot (\mathbf{u} + \mathbf{v})$ **c.** $\|\mathbf{v}\|$ ■

The Angle Between Two Vectors

The **angle between two nonzero vectors** is the angle θ, $0 \le \theta \le \pi$, between their respective standard position vectors, as shown in Figure 6.26. This angle can be found using the dot product.

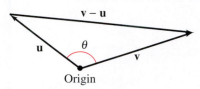

v − u

u θ **v**

Origin

Figure 6.26

Angle Between Two Vectors

If θ is the angle between two nonzero vectors $\mathbf{u}$ and $\mathbf{v}$, then

$$\cos \theta = \frac{\mathbf{u} \cdot \mathbf{v}}{\|\mathbf{u}\| \, \|\mathbf{v}\|}.$$

For a proof of the angle between two vectors, see Proofs in Mathematics on page 335.

EXAMPLE 3   **Finding the Angle Between Two Vectors**

Find the angle θ between $\mathbf{u} = \langle 4, 3 \rangle$ and $\mathbf{v} = \langle 3, 5 \rangle$ (see Figure 6.27).

Solution

$$\cos \theta = \frac{\mathbf{u} \cdot \mathbf{v}}{\|\mathbf{u}\| \, \|\mathbf{v}\|} = \frac{\langle 4, 3 \rangle \cdot \langle 3, 5 \rangle}{\|\langle 4, 3 \rangle\| \, \|\langle 3, 5 \rangle\|} = \frac{4(3) + 3(5)}{\sqrt{4^2 + 3^2} \, \sqrt{3^2 + 5^2}} = \frac{27}{5\sqrt{34}}$$

This implies that the angle between the two vectors is

$$\theta = \cos^{-1} \frac{27}{5\sqrt{34}} \approx 0.3869 \text{ radian.}$$ Use a calculator.

✓ **Checkpoint**  *Audio-video solution in English & Spanish at LarsonPrecalculus.com.*

Find the angle θ between $\mathbf{u} = \langle 2, 1 \rangle$ and $\mathbf{v} = \langle 1, 3 \rangle$. ■

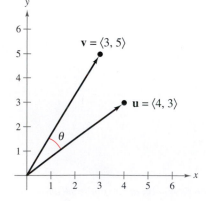

y

v = $\langle 3, 5 \rangle$

u = $\langle 4, 3 \rangle$

θ

x

Figure 6.27

Rewriting the expression for the angle between two vectors in the form

$$\mathbf{u} \cdot \mathbf{v} = \|\mathbf{u}\| \, \|\mathbf{v}\| \cos \theta \qquad \text{Alternative form of dot product}$$

produces an alternative way to calculate the dot product. From this form, you can see that because $\|\mathbf{u}\|$ and $\|\mathbf{v}\|$ are always positive, $\mathbf{u} \cdot \mathbf{v}$ and $\cos \theta$ will always have the same sign. The five possible orientations of two vectors are shown below.

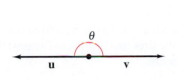

$\theta = \pi$

$\cos \theta = -1$

Opposite direction

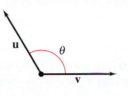

$\dfrac{\pi}{2} < \theta < \pi$

$-1 < \cos \theta < 0$

Obtuse angle

$\theta = \dfrac{\pi}{2}$

$\cos \theta = 0$

90° angle

$0 < \theta < \dfrac{\pi}{2}$

$0 < \cos \theta < 1$

Acute angle

$\theta = 0$

$\cos \theta = 1$

Same direction

Definition of Orthogonal Vectors

The vectors $\mathbf{u}$ and $\mathbf{v}$ are **orthogonal** if and only if $\mathbf{u} \cdot \mathbf{v} = 0$.

The terms *orthogonal* and *perpendicular* mean essentially the same thing—meeting at right angles. Note that the zero vector is orthogonal to every vector $\mathbf{u}$, because $\mathbf{0} \cdot \mathbf{u} = 0$.

▷ **TECHNOLOGY** The graphing utility program, Finding the Angle Between Two Vectors, found on the website for this text at *LarsonPrecalculus.com*, graphs two vectors $\mathbf{u} = \langle a, b \rangle$ and $\mathbf{v} = \langle c, d \rangle$ in standard position and finds the measure of the angle between them. Use the program to verify the solutions for Examples 3 and 4.

EXAMPLE 4 Determining Orthogonal Vectors

Are the vectors $\mathbf{u} = \langle 2, -3 \rangle$ and $\mathbf{v} = \langle 6, 4 \rangle$ orthogonal?

Solution Find the dot product of the two vectors.

$$\mathbf{u} \cdot \mathbf{v} = \langle 2, -3 \rangle \cdot \langle 6, 4 \rangle$$
$$= 2(6) + (-3)(4)$$
$$= 0$$

Because the dot product is 0, the two vectors are orthogonal (see below).

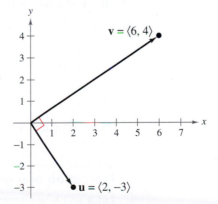

✓ *Checkpoint* ◀))) Audio-video solution in English & Spanish at LarsonPrecalculus.com.

Are the vectors $\mathbf{u} = \langle 6, 10 \rangle$ and $\mathbf{v} = \left\langle -\frac{1}{3}, \frac{1}{5} \right\rangle$ orthogonal?

Finding Vector Components

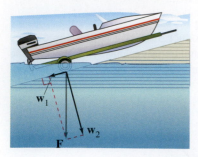

Figure 6.28

You have already seen applications in which two vectors are added to produce a resultant vector. Many applications in physics and engineering pose the reverse problem—decomposing a given vector into the sum of two **vector components.**

Consider a boat on an inclined ramp, as shown in Figure 6.28. The force **F** due to gravity pulls the boat *down* the ramp and *against* the ramp. These two orthogonal forces, $\mathbf{w}_1$ and $\mathbf{w}_2$, are vector components of **F**. That is,

$$\mathbf{F} = \mathbf{w}_1 + \mathbf{w}_2. \qquad \text{Vector components of } \mathbf{F}$$

The negative of component $\mathbf{w}_1$ represents the force needed to keep the boat from rolling down the ramp, whereas $\mathbf{w}_2$ represents the force that the tires must withstand against the ramp. A procedure for finding $\mathbf{w}_1$ and $\mathbf{w}_2$ is shown below.

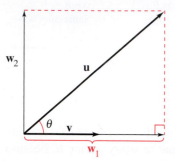

θ is acute.

Definition of Vector Components

Let **u** and **v** be nonzero vectors such that

$$\mathbf{u} = \mathbf{w}_1 + \mathbf{w}_2$$

where $\mathbf{w}_1$ and $\mathbf{w}_2$ are orthogonal and $\mathbf{w}_1$ is parallel to (or a scalar multiple of) **v**, as shown in Figure 6.29. The vectors $\mathbf{w}_1$ and $\mathbf{w}_2$ are called **vector components** of **u**. The vector $\mathbf{w}_1$ is the **projection** of **u** onto **v** and is denoted by

$$\mathbf{w}_1 = \text{proj}_{\mathbf{v}}\mathbf{u}.$$

The vector $\mathbf{w}_2$ is given by

$$\mathbf{w}_2 = \mathbf{u} - \mathbf{w}_1.$$

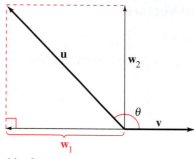

θ is obtuse.

Figure 6.29

From the definition of vector components, you can see that it is easy to find the component $\mathbf{w}_2$ once you have found the projection of **u** onto **v**. To find the projection, you can use the dot product, as follows.

$$\mathbf{u} = \mathbf{w}_1 + \mathbf{w}_2$$

$$\mathbf{u} = c\mathbf{v} + \mathbf{w}_2 \qquad \text{$\mathbf{w}_1$ is a scalar multiple of $\mathbf{v}$.}$$

$$\mathbf{u} \cdot \mathbf{v} = (c\mathbf{v} + \mathbf{w}_2) \cdot \mathbf{v} \qquad \text{Take dot product of each side with $\mathbf{v}$.}$$

$$\mathbf{u} \cdot \mathbf{v} = c\mathbf{v} \cdot \mathbf{v} + \mathbf{w}_2 \cdot \mathbf{v}$$

$$\mathbf{u} \cdot \mathbf{v} = c\|\mathbf{v}\|^2 + 0 \qquad \text{$\mathbf{w}_2$ and $\mathbf{v}$ are orthogonal.}$$

So,

$$c = \frac{\mathbf{u} \cdot \mathbf{v}}{\|\mathbf{v}\|^2}$$

and

$$\mathbf{w}_1 = \text{proj}_{\mathbf{v}}\mathbf{u} = c\mathbf{v} = \left(\frac{\mathbf{u} \cdot \mathbf{v}}{\|\mathbf{v}\|^2}\right)\mathbf{v}.$$

Projection of u onto v

Let **u** and **v** be nonzero vectors. The projection of **u** onto **v** is

$$\text{proj}_{\mathbf{v}}\mathbf{u} = \left(\frac{\mathbf{u} \cdot \mathbf{v}}{\|\mathbf{v}\|^2}\right)\mathbf{v}.$$

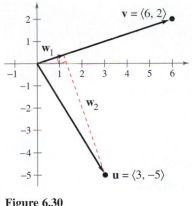

Figure 6.30

EXAMPLE 5 Decomposing a Vector into Components

Find the projection of $\mathbf{u} = \langle 3, -5 \rangle$ onto $\mathbf{v} = \langle 6, 2 \rangle$. Then write $\mathbf{u}$ as the sum of two orthogonal vectors, one of which is $\text{proj}_\mathbf{v}\mathbf{u}$.

Solution The projection of $\mathbf{u}$ onto $\mathbf{v}$ is

$$\mathbf{w}_1 = \text{proj}_\mathbf{v}\mathbf{u} = \left(\frac{\mathbf{u} \cdot \mathbf{v}}{\|\mathbf{v}\|^2}\right)\mathbf{v} = \left(\frac{8}{40}\right)\langle 6, 2 \rangle = \left\langle \frac{6}{5}, \frac{2}{5} \right\rangle$$

as shown in Figure 6.30. The other component, $\mathbf{w}_2$, is

$$\mathbf{w}_2 = \mathbf{u} - \mathbf{w}_1 = \langle 3, -5 \rangle - \left\langle \frac{6}{5}, \frac{2}{5} \right\rangle = \left\langle \frac{9}{5}, -\frac{27}{5} \right\rangle.$$

So,

$$\mathbf{u} = \mathbf{w}_1 + \mathbf{w}_2 = \left\langle \frac{6}{5}, \frac{2}{5} \right\rangle + \left\langle \frac{9}{5}, -\frac{27}{5} \right\rangle = \langle 3, -5 \rangle.$$

✓ *Checkpoint* 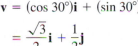 *Audio-video solution in English & Spanish at LarsonPrecalculus.com.*

Find the projection of $\mathbf{u} = \langle 3, 4 \rangle$ onto $\mathbf{v} = \langle 8, 2 \rangle$. Then write $\mathbf{u}$ as the sum of two orthogonal vectors, one of which is $\text{proj}_\mathbf{v}\mathbf{u}$.

EXAMPLE 6 Finding a Force

A 200-pound cart sits on a ramp inclined at 30°, as shown in Figure 6.31. What force is required to keep the cart from rolling down the ramp?

Solution Because the force due to gravity is vertical and downward, you can represent the gravitational force by the vector

$$\mathbf{F} = -200\mathbf{j}. \qquad \textcolor{red}{\text{Force due to gravity}}$$

To find the force required to keep the cart from rolling down the ramp, project $\mathbf{F}$ onto a unit vector $\mathbf{v}$ in the direction of the ramp, as follows.

$$\mathbf{v} = (\cos 30°)\mathbf{i} + (\sin 30°)\mathbf{j}$$

$$= \frac{\sqrt{3}}{2}\mathbf{i} + \frac{1}{2}\mathbf{j} \qquad \textcolor{red}{\text{Unit vector along ramp}}$$

So, the projection of $\mathbf{F}$ onto $\mathbf{v}$ is

$$\mathbf{w}_1 = \text{proj}_\mathbf{v}\mathbf{F}$$

$$= \left(\frac{\mathbf{F} \cdot \mathbf{v}}{\|\mathbf{v}\|^2}\right)\mathbf{v}$$

$$= (\mathbf{F} \cdot \mathbf{v})\mathbf{v}$$

$$= (-200)\left(\frac{1}{2}\right)\mathbf{v}$$

$$= -100\left(\frac{\sqrt{3}}{2}\mathbf{i} + \frac{1}{2}\mathbf{j}\right).$$

The magnitude of this force is 100. So, a force of 100 pounds is required to keep the cart from rolling down the ramp.

✓ *Checkpoint* *Audio-video solution in English & Spanish at LarsonPrecalculus.com.*

Rework Example 6 for a 150-pound cart sitting on a ramp inclined at 15°.

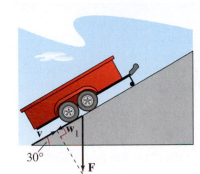

Figure 6.31

Force acts along the line of motion.
Figure 6.32

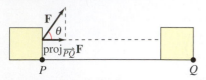

Force acts at angle θ with the line of motion.
Figure 6.33

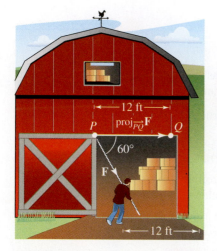

Figure 6.34

Work is done only when an object is moved. It does not matter how much force is applied—if an object does not move, then no work has been done.

Work

The work W done by a *constant* force $\mathbf{F}$ acting along the line of motion of an object is given by

$$W = (\text{magnitude of force})(\text{distance})$$
$$= \|\mathbf{F}\|\,\|\overrightarrow{PQ}\|$$

as shown in Figure 6.32. When the constant force $\mathbf{F}$ is not directed along the line of motion, as shown in Figure 6.33, the work W done by the force is given by

$$W = \|\text{proj}_{\overrightarrow{PQ}}\mathbf{F}\|\,\|\overrightarrow{PQ}\| \qquad \color{red}{\text{Projection form for work}}$$
$$= (\cos\theta)\|\mathbf{F}\|\,\|\overrightarrow{PQ}\| \qquad \color{red}{\|\text{proj}_{\overrightarrow{PQ}}\mathbf{F}\| = (\cos\theta)\|\mathbf{F}\|}$$
$$= \mathbf{F}\cdot\overrightarrow{PQ}. \qquad \color{red}{\text{Dot product form for work}}$$

This notion of work is summarized in the following definition.

Definition of Work

The **work** W done by a constant force $\mathbf{F}$ as its point of application moves along the vector $\overrightarrow{PQ}$ is given by either of the following.

1. $W = \|\text{proj}_{\overrightarrow{PQ}}\mathbf{F}\|\,\|\overrightarrow{PQ}\|$ $\color{red}{\text{Projection form}}$

2. $W = \mathbf{F}\cdot\overrightarrow{PQ}$ $\color{red}{\text{Dot product form}}$

EXAMPLE 7   **Finding Work**

To close a sliding barn door, a person pulls on a rope with a constant force of 50 pounds at a constant angle of 60°, as shown in Figure 6.34. Find the work done in moving the barn door 12 feet to its closed position.

Solution   Using a projection, you can calculate the work as follows.

$$W = \|\text{proj}_{\overrightarrow{PQ}}\mathbf{F}\|\,\|\overrightarrow{PQ}\| = (\cos 60°)\|\mathbf{F}\|\,\|\overrightarrow{PQ}\| = \frac{1}{2}(50)(12) = 300 \text{ foot-pounds}$$

So, the work done is 300 foot-pounds. You can verify this result by finding the vectors $\mathbf{F}$ and $\overrightarrow{PQ}$ and calculating their dot product.

✓ *Checkpoint*  *Audio-video solution in English & Spanish at LarsonPrecalculus.com.*

A wagon is pulled by exerting a force of 35 pounds on a handle that makes a 30° angle with the horizontal. Find the work done in pulling the wagon 40 feet. ■

Summarize   **(Section 6.4)**

1. State the definition of the dot product *(page 309)*. For examples of finding dot products and using the properties of the dot product, see Examples 1 and 2.

2. Describe how to find the angle between two vectors *(page 310)*. For examples involving the angle between two vectors, see Examples 3 and 4.

3. Describe how to decompose a vector into components *(page 312)*. For examples involving vector components, see Examples 5 and 6.

4. State the definition of work *(page 314)*. For an example of finding the work done by a constant force, see Example 7.

6.5 Trigonometric Form of a Complex Number

You can use the trigonometric form of a complex number to perform operations with complex numbers. For instance, the trigonometric forms of complex numbers can be used to find the voltage of an alternating current circuit.

■ Plot complex numbers in the complex plane and find absolute values of complex numbers.
■ Write the trigonometric forms of complex numbers.
■ Multiply and divide complex numbers written in trigonometric form.
■ Use DeMoivre's Theorem to find powers of complex numbers.
■ Find *n*th roots of complex numbers.

The Complex Plane

Just as real numbers can be represented by points on the real number line, you can represent a complex number $z = a + bi$ as the point (a, b) in a coordinate plane (the **complex plane**). The horizontal axis is called the **real axis** and the vertical axis is called the **imaginary axis,** as shown below.

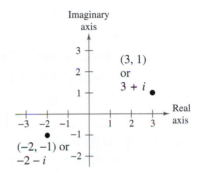

The **absolute value** of the complex number $a + bi$ is defined as the distance between the origin $(0, 0)$ and the point (a, b).

> **Definition of the Absolute Value of a Complex Number**
>
> The **absolute value** of the complex number $z = a + bi$ is
>
> $$|a + bi| = \sqrt{a^2 + b^2}.$$

When the complex number $a + bi$ is a real number (that is, when $b = 0$), this definition agrees with that given for the absolute value of a real number

$$|a + 0i| = \sqrt{a^2 + 0^2}$$
$$= |a|.$$

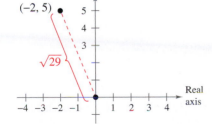

Figure 6.35

EXAMPLE 1   **Finding the Absolute Value of a Complex Number**

Plot $z = -2 + 5i$ and find its absolute value.

Solution   The number is plotted in Figure 6.35. It has an absolute value of

$$|z| = \sqrt{(-2)^2 + 5^2}$$
$$= \sqrt{29}.$$

✓ **Checkpoint** ◀))) *Audio-video solution in English & Spanish at LarsonPrecalculus.com.*

Plot $z = 3 - 4i$ and find its absolute value.

auremar/Shutterstock.com

Trigonometric Form of a Complex Number

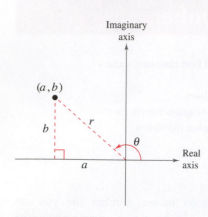

Figure 6.36

In Section 2.4, you learned how to add, subtract, multiply, and divide complex numbers. To work effectively with *powers* and *roots* of complex numbers, it is helpful to write complex numbers in trigonometric form. In Figure 6.36, consider the nonzero complex number

$$a + bi.$$

By letting θ be the angle from the positive real axis (measured counterclockwise) to the line segment connecting the origin and the point (a, b), you can write

$$a = r \cos \theta \quad \text{and} \quad b = r \sin \theta$$

where

$$r = \sqrt{a^2 + b^2}.$$

Consequently, you have

$$a + bi = (r \cos \theta) + (r \sin \theta)i$$

from which you can obtain the **trigonometric form of a complex number.**

Trigonometric Form of a Complex Number

The **trigonometric form** of the complex number $z = a + bi$ is

$$z = r(\cos \theta + i \sin \theta)$$

where $a = r \cos \theta$, $b = r \sin \theta$, $r = \sqrt{a^2 + b^2}$, and $\tan \theta = b/a$. The number r is the **modulus** of z, and θ is called an **argument** of z.

▷

• • REMARK When θ is restricted to the interval $0 \le \theta < 2\pi$, use the following guidelines. When a complex number lies in Quadrant I, $\theta = \arctan(b/a)$. When a complex number lies in Quadrant II or Quadrant III, $\theta = \pi + \arctan(b/a)$. When a complex number lies in Quadrant IV, $\theta = 2\pi + \arctan(b/a)$.

The trigonometric form of a complex number is also called the *polar form*. Because there are infinitely many choices for θ, the trigonometric form of a complex number is not unique. Normally, θ is restricted to the interval $0 \le \theta < 2\pi$, although on occasion it is convenient to use $\theta < 0$.

EXAMPLE 2 **Trigonometric Form of a Complex Number**

Write the complex number $z = -2 - 2\sqrt{3}i$ in trigonometric form.

Solution The absolute value of z is

$$r = \left| -2 - 2\sqrt{3}i \right| = \sqrt{(-2)^2 + \left(-2\sqrt{3}\right)^2} = \sqrt{16} = 4$$

and the argument θ is determined from

$$\tan \theta = \frac{b}{a} = \frac{-2\sqrt{3}}{-2} = \sqrt{3}.$$

Because $z = -2 - 2\sqrt{3}i$ lies in Quadrant III, as shown in Figure 6.37,

$$\theta = \pi + \arctan \sqrt{3} = \pi + \frac{\pi}{3} = \frac{4\pi}{3}.$$

So, the trigonometric form is

$$z = r(\cos \theta + i \sin \theta) = 4\left(\cos \frac{4\pi}{3} + i \sin \frac{4\pi}{3} \right).$$

✓ *Checkpoint* ◀)))) Audio-video solution in English & Spanish at LarsonPrecalculus.com.

Write the complex number $z = 6 - 6i$ in trigonometric form. ■

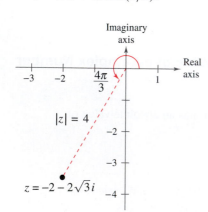

Figure 6.37

▷ **TECHNOLOGY** A graphing utility can be used to convert a complex number in trigonometric (or polar) form to standard form. For specific keystrokes, see the user's manual for your graphing utility.

EXAMPLE 3 **Writing a Complex Number in Standard Form**

Write $z = \sqrt{8}[\cos(-\pi/3) + i \sin(-\pi/3)]$ in standard form $a + bi$.

Solution Because $\cos(-\pi/3) = 1/2$ and $\sin(-\pi/3) = -\sqrt{3}/2$, you can write

$$z = \sqrt{8}\left[\cos\left(-\frac{\pi}{3}\right) + i \sin\left(-\frac{\pi}{3}\right)\right] = 2\sqrt{2}\left(\frac{1}{2} - \frac{\sqrt{3}}{2}i\right) = \sqrt{2} - \sqrt{6}i.$$

✓ **Checkpoint** ◀))) *Audio-video solution in English & Spanish at LarsonPrecalculus.com.*

Write $z = 8[\cos(2\pi/3) + i \sin(2\pi/3)]$ in standard form $a + bi$. ■

Multiplication and Division of Complex Numbers

The trigonometric form adapts nicely to multiplication and division of complex numbers. Suppose you are given two complex numbers $z_1 = r_1(\cos \theta_1 + i \sin \theta_1)$ and $z_2 = r_2(\cos \theta_2 + i \sin \theta_2)$. The product of z_1 and z_2 is

$$z_1 z_2 = r_1 r_2(\cos \theta_1 + i \sin \theta_1)(\cos \theta_2 + i \sin \theta_2)$$

$$= r_1 r_2[(\cos \theta_1 \cos \theta_2 - \sin \theta_1 \sin \theta_2) + i(\sin \theta_1 \cos \theta_2 + \cos \theta_1 \sin \theta_2)].$$

Using the sum and difference formulas for cosine and sine, you can rewrite this equation as

$$z_1 z_2 = r_1 r_2[\cos(\theta_1 + \theta_2) + i \sin(\theta_1 + \theta_2)].$$

This establishes the first part of the following rule. The second part is left for you to verify.

• • **REMARK** Note that this rule says that to *multiply* two complex numbers you multiply moduli and add arguments, whereas to *divide* two complex numbers you divide moduli and subtract arguments.

▷

Product and Quotient of Two Complex Numbers

Let $z_1 = r_1(\cos \theta_1 + i \sin \theta_1)$ and $z_2 = r_2(\cos \theta_2 + i \sin \theta_2)$ be complex numbers.

$$z_1 z_2 = r_1 r_2[\cos(\theta_1 + \theta_2) + i \sin(\theta_1 + \theta_2)] \qquad \text{Product}$$

$$\frac{z_1}{z_2} = \frac{r_1}{r_2}[\cos(\theta_1 - \theta_2) + i \sin(\theta_1 - \theta_2)], \quad z_2 \neq 0 \qquad \text{Quotient}$$

EXAMPLE 4 **Multiplying Complex Numbers**

Find the product $z_1 z_2$ of $z_1 = 3\left(\cos\frac{\pi}{4} + i \sin\frac{\pi}{4}\right)$ and $z_2 = 2\left(\cos\frac{3\pi}{4} + i \sin\frac{3\pi}{4}\right)$.

Solution

$$z_1 z_2 = 3\left(\cos\frac{\pi}{4} + i \sin\frac{\pi}{4}\right) \cdot 2\left(\cos\frac{3\pi}{4} + i \sin\frac{3\pi}{4}\right)$$

$$= 6\left[\cos\left(\frac{\pi}{4} + \frac{3\pi}{4}\right) + i \sin\left(\frac{\pi}{4} + \frac{3\pi}{4}\right)\right]$$

$$= 6(\cos \pi + i \sin \pi)$$

$$= 6[-1 + i(0)]$$

$$= -6$$

✓ **Checkpoint** ◀))) *Audio-video solution in English & Spanish at LarsonPrecalculus.com.*

Find the product $z_1 z_2$ of $z_1 = 2\left(\cos\frac{5\pi}{6} + i \sin\frac{5\pi}{6}\right)$ and $z_2 = 5\left(\cos\frac{7\pi}{6} + i \sin\frac{7\pi}{6}\right)$. ■

EXAMPLE 5 **Multiplying Complex Numbers**

Find the product $z_1 z_2$ of the complex numbers.

$$z_1 = 2\left(\cos\frac{2\pi}{3} + i\sin\frac{2\pi}{3}\right) \qquad z_2 = 8\left(\cos\frac{11\pi}{6} + i\sin\frac{11\pi}{6}\right)$$

Solution

$$z_1 z_2 = 2\left(\cos\frac{2\pi}{3} + i\sin\frac{2\pi}{3}\right) \cdot 8\left(\cos\frac{11\pi}{6} + i\sin\frac{11\pi}{6}\right)$$

$$= 16\left[\cos\left(\frac{2\pi}{3} + \frac{11\pi}{6}\right) + i\sin\left(\frac{2\pi}{3} + \frac{11\pi}{6}\right)\right] \qquad \text{Multiply moduli and add arguments.}$$

$$= 16\left(\cos\frac{5\pi}{2} + i\sin\frac{5\pi}{2}\right)$$

$$= 16\left(\cos\frac{\pi}{2} + i\sin\frac{\pi}{2}\right) \qquad \frac{5\pi}{2} \text{ and } \frac{\pi}{2} \text{ are coterminal.}$$

$$= 16[0 + i(1)]$$

$$= 16i$$

You can check this result by first converting the complex numbers to the standard forms $z_1 = -1 + \sqrt{3}i$ and $z_2 = 4\sqrt{3} - 4i$ and then multiplying algebraically, as in Section 2.4.

$$z_1 z_2 = \left(-1 + \sqrt{3}i\right)\left(4\sqrt{3} - 4i\right)$$

$$= -4\sqrt{3} + 4i + 12i + 4\sqrt{3}$$

$$= 16i$$

✓ *Checkpoint* ◀))) *Audio-video solution in English & Spanish at LarsonPrecalculus.com.*

Find the product $z_1 z_2$ of the complex numbers.

$$z_1 = 3\left(\cos\frac{\pi}{3} + i\sin\frac{\pi}{3}\right) \qquad z_2 = 4\left(\cos\frac{\pi}{6} + i\sin\frac{\pi}{6}\right)$$

▷ **TECHNOLOGY**
Some graphing utilities can multiply and divide complex numbers in trigonometric form. If you have access to such a graphing utility, use it to find $z_1 z_2$ and z_1/z_2 in Examples 5 and 6.

EXAMPLE 6 **Dividing Complex Numbers**

Find the quotient z_1/z_2 of the complex numbers.

$$z_1 = 24(\cos 300° + i\sin 300°) \qquad z_2 = 8(\cos 75° + i\sin 75°)$$

Solution

$$\frac{z_1}{z_2} = \frac{24(\cos 300° + i\sin 300°)}{8(\cos 75° + i\sin 75°)}$$

$$= 3\left[\cos(300° - 75°) + i\sin(300° - 75°)\right] \qquad \text{Divide moduli and subtract arguments.}$$

$$= 3(\cos 225° + i\sin 225°)$$

$$= 3\left[-\frac{\sqrt{2}}{2} + i\left(-\frac{\sqrt{2}}{2}\right)\right]$$

$$= -\frac{3\sqrt{2}}{2} - \frac{3\sqrt{2}}{2}i$$

✓ *Checkpoint* ◀))) *Audio-video solution in English & Spanish at LarsonPrecalculus.com.*

Find the quotient z_1/z_2 of the complex numbers.

$$z_1 = \cos 40° + i\sin 40° \qquad z_2 = \cos 10° + i\sin 10°$$

Powers of Complex Numbers

The trigonometric form of a complex number is used to raise a complex number to a power. To accomplish this, consider repeated use of the multiplication rule.

$$z = r(\cos \theta + i \sin \theta)$$

$$z^2 = r(\cos \theta + i \sin \theta)r(\cos \theta + i \sin \theta) = r^2(\cos 2\theta + i \sin 2\theta)$$

$$z^3 = r^2(\cos 2\theta + i \sin 2\theta)r(\cos \theta + i \sin \theta) = r^3(\cos 3\theta + i \sin 3\theta)$$

$$z^4 = r^4(\cos 4\theta + i \sin 4\theta)$$

$$z^5 = r^5(\cos 5\theta + i \sin 5\theta)$$

$$\vdots$$

This pattern leads to **DeMoivre's Theorem,** which is named after the French mathematician Abraham DeMoivre (1667–1754).

DeMoivre's Theorem

If $z = r(\cos \theta + i \sin \theta)$ is a complex number and n is a positive integer, then

$$z^n = [r(\cos \theta + i \sin \theta)]^n$$

$$= r^n(\cos n\theta + i \sin n\theta).$$

Abraham DeMoivre (1667–1754) is remembered for his work in probability theory and DeMoivre's Theorem. His book *The Doctrine of Chances* (published in 1718) includes the theory of recurring series and the theory of partial fractions.

EXAMPLE 7   **Finding a Power of a Complex Number**

Use DeMoivre's Theorem to find $\left(-1 + \sqrt{3}i\right)^{12}$.

Solution   The absolute value of $z = -1 + \sqrt{3}i$ is

$$r = \left|-1 + \sqrt{3}i\right| = \sqrt{(-1)^2 + \left(\sqrt{3}\right)^2} = 2$$

and the argument θ is given by $\tan \theta = \sqrt{3}/(-1)$. Because $z = -1 + \sqrt{3}i$ lies in Quadrant II,

$$\theta = \pi + \arctan \frac{\sqrt{3}}{-1} = \pi + \left(-\frac{\pi}{3}\right) = \frac{2\pi}{3}.$$

So, the trigonometric form is

$$z = -1 + \sqrt{3}i = 2\left(\cos \frac{2\pi}{3} + i \sin \frac{2\pi}{3}\right).$$

Then, by DeMoivre's Theorem, you have

$$\left(-1 + \sqrt{3}i\right)^{12} = \left[2\left(\cos \frac{2\pi}{3} + i \sin \frac{2\pi}{3}\right)\right]^{12}$$

$$= 2^{12}\left[\cos \frac{12(2\pi)}{3} + i \sin \frac{12(2\pi)}{3}\right]$$

$$= 4096(\cos 8\pi + i \sin 8\pi)$$

$$= 4096(1 + 0)$$

$$= 4096.$$

✓ *Checkpoint*))) *Audio-video solution in English & Spanish at LarsonPrecalculus.com.*

Use DeMoivre's Theorem to find $(-1 - i)^4$.

Alamy

Roots of Complex Numbers

Recall that a consequence of the Fundamental Theorem of Algebra is that a polynomial equation of degree n has n solutions in the complex number system. So, the equation $x^6 = 1$ has six solutions, and in this particular case you can find the six solutions by factoring and using the Quadratic Formula.

$$x^6 - 1 = 0$$

$$(x^3 - 1)(x^3 + 1) = 0$$

$$(x - 1)(x^2 + x + 1)(x + 1)(x^2 - x + 1) = 0$$

Consequently, the solutions are

$$x = \pm 1, \quad x = \frac{-1 \pm \sqrt{3}\,i}{2}, \quad \text{and} \quad x = \frac{1 \pm \sqrt{3}\,i}{2}.$$

Each of these numbers is a sixth root of 1. In general, an **nth root of a complex number** is defined as follows.

Definition of an nth Root of a Complex Number

The complex number $u = a + bi$ is an **nth root** of the complex number z when

$$z = u^n$$

$$= (a + bi)^n.$$

To find a formula for an nth root of a complex number, let u be an nth root of z, where

$$u = s(\cos \beta + i \sin \beta)$$

and

$$z = r(\cos \theta + i \sin \theta).$$

By DeMoivre's Theorem and the fact that $u^n = z$, you have

$$s^n(\cos n\beta + i \sin n\beta) = r(\cos \theta + i \sin \theta).$$

Taking the absolute value of each side of this equation, it follows that $s^n = r$. Substituting back into the previous equation and dividing by r, you get

$$\cos n\beta + i \sin n\beta = \cos \theta + i \sin \theta.$$

So, it follows that

$$\cos n\beta = \cos \theta$$

and

$$\sin n\beta = \sin \theta.$$

Because both sine and cosine have a period of 2π, these last two equations have solutions if and only if the angles differ by a multiple of 2π. Consequently, there must exist an integer k such that

$$n\beta = \theta + 2\pi k$$

$$\beta = \frac{\theta + 2\pi k}{n}.$$

By substituting this value of β into the trigonometric form of u, you get the result stated on the following page.

Finding *n*th Roots of a Complex Number

For a positive integer n, the complex number $z = r(\cos \theta + i \sin \theta)$ has exactly n distinct *n*th roots given by

$$z_k = \sqrt[n]{r}\left(\cos \frac{\theta + 2\pi k}{n} + i \sin \frac{\theta + 2\pi k}{n}\right)$$

where $k = 0, 1, 2, \ldots, n - 1$.

When $k > n - 1$, the roots begin to repeat. For instance, when $k = n$, the angle

$$\frac{\theta + 2\pi n}{n} = \frac{\theta}{n} + 2\pi$$

is coterminal with θ/n, which is also obtained when $k = 0$.

The formula for the *n*th roots of a complex number z has a nice geometrical interpretation, as shown in Figure 6.38. Note that because the *n*th roots of z all have the same magnitude $\sqrt[n]{r}$, they all lie on a circle of radius $\sqrt[n]{r}$ with center at the origin. Furthermore, because successive *n*th roots have arguments that differ by $2\pi/n$, the n roots are equally spaced around the circle.

You have already found the sixth roots of 1 by factoring and using the Quadratic Formula. Example 8 shows how you can solve the same problem with the formula for *n*th roots.

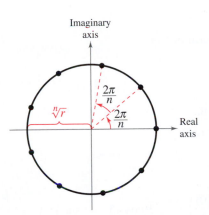

Figure 6.38

EXAMPLE 8   Finding the *n*th Roots of a Real Number

Find all sixth roots of 1.

Solution   First, write 1 in the trigonometric form $z = 1(\cos 0 + i \sin 0)$. Then, by the *n*th root formula, with $n = 6$ and $r = 1$, the roots have the form

$$z_k = \sqrt[6]{1}\left(\cos \frac{0 + 2\pi k}{6} + i \sin \frac{0 + 2\pi k}{6}\right) = \cos \frac{\pi k}{3} + i \sin \frac{\pi k}{3}.$$

So, for $k = 0, 1, 2, 3, 4,$ and 5, the sixth roots are as follows. (See Figure 6.39.)

$$z_0 = \cos 0 + i \sin 0 = 1$$

$$z_1 = \cos \frac{\pi}{3} + i \sin \frac{\pi}{3} = \frac{1}{2} + \frac{\sqrt{3}}{2} i \qquad \text{Increment by } \frac{2\pi}{n} = \frac{2\pi}{6} = \frac{\pi}{3}$$

$$z_2 = \cos \frac{2\pi}{3} + i \sin \frac{2\pi}{3} = -\frac{1}{2} + \frac{\sqrt{3}}{2} i$$

$$z_3 = \cos \pi + i \sin \pi = -1$$

$$z_4 = \cos \frac{4\pi}{3} + i \sin \frac{4\pi}{3} = -\frac{1}{2} - \frac{\sqrt{3}}{2} i$$

$$z_5 = \cos \frac{5\pi}{3} + i \sin \frac{5\pi}{3} = \frac{1}{2} - \frac{\sqrt{3}}{2} i$$

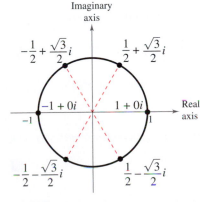

Figure 6.39

✓ **Checkpoint**  Audio-video solution in English & Spanish at LarsonPrecalculus.com.

Find all fourth roots of 1. ■

In Figure 6.39, notice that the roots obtained in Example 8 all have a magnitude of 1 and are equally spaced around the unit circle. Also notice that the complex roots occur in conjugate pairs, as discussed in Section 2.5. The n distinct *n*th roots of 1 are called the **nth roots of unity**.

Finding the *n*th Roots of a Complex Number

Find the three cube roots of $z = -2 + 2i$.

Solution   The absolute value of z is

$$r = |-2 + 2i| = \sqrt{(-2)^2 + 2^2} = \sqrt{8}$$

and the argument θ is given by

$$\tan \theta = \frac{b}{a} = \frac{2}{-2} = -1.$$

Because z lies in Quadrant II, the trigonometric form of z is

$$z = -2 + 2i$$
$$= \sqrt{8}\,(\cos 135° + i \sin 135°). \qquad \color{red}{\theta = \pi + \arctan(-1) = 3\pi/4 = 135°}$$

By the formula for *n*th roots, the cube roots have the form

$$z_k = \sqrt[6]{8}\left(\cos \frac{135° + 360°k}{3} + i \sin \frac{135° + 360°k}{3}\right).$$

Finally, for $k = 0, 1,$ and 2, you obtain the roots

$$z_0 = \sqrt[6]{8}\left(\cos \frac{135° + 360°(0)}{3} + i \sin \frac{135° + 360°(0)}{3}\right) = \sqrt{2}\,(\cos 45° + i \sin 45°)$$
$$= 1 + i$$

$$z_1 = \sqrt[6]{8}\left(\cos \frac{135° + 360°(1)}{3} + i \sin \frac{135° + 360°(1)}{3}\right) = \sqrt{2}\,(\cos 165° + i \sin 165°)$$
$$\approx -1.3660 + 0.3660i$$

$$z_2 = \sqrt[6]{8}\left(\cos \frac{135° + 360°(2)}{3} + i \sin \frac{135° + 360°(2)}{3}\right) = \sqrt{2}\,(\cos 285° + i \sin 285°)$$
$$\approx 0.3660 - 1.3660i.$$

See Figure 6.40.

✓ **Checkpoint** ◄))) *Audio-video solution in English & Spanish at LarsonPrecalculus.com.*

Find the three cube roots of $z = -6 + 6i$. ■

•• REMARK In Example 9, because $r = \sqrt{8}$, it follows that

$$\sqrt[n]{r} = \sqrt[3]{\sqrt{8}}$$
$$= (8^{1/2})^{1/3}$$
$$= 8^{1/6}$$
$$= \sqrt[6]{8}.$$

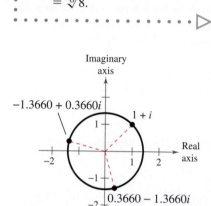

Figure 6.40

Summarize   (Section 6.5)

1. State the definition of the absolute value of a complex number *(page 315)*. For an example of finding the absolute value of a complex number, see Example 1.

2. State the definition of the trigonometric form of a complex number *(page 316)*. For examples of writing complex numbers in trigonometric form and standard form, see Examples 2 and 3.

3. Describe how to multiply and divide complex numbers written in trigonometric form *(page 317)*. For examples of multiplying and dividing complex numbers written in trigonometric form, see Examples 4–6.

4. State DeMoivre's Theorem *(page 319)*. For an example of using DeMoivre's Theorem to find a power of a complex number, see Example 7.

5. Describe how to find the *n*th roots of a complex number *(pages 320 and 321)*. For examples of finding *n*th roots, see Examples 8 and 9.

Chapter Summary

	What Did You Learn?	Explanation/Examples	Review Exercises
Section 6.1	Use the Law of Sines to solve oblique triangles (AAS or ASA) *(p. 290)*.	**Law of Sines** If ABC is a triangle with sides a, b, and c, then $$\frac{a}{\sin A} = \frac{b}{\sin B} = \frac{c}{\sin C}.$$ A is acute. A is obtuse.	1–12
	Use the Law of Sines to solve oblique triangles (SSA) *(p. 292)*.	If two sides and one opposite angle are given, then three possible situations can occur: (1) no such triangle exists (see Example 4), (2) one such triangle exists (see Example 3), or (3) two distinct triangles may satisfy the conditions (see Example 5).	1–12, 31–34
	Find the areas of oblique triangles *(p. 294)*.	The area of any triangle is one-half the product of the lengths of two sides times the sine of their included angle. That is, Area $= \frac{1}{2}bc \sin A = \frac{1}{2}ab \sin C = \frac{1}{2}ac \sin B$.	13–16
	Use the Law of Sines to model and solve real-life problems *(p. 295)*.	You can use the Law of Sines to approximate the total distance of a boat race course. (See Example 7.)	17–20
Section 6.2	Use the Law of Cosines to solve oblique triangles (SSS or SAS) *(p. 296)*.	**Law of Cosines** **Standard Form** **Alternative Form** $a^2 = b^2 + c^2 - 2bc \cos A$ $\cos A = \dfrac{b^2 + c^2 - a^2}{2bc}$ $b^2 = a^2 + c^2 - 2ac \cos B$ $\cos B = \dfrac{a^2 + c^2 - b^2}{2ac}$ $c^2 = a^2 + b^2 - 2ab \cos C$ $\cos C = \dfrac{a^2 + b^2 - c^2}{2ab}$	21–34
	Use the Law of Cosines to model and solve real-life problems *(p. 298)*.	You can use the Law of Cosines to find the distance between the pitcher's mound and first base on a women's softball field. (See Example 3.)	35–38
	Use Heron's Area Formula to find the area of a triangle *(p. 299)*.	**Heron's Area Formula:** Given any triangle with sides of lengths a, b, and c, the area of the triangle is Area $= \sqrt{s(s - a)(s - b)(s - c)}$, where $s = (a + b + c)/2$.	39–42
Section 6.3	Represent vectors as directed line segments *(p. 300)*.		43, 44
	Write the component forms of vectors *(p. 301)*.	The component form of the vector with initial point $P(p_1, p_2)$ and terminal point $Q(q_1, q_2)$ is given by $\overrightarrow{PQ} = \langle q_1 - p_1, q_2 - p_2 \rangle = \langle v_1, v_2 \rangle = \mathbf{v}.$	45–50

What Did You Learn?	**Explanation/Examples**	**Review Exercises**
Section 6.3		
Perform basic vector operations and represent them graphically (*p. 302*).	Let $\mathbf{u} = \langle u_1, u_2 \rangle$ and $\mathbf{v} = \langle v_1, v_2 \rangle$ be vectors and let k be a scalar (a real number). $\mathbf{u} + \mathbf{v} = \langle u_1 + v_1, u_2 + v_2 \rangle \qquad k\mathbf{u} = \langle ku_1, ku_2 \rangle$ $-\mathbf{v} = \langle -v_1, -v_2 \rangle \qquad\qquad \mathbf{u} - \mathbf{v} = \langle u_1 - v_1, u_2 - v_2 \rangle$	51–64
Write vectors as linear combinations of unit vectors (*p. 304*).	The vector sum $\mathbf{v} = \langle v_1, v_2 \rangle = v_1\langle 1, 0 \rangle + v_2\langle 0, 1 \rangle = v_1\mathbf{i} + v_2\mathbf{j}$ is a linear combination of the vectors $\mathbf{i}$ and $\mathbf{j}$.	65–68
Find the direction angles of vectors (*p. 306*).	If $\mathbf{u} = 2\mathbf{i} + 2\mathbf{j}$, then the direction angle is determined from $\tan \theta = 2/2 = 1$. So, $\theta = 45°$.	69–74
Use vectors to model and solve real-life problems (*p. 307*).	You can use vectors to find the resultant speed and direction of an airplane. (See Example 10.)	75–78
Section 6.4		
Find the dot product of two vectors and use the properties of the dot product (*p. 309*).	The dot product of $\mathbf{u} = \langle u_1, u_2 \rangle$ and $\mathbf{v} = \langle v_1, v_2 \rangle$ is $\mathbf{u} \cdot \mathbf{v} = u_1v_1 + u_2v_2$.	79–90
Find the angle between two vectors and determine whether two vectors are orthogonal (*p. 310*).	If θ is the angle between two nonzero vectors $\mathbf{u}$ and $\mathbf{v}$, then $\cos \theta = \dfrac{\mathbf{u} \cdot \mathbf{v}}{\|\mathbf{u}\| \, \|\mathbf{v}\|}.$ Vectors $\mathbf{u}$ and $\mathbf{v}$ are orthogonal if and only if $\mathbf{u} \cdot \mathbf{v} = 0$.	91–98
Write a vector as the sum of two vector components (*p. 312*).	Many applications in physics and engineering require the decomposition of a given vector into the sum of two vector components. (See Example 6.)	99–102
Use vectors to find the work done by a force (*p. 314*).	The work W done by a constant force $\mathbf{F}$ as its point of application moves along the vector $\overrightarrow{PQ}$ is given by either of the following. **1.** $W = \|\text{proj}_{\overrightarrow{PQ}} \mathbf{F}\| \, \|\overrightarrow{PQ}\|$ **2.** $W = \mathbf{F} \cdot \overrightarrow{PQ}$	103–106
Section 6.5		
Plot complex numbers in the complex plane and find absolute values of complex numbers (*p. 315*).	A complex number $z = a + bi$ can be represented as the point (a, b) in the complex plane. The horizontal axis is the real axis and the vertical axis is the imaginary axis. The absolute value of $z = a + bi$ is $|a + bi| = \sqrt{a^2 + b^2}$.	107–112
Write the trigonometric forms of complex numbers (*p. 316*).	The trigonometric form of the complex number $z = a + bi$ is $z = r(\cos \theta + i \sin \theta)$ where $a = r \cos \theta$, $b = r \sin \theta$, $r = \sqrt{a^2 + b^2}$, and $\tan \theta = b/a$.	113–118
Multiply and divide complex numbers written in trigonometric form (*p. 317*).	Let $z_1 = r_1(\cos \theta_1 + i \sin \theta_1)$ and $z_2 = r_2(\cos \theta_2 + i \sin \theta_2)$ be complex numbers. $z_1z_2 = r_1r_2[\cos(\theta_1 + \theta_2) + i \sin(\theta_1 + \theta_2)]$ $z_1/z_2 = (r_1/r_2)[\cos(\theta_1 - \theta_2) + i \sin(\theta_1 - \theta_2)], \quad z_2 \neq 0$	119, 120
Use DeMoivre's Theorem to find powers of complex numbers (*p. 319*).	**DeMoivre's Theorem:** If $z = r(\cos \theta + i \sin \theta)$ is a complex number and n is a positive integer, then $z^n = [r(\cos \theta + i \sin \theta)]^n = r^n(\cos n\theta + i \sin n\theta)$.	121–124
Find nth roots of complex numbers (*p. 320*).	The complex number $u = a + bi$ is an nth root of the complex number z when $z = u^n = (a + bi)^n$.	125–134

Review Exercises
See CalcChat.com for tutorial help and worked-out solutions to odd-numbered exercises.

6.1 Using the Law of Sines In Exercises 1–12, use the Law of Sines to solve (if possible) the triangle. If two solutions exist, find both. Round your answers to two decimal places.

1.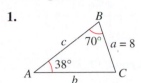

2.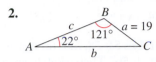

3. $B = 72°$, $C = 82°$, $b = 54$
4. $B = 10°$, $C = 20°$, $c = 33$
5. $A = 16°$, $B = 98°$, $c = 8.4$
6. $A = 95°$, $B = 45°$, $c = 104.8$
7. $A = 24°$, $C = 48°$, $b = 27.5$
8. $B = 64°$, $C = 36°$, $a = 367$
9. $B = 150°$, $b = 30$, $c = 10$
10. $B = 150°$, $a = 10$, $b = 3$
11. $A = 75°$, $a = 51.2$, $b = 33.7$
12. $B = 25°$, $a = 6.2$, $b = 4$

Finding the Area of a Triangle In Exercises 13–16, find the area of the triangle having the indicated angle and sides.

13. $A = 33°$, $b = 7$, $c = 10$
14. $B = 80°$, $a = 4$, $c = 8$
15. $C = 119°$, $a = 18$, $b = 6$
16. $A = 11°$, $b = 22$, $c = 21$

17. Height From a certain distance, the angle of elevation to the top of a building is 17°. At a point 50 meters closer to the building, the angle of elevation is 31°. Approximate the height of the building.

18. Geometry Find the length of the side w of the parallelogram.

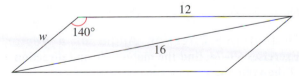

19. Height A tree stands on a hillside of slope 28° from the horizontal. From a point 75 feet down the hill, the angle of elevation to the top of the tree is 45° (see figure). Find the height of the tree.

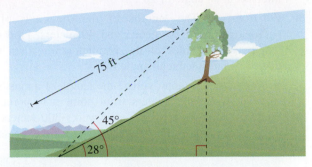

Figure for 19

20. River Width A surveyor finds that a tree on the opposite bank of a river flowing due east has a bearing of N 22° 30′ E from a certain point and a bearing of N 15° W from a point 400 feet downstream. Find the width of the river.

6.2 Using the Law of Cosines In Exercises 21–30, use the Law of Cosines to solve the triangle. Round your answers to two decimal places.

21.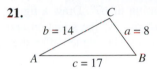

22.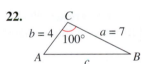

23. $a = 6$, $b = 9$, $c = 14$
24. $a = 75$, $b = 50$, $c = 110$
25. $a = 2.5$, $b = 5.0$, $c = 4.5$
26. $a = 16.4$, $b = 8.8$, $c = 12.2$
27. $B = 108°$, $a = 11$, $c = 11$
28. $B = 150°$, $a = 10$, $c = 20$
29. $C = 43°$, $a = 22.5$, $b = 31.4$
30. $A = 62°$, $b = 11.34$, $c = 19.52$

Solving a Triangle In Exercises 31–34, determine whether the Law of Sines or the Law of Cosines is needed to solve the triangle. Then solve (if possible) the triangle. If two solutions exist, find both. Round your answers to two decimal places.

31. $b = 9$, $c = 13$, $C = 64°$
32. $a = 4$, $c = 5$, $B = 52°$
33. $a = 13$, $b = 15$, $c = 24$
34. $A = 44°$, $B = 31°$, $c = 2.8$

35. Geometry The lengths of the diagonals of a parallelogram are 10 feet and 16 feet. Find the lengths of the sides of the parallelogram when the diagonals intersect at an angle of 28°.

36. Geometry The lengths of the diagonals of a parallelogram are 30 meters and 40 meters. Find the lengths of the sides of the parallelogram when the diagonals intersect at an angle of 34°.

37. Surveying To approximate the length of a marsh, a surveyor walks 425 meters from point *A* to point *B*. Then the surveyor turns 65° and walks 300 meters to point *C* (see figure). Find the length *AC* of the marsh.

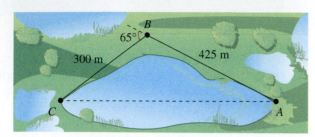

38. Navigation Two planes leave an airport at approximately the same time. One is flying 425 miles per hour at a bearing of 355°, and the other is flying 530 miles per hour at a bearing of 67°. Draw a figure that gives a visual representation of the situation and determine the distance between the planes after they have flown for 2 hours.

Using Heron's Area Formula In Exercises 39–42, use Heron's Area Formula to find the area of the triangle.

39. $a = 3$, $b = 6$, $c = 8$

40. $a = 15$, $b = 8$, $c = 10$

41. $a = 12.3$, $b = 15.8$, $c = 3.7$

42. $a = \dfrac{4}{5}$, $b = \dfrac{3}{4}$, $c = \dfrac{5}{8}$

6.3 **Showing That Two Vectors Are Equivalent** In Exercises 43 and 44, show that u and v are equivalent.

43.

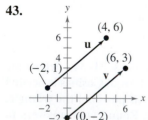

44.

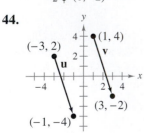

Finding the Component Form of a Vector In Exercises 45–50, find the component form of the vector v satisfying the given conditions.

45.   **46.**

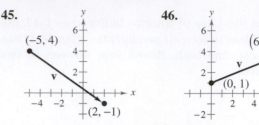

47. Initial point: $(0, 10)$; terminal point: $(7, 3)$

48. Initial point: $(1, 5)$; terminal point: $(15, 9)$

49. $\|\mathbf{v}\| = 8$, $\theta = 120°$ **50.** $\|\mathbf{v}\| = \frac{1}{2}$, $\theta = 225°$

Vector Operations In Exercises 51–58, find (a) u + v, (b) u − v, (c) 4u, and (d) 3v + 5u. Then sketch each resultant vector.

51. $\mathbf{u} = \langle -1, -3 \rangle$, $\mathbf{v} = \langle -3, 6 \rangle$

52. $\mathbf{u} = \langle 4, 5 \rangle$, $\mathbf{v} = \langle 0, -1 \rangle$

53. $\mathbf{u} = \langle -5, 2 \rangle$, $\mathbf{v} = \langle 4, 4 \rangle$

54. $\mathbf{u} = \langle 1, -8 \rangle$, $\mathbf{v} = \langle 3, -2 \rangle$

55. $\mathbf{u} = 2\mathbf{i} - \mathbf{j}$, $\mathbf{v} = 5\mathbf{i} + 3\mathbf{j}$

56. $\mathbf{u} = -7\mathbf{i} - 3\mathbf{j}$, $\mathbf{v} = 4\mathbf{i} - \mathbf{j}$

57. $\mathbf{u} = 4\mathbf{i}$, $\mathbf{v} = -\mathbf{i} + 6\mathbf{j}$ **58.** $\mathbf{u} = -6\mathbf{j}$, $\mathbf{v} = \mathbf{i} + \mathbf{j}$

Vector Operations In Exercises 59–64, find the component form of w and sketch the specified vector operations geometrically, where $\mathbf{u} = 6\mathbf{i} - 5\mathbf{j}$ and $\mathbf{v} = 10\mathbf{i} + 3\mathbf{j}$.

59. $\mathbf{w} = 3\mathbf{v}$ **60.** $\mathbf{w} = \frac{1}{2}\mathbf{v}$

61. $\mathbf{w} = 2\mathbf{u} + \mathbf{v}$ **62.** $\mathbf{w} = 4\mathbf{u} - 5\mathbf{v}$

63. $\mathbf{w} = 5\mathbf{u} - 4\mathbf{v}$ **64.** $\mathbf{w} = -3\mathbf{u} + 2\mathbf{v}$

Writing a Linear Combination of Unit Vectors In Exercises 65–68, the initial and terminal points of a vector are given. Write the vector as a linear combination of the standard unit vectors i and j.

	Initial Point	**Terminal Point**
65.	$(2, 3)$	$(1, 8)$
66.	$(4, -2)$	$(-2, -10)$
67.	$(3, 4)$	$(9, 8)$
68.	$(-2, 7)$	$(5, -9)$

Finding the Direction Angle of a Vector In Exercises 69–74, find the magnitude and direction angle of the vector v.

69. $\mathbf{v} = 7(\cos 60°\mathbf{i} + \sin 60°\mathbf{j})$

70. $\mathbf{v} = 3(\cos 150°\mathbf{i} + \sin 150°\mathbf{j})$

71. $\mathbf{v} = 5\mathbf{i} + 4\mathbf{j}$ **72.** $\mathbf{v} = -4\mathbf{i} + 7\mathbf{j}$

73. $\mathbf{v} = -3\mathbf{i} - 3\mathbf{j}$ **74.** $\mathbf{v} = 8\mathbf{i} - \mathbf{j}$

75. Resultant Force Forces with magnitudes of 85 pounds and 50 pounds act on a single point. The angle between the forces is 15°. Describe the resultant force.

76. Rope Tension A 180-pound weight is supported by two ropes, as shown in the figure. Find the tension in each rope.

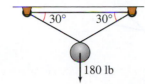

77. Navigation An airplane has an airspeed of 430 miles per hour at a bearing of 135°. The wind velocity is 35 miles per hour in the direction of N 30° E. Find the resultant speed and direction of the airplane.

78. Navigation An airplane has an airspeed of 724 kilometers per hour at a bearing of 30°. The wind velocity is 32 kilometers per hour from the west. Find the resultant speed and direction of the airplane.

6.4 Finding a Dot Product In Exercises 79–82, find the dot product of u and v.

79. $\mathbf{u} = \langle 6, 7 \rangle$
$\mathbf{v} = \langle -3, 9 \rangle$

80. $\mathbf{u} = \langle -7, 12 \rangle$
$\mathbf{v} = \langle -4, -14 \rangle$

81. $\mathbf{u} = 3\mathbf{i} + 7\mathbf{j}$
$\mathbf{v} = 11\mathbf{i} - 5\mathbf{j}$

82. $\mathbf{u} = -7\mathbf{i} + 2\mathbf{j}$
$\mathbf{v} = 16\mathbf{i} - 12\mathbf{j}$

Using Properties of Dot Products In Exercises 83–90, use the vectors $\mathbf{u} = \langle -4, 2 \rangle$ and $\mathbf{v} = \langle 5, 1 \rangle$ to find the indicated quantity. State whether the result is a vector or a scalar.

83. $2\mathbf{u} \cdot \mathbf{u}$

84. $3\mathbf{u} \cdot \mathbf{v}$

85. $4 - \|\mathbf{u}\|$

86. $\|\mathbf{v}\|^2$

87. $\mathbf{u}(\mathbf{u} \cdot \mathbf{v})$

88. $(\mathbf{u} \cdot \mathbf{v})\mathbf{v}$

89. $(\mathbf{u} \cdot \mathbf{u}) - (\mathbf{u} \cdot \mathbf{v})$

90. $(\mathbf{v} \cdot \mathbf{v}) - (\mathbf{v} \cdot \mathbf{u})$

Finding the Angle Between Two Vectors In Exercises 91–94, find the angle θ between the vectors.

91. $\mathbf{u} = \cos \dfrac{7\pi}{4} \mathbf{i} + \sin \dfrac{7\pi}{4} \mathbf{j}$

$\mathbf{v} = \cos \dfrac{5\pi}{6} \mathbf{i} + \sin \dfrac{5\pi}{6} \mathbf{j}$

92. $\mathbf{u} = \cos 45° \mathbf{i} + \sin 45° \mathbf{j}$

$\mathbf{v} = \cos 300° \mathbf{i} + \sin 300° \mathbf{j}$

93. $\mathbf{u} = \langle 2\sqrt{2}, -4 \rangle$, $\mathbf{v} = \langle -\sqrt{2}, 1 \rangle$

94. $\mathbf{u} = \langle 3, \sqrt{3} \rangle$, $\mathbf{v} = \langle 4, 3\sqrt{3} \rangle$

Determining Orthogonal Vectors In Exercises 95–98, determine whether u and v are orthogonal.

95. $\mathbf{u} = \langle -3, 8 \rangle$
$\mathbf{v} = \langle 8, 3 \rangle$

96. $\mathbf{u} = \langle \frac{1}{4}, -\frac{1}{2} \rangle$
$\mathbf{v} = \langle -2, 4 \rangle$

97. $\mathbf{u} = -\mathbf{i}$
$\mathbf{v} = \mathbf{i} + 2\mathbf{j}$

98. $\mathbf{u} = -2\mathbf{i} + \mathbf{j}$
$\mathbf{v} = 3\mathbf{i} + 6\mathbf{j}$

Decomposing a Vector into Components In Exercises 99–102, find the projection of u onto v. Then write u as the sum of two orthogonal vectors, one of which is $\text{proj}_{\mathbf{v}} \mathbf{u}$.

99. $\mathbf{u} = \langle -4, 3 \rangle$, $\mathbf{v} = \langle -8, -2 \rangle$

100. $\mathbf{u} = \langle 5, 6 \rangle$, $\mathbf{v} = \langle 10, 0 \rangle$

101. $\mathbf{u} = \langle 2, 7 \rangle$, $\mathbf{v} = \langle 1, -1 \rangle$

102. $\mathbf{u} = \langle -3, 5 \rangle$, $\mathbf{v} = \langle -5, 2 \rangle$

Work In Exercises 103 and 104, find the work done in moving a particle from P to Q when the magnitude and direction of the force are given by v.

103. $P(5, 3)$, $Q(8, 9)$, $\mathbf{v} = \langle 2, 7 \rangle$

104. $P(-2, -9)$, $Q(-12, 8)$, $\mathbf{v} = 3\mathbf{i} - 6\mathbf{j}$

105. Work Determine the work done (in foot-pounds) by a crane lifting an 18,000-pound truck 48 inches.

106. Work A mover exerts a horizontal force of 25 pounds on a crate as it is pushed up a ramp that is 12 feet long and inclined at an angle of 20° above the horizontal. Find the work done in pushing the crate.

6.5 Finding the Absolute Value of a Complex Number In Exercises 107–112, plot the complex number and find its absolute value.

107. $7i$

108. $-6i$

109. $5 + 3i$

110. $-10 - 4i$

111. $\sqrt{2} - \sqrt{2}i$

112. $-\sqrt{2} + \sqrt{2}i$

Trigonometric Form of a Complex Number In Exercises 113–118, write the complex number in trigonometric form.

113. $4i$

114. -7

115. $7 - 7i$

116. $5 + 12i$

117. $-5 - 12i$

118. $-3\sqrt{3} + 3i$

Multiplying or Dividing Complex Numbers In Exercises 119 and 120, (a) write the two complex numbers in trigonometric form and (b) use the trigonometric forms to find $z_1 z_2$ and z_1/z_2, where $z_2 \neq 0$.

119. $z_1 = 2\sqrt{3} - 2i$, $z_2 = -10i$

120. $z_1 = -3(1 + i)$, $z_2 = 2(\sqrt{3} + i)$

Finding a Power of a Complex Number In Exercises 121–124, use DeMoivre's Theorem to find the indicated power of the complex number. Write the result in standard form.

121. $\left[5\left(\cos\dfrac{\pi}{12} + i \sin\dfrac{\pi}{12}\right)\right]^4$

122. $\left[2\left(\cos\dfrac{4\pi}{15} + i \sin\dfrac{4\pi}{15}\right)\right]^5$

123. $(2 + 3i)^6$

124. $(1 - i)^8$

Finding the *n*th Roots of a Complex Number In Exercises 125–128, (a) use the formula on page 321 to find the indicated roots of the complex number, (b) represent each of the roots graphically, and (c) write each of the roots in standard form.

125. Sixth roots of $-729i$ 126. Fourth roots of $256i$

127. Cube roots of 8   128. Fifth roots of -1024

Solving an Equation In Exercises 129–134, use the formula on page 321 to find all solutions of the equation and represent the solutions graphically.

129. $x^4 + 81 = 0$ 130. $x^5 - 32 = 0$

131. $x^3 + 8i = 0$ 132. $x^4 - 64i = 0$

133. $x^5 + x^3 - x^2 - 1 = 0$

134. $x^5 + 4x^3 - 8x^2 - 32 = 0$

Exploration

True or False? In Exercises 135–137, determine whether the statement is true or false. Justify your answer.

135. The Law of Sines is true when one of the angles in the triangle is a right angle.

136. When the Law of Sines is used, the solution is always unique.

137. $x = \sqrt{3} + i$ is a solution of the equation $x^2 - 8i = 0$.

138. **Law of Sines** State the Law of Sines from memory.

139. **Law of Cosines** State the Law of Cosines from memory.

140. **Reasoning** What characterizes a vector in the plane?

141. **Think About It** Which vectors in the figure appear to be equivalent?

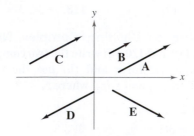

142. **Think About It** The vectors **u** and **v** have the same magnitudes in the two figures. In which figure will the magnitude of the sum be greater? Give a reason for your answer.

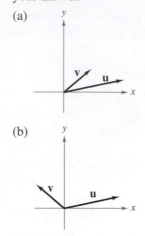

143. **Geometry** Describe geometrically the scalar multiple $k\mathbf{u}$ of the vector **u**, for $k > 0$ and $k < 0$.

144. **Geometry** Describe geometrically the sum of the vectors **u** and **v**.

Graphical Reasoning In Exercises 145–148, use the graph of the roots of a complex number.

(a) **Write each of the roots in trigonometric form.**

(b) **Identify the complex number whose roots are given. Use a graphing utility to verify your results.**

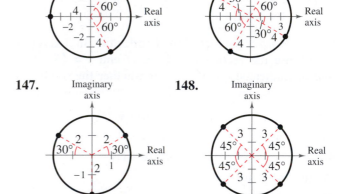

149. **Graphical Reasoning** The figure shows z_1 and z_2. Describe $z_1 z_2$ and z_1/z_2.

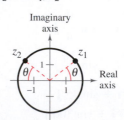

Chapter Test

See CalcChat.com for tutorial help and worked-out solutions to odd-numbered exercises.

Take this test as you would take a test in class. When you are finished, check your work against the answers given in the back of the book.

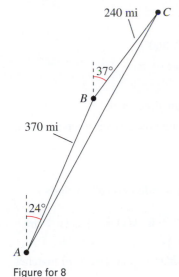

Figure for 8

In Exercises 1–6, determine whether the Law of Sines or the Law of Cosines is needed to solve the triangle. Then solve (if possible) the triangle. If two solutions exist, find both. Round your answers to two decimal places.

1. $A = 24°$, $B = 68°$, $a = 12.2$
2. $B = 110°$, $C = 28°$, $a = 15.6$
3. $A = 24°$, $a = 11.2$, $b = 13.4$
4. $a = 4.0$, $b = 7.3$, $c = 12.4$
5. $B = 100°$, $a = 15$, $b = 23$
6. $C = 121°$, $a = 34$, $b = 55$

7. A triangular parcel of land has borders of lengths 60 meters, 70 meters, and 82 meters. Find the area of the parcel of land.

8. An airplane flies 370 miles from point A to point B with a bearing of 24°. It then flies 240 miles from point B to point C with a bearing of 37° (see figure). Find the distance and bearing from point A to point C.

In Exercises 9 and 10, find the component form of the vector v satisfying the given conditions.

9. Initial point of $\mathbf{v}$: $(-3, 7)$; terminal point of $\mathbf{v}$: $(11, -16)$
10. Magnitude of $\mathbf{v}$: $\|\mathbf{v}\| = 12$; direction of $\mathbf{v}$: $\mathbf{u} = \langle 3, -5 \rangle$

In Exercises 11–14, u = $\langle 2, 7 \rangle$ and v = $\langle -6, 5 \rangle$. Find the resultant vector and sketch its graph.

11. $\mathbf{u} + \mathbf{v}$
12. $\mathbf{u} - \mathbf{v}$
13. $5\mathbf{u} - 3\mathbf{v}$
14. $4\mathbf{u} + 2\mathbf{v}$

15. Find a unit vector in the direction of $\mathbf{u} = \langle 24, -7 \rangle$.

16. Forces with magnitudes of 250 pounds and 130 pounds act on an object at angles of 45° and $-60°$, respectively, with the positive x-axis. Find the direction and magnitude of the resultant of these forces.

17. Find the angle between the vectors $\mathbf{u} = \langle -1, 5 \rangle$ and $\mathbf{v} = \langle 3, -2 \rangle$.

18. Are the vectors $\mathbf{u} = \langle 6, -10 \rangle$ and $\mathbf{v} = \langle 5, 3 \rangle$ orthogonal?

19. Find the projection of $\mathbf{u} = \langle 6, 7 \rangle$ onto $\mathbf{v} = \langle -5, -1 \rangle$. Then write $\mathbf{u}$ as the sum of two orthogonal vectors, one of which is $\text{proj}_\mathbf{v}\mathbf{u}$.

20. A 500-pound motorcycle is headed up a hill inclined at 12°. What force is required to keep the motorcycle from rolling down the hill when stopped at a red light?

21. Write the complex number $z = 4 - 4i$ in trigonometric form.

22. Write the complex number $z = 6(\cos 120° + i \sin 120°)$ in standard form.

In Exercises 23 and 24, use DeMoivre's Theorem to find the indicated power of the complex number. Write the result in standard form.

23. $\left[3\left(\cos \dfrac{7\pi}{6} + i \sin \dfrac{7\pi}{6} \right) \right]^8$

24. $(3 - 3i)^6$

25. Find the fourth roots of $256\left(1 + \sqrt{3}i\right)$.

26. Find all solutions of the equation $x^3 - 27i = 0$ and represent the solutions graphically.

Cumulative Test for Chapters 4–6

See **CalcChat.com** for tutorial help and worked-out solutions to odd-numbered exercises.

Take this test as you would take a test in class. When you are finished, check your work against the answers given in the back of the book.

1. Consider the angle $\theta = -120°$.
   (a) Sketch the angle in standard position.
   (b) Determine a coterminal angle in the interval $[0°, 360°)$.
   (c) Rewrite the angle in radian measure as a multiple of π.
   (d) Find the reference angle θ'.
   (e) Find the exact values of the six trigonometric functions of θ.

2. Convert the angle $\theta = -1.45$ radians to degrees. Round the answer to one decimal place.

3. Find $\cos \theta$ when $\tan \theta = -\frac{21}{20}$ and $\sin \theta < 0$.

In Exercises 4–6, sketch the graph of the function. (Include two full periods.)

4. $f(x) = 3 - 2 \sin \pi x$ 5. $g(x) = \frac{1}{2}\tan\left(x - \frac{\pi}{2}\right)$ 6. $h(x) = -\sec(x + \pi)$

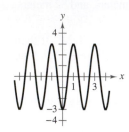

Figure for 7

7. Find a, b, and c such that the graph of the function $h(x) = a \cos(bx + c)$ matches the graph shown in the figure.

8. Sketch the graph of the function $f(x) = \frac{1}{2}x \sin x$ on the interval $-3\pi \leq x \leq 3\pi$.

In Exercises 9 and 10, find the exact value of the expression without using a calculator.

9. $\tan(\arctan 4.9)$ 10. $\tan\left(\arcsin \frac{3}{5}\right)$

11. Write an algebraic expression equivalent to $\sin(\arccos 2x)$.

12. Use the fundamental identities to simplify: $\cos\left(\frac{\pi}{2} - x\right)\csc x$.

13. Subtract and simplify: $\dfrac{\sin \theta - 1}{\cos \theta} - \dfrac{\cos \theta}{\sin \theta - 1}$.

In Exercises 14–16, verify the identity.

14. $\cot^2 \alpha(\sec^2 \alpha - 1) = 1$

15. $\sin(x + y)\sin(x - y) = \sin^2 x - \sin^2 y$

16. $\sin^2 x \cos^2 x = \frac{1}{8}(1 - \cos 4x)$

In Exercises 17 and 18, find all solutions of the equation in the interval $[0, 2\pi)$.

17. $2 \cos^2 \beta - \cos \beta = 0$

18. $3 \tan \theta - \cot \theta = 0$

19. Use the Quadratic Formula to solve the equation in the interval $[0, 2\pi)$: $\sin^2 x + 2 \sin x + 1 = 0$.

20. Given that $\sin u = \frac{12}{13}$, $\cos v = \frac{3}{5}$, and angles u and v are both in Quadrant I, find $\tan(u - v)$.

21. Given that $\tan \theta = \frac{1}{2}$, find the exact value of $\tan(2\theta)$.

22. Given that $\tan \theta = \frac{4}{3}$, find the exact value of $\sin \frac{\theta}{2}$.

23. Write the product $5 \sin \dfrac{3\pi}{4} \cdot \cos \dfrac{7\pi}{4}$ as a sum or difference.

24. Write $\cos 9x - \cos 7x$ as a product.

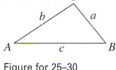

Figure for 25–30

In Exercises 25–30, determine whether the Law of Sines or the Law of Cosines is needed to solve the triangle at the left, then solve the triangle. Round your answers to two decimal places.

25. $A = 30°$, $a = 9$, $b = 8$

26. $A = 30°$, $b = 8$, $c = 10$

27. $A = 30°$, $C = 90°$, $b = 10$

28. $a = 4.7$, $b = 8.1$, $c = 10.3$

29. $A = 45°$, $B = 26°$, $c = 20$

30. $a = 1.2$, $b = 10$, $C = 80°$

31. Two sides of a triangle have lengths 7 inches and 12 inches. Their included angle measures $99°$. Find the area of the triangle.

32. Find the area of a triangle with sides of lengths 30 meters, 41 meters, and 45 meters.

33. Write the vector

$$\mathbf{u} = \langle 7, 8 \rangle$$

as a linear combination of the standard unit vectors $\mathbf{i}$ and $\mathbf{j}$.

34. Find a unit vector in the direction of $\mathbf{v} = \mathbf{i} + \mathbf{j}$.

35. Find $\mathbf{u} \cdot \mathbf{v}$ for $\mathbf{u} = 3\mathbf{i} + 4\mathbf{j}$ and $\mathbf{v} = \mathbf{i} - 2\mathbf{j}$.

36. Find the projection of $\mathbf{u} = \langle 8, -2 \rangle$ onto $\mathbf{v} = \langle 1, 5 \rangle$. Then write $\mathbf{u}$ as the sum of two orthogonal vectors, one of which is $\text{proj}_\mathbf{v}\mathbf{u}$.

37. Write the complex number $-2 + 2i$ in trigonometric form.

38. Find the product of $[4(\cos 30° + i \sin 30°)]$ and $[6(\cos 120° + i \sin 120°)]$. Write the answer in standard form.

39. Find the three cube roots of 1.

40. Find all solutions of the equation $x^5 + 243 = 0$.

41. A ceiling fan with 21-inch blades makes 63 revolutions per minute. Find the angular speed of the fan in radians per minute. Find the linear speed of the tips of the blades in inches per minute.

42. Find the area of the sector of a circle with a radius of 12 yards and a central angle of $105°$.

43. From a point 200 feet from a flagpole, the angles of elevation to the bottom and top of the flag are $16° \, 45'$ and $18°$, respectively. Approximate the height of the flag to the nearest foot.

44. To determine the angle of elevation of a star in the sky, you get the star and the top of the backboard of a basketball hoop that is 5 feet higher than your eyes in your line of vision (see figure). Your horizontal distance from the backboard is 12 feet. What is the angle of elevation of the star?

Figure for 44

45. Find a model for a particle in simple harmonic motion with a displacement (at $t = 0$) of 4 inches, an amplitude of 4 inches, and a period of 8 seconds.

46. An airplane has an airspeed of 500 kilometers per hour at a bearing of $30°$. The wind velocity is 50 kilometers per hour in the direction of N $60°$ E. Find the resultant speed and direction of the airplane.

47. A force of 85 pounds, exerted at an angle of $60°$ with the horizontal, is required to slide an object across a floor. Determine the work done in sliding the object 10 feet.

Proofs in Mathematics ■ ■ ■ ■ ■ ■ ■ ■ ■ ■ ■ ■ ■

LAW OF TANGENTS

Besides the Law of Sines and the Law of Cosines, there is also a Law of Tangents, which was developed by Francois Viète (1540–1603). The Law of Tangents follows from the Law of Sines and the sum-to-product formulas for sine and is defined as follows.

$$\frac{a + b}{a - b} = \frac{\tan[(A + B)/2]}{\tan[(A - B)/2]}$$

The Law of Tangents can be used to solve a triangle when two sides and the included angle are given (SAS). Before calculators were invented, the Law of Tangents was used to solve the SAS case instead of the Law of Cosines because computation with a table of tangent values was easier.

Law of Sines   *(p. 290)*

If ABC is a triangle with sides a, b, and c, then

$$\frac{a}{\sin A} = \frac{b}{\sin B} = \frac{c}{\sin C}.$$

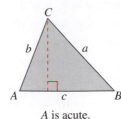

A is acute.

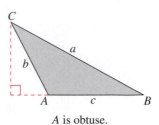

A is obtuse.

Proof

Let h be the altitude of either triangle in the figure above. Then you have

$$\sin A = \frac{h}{b} \qquad \text{or} \qquad h = b \sin A$$

$$\sin B = \frac{h}{a} \qquad \text{or} \qquad h = a \sin B.$$

Equating these two values of h, you have

$$a \sin B = b \sin A \qquad \text{or} \qquad \frac{a}{\sin A} = \frac{b}{\sin B}.$$

Note that $\sin A \neq 0$ and $\sin B \neq 0$ because no angle of a triangle can have a measure of $0°$ or $180°$. In a similar manner, construct an altitude h from vertex B to side AC (extended in the obtuse triangle), as shown at the left. Then you have

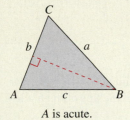

A is acute.

$$\sin A = \frac{h}{c} \qquad \text{or} \qquad h = c \sin A$$

$$\sin C = \frac{h}{a} \qquad \text{or} \qquad h = a \sin C.$$

Equating these two values of h, you have

$$a \sin C = c \sin A \qquad \text{or} \qquad \frac{a}{\sin A} = \frac{c}{\sin C}.$$

By the Transitive Property of Equality, you know that

$$\frac{a}{\sin A} = \frac{b}{\sin B} = \frac{c}{\sin C}.$$

So, the Law of Sines is established. ■

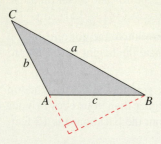

A is obtuse.

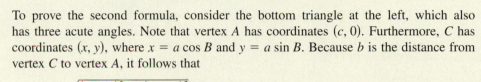

Law of Cosines   *(p. 296)*

Standard Form	**Alternative Form**
$a^2 = b^2 + c^2 - 2bc \cos A$	$\cos A = \dfrac{b^2 + c^2 - a^2}{2bc}$
$b^2 = a^2 + c^2 - 2ac \cos B$	$\cos B = \dfrac{a^2 + c^2 - b^2}{2ac}$
$c^2 = a^2 + b^2 - 2ab \cos C$	$\cos C = \dfrac{a^2 + b^2 - c^2}{2ab}$

Proof

To prove the first formula, consider the top triangle at the left, which has three acute angles. Note that vertex B has coordinates $(c, 0)$. Furthermore, C has coordinates (x, y), where $x = b \cos A$ and $y = b \sin A$. Because a is the distance from vertex C to vertex B, it follows that

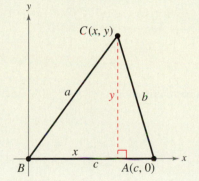

$a = \sqrt{(x - c)^2 + (y - 0)^2}$	Distance Formula
$a^2 = (x - c)^2 + (y - 0)^2$	Square each side.
$a^2 = (b \cos A - c)^2 + (b \sin A)^2$	Substitute for x and y.
$a^2 = b^2 \cos^2 A - 2bc \cos A + c^2 + b^2 \sin^2 A$	Expand.
$a^2 = b^2(\sin^2 A + \cos^2 A) + c^2 - 2bc \cos A$	Factor out b^2.
$a^2 = b^2 + c^2 - 2bc \cos A.$	$\sin^2 A + \cos^2 A = 1$

To prove the second formula, consider the bottom triangle at the left, which also has three acute angles. Note that vertex A has coordinates $(c, 0)$. Furthermore, C has coordinates (x, y), where $x = a \cos B$ and $y = a \sin B$. Because b is the distance from vertex C to vertex A, it follows that

$b = \sqrt{(x - c)^2 + (y - 0)^2}$	Distance Formula
$b^2 = (x - c)^2 + (y - 0)^2$	Square each side.
$b^2 = (a \cos B - c)^2 + (a \sin B)^2$	Substitute for x and y.
$b^2 = a^2 \cos^2 B - 2ac \cos B + c^2 + a^2 \sin^2 B$	Expand.
$b^2 = a^2(\sin^2 B + \cos^2 B) + c^2 - 2ac \cos B$	Factor out a^2.
$b^2 = a^2 + c^2 - 2ac \cos B.$	$\sin^2 B + \cos^2 B = 1$

A similar argument is used to establish the third formula. ◼

Proofs in Mathematics

Heron's Area Formula *(p. 299)*

Given any triangle with sides of lengths a, b, and c, the area of the triangle is

$$\text{Area} = \sqrt{s(s-a)(s-b)(s-c)}$$

where $s = \dfrac{a+b+c}{2}$.

Proof

From Section 6.1, you know that

$$\text{Area} = \frac{1}{2}bc \sin A \qquad\qquad\text{Formula for the area of an oblique triangle}$$

$$(\text{Area})^2 = \frac{1}{4}b^2c^2 \sin^2 A \qquad\qquad\text{Square each side.}$$

$$\text{Area} = \sqrt{\frac{1}{4}b^2c^2 \sin^2 A} \qquad\qquad\text{Take the square root of each side.}$$

$$= \sqrt{\frac{1}{4}b^2c^2(1 - \cos^2 A)} \qquad\qquad\text{Pythagorean identity}$$

$$= \sqrt{\left[\frac{1}{2}bc(1 + \cos A)\right]\left[\frac{1}{2}bc(1 - \cos A)\right]}. \qquad\text{Factor.}$$

Using the Law of Cosines, you can show that

$$\frac{1}{2}bc(1 + \cos A) = \frac{a+b+c}{2} \cdot \frac{-a+b+c}{2}$$

and

$$\frac{1}{2}bc(1 - \cos A) = \frac{a-b+c}{2} \cdot \frac{a+b-c}{2}.$$

Letting $s = (a + b + c)/2$, these two equations can be rewritten as

$$\frac{1}{2}bc(1 + \cos A) = s(s - a)$$

and

$$\frac{1}{2}bc(1 - \cos A) = (s - b)(s - c).$$

By substituting into the last formula for area, you can conclude that

$$\text{Area} = \sqrt{s(s-a)(s-b)(s-c)}.$$

Properties of the Dot Product   *(p. 309)*

Let **u**, **v**, and **w** be vectors in the plane or in space and let c be a scalar.

1. $\mathbf{u} \cdot \mathbf{v} = \mathbf{v} \cdot \mathbf{u}$
2. $\mathbf{0} \cdot \mathbf{v} = 0$
3. $\mathbf{u} \cdot (\mathbf{v} + \mathbf{w}) = \mathbf{u} \cdot \mathbf{v} + \mathbf{u} \cdot \mathbf{w}$
4. $\mathbf{v} \cdot \mathbf{v} = \|\mathbf{v}\|^2$
5. $c(\mathbf{u} \cdot \mathbf{v}) = c\mathbf{u} \cdot \mathbf{v} = \mathbf{u} \cdot c\mathbf{v}$

Proof

Let $\mathbf{u} = \langle u_1, u_2 \rangle$, $\mathbf{v} = \langle v_1, v_2 \rangle$, $\mathbf{w} = \langle w_1, w_2 \rangle$, $\mathbf{0} = \langle 0, 0 \rangle$, and let c be a scalar.

1. $\mathbf{u} \cdot \mathbf{v} = u_1 v_1 + u_2 v_2 = v_1 u_1 + v_2 u_2 = \mathbf{v} \cdot \mathbf{u}$

2. $\mathbf{0} \cdot \mathbf{v} = 0 \cdot v_1 + 0 \cdot v_2 = 0$

3.
$$
\begin{aligned}
\mathbf{u} \cdot (\mathbf{v} + \mathbf{w}) &= \mathbf{u} \cdot \langle v_1 + w_1, v_2 + w_2 \rangle \\
&= u_1(v_1 + w_1) + u_2(v_2 + w_2) \\
&= u_1 v_1 + u_1 w_1 + u_2 v_2 + u_2 w_2 \\
&= (u_1 v_1 + u_2 v_2) + (u_1 w_1 + u_2 w_2) = \mathbf{u} \cdot \mathbf{v} + \mathbf{u} \cdot \mathbf{w}
\end{aligned}
$$

4. $\mathbf{v} \cdot \mathbf{v} = v_1^2 + v_2^2 = \left(\sqrt{v_1^2 + v_2^2} \right)^2 = \|\mathbf{v}\|^2$

5.
$$
\begin{aligned}
c(\mathbf{u} \cdot \mathbf{v}) &= c(\langle u_1, u_2 \rangle \cdot \langle v_1, v_2 \rangle) \\
&= c(u_1 v_1 + u_2 v_2) \\
&= (cu_1)v_1 + (cu_2)v_2 \\
&= \langle cu_1, cu_2 \rangle \cdot \langle v_1, v_2 \rangle \\
&= c\mathbf{u} \cdot \mathbf{v}
\end{aligned}
$$

Angle Between Two Vectors   *(p. 310)*

If θ is the angle between two nonzero vectors **u** and **v**, then $\cos \theta = \dfrac{\mathbf{u} \cdot \mathbf{v}}{\|\mathbf{u}\|\, \|\mathbf{v}\|}$.

Proof

Consider the triangle determined by vectors **u**, **v**, and $\mathbf{v} - \mathbf{u}$, as shown at the left. By the Law of Cosines, you can write

$$\|\mathbf{v} - \mathbf{u}\|^2 = \|\mathbf{u}\|^2 + \|\mathbf{v}\|^2 - 2\|\mathbf{u}\|\,\|\mathbf{v}\| \cos \theta$$

$$(\mathbf{v} - \mathbf{u}) \cdot (\mathbf{v} - \mathbf{u}) = \|\mathbf{u}\|^2 + \|\mathbf{v}\|^2 - 2\|\mathbf{u}\|\,\|\mathbf{v}\| \cos \theta$$

$$(\mathbf{v} - \mathbf{u}) \cdot \mathbf{v} - (\mathbf{v} - \mathbf{u}) \cdot \mathbf{u} = \|\mathbf{u}\|^2 + \|\mathbf{v}\|^2 - 2\|\mathbf{u}\|\,\|\mathbf{v}\| \cos \theta$$

$$\mathbf{v} \cdot \mathbf{v} - \mathbf{u} \cdot \mathbf{v} - \mathbf{v} \cdot \mathbf{u} + \mathbf{u} \cdot \mathbf{u} = \|\mathbf{u}\|^2 + \|\mathbf{v}\|^2 - 2\|\mathbf{u}\|\,\|\mathbf{v}\| \cos \theta$$

$$\|\mathbf{v}\|^2 - 2\mathbf{u} \cdot \mathbf{v} + \|\mathbf{u}\|^2 = \|\mathbf{u}\|^2 + \|\mathbf{v}\|^2 - 2\|\mathbf{u}\|\,\|\mathbf{v}\| \cos \theta$$

$$\cos \theta = \frac{\mathbf{u} \cdot \mathbf{v}}{\|\mathbf{u}\|\,\|\mathbf{v}\|}.$$

P.S. Problem Solving ■ ■ ■ ■ ■ ■ ■ ■ ■ ■ ■ ■ ■

1. Distance In the figure, a beam of light is directed at the blue mirror, reflected to the red mirror, and then reflected back to the blue mirror. Find *PT*, the distance that the light travels from the red mirror back to the blue mirror.

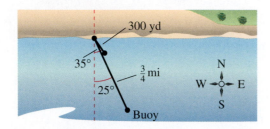

2. Correcting a Course A triathlete sets a course to swim S 25° E from a point on shore to a buoy $\frac{3}{4}$ mile away. After swimming 300 yards through a strong current, the triathlete is off course at a bearing of S 35° E. Find the bearing and distance the triathlete needs to swim to correct her course.

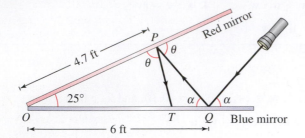

3. Locating Lost Hikers A group of hikers is lost in a national park. Two ranger stations have received an emergency SOS signal from the hikers. Station B is 75 miles due east of station A. The bearing from station A to the signal is S 60° E and the bearing from station B to the signal is S 75° W.

(a) Draw a diagram that gives a visual representation of the problem.

(b) Find the distance from each station to the SOS signal.

(c) A rescue party is in the park 20 miles from station A at a bearing of S 80° E. Find the distance and the bearing the rescue party must travel to reach the lost hikers.

4. Seeding a Courtyard You are seeding a triangular courtyard. One side of the courtyard is 52 feet long and another side is 46 feet long. The angle opposite the 52-foot side is 65°.

(a) Draw a diagram that gives a visual representation of the situation.

(b) How long is the third side of the courtyard?

(c) One bag of grass seed covers an area of 50 square feet. How many bags of grass seed will you need to cover the courtyard?

5. Finding Magnitudes For each pair of vectors, find the following.

(i) $\|\mathbf{u}\|$

(ii) $\|\mathbf{v}\|$

(iii) $\|\mathbf{u} + \mathbf{v}\|$

(iv) $\left\|\dfrac{\mathbf{u}}{\|\mathbf{u}\|}\right\|$

(v) $\left\|\dfrac{\mathbf{v}}{\|\mathbf{v}\|}\right\|$

(vi) $\left\|\dfrac{\mathbf{u} + \mathbf{v}}{\|\mathbf{u} + \mathbf{v}\|}\right\|$

(a) $\mathbf{u} = \langle 1, -1 \rangle$
   $\mathbf{v} = \langle -1, 2 \rangle$

(b) $\mathbf{u} = \langle 0, 1 \rangle$
   $\mathbf{v} = \langle 3, -3 \rangle$

(c) $\mathbf{u} = \left\langle 1, \frac{1}{2} \right\rangle$
   $\mathbf{v} = \langle 2, 3 \rangle$

(d) $\mathbf{u} = \langle 2, -4 \rangle$
   $\mathbf{v} = \langle 5, 5 \rangle$

6. Writing a Vector in Terms of Other Vectors Write the vector **w** in terms of **u** and **v**, given that the terminal point of **w** bisects the line segment (see figure).

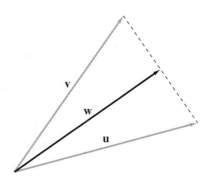

7. Proof Prove that if **u** is orthogonal to **v** and **w**, then **u** is orthogonal to

$$c\mathbf{v} + d\mathbf{w}$$

for any scalars *c* and *d*.

8. Comparing Work Two forces of the same magnitude $\mathbf{F}_1$ and $\mathbf{F}_2$ act at angles θ_1 and θ_2, respectively. Use a diagram to compare the work done by $\mathbf{F}_1$ with the work done by $\mathbf{F}_2$ in moving along the vector *PQ* when

(a) $\theta_1 = -\theta_2$

(b) $\theta_1 = 60°$ and $\theta_2 = 30°$.

9. Skydiving A skydiver is falling at a constant downward velocity of 120 miles per hour. In the figure, vector **u** represents the skydiver's velocity. A steady breeze pushes the skydiver to the east at 40 miles per hour. Vector **v** represents the wind velocity.

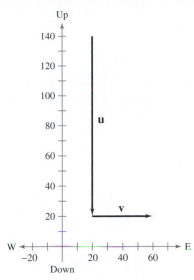

(a) Write the vectors **u** and **v** in component form.

(b) Let

s = **u** + **v**.

Use the figure to sketch **s**. To print an enlarged copy of the graph, go to *MathGraphs.com*.

(c) Find the magnitude of **s**. What information does the magnitude give you about the skydiver's fall?

(d) If there were no wind, then the skydiver would fall in a path perpendicular to the ground. At what angle to the ground is the path of the skydiver when affected by the 40-mile-per-hour wind from due west?

(e) The skydiver is blown to the west at 30 miles per hour. Draw a new figure that gives a visual representation of the problem and find the skydiver's new velocity.

10. Speed and Velocity of an Airplane Four basic forces are in action during flight: weight, lift, thrust, and drag. To fly through the air, an object must overcome its own *weight*. To do this, it must create an upward force called *lift*. To generate lift, a forward motion called *thrust* is needed. The thrust must be great enough to overcome air resistance, which is called *drag*.

For a commercial jet aircraft, a quick climb is important to maximize efficiency because the performance of an aircraft at high altitudes is enhanced. In addition, it is necessary to clear obstacles such as buildings and mountains and to reduce noise in residential areas. In the diagram, the angle θ is called the climb angle. The velocity of the plane can be represented by a vector **v** with a vertical component $\|\mathbf{v}\| \sin \theta$ (called climb speed) and a horizontal component $\|\mathbf{v}\| \cos \theta$, where $\|\mathbf{v}\|$ is the speed of the plane.

When taking off, a pilot must decide how much of the thrust to apply to each component. The more the thrust is applied to the horizontal component, the faster the airplane gains speed. The more the thrust is applied to the vertical component, the quicker the airplane climbs.

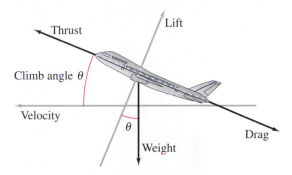

(a) Complete the table for an airplane that has a speed of $\|\mathbf{v}\| = 100$ miles per hour.

θ	0.5°	1.0°	1.5°	2.0°	2.5°	3.0°
$\|\mathbf{v}\| \sin \theta$						
$\|\mathbf{v}\| \cos \theta$						

(b) Does an airplane's speed equal the sum of the vertical and horizontal components of its velocity? If not, how could you find the speed of an airplane whose velocity components were known?

(c) Use the result of part (b) to find the speed of an airplane with the given velocity components.

(i) $\|\mathbf{v}\| \sin \theta = 5.235$ miles per hour
$\|\mathbf{v}\| \cos \theta = 149.909$ miles per hour

(ii) $\|\mathbf{v}\| \sin \theta = 10.463$ miles per hour
$\|\mathbf{v}\| \cos \theta = 149.634$ miles per hour

7 Systems of Equations and Inequalities

Waste Water

Nutrition

Global Positioning System

Fuel Mixture

Traffic Flow

339

Clockwise from top left, tlorna/Shutterstock.com; Kinetic Imagery/Shutterstock.com;
Vlad61/Shutterstock.com; Taras Vyshnya/Shutterstock.com; edobric/Shutterstock.com

7.1 Linear and Nonlinear Systems of Equations

■ Use the method of substitution to solve systems of linear equations in two variables.
■ Use the method of substitution to solve systems of nonlinear equations in two variables.
■ Use a graphical approach to solve systems of equations in two variables.
■ Use systems of equations to model and solve real-life problems.

Graphs of systems of equations can help you solve real-life problems, such as approximating when the consumption of wind energy surpassed the consumption of geothermal energy.

The Method of Substitution

Up to this point in the text, most problems have involved either a function of one variable or a single equation in two variables. However, many problems in science, business, and engineering involve two or more equations in two or more variables. To solve such a problem, you need to find solutions of a **system of equations.** Here is an example of a system of two equations in two unknowns.

$$\begin{cases} 2x + y = 5 & \text{Equation 1} \\ 3x - 2y = 4 & \text{Equation 2} \end{cases}$$

A **solution** of this system is an ordered pair that satisfies each equation in the system. Finding the set of all solutions is called **solving the system of equations.** For instance, the ordered pair $(2, 1)$ is a solution of this system. To check this, substitute 2 for x and 1 for y in *each* equation.

Check (2, 1) in Equation 1 and Equation 2:

$$2x + y = 5 \qquad \text{Write Equation 1.}$$
$$2(2) + 1 \overset{?}{=} 5 \qquad \text{Substitute 2 for } x \text{ and 1 for } y.$$
$$4 + 1 = 5 \qquad \text{Solution checks in Equation 1.} \checkmark$$
$$3x - 2y = 4 \qquad \text{Write Equation 2.}$$
$$3(2) - 2(1) \overset{?}{=} 4 \qquad \text{Substitute 2 for } x \text{ and 1 for } y.$$
$$6 - 2 = 4 \qquad \text{Solution checks in Equation 2.} \checkmark$$

In this chapter, you will study four ways to solve systems of equations, beginning with the **method of substitution.**

Method	Section	Type of System
1. Substitution	7.1	Linear or nonlinear, two variables
2. Graphical method	7.1	Linear or nonlinear, two variables
3. Elimination	7.2	Linear, two variables
4. Gaussian elimination	7.3	Linear, three or more variables

Method of Substitution

1. *Solve* one of the equations for one variable in terms of the other.
2. *Substitute* the expression found in Step 1 into the other equation to obtain an equation in one variable.
3. *Solve* the equation obtained in Step 2.
4. *Back-substitute* the value obtained in Step 3 into the expression obtained in Step 1 to find the value of the other variable.
5. *Check* that the solution satisfies *each* of the original equations.

Systems of equations have a wide variety of real-life applications. For instance, researchers in Italy studying the acoustical noise levels from vehicular traffic at a busy three-way intersection used a system of equations to model the traffic flow at the intersection. To help formulate the system of equations, "operators" stationed themselves at various locations along the intersection and counted the numbers of vehicles that passed them. *(Source: Proceedings of the 11th WSEAS International Conference on Acoustics & Music: Theory & Applications)*

REMARK Because many steps are required to solve a system of equations, it is very easy to make errors in arithmetic. So, you should always check your solution by substituting it into *each* equation in the original system.

EXAMPLE 1   **Solving a System of Equations by Substitution**

Solve the system of equations.

$$\begin{cases} x + y = 4 & \text{Equation 1} \\ x - y = 2 & \text{Equation 2} \end{cases}$$

Solution   Begin by solving for y in Equation 1.

$$y = 4 - x \qquad \text{Solve for } y \text{ in Equation 1.}$$

Next, substitute this expression for y into Equation 2 and solve the resulting single-variable equation for x.

$x - y = 2$	Write Equation 2.
$x - (4 - x) = 2$	Substitute $4 - x$ for y.
$x - 4 + x = 2$	Distributive Property
$2x = 6$	Combine like terms.
$x = 3$	Divide each side by 2.

Finally, solve for y by *back-substituting* $x = 3$ into the equation $y = 4 - x$, to obtain

$y = 4 - x$	Write revised Equation 1.
$y = 4 - 3$	Substitute 3 for x.
$y = 1.$	Solve for y.

The solution is the ordered pair

$$(3, 1).$$

Check this solution as follows.

Check

Substitute $(3, 1)$ into Equation 1:

$x + y = 4$	Write Equation 1.
$3 + 1 \overset{?}{=} 4$	Substitute for x and y.
$4 = 4$	Solution checks in Equation 1. ✓

Substitute $(3, 1)$ into Equation 2:

$x - y = 2$	Write Equation 2.
$3 - 1 \overset{?}{=} 2$	Substitute for x and y.
$2 = 2$	Solution checks in Equation 2. ✓

Because $(3, 1)$ satisfies both equations in the system, it is a solution of the system of equations.

✓ *Checkpoint*))) *Audio-video solution in English & Spanish at LarsonPrecalculus.com.*

Solve the system of equations.

$$\begin{cases} x - \ y = 0 \\ 5x - 3y = 6 \end{cases}$$

The term *back-substitution* implies that you work *backwards*. First you solve for one of the variables, and then you substitute that value *back* into one of the equations in the system to find the value of the other variable.

Taras Vyshnya/Shutterstock.com

EXAMPLE 2 **Solving a System by Substitution**

A total of $12,000 is invested in two funds paying 5% and 3% simple interest. (Recall that the formula for simple interest is $I = Prt$, where P is the principal, r is the annual interest rate, and t is the time.) The yearly interest is $500. How much is invested at each rate?

Solution

Verbal Model:	Amount in 5% fund	+	Amount in 3% fund	=	Total investment

	Interest for 5% fund	+	Interest for 3% fund	=	Total interest

Labels:
Amount in 5% fund = x (dollars)
Interest for 5% fund = $0.05x$ (dollars)
Amount in 3% fund = y (dollars)
Interest for 3% fund = $0.03y$ (dollars)
Total investment = 12,000 (dollars)
Total interest = 500 (dollars)

System:
$$\begin{cases} x + y = 12{,}000 & \text{Equation 1} \\ 0.05x + 0.03y = 500 & \text{Equation 2} \end{cases}$$

• REMARK When using the method of substitution, it does not matter which variable you choose to solve for first. Whether you solve for y first or x first, you will obtain the same solution. You should choose the variable and equation that make your work easier. For instance, in Example 2, solving for x in Equation 1 is easier than solving for x in Equation 2.

▷

To begin, it is convenient to multiply each side of Equation 2 by 100. This eliminates the need to work with decimals.

$100(0.05x + 0.03y) = 100(500)$ Multiply each side of Equation 2 by 100.

$5x + 3y = 50{,}000$ Revised Equation 2

To solve this system, you can solve for x in Equation 1.

$x = 12{,}000 - y$ Revised Equation 1

Then, substitute this expression for x into revised Equation 2 and solve the resulting equation for y.

$5x + 3y = 50{,}000$ Write revised Equation 2.

$5(12{,}000 - y) + 3y = 50{,}000$ Substitute $12{,}000 - y$ for x.

$60{,}000 - 5y + 3y = 50{,}000$ Distributive Property

$-2y = -10{,}000$ Combine like terms.

$y = 5000$ Divide each side by -2.

Next, back-substitute $y = 5000$ to solve for x.

$x = 12{,}000 - y$ Write revised Equation 1.

$x = 12{,}000 - 5000$ Substitute 5000 for y.

$x = 7000$ Subtract.

▷ **TECHNOLOGY** One way to check the answers you obtain in this section is to use a graphing utility. For instance, graph the two equations in Example 2

and find the point of intersection. Does this point agree with the solution obtained at the right?

The solution is (7000, 5000). So, $7000 is invested at 5% and $5000 is invested at 3%. Check this in the original system.

✓ *Checkpoint*  Audio-video solution in English & Spanish at LarsonPrecalculus.com.

A total of $25,000 is invested in two funds paying 6.5% and 8.5% simple interest. The yearly interest is $2000. How much is invested at each rate?

Nonlinear Systems of Equations

The equations in Examples 1 and 2 are linear. The method of substitution can also be used to solve systems in which one or both of the equations are nonlinear.

EXAMPLE 3   Substitution: Two-Solution Case

Solve the system of equations.

$$\begin{cases} 3x^2 + 4x - y = 7 & \text{Equation 1} \\ 2x - y = -1 & \text{Equation 2} \end{cases}$$

Solution   Begin by solving for y in Equation 2 to obtain $y = 2x + 1$. Next, substitute this expression for y into Equation 1 and solve for x.

$$3x^2 + 4x - (2x + 1) = 7 \qquad \text{Substitute } 2x + 1 \text{ for } y \text{ in Equation 1.}$$

$$3x^2 + 2x - 1 = 7 \qquad \text{Simplify.}$$

$$3x^2 + 2x - 8 = 0 \qquad \text{Write in general form.}$$

$$(3x - 4)(x + 2) = 0 \qquad \text{Factor.}$$

$$x = \frac{4}{3}, -2 \qquad \text{Solve for } x.$$

Back-substituting these values of x to solve for the corresponding values of y produces the solutions $\left(\frac{4}{3}, \frac{11}{3}\right)$ and $(-2, -3)$. Check these in the original system.

✓ *Checkpoint* 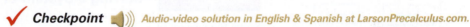 *Audio-video solution in English & Spanish at LarsonPrecalculus.com.*

Solve the system of equations.

$$\begin{cases} -2x + y = 5 \\ x^2 - y + 3x = 1 \end{cases}$$

> **ALGEBRA HELP**   You can review the techniques for factoring in Appendix A.3.

EXAMPLE 4   Substitution: No-Real-Solution Case

Solve the system of equations.

$$\begin{cases} -x + y = 4 & \text{Equation 1} \\ x^2 + y = 3 & \text{Equation 2} \end{cases}$$

Solution   Begin by solving for y in Equation 1 to obtain $y = x + 4$. Next, substitute this expression for y into Equation 2 and solve for x.

$$x^2 + (x + 4) = 3 \qquad \text{Substitute } x + 4 \text{ for } y \text{ in Equation 2.}$$

$$x^2 + x + 1 = 0 \qquad \text{Write in general form.}$$

$$x = \frac{-1 \pm \sqrt{-3}}{2} \qquad \text{Use the Quadratic Formula.}$$

Because the discriminant is negative, the equation

$$x^2 + x + 1 = 0$$

has no (real) solution. So, the original system has no (real) solution.

✓ *Checkpoint* 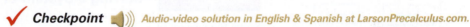 *Audio-video solution in English & Spanish at LarsonPrecalculus.com.*

Solve the system of equations.

$$\begin{cases} 2x - y = -3 \\ 2x^2 + 4x - y^2 = 0 \end{cases}$$

▷ **TECHNOLOGY** Most
graphing utilities have built-in
features that approximate the
point(s) of intersection of two
graphs. Use a graphing utility
to find the points of intersection
of the graphs in Figures 7.1
through 7.3. Be sure to adjust
the viewing window so that you
see all the points of intersection.

▷ **ALGEBRA HELP** You
can review the techniques
for graphing equations in
Section 1.2.

Graphical Approach to Finding Solutions

Notice from Examples 2, 3, and 4 that a system of two equations in two unknowns can have exactly one solution, more than one solution, or no solution. By using a **graphical method,** you can gain insight about the number of solutions and the location(s) of the solution(s) of a system of equations by graphing each of the equations in the same coordinate plane. The solutions of the system correspond to the **points of intersection** of the graphs. For instance, the two equations in Figure 7.1 graph as two lines with a *single point* of intersection; the two equations in Figure 7.2 graph as a parabola and a line with *two points* of intersection; and the two equations in Figure 7.3 graph as a parabola and a line with *no points* of intersection.

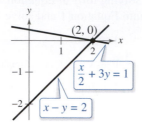

One intersection point
Figure 7.1

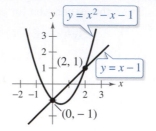

Two intersection points
Figure 7.2

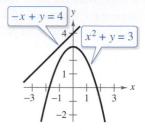

No intersection points
Figure 7.3

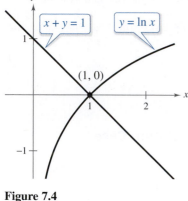

Figure 7.4

EXAMPLE 5 **Solving a System of Equations Graphically**

Solve the system of equations.

$$\begin{cases} y = \ln x & \text{Equation 1} \\ x + y = 1 & \text{Equation 2} \end{cases}$$

Solution There is only one point of intersection of the graphs of the two equations, and $(1, 0)$ is the solution point (see Figure 7.4). Check this solution as follows.

Check (1, 0) in Equation 1:

$y = \ln x$	Write Equation 1.
$0 = \ln 1$	Substitute for x and y.
$0 = 0$	Solution checks in Equation 1. ✓

Check (1, 0) in Equation 2:

$x + y = 1$	Write Equation 2.
$1 + 0 = 1$	Substitute for x and y.
$1 = 1$	Solution checks in Equation 2. ✓

✓ *Checkpoint* 🔊))  *Audio-video solution in English & Spanish at LarsonPrecalculus.com.*

Solve the system of equations.

$$\begin{cases} y = 3 - \log x \\ -2x + y = 1 \end{cases}$$ ∎

Example 5 shows the benefit of a graphical approach to solving systems of equations in two variables. Notice that by trying only the substitution method in Example 5, you would obtain the equation $x + \ln x = 1$. It would be difficult to solve this equation for x using standard algebraic techniques.

Applications

The total cost C of producing x units of a product typically has two components—the initial cost and the cost per unit. When enough units have been sold so that the total revenue R equals the total cost C, the sales are said to have reached the **break-even point.** You will find that the break-even point corresponds to the point of intersection of the cost and revenue curves.

EXAMPLE 6   **Break-Even Analysis**

A shoe company invests \$300,000 in equipment to produce a new line of athletic footwear. Each pair of shoes costs \$5 to produce and sells for \$60. How many pairs of shoes must the company sell to break even?

Algebraic Solution

The total cost of producing x units is

Total cost	=	Cost per unit	·	Number of units	+	Initial cost

$$C = 5x + 300{,}000. \qquad \text{Equation 1}$$

The revenue obtained by selling x units is

Total revenue	=	Price per unit	·	Number of units

$$R = 60x. \qquad \text{Equation 2}$$

Because the break-even point occurs when $R = C$, you have $C = 60x$, and the system of equations to solve is

$$\begin{cases} C = 5x + 300{,}000 \\ C = 60x \end{cases}.$$

Solve by substitution.

$60x = 5x + 300{,}000$ Substitute $60x$ for C in Equation 1.

$55x = 300{,}000$ Subtract $5x$ from each side.

$x \approx 5455$ Divide each side by 55.

So, the company must sell about 5455 pairs of shoes to break even.

Graphical Solution

The total cost of producing x units is

Total cost	=	Cost per unit	·	Number of units	+	Initial cost

$$C = 5x + 300{,}000. \qquad \text{Equation 1}$$

The revenue obtained by selling x units is

Total revenue	=	Price per unit	·	Number of units

$$R = 60x. \qquad \text{Equation 2}$$

Because the break-even point occurs when $R = C$, you have $C = 60x$, and the system of equations to solve is

$$\begin{cases} C = 5x + 300{,}000 \\ C = 60x \end{cases}.$$

Use a graphing utility to graph $y_1 = 5x + 300{,}000$ and $y_2 = 60x$ in the same viewing window.

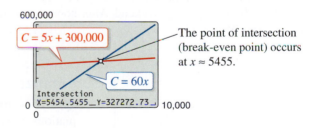

The point of intersection (break-even point) occurs at $x \approx 5455$.

So, the company must sell about 5455 pairs of shoes to break even.

✓ *Checkpoint*  Audio-video solution in English & Spanish at LarsonPrecalculus.com.

In Example 6, each pair of shoes costs \$12 to produce. How many pairs of shoes must the company sell to break even?

Another way to view the solution in Example 6 is to consider the profit function $P = R - C$. The break-even point occurs when the profit is 0, which is the same as saying that $R = C$.

EXAMPLE 7 **Movie Ticket Sales**

The weekly ticket sales for a new comedy movie decreased each week. At the same time, the weekly ticket sales for a new drama movie increased each week. Models that approximate the weekly ticket sales S (in millions of dollars) for the movies are

$$\begin{cases} S = 60 - 8x & \text{Comedy} \\ S = 10 + 4.5x & \text{Drama} \end{cases}$$

where x represents the number of weeks each movie was in theaters, with $x = 0$ corresponding to the opening weekend. After how many weeks will the ticket sales for the two movies be equal?

Algebraic Solution

Because both equations are already solved for S in terms of x, substitute either expression for S into the other equation and solve for x.

$10 + 4.5x = 60 - 8x$ Substitute for S in Equation 1.

$4.5x + 8x = 60 - 10$ Add $8x$ and -10 to each side.

$12.5x = 50$ Combine like terms.

$x = 4$ Divide each side by 12.5.

So, the weekly ticket sales for the two movies will be equal after 4 weeks.

Numerical Solution

You can create a table of values for each model to determine when the ticket sales for the two movies will be equal.

Number of Weeks, x	0	1	2	3	4	5	6
Sales, S (comedy)	60	52	44	36	28	20	12
Sales, S (drama)	10	14.5	19	23.5	28	32.5	37

So, from the above table, the weekly ticket sales for the two movies will be equal after 4 weeks.

✓ *Checkpoint*   *Audio-video solution in English & Spanish at LarsonPrecalculus.com.*

The weekly ticket sales for a new animated movie decreased each week. At the same time, the weekly ticket sales for a new horror movie increased each week. Models that approximate the weekly ticket sales S (in millions of dollars) for the movies are

$$\begin{cases} S = 108 - 9.4x & \text{Animated} \\ S = 16 + 9x & \text{Horror} \end{cases}$$

where x represents the number of weeks each movie was in theaters, with $x = 0$ corresponding to the opening weekend. After how many weeks will the ticket sales for the two movies be equal?

Summarize   **(Section 7.1)**

1. Explain how to use the method of substitution to solve a system of linear equations in two variables *(page 340)*. For examples of using the method of substitution to solve systems of linear equations in two variables, see Examples 1 and 2.

2. Explain how to use the method of substitution to solve a system of nonlinear equations in two variables *(page 343)*. For examples of using the method of substitution to solve systems of nonlinear equations in two variables, see Examples 3 and 4.

3. Explain how to use a graphical approach to solve a system of equations in two variables *(page 344)*. For an example of using a graphical approach to solve a system of equations in two variables, see Example 5.

4. Describe examples of how to use systems of equations to model and solve real-life problems *(pages 345 and 346, Examples 6 and 7)*.

7.2 Two-Variable Linear Systems

■ Use the method of elimination to solve systems of linear equations in two variables.
■ Interpret graphically the numbers of solutions of systems of linear equations in two variables.
■ Use systems of linear equations in two variables to model and solve real-life problems.

The Method of Elimination

In Section 7.1, you studied two methods for solving a system of equations: substitution and graphing. Now, you will study the **method of elimination.** The key step in this method is to obtain, for one of the variables, coefficients that differ only in sign so that *adding* the equations eliminates the variable.

$$3x + 5y = 7 \qquad \text{Equation 1}$$
$$\underline{-3x - 2y = -1} \qquad \text{Equation 2}$$
$$3y = 6 \qquad \text{Add equations.}$$

Note that by adding the two equations, you eliminate the x-terms and obtain a single equation in y. Solving this equation for y produces $y = 2$, which you can then back-substitute into one of the original equations to solve for x.

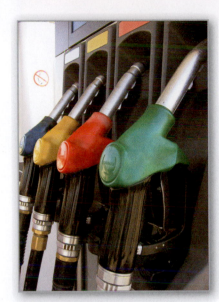

Systems of equations in two variables can help you model and solve real-life problems, such as finding the numbers of gallons of 87- and 92-octane gasoline that must be mixed to obtain 500 gallons of 89-octane gasoline.

EXAMPLE 1   Solving a System of Equations by Elimination

Solve the system of linear equations.

$$\begin{cases} 3x + 2y = 4 & \text{Equation 1} \\ 5x - 2y = 12 & \text{Equation 2} \end{cases}$$

Solution   Because the coefficients of y differ only in sign, eliminate the y-terms by adding the two equations.

$$3x + 2y = 4 \qquad \text{Write Equation 1.}$$
$$\underline{5x - 2y = 12} \qquad \text{Write Equation 2.}$$
$$8x = 16 \qquad \text{Add equations.}$$
$$x = 2 \qquad \text{Solve for } x.$$

Solve for y by back-substituting $x = 2$ into Equation 1.

$$3x + 2y = 4 \qquad \text{Write Equation 1.}$$
$$3(2) + 2y = 4 \qquad \text{Substitute 2 for } x.$$
$$y = -1 \qquad \text{Solve for } y.$$

The solution is $(2, -1)$. Check this in the original system, as follows.

Check

$$3(2) + 2(-1) = 4 \qquad \text{Solution checks in Equation 1. } \checkmark$$
$$5(2) - 2(-1) = 12 \qquad \text{Solution checks in Equation 2. } \checkmark$$

✓ *Checkpoint* ◀))) *Audio-video solution in English & Spanish at LarsonPrecalculus.com.*

Solve the system of linear equations.

$$\begin{cases} 2x + y = 4 \\ 2x - y = -1 \end{cases}$$

Method of Elimination

To use the **method of elimination** to solve a system of two linear equations in x and y, perform the following steps.

1. *Obtain coefficients* for x (or y) that differ only in sign by multiplying all terms of one or both equations by suitably chosen constants.

2. *Add* the equations to eliminate one variable.

3. *Solve* the equation obtained in Step 2.

4. *Back-substitute* the value obtained in Step 3 into either of the original equations and solve for the other variable.

5. *Check* that the solution satisfies *each* of the original equations.

EXAMPLE 2   **Solving a System of Equations by Elimination**

Solve the system of linear equations.

$$\begin{cases} 2x - 4y = -7 & \text{Equation 1} \\ 5x + \ y = -1 & \text{Equation 2} \end{cases}$$

Solution   To obtain coefficients that differ only in sign, multiply Equation 2 by 4.

$$\begin{array}{lll} 2x - 4y = -7 & \Longrightarrow & 2x - 4y = -7 & \text{Write Equation 1.} \\ 5x + \ y = -1 & \Longrightarrow & \underline{20x + 4y = -4} & \text{Multiply Equation 2 by 4.} \\ & & 22x \quad\quad\ = -11 & \text{Add equations.} \\ & & x \quad\quad\quad = -\tfrac{1}{2} & \text{Solve for } x. \end{array}$$

Solve for y by back-substituting $x = -\tfrac{1}{2}$ into Equation 1.

$$\begin{array}{ll} 2x - 4y = -7 & \text{Write Equation 1.} \\ 2\left(-\tfrac{1}{2}\right) - 4y = -7 & \text{Substitute } -\tfrac{1}{2} \text{ for } x. \\ -4y = -6 & \text{Simplify.} \\ y = \tfrac{3}{2} & \text{Solve for } y. \end{array}$$

The solution is $\left(-\tfrac{1}{2}, \tfrac{3}{2}\right)$. Check this in the original system, as follows.

Check

$$\begin{array}{ll} 2x - 4y = -7 & \text{Write Equation 1.} \\ 2\left(-\tfrac{1}{2}\right) - 4\left(\tfrac{3}{2}\right) \overset{?}{=} -7 & \text{Substitute for } x \text{ and } y. \\ -1 - 6 = -7 & \text{Solution checks in Equation 1. } ✔ \\ 5x + y = -1 & \text{Write Equation 2.} \\ 5\left(-\tfrac{1}{2}\right) + \tfrac{3}{2} \overset{?}{=} -1 & \text{Substitute for } x \text{ and } y. \\ -\tfrac{5}{2} + \tfrac{3}{2} = -1 & \text{Solution checks in Equation 2. } ✔ \end{array}$$

✔ ***Checkpoint***  *Audio-video solution in English & Spanish at LarsonPrecalculus.com.*

Solve the system of linear equations.

$$\begin{cases} 2x + 3y = 17 \\ 5x - \ y = 17 \end{cases}$$

In Example 2, the two systems of linear equations (the original system and the system obtained by multiplying by a constant)

$$\begin{cases} 2x - 4y = -7 \\ 5x + y = -1 \end{cases} \quad \text{and} \quad \begin{cases} 2x - 4y = -7 \\ 20x + 4y = -4 \end{cases}$$

are called **equivalent systems** because they have precisely the same solution set. The operations that can be performed on a system of linear equations to produce an equivalent system are (1) interchanging any two equations, (2) multiplying an equation by a nonzero constant, and (3) adding a multiple of one equation to any other equation in the system.

EXAMPLE 3   **Solving the System of Equations by Elimination**

Solve the system of linear equations.

$$\begin{cases} 5x + 3y = 9 \\ 2x - 4y = 14 \end{cases}$$
Equation 1
Equation 2

Algebraic Solution

You can obtain coefficients that differ only in sign by multiplying Equation 1 by 4 and multiplying Equation 2 by 3.

$$5x + 3y = 9 \implies 20x + 12y = 36 \qquad \text{Multiply Equation 1 by 4.}$$
$$\underline{2x - 4y = 14 \implies 6x - 12y = 42} \qquad \text{Multiply Equation 2 by 3.}$$
$$26x = 78 \qquad \text{Add equations.}$$
$$x = 3 \qquad \text{Solve for } x.$$

Solve for y by back-substituting $x = 3$ into Equation 2.

$$2x - 4y = 14 \qquad \text{Write Equation 2.}$$
$$2(3) - 4y = 14 \qquad \text{Substitute 3 for } x.$$
$$-4y = 8 \qquad \text{Simplify.}$$
$$y = -2 \qquad \text{Solve for } y.$$

The solution is

$$(3, -2).$$

Check this in the original system.

Graphical Solution

Solve each equation for y and use a graphing utility to graph the equations in the same viewing window.

$y_1 = 3 - \frac{5}{3}x$

The point of intersection is $(3, -2)$.

$y_2 = -\frac{7}{2} + \frac{1}{2}x$

Intersection
X=3   Y=-2

From the figure, the solution is

$$(3, -2).$$

Check this in the original system.

✓ *Checkpoint* 🔊)) *Audio-video solution in English & Spanish at LarsonPrecalculus.com.*

Solve the system of linear equations.

$$\begin{cases} 3x + 2y = 7 \\ 2x + 5y = 1 \end{cases}$$

■

Check the solution from Example 3, as follows.

$$5(3) + 3(-2) \overset{?}{=} 9 \qquad \text{Substitute 3 for } x \text{ and } -2 \text{ for } y \text{ in Equation 1.}$$
$$15 - 6 = 9 \qquad \text{Solution checks in Equation 1. } ✓$$
$$2(3) - 4(-2) \overset{?}{=} 14 \qquad \text{Substitute 3 for } x \text{ and } -2 \text{ for } y \text{ in Equation 2.}$$
$$6 + 8 = 14 \qquad \text{Solution checks in Equation 2. } ✓$$

Keep in mind that the terminology and methods discussed in this section apply only to systems of *linear* equations.

▷ **TECHNOLOGY** The general solution of the linear system

$$\begin{cases} ax + by = c \\ dx + ey = f \end{cases}$$

is

$$x = \frac{ce - bf}{ae - bd}$$

and

$$y = \frac{af - cd}{ae - bd}.$$

If $ae - bd = 0$, then the system does not have a unique solution. A graphing utility program (called Systems of Linear Equations) for solving such a system can be found at the website for this text at *LarsonPrecalculus.com.* Try using the program to solve the system in Example 4.

Example 4 illustrates a strategy for solving a system of linear equations that has decimal coefficients.

EXAMPLE 4 **A Linear System Having Decimal Coefficients**

Solve the system of linear equations.

$$\begin{cases} 0.02x - 0.05y = -0.38 & \text{Equation 1} \\ 0.03x + 0.04y = 1.04 & \text{Equation 2} \end{cases}$$

Solution Because the coefficients in this system have two decimal places, you can begin by multiplying each equation by 100. This produces a system in which the coefficients are all integers.

$$\begin{cases} 2x - 5y = -38 & \text{Revised Equation 1} \\ 3x + 4y = 104 & \text{Revised Equation 2} \end{cases}$$

Now, to obtain coefficients that differ only in sign, multiply Equation 1 by 3 and multiply Equation 2 by -2.

$$\begin{array}{llll} 2x - 5y = -38 & \Longrightarrow & 6x - 15y = -114 & \text{Multiply Equation 1 by 3.} \\ 3x + 4y = 104 & \Longrightarrow & \underline{-6x - 8y = -208} & \text{Multiply Equation 2 by } -2. \\ & & {-23y = -322} & \text{Add equations.} \end{array}$$

So,

$$y = \frac{-322}{-23}$$

$$= 14.$$

Back-substituting $y = 14$ into revised Equation 2 produces the following.

$$\begin{array}{ll} 3x + 4y = 104 & \text{Write revised Equation 2.} \\ 3x + 4(14) = 104 & \text{Substitute 14 for } y. \\ 3x = 48 & \text{Simplify.} \\ x = 16 & \text{Solve for } x. \end{array}$$

The solution is

$(16, 14)$.

Check this in the original system, as follows.

Check

$$\begin{array}{ll} 0.02x - 0.05y = -0.38 & \text{Write Equation 1.} \\ 0.02(16) - 0.05(14) \stackrel{?}{=} -0.38 & \text{Substitute for } x \text{ and } y. \\ 0.32 - 0.70 = -0.38 & \text{Solution checks in Equation 1. } ✓ \\ 0.03x + 0.04y = 1.04 & \text{Write Equation 2.} \\ 0.03(16) + 0.04(14) \stackrel{?}{=} 1.04 & \text{Substitute for } x \text{ and } y. \\ 0.48 + 0.56 = 1.04 & \text{Solution checks in Equation 2. } ✓ \end{array}$$

✓ *Checkpoint* ◀)) *Audio-video solution in English & Spanish at LarsonPrecalculus.com.*

Solve the system of linear equations.

$$\begin{cases} 0.03x + 0.04y = 0.75 \\ 0.02x + 0.06y = 0.90 \end{cases}$$

Graphical Interpretation of Solutions

It is possible for a *general* system of equations to have exactly one solution, two or more solutions, or no solution. If a system of *linear* equations has two different solutions, then it must have an *infinite* number of solutions.

Graphical Interpretations of Solutions

For a system of two linear equations in two variables, the number of solutions is one of the following.

Number of Solutions	Graphical Interpretation	Slopes of Lines
1. Exactly one solution	The two lines intersect at one point.	The slopes of the two lines are not equal.
2. Infinitely many solutions	The two lines coincide (are identical).	The slopes of the two lines are equal.
3. No solution	The two lines are parallel.	The slopes of the two lines are equal.

A system of linear equations is **consistent** when it has at least one solution. A consistent system with exactly one solution is *independent*, whereas a consistent system with infinitely many solutions is *dependent*. A system is **inconsistent** when it has no solution.

> • • **REMARK**   A comparison of the slopes of two lines gives useful information about the number of solutions of the corresponding system of equations. To solve a system of equations graphically, it helps to begin by writing the equations in slope-intercept form. Try doing this for the systems in Example 5.
> • • • • • • • • • • • • • ▷

EXAMPLE 5   **Recognizing Graphs of Linear Systems**

Match each system of linear equations with its graph. Describe the number of solutions and state whether the system is consistent or inconsistent.

a. $\begin{cases} 2x - 3y = 3 \\ -4x + 6y = 6 \end{cases}$ **b.** $\begin{cases} 2x - 3y = 3 \\ x + 2y = 5 \end{cases}$ **c.** $\begin{cases} 2x - 3y = 3 \\ -4x + 6y = -6 \end{cases}$

i. 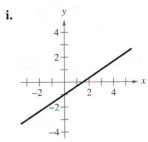   **ii.** 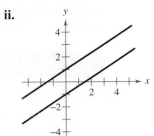   **iii.**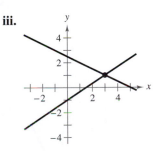

Solution

a. The graph of system (a) is a pair of parallel lines (ii). The lines have no point of intersection, so the system has no solution. The system is inconsistent.

b. The graph of system (b) is a pair of intersecting lines (iii). The lines have one point of intersection, so the system has exactly one solution. The system is consistent.

c. The graph of system (c) is a pair of lines that coincide (i). The lines have infinitely many points of intersection, so the system has infinitely many solutions. The system is consistent.

✓ *Checkpoint*)) *Audio-video solution in English & Spanish at LarsonPrecalculus.com.*

Sketch the graph of the system of linear equations. Then describe the number of solutions and state whether the system is consistent or inconsistent.

$$\begin{cases} -2x + 3y = 6 \\ 4x - 6y = -9 \end{cases}$$

In Examples 6 and 7, note how you can use the method of elimination to determine that a system of linear equations has no solution or infinitely many solutions.

EXAMPLE 6   **No-Solution Case: Method of Elimination**

Solve the system of linear equations.

$$\begin{cases} x - 2y = 3 & \text{Equation 1} \\ -2x + 4y = 1 & \text{Equation 2} \end{cases}$$

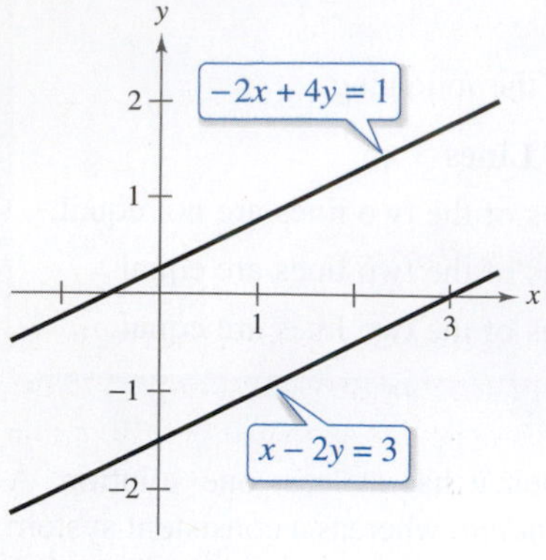

Solution   To obtain coefficients that differ only in sign, multiply Equation 1 by 2.

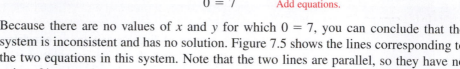

$$
\begin{array}{ll}
x - 2y = 3 \quad\Longrightarrow\quad 2x - 4y = 6 & \text{Multiply Equation 1 by 2.} \\
-2x + 4y = 1 \quad\Longrightarrow\quad \underline{-2x + 4y = 1} & \text{Write Equation 2.} \\
\hphantom{-2x + 4y = 1 \quad\Longrightarrow\quad} 0 = 7 & \text{Add equations.}
\end{array}
$$

Because there are no values of x and y for which $0 = 7$, you can conclude that the system is inconsistent and has no solution. Figure 7.5 shows the lines corresponding to the two equations in this system. Note that the two lines are parallel, so they have no point of intersection.

✔ **Checkpoint** 🔊 *Audio-video solution in English & Spanish at LarsonPrecalculus.com.*

Solve the system of linear equations.

$$\begin{cases} 6x - 5y = 3 \\ -12x + 10y = 5 \end{cases}$$

In Example 6, note that the occurrence of a false statement, such as $0 = 7$, indicates that the system has no solution. In the next example, note that the occurrence of a statement that is true for all values of the variables, such as $0 = 0$, indicates that the system has infinitely many solutions.

EXAMPLE 7   **Many-Solution Case: Method of Elimination**

Solve the system of linear equations.

$$\begin{cases} 2x - y = 1 & \text{Equation 1} \\ 4x - 2y = 2 & \text{Equation 2} \end{cases}$$

Solution   To obtain coefficients that differ only in sign, multiply Equation 1 by -2.

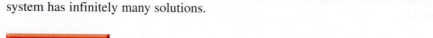

$$
\begin{array}{ll}
2x - y = 1 \quad\Longrightarrow\quad -4x + 2y = -2 & \text{Multiply Equation 1 by } -2. \\
\underline{4x - 2y = 2} \quad\Longrightarrow\quad \underline{4x - 2y = \hphantom{-}2} & \text{Write Equation 2.} \\
\hphantom{4x - 2y = 2 \quad\Longrightarrow\quad} 0 = \hphantom{-}0 & \text{Add equations.}
\end{array}
$$

Because the two equations are equivalent (have the same solution set), the system has infinitely many solutions. The solution set consists of all points (x, y) lying on the line $2x - y = 1$, as shown in Figure 7.6. Letting $x = a$, where a is any real number, the solutions of the system are $(a, 2a - 1)$.

✔ **Checkpoint** 🔊 *Audio-video solution in English & Spanish at LarsonPrecalculus.com.*

Solve the system of linear equations.

$$\begin{cases} \dfrac{1}{2}x - \dfrac{1}{8}y = -\dfrac{3}{8} \\ -4x + y = \hphantom{-}3 \end{cases}$$

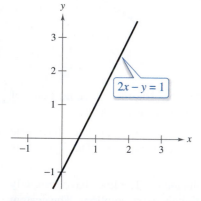

Figure 7.5

Figure 7.6

Applications

At this point, you may be asking the question "How can I tell whether I can solve an application problem using a system of linear equations?" The answer comes from the following considerations.

1. Does the problem involve more than one unknown quantity?

2. Are there two (or more) equations or conditions to be satisfied?

If the answer to one or both of these questions is "yes," then the appropriate model for the problem may be a system of linear equations.

EXAMPLE 8   An Application of a Linear System

Original flight

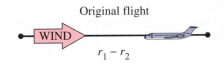

$r_1 - r_2$

Return flight

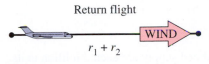

$r_1 + r_2$

Figure 7.7

An airplane flying into a headwind travels the 2000-mile flying distance between Chicopee, Massachusetts, and Salt Lake City, Utah, in 4 hours and 24 minutes. On the return flight, the airplane travels this distance in 4 hours. Find the airspeed of the plane and the speed of the wind, assuming that both remain constant.

Solution   The two unknown quantities are the speeds of the wind and the plane. If r_1 is the speed of the plane and r_2 is the speed of the wind, then

$$r_1 - r_2 = \text{speed of the plane against the wind}$$

$$r_1 + r_2 = \text{speed of the plane with the wind}$$

as shown in Figure 7.7. Using the formula

$$\text{distance} = (\text{rate})(\text{time})$$

for these two speeds, you obtain the following equations.

$$2000 = (r_1 - r_2)\left(4 + \frac{24}{60}\right)$$

$$2000 = (r_1 + r_2)(4)$$

These two equations simplify as follows.

$$\begin{cases} 5000 = 11r_1 - 11r_2 & \text{Equation 1} \\ 500 = r_1 + r_2 & \text{Equation 2} \end{cases}$$

To solve this system by elimination, multiply Equation 2 by 11.

$$
\begin{array}{lll}
5000 = 11r_1 - 11r_2 & \Longrightarrow & 5000 = 11r_1 - 11r_2 \qquad \text{Write Equation 1.} \\
500 = r_1 + r_2 & \Longrightarrow & 5500 = 11r_1 + 11r_2 \qquad \text{Multiply Equation 2 by 11.} \\
\hline
& & 10{,}500 = 22r_1 \qquad\qquad\quad \text{Add equations.}
\end{array}
$$

So,

$$r_1 = \frac{10{,}500}{22} = \frac{5250}{11} \approx 477.27 \text{ miles per hour} \qquad \text{Speed of plane}$$

and

$$r_2 = 500 - \frac{5250}{11} = \frac{250}{11} \approx 22.73 \text{ miles per hour.} \qquad \text{Speed of wind}$$

Check this solution in the original statement of the problem.

✓ *Checkpoint*  *Audio-video solution in English & Spanish at LarsonPrecalculus.com.*

In Example 8, the return flight takes 4 hours and 6 minutes. Find the airspeed of the plane and the speed of the wind, assuming that both remain constant. ■

In a free market, the demands for many products are related to the prices of the products. As the prices decrease, the demands by consumers increase and the amounts that producers are able or willing to supply decrease.

EXAMPLE 9 Finding the Equilibrium Point

The demand and supply equations for a video game console are

$$\begin{cases} p = 180 - 0.00001x & \text{Demand equation} \\ p = 90 + 0.00002x & \text{Supply equation} \end{cases}$$

where p is the price per unit (in dollars) and x is the number of units. Find the equilibrium point for this market. **The equilibrium point** is the price p and number of units x that satisfy both the demand and supply equations.

Solution Because p is written in terms of x, begin by substituting the value of p given in the supply equation into the demand equation.

$$p = 180 - 0.00001x \qquad \text{Write demand equation.}$$

$$90 + 0.00002x = 180 - 0.00001x \qquad \text{Substitute } 90 + 0.00002x \text{ for } p.$$

$$0.00003x = 90 \qquad \text{Combine like terms.}$$

$$x = 3{,}000{,}000 \qquad \text{Solve for } x.$$

So, the equilibrium point occurs when the demand and supply are each 3 million units. (See Figure 7.8.) Obtain the price that corresponds to this x-value by back-substituting $x = 3{,}000{,}000$ into either of the original equations. For instance, back-substituting into the demand equation produces

$$p = 180 - 0.00001(3{,}000{,}000)$$

$$= 180 - 30$$

$$= \$150.$$

The solution is $(3{,}000{,}000, 150)$. Check this by substituting $(3{,}000{,}000, 150)$ into the demand and supply equations.

✓ *Checkpoint* 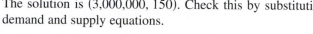 *Audio-video solution in English & Spanish at LarsonPrecalculus.com.*

The demand and supply equations for a flat-screen television are

$$\begin{cases} p = 567 - 0.00002x & \text{Demand equation} \\ p = 492 + 0.00003x & \text{Supply equation} \end{cases}$$

where p is the price per unit (in dollars) and x is the number of units. Find the equilibrium point for this market.

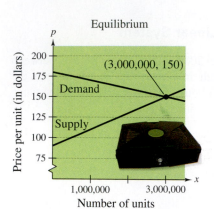

Equilibrium

(3,000,000, 150)

Demand

Supply

1,000,000   3,000,000
Number of units

Price per unit (in dollars)

Figure 7.8

Summarize (Section 7.2)

1. Explain how to use the method of elimination to solve a system of linear equations in two variables *(page 347)*. For examples of using the method of elimination to solve systems of linear equations in two variables, see Examples 1–4.

2. Explain how to interpret graphically the numbers of solutions of systems of linear equations in two variables *(page 351)*. For examples of interpreting graphically the numbers of solutions of systems of linear equations in two variables, see Examples 5–7.

3. Describe examples of how to use systems of linear equations in two variables to model and solve real-life problems *(pages 353 and 354, Examples 8 and 9)*.

7.3 Multivariable Linear Systems

Systems of linear equations in three or more variables can help you model and solve real-life problems. For instance, a system of linear equations in three variables can be used to analyze the reproductive rates of deer in a wildlife preserve.

- **Use back-substitution to solve linear systems in row-echelon form.**
- **Use Gaussian elimination to solve systems of linear equations.**
- **Solve nonsquare systems of linear equations.**
- **Use systems of linear equations in three or more variables to model and solve real-life problems.**

Row-Echelon Form and Back-Substitution

The method of elimination can be applied to a system of linear equations in more than two variables. In fact, this method adapts to computer use for solving linear systems with dozens of variables.

When using the method of elimination to solve a system of linear equations, the goal is to rewrite the system in a form to which back-substitution can be applied. To see how this works, consider the following two systems of linear equations.

$$\begin{cases} x - 2y + 3z = 9 \\ -x + 3y = -4 \\ 2x - 5y + 5z = 17 \end{cases}$$ System of three linear equations in three variables (See Example 3.)

$$\begin{cases} x - 2y + 3z = 9 \\ y + 3z = 5 \\ z = 2 \end{cases}$$ Equivalent system in row-echelon form (See Example 1.)

The second system is in **row-echelon form,** which means that it has a "stair-step" pattern with leading coefficients of 1. After comparing the two systems, it should be clear that it is easier to solve the system in row-echelon form, using back-substitution.

EXAMPLE 1 Using Back-Substitution in Row-Echelon Form

Solve the system of linear equations.

$$\begin{cases} x - 2y + 3z = 9 & \text{Equation 1} \\ y + 3z = 5 & \text{Equation 2} \\ z = 2 & \text{Equation 3} \end{cases}$$

Solution From Equation 3, you know the value of z. To solve for y, back-substitute $z = 2$ into Equation 2 to obtain

$y + 3(2) = 5$ Substitute 2 for z.

$y = -1.$ Solve for y.

Then back-substitute $y = -1$ and $z = 2$ into Equation 1 to obtain

$x - 2(-1) + 3(2) = 9$ Substitute -1 for y and 2 for z.

$x = 1.$ Solve for x.

The solution is $x = 1$, $y = -1$, and $z = 2$, which can be written as the **ordered triple** $(1, -1, 2)$. Check this in the original system of equations.

✓ **Checkpoint** *Audio-video solution in English & Spanish at LarsonPrecalculus.com.*

Solve the system of linear equations.

$$\begin{cases} 2x - y + 5z = 22 \\ y + 3z = 6 \\ z = 3 \end{cases}$$

One of the most influential Chinese mathematics books was the *Chui-chang suan-shu* or *Nine Chapters on the Mathematical Art* (written in approximately 250 B.C.). Chapter Eight of the *Nine Chapters* contained solutions of systems of linear equations such as

$$\begin{cases} 3x + 2y + z = 39 \\ 2x + 3y + z = 34 \\ x + 2y + 3z = 26 \end{cases}$$

using positive and negative numbers. This system was solved using column operations on a matrix. Matrices (plural for matrix) are discussed in the next chapter.

Gaussian Elimination

Two systems of equations are *equivalent* when they have the same solution set. To solve a system that is not in row-echelon form, first convert it to an *equivalent* system that is in row-echelon form by using the following operations.

Operations That Produce Equivalent Systems

Each of the following **row operations** on a system of linear equations produces an *equivalent* system of linear equations.

1. Interchange two equations.

2. Multiply one of the equations by a nonzero constant.

3. Add a multiple of one of the equations to another equation to replace the latter equation.

To see how to use row operations, take another look at the method of elimination, as applied to a system of two linear equations.

EXAMPLE 2   Using Row Operations to Solve a System

Solve the system of linear equations.

$$\begin{cases} 3x - 2y = -1 & \text{Equation 1} \\ x - y = 0 & \text{Equation 2} \end{cases}$$

Solution   There are two strategies that seem reasonable: eliminate the variable x or eliminate the variable y. The following steps show how to eliminate x.

$$\begin{cases} x - y = 0 \\ 3x - 2y = -1 \end{cases}$$ Interchange the two equations in the system.

$$\begin{cases} -3x + 3y = 0 \\ 3x - 2y = -1 \end{cases}$$ Multiply the first equation by -3.

$$\begin{array}{r} -3x + 3y = 0 \\ \underline{3x - 2y = -1} \\ y = -1 \end{array}$$ Add the multiple of the first equation to the second equation to obtain a new second equation.

$$\begin{cases} x - y = 0 \\ y = -1 \end{cases}$$ New system in row-echelon form

Notice in the first step that interchanging rows is a way of obtaining a leading coefficient of 1. Now back-substitute $y = -1$ into Equation 2 and solve for x.

$$x - (-1) = 0$$ Substitute -1 for y.

$$x = -1$$ Solve for x.

The solution is $x = -1$ and $y = -1$, which can be written as the ordered pair $(-1, -1)$.

✓ **Checkpoint** 🔊))) *Audio-video solution in English & Spanish at LarsonPrecalculus.com.*

Solve the system of linear equations.

$$\begin{cases} 2x + y = 3 \\ x + 2y = 3 \end{cases}$$

Rewriting a system of linear equations in row-echelon form usually involves a chain of equivalent systems, which you obtain by using one of the three basic row operations listed on the previous page. This process is called **Gaussian elimination,** after the German mathematician Carl Friedrich Gauss (1777–1855).

EXAMPLE 3 Using Gaussian Elimination to Solve a System

Solve the system of linear equations.

$$\begin{cases} x - 2y + 3z = 9 & \text{Equation 1} \\ -x + 3y \quad\;\; = -4 & \text{Equation 2} \\ 2x - 5y + 5z = 17 & \text{Equation 3} \end{cases}$$

Solution Because the leading coefficient of the first equation is 1, begin by keeping the x in the upper left position and eliminating the other x-terms from the first column.

> **REMARK** Arithmetic errors are common when performing row operations. You should note the operation performed in each step to make checking your work easier.

$$\begin{array}{rl} x - 2y + 3z = 9 & \text{Write Equation 1.} \\ \underline{-x + 3y \qquad\; = -4} & \text{Write Equation 2.} \\ y + 3z = 5 & \text{Add Equation 1 to Equation 2.} \end{array}$$

$$\begin{cases} x - 2y + 3z = 9 \\ \quad\;\; y + 3z = 5 \\ 2x - 5y + 5z = 17 \end{cases}$$

Adding the first equation to the second equation produces a new second equation.

$$\begin{array}{rl} -2x + 4y - 6z = -18 & \text{Multiply Equation 1 by } -2. \\ \underline{2x - 5y + 5z = \quad 17} & \text{Write Equation 3.} \\ -y - z = -1 & \text{Add revised Equation 1 to Equation 3.} \end{array}$$

$$\begin{cases} x - 2y + 3z = 9 \\ \quad\;\; y + 3z = 5 \\ \quad\; -y - z = -1 \end{cases}$$

Adding -2 times the first equation to the third equation produces a new third equation.

Now that you have eliminated all but the x in the upper position of the first column, work on the second column. (You need to eliminate y from the third equation.)

$$\begin{cases} x - 2y + 3z = 9 \\ \quad\;\; y + 3z = 5 \\ \quad\quad\;\; 2z = 4 \end{cases}$$

Adding the second equation to the third equation produces a new third equation.

Finally, you need a coefficient of 1 for z in the third equation.

$$\begin{cases} x - 2y + 3z = 9 \\ \quad\;\; y + 3z = 5 \\ \quad\quad\;\;\; z = 2 \end{cases}$$

Multiplying the third equation by $\frac{1}{2}$ produces a new third equation.

This is the same system in Example 1, and, as in that example, the solution is

$$x = 1, \quad y = -1, \quad \text{and} \quad z = 2.$$

✓ **Checkpoint** ◀))) Audio-video solution in English & Spanish at LarsonPrecalculus.com.

Solve the system of linear equations.

$$\begin{cases} x + y + z = 6 \\ 2x - y + z = 3 \\ 3x + y - z = 2 \end{cases}$$

The next example involves an inconsistent system—one that has no solution. The key to recognizing an inconsistent system is that at some stage in the elimination process, you obtain a false statement such as $0 = -2$.

EXAMPLE 4 **An Inconsistent System**

Solve the system of linear equations.

$$\begin{cases} x - 3y + z = 1 \\ 2x - y - 2z = 2 \\ x + 2y - 3z = -1 \end{cases}$$
Equation 1
Equation 2
Equation 3

Solution

$$\begin{cases} x - 3y + z = 1 \\ 5y - 4z = 0 \\ x + 2y - 3z = -1 \end{cases}$$
Adding -2 times the first equation to the second equation produces a new second equation.

$$\begin{cases} x - 3y + z = 1 \\ 5y - 4z = 0 \\ 5y - 4z = -2 \end{cases}$$
Adding -1 times the first equation to the third equation produces a new third equation.

$$\begin{cases} x - 3y + z = 1 \\ 5y - 4z = 0 \\ 0 = -2 \end{cases}$$
Adding -1 times the second equation to the third equation produces a new third equation.

Because $0 = -2$ is a false statement, this system is inconsistent and has no solution. Moreover, because this system is equivalent to the original system, the original system also has no solution.

✓ **Checkpoint** 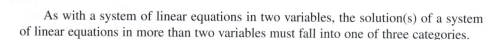 *Audio-video solution in English & Spanish at LarsonPrecalculus.com.*

Solve the system of linear equations.

$$\begin{cases} x + y - 2z = 3 \\ 3x - 2y + 4z = 1 \\ 2x - 3y + 6z = 8 \end{cases}$$

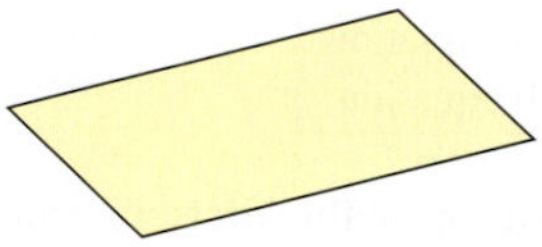

As with a system of linear equations in two variables, the solution(s) of a system of linear equations in more than two variables must fall into one of three categories.

The Number of Solutions of a Linear System

For a system of linear equations, exactly one of the following is true.

1. There is exactly one solution.
2. There are infinitely many solutions.
3. There is no solution.

In Section 7.2, you learned that a system of two linear equations in two variables can be represented graphically as a pair of lines that are intersecting, coincident, or parallel. A system of three linear equations in three variables has a similar graphical representation—it can be represented as three planes in space that intersect in one point (exactly one solution, see Figure 7.9), intersect in a line or a plane (infinitely many solutions, see Figures 7.10 and 7.11), or have no points common to all three planes (no solution, see Figures 7.12 and 7.13).

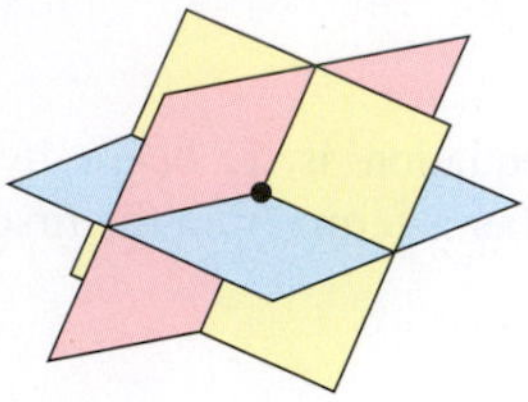

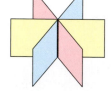

Solution: one point
Figure 7.9

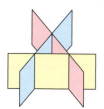

Solution: one line
Figure 7.10

Solution: one plane
Figure 7.11

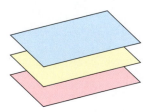

Solution: none
Figure 7.12

Solution: none
Figure 7.13

EXAMPLE 5 **A System with Infinitely Many Solutions**

Solve the system of linear equations.

$$\begin{cases} x + y - 3z = -1 & \text{Equation 1} \\ \quad\quad y - z = 0 & \text{Equation 2} \\ -x + 2y \quad\quad = 1 & \text{Equation 3} \end{cases}$$

Solution

$$\begin{cases} x + y - 3z = -1 \\ \quad\quad y - z = 0 \\ \quad\quad 3y - 3z = 0 \end{cases}$$

> Adding the first equation to the third equation produces a new third equation.

$$\begin{cases} x + y - 3z = -1 \\ \quad\quad y - z = 0 \\ \quad\quad\quad 0 = 0 \end{cases}$$

> Adding -3 times the second equation to the third equation produces a new third equation.

This result means that Equation 3 depends on Equations 1 and 2 in the sense that it gives no additional information about the variables. Because $0 = 0$ is a true statement, this system has infinitely many solutions. However, it is incorrect to say that the solution is "infinite." You must also specify the correct form of the solution. So, the original system is equivalent to the system

$$\begin{cases} x + y - 3z = -1 \\ \quad\quad y - z = 0 \end{cases}.$$

In the second equation, solve for y in terms of z to obtain $y = z$. Back-substituting $y = z$ in the first equation produces $x = 2z - 1$. Finally, letting $z = a$, where a is a real number, the solutions of the given system are all of the form $x = 2a - 1$, $y = a$, and $z = a$. So, every ordered triple of the form

$$(2a - 1, a, a) \qquad\qquad a \text{ is a real number.}$$

is a solution of the system.

✓ *Checkpoint*  Audio-video solution in English & Spanish at *LarsonPrecalculus.com*.

Solve the system of linear equations.

$$\begin{cases} x + 2y - 7z = -4 \\ 2x + 3y + z = 5 \\ 3x + 7y - 36z = -25 \end{cases}$$

In Example 5, there are other ways to write the same infinite set of solutions. For instance, letting $x = b$, the solutions could have been written as

$$\left(b, \tfrac{1}{2}(b + 1), \tfrac{1}{2}(b + 1)\right). \qquad\qquad b \text{ is a real number.}$$

To convince yourself that this description produces the same set of solutions, consider the following.

Substitution	Solution	
$a = 0$	$(2(0) - 1, 0, 0) = (-1, 0, 0)$	Same solution
$b = -1$	$\left(-1, \tfrac{1}{2}(-1 + 1), \tfrac{1}{2}(-1 + 1)\right) = (-1, 0, 0)$	
$a = 1$	$(2(1) - 1, 1, 1) = (1, 1, 1)$	Same solution
$b = 1$	$\left(1, \tfrac{1}{2}(1 + 1), \tfrac{1}{2}(1 + 1)\right) = (1, 1, 1)$	
$a = 2$	$(2(2) - 1, 2, 2) = (3, 2, 2)$	Same solution
$b = 3$	$\left(3, \tfrac{1}{2}(3 + 1), \tfrac{1}{2}(3 + 1)\right) = (3, 2, 2)$	

The Global Positioning System (GPS) is a network of 24 satellites originally developed by the U.S. military as a navigational tool. Civilian applications of GPS receivers include determining directions, locating vessels lost at sea, and monitoring earthquakes. A GPS receiver works by using satellite readings to calculate its location. In a simplified mathematical model, nonsquare systems of three linear equations in four variables (three dimensions and time) determine the coordinates of the receiver as a function of time.

Nonsquare Systems

So far, each system of linear equations you have looked at has been *square*, which means that the number of equations is equal to the number of variables. In a **nonsquare** system, the number of equations differs from the number of variables. A system of linear equations cannot have a unique solution unless there are at least as many equations as there are variables in the system.

EXAMPLE 6 **A System with Fewer Equations than Variables**

Solve the system of linear equations.

$$\begin{cases} x - 2y + z = 2 & \text{Equation 1} \\ 2x - y - z = 1 & \text{Equation 2} \end{cases}$$

Solution Begin by rewriting the system in row-echelon form.

$$\begin{cases} x - 2y + z = 2 \\ 3y - 3z = -3 \end{cases}$$

> Adding -2 times the first equation to the second equation produces a new second equation.

$$\begin{cases} x - 2y + z = 2 \\ y - z = -1 \end{cases}$$

> Multiplying the second equation by $\frac{1}{3}$ produces a new second equation.

Solve for y in terms of z to obtain

$$y = z - 1.$$

Solve for x by back-substituting $y = z - 1$ into Equation 1.

$x - 2y + z = 2$	Write Equation 1.
$x - 2(z - 1) + z = 2$	Substitute $z - 1$ for y.
$x - 2z + 2 + z = 2$	Distributive Property
$x = z$	Solve for x.

Finally, by letting $z = a$, where a is a real number, you have the solution

$$x = a, \quad y = a - 1, \quad \text{and} \quad z = a.$$

So, every ordered triple of the form

$$(a, a - 1, a) \qquad a \text{ is a real number.}$$

is a solution of the system. Because there were originally three variables and only two equations, the system cannot have a unique solution.

✓ **Checkpoint** Audio-video solution in English & Spanish at LarsonPrecalculus.com.

Solve the system of linear equations.

$$\begin{cases} x - y + 4z = 3 \\ 4x - z = 0 \end{cases}$$

In Example 6, try choosing some values of a to obtain different solutions of the system, such as

$$(1, 0, 1), \quad (2, 1, 2), \quad \text{and} \quad (3, 2, 3).$$

Then check each ordered triple in the original system to verify that it is a solution of the system.

Applications

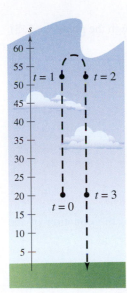

Figure 7.14

EXAMPLE 7   **Vertical Motion**

The height at time t of an object that is moving in a (vertical) line with constant acceleration a is given by the **position equation**

$$s = \tfrac{1}{2}at^2 + v_0 t + s_0$$

where s is the height in feet, a is the acceleration in feet per second squared, t is the time in seconds, v_0 is the initial velocity (at $t = 0$), and s_0 is the initial height. Find the values of a, v_0, and s_0 when $s = 52$ at $t = 1$, $s = 52$ at $t = 2$, and $s = 20$ at $t = 3$, and interpret the result. (See Figure 7.14.)

Solution   By substituting the three values of t and s into the position equation, you can obtain three linear equations in a, v_0, and s_0.

When $t = 1$: $\tfrac{1}{2}a(1)^2 + v_0(1) + s_0 = 52$ $\Longrightarrow$ $a + 2v_0 + 2s_0 = 104$

When $t = 2$: $\tfrac{1}{2}a(2)^2 + v_0(2) + s_0 = 52$ $\Longrightarrow$ $2a + 2v_0 + s_0 = 52$

When $t = 3$: $\tfrac{1}{2}a(3)^2 + v_0(3) + s_0 = 20$ $\Longrightarrow$ $9a + 6v_0 + 2s_0 = 40$

This produces the following system of linear equations.

$$\begin{cases} a + 2v_0 + 2s_0 = 104 \\ 2a + 2v_0 + s_0 = 52 \\ 9a + 6v_0 + 2s_0 = 40 \end{cases}$$

Now solve the system using Gaussian elimination.

$$\begin{cases} a + 2v_0 + 2s_0 = 104 \\ \quad\ -2v_0 - 3s_0 = -156 \\ 9a + 6v_0 + 2s_0 = 40 \end{cases}$$

> Adding -2 times the first equation to the second equation produces a new second equation.

$$\begin{cases} a + 2v_0 + 2s_0 = 104 \\ \quad\ -2v_0 - 3s_0 = -156 \\ \quad\ -12v_0 - 16s_0 = -896 \end{cases}$$

> Adding -9 times the first equation to the third equation produces a new third equation.

$$\begin{cases} a + 2v_0 + 2s_0 = 104 \\ \quad\ -2v_0 - 3s_0 = -156 \\ \quad\quad\ 2s_0 = 40 \end{cases}$$

> Adding -6 times the second equation to the third equation produces a new third equation.

$$\begin{cases} a + 2v_0 + 2s_0 = 104 \\ \quad\ v_0 + \tfrac{3}{2}s_0 = 78 \\ \quad\quad\ s_0 = 20 \end{cases}$$

> Multiplying the second equation by $-\tfrac{1}{2}$ produces a new second equation and multiplying the third equation by $\tfrac{1}{2}$ produces a new third equation.

So, the solution of this system is $a = -32$, $v_0 = 48$, and $s_0 = 20$, which can be written as $(-32, 48, 20)$. This solution results in a position equation of $s = -16t^2 + 48t + 20$ and implies that the object was thrown upward at a velocity of 48 feet per second from a height of 20 feet.

✓ **Checkpoint**   *Audio-video solution in English & Spanish at LarsonPrecalculus.com.*

For the position equation

$$s = \tfrac{1}{2}at^2 + v_0 t + s_0$$

given in Example 7, find the values of a, v_0, and s_0 when $s = 104$ at $t = 1$, $s = 76$ at $t = 2$, and $s = 16$ at $t = 3$, and interpret the result. ■

EXAMPLE 8 Data Analysis: Curve-Fitting

Find a quadratic equation $y = ax^2 + bx + c$ whose graph passes through the points $(-1, 3)$, $(1, 1)$, and $(2, 6)$.

Solution Because the graph of $y = ax^2 + bx + c$ passes through the points $(-1, 3)$, $(1, 1)$, and $(2, 6)$, you can write the following.

When $x = -1, y = 3$: $a(-1)^2 + b(-1) + c = 3$

When $x = 1, y = 1$: $a(1)^2 + b(1) + c = 1$

When $x = 2, y = 6$: $a(2)^2 + b(2) + c = 6$

This produces the following system of linear equations.

$$\begin{cases} a - b + c = 3 & \text{Equation 1} \\ a + b + c = 1 & \text{Equation 2} \\ 4a + 2b + c = 6 & \text{Equation 3} \end{cases}$$

The solution of this system is $a = 2$, $b = -1$, and $c = 0$. So, the equation of the parabola is

$$y = 2x^2 - x$$

as shown below.

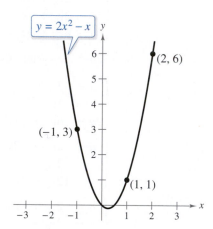

$y = 2x^2 - x$

✓ *Checkpoint* 🔊)) *Audio-video solution in English & Spanish at LarsonPrecalculus.com.*

Find a quadratic equation $y = ax^2 + bx + c$ whose graph passes through the points $(0, 0)$, $(3, -3)$, and $(6, 0)$.

Summarize (Section 7.3)

1. State the definition of row-echelon form *(page 355)*. For an example of solving a linear system in row-echelon form, see Example 1.

2. Describe the process of Gaussian elimination *(pages 356 and 357)*. For examples of using Gaussian elimination to solve systems of linear equations, see Examples 2–5.

3. Explain the difference between a square system of linear equations and a nonsquare system of linear equations *(page 360)*. For an example of solving a nonsquare system of linear equations, see Example 6.

4. Describe examples of how to use systems of linear equations in three or more variables to model and solve real-life problems *(pages 361 and 362, Examples 7 and 8)*.

7.4 Partial Fractions

- ■ Recognize partial fraction decompositions of rational expressions.
- ■ Find partial fraction decompositions of rational expressions.

Introduction

In this section, you will learn to write a rational expression as the sum of two or more simpler rational expressions. For example, the rational expression

$$\frac{x + 7}{x^2 - x - 6}$$

can be written as the sum of two fractions with first-degree denominators. That is,

Partial fraction decomposition
of $\dfrac{x + 7}{x^2 - x - 6}$

$$\frac{x + 7}{x^2 - x - 6} = \overbrace{\underbrace{\frac{2}{x - 3}}_{\substack{\text{Partial} \\ \text{fraction}}} + \underbrace{\frac{-1}{x + 2}}_{\substack{\text{Partial} \\ \text{fraction}}}}.$$

Each fraction on the right side of the equation is a **partial fraction,** and together they make up the **partial fraction decomposition** of the left side.

Partial fractions can help you analyze the behavior of a rational function, such as the predicted cost for a company to remove a percentage of a chemical from its waste water.

▷ **ALGEBRA HELP** You can review how to find the degree of a polynomial (such as $x - 3$ and $x + 2$) in Appendix A.3.

•• **REMARK** Appendix A.4 shows you how to combine expressions such as

$$\frac{1}{x - 2} + \frac{-1}{x + 3} = \frac{5}{(x - 2)(x + 3)}.$$

The method of partial fraction decomposition shows you how to reverse this process and write

$$\frac{5}{(x - 2)(x + 3)} = \frac{1}{x - 2} + \frac{-1}{x + 3}.$$

Decomposition of $N(x)/D(x)$ into Partial Fractions

1. *Divide when improper:* When $N(x)/D(x)$ is an improper fraction [degree of $N(x) \geq$ degree of $D(x)$], divide the denominator into the numerator to obtain

   $$\frac{N(x)}{D(x)} = (\text{polynomial}) + \frac{N_1(x)}{D(x)}$$

   and apply Steps 2, 3, and 4 below to the proper rational expression

   $$\frac{N_1(x)}{D(x)}.$$

   Note that $N_1(x)$ is the remainder from the division of $N(x)$ by $D(x)$.

2. *Factor the denominator:* Completely factor the denominator into factors of the form

   $$(px + q)^m \quad \text{and} \quad (ax^2 + bx + c)^n$$

   where $(ax^2 + bx + c)$ is irreducible.

3. *Linear factors:* For *each* factor of the form $(px + q)^m$, the partial fraction decomposition must include the following sum of m fractions.

   $$\frac{A_1}{(px + q)} + \frac{A_2}{(px + q)^2} + \cdots + \frac{A_m}{(px + q)^m}$$

4. *Quadratic factors:* For *each* factor of the form $(ax^2 + bx + c)^n$, the partial fraction decomposition must include the following sum of n fractions.

   $$\frac{B_1x + C_1}{ax^2 + bx + c} + \frac{B_2x + C_2}{(ax^2 + bx + c)^2} + \cdots + \frac{B_nx + C_n}{(ax^2 + bx + c)^n}$$

Partial Fraction Decomposition

The following examples demonstrate algebraic techniques for determining the constants in the numerators of partial fractions. Note that the techniques vary slightly, depending on the type of factors of the denominator: linear or quadratic, distinct or repeated.

EXAMPLE 1   Distinct Linear Factors

Write the partial fraction decomposition of

$$\frac{x + 7}{x^2 - x - 6}.$$

Solution   The expression is proper, so you should begin by factoring the denominator. Because

$$x^2 - x - 6 = (x - 3)(x + 2)$$

you should include one partial fraction with a constant numerator for each linear factor of the denominator. Write the form of the decomposition as follows.

$$\frac{x + 7}{x^2 - x - 6} = \frac{A}{x - 3} + \frac{B}{x + 2} \qquad \text{Write form of decomposition.}$$

Multiplying each side of this equation by the least common denominator, $(x - 3)(x + 2)$, leads to the **basic equation**

$$x + 7 = A(x + 2) + B(x - 3). \qquad \text{Basic equation}$$

Because this equation is true for all x, substitute any *convenient* values of x that will help determine the constants A and B. Values of x that are especially convenient are those that make the factors $(x + 2)$ and $(x - 3)$ equal to zero. For instance, to solve for B, let $x = -2$. Then

$$-2 + 7 = A(-2 + 2) + B(-2 - 3) \qquad \text{Substitute } -2 \text{ for } x.$$
$$5 = A(0) + B(-5)$$
$$5 = -5B$$
$$-1 = B.$$

To solve for A, let $x = 3$ and obtain

$$3 + 7 = A(3 + 2) + B(3 - 3) \qquad \text{Substitute 3 for } x.$$
$$10 = A(5) + B(0)$$
$$10 = 5A$$
$$2 = A.$$

So, the partial fraction decomposition is

$$\frac{x + 7}{x^2 - x - 6} = \frac{2}{x - 3} + \frac{-1}{x + 2}.$$

Check this result by combining the two partial fractions on the right side of the equation, or by using your graphing utility.

✓ *Checkpoint* ◀))) *Audio-video solution in English & Spanish at LarsonPrecalculus.com.*

Write the partial fraction decomposition of

$$\frac{x + 5}{2x^2 - x - 1}.$$

▷ **TECHNOLOGY** To use a graphing utility to check the decomposition found in Example 1, graph

$$y_1 = \frac{x + 7}{x^2 - x - 6}$$

and

$$y_2 = \frac{2}{x - 3} + \frac{-1}{x + 2}$$

in the same viewing window. The graphs should be identical, as shown below.

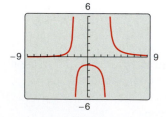

The next example shows how to find the partial fraction decomposition of a rational expression whose denominator has a *repeated* linear factor.

EXAMPLE 2 Repeated Linear Factors

Write the partial fraction decomposition of $\dfrac{x^4 + 2x^3 + 6x^2 + 20x + 6}{x^3 + 2x^2 + x}$.

> **ALGEBRA HELP** You can review long division of polynomials in Section 2.3. You can review factoring of polynomials in Appendix A.3.

Solution This rational expression is improper, so you should begin by dividing the numerator by the denominator to obtain

$$x + \frac{5x^2 + 20x + 6}{x^3 + 2x^2 + x}.$$

Because the denominator of the remainder factors as

$$x^3 + 2x^2 + x = x(x^2 + 2x + 1) = x(x + 1)^2$$

you should include one partial fraction with a constant numerator for each power of x and $(x + 1)$, and write the form of the decomposition as follows.

$$\frac{5x^2 + 20x + 6}{x(x + 1)^2} = \frac{A}{x} + \frac{B}{x + 1} + \frac{C}{(x + 1)^2} \qquad \text{Write form of decomposition.}$$

Multiplying each side by the LCD, $x(x + 1)^2$, leads to the basic equation

$$5x^2 + 20x + 6 = A(x + 1)^2 + Bx(x + 1) + Cx. \qquad \text{Basic equation}$$

> **REMARK** To obtain the basic equation, be sure to multiply *each* fraction in the form of the decomposition by the LCD.

Letting $x = -1$ eliminates the A- and B-terms and yields

$$5(-1)^2 + 20(-1) + 6 = A(-1 + 1)^2 + B(-1)(-1 + 1) + C(-1)$$
$$5 - 20 + 6 = 0 + 0 - C$$
$$C = 9.$$

Letting $x = 0$ eliminates the B- and C-terms and yields

$$5(0)^2 + 20(0) + 6 = A(0 + 1)^2 + B(0)(0 + 1) + C(0)$$
$$6 = A(1) + 0 + 0$$
$$6 = A.$$

At this point, you have exhausted the most convenient values of x, so to find the value of B, use *any other value* of x along with the known values of A and C. So, using $x = 1$, $A = 6$, and $C = 9$,

$$5(1)^2 + 20(1) + 6 = 6(1 + 1)^2 + B(1)(1 + 1) + 9(1)$$
$$31 = 6(4) + 2B + 9$$
$$-2 = 2B$$
$$-1 = B.$$

So, the partial fraction decomposition is

$$\frac{x^4 + 2x^3 + 6x^2 + 20x + 6}{x^3 + 2x^2 + x} = x + \frac{6}{x} + \frac{-1}{x + 1} + \frac{9}{(x + 1)^2}.$$

✓ **Checkpoint** ◀))) Audio-video solution in English & Spanish at LarsonPrecalculus.com.

Write the partial fraction decomposition of

$$\frac{x^4 + x^3 + x + 4}{x^3 + x^2}.$$

The procedure used to solve for the constants in Examples 1 and 2 works well when the factors of the denominator are linear. However, when the denominator contains irreducible quadratic factors, you should use a different procedure, which involves writing the right side of the basic equation in polynomial form, *equating the coefficients* of like terms, and using a system of equations to solve for the coefficients.

EXAMPLE 3   Distinct Linear and Quadratic Factors

Write the partial fraction decomposition of

$$\frac{3x^2 + 4x + 4}{x^3 + 4x}.$$

Solution This expression is proper, so begin by factoring the denominator. Because the denominator factors as

$$x^3 + 4x = x(x^2 + 4)$$

you should include one partial fraction with a constant numerator and one partial fraction with a linear numerator, and write the form of the decomposition as follows.

$$\frac{3x^2 + 4x + 4}{x^3 + 4x} = \frac{A}{x} + \frac{Bx + C}{x^2 + 4}$$ Write form of decomposition.

Multiplying each side by the LCD, $x(x^2 + 4)$, yields the basic equation

$$3x^2 + 4x + 4 = A(x^2 + 4) + (Bx + C)x.$$ Basic equation

Expanding this basic equation and collecting like terms produces

$$3x^2 + 4x + 4 = Ax^2 + 4A + Bx^2 + Cx$$

$$= (A + B)x^2 + Cx + 4A.$$ Polynomial form

Finally, because two polynomials are equal if and only if the coefficients of like terms are equal, equate the coefficients of like terms on opposite sides of the equation.

$$3x^2 + 4x + 4 = (A + B)x^2 + Cx + 4A$$ Equate coefficients of like terms.

Now write the following system of linear equations.

$$\begin{cases} A + B \quad\;\; = 3 & \text{Equation 1} \\ \quad\quad\;\; C = 4 & \text{Equation 2} \\ 4A \quad\quad\; = 4 & \text{Equation 3} \end{cases}$$

From this system, you can see that

$$A = 1 \quad \text{and} \quad C = 4.$$

Moreover, back-substituting $A = 1$ into Equation 1 yields

$$1 + B = 3 \implies B = 2.$$

So, the partial fraction decomposition is

$$\frac{3x^2 + 4x + 4}{x^3 + 4x} = \frac{1}{x} + \frac{2x + 4}{x^2 + 4}.$$

✓ *Checkpoint*  Audio-video solution in English & Spanish at LarsonPrecalculus.com.

Write the partial fraction decomposition of

$$\frac{2x^2 - 5}{x^3 + x}.$$

John Bernoulli (1667–1748), a Swiss mathematician, introduced the method of partial fractions and was instrumental in the early development of calculus. Bernoulli was a professor at the University of Basel and taught many outstanding students, the most famous of whom was Leonhard Euler.

The next example shows how to find the partial fraction decomposition of a rational expression whose denominator has a *repeated* quadratic factor.

EXAMPLE 4   **Repeated Quadratic Factors**

Write the partial fraction decomposition of

$$\frac{8x^3 + 13x}{(x^2 + 2)^2}.$$

Solution   Include one partial fraction with a linear numerator for each power of $(x^2 + 2)$.

$$\frac{8x^3 + 13x}{(x^2 + 2)^2} = \frac{Ax + B}{x^2 + 2} + \frac{Cx + D}{(x^2 + 2)^2}$$ Write form of decomposition.

Multiplying each side by the LCD, $(x^2 + 2)^2$, yields the basic equation

$$8x^3 + 13x = (Ax + B)(x^2 + 2) + Cx + D$$ Basic equation

$$= Ax^3 + 2Ax + Bx^2 + 2B + Cx + D$$

$$= Ax^3 + Bx^2 + (2A + C)x + (2B + D).$$ Polynomial form

Equating coefficients of like terms on opposite sides of the equation

$$8x^3 + 0x^2 + 13x + 0 = Ax^3 + Bx^2 + (2A + C)x + (2B + D)$$

produces the following system of linear equations.

$$\begin{cases} A && = 8 & \text{Equation 1} \\ & B && = 0 & \text{Equation 2} \\ 2A + & & C & = 13 & \text{Equation 3} \\ & 2B + & D & = 0 & \text{Equation 4} \end{cases}$$

Use the values $A = 8$ and $B = 0$ to obtain the following.

$$2(8) + C = 13$$ Substitute 8 for A in Equation 3.

$$C = -3$$

$$2(0) + D = 0$$ Substitute 0 for B in Equation 4.

$$D = 0$$

So, using

$$A = 8, \quad B = 0, \quad C = -3, \quad \text{and} \quad D = 0$$

the partial fraction decomposition is

$$\frac{8x^3 + 13x}{(x^2 + 2)^2} = \frac{8x}{x^2 + 2} + \frac{-3x}{(x^2 + 2)^2}.$$

Check this result by combining the two partial fractions on the right side of the equation, or by using your graphing utility.

✓ **Checkpoint** ◀))) *Audio-video solution in English & Spanish at LarsonPrecalculus.com.*

Write the partial fraction decomposition of

$$\frac{x^3 + 3x^2 - 2x + 7}{(x^2 + 4)^2}.$$

EXAMPLE 5 **Repeated Linear and Quadratic Factors**

Write the partial fraction decomposition of $\dfrac{x+5}{x^2(x^2+1)^2}$.

Solution Include one partial fraction with a constant numerator for each power of x and one partial fraction with a linear numerator for each power of (x^2+1).

$$\frac{x+5}{x^2(x^2+1)^2} = \frac{A}{x} + \frac{B}{x^2} + \frac{Cx+D}{x^2+1} + \frac{Ex+F}{(x^2+1)^2} \qquad \textcolor{red}{\text{Write form of decomposition.}}$$

Multiplying each side by the LCD, $x^2(x^2+1)^2$, yields the basic equation

$$x+5 = Ax(x^2+1)^2 + B(x^2+1)^2 + (Cx+D)x^2(x^2+1) + (Ex+F)x^2 \qquad \textcolor{red}{\substack{\text{Basic}\\\text{equation}}}$$

$$= (A+C)x^5 + (B+D)x^4 + (2A+C+E)x^3 + (2B+D+F)x^2 + Ax + B.$$

Verify, by equating coefficients of like terms on opposite sides of the equation and solving the resulting system of linear equations, that $A=1$, $B=5$, $C=-1$, $D=-5$, $E=-1$, and $F=-5$, and that the partial fraction decomposition is

$$\frac{x+5}{x^2(x^2+1)^2} = \frac{1}{x} + \frac{5}{x^2} - \frac{x+5}{x^2+1} - \frac{x+5}{(x^2+1)^2}.$$

✓ **Checkpoint** *Audio-video solution in English & Spanish at LarsonPrecalculus.com.*

Write the partial fraction decomposition of $\dfrac{4x-8}{x^2(x^2+2)^2}$. ∎

Guidelines for Solving the Basic Equation

Linear Factors

1. Substitute the *zeros* of the distinct linear factors into the basic equation.

2. For repeated linear factors, use the coefficients determined in Step 1 to rewrite the basic equation. Then substitute *other* convenient values of x and solve for the remaining coefficients.

Quadratic Factors

1. Expand the basic equation.

2. Collect terms according to powers of x.

3. Equate the coefficients of like terms to obtain equations involving A, B, C, and so on.

4. Use a system of linear equations to solve for A, B, C, and so on.

Keep in mind that for *improper* rational expressions, you must first divide before applying partial fraction decomposition.

Summarize **(Section 7.4)**

1. Explain what is meant by the partial fraction decomposition of a rational expression *(page 363)*.

2. Explain how to find the partial fraction decomposition of a rational expression *(pages 363–368)*. For examples of finding partial fraction decompositions of rational expressions, see Examples 1–5.

7.5 Systems of Inequalities

- Sketch the graphs of inequalities in two variables.
- Solve systems of inequalities.
- Use systems of inequalities in two variables to model and solve real-life problems.

Systems of inequalities in two variables can help you model and solve real-life problems, such as analyzing the nutritional content of a dietary supplement.

The Graph of an Inequality

The statements

$$3x - 2y < 6 \quad \text{and} \quad 2x^2 + 3y^2 \geq 6$$

are inequalities in two variables. An ordered pair (a, b) is a **solution of an inequality** in x and y when the inequality is true after a and b are substituted for x and y, respectively. The **graph of an inequality** is the collection of all solutions of the inequality. To sketch the graph of an inequality, begin by sketching the graph of the *corresponding equation*. The graph of the equation will usually separate the plane into two or more regions. In each such region, one of the following must be true.

1. *All* points in the region are solutions of the inequality.

2. *No* point in the region is a solution of the inequality.

So, you can determine whether the points in an entire region satisfy the inequality by testing *one* point in the region.

> **Sketching the Graph of an Inequality in Two Variables**
>
> 1. Replace the inequality sign by an equal sign and sketch the graph of the resulting equation. (Use a dashed line for < or > and a solid line for ≤ or ≥.)
>
> 2. Test one point in each of the regions formed by the graph in Step 1. If the point satisfies the inequality, then shade the entire region to denote that every point in the region satisfies the inequality.

• • **REMARK** Be careful when you are sketching the graph of an inequality in two variables. A dashed line means that all points on the line or curve *are not* solutions of the inequality. A solid line means that all points on the line or curve *are* solutions of the inequality.

▷

EXAMPLE 1   Sketching the Graph of an Inequality

Sketch the graph of $y \geq x^2 - 1$.

Solution Begin by graphing the corresponding equation $y = x^2 - 1$, which is a parabola, as shown at the right. Test a point *above* the parabola, such as $(0, 0)$, and a point *below* the parabola, such as $(0, -2)$.

The points that satisfy the inequality are those lying above (or on) the parabola.

$(0, 0)$: $0 \overset{?}{\geq} 0^2 - 1$

$\qquad 0 \geq -1 \qquad \textcolor{red}{(0, 0) \text{ is a solution.}}$

$(0, -2)$: $-2 \overset{?}{\geq} 0^2 - 1$

$\qquad -2 \not\geq -1 \qquad \textcolor{red}{(0, -2) \text{ is not a solution.}}$

$y \geq x^2 - 1$ $y = x^2 - 1$

$(0, 0)$

Test point above parabola   Test point below parabola

$(0, -2)$

✓ **Checkpoint** ◀))) *Audio-video solution in English & Spanish at LarsonPrecalculus.com.*

Sketch the graph of $(x + 2)^2 + (y - 2)^2 < 16$.

The inequality in Example 1 is a nonlinear inequality in two variables. Most of the following examples involve **linear inequalities** such as $ax + by < c$ (where a and b are not both zero). The graph of a linear inequality is a half-plane lying on one side of the line $ax + by = c$.

■ EXAMPLE 2 **Sketching the Graph of a Linear Inequality**

Sketch the graph of each linear inequality.

a. $x > -2$ **b.** $y \leq 3$

Solution

▷ **TECHNOLOGY** A graphing utility can be used to graph an inequality or a system of inequalities. For instance, to graph $y \geq x - 2$, enter $y = x - 2$ and use the *shade* feature of the graphing utility to shade the correct part of the graph. You should obtain the graph shown below. Consult the user's guide for your graphing utility for specific keystrokes.

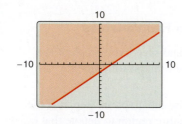

a. The graph of the corresponding equation $x = -2$ is a vertical line. The points that satisfy the inequality $x > -2$ are those lying to the right of this line, as shown in Figure 7.15.

b. The graph of the corresponding equation $y = 3$ is a horizontal line. The points that satisfy the inequality $y \leq 3$ are those lying below (or on) this line, as shown in Figure 7.16.

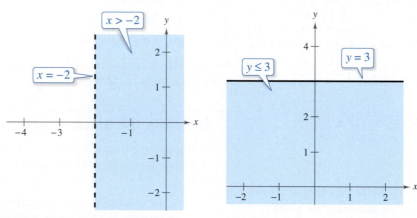

Figure 7.15   Figure 7.16

✓ *Checkpoint*)) *Audio-video solution in English & Spanish at LarsonPrecalculus.com.*

Sketch the graph of $x \geq 3$.

■ EXAMPLE 3 **Sketching the Graph of a Linear Inequality**

Sketch the graph of $x - y < 2$.

Solution The graph of the corresponding equation $x - y = 2$ is a line, as shown in Figure 7.17. Because the origin $(0, 0)$ satisfies the inequality, the graph consists of the half-plane lying above the line. (Try checking a point below the line. Regardless of which point you choose, you will see that it does not satisfy the inequality.)

✓ *Checkpoint*)) *Audio-video solution in English & Spanish at LarsonPrecalculus.com.*

Sketch the graph of $x + y > -2$.

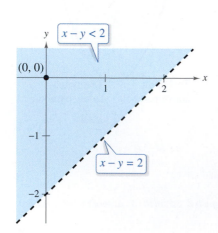

Figure 7.17

To graph a linear inequality, it can help to write the inequality in slope-intercept form. For instance, by writing $x - y < 2$ in the form

$$y > x - 2$$

you can see that the solution points lie *above* the line $x - y = 2$ (or $y = x - 2$), as shown in Figure 7.17.

Systems of Inequalities

Many practical problems in business, science, and engineering involve systems of linear inequalities. A **solution** of a system of inequalities in x and y is a point

$$(x, y)$$

that satisfies each inequality in the system.

To sketch the graph of a system of inequalities in two variables, first sketch the graph of each individual inequality (on the same coordinate system) and then find the region that is *common* to every graph in the system. This region represents the **solution set** of the system. For systems of *linear inequalities,* it is helpful to find the vertices of the solution region.

EXAMPLE 4   Solving a System of Inequalities

Sketch the graph (and label the vertices) of the solution set of the system.

$$\begin{cases} x - y < 2 & \text{Inequality 1} \\ x > -2 & \text{Inequality 2} \\ y \leq 3 & \text{Inequality 3} \end{cases}$$

Solution   The graphs of these inequalities are shown in Figures 7.17, 7.15, and 7.16, respectively, on page 370. The triangular region common to all three graphs can be found by superimposing the graphs on the same coordinate system, as shown below. To find the vertices of the region, solve the three systems of corresponding equations obtained by taking *pairs* of equations representing the boundaries of the individual regions.

Vertex A: $(-2, -4)$

$$\begin{cases} x - y = 2 \\ x = -2 \end{cases}$$

Vertex B: $(5, 3)$

$$\begin{cases} x - y = 2 \\ y = 3 \end{cases}$$

Vertex C: $(-2, 3)$

$$\begin{cases} x = -2 \\ y = 3 \end{cases}$$

REMARK   Using a different colored pencil to shade the solution of each inequality in a system will make identifying the solution of the system of inequalities easier.

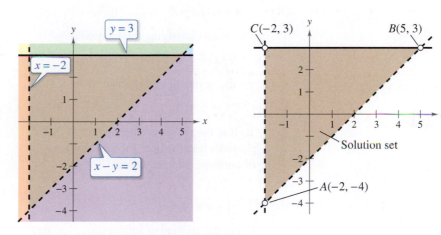

Note that the vertices of the region are represented by open dots. This means that the vertices *are not* solutions of the system of inequalities.

✓ **Checkpoint**  *Audio-video solution in English & Spanish at LarsonPrecalculus.com.*

Sketch the graph (and label the vertices) of the solution set of the system.

$$\begin{cases} x + y \geq 1 \\ -x + y \geq 1 \\ y \leq 2 \end{cases}$$

For the triangular region shown on page 371, each point of intersection of a pair of boundary lines corresponds to a vertex. With more complicated regions, two border lines can sometimes intersect at a point that is not a vertex of the region, as shown below. To keep track of which points of intersection are actually vertices of the region, you should sketch the region and refer to your sketch as you find each point of intersection.

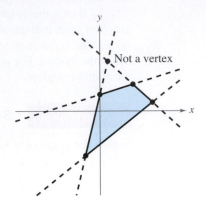

Not a vertex

EXAMPLE 5   **Solving a System of Inequalities**

Sketch the region containing all points that satisfy the system of inequalities.

$$\begin{cases} x^2 - y \le 1 & \text{Inequality 1} \\ -x + y \le 1 & \text{Inequality 2} \end{cases}$$

Solution   As shown in the figure below, the points that satisfy the inequality

$$x^2 - y \le 1 \qquad \text{Inequality 1}$$

are the points lying above (or on) the parabola

$$y = x^2 - 1. \qquad \text{Parabola}$$

The points satisfying the inequality

$$-x + y \le 1 \qquad \text{Inequality 2}$$

are the points lying below (or on) the line

$$y = x + 1. \qquad \text{Line}$$

To find the points of intersection of the parabola and the line, solve the system of corresponding equations.

$$\begin{cases} x^2 - y = 1 \\ -x + y = 1 \end{cases}$$

Using the method of substitution, you can find the solutions to be $(-1, 0)$ and $(2, 3)$. So, the region containing all points that satisfy the system is indicated by the shaded region at the right.

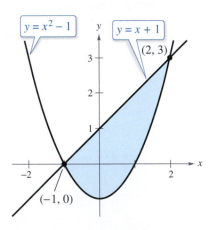

$y = x^2 - 1$ $y = x + 1$

$(2, 3)$

$(-1, 0)$

✓ *Checkpoint* ◀))) *Audio-video solution in English & Spanish at LarsonPrecalculus.com.*

Sketch the region containing all points that satisfy the system of inequalities.

$$\begin{cases} x - y^2 > 0 \\ x + y < 2 \end{cases}$$

When solving a system of inequalities, you should be aware that the system might have no solution *or* its graph might be an unbounded region in the plane. Examples 6 and 7 show these two possibilities.

EXAMPLE 6   **A System with No Solution**

Sketch the solution set of the system of inequalities.

$$\begin{cases} x + y > 3 & \text{Inequality 1} \\ x + y < -1 & \text{Inequality 2} \end{cases}$$

Solution   From the way the system is written, it should be clear that the system has no solution because the quantity $(x + y)$ cannot be both less than -1 and greater than 3. The graph of the inequality $x + y > 3$ is the half-plane lying above the line $x + y = 3$, and the graph of the inequality $x + y < -1$ is the half-plane lying below the line $x + y = -1$, as shown below. These two half-planes have no points in common. So, the system of inequalities has no solution.

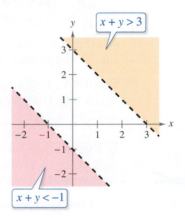

✓ **Checkpoint**   ◀))) *Audio-video solution in English & Spanish at LarsonPrecalculus.com.*

Sketch the solution set of the system of inequalities.

$$\begin{cases} 2x - y < -3 \\ 2x - y > 1 \end{cases}$$

EXAMPLE 7   **An Unbounded Solution Set**

Sketch the solution set of the system of inequalities.

$$\begin{cases} x + y < 3 & \text{Inequality 1} \\ x + 2y > 3 & \text{Inequality 2} \end{cases}$$

Solution   The graph of the inequality $x + y < 3$ is the half-plane that lies below the line $x + y = 3$. The graph of the inequality $x + 2y > 3$ is the half-plane that lies above the line $x + 2y = 3$. The intersection of these two half-planes is an *infinite wedge* that has a vertex at $(3, 0)$, as shown in Figure 7.18. So, the solution set of the system of inequalities is unbounded.

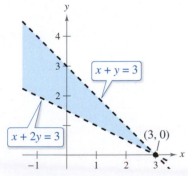

Figure 7.18

✓ **Checkpoint**   ◀))) *Audio-video solution in English & Spanish at LarsonPrecalculus.com.*

Sketch the solution set of the system of inequalities.

$$\begin{cases} x^2 - y < 0 \\ x - y < -2 \end{cases}$$

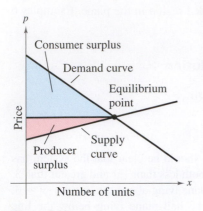

p

Consumer surplus

Demand curve

Equilibrium point

Price

Supply curve

Producer surplus

Number of units

x

Figure 7.19

Applications

Example 9 in Section 7.2 discussed the *equilibrium point* for a system of demand and supply equations. The next example discusses two related concepts that economists call **consumer surplus** and **producer surplus.** As shown in Figure 7.19, the consumer surplus is defined as the area of the region formed by the *demand* curve, the horizontal line passing through the equilibrium point, and the p-axis. Similarly, the producer surplus is defined as the area of the region formed by the *supply* curve, the horizontal line passing through the equilibrium point, and the p-axis. The consumer surplus is a measure of the amount that consumers would have been willing to pay *above what they actually paid,* whereas the producer surplus is a measure of the amount that producers would have been willing to receive *below what they actually received.*

EXAMPLE 8 **Consumer Surplus and Producer Surplus**

The demand and supply equations for a new type of video game console are

$$\begin{cases} p = 180 - 0.00001x & \text{Demand equation} \\ p = 90 + 0.00002x & \text{Supply equation} \end{cases}$$

where p is the price per unit (in dollars) and x is the number of units. Find the consumer surplus and producer surplus for these two equations.

Solution Begin by finding the equilibrium point (when supply and demand are equal) by solving the equation

$$90 + 0.00002x = 180 - 0.00001x.$$

In Example 9 in Section 7.2, you saw that the solution is $x = 3{,}000{,}000$ units, which corresponds to an equilibrium price of $p = \$150$. So, the consumer surplus and producer surplus are the areas of the following triangular regions.

Consumer Surplus	Producer Surplus
$\begin{cases} p \le 180 - 0.00001x \\ p \ge 150 \\ x \ge 0 \end{cases}$	$\begin{cases} p \ge 90 + 0.00002x \\ p \le 150 \\ x \ge 0 \end{cases}$

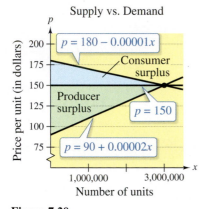

Supply vs. Demand

p

200

$p = 180 - 0.00001x$

175

Consumer surplus

150

Producer surplus

125

$p = 150$

100

75

$p = 90 + 0.00002x$

x

1,000,000   3,000,000

Number of units

Price per unit (in dollars)

Figure 7.20

The consumer and producer surpluses are the areas of the shaded triangles shown in Figure 7.20.

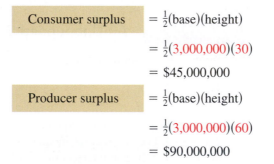

Consumer surplus $= \frac{1}{2}(\text{base})(\text{height})$

$= \frac{1}{2}(3{,}000{,}000)(30)$

$= \$45{,}000{,}000$

Producer surplus $= \frac{1}{2}(\text{base})(\text{height})$

$= \frac{1}{2}(3{,}000{,}000)(60)$

$= \$90{,}000{,}000$

✓ **Checkpoint**  Audio-video solution in English & Spanish at LarsonPrecalculus.com.

The demand and supply equations for a flat-screen television are

$$\begin{cases} p = 567 - 0.00002x & \text{Demand equation} \\ p = 492 + 0.00003x & \text{Supply equation} \end{cases}$$

where p is the price per unit (in dollars) and x is the number of units. Find the consumer surplus and producer surplus for these two equations.

A system of linear inequalities can help you determine amounts of dietary drinks to consume to meet or exceed daily requirements for calories and vitamins.

EXAMPLE 9 **Nutrition**

The liquid portion of a diet is to provide at least 300 calories, 36 units of vitamin A, and 90 units of vitamin C. A cup of dietary drink X provides 60 calories, 12 units of vitamin A, and 10 units of vitamin C. A cup of dietary drink Y provides 60 calories, 6 units of vitamin A, and 30 units of vitamin C. Set up a system of linear inequalities that describes how many cups of each drink should be consumed each day to meet or exceed the minimum daily requirements for calories and vitamins.

Solution Begin by letting x represent the number of cups of dietary drink X and y represent the number of cups of dietary drink Y. To meet or exceed the minimum daily requirements, the following inequalities must be satisfied.

$$\begin{cases} 60x + 60y \geq 300 & \text{Calories} \\ 12x + 6y \geq 36 & \text{Vitamin A} \\ 10x + 30y \geq 90 & \text{Vitamin C} \\ x \geq 0 \\ y \geq 0 \end{cases}$$

The last two inequalities are included because x and y cannot be negative. The graph of this system of inequalities is shown below. (More is said about this application in Example 6 in Section 7.6.)

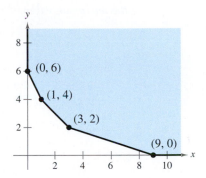

✓ *Checkpoint* *Audio-video solution in English & Spanish at LarsonPrecalculus.com.*

A pet supply company mixes two brands of dry dog food. A bag of brand X contains 8 units of nutrient A, 1 unit of nutrient B, and 2 units of nutrient C. A bag of brand Y contains 2 units of nutrient A, 1 unit of nutrient B, and 7 units of nutrient C. The minimum required amounts of nutrients A, B, and C are 16 units, 5 units, and 20 units, respectively. Set up a system of linear inequalities that describes how many bags of each brand should be mixed to meet or exceed the minimum required amounts of nutrients. ∎

Summarize (Section 7.5)

1. Explain how to sketch the graph of an inequality in two variables *(page 369)*. For examples of sketching the graphs of inequalities in two variables, see Examples 1–3.

2. Explain how to solve a system of inequalities *(page 371)*. For examples of solving systems of inequalities, see Examples 4–7.

3. Describe examples of how to use systems of inequalities in two variables to model and solve real-life problems *(pages 374 and 375, Examples 8 and 9)*.

7.6 Linear Programming

Linear programming is often useful in making real-life decisions, such as determining the optimal acreage and yield for two fruit crops.

■ Solve linear programming problems.
■ Use linear programming to model and solve real-life problems.

Linear Programming: A Graphical Approach

Many applications in business and economics involve a process called **optimization,** in which you find the minimum or maximum value of a quantity. In this section, you will study an optimization strategy called **linear programming.**

A two-dimensional linear programming problem consists of a linear **objective function** and a system of linear inequalities called **constraints.** The objective function gives the quantity to be maximized (or minimized), and the constraints determine the set of **feasible solutions.** For instance, suppose you want to maximize the value of

$$z = ax + by \qquad \text{Objective function}$$

subject to a set of constraints that determines the shaded region shown below. Because every point in the shaded region satisfies each constraint, it is not clear how you should find the point that yields a maximum value of z. Fortunately, it can be shown that when there is an optimal solution, it must occur at one of the vertices. This means that *to find the maximum value of z, test z at each of the vertices.*

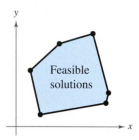

Feasible solutions

Optimal Solution of a Linear Programming Problem

If a linear programming problem has a solution, then it must occur at a vertex of the set of feasible solutions. If there is more than one solution, then at least one solution must occur at such a vertex. In either case, the value of the objective function is unique.

A linear programming problem can include hundreds, and sometimes even thousands, of variables. However, in this section, you will solve linear programming problems that involve only two variables. The guidelines for solving a linear programming problem in two variables are listed below.

Solving a Linear Programming Problem

1. Sketch the region corresponding to the system of constraints. (The points inside or on the boundary of the region are *feasible solutions.*)
2. Find the vertices of the region.
3. Test the objective function at each of the vertices and select the values of the variables that optimize the objective function. For a bounded region, both a minimum and a maximum value will exist. (For an unbounded region, *if* an optimal solution exists, then it will occur at a vertex.)

EXAMPLE 1 **Solving a Linear Programming Problem**

Find the maximum value of

$$z = 3x + 2y \qquad \text{Objective function}$$

subject to the following constraints.

$$\left.\begin{array}{r} x \geq 0 \\ y \geq 0 \\ x + 2y \leq 4 \\ x - y \leq 1 \end{array}\right\} \qquad \text{Constraints}$$

Solution The constraints form the region shown in Figure 7.21. At the four vertices of this region, the objective function has the following values.

At $(0, 0)$: $z = 3(0) + 2(0) = 0$

At $(0, 2)$: $z = 3(0) + 2(2) = 4$

At $(2, 1)$: $z = 3(2) + 2(1) = 8$ Maximum value of z

At $(1, 0)$: $z = 3(1) + 2(0) = 3$

So, the maximum value of z is 8, and this occurs when $x = 2$ and $y = 1$.

✓ **Checkpoint** ◀))) *Audio-video solution in English & Spanish at LarsonPrecalculus.com.*

Find the maximum value of

$$z = 4x + 5y$$

subject to the following constraints.

$$\begin{array}{r} x \geq 0 \\ y \geq 0 \\ x + y \leq 6 \end{array}$$

In Example 1, try testing some of the *interior* points in the region. You will see that the corresponding values of z are less than 8. Here are some examples.

At $(1, 1)$: $z = 3(1) + 2(1) = 5$

At $\left(1, \frac{1}{2}\right)$: $z = 3(1) + 2\left(\frac{1}{2}\right) = 4$

At $\left(\frac{1}{2}, \frac{3}{2}\right)$: $z = 3\left(\frac{1}{2}\right) + 2\left(\frac{3}{2}\right) = \frac{9}{2}$

At $\left(\frac{3}{2}, 1\right)$: $z = 3\left(\frac{3}{2}\right) + 2(1) = \frac{13}{2}$

To see why the maximum value of the objective function in Example 1 must occur at a vertex, consider writing the objective function in slope-intercept form

$$y = -\frac{3}{2}x + \frac{z}{2} \qquad \text{Family of lines}$$

where

$$\frac{z}{2}$$

is the y-intercept of the objective function. This equation represents a family of lines, each of slope $-\frac{3}{2}$. Of these infinitely many lines, you want the one that has the largest z-value while still intersecting the region determined by the constraints. In other words, of all the lines whose slope is $-\frac{3}{2}$, you want the one that has the largest y-intercept *and* intersects the given region, as shown in Figure 7.22. Notice from the graph that such a line will pass through one (or more) of the vertices of the region.

Figure 7.21

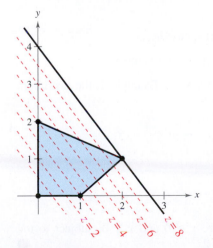

Figure 7.22

The next example shows that the same basic procedure can be used to solve a problem in which the objective function is to be *minimized*.

EXAMPLE 2 **Minimizing an Objective Function**

Find the minimum value of

$$z = 5x + 7y \qquad \text{Objective function}$$

where $x \geq 0$ and $y \geq 0$, subject to the following constraints.

$$\left.\begin{array}{rcl} 2x + 3y &\geq& 6 \\ 3x - y &\leq& 15 \\ -x + y &\leq& 4 \\ 2x + 5y &\leq& 27 \end{array}\right\} \qquad \text{Constraints}$$

Solution Figure 7.23 shows the region bounded by the constraints. By testing the objective function at each vertex, you obtain the following.

At $(0, 2)$: $z = 5(0) + 7(2) = 14$ Minimum value of z

At $(0, 4)$: $z = 5(0) + 7(4) = 28$

At $(1, 5)$: $z = 5(1) + 7(5) = 40$

At $(6, 3)$: $z = 5(6) + 7(3) = 51$

At $(5, 0)$: $z = 5(5) + 7(0) = 25$

At $(3, 0)$: $z = 5(3) + 7(0) = 15$

So, the minimum value of z is 14, and this occurs when $x = 0$ and $y = 2$.

✓ **Checkpoint** Audio-video solution in English & Spanish at LarsonPrecalculus.com.

Find the minimum value of

$$z = 12x + 8y$$

where $x \geq 0$ and $y \geq 0$, subject to the following constraints.

$$\begin{array}{rcl} 5x + 6y &\leq& 420 \\ -x + 6y &\leq& 240 \\ -2x + y &\geq& -100 \end{array}$$

EXAMPLE 3 **Maximizing an Objective Function**

Find the maximum value of

$$z = 5x + 7y \qquad \text{Objective function}$$

where $x \geq 0$ and $y \geq 0$, subject to the constraints given in Example 2.

Solution Using the values of z at the vertices shown in Example 2, the maximum value of z is

$$z = 5(6) + 7(3)$$

$$= 51$$

and occurs when $x = 6$ and $y = 3$.

✓ **Checkpoint** Audio-video solution in English & Spanish at LarsonPrecalculus.com.

Find the maximum value of $z = 12x + 8y$, where $x \geq 0$ and $y \geq 0$, subject to the constraints given in the Checkpoint with Example 2.

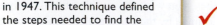

Figure 7.23

George Dantzig (1914–2005) was the first to propose the simplex method, or linear programming, in 1947. This technique defined the steps needed to find the optimal solution to a complex multivariable problem.

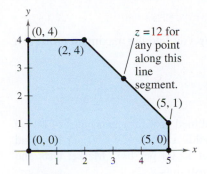

Figure 7.24

It is possible for the maximum (or minimum) value in a linear programming problem to occur at *two* different vertices. For instance, at the vertices of the region shown in Figure 7.24, the objective function

$$z = 2x + 2y \qquad \text{Objective function}$$

has the following values.

At $(0, 0)$: $z = 2(0) + 2(0) = 0$

At $(0, 4)$: $z = 2(0) + 2(4) = 8$

At $(2, 4)$: $z = 2(2) + 2(4) = 12$ Maximum value of z

At $(5, 1)$: $z = 2(5) + 2(1) = 12$ Maximum value of z

At $(5, 0)$: $z = 2(5) + 2(0) = 10$

In this case, the objective function has a maximum value not only at the vertices $(2, 4)$ and $(5, 1)$; it also has a maximum value (of 12) at *any point on the line segment connecting these two vertices*. Note that the objective function in slope-intercept form $y = -x + \frac{1}{2}z$ has the same slope as the line through the vertices $(2, 4)$ and $(5, 1)$.

Some linear programming problems have no optimal solutions. This can occur when the region determined by the constraints is *unbounded*. Example 4 illustrates such a problem.

EXAMPLE 4   **An Unbounded Region**

Find the maximum value of

$$z = 4x + 2y \qquad \text{Objective function}$$

where $x \geq 0$ and $y \geq 0$, subject to the following constraints.

$$\left. \begin{array}{r} x + 2y \geq 4 \\ 3x + y \geq 7 \\ -x + 2y \leq 7 \end{array} \right\} \qquad \text{Constraints}$$

Solution   Figure 7.25 shows the region determined by the constraints. For this unbounded region, there is no maximum value of z. To see this, note that the point $(x, 0)$ lies in the region for all values of $x \geq 4$. Substituting this point into the objective function, you get

$$z = 4(x) + 2(0) = 4x.$$

By choosing x to be large, you can obtain values of z that are as large as you want. So, there is no maximum value of z. However, there *is* a minimum value of z.

At $(1, 4)$: $z = 4(1) + 2(4) = 12$

At $(2, 1)$: $z = 4(2) + 2(1) = 10$ Minimum value of z

At $(4, 0)$: $z = 4(4) + 2(0) = 16$

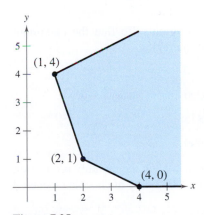

Figure 7.25

So, the minimum value of z is 10, and this occurs when $x = 2$ and $y = 1$.

✓ *Checkpoint*   Audio-video solution in English & Spanish at LarsonPrecalculus.com.

Find the minimum value of

$$z = 3x + 7y$$

where $x \geq 0$ and $y \geq 0$, subject to the following constraints.

$$\begin{array}{r} x + y \geq 8 \\ 3x + 5y \geq 30 \end{array}$$

Applications

Example 5 shows how linear programming can help you find the maximum profit in a business application.

EXAMPLE 5 **Optimal Profit**

A candy manufacturer wants to maximize the combined profit for two types of boxed chocolates. A box of chocolate-covered creams yields a profit of $1.50 per box, and a box of chocolate-covered nuts yields a profit of $2.00 per box. Market tests and available resources have indicated the following constraints.

1. The combined production level should not exceed 1200 boxes per month.

2. The demand for a box of chocolate-covered nuts is no more than half the demand for a box of chocolate-covered creams.

3. The production level for chocolate-covered creams should be less than or equal to 600 boxes plus three times the production level for chocolate-covered nuts.

What is the maximum monthly profit? How many boxes of each type should be produced per month to yield the maximum profit?

Solution   Let x be the number of boxes of chocolate-covered creams and let y be the number of boxes of chocolate-covered nuts. So, the objective function (for the combined profit) is

$$P = 1.5x + 2y. \qquad \text{\color{red}{Objective function}}$$

The three constraints translate into the following linear inequalities.

1. $x + y \le 1200$ $\Longrightarrow$ $x + y \le 1200$

2. $y \le \frac{1}{2}x$ $\Longrightarrow$ $-x + 2y \le 0$

3. $x \le 600 + 3y$ $x - 3y \le 600$

Because neither x nor y can be negative, you also have the two additional constraints of

$$x \ge 0 \quad \text{and} \quad y \ge 0.$$

Figure 7.26 shows the region determined by the constraints. To find the maximum monthly profit, test the values of P at the vertices of the region.

At $(0, 0)$: $\quad P = 1.5(0) \quad + 2(0) \quad = \quad 0$

At $(800, 400)$: $\quad P = 1.5(800) \quad + 2(400) = 2000$ $\qquad$ {\color{red}{Maximum profit}}

At $(1050, 150)$: $\quad P = 1.5(1050) + 2(150) = 1875$

At $(600, 0)$: $\quad P = 1.5(600) \quad + 2(0) \quad = \quad 900$

So, the maximum monthly profit is $2000, and it occurs when the monthly production consists of 800 boxes of chocolate-covered creams and 400 boxes of chocolate-covered nuts.

✓ **Checkpoint**))) *Audio-video solution in English & Spanish at LarsonPrecalculus.com.*

In Example 5, the candy manufacturer improves the production of chocolate-covered creams so that the profit is $2.50 per box. What is the maximum monthly profit? How many boxes of each type should be produced per month to yield the maximum profit? ■

Example 6 shows how linear programming can help you find the optimal cost in a real-life application.

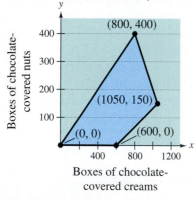

Maximum Monthly Profit

Boxes of chocolate-covered nuts

(800, 400)

(1050, 150)

(0, 0) (600, 0)

Boxes of chocolate-covered creams

Figure 7.26

EXAMPLE 6 **Optimal Cost**

The liquid portion of a diet is to provide at least 300 calories, 36 units of vitamin A, and 90 units of vitamin C. A cup of dietary drink X costs $0.12 and provides 60 calories, 12 units of vitamin A, and 10 units of vitamin C. A cup of dietary drink Y costs $0.15 and provides 60 calories, 6 units of vitamin A, and 30 units of vitamin C. How many cups of each drink should be consumed each day to obtain an optimal cost and still meet the daily requirements?

Solution As in Example 9 in Section 7.5, let x be the number of cups of dietary drink X and let y be the number of cups of dietary drink Y.

$$\left.\begin{array}{lrcl} \text{For calories:} & 60x + 60y &\geq& 300 \\ \text{For vitamin A:} & 12x + 6y &\geq& 36 \\ \text{For vitamin C:} & 10x + 30y &\geq& 90 \\ & x &\geq& 0 \\ & y &\geq& 0 \end{array}\right\} \quad \text{Constraints}$$

The cost C is given by $C = 0.12x + 0.15y$.   Objective function

Figure 7.27 shows the graph of the region corresponding to the constraints. Because you want to incur as little cost as possible, you want to determine the *minimum* cost. To determine the minimum cost, test C at each vertex of the region.

At $(0, 6)$: $C = 0.12(0) + 0.15(6) = 0.90$

At $(1, 4)$: $C = 0.12(1) + 0.15(4) = 0.72$

At $(3, 2)$: $C = 0.12(3) + 0.15(2) = 0.66$ Minimum value of C

At $(9, 0)$: $C = 0.12(9) + 0.15(0) = 1.08$

So, the minimum cost is $0.66 per day, and this occurs when 3 cups of drink X and 2 cups of drink Y are consumed each day.

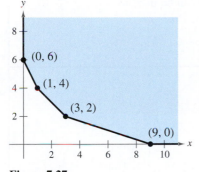

Figure 7.27

✓ *Checkpoint* 🔊)) *Audio-video solution in English & Spanish at LarsonPrecalculus.com.*

A pet supply company mixes two brands of dry dog food. Brand X costs $15 per bag and contains eight units of nutrient A, one unit of nutrient B, and two units of nutrient C. Brand Y costs $30 per bag and contains two units of nutrient A, one unit of nutrient B, and seven units of nutrient C. The minimum required amounts of nutrients A, B, and C are 16 units, 5 units, and 20 units, respectively. How many bags of each brand should be mixed to obtain an optimal cost and still meet the minimum required amounts of nutrients? ◼

Summarize (Section 7.6)

1. State the guidelines for solving a linear programming problem *(page 376)*. For an example of solving a linear programming problem, see Example 1.

2. Explain how to find the minimum and maximum values of an objective function of a linear programming problem *(pages 377 and 378)*. For an example of minimizing an objective function, see Example 2. For examples of maximizing objective functions, see Examples 1 and 3.

3. Describe a situation in which a linear programming problem has no optimal solution *(page 379)*. For an example of solving a linear programming problem that has no optimal solution, see Example 4.

4. Describe real-life examples of how to use linear programming to find the maximum profit in a business application *(page 380, Example 5)* and the optimal cost of diet drinks *(page 381, Example 6)*.

Chapter Summary

	What Did You Learn?	**Explanation/Examples**	**Review Exercises**
Section 7.1	Use the method of substitution to solve systems of linear equations in two variables *(p. 340)*.	**Method of Substitution** 1. *Solve* one of the equations for one variable in terms of the other. 2. *Substitute* the expression found in Step 1 into the other equation to obtain an equation in one variable. 3. *Solve* the equation obtained in Step 2. 4. *Back-substitute* the value obtained in Step 3 into the expression obtained in Step 1 to find the value of the other variable. 5. *Check* that the solution satisfies *each* of the original equations.	1–6
	Use the method of substitution to solve systems of nonlinear equations in two variables *(p. 343)*.	The method of substitution (see steps above) can be used to solve systems in which one or both of the equations are nonlinear. (See Examples 3 and 4.)	7–10
	Use a graphical approach to solve systems of equations in two variables *(p. 344)*.	 One intersection point Two intersection points No intersection points	11–18
	Use systems of equations to model and solve real-life problems *(p. 345)*.	A system of equations can help you find the break-even point for a company. (See Example 6.)	19–22
Section 7.2	Use the method of elimination to solve systems of linear equations in two variables *(p. 347)*.	**Method of Elimination** 1. *Obtain coefficients* for x (or y) that differ only in sign. 2. *Add* the equations to eliminate one variable. 3. *Solve* the equation obtained in Step 2. 4. *Back-substitute* the value obtained in Step 3 into either of the original equations and solve for the other variable. 5. *Check* that the solution satisfies *each* of the original equations.	23–28
	Interpret graphically the numbers of solutions of systems of linear equations in two variables *(p. 351)*.	 Exactly one solution Infinitely many solutions No solution	29–32
	Use systems of linear equations in two variables to model and solve real-life problems *(p. 353)*.	A system of linear equations in two variables can help you find the equilibrium point for a market. (See Example 9.)	33, 34

	What Did You Learn?	**Explanation/Examples**	**Review Exercises**
Section 7.3	Use back-substitution to solve linear systems in row-echelon form (p. 355).	**Row-Echelon Form** $$\begin{cases} x - 2y + 3z = 9 \\ -x + 3y = -4 \\ 2x - 5y + 5z = 17 \end{cases} \implies \begin{cases} x - 2y + 3z = 9 \\ y + 3z = 5 \\ z = 2 \end{cases}$$	35–38
	Use Gaussian elimination to solve systems of linear equations (p. 356).	To produce an equivalent system of linear equations, use row operations by (1) interchanging two equations, (2) multiplying one equation by a nonzero constant, or (3) adding a multiple of one equation to another equation to replace the latter equation.	39–42
	Solve nonsquare systems of linear equations (p. 360).	In a nonsquare system, the number of equations differs from the number of variables. A system of linear equations cannot have a unique solution unless there are at least as many equations as there are variables.	43, 44
	Use systems of linear equations in three or more variables to model and solve real-life problems (p. 361).	A system of linear equations in three variables can help you find the position equation of an object that is moving in a (vertical) line with constant acceleration. (See Example 7.)	45–54
Section 7.4	Recognize partial fraction decompositions of rational expressions (p. 363).	$$\frac{9}{x^3 - 6x^2} = \frac{9}{x^2(x - 6)}$$ $$= \frac{A}{x} + \frac{B}{x^2} + \frac{C}{x - 6}$$	55–58
	Find partial fraction decompositions of rational expressions (p. 364).	The techniques used for determining constants in the numerators of partial fractions vary slightly, depending on the type of factors of the denominator: linear or quadratic, distinct or repeated.	59–66
Section 7.5	Sketch the graphs of inequalities in two variables (p. 369), and solve systems of inequalities (p. 371).	$$\begin{cases} x^2 + y \le 5 \\ x \ge -1 \\ y \ge 0 \end{cases}$$	67–78
	Use systems of inequalities in two variables to model and solve real-life problems (p. 374).	A system of inequalities in two variables can help you find the consumer surplus and producer surplus for given demand and supply equations. (See Example 8.)	79–84
Section 7.6	Solve linear programming problems (p. 376).	To solve a linear programming problem, (1) sketch the region corresponding to the system of constraints, (2) find the vertices of the region, and (3) test the objective function at each of the vertices and select the values of the variables that optimize the objective function.	85–88
	Use linear programming to model and solve real-life problems (p. 380).	Linear programming can help you find the maximum profit in business applications. (See Example 5.)	89, 90

Review Exercises

See **CalcChat.com** for tutorial help and worked-out solutions to odd-numbered exercises.

7.1 Solving a System by Substitution In Exercises 1–10, solve the system by the method of substitution.

1. $\begin{cases} x + y = 2 \\ x - y = 0 \end{cases}$

2. $\begin{cases} 2x - 3y = 3 \\ x - y = 0 \end{cases}$

3. $\begin{cases} 4x - y - 1 = 0 \\ 8x + y - 17 = 0 \end{cases}$

4. $\begin{cases} 10x + 6y + 14 = 0 \\ x + 9y + 7 = 0 \end{cases}$

5. $\begin{cases} 0.5x + y = 0.75 \\ 1.25x - 4.5y = -2.5 \end{cases}$

6. $\begin{cases} -x + \frac{2}{5}y = \frac{3}{5} \\ -x + \frac{1}{5}y = -\frac{4}{5} \end{cases}$

7. $\begin{cases} x^2 - y^2 = 9 \\ x - y = 1 \end{cases}$

8. $\begin{cases} x^2 + y^2 = 169 \\ 3x + 2y = 39 \end{cases}$

9. $\begin{cases} y = 2x^2 \\ y = x^4 - 2x^2 \end{cases}$

10. $\begin{cases} x = y + 3 \\ x = y^2 + 1 \end{cases}$

Solving a System of Equations Graphically In Exercises 11–14, solve the system graphically.

11. $\begin{cases} 2x - y = 10 \\ x + 5y = -6 \end{cases}$

12. $\begin{cases} 8x - 3y = -3 \\ 2x + 5y = 28 \end{cases}$

13. $\begin{cases} y = 2x^2 - 4x + 1 \\ y = x^2 - 4x + 3 \end{cases}$

14. $\begin{cases} y^2 - 2y + x = 0 \\ x + y = 0 \end{cases}$

Solving a System of Equations Graphically In Exercises 15–18, use a graphing utility to solve the systems of equations. Find the solution(s) accurate to two decimal places.

15. $\begin{cases} y = -2e^{-x} \\ 2e^x + y = 0 \end{cases}$

16. $\begin{cases} x^2 + y^2 = 100 \\ 2x - 3y = -12 \end{cases}$

17. $\begin{cases} y = 2 + \log x \\ y = \frac{3}{4}x + 5 \end{cases}$

18. $\begin{cases} y = \ln(x - 1) - 3 \\ y = 4 - \frac{1}{2}x \end{cases}$

19. Body Mass Index Body Mass Index (BMI) is a measure of body fat based on height and weight. The 85th percentile BMI for females, ages 9 to 20, is growing more slowly than that for males of the same age range. Models that represent the 85th percentile BMI for males and females, ages 9 to 20, are

$$\begin{cases} B = 0.77a + 11.7 & \text{Males} \\ B = 0.68a + 13.5 & \text{Females} \end{cases}$$

where B is the BMI $\left(\text{kg/m}^2\right)$ and a represents the age, with $a = 9$ corresponding to 9 years old. Use a graphing utility to determine whether the BMI for males exceeds the BMI for females. *(Source: Centers for Disease Control and Prevention)*

20. Choice of Two Jobs You receive two sales job offers. One company offers an annual salary of $55,000 plus a year-end bonus of 1.5% of your total sales. The other company offers an annual salary of $52,000 plus a year-end bonus of 2% of your total sales. What amount of sales will make the second job offer better? Explain.

21. Geometry The perimeter of a rectangle is 68 feet and its width is $\frac{8}{9}$ times its length. Find the dimensions of the rectangle.

22. Geometry The perimeter of a rectangle is 40 inches. The area of the rectangle is 96 square inches. Find the dimensions of the rectangle.

7.2 Solving a System by Elimination In Exercises 23–28, solve the system by the method of elimination and check any solutions algebraically.

23. $\begin{cases} 2x - y = 2 \\ 6x + 8y = 39 \end{cases}$

24. $\begin{cases} 12x + 42y = -17 \\ 30x - 18y = 19 \end{cases}$

25. $\begin{cases} 3x - 2y = 0 \\ 3x + 2(y + 5) = 10 \end{cases}$

26. $\begin{cases} 7x + 12y = 63 \\ 2x + 3(y + 2) = 21 \end{cases}$

27. $\begin{cases} 1.25x - 2y = 3.5 \\ 5x - 8y = 14 \end{cases}$

28. $\begin{cases} 1.5x + 2.5y = 8.5 \\ 6x + 10y = 24 \end{cases}$

Matching a System with Its Graph In Exercises 29–32, match the system of linear equations with its graph. Describe the number of solutions and state whether the system is consistent or inconsistent. [The graphs are labeled (a), (b), (c), and (d).]

(a)

(b)
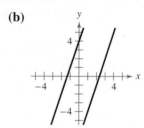

(c)
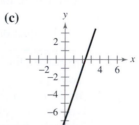

(d)
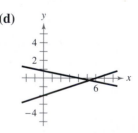

29. $\begin{cases} x + 5y = 4 \\ x - 3y = 6 \end{cases}$

30. $\begin{cases} -3x + y = -7 \\ 9x - 3y = 21 \end{cases}$

31. $\begin{cases} 3x - y = 7 \\ -6x + 2y = 8 \end{cases}$

32. $\begin{cases} 2x - y = -3 \\ x + 5y = 4 \end{cases}$

Finding the Equilibrium Point In Exercises 33 and 34, find the equilibrium point of the demand and supply equations.

Demand	Supply
33. $p = 43 - 0.0002x$	$p = 22 + 0.00001x$
34. $p = 120 - 0.0001x$	$p = 45 + 0.0002x$

7.3 Using Back-Substitution In Exercises 35–38, use back-substitution to solve the system of linear equations.

35. $\begin{cases} x - 4y + 3z = 3 \\ -y + z = -1 \\ z = -5 \end{cases}$

36. $\begin{cases} x - 7y + 8z = 85 \\ y - 9z = -35 \\ z = 3 \end{cases}$

37. $\begin{cases} 4x - 3y - 2z = -65 \\ 8y - 7z = -14 \\ z = 10 \end{cases}$

38. $\begin{cases} 5x - 7z = 9 \\ 3y - 8z = -4 \\ z = -7 \end{cases}$

Using Gaussian Elimination In Exercises 39–42, use Gaussian elimination to solve the system of linear equations and check any solutions algebraically.

39. $\begin{cases} x + 2y + 6z = 4 \\ -3x + 2y - z = -4 \\ 4x + 2z = 16 \end{cases}$

40. $\begin{cases} x - 2y + z = -6 \\ 2x - 3y = -7 \\ -x + 3y - 3z = 11 \end{cases}$

41. $\begin{cases} 2x + 6z = -9 \\ 3x - 2y + 11z = -16 \\ 3x - y + 7z = -11 \end{cases}$

42. $\begin{cases} x + 4w = 1 \\ 3y + z - w = 4 \\ 2y - 3w = 2 \\ 4x - y + 2z = 5 \end{cases}$

Solving a Nonsquare System In Exercises 43 and 44, solve the nonsquare system of equations.

43. $\begin{cases} 5x - 12y + 7z = 16 \\ 3x - 7y + 4z = 9 \end{cases}$

44. $\begin{cases} 2x + 5y - 19z = 34 \\ 3x + 8y - 31z = 54 \end{cases}$

Finding the Equation of a Parabola In Exercises 45 and 46, find the equation of the parabola $y = ax^2 + bx + c$ that passes through the points. Verify your result using a graphing utility.

45.

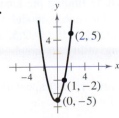

46.

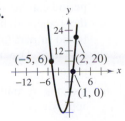

Finding the Equation of a Circle In Exercises 47 and 48, find the equation of the circle

$$x^2 + y^2 + Dx + Ey + F = 0$$

that passes through the points. Verify your result using a graphing utility.

47.

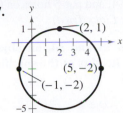

48.

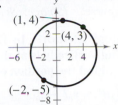

49. **Data Analysis: E-Commerce** The table shows the retail e-commerce sales y (in billions of dollars) in the United States from 2007 through 2009. (*Source: U.S. Census Bureau*)

Year	2007	2008	2009
Retail E-Commerce Sales, y	126.7	141.9	145.2

(a) Find the least squares regression parabola $y = at^2 + bt + c$ for the data, where t represents the year with $t = 7$ corresponding to 2007, by solving the system for a, b, and c.

$$\begin{cases} 3c + 24b + 194a = 413.8 \\ 24c + 194b + 1584a = 3328.9 \\ 194c + 1584b + 13{,}058a = 27{,}051.1 \end{cases}$$

(b) Use a graphing utility to graph the parabola and the data in the same viewing window. How well does the model fit the data?

(c) Use the model to estimate the retail e-commerce sales in the United States in 2015. Does your answer seem reasonable?

50. **Agriculture** A mixture of 6 gallons of chemical A, 8 gallons of chemical B, and 13 gallons of chemical C is required to kill a destructive crop insect. Commercial spray X contains 1, 2, and 2 parts, respectively, of these chemicals. Commercial spray Y contains only chemical C. Commercial spray Z contains chemicals A, B, and C in equal amounts. How much of each type of commercial spray gives the desired mixture?

51. **Investment Analysis** An inheritance of $40,000 was divided among three investments yielding $3500 in interest per year. The interest rates for the three investments were 7%, 9%, and 11%. Find the amount placed in each investment when the second and third were $3000 and $5000 less than the first, respectively.

52. Sports The Old Course at St Andrews Links in St Andrews, Scotland, is one of the oldest golf courses in the world. It is an 18-hole course that consists of par-3 holes, par-4 holes, and par-5 holes. There are seven times as many par-4 holes as par-5 holes, and the sum of the numbers of par-3 and par-5 holes is four. Find the numbers of par-3, par-4, and par-5 holes on the course. *(Source: St Andrews Links Trust)*

Vertical Motion In Exercises 53 and 54, an object moving vertically is at the given heights at the specified times. Find the position equation

$$s = \tfrac{1}{2}at^2 + v_0 t + s_0$$

for the object.

53. At $t = 1$ second, $s = 134$ feet
At $t = 2$ seconds, $s = 86$ feet
At $t = 3$ seconds, $s = 6$ feet

54. At $t = 1$ second, $s = 184$ feet
At $t = 2$ seconds, $s = 116$ feet
At $t = 3$ seconds, $s = 16$ feet

7.4 **Writing the Form of the Decomposition** In Exercises 55–58, write the form of the partial fraction decomposition of the rational expression. Do not solve for the constants.

55. $\dfrac{3}{x^2 + 20x}$ **56.** $\dfrac{x - 8}{x^2 - 3x - 28}$

57. $\dfrac{3x - 4}{x^3 - 5x^2}$ **58.** $\dfrac{x - 2}{x(x^2 + 2)^2}$

Writing the Partial Fraction Decomposition In Exercises 59–66, write the partial fraction decomposition of the rational expression. Check your result algebraically.

59. $\dfrac{4 - x}{x^2 + 6x + 8}$ **60.** $\dfrac{-x}{x^2 + 3x + 2}$

61. $\dfrac{x^2}{x^2 + 2x - 15}$ **62.** $\dfrac{9}{x^2 - 9}$

63. $\dfrac{x^2 + 2x}{x^3 - x^2 + x - 1}$ **64.** $\dfrac{4x}{3(x - 1)^2}$

65. $\dfrac{3x^2 + 4x}{(x^2 + 1)^2}$ **66.** $\dfrac{4x^2}{(x - 1)(x^2 + 1)}$

7.5 **Graphing an Inequality** In Exercises 67–72, sketch the graph of the inequality.

67. $y \le 5 - \tfrac{1}{2}x$ **68.** $3y - x \ge 7$

69. $y - 4x^2 > -1$ **70.** $y \le \dfrac{3}{x^2 + 2}$

71. $(x - 1)^2 + (y - 3)^2 < 16$
72. $x^2 + (y + 5)^2 > 1$

Solving a System of Inequalities In Exercises 73–78, sketch the graph (and label the vertices) of the solution set of the system of inequalities.

73. $\begin{cases} x + 2y \le 160 \\ 3x + y \le 180 \\ x \ge 0 \\ y \ge 0 \end{cases}$ **74.** $\begin{cases} 3x + 2y \ge 24 \\ x + 2y \ge 12 \\ 2 \le x \le 15 \\ y \le 15 \end{cases}$

75. $\begin{cases} y < x + 1 \\ y > x^2 - 1 \end{cases}$ **76.** $\begin{cases} y \le 6 - 2x - x^2 \\ y \ge x + 6 \end{cases}$

77. $\begin{cases} 2x - 3y \ge 0 \\ 2x - y \le 8 \\ y \ge 0 \end{cases}$ **78.** $\begin{cases} x^2 + y^2 \le 9 \\ (x - 3)^2 + y^2 \le 9 \end{cases}$

Consumer Surplus and Producer Surplus In Exercises 79 and 80, (a) graph the systems representing the consumer surplus and producer surplus for the supply and demand equations and (b) find the consumer surplus and producer surplus.

	Demand	Supply
79.	$p = 160 - 0.0001x$	$p = 70 + 0.0002x$
80.	$p = 130 - 0.0002x$	$p = 30 + 0.0003x$

81. Geometry Write a system of inequalities to describe the region of a rectangle with vertices at $(3, 1)$, $(7, 1)$, $(7, 10)$, and $(3, 10)$.

82. Data Analysis: Sales The table shows the sales y (in millions of dollars) for Aéropostale, Inc., from 2003 through 2010. *(Source: Aéropostale, Inc.)*

Year	Sales, y
2003	734.9
2004	964.2
2005	1204.3
2006	1413.2
2007	1590.9
2008	1885.5
2009	2230.1
2010	2400.4

Spreadsheet at LarsonPrecalculus.com

(a) Use the *regression* feature of a graphing utility to find a linear model for the data. Let t represent the year, with $t = 3$ corresponding to 2003.

(b) The total sales for Aéropostale during this eight-year period can be approximated by finding the area of the trapezoid bounded by the linear model you found in part (a) and the lines $y = 0$, $t = 2.5$, and $t = 10.5$. Use the graphing utility to graph this region.

(c) Use the formula for the area of a trapezoid to approximate the total sales for Aéropostale.

83. Inventory Costs A warehouse operator has 24,000 square feet of floor space in which to store two products. Each unit of product I requires 20 square feet of floor space and costs $12 per day to store. Each unit of product II requires 30 square feet of floor space and costs $8 per day to store. The total storage cost per day cannot exceed $12,400. Find and graph a system of inequalities describing all possible inventory levels.

84. Nutrition A dietitian designs a special dietary supplement using two different foods. Each ounce of food X contains 12 units of calcium, 10 units of iron, and 20 units of vitamin B. Each ounce of food Y contains 15 units of calcium, 20 units of iron, and 12 units of vitamin B. The minimum daily requirements of the diet are 300 units of calcium, 280 units of iron, and 300 units of vitamin B.

(a) Write a system of inequalities describing the different amounts of food X and food Y that can be used.

(b) Sketch a graph of the region in part (a).

(c) Find two solutions to the system and interpret their meanings in the context of the problem.

7.6 Solving a Linear Programming Problem In Exercises 85–88, sketch the region determined by the constraints. Then find the minimum and maximum values of the objective function (if possible) and where they occur, subject to the indicated constraints.

85. Objective function:

$z = 3x + 4y$

Constraints:

$x \geq 0$
$y \geq 0$
$2x + 5y \leq 50$
$4x + y \leq 28$

86. Objective function:

$z = 10x + 7y$

Constraints:

$x \geq 0$
$y \geq 0$
$2x + y \geq 100$
$x + y \geq 75$

87. Objective function:

$z = 1.75x + 2.25y$

Constraints:

$x \geq 0$
$y \geq 0$
$2x + y \geq 25$
$3x + 2y \geq 45$

88. Objective function:

$z = 50x + 70y$

Constraints:

$x \geq 0$
$y \geq 0$
$x + 2y \leq 1500$
$5x + 2y \leq 3500$

89. Optimal Revenue A student is working part time as a hairdresser to pay college expenses. The student may work no more than 24 hours per week. Haircuts cost $25 and require an average of 20 minutes, and permanents cost $70 and require an average of 1 hour and 10 minutes. How many haircuts and/or permanents will yield an optimal revenue? What is the optimal revenue?

90. Optimal Profit A manufacturer produces two models of bicycles. The table shows the times (in hours) required for assembling, painting, and packaging each model.

Process	Hours, Model A	Hours, Model B
Assembling	2	2.5
Painting	4	1
Packaging	1	0.75

The total times available for assembling, painting, and packaging are 4000 hours, 4800 hours, and 1500 hours, respectively. The profits per unit are $45 for model A and $50 for model B. What is the optimal production level for each model? What is the optimal profit?

Exploration

True or False? In Exercises 91 and 92, determine whether the statement is true or false. Justify your answer.

91. The system

$$\begin{cases} y \leq 5 \\ y \geq -2 \\ y \leq \frac{7}{2}x - 9 \\ y \leq -\frac{7}{2}x + 26 \end{cases}$$

represents the region covered by an isosceles trapezoid.

92. It is possible for an objective function of a linear programming problem to have exactly 10 maximum value points.

Finding a System of Linear Equations In Exercises 93–96, find a system of linear equations having the ordered pair as a solution. (There are many correct answers.)

93. $(-8, 10)$ **94.** $(5, -4)$

95. $\left(\frac{4}{3}, 3\right)$ **96.** $\left(-2, \frac{11}{5}\right)$

Finding a System of Linear Equations In Exercises 97–100, find a system of linear equations having the ordered triple as a solution. (There are many correct answers.)

97. $(4, -1, 3)$ **98.** $(-3, 5, 6)$

99. $\left(5, \frac{3}{2}, 2\right)$ **100.** $\left(-\frac{1}{2}, -2, -\frac{3}{4}\right)$

101. Writing Explain what is meant by an inconsistent system of linear equations.

102. Graphical Reasoning How can you tell graphically that a system of linear equations in two variables has no solution? Give an example.

Chapter Test

See CalcChat.com for tutorial help and worked-out solutions to odd-numbered exercises.

Take this test as you would take a test in class. When you are finished, check your work against the answers given in the back of the book.

In Exercises 1–3, solve the system by the method of substitution.

1. $\begin{cases} x + y = -9 \\ 5x - 8y = 20 \end{cases}$

2. $\begin{cases} y = x - 1 \\ y = (x - 1)^3 \end{cases}$

3. $\begin{cases} 2x - y^2 = 0 \\ x - y = 4 \end{cases}$

In Exercises 4–6, solve the system graphically.

4. $\begin{cases} 3x - 6y = 0 \\ 3x + 6y = 18 \end{cases}$

5. $\begin{cases} y = 9 - x^2 \\ y = x + 3 \end{cases}$

6. $\begin{cases} y - \ln x = 12 \\ 7x - 2y + 11 = -6 \end{cases}$

In Exercises 7 and 8, solve the system by the method of elimination.

7. $\begin{cases} 3x + 4y = -26 \\ 7x - 5y = 11 \end{cases}$

8. $\begin{cases} 1.4x - y = 17 \\ 0.8x + 6y = -10 \end{cases}$

In Exercises 9 and 10, solve the system of linear equations and check any solutions algebraically.

9. $\begin{cases} x - 2y + 3z = 11 \\ 2x - z = 3 \\ 3y + z = -8 \end{cases}$

10. $\begin{cases} 3x + 2y + z = 17 \\ -x + y + z = 4 \\ x - y - z = 3 \end{cases}$

In Exercises 11–14, write the partial fraction decomposition of the rational expression. Check your result algebraically.

11. $\dfrac{2x + 5}{x^2 - x - 2}$

12. $\dfrac{3x^2 - 2x + 4}{x^2(2 - x)}$

13. $\dfrac{x^2 + 5}{x^3 - x}$

14. $\dfrac{x^2 - 4}{x^3 + 2x}$

In Exercises 15–17, sketch the graph (and label the vertices) of the solution set of the system of inequalities.

15. $\begin{cases} 2x + y \le 4 \\ 2x - y \ge 0 \\ x \ge 0 \end{cases}$

16. $\begin{cases} y < -x^2 + x + 4 \\ y > 4x \end{cases}$

17. $\begin{cases} x^2 + y^2 \le 36 \\ x \ge 2 \\ y \ge -4 \end{cases}$

18. Find the minimum and maximum values of the objective function $z = 20x + 12y$ and where they occur, subject to the following constraints.

$$\left. \begin{array}{r} x \ge 0 \\ y \ge 0 \\ x + 4y \le 32 \\ 3x + 2y \le 36 \end{array} \right\} \quad \text{Constraints}$$

19. A total of $50,000 is invested in two funds that pay 4% and 5.5% simple interest. The yearly interest is $2390. How much is invested at each rate?

20. Find the equation of the parabola $y = ax^2 + bx + c$ that passes through the points $(0, 6)$, $(-2, 2)$, and $\left(3, \frac{9}{2}\right)$.

21. A manufacturer produces two models of television stands. The table at the left shows the times (in hours) required for assembling, staining, and packaging the two models. The total times available for assembling, staining, and packaging are 4000 hours, 8950 hours, and 2650 hours, respectively. The profits per unit are $30 for model I and $40 for model II. What is the optimal inventory level for each model? What is the optimal profit?

	Model I	Model II
Assembling	0.5	0.75
Staining	2.0	1.5
Packaging	0.5	0.5

Table for 21

Proofs in Mathematics ■ ■ ■ ■ ■ ■ ■ ■ ■ ■ ■ ■ ■

An **indirect proof** can be useful in proving statements of the form "p implies q." Recall that the conditional statement $p \to q$ is false only when p is true and q is false. To prove a conditional statement indirectly, assume that p is true and q is false. If this assumption leads to an impossibility, then you have proved that the conditional statement is true. An indirect proof is also called a **proof by contradiction.**

To use an indirect proof to prove the following conditional statement,

"If a is a positive integer and a^2 is divisible by 2, then a is divisible by 2,"

proceed as follows. First, assume that p, "a is a positive integer and a^2 is divisible by 2," is true and q, "a is divisible by 2," is false. This means that a is not divisible by 2. If so, then a is odd and can be written as $a = 2n + 1$, where n is an integer.

$$a = 2n + 1 \qquad \text{Definition of an odd integer}$$

$$a^2 = 4n^2 + 4n + 1 \qquad \text{Square each side.}$$

$$a^2 = 2(2n^2 + 2n) + 1 \qquad \text{Distributive Property}$$

So, by the definition of an odd integer, a^2 is odd. This contradicts the assumption, and you can conclude that a is divisible by 2.

EXAMPLE   **Using an Indirect Proof**

Use an indirect proof to prove that $\sqrt{2}$ is an irrational number.

Solution   Begin by assuming that $\sqrt{2}$ is *not* an irrational number. Then $\sqrt{2}$ can be written as the quotient of two integers a and b $(b \neq 0)$ that have no common factors.

$$\sqrt{2} = \frac{a}{b} \qquad \text{Assume that } \sqrt{2} \text{ is a rational number.}$$

$$2 = \frac{a^2}{b^2} \qquad \text{Square each side.}$$

$$2b^2 = a^2 \qquad \text{Multiply each side by } b^2.$$

This implies that 2 is a factor of a^2. So, 2 is also a factor of a, and a can be written as $2c$, where c is an integer.

$$2b^2 = (2c)^2 \qquad \text{Substitute } 2c \text{ for } a.$$

$$2b^2 = 4c^2 \qquad \text{Simplify.}$$

$$b^2 = 2c^2 \qquad \text{Divide each side by 2.}$$

This implies that 2 is a factor of b^2 and also a factor of b. So, 2 is a factor of both a and b. This contradicts the assumption that a and b have no common factors. So, you can conclude that $\sqrt{2}$ is an irrational number. ■

P.S. Problem Solving

1. **Geometry** A theorem from geometry states that if a triangle is inscribed in a circle such that one side of the triangle is a diameter of the circle, then the triangle is a right triangle. Show that this theorem is true for the circle

$$x^2 + y^2 = 100$$

and the triangle formed by the lines

$y = 0$, $y = \frac{1}{2}x + 5$, and $y = -2x + 20$.

2. **Finding Values of Constants** Find values of k_1 and k_2 such that the system of equations has an infinite number of solutions.

$$\begin{cases} 3x - 5y = 8 \\ 2x + k_1y = k_2 \end{cases}$$

3. **Finding Conditions on Constants** Consider the following system of linear equations in x and y.

$$\begin{cases} ax + by = e \\ cx + dy = f \end{cases}$$

Under what condition(s) will the system have exactly one solution?

4. **Finding Values of Constants** Find values of a, b, and c (if possible) such that the system of linear equations has (a) a unique solution, (b) no solution, and (c) an infinite number of solutions.

$$\begin{cases} x + y \quad\;\; = 2 \\ \quad\; y + z = 2 \\ x \quad\;\; + z = 2 \\ ax + by + cz = 0 \end{cases}$$

5. **Graphical Analysis** Graph the lines determined by each system of linear equations. Then use Gaussian elimination to solve each system. At each step of the elimination process, graph the corresponding lines. What do you observe?

   (a) $\begin{cases} x - 4y = -3 \\ 5x - 6y = 13 \end{cases}$

   (b) $\begin{cases} 2x - 3y = 7 \\ -4x + 6y = -14 \end{cases}$

6. **Maximum Numbers of Solutions** A system of two equations in two unknowns has a finite number of solutions. Determine the maximum number of solutions of the system satisfying each condition.

   (a) Both equations are linear.

   (b) One equation is linear and the other is quadratic.

   (c) Both equations are quadratic.

7. **Vietnam Veterans Memorial** The Vietnam Veterans Memorial (or "The Wall") in Washington, D.C., was designed by Maya Ying Lin when she was a student at Yale University. This monument has two vertical, triangular sections of black granite with a common side (see figure). The bottom of each section is level with the ground. The tops of the two sections can be approximately modeled by the equations

$$-2x + 50y = 505$$

and

$$2x + 50y = 505$$

when the x-axis is superimposed at the base of the wall. Each unit in the coordinate system represents 1 foot. How high is the memorial at the point where the two sections meet? How long is each section?

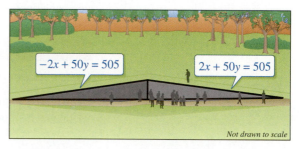

Not drawn to scale

8. **Finding Atomic Weights** Weights of atoms and molecules are measured in atomic mass units (u). A molecule of C_2H_6 (ethane) is made up of two carbon atoms and six hydrogen atoms and weighs 30.070 u. A molecule of C_3H_8 (propane) is made up of three carbon atoms and eight hydrogen atoms and weighs 44.097 u. Find the weights of a carbon atom and a hydrogen atom.

9. **DVD Connector Cables** Connecting a DVD player to a television set requires a cable with special connectors at both ends. You buy a six-foot cable for $15.50 and a three-foot cable for $10.25. Assuming that the cost of a cable is the sum of the cost of the two connectors and the cost of the cable itself, what is the cost of a four-foot cable? Explain your reasoning.

10. **Distance** A hotel 35 miles from an airport runs a shuttle service to and from the airport. The 9:00 A.M. bus leaves for the airport traveling at 30 miles per hour. The 9:15 A.M. bus leaves for the airport traveling at 40 miles per hour. Write a system of linear equations that represents distance as a function of time for the buses. Graph and solve the system. How far from the airport will the 9:15 A.M. bus catch up to the 9:00 A.M. bus?

11. Systems with Rational Expressions Solve each system of equations by letting $X = 1/x$, $Y = 1/y$, and $Z = 1/z$.

(a) $\begin{cases} \dfrac{12}{x} - \dfrac{12}{y} = 7 \\ \dfrac{3}{x} + \dfrac{4}{y} = 0 \end{cases}$

(b) $\begin{cases} \dfrac{2}{x} + \dfrac{1}{y} - \dfrac{3}{z} = 4 \\ \dfrac{4}{x} \quad + \dfrac{2}{z} = 10 \\ -\dfrac{2}{x} + \dfrac{3}{y} - \dfrac{13}{z} = -8 \end{cases}$

12. Finding Values of Constants What values should be given to a, b, and c so that the following linear system has $(-1, 2, -3)$ as its only solution?

$\begin{cases} x + 2y - 3z = a & \text{Equation 1} \\ -x - y + z = b & \text{Equation 2} \\ 2x + 3y - 2z = c & \text{Equation 3} \end{cases}$

13. System of Linear Equations The following system has one solution: $x = 1$, $y = -1$, and $z = 2$.

$\begin{cases} 4x - 2y + 5z = 16 & \text{Equation 1} \\ x + y = 0 & \text{Equation 2} \\ -x - 3y + 2z = 6 & \text{Equation 3} \end{cases}$

Solve the system given by (a) Equation 1 and Equation 2, (b) Equation 1 and Equation 3, and (c) Equation 2 and Equation 3. (d) How many solutions does each of these systems have?

14. System of Linear Equations Solve the system of linear equations algebraically.

$\begin{cases} x_1 - x_2 + 2x_3 + 2x_4 + 6x_5 = 6 \\ 3x_1 - 2x_2 + 4x_3 + 4x_4 + 12x_5 = 14 \\ - x_2 - x_3 - x_4 - 3x_5 = -3 \\ 2x_1 - 2x_2 + 4x_3 + 5x_4 + 15x_5 = 10 \\ 2x_1 - 2x_2 + 4x_3 + 4x_4 + 13x_5 = 13 \end{cases}$

15. Biology Each day, an average adult moose can process about 32 kilograms of terrestrial vegetation (twigs and leaves) and aquatic vegetation. From this food, it needs to obtain about 1.9 grams of sodium and 11,000 calories of energy. Aquatic vegetation has about 0.15 gram of sodium per kilogram and about 193 calories of energy per kilogram, whereas terrestrial vegetation has minimal sodium and about four times as much energy as aquatic vegetation. Write and graph a system of inequalities that describes the amounts t and a of terrestrial and aquatic vegetation, respectively, for the daily diet of an average adult moose. *(Source: Biology by Numbers)*

16. Height and Weight For a healthy person who is 4 feet 10 inches tall, the recommended minimum weight is about 91 pounds and increases by about 3.6 pounds for each additional inch of height. The recommended maximum weight is about 115 pounds and increases by about 4.5 pounds for each additional inch of height. *(Source: National Institutes of Health)*

(a) Let x be the number of inches by which a person's height exceeds 4 feet 10 inches and let y be the person's weight in pounds. Write a system of inequalities that describes the possible values of x and y for a healthy person.

(b) Use a graphing utility to graph the system of inequalities from part (a).

(c) What is the recommended weight range for a healthy person who is 6 feet tall?

17. Cholesterol Cholesterol in human blood is necessary, but too much can lead to health problems. There are three main types of cholesterol: HDL (high-density lipoproteins), LDL (low-density lipoproteins), and VLDL (very low-density lipoproteins). HDL is considered "good" cholesterol; LDL and VLDL are considered "bad" cholesterol.

A standard fasting cholesterol blood test measures total cholesterol, HDL cholesterol, and triglycerides. These numbers are used to estimate LDL and VLDL, which are difficult to measure directly. It is recommended that your combined LDL/VLDL cholesterol level be less than 130 milligrams per deciliter, your HDL cholesterol level be at least 40 milligrams per deciliter, and your total cholesterol level be no more than 200 milligrams per deciliter. *(Source: American Heart Association)*

(a) Write a system of linear inequalities for the recommended cholesterol levels. Let x represent the HDL cholesterol level, and let y represent the combined LDL/VLDL cholesterol level.

(b) Graph the system of inequalities from part (a). Label any vertices of the solution region.

(c) Are the following cholesterol levels within recommendations? Explain your reasoning.

LDL/VLDL: 120 milligrams per deciliter

HDL: 90 milligrams per deciliter

Total: 210 milligrams per deciliter

(d) Give an example of cholesterol levels in which the LDL/VLDL cholesterol level is too high but the HDL cholesterol level is acceptable.

(e) Another recommendation is that the ratio of total cholesterol to HDL cholesterol be less than 4 (that is, less than 4 to 1). Find a point in the solution region from part (b) that meets this recommendation, and explain why it meets the recommendation.

391

8 Matrices and Determinants

Sudoku

Data Encryption

Beam Deflection

Flight Crew Scheduling

Health and Wellness

393

8.1 Matrices and Systems of Equations

You can use matrices to solve systems of linear equations in two or more variables. For instance, a matrix can be used to find a model for the numbers of new cases of a waterborne disease in a small city.

- ■ Write matrices and identify their orders.
- ■ Perform elementary row operations on matrices.
- ■ Use matrices and Gaussian elimination to solve systems of linear equations.
- ■ Use matrices and Gauss-Jordan elimination to solve systems of linear equations.

Matrices

In this section, you will study a streamlined technique for solving systems of linear equations. This technique involves the use of a rectangular array of real numbers called a **matrix**. The plural of matrix is *matrices*.

Definition of Matrix

If m and n are positive integers, then an $m \times n$ (read "m by n") matrix is a rectangular array

$$
\begin{array}{c}
\quad\text{Column 1}\ \text{Column 2}\ \text{Column 3}\ \ldots\ \text{Column } n \\
\begin{array}{c}
\text{Row 1} \\
\text{Row 2} \\
\text{Row 3} \\
\vdots \\
\text{Row } m
\end{array}
\begin{bmatrix}
a_{11} & a_{12} & a_{13} & \cdots & a_{1n} \\
a_{21} & a_{22} & a_{23} & \cdots & a_{2n} \\
a_{31} & a_{32} & a_{33} & \cdots & a_{3n} \\
\vdots & \vdots & \vdots & & \vdots \\
a_{m1} & a_{m2} & a_{m3} & \cdots & a_{mn}
\end{bmatrix}
\end{array}
$$

in which each **entry** a_{ij} of the matrix is a number. An $m \times n$ matrix has m rows and n columns. Matrices are usually denoted by capital letters.

The entry in the ith row and jth column is denoted by the *double subscript* notation a_{ij}. For instance, a_{23} refers to the entry in the second row, third column. A matrix that has only one row is called a **row matrix**, and a matrix that has only one column is called a **column matrix**. A matrix having m rows and n columns is said to be of **order** $m \times n$. If $m = n$, then the matrix is **square** of order $m \times m$ (or $n \times n$). For a square matrix, the entries $a_{11}, a_{22}, a_{33}, \ldots$ are the **main diagonal** entries.

EXAMPLE 1 Orders of Matrices

Determine the order of each matrix.

a. $\begin{bmatrix} 2 \end{bmatrix}$ **b.** $\begin{bmatrix} 1 & -3 & 0 & \frac{1}{2} \end{bmatrix}$ **c.** $\begin{bmatrix} 0 & 0 \\ 0 & 0 \end{bmatrix}$ **d.** $\begin{bmatrix} 5 & 0 \\ 2 & -2 \\ -7 & 4 \end{bmatrix}$

Solution

a. This matrix has *one* row and *one* column. The order of the matrix is 1×1.

b. This matrix has *one* row and *four* columns. The order of the matrix is 1×4.

c. This matrix has *two* rows and *two* columns. The order of the matrix is 2×2.

d. This matrix has *three* rows and *two* columns. The order of the matrix is 3×2.

✓ *Checkpoint*  *Audio-video solution in English & Spanish at LarsonPrecalculus.com.*

Determine the order of the matrix $\begin{bmatrix} 14 & 7 & 10 \\ -2 & -3 & -8 \end{bmatrix}$.

A matrix derived from a system of linear equations (each written in standard form with the constant term on the right) is the **augmented matrix** of the system. Moreover, the matrix derived from the coefficients of the system (but not including the constant terms) is the **coefficient matrix** of the system.

$$\text{System:} \begin{cases} x - 4y + 3z = 5 \\ -x + 3y - z = -3 \\ 2x - 4z = 6 \end{cases}$$

$$\text{Augmented Matrix:} \begin{bmatrix} 1 & -4 & 3 & \vdots & 5 \\ -1 & 3 & -1 & \vdots & -3 \\ 2 & 0 & -4 & \vdots & 6 \end{bmatrix} \qquad \text{Coefficient Matrix:} \begin{bmatrix} 1 & -4 & 3 \\ -1 & 3 & -1 \\ 2 & 0 & -4 \end{bmatrix}$$

▷
•• REMARK The vertical dots in an augmented matrix separate the coefficients of the linear system from the constant terms.

Note the use of 0 for the coefficient of the missing y-variable in the third equation, and also note the fourth column of constant terms in the augmented matrix.

When forming either the coefficient matrix or the augmented matrix of a system, you should begin by vertically aligning the variables in the equations and using zeros for the coefficients of the missing variables.

EXAMPLE 2 Writing an Augmented Matrix

Write the augmented matrix for the system of linear equations.

$$\begin{cases} x + 3y - w = 9 \\ -y + 4z + 2w = -2 \\ x - 5z - 6w = 0 \\ 2x + 4y - 3z = 4 \end{cases}$$

What is the order of the augmented matrix?

Solution

Begin by rewriting the linear system and aligning the variables.

$$\begin{cases} x + 3y \quad\quad - w = 9 \\ \quad -y + 4z + 2w = -2 \\ x \quad\quad - 5z - 6w = 0 \\ 2x + 4y - 3z \quad\quad = 4 \end{cases}$$

Next, use the coefficients and constant terms as the matrix entries. Include zeros for the coefficients of the missing variables.

$$\begin{matrix} R_1 \\ R_2 \\ R_3 \\ R_4 \end{matrix} \begin{bmatrix} 1 & 3 & 0 & -1 & \vdots & 9 \\ 0 & -1 & 4 & 2 & \vdots & -2 \\ 1 & 0 & -5 & -6 & \vdots & 0 \\ 2 & 4 & -3 & 0 & \vdots & 4 \end{bmatrix}$$

The augmented matrix has four rows and five columns, so it is a 4×5 matrix. The notation R_n is used to designate each row in the matrix. For example, Row 1 is represented by R_1.

✓ *Checkpoint* ◁)) *Audio-video solution in English & Spanish at LarsonPrecalculus.com.*

Write the augmented matrix for the system of linear equations.

$$\begin{cases} x + y + z = 2 \\ 2x - y + 3z = -1 \\ -x + 2y - z = 4 \end{cases}$$

■

Elementary Row Operations

In Section 7.3, you studied three operations that can be used on a system of linear equations to produce an equivalent system.

1. Interchange two equations.

2. Multiply an equation by a nonzero constant.

3. Add a multiple of an equation to another equation.

In matrix terminology, these three operations correspond to **elementary row operations.** An elementary row operation on an augmented matrix of a given system of linear equations produces a new augmented matrix corresponding to a new (but equivalent) system of linear equations. Two matrices are **row-equivalent** when one can be obtained from the other by a sequence of elementary row operations.

> **REMARK** Although elementary row operations are simple to perform, they involve a lot of arithmetic. Because it is easy to make a mistake, you should get in the habit of noting the elementary row operations performed in each step so that you can go back and check your work.

Elementary Row Operations

Operation	Notation
1. Interchange two rows.	$R_a \leftrightarrow R_b$
2. Multiply a row by a nonzero constant.	$cR_a \quad (c \neq 0)$
3. Add a multiple of a row to another row.	$cR_a + R_b$

EXAMPLE 3 **Elementary Row Operations**

a. Interchange the first and second rows of the original matrix.

Original Matrix

$$\begin{bmatrix} 0 & 1 & 3 & 4 \\ -1 & 2 & 0 & 3 \\ 2 & -3 & 4 & 1 \end{bmatrix}$$

New Row-Equivalent Matrix

$$\begin{matrix} R_2 \\ R_1 \end{matrix} \begin{bmatrix} -1 & 2 & 0 & 3 \\ 0 & 1 & 3 & 4 \\ 2 & -3 & 4 & 1 \end{bmatrix}$$

b. Multiply the first row of the original matrix by $\frac{1}{2}$.

Original Matrix

$$\begin{bmatrix} 2 & -4 & 6 & -2 \\ 1 & 3 & -3 & 0 \\ 5 & -2 & 1 & 2 \end{bmatrix}$$

New Row-Equivalent Matrix

$$\tfrac{1}{2}R_1 \rightarrow \begin{bmatrix} 1 & -2 & 3 & -1 \\ 1 & 3 & -3 & 0 \\ 5 & -2 & 1 & 2 \end{bmatrix}$$

> **TECHNOLOGY** Most graphing utilities can perform elementary row operations on matrices. Consult the user's guide for your graphing utility for specific keystrokes.
> After performing a row operation, the new row-equivalent matrix that is displayed on your graphing utility is stored in the *answer* variable. You should use the *answer* variable and not the original matrix for subsequent row operations.

c. Add -2 times the first row of the original matrix to the third row.

Original Matrix

$$\begin{bmatrix} 1 & 2 & -4 & 3 \\ 0 & 3 & -2 & -1 \\ 2 & 1 & 5 & -2 \end{bmatrix}$$

New Row-Equivalent Matrix

$$\begin{matrix} \\ \\ -2R_1 + R_3 \rightarrow \end{matrix} \begin{bmatrix} 1 & 2 & -4 & 3 \\ 0 & 3 & -2 & -1 \\ 0 & -3 & 13 & -8 \end{bmatrix}$$

Note that the elementary row operation is written beside the row that is *changed*.

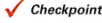 **Checkpoint**  Audio-video solution in English & Spanish at LarsonPrecalculus.com.

Identify the elementary row operation being performed to obtain the new row-equivalent matrix.

Original Matrix

$$\begin{bmatrix} 1 & 0 & 2 \\ 3 & 1 & 7 \\ 2 & -6 & 14 \end{bmatrix}$$

New Row-Equivalent Matrix

$$\begin{bmatrix} 1 & 0 & 2 \\ 0 & 1 & 1 \\ 2 & -6 & 14 \end{bmatrix}$$

Gaussian Elimination with Back-Substitution

In Example 3 in Section 7.3, you used Gaussian elimination with back-substitution to solve a system of linear equations. The next example demonstrates the matrix version of Gaussian elimination. The two methods are essentially the same. The basic difference is that with matrices you do not need to keep writing the variables.

EXAMPLE 4 **Comparing Linear Systems and Matrix Operations**

Linear System

$$\begin{cases} x - 2y + 3z = 9 \\ -x + 3y = -4 \\ 2x - 5y + 5z = 17 \end{cases}$$

Associated Augmented Matrix

$$\begin{bmatrix} 1 & -2 & 3 & \vdots & 9 \\ -1 & 3 & 0 & \vdots & -4 \\ 2 & -5 & 5 & \vdots & 17 \end{bmatrix}$$

Add the first equation to the second equation.

$$\begin{cases} x - 2y + 3z = 9 \\ y + 3z = 5 \\ 2x - 5y + 5z = 17 \end{cases}$$

Add the first row to the second row $(R_1 + R_2)$.

$$R_1 + R_2 \rightarrow \begin{bmatrix} 1 & -2 & 3 & \vdots & 9 \\ 0 & 1 & 3 & \vdots & 5 \\ 2 & -5 & 5 & \vdots & 17 \end{bmatrix}$$

Add -2 times the first equation to the third equation.

$$\begin{cases} x - 2y + 3z = 9 \\ y + 3z = 5 \\ -y - z = -1 \end{cases}$$

Add -2 times the first row to the third row $(-2R_1 + R_3)$.

$$-2R_1 + R_3 \rightarrow \begin{bmatrix} 1 & -2 & 3 & \vdots & 9 \\ 0 & 1 & 3 & \vdots & 5 \\ 0 & -1 & -1 & \vdots & -1 \end{bmatrix}$$

Add the second equation to the third equation.

$$\begin{cases} x - 2y + 3z = 9 \\ y + 3z = 5 \\ 2z = 4 \end{cases}$$

Add the second row to the third row $(R_2 + R_3)$.

$$R_2 + R_3 \rightarrow \begin{bmatrix} 1 & -2 & 3 & \vdots & 9 \\ 0 & 1 & 3 & \vdots & 5 \\ 0 & 0 & 2 & \vdots & 4 \end{bmatrix}$$

Multiply the third equation by $\frac{1}{2}$.

$$\begin{cases} x - 2y + 3z = 9 \\ y + 3z = 5 \\ z = 2 \end{cases}$$

Multiply the third row by $\frac{1}{2}$ $\left(\frac{1}{2}R_3\right)$.

$$\frac{1}{2}R_3 \rightarrow \begin{bmatrix} 1 & -2 & 3 & \vdots & 9 \\ 0 & 1 & 3 & \vdots & 5 \\ 0 & 0 & 1 & \vdots & 2 \end{bmatrix}$$

> **• • REMARK** Remember that you should check a solution by substituting the values of x, y, and z into each equation of the original system. For example, you can check the solution to Example 4 as follows.
>
> Equation 1:
> $1 - 2(-1) + 3(2) = 9$ ✓
>
> Equation 2:
> $-1 + 3(-1) = -4$ ✓
>
> Equation 3:
> $2(1) - 5(-1) + 5(2) = 17$ ✓

At this point, you can use back-substitution to find x and y.

$$y + 3(2) = 5 \qquad \text{Substitute 2 for } z.$$
$$y = -1 \qquad \text{Solve for } y.$$
$$x - 2(-1) + 3(2) = 9 \qquad \text{Substitute } -1 \text{ for } y \text{ and 2 for } z.$$
$$x = 1 \qquad \text{Solve for } x.$$

The solution is $x = 1$, $y = -1$, and $z = 2$.

✓ **Checkpoint** 🔊)) *Audio-video solution in English & Spanish at LarsonPrecalculus.com.*

Compare solving the linear system below to solving it using its associated augmented matrix.

$$\begin{cases} 2x + y - z = -3 \\ 4x - 2y + 2z = -2 \\ -6x + 5y + 4z = 10 \end{cases}$$

The last matrix in Example 4 is said to be in **row-echelon form.** The term *echelon* refers to the stair-step pattern formed by the nonzero entries of the matrix. To be in this form, a matrix must have the following properties.

Row-Echelon Form and Reduced Row-Echelon Form

A matrix in **row-echelon form** has the following properties.

1. Any rows consisting entirely of zeros occur at the bottom of the matrix.
2. For each row that does not consist entirely of zeros, the first nonzero entry is 1 (called a **leading 1**).
3. For two successive (nonzero) rows, the leading 1 in the higher row is farther to the left than the leading 1 in the lower row.

A matrix in *row-echelon form* is in **reduced row-echelon form** when every column that has a leading 1 has zeros in every position above and below its leading 1.

It is worth noting that the row-echelon form of a matrix is not unique. That is, two different sequences of elementary row operations may yield different row-echelon forms. However, the *reduced* row-echelon form of a given matrix is unique.

EXAMPLE 5   Row-Echelon Form

Determine whether each matrix is in row-echelon form. If it is, determine whether the matrix is in reduced row-echelon form.

a. $\begin{bmatrix} 1 & 2 & -1 & 4 \\ 0 & 1 & 0 & 3 \\ 0 & 0 & 1 & -2 \end{bmatrix}$ b. $\begin{bmatrix} 1 & 2 & -1 & 2 \\ 0 & 0 & 0 & 0 \\ 0 & 1 & 2 & -4 \end{bmatrix}$

c. $\begin{bmatrix} 1 & -5 & 2 & -1 & 3 \\ 0 & 0 & 1 & 3 & -2 \\ 0 & 0 & 0 & 1 & 4 \\ 0 & 0 & 0 & 0 & 1 \end{bmatrix}$ d. $\begin{bmatrix} 1 & 0 & 0 & -1 \\ 0 & 1 & 0 & 2 \\ 0 & 0 & 1 & 3 \\ 0 & 0 & 0 & 0 \end{bmatrix}$

e. $\begin{bmatrix} 1 & 2 & -3 & 4 \\ 0 & 2 & 1 & -1 \\ 0 & 0 & 1 & -3 \end{bmatrix}$ f. $\begin{bmatrix} 0 & 1 & 0 & 5 \\ 0 & 0 & 1 & 3 \\ 0 & 0 & 0 & 0 \end{bmatrix}$

Solution   The matrices in (a), (c), (d), and (f) are in row-echelon form. The matrices in (d) and (f) are in *reduced* row-echelon form because every column that has a leading 1 has zeros in every position above and below its leading 1. The matrix in (b) is not in row-echelon form because a row of all zeros does not occur at the bottom of the matrix. The matrix in (e) is not in row-echelon form because the first nonzero entry in Row 2 is not a leading 1.

✔ *Checkpoint*   *Audio-video solution in English & Spanish at LarsonPrecalculus.com.*

Determine whether the matrix is in row-echelon form. If it is, determine whether it is in reduced row-echelon form.

$$\begin{bmatrix} 1 & 0 & -2 & 4 \\ 0 & 1 & 11 & 3 \\ 0 & 0 & 0 & 0 \end{bmatrix}$$

■

Every matrix is row-equivalent to a matrix in row-echelon form. For instance, in Example 5, you can change the matrix in part (e) to row-echelon form by multiplying its second row by $\frac{1}{2}$.

Gaussian elimination with back-substitution works well for solving systems of linear equations by hand or with a computer. For this algorithm, the order in which the elementary row operations are performed is important. You should operate from left to right by columns, using elementary row operations to obtain zeros in all entries directly below the leading 1's.

EXAMPLE 6 Gaussian Elimination with Back-Substitution

Solve the system
$$\begin{cases} \quad\; y + \;z - 2w = -3 \\ x + 2y - \;z \qquad\;\; = \;\;2 \\ 2x + 4y + \;z - 3w = -2 \\ x - 4y - 7z - \;w = -19 \end{cases}.$$

Solution

$$\begin{bmatrix} 0 & 1 & 1 & -2 & \vdots & -3 \\ 1 & 2 & -1 & 0 & \vdots & 2 \\ 2 & 4 & 1 & -3 & \vdots & -2 \\ 1 & -4 & -7 & -1 & \vdots & -19 \end{bmatrix}$$ Write augmented matrix.

$$\begin{matrix} R_2 \\ R_1 \end{matrix} \begin{bmatrix} 1 & 2 & -1 & 0 & \vdots & 2 \\ 0 & 1 & 1 & -2 & \vdots & -3 \\ 2 & 4 & 1 & -3 & \vdots & -2 \\ 1 & -4 & -7 & -1 & \vdots & -19 \end{bmatrix}$$ Interchange R_1 and R_2 so first column has leading 1 in upper left corner.

$$\begin{matrix} \\ \\ -2R_1 + R_3 \to \\ -R_1 + R_4 \to \end{matrix} \begin{bmatrix} 1 & 2 & -1 & 0 & \vdots & 2 \\ 0 & 1 & 1 & -2 & \vdots & -3 \\ 0 & 0 & 3 & -3 & \vdots & -6 \\ 0 & -6 & -6 & -1 & \vdots & -21 \end{bmatrix}$$ Perform operations on R_3 and R_4 so first column has zeros below its leading 1.

$$\begin{matrix} \\ \\ \\ 6R_2 + R_4 \to \end{matrix} \begin{bmatrix} 1 & 2 & -1 & 0 & \vdots & 2 \\ 0 & 1 & 1 & -2 & \vdots & -3 \\ 0 & 0 & 3 & -3 & \vdots & -6 \\ 0 & 0 & 0 & -13 & \vdots & -39 \end{bmatrix}$$ Perform operations on R_4 so second column has zeros below its leading 1.

$$\begin{matrix} \\ \\ \frac{1}{3}R_3 \to \\ -\frac{1}{13}R_4 \to \end{matrix} \begin{bmatrix} 1 & 2 & -1 & 0 & \vdots & 2 \\ 0 & 1 & 1 & -2 & \vdots & -3 \\ 0 & 0 & 1 & -1 & \vdots & -2 \\ 0 & 0 & 0 & 1 & \vdots & 3 \end{bmatrix}$$ Perform operations on R_3 and R_4 so third and fourth columns have leading 1's.

The matrix is now in row-echelon form, and the corresponding system is

$$\begin{cases} x + 2y - z \qquad\;\; = \;\;2 \\ \quad\; y + z - 2w = -3 \\ \qquad\;\; z - \;w = -2 \\ \qquad\qquad\;\; w = \;\;3 \end{cases}.$$

Using back-substitution, you can determine that the solution is

$$x = -1, \quad y = 2, \quad z = 1, \quad \text{and} \quad w = 3.$$

✓ *Checkpoint* �))) *Audio-video solution in English & Spanish at LarsonPrecalculus.com.*

Solve the system
$$\begin{cases} -3x + 5y + 3z = -19 \\ 3x + 4y + 4z = \;\;8 \\ 4x - 8y - 6z = \;\;26 \end{cases}.$$

The procedure for using Gaussian elimination with back-substitution is summarized below.

Gaussian Elimination with Back-Substitution

1. Write the augmented matrix of the system of linear equations.
2. Use elementary row operations to rewrite the augmented matrix in row-echelon form.
3. Write the system of linear equations corresponding to the matrix in row-echelon form and use back-substitution to find the solution.

When solving a system of linear equations, remember that it is possible for the system to have no solution. If, in the elimination process, you obtain a row of all zeros except for the last entry, then the system has no solution, or is *inconsistent*.

EXAMPLE 7 **A System with No Solution**

Solve the system $\begin{cases} x - y + 2z = 4 \\ x \qquad + z = 6 \\ 2x - 3y + 5z = 4 \\ 3x + 2y - z = 1 \end{cases}$.

Solution

$$\left[\begin{array}{ccc:c} 1 & -1 & 2 & 4 \\ 1 & 0 & 1 & 6 \\ 2 & -3 & 5 & 4 \\ 3 & 2 & -1 & 1 \end{array}\right]$$ Write augmented matrix.

$$\begin{array}{c} \\ -R_1 + R_2 \rightarrow \\ -2R_1 + R_3 \rightarrow \\ -3R_1 + R_4 \rightarrow \end{array}\left[\begin{array}{ccc:c} 1 & -1 & 2 & 4 \\ 0 & 1 & -1 & 2 \\ 0 & -1 & 1 & -4 \\ 0 & 5 & -7 & -11 \end{array}\right]$$ Perform row operations.

$$\begin{array}{c} \\ \\ R_2 + R_3 \rightarrow \\ \\ \end{array}\left[\begin{array}{ccc:c} 1 & -1 & 2 & 4 \\ 0 & 1 & -1 & 2 \\ 0 & 0 & 0 & -2 \\ 0 & 5 & -7 & -11 \end{array}\right]$$ Perform row operations.

Note that the third row of this matrix consists entirely of zeros except for the last entry. This means that the original system of linear equations is inconsistent. You can see why this is true by converting back to a system of linear equations.

$$\begin{cases} x - y + 2z = 4 \\ \quad y - z = 2 \\ \quad\quad\quad 0 = -2 \\ \quad 5y - 7z = -11 \end{cases}$$

Because the third equation is not possible, the system has no solution.

✓ *Checkpoint* ◄)) *Audio-video solution in English & Spanish at LarsonPrecalculus.com.*

Solve the system $\begin{cases} x + y + z = 1 \\ x + 2y + 2z = 2 \\ x - y - z = 1 \end{cases}$.

Gauss-Jordan Elimination

With Gaussian elimination, elementary row operations are applied to a matrix to obtain a (row-equivalent) row-echelon form of the matrix. A second method of elimination, called **Gauss-Jordan elimination,** after Carl Friedrich Gauss and Wilhelm Jordan (1842–1899), continues the reduction process until a *reduced* row-echelon form is obtained. This procedure is demonstrated in Example 8.

EXAMPLE 8   **Gauss-Jordan Elimination**

▷ **TECHNOLOGY** For a demonstration of a graphical approach to Gauss-Jordan elimination on a 2×3 matrix, see the Visualizing Row Operations Program available for several models of graphing calculators at the website for this text at *LarsonPrecalculus.com*.

Use Gauss-Jordan elimination to solve the system $\begin{cases} x - 2y + 3z = 9 \\ -x + 3y = -4. \\ 2x - 5y + 5z = 17 \end{cases}$

Solution   In Example 4, Gaussian elimination was used to obtain the row-echelon form of the linear system above.

$$\begin{bmatrix} 1 & -2 & 3 & \vdots & 9 \\ 0 & 1 & 3 & \vdots & 5 \\ 0 & 0 & 1 & \vdots & 2 \end{bmatrix}$$

Now, apply elementary row operations until you obtain zeros above each of the leading 1's, as follows.

$$\begin{matrix} 2R_2 + R_1 \rightarrow \\ \\ \end{matrix} \begin{bmatrix} 1 & 0 & 9 & \vdots & 19 \\ 0 & 1 & 3 & \vdots & 5 \\ 0 & 0 & 1 & \vdots & 2 \end{bmatrix}$$

Perform operations on R_1 so second column has a zero above its leading 1.

$$\begin{matrix} -9R_3 + R_1 \rightarrow \\ -3R_3 + R_2 \rightarrow \\ \end{matrix} \begin{bmatrix} 1 & 0 & 0 & \vdots & 1 \\ 0 & 1 & 0 & \vdots & -1 \\ 0 & 0 & 1 & \vdots & 2 \end{bmatrix}$$

Perform operations on R_1 and R_2 so third column has zeros above its leading 1.

•• **REMARK** The advantage of using Gauss-Jordan elimination to solve a system of linear equations is that the solution of the system is easily found without using back-substitution, as illustrated in Example 8.
···············▷

The matrix is now in reduced row-echelon form. Converting back to a system of linear equations, you have

$$\begin{cases} x = 1 \\ y = -1. \\ z = 2 \end{cases}$$

Now you can simply read the solution, $x = 1$, $y = -1$, and $z = 2$, which can be written as the ordered triple $(1, -1, 2)$.

✓ *Checkpoint*  *Audio-video solution in English & Spanish at LarsonPrecalculus.com.*

Use Gauss-Jordan elimination to solve the system $\begin{cases} -3x + 7y + 2z = 1 \\ -5x + 3y - 5z = -8. \\ 2x - 2y - 3z = 15 \end{cases}$ ■

The elimination procedures described in this section sometimes result in fractional coefficients. For instance, in the elimination procedure for the system

$$\begin{cases} 2x - 5y + 5z = 17 \\ 3x - 2y + 3z = 11 \\ -3x + 3y = -6 \end{cases}$$

you may be inclined to multiply the first row by $\frac{1}{2}$ to produce a leading 1, which will result in working with fractional coefficients. You can sometimes avoid fractions by judiciously choosing the order in which you apply elementary row operations.

EXAMPLE 9 **A System with an Infinite Number of Solutions**

Solve the system $\begin{cases} 2x + 4y - 2z = 0 \\ 3x + 5y = 1 \end{cases}$.

Solution

$$\begin{bmatrix} 2 & 4 & -2 & \vdots & 0 \\ 3 & 5 & 0 & \vdots & 1 \end{bmatrix}$$

$$\tfrac{1}{2}R_1 \rightarrow \begin{bmatrix} 1 & 2 & -1 & \vdots & 0 \\ 3 & 5 & 0 & \vdots & 1 \end{bmatrix}$$

$$-3R_1 + R_2 \rightarrow \begin{bmatrix} 1 & 2 & -1 & \vdots & 0 \\ 0 & -1 & 3 & \vdots & 1 \end{bmatrix}$$

$$-R_2 \rightarrow \begin{bmatrix} 1 & 2 & -1 & \vdots & 0 \\ 0 & 1 & -3 & \vdots & -1 \end{bmatrix}$$

$$-2R_2 + R_1 \rightarrow \begin{bmatrix} 1 & 0 & 5 & \vdots & 2 \\ 0 & 1 & -3 & \vdots & -1 \end{bmatrix}$$

The corresponding system of equations is

$$\begin{cases} x + 5z = 2 \\ y - 3z = -1 \end{cases}.$$

Solving for x and y in terms of z, you have

$$x = -5z + 2 \quad \text{and} \quad y = 3z - 1.$$

To write a solution of the system that does not use any of the three variables of the system, let a represent any real number and let $z = a$. Substituting a for z in the equations for x and y, you have

$$x = -5z + 2 = -5a + 2 \quad \text{and} \quad y = 3z - 1 = 3a - 1.$$

So, the solution set can be written as an ordered triple of the form

$$(-5a + 2, 3a - 1, a)$$

where a is any real number. Remember that a solution set of this form represents an infinite number of solutions. Try substituting values for a to obtain a few solutions. Then check each solution in the original system of equations.

 Checkpoint 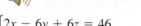 Audio-video solution in English & Spanish at LarsonPrecalculus.com.

Solve the system $\begin{cases} 2x - 6y + 6z = 46 \\ 2x - 3y = 31 \end{cases}$.

Summarize (Section 8.1)

1. State the definition of a matrix *(page 394)*. For examples of writing matrices and determining their orders, see Examples 1 and 2.

2. List the elementary row operations *(page 396)*. For examples of performing elementary row operations, see Examples 3 and 4.

3. State the definitions of row-echelon form and reduced row-echelon form *(page 398)*. For an example of matrices in these forms, see Example 5.

4. Describe Gaussian elimination with back-substitution *(pages 399 and 400)*. For examples of using this procedure, see Examples 6 and 7.

5. Describe Gauss-Jordan elimination *(page 401)*. For examples of using this procedure, see Examples 8 and 9.

8.2   Operations with Matrices

You can use matrix operations to model and solve real-life problems. For instance, matrix multiplication can be used to determine the numbers of calories burned by individuals of different body weights while performing different types of exercises.

- Decide whether two matrices are equal.
- Add and subtract matrices, and multiply matrices by scalars.
- Multiply two matrices.
- Use matrix operations to model and solve real-life problems.

Equality of Matrices

In Section 8.1, you used matrices to solve systems of linear equations. There is a rich mathematical theory of matrices, and its applications are numerous. This section and the next two introduce some fundamentals of matrix theory. It is standard mathematical convention to represent matrices in any of the following three ways.

Representation of Matrices

1. A matrix can be denoted by an uppercase letter such as A, B, or C.

2. A matrix can be denoted by a representative element enclosed in brackets, such as $[a_{ij}]$, $[b_{ij}]$, or $[c_{ij}]$.

3. A matrix can be denoted by a rectangular array of numbers such as

$$A = [a_{ij}] = \begin{bmatrix} a_{11} & a_{12} & a_{13} & \cdots & a_{1n} \\ a_{21} & a_{22} & a_{23} & \cdots & a_{2n} \\ a_{31} & a_{32} & a_{33} & \cdots & a_{3n} \\ \vdots & \vdots & \vdots & & \vdots \\ a_{m1} & a_{m2} & a_{m3} & \cdots & a_{mn} \end{bmatrix}.$$

Two matrices $A = [a_{ij}]$ and $B = [b_{ij}]$ are **equal** when they have the same order $(m \times n)$ and $a_{ij} = b_{ij}$ for $1 \le i \le m$ and $1 \le j \le n$. In other words, two matrices are equal when their corresponding entries are equal.

EXAMPLE 1   Equality of Matrices

Solve for a_{11}, a_{12}, a_{21}, and a_{22} in the following matrix equation.

$$\begin{bmatrix} a_{11} & a_{12} \\ a_{21} & a_{22} \end{bmatrix} = \begin{bmatrix} 2 & -1 \\ -3 & 0 \end{bmatrix}$$

Solution   Because two matrices are equal when their corresponding entries are equal, you can conclude that $a_{11} = 2$, $a_{12} = -1$, $a_{21} = -3$, and $a_{22} = 0$.

✓ **Checkpoint**  *Audio-video solution in English & Spanish at LarsonPrecalculus.com.*

Solve for a_{11}, a_{12}, a_{21}, and a_{22} in the following matrix equation.

$$\begin{bmatrix} a_{11} & a_{12} \\ a_{21} & a_{22} \end{bmatrix} = \begin{bmatrix} 6 & 3 \\ -2 & 4 \end{bmatrix}$$

Be sure you see that for two matrices to be equal, they must have the same order *and* their corresponding entries must be equal. For instance,

$$\begin{bmatrix} 2 & -1 \\ \sqrt{4} & \frac{1}{2} \end{bmatrix} = \begin{bmatrix} 2 & -1 \\ 2 & 0.5 \end{bmatrix} \quad \text{but} \quad \begin{bmatrix} 2 & -1 & 0 \\ 3 & 4 & 0 \end{bmatrix} \ne \begin{bmatrix} 2 & -1 \\ 3 & 4 \end{bmatrix}.$$

Matrix Addition and Scalar Multiplication

In this section, three basic matrix operations will be covered. The first two are matrix addition and scalar multiplication. With matrix addition, you can add two matrices (of the same order) by adding their corresponding entries.

Definition of Matrix Addition

If $A = [a_{ij}]$ and $B = [b_{ij}]$ are matrices of order $m \times n$, then their sum is the $m \times n$ matrix given by

$$A + B = [a_{ij} + b_{ij}].$$

The sum of two matrices of different orders is undefined.

EXAMPLE 2   **Addition of Matrices**

a. $\begin{bmatrix} -1 & 2 \\ 0 & 1 \end{bmatrix} + \begin{bmatrix} 1 & 3 \\ -1 & 2 \end{bmatrix} = \begin{bmatrix} -1+1 & 2+3 \\ 0+(-1) & 1+2 \end{bmatrix} = \begin{bmatrix} 0 & 5 \\ -1 & 3 \end{bmatrix}$

b. $\begin{bmatrix} 0 & 1 & -2 \\ 1 & 2 & 3 \end{bmatrix} + \begin{bmatrix} 0 & 0 & 0 \\ 0 & 0 & 0 \end{bmatrix} = \begin{bmatrix} 0 & 1 & -2 \\ 1 & 2 & 3 \end{bmatrix}$

c. $\begin{bmatrix} 1 \\ -3 \\ -2 \end{bmatrix} + \begin{bmatrix} -1 \\ 3 \\ 2 \end{bmatrix} = \begin{bmatrix} 0 \\ 0 \\ 0 \end{bmatrix}$

d. The sum of

$$A = \begin{bmatrix} 2 & 1 & 0 \\ 4 & 0 & -1 \\ 3 & -2 & 2 \end{bmatrix}$$

and

$$B = \begin{bmatrix} 0 & 1 \\ -1 & 3 \\ 2 & 4 \end{bmatrix}$$

is undefined because A is of order 3×3 and B is of order 3×2.

✓ **Checkpoint** *Audio-video solution in English & Spanish at LarsonPrecalculus.com.*

Evaluate the expression.

$$\begin{bmatrix} 4 & -1 \\ 2 & -3 \end{bmatrix} + \begin{bmatrix} 2 & -1 \\ 0 & 6 \end{bmatrix}$$

In operations with matrices, numbers are usually referred to as **scalars.** In this text, scalars will always be real numbers. You can multiply a matrix A by a scalar c by multiplying each entry in A by c.

Definition of Scalar Multiplication

If $A = [a_{ij}]$ is an $m \times n$ matrix and c is a scalar, then the **scalar multiple** of A by c is the $m \times n$ matrix given by

$$cA = [ca_{ij}].$$

The symbol $-A$ represents the negation of A, which is the scalar product $(-1)A$. Moreover, if A and B are of the same order, then $A - B$ represents the sum of A and $(-1)B$. That is,

$$A - B = A + (-1)B. \qquad \text{Subtraction of matrices}$$

EXAMPLE 3   Scalar Multiplication and Matrix Subtraction

For the following matrices, find (a) $3A$, (b) $-B$, and (c) $3A - B$.

$$A = \begin{bmatrix} 2 & 2 & 4 \\ -3 & 0 & -1 \\ 2 & 1 & 2 \end{bmatrix} \quad \text{and} \quad B = \begin{bmatrix} 2 & 0 & 0 \\ 1 & -4 & 3 \\ -1 & 3 & 2 \end{bmatrix}$$

Solution

a. $3A = 3 \begin{bmatrix} 2 & 2 & 4 \\ -3 & 0 & -1 \\ 2 & 1 & 2 \end{bmatrix}$ Scalar multiplication

$$= \begin{bmatrix} 3(2) & 3(2) & 3(4) \\ 3(-3) & 3(0) & 3(-1) \\ 3(2) & 3(1) & 3(2) \end{bmatrix} \qquad \text{Multiply each entry by 3.}$$

$$= \begin{bmatrix} 6 & 6 & 12 \\ -9 & 0 & -3 \\ 6 & 3 & 6 \end{bmatrix} \qquad \text{Simplify.}$$

b. $-B = (-1) \begin{bmatrix} 2 & 0 & 0 \\ 1 & -4 & 3 \\ -1 & 3 & 2 \end{bmatrix}$ Definition of negation

$$= \begin{bmatrix} -2 & 0 & 0 \\ -1 & 4 & -3 \\ 1 & -3 & -2 \end{bmatrix} \qquad \text{Multiply each entry by } -1.$$

c. $3A - B = \begin{bmatrix} 6 & 6 & 12 \\ -9 & 0 & -3 \\ 6 & 3 & 6 \end{bmatrix} - \begin{bmatrix} 2 & 0 & 0 \\ 1 & -4 & 3 \\ -1 & 3 & 2 \end{bmatrix}$ Perform scalar multiplication first.

REMARK The order of operations for matrix expressions is similar to that for real numbers. In particular, you perform scalar multiplication before matrix addition and subtraction, as shown in Example 3(c).

$$= \begin{bmatrix} 4 & 6 & 12 \\ -10 & 4 & -6 \\ 7 & 0 & 4 \end{bmatrix} \qquad \text{Subtract corresponding entries.}$$

✓ **Checkpoint** 🔊 *Audio-video solution in English & Spanish at LarsonPrecalculus.com.*

For the following matrices, find (a) $A + B$, (b) $A - B$, (c) $3A$, and (d) $3A - 2B$.

$$A = \begin{bmatrix} 4 & -1 \\ 0 & 4 \\ -3 & 8 \end{bmatrix}, \quad B = \begin{bmatrix} 0 & 4 \\ -1 & 3 \\ 1 & 7 \end{bmatrix}$$

It is often convenient to rewrite the scalar multiple cA by factoring c out of every entry in the matrix. For instance, in the following example, the scalar $\frac{1}{2}$ has been factored out of the matrix.

$$\begin{bmatrix} \frac{1}{2} & -\frac{3}{2} \\ \frac{5}{2} & \frac{1}{2} \end{bmatrix} = \begin{bmatrix} \frac{1}{2}(1) & \frac{1}{2}(-3) \\ \frac{1}{2}(5) & \frac{1}{2}(1) \end{bmatrix} = \frac{1}{2} \begin{bmatrix} 1 & -3 \\ 5 & 1 \end{bmatrix}$$

▷ **ALGEBRA HELP**   You can
review the properties of addition
and multiplication of real
numbers (and other properties of
real numbers) in Appendix A.1.

The properties of matrix addition and scalar multiplication are similar to those of addition and multiplication of real numbers.

Properties of Matrix Addition and Scalar Multiplication

Let A, B, and C be $m \times n$ matrices and let c and d be scalars.

1. $A + B = B + A$ Commutative Property of Matrix Addition

2. $A + (B + C) = (A + B) + C$ Associative Property of Matrix Addition

3. $(cd)A = c(dA)$ Associative Property of Scalar Multiplication

4. $1A = A$ Scalar Identity Property

5. $c(A + B) = cA + cB$ Distributive Property

6. $(c + d)A = cA + dA$ Distributive Property

Note that the Associative Property of Matrix Addition allows you to write expressions such as $A + B + C$ without ambiguity because the same sum occurs no matter how the matrices are grouped. This same reasoning applies to sums of four or more matrices.

▷ **TECHNOLOGY**   Most
graphing utilities have the
capability of performing matrix
operations. Consult the user's
guide for your graphing utility
for specific keystrokes. Try
using a graphing utility to find
the sum of the matrices

$$A = \begin{bmatrix} 2 & -3 \\ -1 & 0 \end{bmatrix}$$

and

$$B = \begin{bmatrix} -1 & 4 \\ 2 & -5 \end{bmatrix}.$$

EXAMPLE 4   **Addition of More than Two Matrices**

By adding corresponding entries, you obtain the following sum of four matrices.

$$\begin{bmatrix} 1 \\ 2 \\ -3 \end{bmatrix} + \begin{bmatrix} -1 \\ -1 \\ 2 \end{bmatrix} + \begin{bmatrix} 0 \\ 1 \\ 4 \end{bmatrix} + \begin{bmatrix} 2 \\ -3 \\ -2 \end{bmatrix} = \begin{bmatrix} 2 \\ -1 \\ 1 \end{bmatrix}$$

✓ **Checkpoint** ◀))) *Audio-video solution in English & Spanish at LarsonPrecalculus.com.*

Evaluate the expression.

$$\begin{bmatrix} 3 & -8 \\ 0 & 2 \end{bmatrix} + \begin{bmatrix} -2 & 3 \\ 6 & -5 \end{bmatrix} + \begin{bmatrix} 0 & 7 \\ 4 & -1 \end{bmatrix}$$

EXAMPLE 5   **Using the Distributive Property**

$$3\left(\begin{bmatrix} -2 & 0 \\ 4 & 1 \end{bmatrix} + \begin{bmatrix} 4 & -2 \\ 3 & 7 \end{bmatrix} \right) = 3\begin{bmatrix} -2 & 0 \\ 4 & 1 \end{bmatrix} + 3\begin{bmatrix} 4 & -2 \\ 3 & 7 \end{bmatrix}$$

$$= \begin{bmatrix} -6 & 0 \\ 12 & 3 \end{bmatrix} + \begin{bmatrix} 12 & -6 \\ 9 & 21 \end{bmatrix}$$

$$= \begin{bmatrix} 6 & -6 \\ 21 & 24 \end{bmatrix}$$

✓ **Checkpoint** ◀))) *Audio-video solution in English & Spanish at LarsonPrecalculus.com.*

Evaluate the expression using the Distributive Property.

$$2\left(\begin{bmatrix} 1 & 3 \\ -2 & 2 \end{bmatrix} + \begin{bmatrix} -4 & 0 \\ -3 & 1 \end{bmatrix} \right)$$

In Example 5, you could add the two matrices first and then multiply the resulting matrix by 3. The result would be the same.

One important property of addition of real numbers is that the number 0 is the additive identity. That is, $c + 0 = c$ for any real number c. For matrices, a similar property holds. That is, if A is an $m \times n$ matrix and O is the $m \times n$ **zero matrix** consisting entirely of zeros, then $A + O = A$.

In other words, O is the **additive identity** for the set of all $m \times n$ matrices. For example, the following matrices are the additive identities for the sets of all 2×3 and 2×2 matrices.

$$O = \begin{bmatrix} 0 & 0 & 0 \\ 0 & 0 & 0 \end{bmatrix} \quad \text{and} \quad O = \begin{bmatrix} 0 & 0 \\ 0 & 0 \end{bmatrix}.$$

2×3 zero matrix 2×2 zero matrix

The algebra of real numbers and the algebra of matrices have many similarities. For example, compare the following solutions.

Real Numbers (Solve for x.)	**$m \times n$ Matrices** (Solve for X.)
$x + a = b$	$X + A = B$
$x + a + (-a) = b + (-a)$	$X + A + (-A) = B + (-A)$
$x + 0 = b - a$	$X + O = B - A$
$x = b - a$	$X = B - A$

> •• **REMARK** Remember that matrices are denoted by capital letters. So, when you solve for X, you are solving for a *matrix* that makes the matrix equation true.

The algebra of real numbers and the algebra of matrices also have important differences (see Example 10).

EXAMPLE 6   Solving a Matrix Equation

Solve for X in the equation $3X + A = B$, where

$$A = \begin{bmatrix} 1 & -2 \\ 0 & 3 \end{bmatrix} \quad \text{and} \quad B = \begin{bmatrix} -3 & 4 \\ 2 & 1 \end{bmatrix}.$$

Solution   Begin by solving the matrix equation for X to obtain

$$3X + A = B$$
$$3X = B - A$$
$$X = \frac{1}{3}(B - A).$$

Now, using the matrices A and B, you have

$$X = \frac{1}{3}\left(\begin{bmatrix} -3 & 4 \\ 2 & 1 \end{bmatrix} - \begin{bmatrix} 1 & -2 \\ 0 & 3 \end{bmatrix} \right) \qquad \text{Substitute the matrices.}$$

$$= \frac{1}{3}\begin{bmatrix} -4 & 6 \\ 2 & -2 \end{bmatrix} \qquad \text{Subtract matrix } A \text{ from matrix } B.$$

$$= \begin{bmatrix} -\dfrac{4}{3} & 2 \\ \dfrac{2}{3} & -\dfrac{2}{3} \end{bmatrix}. \qquad \text{Multiply the resulting matrix by } \frac{1}{3}.$$

✓ **Checkpoint** ◀))   Audio-video solution in English & Spanish at LarsonPrecalculus.com.

Solve for X in the equation $2X - A = B$, where

$$A = \begin{bmatrix} 6 & 1 \\ 0 & 3 \end{bmatrix} \quad \text{and} \quad B = \begin{bmatrix} 4 & -1 \\ -2 & 5 \end{bmatrix}.$$

Matrix Multiplication

Another basic matrix operation is **matrix multiplication.** At first glance, the definition may seem unusual. You will see later, however, that this definition of the product of two matrices has many practical applications.

> ### Definition of Matrix Multiplication
>
> If $A = [a_{ij}]$ is an $m \times n$ matrix and $B = [b_{ij}]$ is an $n \times p$ matrix, then the product AB is an $m \times p$ matrix
>
> $$AB = [c_{ij}]$$
>
> where $c_{ij} = a_{i1}b_{1j} + a_{i2}b_{2j} + a_{i3}b_{3j} + \cdots + a_{in}b_{nj}.$

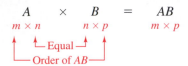

$$
\begin{array}{ccccc}
A & \times & B & = & AB \\
m \times n & & n \times p & & m \times p
\end{array}
$$

Equal
Order of AB

The definition of matrix multiplication indicates a *row-by-column* multiplication, where the entry in the ith row and jth column of the product AB is obtained by multiplying the entries in the ith row of A by the corresponding entries in the jth column of B and then adding the results. So, for the product of two matrices to be defined, the number of columns of the first matrix must equal the number of rows of the second matrix. That is, the middle two indices must be the same. The outside two indices give the order of the product, as shown at the left. The general pattern for matrix multiplication is as follows.

$$
\begin{bmatrix}
a_{11} & a_{12} & a_{13} & \cdots & a_{1n} \\
a_{21} & a_{22} & a_{23} & \cdots & a_{2n} \\
a_{31} & a_{32} & a_{33} & \cdots & a_{3n} \\
\vdots & \vdots & \vdots & & \vdots \\
a_{i1} & a_{i2} & a_{i3} & \cdots & a_{in} \\
\vdots & \vdots & \vdots & & \vdots \\
a_{m1} & a_{m2} & a_{m3} & \cdots & a_{mn}
\end{bmatrix}
\begin{bmatrix}
b_{11} & b_{12} & \cdots & b_{1j} & \cdots & b_{1p} \\
b_{21} & b_{22} & \cdots & b_{2j} & \cdots & b_{2p} \\
b_{31} & b_{32} & \cdots & b_{3j} & \cdots & b_{3p} \\
\vdots & \vdots & & \vdots & & \vdots \\
b_{n1} & b_{n2} & \cdots & b_{nj} & \cdots & b_{np}
\end{bmatrix}
=
\begin{bmatrix}
c_{11} & c_{12} & \cdots & c_{1j} & \cdots & c_{1p} \\
c_{21} & c_{22} & \cdots & c_{2j} & \cdots & c_{2p} \\
 & & & \vdots & & \vdots \\
c_{i1} & c_{i2} & \cdots & c_{ij} & \cdots & c_{ip} \\
 & & & \vdots & & \vdots \\
c_{m1} & c_{m2} & \cdots & c_{mj} & \cdots & c_{mp}
\end{bmatrix}
$$

$$a_{i1}b_{1j} + a_{i2}b_{2j} + a_{i3}b_{3j} + \cdots + a_{in}b_{nj} = c_{ij}$$

EXAMPLE 7   **Finding the Product of Two Matrices**

Find the product AB using $A = \begin{bmatrix} -1 & 3 \\ 4 & -2 \\ 5 & 0 \end{bmatrix}$ and $B = \begin{bmatrix} -3 & 2 \\ -4 & 1 \end{bmatrix}$.

Solution   To find the entries of the product, multiply each row of A by each column of B.

$$
AB = \begin{bmatrix} -1 & 3 \\ 4 & -2 \\ 5 & 0 \end{bmatrix} \begin{bmatrix} -3 & 2 \\ -4 & 1 \end{bmatrix}
$$

$$
= \begin{bmatrix}
(-1)(-3) + (3)(-4) & (-1)(2) + (3)(1) \\
(4)(-3) + (-2)(-4) & (4)(2) + (-2)(1) \\
(5)(-3) + (0)(-4) & (5)(2) + (0)(1)
\end{bmatrix}
$$

$$
= \begin{bmatrix} -9 & 1 \\ -4 & 6 \\ -15 & 10 \end{bmatrix}
$$

• • REMARK In Example 7, the product AB is defined because the number of columns of A is equal to the number of rows of B. Also, note that the product AB has order 3×2.

✓ *Checkpoint*))) *Audio-video solution in English & Spanish at LarsonPrecalculus.com.*

Find the product AB using $A = \begin{bmatrix} -1 & 4 \\ 2 & 0 \\ 1 & 2 \end{bmatrix}$ and $B = \begin{bmatrix} 1 & -2 \\ 0 & 7 \end{bmatrix}$.

EXAMPLE 8   **Finding the Product of Two Matrices**

Find the product AB using $A = \begin{bmatrix} 1 & 0 & 3 \\ 2 & -1 & -2 \end{bmatrix}$ and $B = \begin{bmatrix} -2 & 4 \\ 1 & 0 \\ -1 & 1 \end{bmatrix}$.

Solution   Note that the order of A is 2×3 and the order of B is 3×2. So, the product AB has order 2×2.

$$AB = \begin{bmatrix} 1 & 0 & 3 \\ 2 & -1 & -2 \end{bmatrix}\begin{bmatrix} -2 & 4 \\ 1 & 0 \\ -1 & 1 \end{bmatrix}$$

$$= \begin{bmatrix} 1(-2) + & 0(1) + & 3(-1) & 1(4) + & 0(0) + & 3(1) \\ 2(-2) + & (-1)(1) + & (-2)(-1) & 2(4) + & (-1)(0) + & (-2)(1) \end{bmatrix}$$

$$= \begin{bmatrix} -5 & 7 \\ -3 & 6 \end{bmatrix}$$

✓ **Checkpoint** 🔊)) Audio-video solution in English & Spanish at LarsonPrecalculus.com.

Find the product AB using $A = \begin{bmatrix} 0 & 4 & -3 \\ 2 & 1 & 7 \\ 3 & -2 & 1 \end{bmatrix}$ and $B = \begin{bmatrix} -2 & 0 \\ 0 & -4 \\ 1 & 2 \end{bmatrix}$.

EXAMPLE 9   **Patterns in Matrix Multiplication**

a. $\underset{2 \times 2}{\begin{bmatrix} 3 & 4 \\ -2 & 5 \end{bmatrix}} \underset{2 \times 2}{\begin{bmatrix} 1 & 0 \\ 0 & 1 \end{bmatrix}} = \underset{2 \times 2}{\begin{bmatrix} 3 & 4 \\ -2 & 5 \end{bmatrix}}$

b. $\underset{3 \times 3}{\begin{bmatrix} 6 & 2 & 0 \\ 3 & -1 & 2 \\ 1 & 4 & 6 \end{bmatrix}} \underset{3 \times 1}{\begin{bmatrix} 1 \\ 2 \\ -3 \end{bmatrix}} = \underset{3 \times 1}{\begin{bmatrix} 10 \\ -5 \\ -9 \end{bmatrix}}$

c. The product AB for the following matrices is not defined.

$$A = \underset{3 \times 2}{\begin{bmatrix} -2 & 1 \\ 1 & -3 \\ 1 & 4 \end{bmatrix}} \quad \text{and} \quad B = \underset{3 \times 4}{\begin{bmatrix} -2 & 3 & 1 & 4 \\ 0 & 1 & -1 & 2 \\ 2 & -1 & 0 & 1 \end{bmatrix}}$$

✓ **Checkpoint** 🔊)) Audio-video solution in English & Spanish at LarsonPrecalculus.com.

Find, if possible, the product AB using $A = \begin{bmatrix} 3 & 1 & 2 \\ 7 & 0 & -2 \end{bmatrix}$ and $B = \begin{bmatrix} 6 & 4 \\ 2 & -1 \end{bmatrix}$.

▷ **REMARK** In Example 10, note that the two products are different. Even when both AB and BA are defined, matrix multiplication is not, in general, commutative. That is, for most matrices, $AB \neq BA$. This is one way in which the algebra of real numbers and the algebra of matrices differ.

EXAMPLE 10   **Patterns in Matrix Multiplication**

a. $\underset{1 \times 3}{\begin{bmatrix} 1 & -2 & -3 \end{bmatrix}} \underset{3 \times 1}{\begin{bmatrix} 2 \\ -1 \\ 1 \end{bmatrix}} = \underset{1 \times 1}{\begin{bmatrix} 1 \end{bmatrix}}$ b. $\underset{3 \times 1}{\begin{bmatrix} 2 \\ -1 \\ 1 \end{bmatrix}} \underset{1 \times 3}{\begin{bmatrix} 1 & -2 & -3 \end{bmatrix}} = \underset{3 \times 3}{\begin{bmatrix} 2 & -4 & -6 \\ -1 & 2 & 3 \\ 1 & -2 & -3 \end{bmatrix}}$

✓ **Checkpoint** 🔊)) Audio-video solution in English & Spanish at LarsonPrecalculus.com.

Find AB and BA using $A = \begin{bmatrix} 3 & -1 \end{bmatrix}$ and $B = \begin{bmatrix} 1 \\ -3 \end{bmatrix}$.

EXAMPLE 11   **Squaring a Matrix**

Find A^2, where $A = \begin{bmatrix} 3 & 1 \\ -1 & 2 \end{bmatrix}$. (*Note:* $A^2 = AA$.)

Solution

$$A^2 = AA = \begin{bmatrix} 3 & 1 \\ -1 & 2 \end{bmatrix}\begin{bmatrix} 3 & 1 \\ -1 & 2 \end{bmatrix} = \begin{bmatrix} 3(3) + 1(-1) & 3(1) + 1(2) \\ -1(3) + 2(-1) & -1(1) + 2(2) \end{bmatrix}$$

$$= \begin{bmatrix} 8 & 5 \\ -5 & 3 \end{bmatrix}$$

✓ *Checkpoint*))) Audio-video solution in English & Spanish at LarsonPrecalculus.com.

Find A^2, where $A = \begin{bmatrix} 2 & 1 \\ 3 & -2 \end{bmatrix}$.

Properties of Matrix Multiplication

Let A, B, and C be matrices and let c be a scalar.

1. $A(BC) = (AB)C$ Associative Property of Matrix Multiplication

2. $A(B + C) = AB + AC$ Distributive Property

3. $(A + B)C = AC + BC$ Distributive Property

4. $c(AB) = (cA)B = A(cB)$ Associative Property of Scalar Multiplication

---

Definition of Identity Matrix

The $n \times n$ matrix that consists of 1's on its main diagonal and 0's elsewhere is called the **identity matrix of order** $n \times n$ and is denoted by

$$I_n = \begin{bmatrix} 1 & 0 & 0 & \cdots & 0 \\ 0 & 1 & 0 & \cdots & 0 \\ 0 & 0 & 1 & \cdots & 0 \\ \vdots & \vdots & \vdots & & \vdots \\ 0 & 0 & 0 & \cdots & 1 \end{bmatrix}.$$ Identity matrix

Note that an identity matrix must be *square*. When the order is understood to be $n \times n$, you can denote I_n simply by I.

If A is an $n \times n$ matrix, then the identity matrix has the property that $AI_n = A$ and $I_nA = A$. For example,

$$\begin{bmatrix} 3 & -2 & 5 \\ 1 & 0 & 4 \\ -1 & 2 & -3 \end{bmatrix}\begin{bmatrix} 1 & 0 & 0 \\ 0 & 1 & 0 \\ 0 & 0 & 1 \end{bmatrix} = \begin{bmatrix} 3 & -2 & 5 \\ 1 & 0 & 4 \\ -1 & 2 & -3 \end{bmatrix}$$ $AI = A$

and

$$\begin{bmatrix} 1 & 0 & 0 \\ 0 & 1 & 0 \\ 0 & 0 & 1 \end{bmatrix}\begin{bmatrix} 3 & -2 & 5 \\ 1 & 0 & 4 \\ -1 & 2 & -3 \end{bmatrix} = \begin{bmatrix} 3 & -2 & 5 \\ 1 & 0 & 4 \\ -1 & 2 & -3 \end{bmatrix}.$$ $IA = A$

Applications

Matrix multiplication can be used to represent a system of linear equations. Note how the system below can be written as the matrix equation $AX = B$, where A is the *coefficient matrix* of the system and X and B are column matrices. The column matrix B is also called a *constant matrix*. Its entries are the constant terms in the system of equations.

System

$$\begin{cases} a_{11}x_1 + a_{12}x_2 + a_{13}x_3 = b_1 \\ a_{21}x_1 + a_{22}x_2 + a_{23}x_3 = b_2 \\ a_{31}x_1 + a_{32}x_2 + a_{33}x_3 = b_3 \end{cases}$$

Matrix Equation $AX = B$

$$\underset{A}{\begin{bmatrix} a_{11} & a_{12} & a_{13} \\ a_{21} & a_{22} & a_{23} \\ a_{31} & a_{32} & a_{33} \end{bmatrix}} \underset{X}{\begin{bmatrix} x_1 \\ x_2 \\ x_3 \end{bmatrix}} = \underset{B}{\begin{bmatrix} b_1 \\ b_2 \\ b_3 \end{bmatrix}}$$

In Example 12, $[A \vdots B]$ represents the augmented matrix formed when matrix B is *adjoined* to matrix A. Also, $[I \vdots X]$ represents the reduced row-echelon form of the augmented matrix that yields the solution of the system.

Many real-life applications of linear systems involve enormous numbers of equations and variables. For example, a flight crew scheduling problem for American Airlines required the manipulation of matrices with 837 rows and 12,753,313 columns. *(Source: Very Large-Scale Linear Programming. A Case Study in Combining Interior Point and Simplex Methods, Bixby, Robert E., et al., Operations Research, 40, no. 5)*

EXAMPLE 12   Solving a System of Linear Equations

For the system of linear equations, (a) write the system as a matrix equation, $AX = B$, and (b) use Gauss-Jordan elimination on $[A \vdots B]$ to solve for the matrix X.

$$\begin{cases} x_1 - 2x_2 + x_3 = -4 \\ \quad\quad x_2 + 2x_3 = 4 \\ 2x_1 + 3x_2 - 2x_3 = 2 \end{cases}$$

Solution

a. In matrix form, $AX = B$, the system can be written as follows.

$$\begin{bmatrix} 1 & -2 & 1 \\ 0 & 1 & 2 \\ 2 & 3 & -2 \end{bmatrix} \begin{bmatrix} x_1 \\ x_2 \\ x_3 \end{bmatrix} = \begin{bmatrix} -4 \\ 4 \\ 2 \end{bmatrix}$$

b. The augmented matrix is formed by adjoining matrix B to matrix A.

$$[A \vdots B] = \begin{bmatrix} 1 & -2 & 1 & \vdots & -4 \\ 0 & 1 & 2 & \vdots & 4 \\ 2 & 3 & -2 & \vdots & 2 \end{bmatrix}$$

Using Gauss-Jordan elimination, you can rewrite this matrix as

$$[I \vdots X] = \begin{bmatrix} 1 & 0 & 0 & \vdots & -1 \\ 0 & 1 & 0 & \vdots & 2 \\ 0 & 0 & 1 & \vdots & 1 \end{bmatrix}.$$

So, the solution of the matrix equation is

$$X = \begin{bmatrix} x_1 \\ x_2 \\ x_3 \end{bmatrix} = \begin{bmatrix} -1 \\ 2 \\ 1 \end{bmatrix}.$$

✓ *Checkpoint* *Audio-video solution in English & Spanish at LarsonPrecalculus.com.*

For the system of linear equations, (a) write the system as a matrix equation, $AX = B$, and (b) use Gauss-Jordan elimination on $[A \vdots B]$ to solve for the matrix X.

$$\begin{cases} -2x_1 - 3x_2 = -4 \\ \quad 6x_1 + x_2 = -36 \end{cases}$$

EXAMPLE 13 **Softball Team Expenses**

Two softball teams submit equipment lists to their sponsors.

Equipment	Women's Team	Men's Team
Bats	12	15
Balls	45	38
Gloves	15	17

Each bat costs $80, each ball costs $6, and each glove costs $60. Use matrices to find the total cost of equipment for each team.

Solution The equipment lists E and the costs per item C can be written in matrix form as

$$E = \begin{bmatrix} 12 & 15 \\ 45 & 38 \\ 15 & 17 \end{bmatrix}$$

and

$$C = \begin{bmatrix} 80 & 6 & 60 \end{bmatrix}.$$

The total cost of equipment for each team is given by the product

$$CE = \begin{bmatrix} 80 & 6 & 60 \end{bmatrix} \begin{bmatrix} 12 & 15 \\ 45 & 38 \\ 15 & 17 \end{bmatrix}$$

$$= \begin{bmatrix} 80(12) + 6(45) + 60(15) & 80(15) + 6(38) + 60(17) \end{bmatrix}$$

$$= \begin{bmatrix} 2130 & 2448 \end{bmatrix}.$$

So, the total cost of equipment for the women's team is $2130, and the total cost of equipment for the men's team is $2448.

 Checkpoint ◄))) *Audio-video solution in English & Spanish at LarsonPrecalculus.com.*

Repeat Example 13 when each bat costs $100, each ball costs $7, and each glove costs $65. ■

▸ **• • REMARK** Notice in Example 13 that you cannot find the total cost using the product EC because EC is not defined. That is, the number of columns of E (2 columns) does not equal the number of rows of C (1 row).

Summarize *(Section 8.2)*

1. State the conditions under which two matrices are equal *(page 403)*. For an example of the equality of matrices, see Example 1.

2. State the definition of matrix addition *(page 404)*. For an example of matrix addition, see Example 2.

3. State the definition of scalar multiplication *(page 404)*. For an example of scalar multiplication, see Example 3.

4. List the properties of matrix addition and scalar multiplication *(page 406)*. For examples of using these properties, see Examples 4, 5, and 6.

5. State the definition of matrix multiplication *(page 408)*. For examples of matrix multiplication, see Examples 7–11.

6. Describe applications of matrix multiplication *(pages 411 and 412, Examples 12 and 13)*.

8.3 The Inverse of a Square Matrix

You can use inverse matrices to model and solve real-life problems, such as finding the currents in a circuit.

- Verify that two matrices are inverses of each other.
- Use Gauss-Jordan elimination to find the inverses of matrices.
- Use a formula to find the inverses of 2 × 2 matrices.
- Use inverse matrices to solve systems of linear equations.

The Inverse of a Matrix

This section further develops the algebra of matrices. To begin, consider the real number equation $ax = b$. To solve this equation for x, multiply each side of the equation by a^{-1} (provided that $a \neq 0$).

$$ax = b$$
$$(a^{-1}a)x = a^{-1}b$$
$$(1)x = a^{-1}b$$
$$x = a^{-1}b$$

The number a^{-1} is called the *multiplicative inverse* of a because $a^{-1}a = 1$. The definition of the multiplicative **inverse of a matrix** is similar.

Definition of the Inverse of a Square Matrix

Let A be an $n \times n$ matrix and let I_n be the $n \times n$ identity matrix. If there exists a matrix A^{-1} such that

$$AA^{-1} = I_n = A^{-1}A$$

then A^{-1} is called the **inverse** of A. The symbol A^{-1} is read "A inverse."

EXAMPLE 1   **The Inverse of a Matrix**

Show that B is the inverse of A, where

$$A = \begin{bmatrix} -1 & 2 \\ -1 & 1 \end{bmatrix} \quad \text{and} \quad B = \begin{bmatrix} 1 & -2 \\ 1 & -1 \end{bmatrix}.$$

Solution   To show that B is the inverse of A, show that $AB = I = BA$, as follows.

$$AB = \begin{bmatrix} -1 & 2 \\ -1 & 1 \end{bmatrix}\begin{bmatrix} 1 & -2 \\ 1 & -1 \end{bmatrix} = \begin{bmatrix} -1+2 & 2-2 \\ -1+1 & 2-1 \end{bmatrix} = \begin{bmatrix} 1 & 0 \\ 0 & 1 \end{bmatrix}$$

$$BA = \begin{bmatrix} 1 & -2 \\ 1 & -1 \end{bmatrix}\begin{bmatrix} -1 & 2 \\ -1 & 1 \end{bmatrix} = \begin{bmatrix} -1+2 & 2-2 \\ -1+1 & 2-1 \end{bmatrix} = \begin{bmatrix} 1 & 0 \\ 0 & 1 \end{bmatrix}$$

As you can see, $AB = I = BA$. This is an example of a square matrix that has an inverse. Note that not all square matrices have inverses.

✓ *Checkpoint*  *Audio-video solution in English & Spanish at LarsonPrecalculus.com.*

Show that B is the inverse of A, where

$$A = \begin{bmatrix} 2 & -1 \\ -3 & 1 \end{bmatrix} \quad \text{and} \quad B = \begin{bmatrix} -1 & -1 \\ -3 & -2 \end{bmatrix}.$$

Recall that it is not always true that $AB = BA$, even when both products are defined. However, if A and B are both square matrices and $AB = I_n$, then it can be shown that $BA = I_n$. So, in Example 1, you need only to check that $AB = I_2$.

Ingrid Prats/Shutterstock.com

One real-life application of inverse matrices is in the study of beam deflection. In a simply supported elastic beam subjected to multiple forces, deflection **d** is related to force **w** by the matrix equation

d = F**w**

where F is a *flexibility matrix* whose entries depend on the material of the beam. The inverse of the flexibility matrix, F^{-1}, is called the *stiffness matrix*.

Finding Inverse Matrices

If a matrix A has an inverse, then A is called **invertible** (or **nonsingular**); otherwise, A is called **singular**. A nonsquare matrix cannot have an inverse. To see this, note that when A is of order $m \times n$ and B is of order $n \times m$ (where $m \neq n$), the products AB and BA are of different orders and so cannot be equal to each other. Not all square matrices have inverses (see the matrix at the bottom of page 416). When, however, a matrix does have an inverse, that inverse is unique. Example 2 shows how to use a system of equations to find the inverse of a matrix.

EXAMPLE 2   **Finding the Inverse of a Matrix**

Find the inverse of

$$A = \begin{bmatrix} 1 & 4 \\ -1 & -3 \end{bmatrix}.$$

Solution   To find the inverse of A, solve the matrix equation $AX = I$ for X.

$$\underset{A}{\begin{bmatrix} 1 & 4 \\ -1 & -3 \end{bmatrix}} \underset{X}{\begin{bmatrix} x_{11} & x_{12} \\ x_{21} & x_{22} \end{bmatrix}} = \underset{I}{\begin{bmatrix} 1 & 0 \\ 0 & 1 \end{bmatrix}}$$ Write matrix equation.

$$\begin{bmatrix} x_{11} + 4x_{21} & x_{12} + 4x_{22} \\ -x_{11} - 3x_{21} & -x_{12} - 3x_{22} \end{bmatrix} = \begin{bmatrix} 1 & 0 \\ 0 & 1 \end{bmatrix}$$ Multiply.

Equating corresponding entries, you obtain two systems of linear equations.

$$\begin{cases} x_{11} + 4x_{21} = 1 \\ -x_{11} - 3x_{21} = 0 \end{cases} \qquad \begin{cases} x_{12} + 4x_{22} = 0 \\ -x_{12} - 3x_{22} = 1 \end{cases}$$

Solve the first system using elementary row operations to determine that

$$x_{11} = -3 \quad \text{and} \quad x_{21} = 1.$$

From the second system, you can determine that

$$x_{12} = -4 \quad \text{and} \quad x_{22} = 1.$$

So, the inverse of A is

$$X = A^{-1}$$
$$= \begin{bmatrix} -3 & -4 \\ 1 & 1 \end{bmatrix}.$$

You can use matrix multiplication to check this result.

Check

$$AA^{-1} = \begin{bmatrix} 1 & 4 \\ -1 & -3 \end{bmatrix} \begin{bmatrix} -3 & -4 \\ 1 & 1 \end{bmatrix} = \begin{bmatrix} 1 & 0 \\ 0 & 1 \end{bmatrix} \checkmark$$

$$A^{-1}A = \begin{bmatrix} -3 & -4 \\ 1 & 1 \end{bmatrix} \begin{bmatrix} 1 & 4 \\ -1 & -3 \end{bmatrix} = \begin{bmatrix} 1 & 0 \\ 0 & 1 \end{bmatrix} \checkmark$$

✔ *Checkpoint*    *Audio-video solution in English & Spanish at LarsonPrecalculus.com.*

Find the inverse of

$$A = \begin{bmatrix} 1 & -2 \\ -1 & 3 \end{bmatrix}.$$

In Example 2, note that the two systems of linear equations have the *same coefficient matrix A.* Rather than solve the two systems represented by

$$\begin{bmatrix} 1 & 4 & \vdots & 1 \\ -1 & -3 & \vdots & 0 \end{bmatrix}$$

and

$$\begin{bmatrix} 1 & 4 & \vdots & 0 \\ -1 & -3 & \vdots & 1 \end{bmatrix}$$

separately, you can solve them *simultaneously* by *adjoining* the identity matrix to the coefficient matrix to obtain

$$\begin{matrix} A & & & I & \\ \begin{bmatrix} 1 & 4 & \vdots & 1 & 0 \\ -1 & -3 & \vdots & 0 & 1 \end{bmatrix}. \end{matrix}$$

This "doubly augmented" matrix can be represented as

$$[A \vdots I].$$

By applying Gauss-Jordan elimination to this matrix, you can solve *both* systems with a single elimination process.

$$\begin{bmatrix} 1 & 4 & \vdots & 1 & 0 \\ -1 & -3 & \vdots & 0 & 1 \end{bmatrix}$$

$$R_1 + R_2 \rightarrow \begin{bmatrix} 1 & 4 & \vdots & 1 & 0 \\ 0 & 1 & \vdots & 1 & 1 \end{bmatrix}$$

$$-4R_2 + R_1 \rightarrow \begin{bmatrix} 1 & 0 & \vdots & -3 & -4 \\ 0 & 1 & \vdots & 1 & 1 \end{bmatrix}$$

So, from the "doubly augmented" matrix $[A \vdots I]$, you obtain the matrix $[I \vdots A^{-1}]$.

$$\begin{matrix} A & & I & & & I & & A^{-1} \\ \begin{bmatrix} 1 & 4 & \vdots & 1 & 0 \\ -1 & -3 & \vdots & 0 & 1 \end{bmatrix} \Rightarrow \begin{bmatrix} 1 & 0 & \vdots & -3 & -4 \\ 0 & 1 & \vdots & 1 & 1 \end{bmatrix} \end{matrix}$$

This procedure (or algorithm) works for any square matrix that has an inverse.

▷ **TECHNOLOGY** Most graphing utilities can find the inverse of a square matrix. To do so, you may have to use the inverse key $\boxed{x^{-1}}$. Consult the user's guide for your graphing utility for specific keystrokes.

Finding an Inverse Matrix

Let A be a square matrix of order $n \times n$.

1. Write the $n \times 2n$ matrix that consists of the given matrix A on the left and the $n \times n$ identity matrix I on the right to obtain

   $$[A \vdots I].$$

2. If possible, row reduce A to I using elementary row operations on the *entire* matrix

   $$[A \vdots I].$$

   The result will be the matrix

   $$[I \vdots A^{-1}].$$

   If this is not possible, then A is not invertible.

3. Check your work by multiplying to see that

   $$AA^{-1} = I = A^{-1}A.$$

EXAMPLE 3   **Finding the Inverse of a Matrix**

Find the inverse of

$$A = \begin{bmatrix} 1 & -1 & 0 \\ 1 & 0 & -1 \\ 6 & -2 & -3 \end{bmatrix}.$$

Solution   Begin by adjoining the identity matrix to A to form the matrix

$$[A \ \vdots \ I] = \begin{bmatrix} 1 & -1 & 0 & \vdots & 1 & 0 & 0 \\ 1 & 0 & -1 & \vdots & 0 & 1 & 0 \\ 6 & -2 & -3 & \vdots & 0 & 0 & 1 \end{bmatrix}.$$

Use elementary row operations to obtain the form $[I \ \vdots \ A^{-1}]$, as follows.

$$\begin{matrix} \\ -R_1 + R_2 \rightarrow \\ -6R_1 + R_3 \rightarrow \end{matrix} \begin{bmatrix} 1 & -1 & 0 & \vdots & 1 & 0 & 0 \\ 0 & 1 & -1 & \vdots & -1 & 1 & 0 \\ 0 & 4 & -3 & \vdots & -6 & 0 & 1 \end{bmatrix}$$

$$\begin{matrix} R_2 + R_1 \rightarrow \\ \\ -4R_2 + R_3 \rightarrow \end{matrix} \begin{bmatrix} 1 & 0 & -1 & \vdots & 0 & 1 & 0 \\ 0 & 1 & -1 & \vdots & -1 & 1 & 0 \\ 0 & 0 & 1 & \vdots & -2 & -4 & 1 \end{bmatrix}$$

$$\begin{matrix} R_3 + R_1 \rightarrow \\ R_3 + R_2 \rightarrow \\ \\ \end{matrix} \begin{bmatrix} 1 & 0 & 0 & \vdots & -2 & -3 & 1 \\ 0 & 1 & 0 & \vdots & -3 & -3 & 1 \\ 0 & 0 & 1 & \vdots & -2 & -4 & 1 \end{bmatrix} = [I \ \vdots \ A^{-1}]$$

So, the matrix A is invertible and its inverse is $A^{-1} = \begin{bmatrix} -2 & -3 & 1 \\ -3 & -3 & 1 \\ -2 & -4 & 1 \end{bmatrix}.$

Confirm this result by multiplying A and A^{-1} to obtain I, as follows.

Check

$$AA^{-1} = \begin{bmatrix} 1 & -1 & 0 \\ 1 & 0 & -1 \\ 6 & -2 & -3 \end{bmatrix} \begin{bmatrix} -2 & -3 & 1 \\ -3 & -3 & 1 \\ -2 & -4 & 1 \end{bmatrix} = \begin{bmatrix} 1 & 0 & 0 \\ 0 & 1 & 0 \\ 0 & 0 & 1 \end{bmatrix} = I$$

> **• • REMARK** Be sure to check your solution because it is easy to make algebraic errors when using elementary row operations.

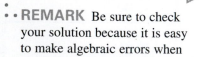 **Checkpoint**   *Audio-video solution in English & Spanish at LarsonPrecalculus.com.*

Find the inverse of

$$A = \begin{bmatrix} 1 & -2 & -1 \\ 0 & -1 & 2 \\ 1 & -2 & 0 \end{bmatrix}.$$

The process shown in Example 3 applies to any $n \times n$ matrix A. When using this algorithm, if the matrix A does not reduce to the identity matrix, then A does not have an inverse. For instance, the following matrix has no inverse.

$$A = \begin{bmatrix} 1 & 2 & 0 \\ 3 & -1 & 2 \\ -2 & 3 & -2 \end{bmatrix}$$

To confirm that matrix A above has no inverse, adjoin the identity matrix to A to form $[A \ \vdots \ I]$ and perform elementary row operations on the matrix. After doing so, you will see that it is impossible to obtain the identity matrix I on the left. So, A is not invertible.

The Inverse of a 2 × 2 Matrix

Using Gauss-Jordan elimination to find the inverse of a matrix works well (even as a computer technique) for matrices of order 3 × 3 or greater. For 2 × 2 matrices, however, many people prefer to use a formula for the inverse rather than Gauss-Jordan elimination. This simple formula, which works *only* for 2 × 2 matrices, is explained as follows. If A is a 2 × 2 matrix given by

$$A = \begin{bmatrix} a & b \\ c & d \end{bmatrix}$$

then A is invertible if and only if $ad - bc \neq 0$. Moreover, if $ad - bc \neq 0$, then the inverse is given by

$$A^{-1} = \frac{1}{ad - bc}\begin{bmatrix} d & -b \\ -c & a \end{bmatrix}.$$ Formula for the inverse of a 2 × 2 matrix

The denominator $ad - bc$ is called the **determinant** of the 2 × 2 matrix A. You will study determinants in the next section.

EXAMPLE 4   Finding the Inverse of a 2 × 2 Matrix

If possible, find the inverse of each matrix.

a. $A = \begin{bmatrix} 3 & -1 \\ -2 & 2 \end{bmatrix}$ **b.** $B = \begin{bmatrix} 3 & -1 \\ -6 & 2 \end{bmatrix}$

Solution

a. For the matrix A, apply the formula for the inverse of a 2 × 2 matrix to obtain

$$ad - bc = 3(2) - (-1)(-2) = 4.$$

Because this quantity is not zero, the matrix is invertible. The inverse is formed by interchanging the entries on the main diagonal, changing the signs of the other two entries, and multiplying by the scalar $\frac{1}{4}$, as follows.

$$A^{-1} = \frac{1}{ad - bc}\begin{bmatrix} d & -b \\ -c & a \end{bmatrix}$$ Formula for the inverse of a 2 × 2 matrix

$$= \frac{1}{4}\begin{bmatrix} 2 & 1 \\ 2 & 3 \end{bmatrix}$$ Substitute for a, b, c, d, and the determinant.

$$= \begin{bmatrix} \frac{1}{4}(2) & \frac{1}{4}(1) \\ \frac{1}{4}(2) & \frac{1}{4}(3) \end{bmatrix}$$ Multiply by the scalar $\frac{1}{4}$.

$$= \begin{bmatrix} \frac{1}{2} & \frac{1}{4} \\ \frac{1}{2} & \frac{3}{4} \end{bmatrix}$$ Simplify.

b. For the matrix B, you have

$$ad - bc = 3(2) - (-1)(-6) = 0.$$

Because $ad - bc = 0$, B is not invertible.

✓ **Checkpoint** ◀))) *Audio-video solution in English & Spanish at LarsonPrecalculus.com.*

If possible, find the inverse of $A = \begin{bmatrix} 5 & -1 \\ 3 & 4 \end{bmatrix}$.

Systems of Linear Equations

You know that a system of linear equations can have exactly one solution, infinitely many solutions, or no solution. If the coefficient matrix A of a *square* system (a system that has the same number of equations as variables) is invertible, then the system has a unique solution, which is defined as follows.

> **A System of Equations with a Unique Solution**
>
> If A is an invertible matrix, then the system of linear equations represented by $AX = B$ has a unique solution given by $X = A^{-1}B$.

EXAMPLE 5 **Solving a System Using an Inverse Matrix**

Use an inverse matrix to solve the system

$$\begin{cases} x + y + z = 10{,}000 \\ 0.06x + 0.075y + 0.095z = 730. \\ x - 2z = 0 \end{cases}$$

▷ **TECHNOLOGY** To solve a system of equations with a graphing utility, enter the matrices A and B in the matrix editor. Then, using the inverse key, solve for X.

A [x⁻¹] B [ENTER]

The screen will display the solution, matrix X.

Solution Begin by writing the system in the matrix form $AX = B$.

$$\begin{bmatrix} 1 & 1 & 1 \\ 0.06 & 0.075 & 0.095 \\ 1 & 0 & -2 \end{bmatrix} \begin{bmatrix} x \\ y \\ z \end{bmatrix} = \begin{bmatrix} 10{,}000 \\ 730 \\ 0 \end{bmatrix}$$

Then, use Gauss-Jordan elimination to find A^{-1}.

$$A^{-1} = \begin{bmatrix} 15 & -200 & -2 \\ -21.5 & 300 & 3.5 \\ 7.5 & -100 & -1.5 \end{bmatrix}$$

Finally, multiply B by A^{-1} on the left to obtain the solution.

$$X = A^{-1}B = \begin{bmatrix} 15 & -200 & -2 \\ -21.5 & 300 & 3.5 \\ 7.5 & -100 & -1.5 \end{bmatrix} \begin{bmatrix} 10{,}000 \\ 730 \\ 0 \end{bmatrix} = \begin{bmatrix} 4000 \\ 4000 \\ 2000 \end{bmatrix}$$

The solution of the system is $x = 4000$, $y = 4000$, and $z = 2000$.

✓ **Checkpoint** ◀)) *Audio-video solution in English & Spanish at LarsonPrecalculus.com.*

Use an inverse matrix to solve the system $\begin{cases} 2x + 3y + z = -1 \\ 3x + 3y + z = 1. \\ 2x + 4y + z = -2 \end{cases}$

Summarize (Section 8.3)

1. State the definition of the inverse of a square matrix *(page 413)*. For an example of how to show that a matrix is the inverse of another matrix, see Example 1.

2. Explain how to find an inverse matrix *(pages 414 and 415)*. For examples of finding inverse matrices, see Examples 2 and 3.

3. Give the formula for finding the inverse of a 2×2 matrix *(page 417)*. For an example of using this formula, see Example 4.

4. Describe how to use an inverse matrix to solve a system of linear equations *(page 418)*. For an example of using an inverse matrix to solve a system of linear equations, see Example 5.

8.4 The Determinant of a Square Matrix

Determinants are often used in other branches of mathematics. For instance, some types of determinants are useful when changes of variables are made in calculus.

■ Find the determinants of 2×2 matrices.
■ Find minors and cofactors of square matrices.
■ Find the determinants of square matrices.

The Determinant of a 2×2 Matrix

Every *square* matrix can be associated with a real number called its **determinant.** Determinants have many uses, and several will be discussed in this and the next section. Historically, the use of determinants arose from special number patterns that occur when systems of linear equations are solved. For instance, the system

$$\begin{cases} a_1 x + b_1 y = c_1 \\ a_2 x + b_2 y = c_2 \end{cases}$$

has a solution

$$x = \frac{c_1 b_2 - c_2 b_1}{a_1 b_2 - a_2 b_1} \quad \text{and} \quad y = \frac{a_1 c_2 - a_2 c_1}{a_1 b_2 - a_2 b_1}$$

provided that $a_1 b_2 - a_2 b_1 \neq 0$. Note that the denominators of the two fractions are the same. This denominator is called the *determinant* of the coefficient matrix of the system.

Coefficient Matrix   **Determinant**

$$A = \begin{bmatrix} a_1 & b_1 \\ a_2 & b_2 \end{bmatrix} \qquad \det(A) = a_1 b_2 - a_2 b_1$$

The determinant of the matrix A can also be denoted by vertical bars on both sides of the matrix, as indicated in the following definition.

Definition of the Determinant of a 2×2 Matrix

The **determinant** of the matrix

$$A = \begin{bmatrix} a_1 & b_1 \\ a_2 & b_2 \end{bmatrix}$$

is given by

$$\det(A) = |A| = \begin{vmatrix} a_1 & b_1 \\ a_2 & b_2 \end{vmatrix} = a_1 b_2 - a_2 b_1.$$

In this text, $\det(A)$ and $|A|$ are used interchangeably to represent the determinant of A. Although vertical bars are also used to denote the absolute value of a real number, the context will show which use is intended.

A convenient method for remembering the formula for the determinant of a 2×2 matrix is shown in the following diagram.

$$\det(A) = \begin{vmatrix} a_1 & b_1 \\ a_2 & b_2 \end{vmatrix} = a_1 b_2 - a_2 b_1$$

Note that the determinant is the difference of the products of the two diagonals of the matrix.

In Example 1, you will see that the determinant of a matrix can be positive, zero, or negative.

> **EXAMPLE 1** **The Determinant of a 2 × 2 Matrix**

Find the determinant of each matrix.

a. $A = \begin{bmatrix} 2 & -3 \\ 1 & 2 \end{bmatrix}$

b. $B = \begin{bmatrix} 2 & 1 \\ 4 & 2 \end{bmatrix}$

c. $C = \begin{bmatrix} 0 & \frac{3}{2} \\ 2 & 4 \end{bmatrix}$

Solution

a. $\det(A) = \begin{vmatrix} 2 & -3 \\ 1 & 2 \end{vmatrix}$

$= 2(2) - 1(-3)$

$= 4 + 3$

$= 7$

b. $\det(B) = \begin{vmatrix} 2 & 1 \\ 4 & 2 \end{vmatrix}$

$= 2(2) - 4(1)$

$= 4 - 4$

$= 0$

c. $\det(C) = \begin{vmatrix} 0 & \frac{3}{2} \\ 2 & 4 \end{vmatrix}$

$= 0(4) - 2\left(\frac{3}{2}\right)$

$= 0 - 3$

$= -3$

✓ **Checkpoint** ◀)) *Audio-video solution in English & Spanish at LarsonPrecalculus.com.*

Find the determinant of each matrix.

a. $A = \begin{bmatrix} 1 & 2 \\ 3 & -1 \end{bmatrix}$ **b.** $B = \begin{bmatrix} 5 & 0 \\ -4 & 2 \end{bmatrix}$ **c.** $C = \begin{bmatrix} 3 & 6 \\ 2 & 4 \end{bmatrix}$ ■

The determinant of a matrix of order 1×1 is defined simply as the entry of the matrix. For instance, if $A = [-2]$, then $\det(A) = -2$.

▷ **TECHNOLOGY** Most graphing utilities can evaluate the determinant of a matrix. For instance, you can evaluate the determinant of

$$A = \begin{bmatrix} 2 & -3 \\ 1 & 2 \end{bmatrix}$$

by entering the matrix as $[A]$ and then choosing the *determinant* feature. The result should be 7, as in Example 1(a). Try evaluating the determinants of other matrices. Consult the user's guide for your graphing utility for specific keystrokes.

Minors and Cofactors

To define the determinant of a square matrix of order 3×3 or higher, it is convenient to introduce the concepts of **minors** and **cofactors**.

Sign Pattern for Cofactors

$$\begin{bmatrix} + & - & + \\ - & + & - \\ + & - & + \end{bmatrix}$$
3×3 matrix

$$\begin{bmatrix} + & - & + & - \\ - & + & - & + \\ + & - & + & - \\ - & + & - & + \end{bmatrix}$$
4×4 matrix

$$\begin{bmatrix} + & - & + & - & + & \cdots \\ - & + & - & + & - & \cdots \\ + & - & + & - & + & \cdots \\ - & + & - & + & - & \cdots \\ + & - & + & - & + & \cdots \\ \vdots & \vdots & \vdots & \vdots & \vdots & \end{bmatrix}$$
$n \times n$ matrix

Minors and Cofactors of a Square Matrix

If A is a square matrix, then the **minor** M_{ij} of the entry a_{ij} is the determinant of the matrix obtained by deleting the ith row and jth column of A. The **cofactor** C_{ij} of the entry a_{ij} is

$$C_{ij} = (-1)^{i+j} M_{ij}.$$

In the sign pattern for cofactors at the left, notice that *odd* positions (where $i + j$ is odd) have negative signs and *even* positions (where $i + j$ is even) have positive signs.

EXAMPLE 2   Finding the Minors and Cofactors of a Matrix

Find all the minors and cofactors of

$$A = \begin{bmatrix} 0 & 2 & 1 \\ 3 & -1 & 2 \\ 4 & 0 & 1 \end{bmatrix}.$$

Solution To find the minor M_{11}, delete the first row and first column of A and evaluate the determinant of the resulting matrix.

$$\begin{bmatrix} 0 & 2 & 1 \\ 3 & -1 & 2 \\ 4 & 0 & 1 \end{bmatrix}, \quad M_{11} = \begin{vmatrix} -1 & 2 \\ 0 & 1 \end{vmatrix} = -1(1) - 0(2) = -1$$

Similarly, to find M_{12}, delete the first row and second column.

$$\begin{bmatrix} 0 & 2 & 1 \\ 3 & -1 & 2 \\ 4 & 0 & 1 \end{bmatrix}, \quad M_{12} = \begin{vmatrix} 3 & 2 \\ 4 & 1 \end{vmatrix} = 3(1) - 4(2) = -5$$

Continuing this pattern, you obtain the minors.

$$M_{11} = -1 \qquad M_{12} = -5 \qquad M_{13} = 4$$
$$M_{21} = 2 \qquad M_{22} = -4 \qquad M_{23} = -8$$
$$M_{31} = 5 \qquad M_{32} = -3 \qquad M_{33} = -6$$

Now, to find the cofactors, combine these minors with the checkerboard pattern of signs for a 3×3 matrix shown at the upper left.

$$C_{11} = -1 \qquad C_{12} = 5 \qquad C_{13} = 4$$
$$C_{21} = -2 \qquad C_{22} = -4 \qquad C_{23} = 8$$
$$C_{31} = 5 \qquad C_{32} = 3 \qquad C_{33} = -6$$

✓ **Checkpoint** ◀))) *Audio-video solution in English & Spanish at LarsonPrecalculus.com.*

Find all the minors and cofactors of

$$A = \begin{bmatrix} 1 & 2 & 3 \\ 0 & -1 & 5 \\ 2 & 1 & 4 \end{bmatrix}.$$

The Determinant of a Square Matrix

The definition below is called *inductive* because it uses determinants of matrices of order $n - 1$ to define determinants of matrices of order n.

Determinant of a Square Matrix

If A is a square matrix (of order 2×2 or greater), then the determinant of A is the sum of the entries in any row (or column) of A multiplied by their respective cofactors. For instance, expanding along the first row yields

$$|A| = a_{11}C_{11} + a_{12}C_{12} + \cdots + a_{1n}C_{1n}.$$

Applying this definition to find a determinant is called **expanding by cofactors.**

Try checking that for a 2×2 matrix

$$A = \begin{bmatrix} a_1 & b_1 \\ a_2 & b_2 \end{bmatrix}$$

this definition of the determinant yields

$$|A| = a_1 b_2 - a_2 b_1$$

as previously defined.

EXAMPLE 3 **The Determinant of a 3 × 3 Matrix**

Find the determinant of $A = \begin{bmatrix} 0 & 2 & 1 \\ 3 & -1 & 2 \\ 4 & 0 & 1 \end{bmatrix}$.

Solution Note that this is the same matrix that was given in Example 2. There you found the cofactors of the entries in the first row to be

$$C_{11} = -1, \quad C_{12} = 5, \quad \text{and} \quad C_{13} = 4.$$

So, by the definition of a determinant, you have

$$|A| = a_{11}C_{11} + a_{12}C_{12} + a_{13}C_{13} \qquad \text{First-row expansion}$$

$$= 0(-1) + 2(5) + 1(4)$$

$$= 14.$$

✓ *Checkpoint* 🔊 *Audio-video solution in English & Spanish at LarsonPrecalculus.com.*

Find the determinant of $A = \begin{bmatrix} 3 & 4 & -2 \\ 3 & 5 & 0 \\ -1 & 4 & 1 \end{bmatrix}$.

In Example 3, the determinant was found by expanding by cofactors in the first row. You could have used any row or column. For instance, you could have expanded along the second row to obtain

$$|A| = a_{21}C_{21} + a_{22}C_{22} + a_{23}C_{23} \qquad \text{Second-row expansion}$$

$$= 3(-2) + (-1)(-4) + 2(8)$$

$$= 14.$$

When expanding by cofactors, you do not need to find cofactors of zero entries, because zero times its cofactor is zero. So, the row (or column) containing the most zeros is usually the best choice for expansion by cofactors. This is demonstrated in the next example.

EXAMPLE 4   **The Determinant of a 4 × 4 Matrix**

Find the determinant of $A = \begin{bmatrix} 1 & -2 & 3 & 0 \\ -1 & 1 & 0 & 2 \\ 0 & 2 & 0 & 3 \\ 3 & 4 & 0 & 2 \end{bmatrix}$.

Solution   After inspecting this matrix, you can see that three of the entries in the third column are zeros. So, you can eliminate some of the work in the expansion by using the third column.

$$|A| = 3(C_{13}) + 0(C_{23}) + 0(C_{33}) + 0(C_{43})$$

Because C_{23}, C_{33}, and C_{43} have zero coefficients, you need only find the cofactor C_{13}. To do this, delete the first row and third column of A and evaluate the determinant of the resulting matrix.

$$C_{13} = (-1)^{1+3} \begin{vmatrix} -1 & 1 & 2 \\ 0 & 2 & 3 \\ 3 & 4 & 2 \end{vmatrix} \qquad \text{Delete 1st row and 3rd column.}$$

$$= \begin{vmatrix} -1 & 1 & 2 \\ 0 & 2 & 3 \\ 3 & 4 & 2 \end{vmatrix} \qquad \text{Simplify.}$$

Expanding by cofactors in the second row yields

$$C_{13} = 0(-1)^3 \begin{vmatrix} 1 & 2 \\ 4 & 2 \end{vmatrix} + 2(-1)^4 \begin{vmatrix} -1 & 2 \\ 3 & 2 \end{vmatrix} + 3(-1)^5 \begin{vmatrix} -1 & 1 \\ 3 & 4 \end{vmatrix}$$

$$= 0 + 2(1)(-8) + 3(-1)(-7)$$

$$= 5.$$

So, $|A| = 3C_{13} = 3(5) = 15$.

✓ **Checkpoint** Audio-video solution in English & Spanish at LarsonPrecalculus.com.

Find the determinant of $A = \begin{bmatrix} 2 & 6 & -4 & 2 \\ 2 & -2 & 3 & 6 \\ 1 & 5 & 0 & 1 \\ 3 & 1 & 0 & -5 \end{bmatrix}$.

In the number-placement puzzle Sudoku, the object is to fill out a partially completed 9 × 9 grid of boxes with numbers from 1 to 9 so that each column, row, and 3 × 3 sub-grid contains each of these numbers without repetition. For a completed Sudoku grid to be valid, no two rows (or columns) will have the numbers in the same order. If this should happen in a row or column, then the determinant of the matrix formed by the numbers in the grid will be zero.

Summarize   (Section 8.4)

1. State the definition of the determinant of a 2 × 2 matrix *(page 419)*. For an example of finding the determinants of 2 × 2 matrices, see Example 1.

2. State the definitions of minors and cofactors of a square matrix *(page 421)*. For an example of finding the minors and cofactors of a square matrix, see Example 2.

3. State the definition of the determinant of a square matrix using expansion by cofactors *(page 422)*. For examples of finding determinants using this definition, see Examples 3 and 4.

8.5 Applications of Matrices and Determinants

- ■ Use Cramer's Rule to solve systems of linear equations.
- ■ Use determinants to find the areas of triangles.
- ■ Use a determinant to test for collinear points and find an equation of a line passing through two points.
- ■ Use matrices to encode and decode messages.

Cramer's Rule

So far, you have studied three methods for solving a system of linear equations: substitution, elimination with equations, and elimination with matrices. In this section, you will study one more method, **Cramer's Rule,** named after Gabriel Cramer (1704–1752). This rule uses determinants to write the solution of a system of linear equations. To see how Cramer's Rule works, take another look at the solution described at the beginning of Section 8.4. There, it was pointed out that the system

$$\begin{cases} a_1x + b_1y = c_1 \\ a_2x + b_2y = c_2 \end{cases}$$

has a solution

$$x = \frac{c_1b_2 - c_2b_1}{a_1b_2 - a_2b_1} \quad \text{and} \quad y = \frac{a_1c_2 - a_2c_1}{a_1b_2 - a_2b_1}$$

provided that

$$a_1b_2 - a_2b_1 \neq 0.$$

Each numerator and denominator in this solution can be expressed as a determinant, as follows.

$$x = \frac{c_1b_2 - c_2b_1}{a_1b_2 - a_2b_1} = \frac{\begin{vmatrix} c_1 & b_1 \\ c_2 & b_2 \end{vmatrix}}{\begin{vmatrix} a_1 & b_1 \\ a_2 & b_2 \end{vmatrix}} \qquad y = \frac{a_1c_2 - a_2c_1}{a_1b_2 - a_2b_1} = \frac{\begin{vmatrix} a_1 & c_1 \\ a_2 & c_2 \end{vmatrix}}{\begin{vmatrix} a_1 & b_1 \\ a_2 & b_2 \end{vmatrix}}$$

Relative to the original system, the denominators for x and y are simply the determinant of the *coefficient* matrix of the system. This determinant is denoted by D. The numerators for x and y are denoted by D_x and D_y, respectively. They are formed by using the column of constants as replacements for the coefficients of x and y, as follows.

Coefficient Matrix	D	D_x	D_y
$\begin{bmatrix} a_1 & b_1 \\ a_2 & b_2 \end{bmatrix}$	$\begin{vmatrix} a_1 & b_1 \\ a_2 & b_2 \end{vmatrix}$	$\begin{vmatrix} c_1 & b_1 \\ c_2 & b_2 \end{vmatrix}$	$\begin{vmatrix} a_1 & c_1 \\ a_2 & c_2 \end{vmatrix}$

For example, given the system

$$\begin{cases} 2x - 5y = 3 \\ -4x + 3y = 8 \end{cases}$$

the coefficient matrix, D, D_x, and D_y are as follows.

Coefficient Matrix	D	D_x	D_y
$\begin{bmatrix} 2 & -5 \\ -4 & 3 \end{bmatrix}$	$\begin{vmatrix} 2 & -5 \\ -4 & 3 \end{vmatrix}$	$\begin{vmatrix} 3 & -5 \\ 8 & 3 \end{vmatrix}$	$\begin{vmatrix} 2 & 3 \\ -4 & 8 \end{vmatrix}$

You can use determinants to solve real-life problems, such as finding the area of a region of forest infested with gypsy moths.

Martynova Anna/Shutterstock.com

Cramer's Rule generalizes easily to systems of n equations in n variables. The value of each variable is given as the quotient of two determinants. The denominator is the determinant of the coefficient matrix, and the numerator is the determinant of the matrix formed by replacing the column corresponding to the variable (being solved for) with the column representing the constants. For instance, the solution for x_3 in the following system is shown.

$$\begin{cases} a_{11}x_1 + a_{12}x_2 + a_{13}x_3 = b_1 \\ a_{21}x_1 + a_{22}x_2 + a_{23}x_3 = b_2 \\ a_{31}x_1 + a_{32}x_2 + a_{33}x_3 = b_3 \end{cases} \qquad x_3 = \frac{|A_3|}{|A|} = \frac{\begin{vmatrix} a_{11} & a_{12} & b_1 \\ a_{21} & a_{22} & b_2 \\ a_{31} & a_{32} & b_3 \end{vmatrix}}{\begin{vmatrix} a_{11} & a_{12} & a_{13} \\ a_{21} & a_{22} & a_{23} \\ a_{31} & a_{32} & a_{33} \end{vmatrix}}$$

Cramer's Rule

If a system of n linear equations in n variables has a coefficient matrix A with a nonzero determinant $|A|$, then the solution of the system is

$$x_1 = \frac{|A_1|}{|A|}, \quad x_2 = \frac{|A_2|}{|A|}, \quad \ldots \quad , \quad x_n = \frac{|A_n|}{|A|}$$

where the ith column of A_i is the column of constants in the system of equations. If the determinant of the coefficient matrix is zero, then the system has either no solution or infinitely many solutions.

EXAMPLE 1   **Using Cramer's Rule for a 2 × 2 System**

Use Cramer's Rule to solve the system

$$\begin{cases} 4x - 2y = 10 \\ 3x - 5y = 11 \end{cases}.$$

Solution   To begin, find the determinant of the coefficient matrix.

$$D = \begin{vmatrix} 4 & -2 \\ 3 & -5 \end{vmatrix} = -20 - (-6) = -14$$

Because this determinant is not zero, you can apply Cramer's Rule.

$$x = \frac{D_x}{D} = \frac{\begin{vmatrix} 10 & -2 \\ 11 & -5 \end{vmatrix}}{-14} = \frac{-50 - (-22)}{-14} = \frac{-28}{-14} = 2$$

$$y = \frac{D_y}{D} = \frac{\begin{vmatrix} 4 & 10 \\ 3 & 11 \end{vmatrix}}{-14} = \frac{44 - 30}{-14} = \frac{14}{-14} = -1$$

So, the solution is $x = 2$ and $y = -1$. Check this in the original system.

✓ *Checkpoint* 🔊))) *Audio-video solution in English & Spanish at LarsonPrecalculus.com.*

Use Cramer's Rule to solve the system

$$\begin{cases} 3x + 4y = 1 \\ 5x + 3y = 9 \end{cases}.$$

EXAMPLE 2 **Using Cramer's Rule for a 3 × 3 System**

Use Cramer's Rule to solve the system $\begin{cases} -x + 2y - 3z = 1 \\ 2x \quad\quad + z = 0. \\ 3x - 4y + 4z = 2 \end{cases}$

Solution To find the determinant of the coefficient matrix

$$\begin{bmatrix} -1 & 2 & -3 \\ 2 & 0 & 1 \\ 3 & -4 & 4 \end{bmatrix}$$

expand along the second row, as follows.

$$D = 2(-1)^3 \begin{vmatrix} 2 & -3 \\ -4 & 4 \end{vmatrix} + 0(-1)^4 \begin{vmatrix} -1 & -3 \\ 3 & 4 \end{vmatrix} + 1(-1)^5 \begin{vmatrix} -1 & 2 \\ 3 & -4 \end{vmatrix}$$

$$= -2(-4) + 0 - 1(-2)$$

$$= 10$$

Because this determinant is not zero, you can apply Cramer's Rule.

$$x = \frac{D_x}{D} = \frac{\begin{vmatrix} 1 & 2 & -3 \\ 0 & 0 & 1 \\ 2 & -4 & 4 \end{vmatrix}}{10} = \frac{8}{10} = \frac{4}{5}$$

$$y = \frac{D_y}{D} = \frac{\begin{vmatrix} -1 & 1 & -3 \\ 2 & 0 & 1 \\ 3 & 2 & 4 \end{vmatrix}}{10} = \frac{-15}{10} = -\frac{3}{2}$$

$$z = \frac{D_z}{D} = \frac{\begin{vmatrix} -1 & 2 & 1 \\ 2 & 0 & 0 \\ 3 & -4 & 2 \end{vmatrix}}{10} = \frac{-16}{10} = -\frac{8}{5}$$

The solution is $\left(\frac{4}{5}, -\frac{3}{2}, -\frac{8}{5}\right)$. Check this in the original system, as follows.

Check

$$-\left(\tfrac{4}{5}\right) + 2\left(-\tfrac{3}{2}\right) - 3\left(-\tfrac{8}{5}\right) \overset{?}{=} 1 \qquad \text{Substitute into Equation 1.}$$

$$-\tfrac{4}{5} - 3 + \tfrac{24}{5} = 1 \qquad \text{Equation 1 checks. } \checkmark$$

$$2\left(\tfrac{4}{5}\right) + \left(-\tfrac{8}{5}\right) \overset{?}{=} 0 \qquad \text{Substitute into Equation 2.}$$

$$\tfrac{8}{5} - \tfrac{8}{5} = 0 \qquad \text{Equation 2 checks. } \checkmark$$

$$3\left(\tfrac{4}{5}\right) - 4\left(-\tfrac{3}{2}\right) + 4\left(-\tfrac{8}{5}\right) \overset{?}{=} 2 \qquad \text{Substitute into Equation 3.}$$

$$\tfrac{12}{5} + 6 - \tfrac{32}{5} = 2 \qquad \text{Equation 3 checks. } \checkmark$$

✓ **Checkpoint** ◀))) *Audio-video solution in English & Spanish at LarsonPrecalculus.com.*

Use Cramer's Rule to solve the system $\begin{cases} 4x - y + z = 12 \\ 2x + 2y + 3z = 1. \\ 5x - 2y + 6z = 22 \end{cases}$

Remember that Cramer's Rule does not apply when the determinant of the coefficient matrix is zero. This would create division by zero, which is undefined.

Area of a Triangle

Another application of matrices and determinants is finding the area of a triangle whose vertices are given as points in a coordinate plane.

Area of a Triangle

The area of a triangle with vertices (x_1, y_1), (x_2, y_2), and (x_3, y_3) is

$$\text{Area} = \pm \frac{1}{2} \begin{vmatrix} x_1 & y_1 & 1 \\ x_2 & y_2 & 1 \\ x_3 & y_3 & 1 \end{vmatrix}$$

where the symbol $\pm$ indicates that the appropriate sign should be chosen to yield a positive area.

EXAMPLE 3   **Finding the Area of a Triangle**

Find the area of a triangle whose vertices are $(1, 0)$, $(2, 2)$, and $(4, 3)$, as shown below.

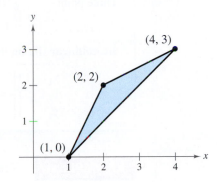

Solution   Let $(x_1, y_1) = (1, 0)$, $(x_2, y_2) = (2, 2)$, and $(x_3, y_3) = (4, 3)$. Then, to find the area of the triangle, evaluate the determinant.

$$\begin{vmatrix} x_1 & y_1 & 1 \\ x_2 & y_2 & 1 \\ x_3 & y_3 & 1 \end{vmatrix} = \begin{vmatrix} 1 & 0 & 1 \\ 2 & 2 & 1 \\ 4 & 3 & 1 \end{vmatrix}$$

$$= 1(-1)^2 \begin{vmatrix} 2 & 1 \\ 3 & 1 \end{vmatrix} + 0(-1)^3 \begin{vmatrix} 2 & 1 \\ 4 & 1 \end{vmatrix} + 1(-1)^4 \begin{vmatrix} 2 & 2 \\ 4 & 3 \end{vmatrix}$$

$$= 1(-1) + 0 + 1(-2)$$

$$= -3$$

Using this value, you can conclude that the area of the triangle is

$$\text{Area} = -\frac{1}{2} \begin{vmatrix} 1 & 0 & 1 \\ 2 & 2 & 1 \\ 4 & 3 & 1 \end{vmatrix} \qquad \text{\color{red}{Choose $(-)$ so that the area is positive.}}$$

$$= -\frac{1}{2}(-3)$$

$$= \frac{3}{2} \text{ square units.}$$

✓ *Checkpoint* ◄))) *Audio-video solution in English & Spanish at LarsonPrecalculus.com.*

Find the area of a triangle whose vertices are $(0, 0)$, $(4, 1)$, and $(2, 5)$. ∎

Lines in a Plane

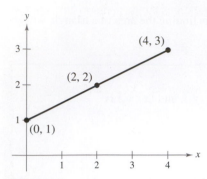

Figure 8.1

Suppose the three points in Example 3 had been on the same line. What would have happened had the area formula been applied to three such points? The answer is that the determinant would have been zero. Consider, for instance, the three collinear points $(0, 1)$, $(2, 2)$, and $(4, 3)$, as shown in Figure 8.1. The area of the "triangle" that has these three points as vertices is

$$\frac{1}{2}\begin{vmatrix} 0 & 1 & 1 \\ 2 & 2 & 1 \\ 4 & 3 & 1 \end{vmatrix} = \frac{1}{2}\left[0(-1)^2\begin{vmatrix} 2 & 1 \\ 3 & 1 \end{vmatrix} + 1(-1)^3\begin{vmatrix} 2 & 1 \\ 4 & 1 \end{vmatrix} + 1(-1)^4\begin{vmatrix} 2 & 2 \\ 4 & 3 \end{vmatrix} \right]$$

$$= \frac{1}{2}[0 - 1(-2) + 1(-2)]$$

$$= 0.$$

This result is generalized as follows.

Test for Collinear Points

Three points

$$(x_1, y_1), \quad (x_2, y_2), \quad \text{and} \quad (x_3, y_3)$$

are **collinear** (lie on the same line) if and only if

$$\begin{vmatrix} x_1 & y_1 & 1 \\ x_2 & y_2 & 1 \\ x_3 & y_3 & 1 \end{vmatrix} = 0.$$

EXAMPLE 4   **Testing for Collinear Points**

Determine whether the points $(-2, -2)$, $(1, 1)$, and $(7, 5)$ are collinear. (See Figure 8.2.)

Solution   Letting $(x_1, y_1) = (-2, -2)$, $(x_2, y_2) = (1, 1)$, and $(x_3, y_3) = (7, 5)$, you have

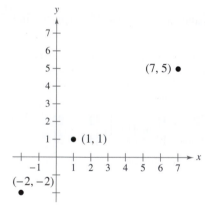

Figure 8.2

$$\begin{vmatrix} x_1 & y_1 & 1 \\ x_2 & y_2 & 1 \\ x_3 & y_3 & 1 \end{vmatrix} = \begin{vmatrix} -2 & -2 & 1 \\ 1 & 1 & 1 \\ 7 & 5 & 1 \end{vmatrix}$$

$$= -2(-1)^2\begin{vmatrix} 1 & 1 \\ 5 & 1 \end{vmatrix} + (-2)(-1)^3\begin{vmatrix} 1 & 1 \\ 7 & 1 \end{vmatrix} + 1(-1)^4\begin{vmatrix} 1 & 1 \\ 7 & 5 \end{vmatrix}$$

$$= -2(-4) + 2(-6) + 1(-2)$$

$$= -6.$$

Because the value of this determinant is *not* zero, you can conclude that the three points do not lie on the same line. Moreover, the area of the triangle with vertices at these points is $\left(-\frac{1}{2}\right)(-6) = 3$ square units.

✓ *Checkpoint* ◀)) *Audio-video solution in English & Spanish at LarsonPrecalculus.com.*

Determine whether the points

$$(-2, 4), \quad (3, -1), \quad \text{and} \quad (6, -4)$$

are collinear.

The test for collinear points can be adapted for another use. Given two points on a rectangular coordinate system, you can find an equation of the line passing through the two points.

Two-Point Form of the Equation of a Line

An equation of the line passing through the distinct points (x_1, y_1) and (x_2, y_2) is given by

$$\begin{vmatrix} x & y & 1 \\ x_1 & y_1 & 1 \\ x_2 & y_2 & 1 \end{vmatrix} = 0.$$

EXAMPLE 5   **Finding an Equation of a Line**

Find an equation of the line passing through the two points $(2, 4)$ and $(-1, 3)$, as shown below.

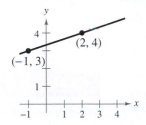

Solution   Let $(x_1, y_1) = (2, 4)$ and $(x_2, y_2) = (-1, 3)$. Applying the determinant formula for the equation of a line produces

$$\begin{vmatrix} x & y & 1 \\ 2 & 4 & 1 \\ -1 & 3 & 1 \end{vmatrix} = 0.$$

To evaluate this determinant, you can expand by cofactors along the first row to find an equation of the line.

$$x(-1)^2 \begin{vmatrix} 4 & 1 \\ 3 & 1 \end{vmatrix} + y(-1)^3 \begin{vmatrix} 2 & 1 \\ -1 & 1 \end{vmatrix} + 1(-1)^4 \begin{vmatrix} 2 & 4 \\ -1 & 3 \end{vmatrix} = 0$$

$$x(1) - y(3) + (1)(10) = 0$$

$$x - 3y + 10 = 0$$

✓ *Checkpoint* 🔊)) *Audio-video solution in English & Spanish at LarsonPrecalculus.com.*

Find an equation of the line passing through the two points $(-3, -1)$ and $(3, 5)$. ∎

Note that this method of finding the equation of a line works for all lines, including horizontal and vertical lines. For instance, the equation of the vertical line passing through $(2, 0)$ and $(2, 2)$ is

$$\begin{vmatrix} x & y & 1 \\ 2 & 0 & 1 \\ 2 & 2 & 1 \end{vmatrix} = 0$$

$$4 - 2x = 0$$

$$x = 2.$$

Cryptography

A **cryptogram** is a message written according to a secret code. (The Greek word *kryptos* means "hidden.") Matrix multiplication can be used to encode and decode messages. To begin, you need to assign a number to each letter in the alphabet (with 0 assigned to a blank space), as follows.

0 = _	9 = I	18 = R
1 = A	10 = J	19 = S
2 = B	11 = K	20 = T
3 = C	12 = L	21 = U
4 = D	13 = M	22 = V
5 = E	14 = N	23 = W
6 = F	15 = O	24 = X
7 = G	16 = P	25 = Y
8 = H	17 = Q	26 = Z

Then the message is converted to numbers and partitioned into **uncoded row matrices,** each having n entries, as demonstrated in Example 6.

Because of the heavy use of the Internet to conduct business, Internet security is of the utmost importance. If a malicious party should receive confidential information such as passwords, personal identification numbers, credit card numbers, social security numbers, bank account details, or corporate secrets, the effects can be damaging. To protect the confidentiality and integrity of such information, the most popular forms of Internet security use data *encryption,* the process of encoding information so that the only way to decode it, apart from a brute force "exhaustion attack," is to use a *key.* Data encryption technology uses algorithms based on the material presented here, but on a much more sophisticated level, to prevent malicious parties from discovering the key.

EXAMPLE 6   **Forming Uncoded Row Matrices**

Write the uncoded row matrices of order 1×3 for the message

   MEET ME MONDAY.

Solution   Partitioning the message (including blank spaces, but ignoring punctuation) into groups of three produces the following uncoded row matrices.

$$[13 \quad 5 \quad 5] \quad [20 \quad 0 \quad 13] \quad [5 \quad 0 \quad 13] \quad [15 \quad 14 \quad 4] \quad [1 \quad 25 \quad 0]$$
$$\text{M} \quad \text{E} \quad \text{E} \quad \text{T} \quad \quad \text{M} \quad \text{E} \quad \quad \text{M} \quad \text{O} \quad \text{N} \quad \text{D} \quad \text{A} \quad \text{Y}$$

Note that a blank space is used to fill out the last uncoded row matrix.

✓ *Checkpoint*  *Audio-video solution in English & Spanish at LarsonPrecalculus.com.*

Write the uncoded row matrices of order 1×3 for the message

   OWLS ARE NOCTURNAL.

To encode a message, use the techniques demonstrated in Section 8.3 to choose an $n \times n$ invertible matrix such as

$$A = \begin{bmatrix} 1 & -2 & 2 \\ -1 & 1 & 3 \\ 1 & -1 & -4 \end{bmatrix}$$

and multiply the uncoded row matrices by A (on the right) to obtain **coded row matrices.** Here is an example.

Uncoded Matrix	**Encoding Matrix A**	**Coded Matrix**
$[13 \quad 5 \quad 5]$	$\begin{bmatrix} 1 & -2 & 2 \\ -1 & 1 & 3 \\ 1 & -1 & -4 \end{bmatrix} =$	$[13 \quad -26 \quad 21]$

EXAMPLE 7   **Encoding a Message**

Use the following invertible matrix to encode the message MEET ME MONDAY.

$$A = \begin{bmatrix} 1 & -2 & 2 \\ -1 & 1 & 3 \\ 1 & -1 & -4 \end{bmatrix}$$

Solution   The coded row matrices are obtained by multiplying each of the uncoded row matrices found in Example 6 by the matrix A, as follows.

Uncoded Matrix   Encoding Matrix A Coded Matrix

$$\begin{bmatrix} 13 & 5 & 5 \end{bmatrix} \begin{bmatrix} 1 & -2 & 2 \\ -1 & 1 & 3 \\ 1 & -1 & -4 \end{bmatrix} = \begin{bmatrix} 13 & -26 & 21 \end{bmatrix}$$

$$\begin{bmatrix} 20 & 0 & 13 \end{bmatrix} \begin{bmatrix} 1 & -2 & 2 \\ -1 & 1 & 3 \\ 1 & -1 & -4 \end{bmatrix} = \begin{bmatrix} 33 & -53 & -12 \end{bmatrix}$$

$$\begin{bmatrix} 5 & 0 & 13 \end{bmatrix} \begin{bmatrix} 1 & -2 & 2 \\ -1 & 1 & 3 \\ 1 & -1 & -4 \end{bmatrix} = \begin{bmatrix} 18 & -23 & -42 \end{bmatrix}$$

$$\begin{bmatrix} 15 & 14 & 4 \end{bmatrix} \begin{bmatrix} 1 & -2 & 2 \\ -1 & 1 & 3 \\ 1 & -1 & -4 \end{bmatrix} = \begin{bmatrix} 5 & -20 & 56 \end{bmatrix}$$

$$\begin{bmatrix} 1 & 25 & 0 \end{bmatrix} \begin{bmatrix} 1 & -2 & 2 \\ -1 & 1 & 3 \\ 1 & -1 & -4 \end{bmatrix} = \begin{bmatrix} -24 & 23 & 77 \end{bmatrix}$$

So, the sequence of coded row matrices is

$$\begin{bmatrix} 13 & -26 & 21 \end{bmatrix} \begin{bmatrix} 33 & -53 & -12 \end{bmatrix} \begin{bmatrix} 18 & -23 & -42 \end{bmatrix} \begin{bmatrix} 5 & -20 & 56 \end{bmatrix} \begin{bmatrix} -24 & 23 & 77 \end{bmatrix}.$$

Finally, removing the matrix notation produces the following cryptogram.

$$13 \ -26 \ 21 \ 33 \ -53 \ -12 \ 18 \ -23 \ -42 \ 5 \ -20 \ 56 \ -24 \ 23 \ 77$$

✓ *Checkpoint*   *Audio-video solution in English & Spanish at LarsonPrecalculus.com.*

Use the following invertible matrix to encode the message OWLS ARE NOCTURNAL.

$$A = \begin{bmatrix} 1 & -1 & 0 \\ 1 & 0 & -1 \\ 6 & -2 & -3 \end{bmatrix}$$

For those who do not know the encoding matrix A, decoding the cryptogram found in Example 7 is difficult. But for an authorized receiver who knows the encoding matrix A, decoding is simple. The receiver just needs to multiply the coded row matrices by A^{-1} (on the right) to retrieve the uncoded row matrices. Here is an example.

$$\underbrace{\begin{bmatrix} 13 & -26 & 21 \end{bmatrix}}_{\text{Coded}} \underbrace{\begin{bmatrix} -1 & -10 & -8 \\ -1 & -6 & -5 \\ 0 & -1 & -1 \end{bmatrix}}_{A^{-1}} = \underbrace{\begin{bmatrix} 13 & 5 & 5 \end{bmatrix}}_{\text{Uncoded}}$$

EXAMPLE 8 **Decoding a Message**

Use the inverse of the matrix A in Example 7 to decode the cryptogram

$$13 \ -26 \ 21 \ 33 \ -53 \ -12 \ 18 \ -23 \ -42 \ 5 \ -20 \ 56 \ -24 \ 23 \ 77.$$

Solution First find the decoding matrix A^{-1} by using the techniques demonstrated in Section 8.3. Then partition the message into groups of three to form the coded row matrices. Finally, multiply each coded row matrix by A^{-1} (on the right).

Coded Matrix	Decoding Matrix A^{-1}	Decoded Matrix

$$\begin{bmatrix} 13 & -26 & 21 \end{bmatrix} \begin{bmatrix} -1 & -10 & -8 \\ -1 & -6 & -5 \\ 0 & -1 & -1 \end{bmatrix} = \begin{bmatrix} 13 & 5 & 5 \end{bmatrix}$$

$$\begin{bmatrix} 33 & -53 & -12 \end{bmatrix} \begin{bmatrix} -1 & -10 & -8 \\ -1 & -6 & -5 \\ 0 & -1 & -1 \end{bmatrix} = \begin{bmatrix} 20 & 0 & 13 \end{bmatrix}$$

$$\begin{bmatrix} 18 & -23 & -42 \end{bmatrix} \begin{bmatrix} -1 & -10 & -8 \\ -1 & -6 & -5 \\ 0 & -1 & -1 \end{bmatrix} = \begin{bmatrix} 5 & 0 & 13 \end{bmatrix}$$

$$\begin{bmatrix} 5 & -20 & 56 \end{bmatrix} \begin{bmatrix} -1 & -10 & -8 \\ -1 & -6 & -5 \\ 0 & -1 & -1 \end{bmatrix} = \begin{bmatrix} 15 & 14 & 4 \end{bmatrix}$$

$$\begin{bmatrix} -24 & 23 & 77 \end{bmatrix} \begin{bmatrix} -1 & -10 & -8 \\ -1 & -6 & -5 \\ 0 & -1 & -1 \end{bmatrix} = \begin{bmatrix} 1 & 25 & 0 \end{bmatrix}$$

So, the message is as follows.

$$\begin{bmatrix} 13 & 5 & 5 \end{bmatrix} \ \begin{bmatrix} 20 & 0 & 13 \end{bmatrix} \ \begin{bmatrix} 5 & 0 & 13 \end{bmatrix} \ \begin{bmatrix} 15 & 14 & 4 \end{bmatrix} \ \begin{bmatrix} 1 & 25 & 0 \end{bmatrix}$$

M E E T   M E   M O N D A Y

✓ *Checkpoint* 🔊))) *Audio-video solution in English & Spanish at LarsonPrecalculus.com.*

Use the inverse of the matrix A in the Checkpoint with Example 7 to decode the cryptogram

$$110 \ -39 \ -59 \ 25 \ -21 \ -3 \ 23 \ -18 \ -5 \ 47 \ -20 \ -24$$
$$149 \ -56 \ -75 \ 87 \ -38 \ -37.$$

Summarize (Section 8.5)

1. State Cramer's Rule *(page 425)*. For examples of using Cramer's Rule to solve systems of linear equations, see Examples 1 and 2.

2. State the formula for finding the area of a triangle using a determinant *(page 427)*. For an example of finding the area of a triangle, see Example 3.

3. State the test for collinear points *(page 428)*. For an example of testing for collinear points, see Example 4.

4. Explain how to use a determinant to find an equation of a line *(page 429)*. For an example of finding an equation of a line, see Example 5.

5. Describe how to use matrices to encode and decode a message *(pages 430–432, Examples 6–8)*.

Chapter Summary

What Did You Learn?	Explanation/Examples	Review Exercises

Section 8.1

What Did You Learn?	Explanation/Examples	Review Exercises
Write matrices and identify their orders (p. 394).	$\begin{bmatrix} -1 & 1 \\ 4 & 7 \end{bmatrix}$ $\quad$ $[-2 \ \ 3 \ \ 0]$ $\quad$ $\begin{bmatrix} 4 & -3 \\ 5 & 0 \\ -2 & 1 \end{bmatrix}$ $\quad$ $\begin{bmatrix} 8 \\ -8 \end{bmatrix}$ $\quad \ 2 \times 2 \qquad \quad 1 \times 3 \qquad \qquad \ 3 \times 2 \qquad \ 2 \times 1$	1–8
Perform elementary row operations on matrices (p. 396).	**Elementary Row Operations** 1. Interchange two rows. 2. Multiply a row by a nonzero constant. 3. Add a multiple of a row to another row.	9, 10
Use matrices and Gaussian elimination to solve systems of linear equations (p. 397).	**Gaussian Elimination with Back-Substitution** 1. Write the augmented matrix of the system of linear equations. 2. Use elementary row operations to rewrite the augmented matrix in row-echelon form. 3. Write the system of linear equations corresponding to the matrix in row-echelon form and use back-substitution to find the solution.	11–28
Use matrices and Gauss-Jordan elimination to solve systems of linear equations (p. 401).	Gauss-Jordan elimination continues the reduction process on a matrix in row-echelon form until a *reduced* row-echelon form is obtained. (See Example 8.)	29–36

Section 8.2

What Did You Learn?	Explanation/Examples	Review Exercises
Decide whether two matrices are equal (p. 402).	Two matrices are equal when their corresponding entries are equal.	37–40
Add and subtract matrices, and multiply matrices by scalars (p. 404).	**Definition of Matrix Addition** If $A = [a_{ij}]$ and $B = [b_{ij}]$ are matrices of order $m \times n$, then their sum is the $m \times n$ matrix given by $A + B = [a_{ij} + b_{ij}]$. **Definition of Scalar Multiplication** If $A = [a_{ij}]$ is an $m \times n$ matrix and c is a scalar, then the scalar multiple of A by c is the $m \times n$ matrix given by $cA = [ca_{ij}]$.	41–54
Multiply two matrices (p. 408).	**Definition of Matrix Multiplication** If $A = [a_{ij}]$ is an $m \times n$ matrix and $B = [b_{ij}]$ is an $n \times p$ matrix, then the product AB is an $m \times p$ matrix $AB = [c_{ij}]$ where $c_{ij} = a_{i1}b_{1j} + a_{i2}b_{2j} + a_{i3}b_{3j} + \cdots + a_{in}b_{nj}$.	55–66
Use matrix operations to model and solve real-life problems (p. 411).	You can use matrix operations to find the total cost of equipment for two softball teams. (See Example 13.)	67–70

Section 8.3

What Did You Learn?	Explanation/Examples	Review Exercises
Verify that two matrices are inverses of each other (p. 413).	**Definition of the Inverse of a Square Matrix** Let A be an $n \times n$ matrix and let I_n be the $n \times n$ identity matrix. If there exists a matrix A^{-1} such that $AA^{-1} = I_n = A^{-1}A$ then A^{-1} is the inverse of A.	71–74

	What Did You Learn?	**Explanation/Examples**	**Review Exercises**
Section 8.3	Use Gauss-Jordan elimination to find the inverses of matrices *(p. 414)*.	**Finding an Inverse Matrix** Let A be a square matrix of order $n \times n$. 1. Write the $n \times 2n$ matrix that consists of the given matrix A on the left and the $n \times n$ identity matrix I on the right to obtain $[A \ \vdots \ I]$. 2. If possible, row reduce A to I using elementary row operations on the *entire* matrix $[A \ \vdots \ I]$. The result will be the matrix $[I \ \vdots \ A^{-1}]$. If this is not possible, then A is not invertible. 3. Check your work by multiplying to see that $AA^{-1} = I = A^{-1}A$.	75–82
	Use a formula to find the inverses of 2×2 matrices *(p. 417)*.	If $A = \begin{bmatrix} a & b \\ c & d \end{bmatrix}$ and $ad - bc \neq 0$, then $A^{-1} = \dfrac{1}{ad - bc}\begin{bmatrix} d & -b \\ -c & a \end{bmatrix}$.	83–90
	Use inverse matrices to solve systems of linear equations *(p. 418)*.	If A is an invertible matrix, then the system of linear equations represented by $AX = B$ has a unique solution given by $X = A^{-1}B$.	91–108
Section 8.4	Find the determinants of 2×2 matrices *(p. 419)*.	The determinant of the matrix $A = \begin{bmatrix} a_1 & b_1 \\ a_2 & b_2 \end{bmatrix}$ is given by $\det(A) = |A| = \begin{vmatrix} a_1 & b_1 \\ a_2 & b_2 \end{vmatrix} = a_1 b_2 - a_2 b_1$.	109–112
	Find minors and cofactors of square matrices *(p. 421)*.	If A is a square matrix, then the minor M_{ij} of the entry a_{ij} is the determinant of the matrix obtained by deleting the ith row and jth column of A. The cofactor C_{ij} of the entry a_{ij} is $C_{ij} = (-1)^{i+j}M_{ij}$.	113–116
	Find the determinants of square matrices *(p. 422)*.	If A is a square matrix (of order 2×2 or greater), then the determinant of A is the sum of the entries in any row (or column) of A multiplied by their respective cofactors.	117–126
Section 8.5	Use Cramer's Rule to solve systems of linear equations *(p. 424)*.	Cramer's Rule uses determinants to write the solution of a system of linear equations.	127–130
	Use determinants to find the areas of triangles *(p. 427)*.	The area of a triangle with vertices (x_1, y_1), (x_2, y_2), and (x_3, y_3) is $\text{Area} = \pm\dfrac{1}{2}\begin{vmatrix} x_1 & y_1 & 1 \\ x_2 & y_2 & 1 \\ x_3 & y_3 & 1 \end{vmatrix}$ where the symbol $\pm$ indicates that the appropriate sign should be chosen to yield a positive area.	131–134
	Use a determinant to test for collinear points and find an equation of a line passing through two points *(p. 428)*.	Three points (x_1, y_1), (x_2, y_2), and (x_3, y_3) are collinear (lie on the same line) if and only if $\begin{vmatrix} x_1 & y_1 & 1 \\ x_2 & y_2 & 1 \\ x_3 & y_3 & 1 \end{vmatrix} = 0.$	135–140
	Use matrices to encode and decode messages *(p. 430)*.	You can use the inverse of a matrix to decode a cryptogram. (See Example 8.)	141–144

See **CalcChat.com** for tutorial help and worked-out solutions to odd-numbered exercises.

Review Exercises

8.1 **Order of a Matrix**   In Exercises 1–4, determine the order of the matrix.

1. $\begin{bmatrix} -4 \\ 0 \\ 5 \end{bmatrix}$

2. $\begin{bmatrix} 3 & -1 & 0 & 6 \\ -2 & 7 & 1 & 4 \end{bmatrix}$

3. $[3]$

4. $\begin{bmatrix} 6 & 2 & -5 & 8 & 0 \end{bmatrix}$

Writing an Augmented Matrix   In Exercises 5 and 6, write the augmented matrix for the system of linear equations.

5. $\begin{cases} 3x - 10y = 15 \\ 5x + 4y = 22 \end{cases}$

6. $\begin{cases} 8x - 7y + 4z = 12 \\ 3x - 5y + 2z = 20 \end{cases}$

Writing a System of Equations   In Exercises 7 and 8, write the system of linear equations represented by the augmented matrix. (Use variables x, y, z, and w, if applicable.)

7. $\begin{bmatrix} 5 & 1 & 7 & \vdots & -9 \\ 4 & 2 & 0 & \vdots & 10 \\ 9 & 4 & 2 & \vdots & 3 \end{bmatrix}$

8. $\begin{bmatrix} 13 & 16 & 7 & 3 & \vdots & 2 \\ 4 & 10 & -4 & 3 & \vdots & -1 \end{bmatrix}$

Writing a Matrix in Row-Echelon Form   In Exercises 9 and 10, write the matrix in row-echelon form. (Remember that the row-echelon form of a matrix is not unique.)

9. $\begin{bmatrix} 0 & 1 & 1 \\ 1 & 2 & 3 \\ 2 & 2 & 2 \end{bmatrix}$

10. $\begin{bmatrix} 4 & 8 & 16 \\ 3 & -1 & 2 \\ -2 & 10 & 12 \end{bmatrix}$

Using Back-Substitution   In Exercises 11–14, write the system of linear equations represented by the augmented matrix. Then use back-substitution to solve the system. (Use variables x, y, and z, if applicable.)

11. $\begin{bmatrix} 1 & 2 & 3 & \vdots & 9 \\ 0 & 1 & -2 & \vdots & 2 \\ 0 & 0 & 1 & \vdots & 0 \end{bmatrix}$

12. $\begin{bmatrix} 1 & 3 & -9 & \vdots & 4 \\ 0 & 1 & -1 & \vdots & 10 \\ 0 & 0 & 1 & \vdots & -2 \end{bmatrix}$

13. $\begin{bmatrix} 1 & -5 & 4 & \vdots & 1 \\ 0 & 1 & 2 & \vdots & 3 \\ 0 & 0 & 1 & \vdots & 4 \end{bmatrix}$

14. $\begin{bmatrix} 1 & -8 & 0 & \vdots & -2 \\ 0 & 1 & -1 & \vdots & -7 \\ 0 & 0 & 1 & \vdots & 1 \end{bmatrix}$

Gaussian Elimination with Back-Substitution   In Exercises 15–28, use matrices to solve the system of equations (if possible). Use Gaussian elimination with back-substitution.

15. $\begin{cases} 5x + 4y = 2 \\ -x + y = -22 \end{cases}$

16. $\begin{cases} 2x - 5y = 2 \\ 3x - 7y = 1 \end{cases}$

17. $\begin{cases} 0.3x - 0.1y = -0.13 \\ 0.2x - 0.3y = -0.25 \end{cases}$

18. $\begin{cases} 0.2x - 0.1y = 0.07 \\ 0.4x - 0.5y = -0.01 \end{cases}$

19. $\begin{cases} -x + 2y = 3 \\ 2x - 4y = 6 \end{cases}$

20. $\begin{cases} -x + 2y = 3 \\ 2x - 4y = -6 \end{cases}$

21. $\begin{cases} x - 2y + z = 7 \\ 2x + y - 2z = -4 \\ -x + 3y + 2z = -3 \end{cases}$

22. $\begin{cases} x - 2y + z = 4 \\ 2x + y - 2z = -24 \\ -x + 3y + 2z = 20 \end{cases}$

23. $\begin{cases} 2x + y + 2z = 4 \\ 2x + 2y = 5 \\ 2x - y + 6z = 2 \end{cases}$

24. $\begin{cases} x + 2y + 6z = 1 \\ 2x + 5y + 15z = 4 \\ 3x + y + 3z = -6 \end{cases}$

25. $\begin{cases} 2x + 3y + z = 10 \\ 2x - 3y - 3z = 22 \\ 4x - 2y + 3z = -2 \end{cases}$

26. $\begin{cases} 2x + 3y + 3z = 3 \\ 6x + 6y + 12z = 13 \\ 12x + 9y - z = 2 \end{cases}$

27. $\begin{cases} 2x + y + z = 6 \\ -2y + 3z - w = 9 \\ 3x + 3y - 2z - 2w = -11 \\ x + z + 3w = 14 \end{cases}$

28. $\begin{cases} x + 2y + w = 3 \\ -3y + 3z = 0 \\ 4x + 4y + z + 2w = 0 \\ 2x + z = 3 \end{cases}$

Gauss-Jordan Elimination   In Exercises 29–34, use matrices to solve the system of equations (if possible). Use Gauss-Jordan elimination.

29. $\begin{cases} x + 2y - z = 3 \\ x - y - z = -3 \\ 2x + y + 3z = 10 \end{cases}$

30. $\begin{cases} x - 3y + z = 2 \\ 3x - y - z = -6 \\ -x + y - 3z = -2 \end{cases}$

31. $\begin{cases} -x + y + 2z = 1 \\ 2x + 3y + z = -2 \\ 5x + 4y + 2z = 4 \end{cases}$

32. $\begin{cases} 4x + 4y + 4z = 5 \\ 4x - 2y - 8z = 1 \\ 5x + 3y + 8z = 6 \end{cases}$

33. $\begin{cases} 2x - y + 9z = -8 \\ -x - 3y + 4z = -15 \\ 5x + 2y - z = 17 \end{cases}$

34. $\begin{cases} -3x + y + 7z = -20 \\ 5x - 2y - z = 34 \\ -x + y + 4z = -8 \end{cases}$

Using a Graphing Utility In Exercises 35 and 36, use the matrix capabilities of a graphing utility to write the augmented matrix corresponding to the system of equations in reduced row-echelon form. Then solve the system.

35. $\begin{cases} 3x - y + 5z - 2w = -44 \\ x + 6y + 4z - w = 1 \\ 5x - y + z + 3w = -15 \\ 4y - z - 8w = 58 \end{cases}$

36. $\begin{cases} 4x + 12y + 2z = 20 \\ x + 6y + 4z = 12 \\ x + 6y + z = 8 \\ -2x - 10y - 2z = -10 \end{cases}$

8.2 Equality of Matrices In Exercises 37–40, find x and y.

37. $\begin{bmatrix} -1 & x \\ y & 9 \end{bmatrix} = \begin{bmatrix} -1 & 12 \\ -7 & 9 \end{bmatrix}$

38. $\begin{bmatrix} -1 & 0 \\ x & 5 \\ -4 & y \end{bmatrix} = \begin{bmatrix} -1 & 0 \\ 8 & 5 \\ -4 & 0 \end{bmatrix}$

39. $\begin{bmatrix} x+3 & -4 & 4y \\ 0 & -3 & 2 \\ -2 & y+5 & 6x \end{bmatrix} = \begin{bmatrix} 5x-1 & -4 & 44 \\ 0 & -3 & 2 \\ -2 & 16 & 6 \end{bmatrix}$

40. $\begin{bmatrix} -9 & 4 & 2 & -5 \\ 0 & -3 & 7 & -4 \\ 6 & -1 & 1 & 0 \end{bmatrix} = \begin{bmatrix} -9 & 4 & x-10 & -5 \\ 0 & -3 & 7 & 2y \\ \frac{1}{2}x & -1 & 1 & 0 \end{bmatrix}$

Operations with Matrices In Exercises 41–44, if possible, find (a) $A + B$, (b) $A - B$, (c) $4A$, and (d) $A + 3B$.

41. $A = \begin{bmatrix} 2 & -2 \\ 3 & 5 \end{bmatrix}$, $B = \begin{bmatrix} -3 & 10 \\ 12 & 8 \end{bmatrix}$

42. $A = \begin{bmatrix} 4 & 3 \\ -6 & 1 \\ 10 & 1 \end{bmatrix}$, $B = \begin{bmatrix} 3 & 11 \\ 15 & 25 \\ 20 & 29 \end{bmatrix}$

43. $A = \begin{bmatrix} 5 & 4 \\ -7 & 2 \\ 11 & 2 \end{bmatrix}$, $B = \begin{bmatrix} 0 & 3 \\ 4 & 12 \\ 20 & 40 \end{bmatrix}$

44. $A = \begin{bmatrix} 6 & -5 & 7 \end{bmatrix}$, $B = \begin{bmatrix} -1 \\ 4 \\ 8 \end{bmatrix}$

Evaluating an Expression In Exercises 45–50, evaluate the expression. If it is not possible, explain why.

45. $\begin{bmatrix} 7 & 3 \\ -1 & 5 \end{bmatrix} + \begin{bmatrix} 10 & -20 \\ 14 & -3 \end{bmatrix}$

46. $\begin{bmatrix} -11 & 16 & 19 \\ -7 & -2 & 1 \end{bmatrix} - \begin{bmatrix} 6 & 0 \\ 8 & -4 \\ -2 & 10 \end{bmatrix}$

47. $-2\begin{bmatrix} 1 & 2 \\ 5 & -4 \\ 6 & 0 \end{bmatrix} + 8\begin{bmatrix} 7 & 1 \\ 1 & 2 \\ 1 & 4 \end{bmatrix}$

48. $-\begin{bmatrix} 8 & -1 & 8 \\ -2 & 4 & 12 \\ 0 & -6 & 0 \end{bmatrix} - 5\begin{bmatrix} -2 & 0 & -4 \\ 3 & -1 & 1 \\ 6 & 12 & -8 \end{bmatrix}$

49. $3\begin{bmatrix} 8 & -2 & 5 \\ 1 & 3 & -1 \end{bmatrix} + 6\begin{bmatrix} 4 & -2 & -3 \\ 2 & 7 & 6 \end{bmatrix}$

50. $-5\begin{bmatrix} 2 & 0 \\ 7 & -2 \\ 8 & 2 \end{bmatrix} + 4\begin{bmatrix} 4 & -2 \\ 6 & 11 \\ -1 & 3 \end{bmatrix}$

Solving a Matrix Equation In Exercises 51–54, solve for X in the equation, where

$$A = \begin{bmatrix} -4 & 0 \\ 1 & -5 \\ -3 & 2 \end{bmatrix} \quad \text{and} \quad B = \begin{bmatrix} 1 & 2 \\ -2 & 1 \\ 4 & 4 \end{bmatrix}.$$

51. $X = 2A - 3B$

52. $6X = 4A + 3B$

53. $3X + 2A = B$

54. $2A - 5B = 3X$

Finding the Product of Two Matrices In Exercises 55–58, find AB, if possible.

55. $A = \begin{bmatrix} 2 & -2 \\ 3 & 5 \end{bmatrix}$, $B = \begin{bmatrix} -3 & 10 \\ 12 & 8 \end{bmatrix}$

56. $A = \begin{bmatrix} 5 & 4 \\ -7 & 2 \\ 11 & 2 \end{bmatrix}$, $B = \begin{bmatrix} 4 & 12 \\ 20 & 40 \\ 15 & 30 \end{bmatrix}$

57. $A = \begin{bmatrix} 5 & 4 \\ -7 & 2 \\ 11 & 2 \end{bmatrix}$, $B = \begin{bmatrix} 4 & 12 \\ 20 & 40 \end{bmatrix}$

58. $A = \begin{bmatrix} 6 & -5 & 7 \end{bmatrix}$, $B = \begin{bmatrix} -1 \\ 4 \\ 8 \end{bmatrix}$

Evaluating an Expression In Exercises 59–62, evaluate the expression. If it is not possible, explain why.

59. $\begin{bmatrix} 1 & 2 \\ 5 & -4 \\ 6 & 0 \end{bmatrix}\begin{bmatrix} 6 & -2 & 8 \\ 4 & 0 & 0 \end{bmatrix}$

60. $\begin{bmatrix} 1 & 5 & 6 \\ 2 & -4 & 0 \end{bmatrix}\begin{bmatrix} 6 & -2 & 8 \\ 4 & 0 & 0 \end{bmatrix}$

61. $\begin{bmatrix} 1 & 5 & 6 \\ 2 & -4 & 0 \end{bmatrix}\begin{bmatrix} 6 & 4 \\ -2 & 0 \\ 8 & 0 \end{bmatrix}$

62. $\begin{bmatrix} 1 & 3 & 2 \\ 0 & 2 & -4 \\ 0 & 0 & 3 \end{bmatrix}\begin{bmatrix} 4 & -3 & 2 \\ 0 & 3 & -1 \\ 0 & 0 & 2 \end{bmatrix}$

Matrix Multiplication In Exercises 63–66, use the matrix capabilities of a graphing utility to find the product, if possible.

63. $\begin{bmatrix} 4 & 1 \\ 11 & -7 \\ 12 & 3 \end{bmatrix} \begin{bmatrix} 3 & -5 & 6 \\ 2 & -2 & -2 \end{bmatrix}$

64. $\begin{bmatrix} -2 & 3 & 10 \\ 4 & -2 & 2 \end{bmatrix} \begin{bmatrix} 1 & 1 \\ -5 & 2 \\ 3 & 2 \end{bmatrix}$

65. $\begin{bmatrix} 1 & 2 & -1 \\ 0 & 4 & -2 \\ 1 & 1 & 3 \end{bmatrix} \begin{bmatrix} 1 & -1 & 2 \end{bmatrix}$

66. $\begin{bmatrix} 4 & -2 & 6 \end{bmatrix} \begin{bmatrix} -2 & 1 \\ 0 & -3 \\ 2 & 0 \end{bmatrix}$

67. **Manufacturing** A tire corporation has three factories, each of which manufactures two models of tires. The production levels are represented by A.

$$A = \begin{array}{c} \\ \begin{bmatrix} 80 & 120 & 140 \\ 40 & 100 & 80 \end{bmatrix} \begin{array}{l} A \\ B \end{array} \Big\} \text{Model} \end{array}$$

with Factory columns labeled 1, 2, 3.

Find the production levels when production is decreased by 5%.

68. **Manufacturing** A power tool company has four manufacturing plants, each of which produces three types of cordless power tools. The production levels are represented by A.

$$A = \begin{bmatrix} 80 & 70 & 90 & 40 \\ 50 & 30 & 80 & 20 \\ 90 & 60 & 100 & 50 \end{bmatrix} \begin{array}{l} A \\ B \\ C \end{array} \Big\} \text{Type}$$

with Plant columns labeled 1, 2, 3, 4.

Find the production levels when production is increased by 20%.

69. **Manufacturing** An electronics manufacturing company produces three different models of headphones that are shipped to two warehouses. The shipment levels are represented by A.

$$A = \begin{bmatrix} 8200 & 7400 \\ 6500 & 9800 \\ 5400 & 4800 \end{bmatrix} \begin{array}{l} A \\ B \\ C \end{array} \Big\} \text{Model}$$

with Warehouse columns labeled 1, 2.

The prices per unit are represented by the matrix

$B = [\$79.99 \quad \$109.95 \quad \$189.99]$.

Compute BA and interpret the result.

70. **Cell Phone Charges** The pay-as-you-go charges (in dollars per minute) of two cellular telephone companies for calls inside the coverage area, regional roaming calls, and calls outside the coverage area are represented by C.

$$C = \begin{bmatrix} 0.07 & 0.095 \\ 0.10 & 0.08 \\ 0.28 & 0.25 \end{bmatrix} \begin{array}{l} \text{Inside} \\ \text{Regional roaming} \\ \text{Outside} \end{array} \Big\} \text{Coverage area}$$

with Company columns labeled A, B.

The numbers of minutes you plan to use in the coverage areas per month are represented by the matrix

$T = [120 \quad 80 \quad 20]$.

Compute TC and interpret the result.

8.3 The Inverse of a Matrix In Exercises 71–74, show that B is the inverse of A.

71. $A = \begin{bmatrix} -4 & -1 \\ 7 & 2 \end{bmatrix}$, $B = \begin{bmatrix} -2 & -1 \\ 7 & 4 \end{bmatrix}$

72. $A = \begin{bmatrix} 5 & -1 \\ 11 & -2 \end{bmatrix}$, $B = \begin{bmatrix} -2 & 1 \\ -11 & 5 \end{bmatrix}$

73. $A = \begin{bmatrix} 1 & 1 & 0 \\ 1 & 0 & 1 \\ 6 & 2 & 3 \end{bmatrix}$, $B = \begin{bmatrix} -2 & -3 & 1 \\ 3 & 3 & -1 \\ 2 & 4 & -1 \end{bmatrix}$

74. $A = \begin{bmatrix} 1 & -1 & 0 \\ -1 & 0 & -1 \\ 8 & -4 & 2 \end{bmatrix}$, $B = \begin{bmatrix} -2 & 1 & \frac{1}{2} \\ -3 & 1 & \frac{1}{2} \\ 2 & -2 & -\frac{1}{2} \end{bmatrix}$

Finding the Inverse of a Matrix In Exercises 75–78, find the inverse of the matrix (if it exists).

75. $\begin{bmatrix} -6 & 5 \\ -5 & 4 \end{bmatrix}$

76. $\begin{bmatrix} -3 & -5 \\ 2 & 3 \end{bmatrix}$

77. $\begin{bmatrix} 2 & 0 & 3 \\ -1 & 1 & 1 \\ 2 & -2 & 1 \end{bmatrix}$

78. $\begin{bmatrix} 0 & -2 & 1 \\ -5 & -2 & -3 \\ 7 & 3 & 4 \end{bmatrix}$

Finding the Inverse of a Matrix In Exercises 79–82, use the matrix capabilities of a graphing utility to find the inverse of the matrix (if it exists).

79. $\begin{bmatrix} -1 & -2 & -2 \\ 3 & 7 & 9 \\ 1 & 4 & 7 \end{bmatrix}$

80. $\begin{bmatrix} 1 & 4 & 6 \\ 2 & -3 & 1 \\ -1 & 18 & 16 \end{bmatrix}$

81. $\begin{bmatrix} 1 & 3 & 1 & 6 \\ 4 & 4 & 2 & 6 \\ 3 & 4 & 1 & 2 \\ -1 & 2 & -1 & -2 \end{bmatrix}$

82. $\begin{bmatrix} 8 & 0 & 2 & 8 \\ 4 & -2 & 0 & -2 \\ 1 & 2 & 1 & 4 \\ -1 & 4 & 1 & 1 \end{bmatrix}$

Finding the Inverse of a 2 × 2 Matrix   In Exercises 83–90, use the formula on page 417 to find the inverse of the 2 × 2 matrix (if it exists).

83. $\begin{bmatrix} -7 & 2 \\ -8 & 2 \end{bmatrix}$ 84. $\begin{bmatrix} 10 & 4 \\ 7 & 3 \end{bmatrix}$

85. $\begin{bmatrix} -12 & 6 \\ 10 & -5 \end{bmatrix}$ 86. $\begin{bmatrix} -18 & -15 \\ -6 & -5 \end{bmatrix}$

87. $\begin{bmatrix} -\frac{1}{2} & 20 \\ \frac{3}{10} & -6 \end{bmatrix}$ 88. $\begin{bmatrix} -\frac{3}{4} & \frac{5}{2} \\ -\frac{4}{5} & -\frac{8}{3} \end{bmatrix}$

89. $\begin{bmatrix} 0.5 & 0.1 \\ -0.2 & -0.4 \end{bmatrix}$ 90. $\begin{bmatrix} 1.6 & -3.2 \\ 1.2 & -2.4 \end{bmatrix}$

Solving a System Using an Inverse Matrix   In Exercises 91–102, use an inverse matrix to solve (if possible) the system of linear equations.

91. $\begin{cases} -x + 4y = 8 \\ 2x - 7y = -5 \end{cases}$ 92. $\begin{cases} 5x - y = 13 \\ -9x + 2y = -24 \end{cases}$

93. $\begin{cases} -3x + 10y = 8 \\ 5x - 17y = -13 \end{cases}$ 94. $\begin{cases} 4x - 2y = -10 \\ -19x + 9y = 47 \end{cases}$

95. $\begin{cases} \frac{1}{2}x + \frac{1}{3}y = 2 \\ -3x + 2y = 0 \end{cases}$ 96. $\begin{cases} -\frac{5}{6}x + \frac{3}{8}y = -2 \\ 4x - 3y = 0 \end{cases}$

97. $\begin{cases} 0.3x + 0.7y = 10.2 \\ 0.4x + 0.6y = 7.6 \end{cases}$ 98. $\begin{cases} 3.5x - 4.5y = 8 \\ 2.5x - 7.5y = 25 \end{cases}$

99. $\begin{cases} 3x + 2y - z = 6 \\ x - y + 2z = -1 \\ 5x + y + z = 7 \end{cases}$ 100. $\begin{cases} 4x + 5y - 6z = -6 \\ 3x + 2y + 2z = 8 \\ 2x + y + z = 3 \end{cases}$

101. $\begin{cases} -2x + y + 2z = -13 \\ -x - 4y + z = -11 \\ -y - z = 0 \end{cases}$

102. $\begin{cases} 3x - y + 5z = -14 \\ -x + y + 6z = 8 \\ -8x + 4y - z = 44 \end{cases}$

Using a Graphing Utility   In Exercises 103–108, use the matrix capabilities of a graphing utility to solve (if possible) the system of linear equations.

103. $\begin{cases} x + 2y = -1 \\ 3x + 4y = -5 \end{cases}$ 104. $\begin{cases} x + 3y = 23 \\ -6x + 2y = -18 \end{cases}$

105. $\begin{cases} \frac{6}{5}x - \frac{4}{7}y = \frac{6}{5} \\ -\frac{12}{5}x + \frac{12}{7}y = -\frac{17}{5} \end{cases}$ 106. $\begin{cases} 5x + 10y = 7 \\ 2x + y = -98 \end{cases}$

107. $\begin{cases} -3x - 3y - 4z = 2 \\ y + z = -1 \\ 4x + 3y + 4z = -1 \end{cases}$

108. $\begin{cases} x - 3y - 2z = 8 \\ -2x + 7y + 3z = -19 \\ x - y - 3z = 3 \end{cases}$

8.4 Finding the Determinant of a Matrix   In Exercises 109–112, find the determinant of the matrix.

109. $\begin{bmatrix} 8 & 5 \\ 2 & -4 \end{bmatrix}$ 110. $\begin{bmatrix} -9 & 11 \\ 7 & -4 \end{bmatrix}$

111. $\begin{bmatrix} 50 & -30 \\ 10 & 5 \end{bmatrix}$ 112. $\begin{bmatrix} 14 & -24 \\ 12 & -15 \end{bmatrix}$

Finding the Minors and Cofactors of a Matrix   In Exercises 113–116, find all the (a) minors and (b) cofactors of the matrix.

113. $\begin{bmatrix} 2 & -1 \\ 7 & 4 \end{bmatrix}$ 114. $\begin{bmatrix} 3 & 6 \\ 5 & -4 \end{bmatrix}$

115. $\begin{bmatrix} 3 & 2 & -1 \\ -2 & 5 & 0 \\ 1 & 8 & 6 \end{bmatrix}$ 116. $\begin{bmatrix} 8 & 3 & 4 \\ 6 & 5 & -9 \\ -4 & 1 & 2 \end{bmatrix}$

Finding the Determinant of a Matrix   In Exercises 117–126, find the determinant of the matrix. Expand by cofactors using the row or column that appears to make the computations easiest.

117. $\begin{bmatrix} -2 & 0 & 0 \\ 2 & -1 & 0 \\ -1 & 1 & -3 \end{bmatrix}$ 118. $\begin{bmatrix} 0 & 1 & -2 \\ 0 & 1 & 2 \\ -1 & -1 & 3 \end{bmatrix}$

119. $\begin{vmatrix} 4 & 1 & -1 \\ 2 & 3 & 2 \\ 1 & -1 & 0 \end{vmatrix}$ 120. $\begin{bmatrix} -1 & -2 & 1 \\ 2 & 3 & 0 \\ -5 & -1 & 3 \end{bmatrix}$

121. $\begin{vmatrix} -2 & 4 & 1 \\ -6 & 0 & 2 \\ 5 & 3 & 4 \end{vmatrix}$ 122. $\begin{vmatrix} 1 & 1 & 4 \\ -4 & 1 & 2 \\ 0 & 1 & -1 \end{vmatrix}$

123. $\begin{bmatrix} 1 & 2 & -1 & 0 \\ 1 & 2 & -4 & 1 \\ 2 & -4 & -3 & 1 \\ 2 & 0 & 0 & 0 \end{bmatrix}$ 124. $\begin{bmatrix} 1 & -2 & 1 & 2 \\ 4 & 1 & 4 & 1 \\ 2 & 3 & 3 & 0 \\ 0 & -2 & -4 & 2 \end{bmatrix}$

125. $\begin{bmatrix} 3 & 0 & -4 & 0 \\ 0 & 8 & 1 & 2 \\ 6 & 1 & 8 & 2 \\ 0 & 3 & -4 & 1 \end{bmatrix}$ 126. $\begin{bmatrix} -5 & 6 & 0 & 0 \\ 0 & 1 & -1 & 2 \\ -3 & 4 & -5 & 1 \\ 1 & 6 & 0 & 3 \end{bmatrix}$

8.5 Using Cramer's Rule   In Exercises 127–130, use Cramer's Rule to solve (if possible) the system of equations.

127. $\begin{cases} 5x - 2y = 6 \\ -11x + 3y = -23 \end{cases}$ 128. $\begin{cases} 3x + 8y = -7 \\ 9x - 5y = 37 \end{cases}$

129. $\begin{cases} -2x + 3y - 5z = -11 \\ 4x - y + z = -3 \\ -x - 4y + 6z = 15 \end{cases}$

130. $\begin{cases} 5x - 2y + z = 15 \\ 3x - 3y - z = -7 \\ 2x - y - 7z = -3 \end{cases}$

Finding the Area of a Triangle In Exercises 131–134, use a determinant to find the area of the triangle with the given vertices.

131.
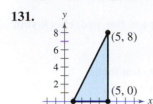

132.
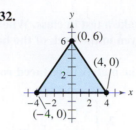

133.
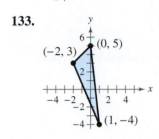

134.
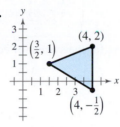

Testing for Collinear Points In Exercises 135 and 136, use a determinant to determine whether the points are collinear.

135. $(-1, 7), (3, -9), (-3, 15)$

136. $(0, -5), (-2, -6), (8, -1)$

Finding an Equation of a Line In Exercises 137–140, use a determinant to find an equation of the line passing through the points.

137. $(-4, 0), (4, 4)$ **138.** $(2, 5), (6, -1)$
139. $\left(-\frac{5}{2}, 3\right), \left(\frac{7}{2}, 1\right)$ **140.** $(-0.8, 0.2), (0.7, 3.2)$

Encoding a Message In Exercises 141 and 142, (a) write the uncoded 1×3 row matrices for the message and then (b) encode the message using the encoding matrix.

Message	Encoding Matrix
141. LOOK OUT BELOW	$\begin{bmatrix} 2 & -2 & 0 \\ 3 & 0 & -3 \\ -6 & 2 & 3 \end{bmatrix}$
142. HEAD DUE WEST	$\begin{bmatrix} 1 & 2 & 2 \\ 3 & 7 & 9 \\ -1 & -4 & -7 \end{bmatrix}$

Decoding a Message In Exercises 143 and 144, decode the cryptogram by using the inverse of the matrix

$$A = \begin{bmatrix} -5 & 4 & -3 \\ 10 & -7 & 6 \\ 8 & -6 & 5 \end{bmatrix}.$$

143. $-5 \;\; 11 \;\; -2 \;\; 370 \;\; -265 \;\; 225 \;\; -57 \;\; 48 \;\; -33 \;\; 32$
$-15 \;\; 20 \;\; 245 \;\; -171 \;\; 147$

144. $145 \;\; -105 \;\; 92 \;\; 264 \;\; -188 \;\; 160 \;\; 23 \;\; -16 \;\; 15$
$129 \;\; -84 \;\; 78 \;\; -9 \;\; 8 \;\; -5 \;\; 159 \;\; -118 \;\; 100 \;\; 219$
$-152 \;\; 133 \;\; 370 \;\; -265 \;\; 225 \;\; -105 \;\; 84 \;\; -63$

Exploration

True or False? In Exercises 145 and 146, determine whether the statement is true or false. Justify your answer.

145. It is possible to find the determinant of a 4×5 matrix.

146.
$$\begin{vmatrix} a_{11} & a_{12} & a_{13} \\ a_{21} & a_{22} & a_{23} \\ a_{31} + c_1 & a_{32} + c_2 & a_{33} + c_3 \end{vmatrix} =$$
$$\begin{vmatrix} a_{11} & a_{12} & a_{13} \\ a_{21} & a_{22} & a_{23} \\ a_{31} & a_{32} & a_{33} \end{vmatrix} + \begin{vmatrix} a_{11} & a_{12} & a_{13} \\ a_{21} & a_{22} & a_{23} \\ c_1 & c_2 & c_3 \end{vmatrix}$$

147. Using a Graphing Utility Use the matrix capabilities of a graphing utility to find the inverse of the matrix

$$A = \begin{bmatrix} 1 & -3 \\ -2 & 6 \end{bmatrix}.$$

What message appears on the screen? Why does the graphing utility display this message?

148. Invertible Matrices Under what conditions does a matrix have an inverse?

149. Writing What is meant by the cofactor of an entry of a matrix? How are cofactors used to find the determinant of the matrix?

150. Think About It Three people were asked to solve a system of equations using an augmented matrix. Each person reduced the matrix to row-echelon form. The reduced matrices were

$$\begin{bmatrix} 1 & 2 & \vdots & 3 \\ 0 & 1 & \vdots & 1 \end{bmatrix},$$

$$\begin{bmatrix} 1 & 0 & \vdots & 1 \\ 0 & 1 & \vdots & 1 \end{bmatrix},$$

and

$$\begin{bmatrix} 1 & 2 & \vdots & 3 \\ 0 & 0 & \vdots & 0 \end{bmatrix}.$$

Can all three be right? Explain.

151. Think About It Describe the row-echelon form of an augmented matrix that corresponds to a system of linear equations that has a unique solution.

152. Solving an Equation Solve the equation for λ.

$$\begin{vmatrix} 2 - \lambda & 5 \\ 3 & -8 - \lambda \end{vmatrix} = 0$$

Chapter Test

See CalcChat.com for tutorial help and worked-out solutions to odd-numbered exercises.

Take this test as you would take a test in class. When you are finished, check your work against the answers given in the back of the book.

In Exercises 1 and 2, write the matrix in reduced row-echelon form.

1. $\begin{bmatrix} 1 & -1 & 5 \\ 6 & 2 & 3 \\ 5 & 3 & -3 \end{bmatrix}$

2. $\begin{bmatrix} 1 & 0 & -1 & 2 \\ -1 & 1 & 1 & -3 \\ 1 & 1 & -1 & 1 \\ 3 & 2 & -3 & 4 \end{bmatrix}$

3. Write the augmented matrix for the system of equations and solve the system.

$$\begin{cases} 4x + 3y - 2z = 14 \\ -x - y + 2z = -5 \\ 3x + y - 4z = 8 \end{cases}$$

4. Find (a) $A - B$, (b) $3A$, (c) $3A - 2B$, and (d) AB (if possible).

$$A = \begin{bmatrix} 6 & 5 \\ -5 & -5 \end{bmatrix}, \quad B = \begin{bmatrix} 5 & 0 \\ -5 & -1 \end{bmatrix}$$

In Exercises 5 and 6, find the inverse of the matrix (if it exists).

5. $\begin{bmatrix} -4 & 3 \\ 5 & -2 \end{bmatrix}$

6. $\begin{bmatrix} -2 & 4 & -6 \\ 2 & 1 & 0 \\ 4 & -2 & 5 \end{bmatrix}$

7. Use the result of Exercise 5 to solve the system.

$$\begin{cases} -4x + 3y = 6 \\ 5x - 2y = 24 \end{cases}$$

In Exercises 8–10, find the determinant of the matrix.

8. $\begin{bmatrix} -6 & 4 \\ 10 & 12 \end{bmatrix}$

9. $\begin{bmatrix} \frac{5}{2} & \frac{13}{4} \\ -8 & \frac{6}{5} \end{bmatrix}$

10. $\begin{bmatrix} 6 & -7 & 2 \\ 3 & -2 & 0 \\ 1 & 5 & 1 \end{bmatrix}$

In Exercises 11 and 12, use Cramer's Rule to solve (if possible) the system of equations.

11. $\begin{cases} 7x + 6y = 9 \\ -2x - 11y = -49 \end{cases}$

12. $\begin{cases} 6x - y + 2z = -4 \\ -2x + 3y - z = 10 \\ 4x - 4y + z = -18 \end{cases}$

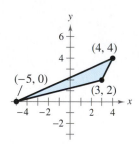

Figure for 13

13. Use a determinant to find the area of the triangle at the left.

14. Write the uncoded 1×3 row matrices for the message KNOCK ON WOOD. Then encode the message using the encoding matrix A below.

$$A = \begin{bmatrix} 1 & -1 & 0 \\ 1 & 0 & -1 \\ 6 & -2 & -3 \end{bmatrix}$$

15. One hundred liters of a 50% solution is obtained by mixing a 60% solution with a 20% solution. How many liters of each solution must be used to obtain the desired mixture?

Proofs in Mathematics ■ ■ ■ ■ ■ ■ ■ ■ ■ ■ ■ ■ ■ ■

A proof without words is a picture or diagram that gives a visual understanding of why a theorem or statement is true. It can also provide a starting point for writing a formal proof. The following proof shows that a 2×2 determinant is the area of a parallelogram.

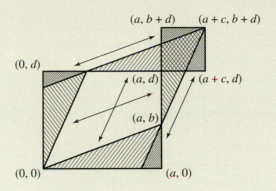

$$\begin{vmatrix} a & b \\ c & d \end{vmatrix} = ad - bc = \|\square\| - \|\square\| = \|\square\|$$

The following is a color-coded version of the proof along with a brief explanation of why this proof works.

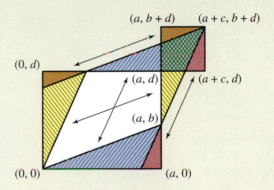

$$\begin{vmatrix} a & b \\ c & d \end{vmatrix} = ad - bc = \|\square\| - \|\square\| = \|\square\|$$

Area of $\square$ = Area of orange $\triangle$ + Area of yellow $\triangle$ + Area of blue $\triangle$ + Area of pink $\triangle$ + Area of white quadrilateral

Area of $\square$ = Area of orange $\triangle$ + Area of pink $\triangle$ + Area of green quadrilateral

Area of $\square$ = Area of white quadrilateral + Area of blue $\triangle$ + Area of yellow $\triangle$ − Area of green quadrilateral

= Area of $\square$ − Area of $\square$ ■

From "Proof Without Words" by Solomon W. Golomb, *Mathematics Magazine,* March 1985, Vol. 58, No. 2, pg. 107. Reprinted with permission.

P.S. Problem Solving

1. **Multiplying by a Transformation Matrix** The columns of matrix T show the coordinates of the vertices of a triangle. Matrix A is a transformation matrix.

$$A = \begin{bmatrix} 0 & -1 \\ 1 & 0 \end{bmatrix} \quad T = \begin{bmatrix} 1 & 2 & 3 \\ 1 & 4 & 2 \end{bmatrix}$$

(a) Find AT and AAT. Then sketch the original triangle and the two images of the triangle. What transformation does A represent?

(b) Given the triangle determined by AAT, describe the transformation that produces the triangle determined by AT and then the triangle determined by T.

2. **Population** The matrices show the number of people (in thousands) who lived in each region of the United States in 2000 and the number of people (in thousands) who lived in each region in 2010. The regional populations are separated into three age groups. *(Source: U.S. Census Bureau)*

2000

	0–17	18–64	65+
Northeast	13,048	33,174	7,372
Midwest	16,648	39,486	8,259
South	25,567	62,232	12,438
West	17,031	39,244	6,922

2010

	0–17	18–64	65+
Northeast	12,333	35,179	7,805
Midwest	16,128	41,777	9,022
South	27,789	71,873	14,894
West	17,931	45,467	8,547

(a) The total population in 2000 was approximately 281,422,000 and the total population in 2010 was approximately 308,746,000. Rewrite the matrices to give the information as percents of the total population.

(b) Write a matrix that gives the change in the percent of the population in each region and age group from 2000 to 2010.

(c) Based on the result of part (b), which region(s) and age group(s) had percents that decreased from 2000 to 2010?

3. **Determining Whether Matrices are Idempotent** Determine whether the matrix is idempotent. A square matrix is **idempotent** when $A^2 = A$.

(a) $\begin{bmatrix} 1 & 0 \\ 0 & 0 \end{bmatrix}$ (b) $\begin{bmatrix} 0 & 1 \\ 1 & 0 \end{bmatrix}$

(c) $\begin{bmatrix} 2 & 3 \\ -1 & -2 \end{bmatrix}$ (d) $\begin{bmatrix} 2 & 3 \\ 1 & 2 \end{bmatrix}$

4. **Quadratic Matrix Equation** Let $A = \begin{bmatrix} 1 & 2 \\ -2 & 1 \end{bmatrix}$.

(a) Show that $A^2 - 2A + 5I = O$, where I is the identity matrix of order 2×2.

(b) Show that $A^{-1} = \frac{1}{5}(2I - A)$.

(c) Show in general that for any square matrix satisfying

$A^2 - 2A + 5I = O$

the inverse of A is given by

$A^{-1} = \frac{1}{5}(2I - A)$.

5. **Satellite Television** Two competing companies offer satellite television to a city with 100,000 households. Gold Satellite System has 25,000 subscribers and Galaxy Satellite Network has 30,000 subscribers. (The other 45,000 households do not subscribe.) The percent changes in satellite subscriptions each year are shown in the matrix below.

Percent Changes	From Gold	From Galaxy	From Nonsubscriber
To Gold	0.70	0.15	0.15
To Galaxy	0.20	0.80	0.15
To Nonsubscriber	0.10	0.05	0.70

(a) Find the number of subscribers each company will have in 1 year using matrix multiplication. Explain how you obtained your answer.

(b) Find the number of subscribers each company will have in 2 years using matrix multiplication. Explain how you obtained your answer.

(c) Find the number of subscribers each company will have in 3 years using matrix multiplication. Explain how you obtained your answer.

(d) What is happening to the number of subscribers to each company? What is happening to the number of nonsubscribers?

6. **The Transpose of a Matrix** The **transpose** of a matrix, denoted A^T, is formed by writing its columns as rows. Find the transpose of each matrix and verify that $(AB)^T = B^TA^T$.

$$A = \begin{bmatrix} -1 & 1 & -2 \\ 2 & 0 & 1 \end{bmatrix}, \quad B = \begin{bmatrix} -3 & 0 \\ 1 & 2 \\ 1 & -1 \end{bmatrix}$$

7. **Finding a Value** Find x such that the matrix is equal to its own inverse.

$$A = \begin{bmatrix} 3 & x \\ -2 & -3 \end{bmatrix}$$

8. Finding a Value Find x such that the matrix is singular.

$$A = \begin{bmatrix} 4 & x \\ -2 & -3 \end{bmatrix}$$

9. Finding a Matrix Find a singular 2×2 matrix satisfying $A^2 = A$.

10. Verifying an Equation Verify the following equation.

$$\begin{vmatrix} 1 & 1 & 1 \\ a & b & c \\ a^2 & b^2 & c^2 \end{vmatrix} = (a - b)(b - c)(c - a)$$

11. Verifying an Equation Verify the following equation.

$$\begin{vmatrix} 1 & 1 & 1 \\ a & b & c \\ a^3 & b^3 & c^3 \end{vmatrix} = (a - b)(b - c)(c - a)(a + b + c)$$

12. Verifying an Equation Verify the following equation.

$$\begin{vmatrix} x & 0 & c \\ -1 & x & b \\ 0 & -1 & a \end{vmatrix} = ax^2 + bx + c$$

13. Finding a Determinant Use the equation given in Exercise 12 as a model to find a determinant that is equal to $ax^3 + bx^2 + cx + d$.

14. Finding Atomic Masses The atomic masses of three compounds are shown in the table. Use a linear system and Cramer's Rule to find the atomic masses of sulfur (S), nitrogen (N), and fluorine (F).

Compound	Formula	Atomic Mass
Tetrasulfur tetranitride	S_4N_4	184
Sulfur hexafluoride	SF_6	146
Dinitrogen tetrafluoride	N_2F_4	104

15. Finding the Costs of Items A walkway lighting package includes a transformer, a certain length of wire, and a certain number of lights on the wire. The price of each lighting package depends on the length of wire and the number of lights on the wire. Use the following information to find the cost of a transformer, the cost per foot of wire, and the cost of a light. Assume that the cost of each item is the same in each lighting package.

- A package that contains a transformer, 25 feet of wire, and 5 lights costs \$20.
- A package that contains a transformer, 50 feet of wire, and 15 lights costs \$35.
- A package that contains a transformer, 100 feet of wire, and 20 lights costs \$50.

16. Decoding a Message Use the inverse of matrix A to decode the cryptogram.

$$A = \begin{bmatrix} 1 & -2 & 2 \\ 1 & 1 & -3 \\ 1 & -1 & 4 \end{bmatrix}$$

23	13	-34	31	-34	63	25	-17	61
24	14	-37	41	-17	-8	20	-29	40
38	-56	116	13	-11	1	22	-3	-6
41	-53	85	28	-32	16			

17. Decoding a Message A code breaker intercepted the encoded message below.

45 -35 38 -30 18 -18 35 -30 81 -60
42 -28 75 -55 2 -2 22 -21 15 -10

Let $A^{-1} = \begin{bmatrix} w & x \\ y & z \end{bmatrix}$.

(a) You know that $\begin{bmatrix} 45 & -35 \end{bmatrix} A^{-1} = \begin{bmatrix} 10 & 15 \end{bmatrix}$ and that $\begin{bmatrix} 38 & -30 \end{bmatrix} A^{-1} = \begin{bmatrix} 8 & 14 \end{bmatrix}$, where A^{-1} is the inverse of the encoding matrix A. Write and solve two systems of equations to find w, x, y, and z.

(b) Decode the message.

18. Conjecture Let

$$A = \begin{bmatrix} 6 & 4 & 1 \\ 0 & 2 & 3 \\ 1 & 1 & 2 \end{bmatrix}.$$

Use a graphing utility to find A^{-1}. Compare $|A^{-1}|$ with $|A|$. Make a conjecture about the determinant of the inverse of a matrix.

19. Finding the Determinant of a Matrix Let A be an $n \times n$ matrix each of whose rows adds up to zero. Find $|A|$.

20. Conjecture Consider matrices of the form

$$A = \begin{bmatrix} 0 & a_{12} & a_{13} & a_{14} & \cdots & a_{1n} \\ 0 & 0 & a_{23} & a_{24} & \cdots & a_{2n} \\ 0 & 0 & 0 & a_{34} & \cdots & a_{3n} \\ \vdots & \vdots & \vdots & \vdots & \cdots & \vdots \\ 0 & 0 & 0 & 0 & \cdots & a_{(n-1)n} \\ 0 & 0 & 0 & 0 & \cdots & 0 \end{bmatrix}.$$

(a) Write a 2×2 matrix and a 3×3 matrix in the form of A.

(b) Use a graphing utility to raise each of the matrices to higher powers. Describe the result.

(c) Use the result of part (b) to make a conjecture about powers of A when A is a 4×4 matrix. Use the graphing utility to test your conjecture.

(d) Use the results of parts (b) and (c) to make a conjecture about powers of A when A is an $n \times n$ matrix.

9 Sequences, Series, and Probability

Horse Racing

Tossing Dice

Electricity

Dominoes

AIDS Cases

445

9.1 Sequences and Series

■ Use sequence notation to write the terms of sequences.
■ Use factorial notation.
■ Use summation notation to write sums.
■ Find the sums of series.
■ Use sequences and series to model and solve real-life problems.

Sequences

In mathematics, the word *sequence* is used in much the same way as in ordinary English. Saying that a collection is listed in *sequence* means that it is ordered so that it has a first member, a second member, a third member, and so on. Two examples are $1, 2, 3, 4, \ldots$ and $1, 3, 5, 7, \ldots$.

Mathematically, you can think of a sequence as a *function* whose domain is the set of positive integers. Rather than using function notation, however, sequences are usually written using subscript notation, as indicated in the following definition.

Sequences and series can help you model real-life problems, such as the numbers of AIDS cases reported from 2003 through 2010.

Definition of Sequence

An **infinite sequence** is a function whose domain is the set of positive integers. The function values

$$a_1, a_2, a_3, a_4, \ldots, a_n, \ldots$$

are the **terms** of the sequence. When the domain of the function consists of the first n positive integers only, the sequence is a **finite sequence.**

On occasion, it is convenient to begin subscripting a sequence with 0 instead of 1 so that the terms of the sequence become

$$a_0, a_1, a_2, a_3, \ldots.$$

When this is the case, the domain includes 0.

▷

REMARK The subscripts of a sequence make up the domain of the sequence and serve to identify the locations of terms within the sequence. For example, a_4 is the fourth term of the sequence, and a_n is the nth term of the sequence. Any variable can be a subscript. The most commonly used variable subscripts in sequence and series notation are $i, j, k,$ and n.

EXAMPLE 1   Writing the Terms of a Sequence

a. The first four terms of the sequence given by $a_n = 3n - 2$ are

$a_1 = 3(1) - 2 = 1$ 1st term

$a_2 = 3(2) - 2 = 4$ 2nd term

$a_3 = 3(3) - 2 = 7$ 3rd term

$a_4 = 3(4) - 2 = 10.$ 4th term

b. The first four terms of the sequence given by $a_n = 3 + (-1)^n$ are

$a_1 = 3 + (-1)^1 = 3 - 1 = 2$ 1st term

$a_2 = 3 + (-1)^2 = 3 + 1 = 4$ 2nd term

$a_3 = 3 + (-1)^3 = 3 - 1 = 2$ 3rd term

$a_4 = 3 + (-1)^4 = 3 + 1 = 4.$ 4th term

✓ *Checkpoint*) *Audio-video solution in English & Spanish at LarsonPrecalculus.com.*

Write the first four terms of the sequence given by $a_n = 2n + 1$.

•• **REMARK**   Write the first
four terms of the sequence
whose nth term is
$$a_n = \frac{(-1)^{n+1}}{2n+1}.$$
Are they the same as the first
four terms of the sequence in
Example 2? If not, then how do
they differ?
⋯⋯⋯⋯⋯⋯⋯⋯▷

EXAMPLE 2   **A Sequence Whose Terms Alternate in Sign**

Write the first four terms of the sequence given by $a_n = \dfrac{(-1)^n}{2n+1}$.

Solution   The first four terms of the sequence are as follows.

$$a_1 = \frac{(-1)^1}{2(1)+1} = \frac{-1}{2+1} = -\frac{1}{3} \qquad \text{1st term}$$

$$a_2 = \frac{(-1)^2}{2(2)+1} = \frac{1}{4+1} = \frac{1}{5} \qquad \text{2nd term}$$

$$a_3 = \frac{(-1)^3}{2(3)+1} = \frac{-1}{6+1} = -\frac{1}{7} \qquad \text{3rd term}$$

$$a_4 = \frac{(-1)^4}{2(4)+1} = \frac{1}{8+1} = \frac{1}{9} \qquad \text{4th term}$$

✓ *Checkpoint* 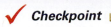 *Audio-video solution in English & Spanish at LarsonPrecalculus.com.*

Write the first four terms of the sequence given by $a_n = \dfrac{2+(-1)^n}{n}$.   ■

Simply listing the first few terms is not sufficient to define a unique sequence—the nth term *must be given*. To see this, consider the following sequences, both of which have the same first three terms.

$$\frac{1}{2}, \frac{1}{4}, \frac{1}{8}, \frac{1}{16}, \cdots, \frac{1}{2^n}, \cdots$$

$$\frac{1}{2}, \frac{1}{4}, \frac{1}{8}, \frac{1}{15}, \cdots, \frac{6}{(n+1)(n^2-n+6)}, \cdots$$

▷ **TECHNOLOGY**
To graph a sequence using a
graphing utility, set the mode
to *sequence* and *dot* and enter
the sequence. The graph of the
sequence in Example 3(a) is
below. To identify the terms,
use the *trace* feature or *value*
feature.

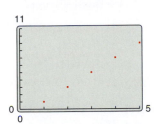

EXAMPLE 3   **Finding the nth Term of a Sequence**

Write an expression for the apparent nth term (a_n) of each sequence.

a. $1, 3, 5, 7, \ldots$ **b.** $2, -5, 10, -17, \ldots$

Solution

a. n: 1 2 3 4 … n
      Terms: 1 3 5 7 … a_n

Apparent pattern: Each term is 1 less than twice n, which implies that

$$a_n = 2n - 1.$$

b. n: 1   2   3   4 … n
      Terms: 2 −5 10 −17 … a_n

Apparent pattern: The absolute value of each term is 1 more than the square of n, and the terms have alternating signs, with those in the even positions being negative. This implies that

$$a_n = (-1)^{n+1}(n^2 + 1).$$

✓ *Checkpoint* 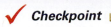 *Audio-video solution in English & Spanish at LarsonPrecalculus.com.*

Write an expression for the apparent nth term (a_n) of each sequence.

a. $1, 5, 9, 13, \ldots$ **b.** $2, -4, 6, -8, \ldots$ ■

Some sequences are defined **recursively.** To define a sequence recursively, you need to be given one or more of the first few terms. All other terms of the sequence are then defined using previous terms.

EXAMPLE 4 **A Recursive Sequence**

Write the first five terms of the sequence defined recursively as

$$a_1 = 3, \quad a_k = 2a_{k-1} + 1, \quad \text{where } k \geq 2.$$

Solution

$a_1 = 3$	1st term is given.
$a_2 = 2a_{2-1} + 1 = 2a_1 + 1 = 2(3) + 1 = 7$	Use recursion formula.
$a_3 = 2a_{3-1} + 1 = 2a_2 + 1 = 2(7) + 1 = 15$	Use recursion formula.
$a_4 = 2a_{4-1} + 1 = 2a_3 + 1 = 2(15) + 1 = 31$	Use recursion formula.
$a_5 = 2a_{5-1} + 1 = 2a_4 + 1 = 2(31) + 1 = 63$	Use recursion formula.

✓ *Checkpoint* ◀))) *Audio-video solution in English & Spanish at LarsonPrecalculus.com.*

Write the first five terms of the sequence defined recursively as

$$a_1 = 6, \quad a_{k+1} = a_k + 1, \quad \text{where} \quad k \geq 2.$$

In the next example, you will study a well-known recursive sequence, the Fibonacci sequence.

EXAMPLE 5 **The Fibonacci Sequence: A Recursive Sequence**

The Fibonacci sequence is defined recursively, as follows.

$$a_0 = 1, \quad a_1 = 1, \quad a_k = a_{k-2} + a_{k-1}, \quad \text{where} \quad k \geq 2$$

Write the first six terms of this sequence.

Solution

$a_0 = 1$	0th term is given.
$a_1 = 1$	1st term is given.
$a_2 = a_{2-2} + a_{2-1} = a_0 + a_1 = 1 + 1 = 2$	Use recursion formula.
$a_3 = a_{3-2} + a_{3-1} = a_1 + a_2 = 1 + 2 = 3$	Use recursion formula.
$a_4 = a_{4-2} + a_{4-1} = a_2 + a_3 = 2 + 3 = 5$	Use recursion formula.
$a_5 = a_{5-2} + a_{5-1} = a_3 + a_4 = 3 + 5 = 8$	Use recursion formula.

✓ *Checkpoint* ◀))) *Audio-video solution in English & Spanish at LarsonPrecalculus.com.*

Write the first five terms of the sequence defined recursively as

$$a_0 = 1, \quad a_1 = 3, \quad a_k = a_{k-2} + a_{k-1}, \quad \text{where} \quad k \geq 2.$$

Factorial Notation

Some very important sequences in mathematics involve terms that are defined with special types of products called **factorials.**

Definition of Factorial

If n is a positive integer, then **n factorial** is defined as

$$n! = 1 \cdot 2 \cdot 3 \cdot 4 \cdots (n-1) \cdot n.$$

As a special case, zero factorial is defined as $0! = 1$.

•• **REMARK** The value of n does not have to be very large before the value of $n!$ becomes extremely large. For instance, $10! = 3{,}628{,}800$.

Notice that $0! = 1$ and $1! = 1$. Here are some other values of $n!$.

$$2! = 1 \cdot 2 = 2 \qquad 3! = 1 \cdot 2 \cdot 3 = 6 \qquad 4! = 1 \cdot 2 \cdot 3 \cdot 4 = 24$$

Factorials follow the same conventions for order of operations as do exponents. For instance,

$$2n! = 2(n!) = 2(1 \cdot 2 \cdot 3 \cdot 4 \cdots n), \quad \text{whereas} \quad (2n)! = 1 \cdot 2 \cdot 3 \cdot 4 \cdots 2n.$$

EXAMPLE 6 **Writing the Terms of a Sequence Involving Factorials**

Write the first five terms of the sequence given by $a_n = \dfrac{2^n}{n!}$. Begin with $n = 0$.

Algebraic Solution

$a_0 = \dfrac{2^0}{0!} = \dfrac{1}{1} = 1$	0th term
$a_1 = \dfrac{2^1}{1!} = \dfrac{2}{1} = 2$	1st term
$a_2 = \dfrac{2^2}{2!} = \dfrac{4}{2} = 2$	2nd term
$a_3 = \dfrac{2^3}{3!} = \dfrac{8}{6} = \dfrac{4}{3}$	3rd term
$a_4 = \dfrac{2^4}{4!} = \dfrac{16}{24} = \dfrac{2}{3}$	4th term

Graphical Solution

Using a graphing utility set to *dot* and *sequence* modes, enter the sequence. Next, graph the sequence. You can estimate the first five terms of the sequence as follows.

$$u_0 = 1$$
$$u_1 = 2$$
$$u_2 = 2$$
$$u_3 \approx 1.333 \approx \tfrac{4}{3}$$
$$u_4 \approx 0.667 \approx \tfrac{2}{3}$$

Use the *trace* feature to approximate the first five terms.

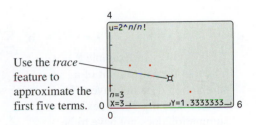

✓ **Checkpoint** ◄))) *Audio-video solution in English & Spanish at LarsonPrecalculus.com.*

Write the first five terms of the sequence given by $a_n = \dfrac{3^n + 1}{n!}$. Begin with $n = 0$.

When working with fractions involving factorials, you will often be able to reduce the fractions to simplify the computations.

EXAMPLE 7 **Simplifying Factorial Expressions**

▷ **ALGEBRA HELP** You can also simplify the expression in Example 7(a) as follows.

$$\frac{8!}{2! \cdot 6!} = \frac{8 \cdot 7 \cdot 6!}{2! \cdot 6!}$$

$$= \frac{8 \cdot 7}{2 \cdot 1} = 28$$

a. $\dfrac{8!}{2! \cdot 6!} = \dfrac{1 \cdot 2 \cdot 3 \cdot 4 \cdot 5 \cdot 6 \cdot 7 \cdot 8}{1 \cdot 2 \cdot 1 \cdot 2 \cdot 3 \cdot 4 \cdot 5 \cdot 6} = \dfrac{7 \cdot 8}{2} = 28$

b. $\dfrac{n!}{(n-1)!} = \dfrac{1 \cdot 2 \cdot 3 \cdots (n-1) \cdot n}{1 \cdot 2 \cdot 3 \cdots (n-1)} = n$

✓ **Checkpoint** ◄))) *Audio-video solution in English & Spanish at LarsonPrecalculus.com.*

Simplify the factorial expression $\dfrac{4!(n+1)!}{3!n!}$.

▷ **TECHNOLOGY** Most graphing utilities can sum the first n terms of a sequence. Check your user's guide for a *sum sequence* feature or a *series* feature.

Summation Notation

A convenient notation for the sum of the terms of a finite sequence is called **summation notation** or **sigma notation**. It involves the use of the uppercase Greek letter sigma, written as Σ.

▷ **REMARK** Summation notation is an instruction to add the terms of a sequence. From the definition at the right, the upper limit of summation tells you the last term of the sum. Summation notation helps you generate the terms of the sequence prior to finding the sum.

> **Definition of Summation Notation**
>
> The sum of the first n terms of a sequence is represented by
>
> $$\sum_{i=1}^{n} a_i = a_1 + a_2 + a_3 + a_4 + \cdots + a_n$$
>
> where i is called the **index of summation**, n is the **upper limit of summation**, and 1 is the **lower limit of summation**.

▷ **REMARK** In Example 8, note that the lower limit of a summation does not have to be 1. Also note that the index of summation does not have to be the letter i. For instance, in part (b), the letter k is the index of summation.

EXAMPLE 8   **Summation Notation for a Sum**

a. $\displaystyle\sum_{i=1}^{5} 3i = 3(1) + 3(2) + 3(3) + 3(4) + 3(5)$

$= 45$

b. $\displaystyle\sum_{k=3}^{6} (1 + k^2) = (1 + 3^2) + (1 + 4^2) + (1 + 5^2) + (1 + 6^2)$

$= 10 + 17 + 26 + 37$

$= 90$

c. $\displaystyle\sum_{i=0}^{8} \frac{1}{i!} = \frac{1}{0!} + \frac{1}{1!} + \frac{1}{2!} + \frac{1}{3!} + \frac{1}{4!} + \frac{1}{5!} + \frac{1}{6!} + \frac{1}{7!} + \frac{1}{8!}$

$= 1 + 1 + \dfrac{1}{2} + \dfrac{1}{6} + \dfrac{1}{24} + \dfrac{1}{120} + \dfrac{1}{720} + \dfrac{1}{5040} + \dfrac{1}{40{,}320}$

≈ 2.71828

For this summation, note that the sum is very close to the irrational number

$e \approx 2.718281828.$

It can be shown that as more terms of the sequence whose nth term is $1/n!$ are added, the sum becomes closer and closer to e.

✓ *Checkpoint* *Audio-video solution in English & Spanish at LarsonPrecalculus.com.*

Find the sum $\displaystyle\sum_{i=1}^{4} (4i + 1)$.

> **Properties of Sums**
>
> **1.** $\displaystyle\sum_{i=1}^{n} c = cn$, c is a constant.   **2.** $\displaystyle\sum_{i=1}^{n} ca_i = c\sum_{i=1}^{n} a_i$, c is a constant.
>
> **3.** $\displaystyle\sum_{i=1}^{n} (a_i + b_i) = \sum_{i=1}^{n} a_i + \sum_{i=1}^{n} b_i$ **4.** $\displaystyle\sum_{i=1}^{n} (a_i - b_i) = \sum_{i=1}^{n} a_i - \sum_{i=1}^{n} b_i$

For proofs of these properties, see Proofs in Mathematics on page 501.

Series

Many applications involve the sum of the terms of a finite or infinite sequence. Such a sum is called a **series**.

Definition of Series

Consider the infinite sequence $a_1, a_2, a_3, \ldots, a_i, \ldots$.

1. The sum of the first n terms of the sequence is called a **finite series** or the **nth partial sum** of the sequence and is denoted by

$$a_1 + a_2 + a_3 + \cdots + a_n = \sum_{i=1}^{n} a_i.$$

2. The sum of all the terms of the infinite sequence is called an **infinite series** and is denoted by

$$a_1 + a_2 + a_3 + \cdots + a_i + \cdots = \sum_{i=1}^{\infty} a_i.$$

EXAMPLE 9   **Finding the Sum of a Series**

For the series

$$\sum_{i=1}^{\infty} \frac{3}{10^i}$$

find (a) the third partial sum and (b) the sum.

Solution

a. The third partial sum is

$$\sum_{i=1}^{3} \frac{3}{10^i} = \frac{3}{10^1} + \frac{3}{10^2} + \frac{3}{10^3}$$
$$= 0.3 + 0.03 + 0.003$$
$$= 0.333.$$

b. The sum of the series is

$$\sum_{i=1}^{\infty} \frac{3}{10^i} = \frac{3}{10^1} + \frac{3}{10^2} + \frac{3}{10^3} + \frac{3}{10^4} + \frac{3}{10^5} + \cdots$$
$$= 0.3 + 0.03 + 0.003 + 0.0003 + 0.00003 + \cdots$$
$$= 0.33333\ldots$$
$$= \frac{1}{3}.$$

✓ **Checkpoint** ◀))) *Audio-video solution in English & Spanish at LarsonPrecalculus.com.*

For the series

$$\sum_{i=1}^{\infty} \frac{5}{10^i}$$

find (a) the fourth partial sum and (b) the sum.

Notice in Example 9(b) that the sum of an infinite series can be a finite number.

Application

Sequences have many applications in business and science. Example 10 illustrates such an application.

EXAMPLE 10 **Compound Interest**

An investor deposits $5000 in an account that earns 3% interest compounded quarterly. The balance in the account after n quarters is given by

$$A_n = 5000\left(1 + \frac{0.03}{4}\right)^n, \quad n = 0, 1, 2, \dots .$$

a. Write the first three terms of the sequence.

b. Find the balance in the account after 10 years by computing the 40th term of the sequence.

Solution

a. The first three terms of the sequence are as follows.

$$A_0 = 5000\left(1 + \frac{0.03}{4}\right)^0 = \$5000.00 \qquad \text{Original deposit}$$

$$A_1 = 5000\left(1 + \frac{0.03}{4}\right)^1 = \$5037.50 \qquad \text{First-quarter balance}$$

$$A_2 = 5000\left(1 + \frac{0.03}{4}\right)^2 \approx \$5075.28 \qquad \text{Second-quarter balance}$$

b. The 40th term of the sequence is

$$A_{40} = 5000\left(1 + \frac{0.03}{4}\right)^{40} \approx \$6741.74. \qquad \text{Ten-year balance}$$

✓ *Checkpoint*))) *Audio-video solution in English & Spanish at LarsonPrecalculus.com.*

An investor deposits $1000 in an account that earns 3% interest compounded monthly. The balance in the account after n months is given by

$$A_n = 1000\left(1 + \frac{0.03}{12}\right)^n, \quad n = 0, 1, 2, \dots .$$

a. Write the first three terms of the sequence.

b. Find the balance in the account after four years by computing the 48th term of the sequence.

Summarize   (Section 9.1)

1. State the definition of a sequence (*page 446*). For examples of writing the terms of sequences, see Examples 1–5.

2. State the definition of a factorial (*page 449*). For examples of using factorial notation, see Examples 6 and 7.

3. State the definition of summation notation (*page 450*). For an example of using summation notation, see Example 8.

4. State the definition of a series (*page 451*). For an example of finding the sum of a series, see Example 9.

5. Describe an example of how to use a sequence to model and solve a real-life problem (*page 452, Example 10*).

9.2 Arithmetic Sequences and Partial Sums

- Recognize, write, and find the *n*th terms of arithmetic sequences.
- Find *n*th partial sums of arithmetic sequences.
- Use arithmetic sequences to model and solve real-life problems.

Arithmetic Sequences

A sequence whose consecutive terms have a common difference is called an **arithmetic sequence.**

Arithmetic sequences have many real-life applications. For instance, an arithmetic sequence can be used to determine how far an object falls in 7 seconds from the top of the Willis Tower in Chicago.

Definition of Arithmetic Sequence

A sequence is **arithmetic** when the differences between consecutive terms are the same. So, the sequence

$$a_1, a_2, a_3, a_4, \ldots, a_n, \ldots$$

is arithmetic when there is a number d such that

$$a_2 - a_1 = a_3 - a_2 = a_4 - a_3 = \cdots = d.$$

The number d is the **common difference** of the arithmetic sequence.

EXAMPLE 1 **Examples of Arithmetic Sequences**

a. The sequence whose *n*th term is $4n + 3$ is arithmetic. For this sequence, the common difference between consecutive terms is 4.

$$7, 11, 15, 19, \ldots, 4n + 3, \ldots \qquad \text{Begin with } n = 1.$$

$$11 - 7 = 4$$

b. The sequence whose *n*th term is $7 - 5n$ is arithmetic. For this sequence, the common difference between consecutive terms is -5.

$$2, -3, -8, -13, \ldots, 7 - 5n, \ldots \qquad \text{Begin with } n = 1.$$

$$-3 - 2 = -5$$

c. The sequence whose *n*th term is $\frac{1}{4}(n + 3)$ is arithmetic. For this sequence, the common difference between consecutive terms is $\frac{1}{4}$.

$$1, \frac{5}{4}, \frac{3}{2}, \frac{7}{4}, \ldots, \frac{n + 3}{4}, \ldots \qquad \text{Begin with } n = 1.$$

$$\tfrac{5}{4} - 1 = \tfrac{1}{4}$$

✓ **Checkpoint** 🔊))) *Audio-video solution in English & Spanish at LarsonPrecalculus.com.*

Write the first four terms of the arithmetic sequence whose *n*th term is $3n - 1$. Then find the common difference between consecutive terms. ■

The sequence $1, 4, 9, 16, \ldots$, whose *n*th term is n^2, is *not* arithmetic. The difference between the first two terms is

$$a_2 - a_1 = 4 - 1 = 3$$

but the difference between the second and third terms is

$$a_3 - a_2 = 9 - 4 = 5.$$

Eugene Moerman/Shutterstock.com

The nth term of an arithmetic sequence can be derived from the following pattern.

$a_1 = a_1$ 1st term

$a_2 = a_1 + d$ 2nd term

$a_3 = a_1 + 2d$ 3rd term

$a_4 = a_1 + 3d$ 4th term

$a_5 = a_1 + 4d$ 5th term

1 less

$\vdots$

$a_n = a_1 + (n - 1)d$ nth term

1 less

The following definition summarizes this result.

The nth Term of an Arithmetic Sequence

The nth term of an arithmetic sequence has the form

$$a_n = a_1 + (n - 1)d$$

where d is the common difference between consecutive terms of the sequence and a_1 is the first term.

EXAMPLE 2   **Finding the nth Term**

Find a formula for the nth term of the arithmetic sequence whose common difference is 3 and whose first term is 2.

Solution   You know that the formula for the nth term is of the form $a_n = a_1 + (n - 1)d$. Moreover, because the common difference is $d = 3$ and the first term is $a_1 = 2$, the formula must have the form

$$a_n = 2 + 3(n - 1).$$ Substitute 2 for a_1 and 3 for d.

So, the formula for the nth term is $a_n = 3n - 1$. The sequence therefore has the following form.

$$2, 5, 8, 11, 14, \ldots, 3n - 1, \ldots$$

The figure below shows a graph of the first 15 terms of the sequence. Notice that the points lie on a line. This makes sense because a_n is a linear function of n. In other words, the terms "arithmetic" and "linear" are closely connected.

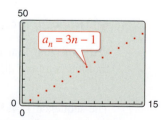

✓ Checkpoint   🔊)) *Audio-video solution in English & Spanish at LarsonPrecalculus.com.*

Find a formula for the nth term of the arithmetic sequence whose common difference is 5 and whose first term is -1.

• REMARK You can find a_1 in Example 3 by using the nth term of an arithmetic sequence, as follows.

$$a_n = a_1 + (n-1)d$$
$$a_4 = a_1 + (4-1)d$$
$$20 = a_1 + (4-1)5$$
$$20 = a_1 + 15$$
$$5 = a_1$$

$\triangleright$

EXAMPLE 3	**Writing the Terms of an Arithmetic Sequence**

The fourth term of an arithmetic sequence is 20, and the 13th term is 65. Write the first 11 terms of this sequence.

Solution You know that $a_4 = 20$ and $a_{13} = 65$. So, you must add the common difference d nine times to the fourth term to obtain the 13th term. Therefore, the fourth and 13th terms of the sequence are related by

$$a_{13} = a_4 + 9d. \qquad \text{\color{red}{a_4 and a_{13} are nine terms apart.}}$$

Using $a_4 = 20$ and $a_{13} = 65$, solve for d to find that $d = 5$, which implies that the sequence is as follows.

a_1	a_2	a_3	a_4	a_5	a_6	a_7	a_8	a_9	a_{10}	a_{11}	$\cdots$
5	10	15	20	25	30	35	40	45	50	55	$\cdots$

✓ **Checkpoint**  *Audio-video solution in English & Spanish at LarsonPrecalculus.com.*

The eighth term of an arithmetic sequence is 25, and the 12th term is 41. Write the first 11 terms of this sequence. ∎

When you know the nth term of an arithmetic sequence *and* you know the common difference of the sequence, you can find the $(n+1)$th term by using the *recursion formula*

$$a_{n+1} = a_n + d. \qquad \text{\color{red}{Recursion formula}}$$

With this formula, you can find any term of an arithmetic sequence, *provided* that you know the preceding term. For instance, when you know the first term, you can find the second term. Then, knowing the second term, you can find the third term, and so on.

EXAMPLE 4	**Using a Recursion Formula**

Find the ninth term of the arithmetic sequence that begins with 2 and 9.

Solution For this sequence, the common difference is

$$d = 9 - 2 = 7.$$

There are two ways to find the ninth term. One way is to write the first nine terms (by repeatedly adding 7).

$$2, 9, 16, 23, 30, 37, 44, 51, 58$$

Another way to find the ninth term is to first find a formula for the nth term. Because the common difference is $d = 7$ and the first term is $a_1 = 2$, the formula must have the form

$$a_n = 2 + 7(n-1). \qquad \text{\color{red}{Substitute 2 for a_1 and 7 for d.}}$$

Therefore, a formula for the nth term is

$$a_n = 7n - 5$$

which implies that the ninth term is

$$a_9 = 7(9) - 5$$
$$= 58.$$

✓ **Checkpoint**  *Audio-video solution in English & Spanish at LarsonPrecalculus.com.*

Find the tenth term of the arithmetic sequence that begins with 7 and 15. ∎

The Sum of a Finite Arithmetic Sequence

• • **REMARK**   Note that this
formula works only for
arithmetic sequences.

There is a formula for the *sum* of a finite arithmetic sequence.

> ### The Sum of a Finite Arithmetic Sequence
>
> The sum of a finite arithmetic sequence with n terms is $S_n = \dfrac{n}{2}(a_1 + a_n)$.

For a proof of this formula for the sum of a finite arithmetic sequence, see Proofs in Mathematics on page 502.

EXAMPLE 5   Sum of a Finite Arithmetic Sequence

Find the sum: $1 + 3 + 5 + 7 + 9 + 11 + 13 + 15 + 17 + 19$.

Solution   To begin, notice that the sequence is arithmetic (with a common difference of 2). Moreover, the sequence has 10 terms. So, the sum of the sequence is

$$S_n = \frac{n}{2}(a_1 + a_n) \qquad \text{Sum of a finite arithmetic sequence}$$

$$= \frac{10}{2}(1 + 19) \qquad \text{Substitute 10 for } n, \text{ 1 for } a_1, \text{ and 19 for } a_n.$$

$$= 100. \qquad \text{Simplify.}$$

✓ **Checkpoint** ◀))) *Audio-video solution in English & Spanish at LarsonPrecalculus.com.*

Find the sum: $40 + 37 + 34 + 31 + 28 + 25 + 22$.

EXAMPLE 6   Sum of a Finite Arithmetic Sequence

Find the sum of the integers (a) from 1 to 100 and (b) from 1 to N.

Solution

a. The integers from 1 to 100 form an arithmetic sequence that has 100 terms. So, you can use the formula for the sum of a finite arithmetic sequence, as follows.

$$S_n = 1 + 2 + 3 + 4 + 5 + 6 + \cdots + 99 + 100$$

$$= \frac{n}{2}(a_1 + a_n) \qquad \text{Sum of a finite arithmetic sequence}$$

$$= \frac{100}{2}(1 + 100) \qquad \text{Substitute 100 for } n, \text{ 1 for } a_1, \text{ and 100 for } a_n.$$

$$= 5050 \qquad \text{Simplify.}$$

b. $S_n = 1 + 2 + 3 + 4 + \cdots + N$

$$= \frac{n}{2}(a_1 + a_n) \qquad \text{Sum of a finite arithmetic sequence}$$

$$= \frac{N}{2}(1 + N) \qquad \text{Substitute } N \text{ for } n, \text{ 1 for } a_1, \text{ and } N \text{ for } a_n.$$

✓ **Checkpoint** ◀))) *Audio-video solution in English & Spanish at LarsonPrecalculus.com.*

Find the sum of the integers (a) from 1 to 35 and (b) from 1 to $2N$.

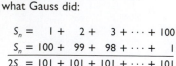

A teacher of Carl Friedrich Gauss (1777–1855) asked him to add all the integers from 1 to 100. When Gauss returned with the correct answer after only a few moments, the teacher could only look at him in astounded silence. This is what Gauss did:

$$\begin{aligned} S_n &= 1 + 2 + 3 + \cdots + 100 \\ S_n &= 100 + 99 + 98 + \cdots + 1 \\ \hline 2S_n &= 101 + 101 + 101 + \cdots + 101 \end{aligned}$$

$$S_n = \frac{100 \times 101}{2} = 5050$$

Bettmann/Corbis

The sum of the first n terms of an infinite sequence is the *nth partial sum*. The nth partial sum can be found by using the formula for the sum of a finite arithmetic sequence.

EXAMPLE 7 **Partial Sum of an Arithmetic Sequence**

Find the 150th partial sum of the arithmetic sequence

5, 16, 27, 38, 49, ...

Solution For this arithmetic sequence, $a_1 = 5$ and $d = 16 - 5 = 11$. So,

$$a_n = 5 + 11(n - 1)$$

and the nth term is

$$a_n = 11n - 6.$$

Therefore, $a_{150} = 11(150) - 6 = 1644$, and the sum of the first 150 terms is

$$S_{150} = \frac{n}{2}(a_1 + a_{150}) \qquad \text{\textit{n}th partial sum formula}$$

$$= \frac{150}{2}(5 + 1644) \qquad \text{Substitute 150 for \textit{n}, 5 for } a_1, \text{ and 1644 for } a_{150}.$$

$$= 75(1649) \qquad \text{Simplify.}$$

$$= 123{,}675. \qquad \text{\textit{n}th partial sum}$$

✓ **Checkpoint**  Audio-video solution in English & Spanish at LarsonPrecalculus.com.

Find the 120th partial sum of the arithmetic sequence

6, 12, 18, 24, 30, ...

EXAMPLE 8 **Partial Sum of an Arithmetic Sequence**

Find the 16th partial sum of the arithmetic sequence

100, 95, 90, 85, 80, ...

Solution For this arithmetic sequence, $a_1 = 100$ and $d = 95 - 100 = -5$. So,

$$a_n = 100 + (-5)(n - 1)$$

and the nth term is

$$a_n = -5n + 105.$$

Therefore, $a_{16} = -5(16) + 105 = 25$, and the sum of the first 16 terms is

$$S_{16} = \frac{n}{2}(a_1 + a_{16}) \qquad \text{\textit{n}th partial sum formula}$$

$$= \frac{16}{2}(100 + 25) \qquad \text{Substitute 16 for \textit{n}, 100 for } a_1, \text{ and 25 for } a_{16}.$$

$$= 8(125) \qquad \text{Simplify.}$$

$$= 1000. \qquad \text{\textit{n}th partial sum}$$

✓ **Checkpoint**  Audio-video solution in English & Spanish at LarsonPrecalculus.com.

Find the 30th partial sum of the arithmetic sequence

78, 76, 74, 72, 70, ...

Application

EXAMPLE 9 **Total Sales**

A small business sells \$10,000 worth of skin care products during its first year. The owner of the business has set a goal of increasing annual sales by \$7500 each year for 9 years. Assuming that this goal is met, find the total sales during the first 10 years this business is in operation.

Solution The annual sales form an arithmetic sequence in which

$$a_1 = 10,000 \quad \text{and} \quad d = 7500.$$

So,

$$a_n = 10,000 + 7500(n - 1)$$

and the nth term of the sequence is

$$a_n = 7500n + 2500.$$

Therefore, the 10th term of the sequence is

$$a_{10} = 7500(10) + 2500$$

$$= 77,500. \qquad \text{See figure.}$$

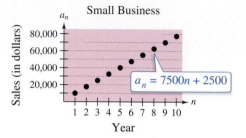

Small Business

The sum of the first 10 terms of the sequence is

$$S_{10} = \frac{n}{2}(a_1 + a_{10}) \qquad \text{\textcolor{red}{\textit{n}th partial sum formula}}$$

$$= \frac{10}{2}(10,000 + 77,500) \qquad \text{\textcolor{red}{Substitute 10 for } n\text{, 10,000 for } a_1\text{, and 77,500 for } a_{10}.}$$

$$= 5(87,500) \qquad \text{\textcolor{red}{Simplify.}}$$

$$= 437,500. \qquad \text{\textcolor{red}{Multiply.}}$$

So, the total sales for the first 10 years will be \$437,500.

✓ **Checkpoint**  *Audio-video solution in English & Spanish at LarsonPrecalculus.com.*

A company sells \$160,000 worth of printing paper during its first year. The sales manager has set a goal of increasing annual sales of printing paper by \$20,000 each year for 9 years. Assuming that this goal is met, find the total sales of printing paper during the first 10 years this company is in operation. ∎

Summarize   (Section 9.2)

1. State the definition of an arithmetic sequence *(page 453)*, and state the formula for the nth term of an arithmetic sequence *(page 454)*. For examples of recognizing, writing, and finding the nth terms of arithmetic sequences, see Examples 1–4.

2. State the formula for the sum of a finite arithmetic sequence and explain how to use it to find an nth partial sum *(pages 456 and 457)*. For examples of finding sums of arithmetic sequences, see Examples 5–8.

3. Describe an example of how to use an arithmetic sequence to model and solve a real-life problem *(page 458, Example 9)*.

9.3 Geometric Sequences and Series

- Recognize, write, and find the *n*th terms of geometric sequences.
- Find the sum of a finite geometric sequence.
- Find the sum of an infinite geometric series.
- Use geometric sequences to model and solve real-life problems.

Geometric Sequences

In Section 9.2, you learned that a sequence whose consecutive terms have a common *difference* is an arithmetic sequence. In this section, you will study another important type of sequence called a **geometric sequence.** Consecutive terms of a geometric sequence have a common *ratio.*

Geometric sequences can help you model and solve real-life problems. For instance, a geometric sequence can be used to model the populations of China from 2004 through 2010.

> **Definition of Geometric Sequence**
>
> A sequence is **geometric** when the ratios of consecutive terms are the same. So, the sequence $a_1, a_2, a_3, a_4, \ldots, a_n, \ldots$ is geometric when there is a number r such that
>
> $$\frac{a_2}{a_1} = \frac{a_3}{a_2} = \frac{a_4}{a_3} = \cdots = r, \quad r \neq 0.$$
>
> The number r is the **common ratio** of the geometric sequence.

REMARK Be sure you understand that a sequence such as $1, 4, 9, 16, \ldots$, whose *n*th term is n^2, is *not* geometric. The ratio of the second term to the first term is

$$\frac{a_2}{a_1} = \frac{4}{1} = 4$$

but the ratio of the third term to the second term is

$$\frac{a_3}{a_2} = \frac{9}{4}.$$

EXAMPLE 1 **Examples of Geometric Sequences**

a. The sequence whose *n*th term is 2^n is geometric. For this sequence, the common ratio of consecutive terms is 2.

$$2, 4, 8, 16, \ldots, 2^n, \ldots \qquad \text{Begin with } n = 1.$$

$$\frac{4}{2} = 2$$

b. The sequence whose *n*th term is $4(3^n)$ is geometric. For this sequence, the common ratio of consecutive terms is 3.

$$12, 36, 108, 324, \ldots, 4(3^n), \ldots \qquad \text{Begin with } n = 1.$$

$$\frac{36}{12} = 3$$

c. The sequence whose *n*th term is $\left(-\frac{1}{3}\right)^n$ is geometric. For this sequence, the common ratio of consecutive terms is $-\frac{1}{3}$.

$$-\frac{1}{3}, \frac{1}{9}, -\frac{1}{27}, \frac{1}{81}, \ldots, \left(-\frac{1}{3}\right)^n, \ldots \qquad \text{Begin with } n = 1.$$

$$\frac{1/9}{-1/3} = -\frac{1}{3}$$

✓ *Checkpoint* 🔊))) *Audio-video solution in English & Spanish at LarsonPrecalculus.com.*

Write the first four terms of the geometric sequence whose *n*th term is $6(-2)^n$. Then find the common ratio of the consecutive terms.

In Example 1, notice that each of the geometric sequences has an *n*th term that is of the form ar^n, where the common ratio of the sequence is r. A geometric sequence may be thought of as an exponential function whose domain is the set of natural numbers.

The nth Term of a Geometric Sequence

The nth term of a geometric sequence has the form

$$a_n = a_1 r^{n-1}$$

where r is the common ratio of consecutive terms of the sequence. So, every geometric sequence can be written in the following form.

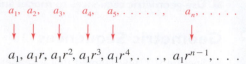

$$a_1, \; a_1 r, \; a_1 r^2, \; a_1 r^3, \; a_1 r^4, \ldots, \; a_1 r^{n-1}, \ldots$$

When you know the nth term of a geometric sequence, you can find the $(n + 1)$th term by multiplying by r. That is, $a_{n+1} = a_n r$.

EXAMPLE 2   **Writing the Terms of a Geometric Sequence**

Write the first five terms of the geometric sequence whose first term is $a_1 = 3$ and whose common ratio is $r = 2$. Then graph the terms on a set of coordinate axes.

Solution   Starting with 3, repeatedly multiply by 2 to obtain the following.

$a_1 = 3$	1st term	$a_4 = 3(2^3) = 24$ 4th term
$a_2 = 3(2^1) = 6$	2nd term	$a_5 = 3(2^4) = 48$ 5th term
$a_3 = 3(2^2) = 12$	3rd term	

Figure 9.1 shows the first five terms of this geometric sequence.

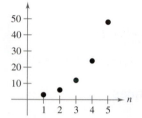

Figure 9.1

✓ **Checkpoint**   *Audio-video solution in English & Spanish at LarsonPrecalculus.com.*

Write the first five terms of the geometric sequence whose first term is $a_1 = 2$ and whose common ratio is $r = 4$. Then graph the terms on a set of coordinate axes.

EXAMPLE 3   **Finding a Term of a Geometric Sequence**

Find the 15th term of the geometric sequence whose first term is 20 and whose common ratio is 1.05.

Algebraic Solution

$$a_n = a_1 r^{n-1}$$
Formula for nth term of a geometric sequence

$$a_{15} = 20(1.05)^{15-1}$$
Substitute 20 for a_1, 1.05 for r, and 15 for n.

$$\approx 39.60$$
Use a calculator.

Numerical Solution

For this sequence, $r = 1.05$ and $a_1 = 20$. So, $a_n = 20(1.05)^{n-1}$. Use a graphing utility to create a table that shows the terms of the sequence.

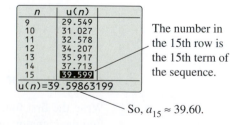

n	$u(n)$
9	29.549
10	31.027
11	32.578
12	34.207
13	35.917
14	37.713
15	39.599

$u(n)=39.59863199$

The number in the 15th row is the 15th term of the sequence.

So, $a_{15} \approx 39.60$.

✓ **Checkpoint** 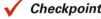 *Audio-video solution in English & Spanish at LarsonPrecalculus.com.*

Find the 12th term of the geometric sequence whose first term is 14 and whose common ratio is 1.2.

EXAMPLE 4   **Finding a Term of a Geometric Sequence**

Find the 12th term of the geometric sequence

$$5, 15, 45, \ldots.$$

Solution   The common ratio of this sequence is $r = 15/5 = 3$. Because the first term is $a_1 = 5$, the 12th term ($n = 12$) is

$$a_n = a_1 r^{n-1} \qquad \text{Formula for } n\text{th term of a geometric sequence}$$

$$a_{12} = 5(3)^{12-1} \qquad \text{Substitute 5 for } a_1, 3 \text{ for } r, \text{ and 12 for } n.$$

$$= 5(177{,}147) \qquad \text{Use a calculator.}$$

$$= 885{,}735. \qquad \text{Multiply.}$$

✓ **Checkpoint**)))   *Audio-video solution in English & Spanish at LarsonPrecalculus.com.*

Find the 12th term of the geometric sequence

$$4, 20, 100, \ldots.$$

When you know *any* two terms of a geometric sequence, you can use that information to find *any other* term of the sequence.

EXAMPLE 5   **Finding a Term of a Geometric Sequence**

▷ **ALGEBRA HELP**   Remember that r is the common ratio of consecutive terms of a geometric sequence. So, in Example 5

$$a_{10} = a_1 r^9$$

$$= a_1 \cdot r \cdot r \cdot r \cdot r^6$$

$$= a_1 \cdot \frac{a_2}{a_1} \cdot \frac{a_3}{a_2} \cdot \frac{a_4}{a_3} \cdot r^6$$

$$= a_4 r^6.$$

The fourth term of a geometric sequence is 125, and the 10th term is 125/64. Find the 14th term. (Assume that the terms of the sequence are positive.)

Solution   The 10th term is related to the fourth term by the equation

$$a_{10} = a_4 r^6 \qquad \text{Multiply fourth term by } r^{10-4}.$$

Because $a_{10} = 125/64$ and $a_4 = 125$, you can solve for r as follows.

$$\frac{125}{64} = 125r^6 \qquad \text{Substitute } \tfrac{125}{64} \text{ for } a_{10} \text{ and 125 for } a_4.$$

$$\frac{1}{64} = r^6 \qquad \text{Divide each side by 125.}$$

$$\frac{1}{2} = r \qquad \text{Take the sixth root of each side.}$$

You can obtain the 14th term by multiplying the 10th term by r^4.

$$a_{14} = a_{10} r^4 \qquad \text{Multiply the 10th term by } r^{14-10}.$$

$$= \frac{125}{64}\left(\frac{1}{2}\right)^4 \qquad \text{Substitute } \tfrac{125}{64} \text{ for } a_{10} \text{ and } \tfrac{1}{2} \text{ for } r.$$

$$= \frac{125}{64}\left(\frac{1}{16}\right) \qquad \text{Evaluate power.}$$

$$= \frac{125}{1024} \qquad \text{Multiply fractions.}$$

✓ **Checkpoint**)))   *Audio-video solution in English & Spanish at LarsonPrecalculus.com.*

The second term of a geometric sequence is 6, and the fifth term is 81/4. Find the eighth term. (Assume that the terms of the sequence are positive.)

The Sum of a Finite Geometric Sequence

The formula for the sum of a *finite* geometric sequence is as follows.

> **The Sum of a Finite Geometric Sequence**
>
> The sum of the finite geometric sequence
>
> $$a_1, \ a_1 r, \ a_1 r^2, \ a_1 r^3, \ a_1 r^4, \ldots, a_1 r^{n-1}$$
>
> with common ratio $r \neq 1$ is given by $S_n = \displaystyle\sum_{i=1}^{n} a_1 r^{i-1} = a_1 \left(\dfrac{1 - r^n}{1 - r} \right)$.

For a proof of this formula for the sum of a finite geometric sequence, see Proofs in Mathematics on page 502.

EXAMPLE 6 **Sum of a Finite Geometric Sequence**

Find the sum $\displaystyle\sum_{i=1}^{12} 4(0.3)^{i-1}$.

Solution You have

$$\sum_{i=1}^{12} 4(0.3)^{i-1} = 4(0.3)^0 + 4(0.3)^1 + 4(0.3)^2 + \cdots + 4(0.3)^{11}.$$

Now, $a_1 = 4$, $r = 0.3$, and $n = 12$, so applying the formula for the sum of a finite geometric sequence, you obtain

$$S_n = a_1 \left(\frac{1 - r^n}{1 - r} \right)$$
Sum of a finite geometric sequence

$$\sum_{i=1}^{12} 4(0.3)^{i-1} = 4 \left[\frac{1 - (0.3)^{12}}{1 - 0.3} \right]$$
Substitute 4 for a_1, 0.3 for r, and 12 for n.

$$\approx 5.714.$$
Use a calculator.

✓ **Checkpoint** 🔊))) *Audio-video solution in English & Spanish at LarsonPrecalculus.com.*

Find the sum $\displaystyle\sum_{i=1}^{10} 2(0.25)^{i-1}$.

When using the formula for the sum of a finite geometric sequence, be careful to check that the sum is of the form

$$\sum_{i=1}^{n} a_1 r^{i-1}.$$
Exponent for r is $i - 1$.

For a sum that is not of this form, you must adjust the formula. For instance, if the sum in Example 6 were $\displaystyle\sum_{i=1}^{12} 4(0.3)^i$, then you would evaluate the sum as follows.

$$\sum_{i=1}^{12} 4(0.3)^i = 4(0.3) + 4(0.3)^2 + 4(0.3)^3 + \cdots + 4(0.3)^{12}$$

$$= 4(0.3) + [4(0.3)](0.3) + [4(0.3)](0.3)^2 + \cdots + [4(0.3)](0.3)^{11}$$

$$= 4(0.3) \left[\frac{1 - (0.3)^{12}}{1 - 0.3} \right]$$
$a_1 = 4(0.3)$, $r = 0.3$, $n = 12$

$$\approx 1.714$$

Geometric Series

The summation of the terms of an infinite geometric *sequence* is called an **infinite geometric series** or simply a **geometric series.**

The formula for the sum of a *finite* geometric *sequence* can, depending on the value of r, be extended to produce a formula for the sum of an *infinite* geometric *series.* Specifically, if the common ratio r has the property that $|r| < 1$, then it can be shown that r^n approaches zero as n increases without bound. Consequently,

$$a_1\left(\frac{1 - r^n}{1 - r}\right) \longrightarrow a_1\left(\frac{1 - 0}{1 - r}\right) \quad \text{as} \quad n \longrightarrow \infty.$$

The following summarizes this result.

> **The Sum of an Infinite Geometric Series**
>
> If $|r| < 1$, then the infinite geometric series
>
> $$a_1 + a_1 r + a_1 r^2 + a_1 r^3 + \cdots + a_1 r^{n-1} + \cdots$$
>
> has the sum
>
> $$S = \sum_{i=0}^{\infty} a_1 r^i = \frac{a_1}{1 - r}.$$

Note that when $|r| \geq 1$, the series does not have a sum.

EXAMPLE 7   **Finding the Sum of an Infinite Geometric Series**

Find each sum.

a. $\displaystyle\sum_{n=0}^{\infty} 4(0.6)^n$

b. $3 + 0.3 + 0.03 + 0.003 + \cdots$

Solution

a. $\displaystyle\sum_{n=0}^{\infty} 4(0.6)^n = 4 + 4(0.6) + 4(0.6)^2 + 4(0.6)^3 + \cdots + 4(0.6)^n + \cdots$

$$= \frac{4}{1 - 0.6} \qquad\qquad \frac{a_1}{1 - r}$$

$$= 10$$

b. $3 + 0.3 + 0.03 + 0.003 + \cdots = 3 + 3(0.1) + 3(0.1)^2 + 3(0.1)^3 + \cdots$

$$= \frac{3}{1 - 0.1} \qquad\qquad \frac{a_1}{1 - r}$$

$$= \frac{10}{3}$$

$$\approx 3.33$$

✓ **Checkpoint** ◀))) *Audio-video solution in English & Spanish at LarsonPrecalculus.com.*

Find each sum.

a. $\displaystyle\sum_{n=0}^{\infty} 5(0.5)^n$

b. $5 + 1 + 0.2 + 0.04 + \cdots$

Application

EXAMPLE 8 Increasing Annuity

An investor deposits \$50 on the first day of each month in an account that pays 3% interest, compounded monthly. What is the balance at the end of 2 years? (This type of investment plan is called an **increasing annuity.**)

Solution To find the balance in the account after 24 months, consider each of the 24 deposits separately. The first deposit will gain interest for 24 months, and its balance will be

$$A_{24} = 50\left(1 + \frac{0.03}{12}\right)^{24}$$
$$= 50(1.0025)^{24}.$$

The second deposit will gain interest for 23 months, and its balance will be

$$A_{23} = 50\left(1 + \frac{0.03}{12}\right)^{23}$$
$$= 50(1.0025)^{23}.$$

The last deposit will gain interest for only 1 month, and its balance will be

$$A_1 = 50\left(1 + \frac{0.03}{12}\right)^{1}$$
$$= 50(1.0025).$$

The total balance in the annuity will be the sum of the balances of the 24 deposits. Using the formula for the sum of a finite geometric sequence, with $A_1 = 50(1.0025)$, $r = 1.0025$, and $n = 24$, you have

$$S_n = A_1\left(\frac{1 - r^n}{1 - r}\right) \qquad \text{Sum of a finite geometric sequence}$$

$$S_{24} = 50(1.0025)\left[\frac{1 - (1.0025)^{24}}{1 - 1.0025}\right] \qquad \begin{array}{l}\text{Substitute } 50(1.0025) \text{ for } A_1,\\ 1.0025 \text{ for } r, \text{ and } 24 \text{ for } n.\end{array}$$

$$\approx \$1238.23. \qquad \text{Use a calculator.}$$

✓ **Checkpoint** *Audio-video solution in English & Spanish at LarsonPrecalculus.com.*

An investor deposits \$70 on the first day of each month in an account that pays 2% interest, compounded monthly. What is the balance at the end of 4 years? ▪

> **REMARK** Recall from Section 3.1 that the formula for compound interest (for n compoundings per year) is
>
> $$A = P\left(1 + \frac{r}{n}\right)^{nt}.$$
>
> So, in Example 8, \$50 is the principal P, 0.03 is the annual interest rate r, 12 is the number n of compoundings per year, and 2 is the time t in years. When you substitute these values into the formula, you obtain
>
> $$A = 50\left(1 + \frac{0.03}{12}\right)^{12(2)}$$
> $$= 50\left(1 + \frac{0.03}{12}\right)^{24}.$$

Summarize (Section 9.3)

1. State the definition of a geometric sequence *(page 459)* and state the formula for the nth term of a geometric sequence *(page 460)*. For examples of recognizing, writing, and finding the nth terms of geometric sequences, see Examples 1–5.

2. State the formula for the sum of a finite geometric sequence *(page 462)*. For an example of finding the sum of a finite geometric sequence, see Example 6.

3. State the formula for the sum of an infinite geometric series *(page 463)*. For an example of finding the sums of infinite geometric series, see Example 7.

4. Describe an example of how to use a geometric series to model and solve a real-life problem *(page 464, Example 8)*.

9.4   Mathematical Induction

- Use mathematical induction to prove statements involving a positive integer n.
- Use pattern recognition and mathematical induction to write a formula for the nth term of a sequence.
- Find the sums of powers of integers.
- Find finite differences of sequences.

Finite differences can help you determine what type of model to use to represent a sequence. For instance, finite differences can be used to find a model that represents the numbers of Alaskan residents from 2005 through 2010.

Introduction

In this section, you will study a form of mathematical proof called **mathematical induction.** It is important that you see clearly the logical need for it, so take a closer look at the problem discussed in Example 5 in Section 9.2.

$$S_1 = 1 = 1^2$$

$$S_2 = 1 + 3 = 2^2$$

$$S_3 = 1 + 3 + 5 = 3^2$$

$$S_4 = 1 + 3 + 5 + 7 = 4^2$$

$$S_5 = 1 + 3 + 5 + 7 + 9 = 5^2$$

$$S_6 = 1 + 3 + 5 + 7 + 9 + 11 = 6^2$$

Judging from the pattern formed by these first six sums, it appears that the sum of the first n odd integers is

$$S_n = 1 + 3 + 5 + 7 + 9 + 11 + \cdot \cdot \cdot + (2n - 1) = n^2.$$

Although this particular formula *is* valid, it is important for you to see that recognizing a pattern and then simply *jumping to the conclusion* that the pattern must be true for all values of n is *not* a logically valid method of proof. There are many examples in which a pattern appears to be developing for small values of n, but then at some point the pattern fails. One of the most famous cases of this was the conjecture by the French mathematician Pierre de Fermat (1601–1665), who speculated that all numbers of the form

$$F_n = 2^{2^n} + 1, \quad n = 0, 1, 2, \ldots$$

are prime. For $n = 0, 1, 2, 3,$ and 4, the conjecture is true.

$$F_0 = 3$$

$$F_1 = 5$$

$$F_2 = 17$$

$$F_3 = 257$$

$$F_4 = 65,537$$

The size of the next Fermat number ($F_5 = 4,294,967,297$) is so great that it was difficult for Fermat to determine whether it was prime or not. However, another well-known mathematician, Leonhard Euler (1707–1783), later found the factorization

$$F_5 = 4,294,967,297$$

$$= 641(6,700,417)$$

which proved that F_5 is not prime and therefore Fermat's conjecture was false.

Just because a rule, pattern, or formula seems to work for several values of n, you cannot simply decide that it is valid for all values of n without going through a *legitimate proof.* Mathematical induction is one method of proof.

· · · · · · · · · · · · · · ▷
·
·· **REMARK** It is important to recognize that in order to prove a statement by induction, both parts of the Principle of Mathematical Induction are necessary.

The Principle of Mathematical Induction

Let P_n be a statement involving the positive integer n. If

1. P_1 is true, and
2. for every positive integer k, the truth of P_k implies the truth of P_{k+1}

then the statement P_n must be true for all positive integers n.

To apply the Principle of Mathematical Induction, you need to be able to determine the statement P_{k+1} for a given statement P_k. To determine P_{k+1}, substitute the quantity $k + 1$ for k in the statement P_k.

EXAMPLE 1 **A Preliminary Example**

Find the statement P_{k+1} for each given statement P_k.

a. P_k: $S_k = \dfrac{k^2(k + 1)^2}{4}$

b. P_k: $S_k = 1 + 5 + 9 + \cdots + [4(k - 1) - 3] + (4k - 3)$

c. P_k: $k + 3 < 5k^2$

d. P_k: $3^k \geq 2k + 1$

Solution

a. P_{k+1}: $S_{k+1} = \dfrac{(k + 1)^2(k + 1 + 1)^2}{4}$ Replace k with $k + 1$.

$= \dfrac{(k + 1)^2(k + 2)^2}{4}$ Simplify.

b. P_{k+1}: $S_{k+1} = 1 + 5 + 9 + \cdots + \{4[(k + 1) - 1] - 3\} + [4(k + 1) - 3]$

$= 1 + 5 + 9 + \cdots + (4k - 3) + (4k + 1)$

c. P_{k+1}: $(k + 1) + 3 < 5(k + 1)^2$

$k + 4 < 5(k^2 + 2k + 1)$

d. P_{k+1}: $3^{k+1} \geq 2(k + 1) + 1$

$3^{k+1} \geq 2k + 3$

✓ *Checkpoint* *Audio-video solution in English & Spanish at LarsonPrecalculus.com.*

Find the statement P_{k+1} for each given statement P_k.

a. P_k: $S_k = \dfrac{6}{k(k + 3)}$ **b.** P_k: $k + 2 \leq 3(k - 1)^2$ **c.** P_k: $2^{4k-2} + 1 > 5k$ ◼

A well-known illustration used to explain why the Principle of Mathematical Induction works is that of an unending line of dominoes. If the line actually contains infinitely many dominoes, then it is clear that you could not knock the entire line down by knocking down only *one domino* at a time. However, suppose it were true that each domino would knock down the next one as it fell. Then you could knock them all down by pushing the first one and starting a chain reaction. Mathematical induction works in the same way. If the truth of P_k implies the truth of P_{k+1} and if P_1 is true, then the chain reaction proceeds as follows: P_1 implies P_2, P_2 implies P_3, P_3 implies P_4, and so on.

An unending line of dominoes can illustrate why the Principle of Mathematical Induction works.

When using mathematical induction to prove a *summation* formula (such as the one in Example 2), it is helpful to think of S_{k+1} as $S_{k+1} = S_k + a_{k+1}$, where a_{k+1} is the $(k + 1)$th term of the original sum.

EXAMPLE 2   Using Mathematical Induction

Use mathematical induction to prove the formula

$$S_n = 1 + 3 + 5 + 7 + \cdots + (2n - 1)$$

$$= n^2$$

for all integers $n \geq 1$.

Solution   Mathematical induction consists of two distinct parts.

1. First, you must show that the formula is true when $n = 1$. When $n = 1$, the formula is valid, because

$$S_1 = 1$$

$$= 1^2.$$

2. The second part of mathematical induction has two steps. The first step is to *assume* that the formula is valid for some integer k. The second step is to use this assumption to prove that the formula is valid for the *next* integer, $k + 1$. Assuming that the formula

$$S_k = 1 + 3 + 5 + 7 + \cdots + (2k - 1)$$

$$= k^2$$

is true, you must show that the formula $S_{k+1} = (k + 1)^2$ is true.

$$S_{k+1} = 1 + 3 + 5 + 7 + \cdots + (2k - 1)$$
$$+ [2(k + 1) - 1] \qquad \text{\small $S_{k+1} = S_k + a_{k+1}$}$$
$$= [1 + 3 + 5 + 7 + \cdots + (2k - 1)] + (2k + 2 - 1)$$
$$= S_k + (2k + 1) \qquad \text{\small Group terms to form S_k.}$$
$$= k^2 + 2k + 1 \qquad \text{\small Substitute k^2 for S_k.}$$
$$= (k + 1)^2$$

Combining the results of parts (1) and (2), you can conclude by mathematical induction that the formula is valid for all integers $n \geq 1$.

✓ **Checkpoint**  *Audio-video solution in English & Spanish at LarsonPrecalculus.com.*

Use mathematical induction to prove the formula

$$S_n = 5 + 7 + 9 + 11 + \cdots + (2n + 3) = n(n + 4)$$

for all integers $n \geq 1$.

It occasionally happens that a statement involving natural numbers is not true for the first $k - 1$ positive integers but is true for all values of $n \geq k$. In these instances, you use a slight variation of the Principle of Mathematical Induction in which you verify P_k rather than P_1. This variation is called the *Extended Principle of Mathematical Induction*. To see the validity of this, note in the unending line of dominoes discussed on page 466 that all but the first $k - 1$ dominoes can be knocked down by knocking over the kth domino. This suggests that you can prove a statement P_n to be true for $n \geq k$ by showing that P_k is true and that P_k implies P_{k+1}.

EXAMPLE 3 **Using Mathematical Induction**

Use mathematical induction to prove the formula

$$S_n = 1^2 + 2^2 + 3^2 + 4^2 + \cdots + n^2$$

$$= \frac{n(n + 1)(2n + 1)}{6}$$

for all integers $n \geq 1$.

Solution

1. When $n = 1$, the formula is valid, because

$$S_1 = 1^2$$

$$= \frac{1(2)(3)}{6}.$$

2. Assuming that

$$S_k = 1^2 + 2^2 + 3^2 + 4^2 + \cdots + k^2 \qquad a_k = k^2$$

$$= \frac{k(k + 1)(2k + 1)}{6}$$

you must show that

$$S_{k+1} = \frac{(k + 1)(k + 1 + 1)[2(k + 1) + 1]}{6}$$

$$= \frac{(k + 1)(k + 2)(2k + 3)}{6}.$$

To do this, write the following.

$$S_{k+1} = S_k + a_{k+1}$$

$$= (1^2 + 2^2 + 3^2 + 4^2 + \cdots + k^2) + (k + 1)^2 \qquad \text{Substitute for } S_k, \ a_{k+1} = (k + 1)^2.$$

$$= \frac{k(k + 1)(2k + 1)}{6} + (k + 1)^2 \qquad \text{By assumption}$$

$$= \frac{k(k + 1)(2k + 1) + 6(k + 1)^2}{6} \qquad \text{Combine fractions.}$$

$$= \frac{(k + 1)[k(2k + 1) + 6(k + 1)]}{6} \qquad \text{Factor.}$$

$$= \frac{(k + 1)(2k^2 + 7k + 6)}{6} \qquad \text{Simplify.}$$

$$= \frac{(k + 1)(k + 2)(2k + 3)}{6} \qquad S_k \text{ implies } S_{k+1}.$$

> **REMARK** Remember that when adding rational expressions, you must first find a common denominator. Example 3 uses the *least* common denominator of 6.

Combining the results of parts (1) and (2), you can conclude by mathematical induction that the formula is valid for all positive integers n.

✓ **Checkpoint** ◀)) *Audio-video solution in English & Spanish at LarsonPrecalculus.com.*

Use mathematical induction to prove the formula

$$S_n = 1(1 - 1) + 2(2 - 1) + 3(3 - 1) + \cdots + n(n - 1) = \frac{n(n - 1)(n + 1)}{3}$$

for all integers $n \geq 1$.

When proving a formula using mathematical induction, the only statement that you *need* to verify is P_1. As a check, however, it is a good idea to try verifying some of the other statements. For instance, in Example 3, try verifying P_2 and P_3.

EXAMPLE 4   Proving an Inequality

Prove that

$$n < 2^n$$

for all positive integers n.

REMARK To check a result that you have proved by mathematical induction, it helps to list the statement for several values of n. For instance, in Example 4, you could list

$$1 < 2^1 = 2, \quad 2 < 2^2 = 4,$$
$$3 < 2^3 = 8, \quad 4 < 2^4 = 16,$$
$$5 < 2^5 = 32, \quad 6 < 2^6 = 64.$$

From this list, your intuition confirms that the statement $n < 2^n$ is reasonable.

Solution

1. For $n = 1$ and $n = 2$, the statement is true because

$$1 < 2^1 \quad \text{and} \quad 2 < 2^2.$$

2. Assuming that

$$k < 2^k$$

you need to show that $k + 1 < 2^{k+1}$. For $n = k$, you have

$$2^{k+1} = 2(2^k) > 2(k) = 2k. \qquad \textcolor{red}{\text{By assumption}}$$

Because $2k = k + k > k + 1$ for all $k > 1$, it follows that

$$2^{k+1} > 2k > k + 1 \quad \text{or} \quad k + 1 < 2^{k+1}.$$

Combining the results of parts (1) and (2), you can conclude by mathematical induction that $n < 2^n$ for all integers $n \geq 1$.

✓ **Checkpoint**   *Audio-video solution in English & Spanish at LarsonPrecalculus.com.*

Prove that

$$n! \geq n$$

for all positive integers n.

EXAMPLE 5   Proving a Factor

Prove that 3 is a factor of $4^n - 1$ for all positive integers n.

Solution

1. For $n = 1$, the statement is true because

$$4^1 - 1 = 3.$$

So, 3 is a factor.

2. Assuming that 3 is a factor of $4^k - 1$, you must show that 3 is a factor of $4^{k+1} - 1$. To do this, write the following.

$$4^{k+1} - 1 = 4^{k+1} - 4^k + 4^k - 1 \qquad \textcolor{red}{\text{Subtract and add } 4^k.}$$
$$= 4^k(4 - 1) + (4^k - 1) \qquad \textcolor{red}{\text{Regroup terms.}}$$
$$= 4^k \cdot 3 + (4^k - 1) \qquad \textcolor{red}{\text{Simplify.}}$$

Because 3 is a factor of $4^k \cdot 3$ and 3 is also a factor of $4^k - 1$, it follows that 3 is a factor of $4^{k+1} - 1$. Combining the results of parts (1) and (2), you can conclude by mathematical induction that 3 is a factor of $4^n - 1$ for all positive integers n.

✓ **Checkpoint**   *Audio-video solution in English & Spanish at LarsonPrecalculus.com.*

Prove that 2 is a factor of $3^n + 1$ for all positive integers n.

Pattern Recognition

Although choosing a formula on the basis of a few observations does *not* guarantee the validity of the formula, pattern recognition *is* important. Once you have a pattern or formula that you think works, try using mathematical induction to prove your formula.

> **Finding a Formula for the *n*th Term of a Sequence**
>
> To find a formula for the *n*th term of a sequence, consider these guidelines.
>
> **1.** Calculate the first several terms of the sequence. It is often a good idea to write the terms in both simplified and factored forms.
>
> **2.** Try to find a recognizable pattern for the terms and write a formula for the *n*th term of the sequence. This is your *hypothesis* or *conjecture*. You might try computing one or two more terms in the sequence to test your hypothesis.
>
> **3.** Use mathematical induction to prove your hypothesis.

EXAMPLE 6 **Finding a Formula for a Finite Sum**

Find a formula for the finite sum and prove its validity.

$$\frac{1}{1 \cdot 2} + \frac{1}{2 \cdot 3} + \frac{1}{3 \cdot 4} + \frac{1}{4 \cdot 5} + \cdots + \frac{1}{n(n+1)}$$

Solution Begin by writing the first few sums.

$$S_1 = \frac{1}{1 \cdot 2} = \frac{1}{2} = \frac{1}{1+1}$$

$$S_2 = \frac{1}{1 \cdot 2} + \frac{1}{2 \cdot 3} = \frac{4}{6} = \frac{2}{3} = \frac{2}{2+1}$$

$$S_3 = \frac{1}{1 \cdot 2} + \frac{1}{2 \cdot 3} + \frac{1}{3 \cdot 4} = \frac{9}{12} = \frac{3}{4} = \frac{3}{3+1}$$

From this sequence, it appears that the formula for the *k*th sum is

$$S_k = \frac{1}{1 \cdot 2} + \frac{1}{2 \cdot 3} + \frac{1}{3 \cdot 4} + \frac{1}{4 \cdot 5} + \cdots + \frac{1}{k(k+1)} = \frac{k}{k+1}.$$

To prove the validity of this hypothesis, use mathematical induction. Note that you have already verified the formula for $n = 1$, so begin by assuming that the formula is valid for $n = k$ and trying to show that it is valid for $n = k + 1$.

$$S_{k+1} = \left[\frac{1}{1 \cdot 2} + \frac{1}{2 \cdot 3} + \frac{1}{3 \cdot 4} + \frac{1}{4 \cdot 5} + \cdots + \frac{1}{k(k+1)} \right] + \frac{1}{(k+1)(k+2)}$$

$$= \frac{k}{k+1} + \frac{1}{(k+1)(k+2)} \qquad \text{By assumption}$$

$$= \frac{k(k+2) + 1}{(k+1)(k+2)} = \frac{k^2 + 2k + 1}{(k+1)(k+2)} = \frac{(k+1)^2}{(k+1)(k+2)} = \frac{k+1}{k+2}$$

So, by mathematical induction the hypothesis is valid.

✓ *Checkpoint* ◀))) *Audio-video solution in English & Spanish at LarsonPrecalculus.com.*

Find a formula for the finite sum and prove its validity.

$$3 + 7 + 11 + 15 + \cdots + 4n - 1$$

Sums of Powers of Integers

The formula in Example 3 is one of a collection of useful summation formulas. This and other formulas dealing with the sums of various powers of the first n positive integers are as follows.

Sums of Powers of Integers

1. $1 + 2 + 3 + 4 + \cdots + n = \dfrac{n(n + 1)}{2}$

2. $1^2 + 2^2 + 3^2 + 4^2 + \cdots + n^2 = \dfrac{n(n + 1)(2n + 1)}{6}$

3. $1^3 + 2^3 + 3^3 + 4^3 + \cdots + n^3 = \dfrac{n^2(n + 1)^2}{4}$

4. $1^4 + 2^4 + 3^4 + 4^4 + \cdots + n^4 = \dfrac{n(n + 1)(2n + 1)(3n^2 + 3n - 1)}{30}$

5. $1^5 + 2^5 + 3^5 + 4^5 + \cdots + n^5 = \dfrac{n^2(n + 1)^2(2n^2 + 2n - 1)}{12}$

EXAMPLE 7   **Finding Sums**

Find each sum.

a. $\displaystyle\sum_{i=1}^{7} i^3 = 1^3 + 2^3 + 3^3 + 4^3 + 5^3 + 6^3 + 7^3$ **b.** $\displaystyle\sum_{i=1}^{4}(6i - 4i^2)$

Solution

a. Using the formula for the sum of the cubes of the first n positive integers, you obtain

$$\sum_{i=1}^{7} i^3 = 1^3 + 2^3 + 3^3 + 4^3 + 5^3 + 6^3 + 7^3$$

$$= \frac{7^2(7 + 1)^2}{4} \qquad\qquad\qquad\text{Formula 3}$$

$$= \frac{49(64)}{4}$$

$$= 784.$$

b. $\displaystyle\sum_{i=1}^{4}(6i - 4i^2) = \sum_{i=1}^{4} 6i - \sum_{i=1}^{4} 4i^2$

$$= 6\sum_{i=1}^{4} i - 4\sum_{i=1}^{4} i^2$$

$$= 6\left[\frac{4(4 + 1)}{2}\right] - 4\left[\frac{4(4 + 1)(2 \cdot 4 + 1)}{6}\right] \qquad\text{Formulas 1 and 2}$$

$$= 6(10) - 4(30)$$

$$= -60$$

✓ *Checkpoint*))) Audio-video solution in English & Spanish at LarsonPrecalculus.com.

Find each sum.

a. $\displaystyle\sum_{i=1}^{20} i$ **b.** $\displaystyle\sum_{i=1}^{5}(2i^2 + 3i^3)$

Finite Differences

The **first differences** of a sequence are found by subtracting consecutive terms. The **second differences** are found by subtracting consecutive first differences. The first and second differences of the sequence 3, 5, 8, 12, 17, 23, . . . are as follows.

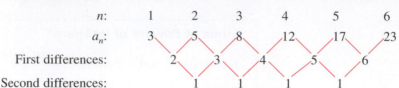

For this sequence, the second differences are all the same. When this happens, the sequence has a perfect *quadratic* model. When the first differences are all the same, the sequence has a perfect *linear* model. That is, it is arithmetic.

EXAMPLE 8 **Finding a Quadratic Model**

Find the quadratic model for the sequence

3, 5, 8, 12, 17, 23, ...

Solution You know from the second differences shown above that the model is quadratic and has the form $a_n = an^2 + bn + c$. By substituting 1, 2, and 3 for n, you can obtain a system of three linear equations in three variables.

$a_1 = a(1)^2 + b(1) + c = 3$ Substitute 1 for *n*.

$a_2 = a(2)^2 + b(2) + c = 5$ Substitute 2 for *n*.

$a_3 = a(3)^2 + b(3) + c = 8$ Substitute 3 for *n*.

You now have a system of three equations in a, b, and c.

$$\begin{cases} a + b + c = 3 & \text{\color{red}Equation 1} \\ 4a + 2b + c = 5 & \text{\color{red}Equation 2} \\ 9a + 3b + c = 8 & \text{\color{red}Equation 3} \end{cases}$$

Using the techniques discussed in Chapter 7, you can find the solution to be $a = \frac{1}{2}$, $b = \frac{1}{2}$, and $c = 2$. So, the quadratic model is $a_n = \frac{1}{2}n^2 + \frac{1}{2}n + 2$. Try checking the values of a_1, a_2, and a_3.

✓ *Checkpoint*  *Audio-video solution in English & Spanish at LarsonPrecalculus.com.*

Find a quadratic model for the sequence -2, 0, 4, 10, 18, 28, ...

Summarize (Section 9.4)

1. State the Principle of Mathematical Induction *(page 466)*. For examples of using mathematical induction to prove statements involving a positive integer *n*, see Examples 2–5.

2. Explain how to use pattern recognition and mathematical induction to write a formula for the *n*th term of a sequence *(page 470)*. For an example of using pattern recognition and mathematical induction to write a formula for the *n*th term of a sequence, see Example 6.

3. State the formulas for the sums of powers of integers *(page 471)*. For an example of finding sums of powers of integers, see Example 7.

4. Explain how to find finite differences of sequences *(page 472)*. For an example of using finite differences to find a quadratic model, see Example 8.

9.5   The Binomial Theorem

- ■ Use the Binomial Theorem to calculate binomial coefficients.
- ■ Use Pascal's Triangle to calculate binomial coefficients.
- ■ Use binomial coefficients to write binomial expansions.

Binomial Coefficients

Recall that a *binomial* is a polynomial that has two terms. In this section, you will study a formula that provides a quick method of raising a binomial to a power. To begin, look at the expansion of

$$(x + y)^n$$

for several values of n.

$$(x + y)^0 = 1$$

$$(x + y)^1 = x + y$$

$$(x + y)^2 = x^2 + 2xy + y^2$$

$$(x + y)^3 = x^3 + 3x^2y + 3xy^2 + y^3$$

$$(x + y)^4 = x^4 + 4x^3y + 6x^2y^2 + 4xy^3 + y^4$$

$$(x + y)^5 = x^5 + 5x^4y + 10x^3y^2 + 10x^2y^3 + 5xy^4 + y^5$$

There are several observations you can make about these expansions.

1. In each expansion, there are $n + 1$ terms.

2. In each expansion, x and y have symmetrical roles. The powers of x decrease by 1 in successive terms, whereas the powers of y increase by 1.

3. The sum of the powers of each term is n. For instance, in the expansion of $(x + y)^5$, the sum of the powers of each term is 5.

4 + 1 = 5   3 + 2 = 5

$$(x + y)^5 = x^5 + 5x^4y^1 + 10x^3y^2 + 10x^2y^3 + 5x^1y^4 + y^5$$

4. The coefficients increase and then decrease in a symmetric pattern.

The coefficients of a binomial expansion are called **binomial coefficients.** To find them, you can use the **Binomial Theorem.**

The Binomial Theorem

In the expansion of $(x + y)^n$

$$(x + y)^n = x^n + nx^{n-1}y + \cdots + {}_nC_r\, x^{n-r}y^r + \cdots + nxy^{n-1} + y^n$$

the coefficient of $x^{n-r}y^r$ is

$${}_nC_r = \frac{n!}{(n-r)!r!}.$$

The symbol $\binom{n}{r}$ is often used in place of ${}_nC_r$ to denote binomial coefficients.

For a proof of the Binomial Theorem, see Proofs in Mathematics on page 503.

Binomial coefficients can help you model and solve real-life problems. For instance, binomial coefficients can be used to write the expansion of a model that represents the average prices of residential electricity in the United States from 2003 through 2010.

▷ **TECHNOLOGY**

Most graphing calculators have the capability of evaluating $_nC_r$. If yours does, then evaluate $_8C_5$. You should get an answer of 56.

EXAMPLE 1   **Finding Binomial Coefficients**

Find each binomial coefficient.

a. $_8C_2$ **b.** $\binom{10}{3}$ **c.** $_7C_0$ **d.** $\binom{8}{8}$

Solution

a. $_8C_2 = \dfrac{8!}{6! \cdot 2!} = \dfrac{(8 \cdot 7) \cdot 6!}{6! \cdot 2!} = \dfrac{8 \cdot 7}{2 \cdot 1} = 28$

b. $\binom{10}{3} = \dfrac{10!}{7! \cdot 3!} = \dfrac{(10 \cdot 9 \cdot 8) \cdot 7!}{7! \cdot 3!} = \dfrac{10 \cdot 9 \cdot 8}{3 \cdot 2 \cdot 1} = 120$

c. $_7C_0 = \dfrac{7!}{7! \cdot 0!} = 1$ **d.** $\binom{8}{8} = \dfrac{8!}{0! \cdot 8!} = 1$

✓ *Checkpoint* ◀))) Audio-video solution in English & Spanish at LarsonPrecalculus.com.

Find each binomial coefficient.

a. $\binom{11}{5}$ **b.** $_9C_2$ **c.** $\binom{5}{0}$ **d.** $_{15}C_{15}$ ■

When $r \neq 0$ and $r \neq n$, as in parts (a) and (b) above, there is a pattern for evaluating binomial coefficients that works because there will always be factorial terms that divide out from the expression.

$$_8C_2 = \frac{\overbrace{8 \cdot 7}^{\text{2 factors}}}{\underbrace{2 \cdot 1}_{\text{2 factors}}} \quad \text{and} \quad \binom{10}{3} = \frac{\overbrace{10 \cdot 9 \cdot 8}^{\text{3 factors}}}{\underbrace{3 \cdot 2 \cdot 1}_{\text{3 factors}}}$$

EXAMPLE 2   **Finding Binomial Coefficients**

a. $_7C_3 = \dfrac{7 \cdot 6 \cdot 5}{3 \cdot 2 \cdot 1} = 35$

b. $\binom{7}{4} = \dfrac{7 \cdot 6 \cdot 5 \cdot 4}{4 \cdot 3 \cdot 2 \cdot 1} = 35$

c. $_{12}C_1 = \dfrac{12}{1} = 12$

d. $\binom{12}{11} = \dfrac{12!}{1! \cdot 11!} = \dfrac{(12) \cdot 11!}{1! \cdot 11!} = \dfrac{12}{1} = 12$

✓ *Checkpoint* ◀))) Audio-video solution in English & Spanish at LarsonPrecalculus.com.

Find each binomial coefficient.

a. $_7C_5$ **b.** $\binom{7}{2}$ **c.** $_{14}C_{13}$ **d.** $\binom{14}{1}$ ■

It is not a coincidence that the results in parts (a) and (b) of Example 2 are the same and that the results in parts (c) and (d) are the same. In general, it is true that $_nC_r = {_nC_{n-r}}$. This shows the symmetric property of binomial coefficients identified earlier.

Pascal's Triangle

There is a convenient way to remember the pattern for binomial coefficients. By arranging the coefficients in a triangular pattern, you obtain the following array, which is called **Pascal's Triangle.** This triangle is named after the famous French mathematician Blaise Pascal (1623–1662).

$$
\begin{array}{ccccccccccccccc}
& & & & & & & 1 & & & & & & & \\
& & & & & & 1 & & 1 & & & & & & \\
& & & & & 1 & & 2 & & 1 & & & & & \\
& & & & 1 & & 3 & & 3 & & 1 & & & & \\
& & & 1 & & 4 & & 6 & & 4 & & 1 & & & \quad 4 + 6 = 10 \\
& & 1 & & 5 & & 10 & & 10 & & 5 & & 1 & & \\
& 1 & & 6 & & 15 & & 20 & & 15 & & 6 & & 1 & \\
1 & & 7 & & 21 & & 35 & & 35 & & 21 & & 7 & & 1 \quad 15 + 6 = 21
\end{array}
$$

The first and last numbers in each row of Pascal's Triangle are 1. Every other number in each row is formed by adding the two numbers immediately above the number. Pascal noticed that numbers in this triangle are precisely the same numbers that are the coefficients of binomial expansions, as follows.

$$(x + y)^0 = 1 \qquad \text{0th row}$$

$$(x + y)^1 = 1x + 1y \qquad \text{1st row}$$

$$(x + y)^2 = 1x^2 + 2xy + 1y^2 \qquad \text{2nd row}$$

$$(x + y)^3 = 1x^3 + 3x^2y + 3xy^2 + 1y^3 \qquad \text{3rd row}$$

$$(x + y)^4 = 1x^4 + 4x^3y + 6x^2y^2 + 4xy^3 + 1y^4 \qquad \vdots$$

$$(x + y)^5 = 1x^5 + 5x^4y + 10x^3y^2 + 10x^2y^3 + 5xy^4 + 1y^5$$

$$(x + y)^6 = 1x^6 + 6x^5y + 15x^4y^2 + 20x^3y^3 + 15x^2y^4 + 6xy^5 + 1y^6$$

$$(x + y)^7 = 1x^7 + 7x^6y + 21x^5y^2 + 35x^4y^3 + 35x^3y^4 + 21x^2y^5 + 7xy^6 + 1y^7$$

The top row in Pascal's Triangle is called the *zeroth row* because it corresponds to the binomial expansion $(x + y)^0 = 1$. Similarly, the next row is called the *first row* because it corresponds to the binomial expansion

$$(x + y)^1 = 1(x) + 1(y).$$

In general, the *nth row* in Pascal's Triangle gives the coefficients of $(x + y)^n$.

EXAMPLE 3   **Using Pascal's Triangle**

Use the seventh row of Pascal's Triangle to find the binomial coefficients.

$$_8C_0, \, _8C_1, \, _8C_2, \, _8C_3, \, _8C_4, \, _8C_5, \, _8C_6, \, _8C_7, \, _8C_8$$

Solution

$$
\begin{array}{ccccccccccccccccc}
& 1 & & 7 & & 21 & & 35 & & 35 & & 21 & & 7 & & 1 & \\
1 & & 8 & & 28 & & 56 & & 70 & & 56 & & 28 & & 8 & & 1 \\
_8C_0 & & _8C_1 & & _8C_2 & & _8C_3 & & _8C_4 & & _8C_5 & & _8C_6 & & _8C_7 & & _8C_8
\end{array}
$$

✓ *Checkpoint* 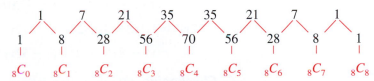 *Audio-video solution in English & Spanish at LarsonPrecalculus.com.*

Use the eighth row of Pascal's Triangle to find the binomial coefficients.

$$_9C_0, \, _9C_1, \, _9C_2, \, _9C_3, \, _9C_4, \, _9C_5, \, _9C_6, \, _9C_7, \, _9C_8, \, _9C_9$$

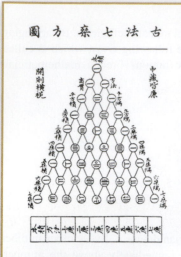

PRECIOUS MIRROR OF THE FOUR ELEMENTS

Eastern cultures were familiar with "Pascal's" Triangle and forms of the Binomial Theorem prior to the Western "discovery" of the theorem. A Chinese text entitled *Precious Mirror of the Four Elements* contains a triangle of binomial expansions through the eighth power.

Binomial Expansions

As mentioned at the beginning of this section, when you write the coefficients for a binomial that is raised to a power, you are **expanding a binomial.** The formula for binomial coefficients and Pascal's Triangle give you a relatively easy way to expand binomials, as demonstrated in the next four examples.

EXAMPLE 4   **Expanding a Binomial**

Write the expansion of the expression

$(x + 1)^3$.

Solution   The binomial coefficients from the third row of Pascal's Triangle are

1, 3, 3, 1.

So, the expansion is as follows.

$$(x + 1)^3 = (1)x^3 + (3)x^2(1) + (3)x(1^2) + (1)(1^3)$$
$$= x^3 + 3x^2 + 3x + 1$$

✓ **Checkpoint** 🔊))) *Audio-video solution in English & Spanish at LarsonPrecalculus.com.*

Write the expansion of the expression

$(x + 2)^4$.

To expand binomials representing *differences* rather than sums, you alternate signs. Here are two examples.

$$(x - 1)^3 = x^3 - 3x^2 + 3x - 1$$
$$(x - 1)^4 = x^4 - 4x^3 + 6x^2 - 4x + 1$$

▷ **ALGEBRA HELP** The solutions to Example 5 use the property of exponents

$(ab)^m = a^m b^m$.

For instance, in Example 5(a),

$(2x)^4 = 2^4 x^4 = 16x^4$.

You can review properties of exponents in Appendix A.2.

EXAMPLE 5   **Expanding a Binomial**

Write the expansion of each expression.

a. $(2x - 3)^4$

b. $(x - 2y)^4$

Solution   The binomial coefficients from the fourth row of Pascal's Triangle are

1, 4, 6, 4, 1.

So, the expansions are as follows.

a. $(2x - 3)^4 = (1)(2x)^4 - (4)(2x)^3(3) + (6)(2x)^2(3^2) - (4)(2x)(3^3) + (1)(3^4)$
$$= 16x^4 - 96x^3 + 216x^2 - 216x + 81$$

b. $(x - 2y)^4 = (1)x^4 - (4)x^3(2y) + (6)x^2(2y)^2 - (4)x(2y)^3 + (1)(2y)^4$
$$= x^4 - 8x^3y + 24x^2y^2 - 32xy^3 + 16y^4$$

✓ **Checkpoint** 🔊))) *Audio-video solution in English & Spanish at LarsonPrecalculus.com.*

Write the expansion of each expression.

a. $(y - 2)^4$

b. $(2x - y)^5$

▷ **TECHNOLOGY** Use a graphing utility to check the expansion in Example 6. Graph the original binomial expression and the expansion in the same viewing window. The graphs should coincide, as shown below.

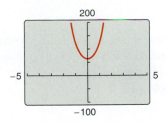

EXAMPLE 6   **Expanding a Binomial**

Write the expansion of $(x^2 + 4)^3$.

Solution   Use the third row of Pascal's Triangle, as follows.

$$(x^2 + 4)^3 = (1)(x^2)^3 + (3)(x^2)^2(4) + (3)x^2(4^2) + (1)(4^3)$$

$$= x^6 + 12x^4 + 48x^2 + 64$$

✓ **Checkpoint** *Audio-video solution in English & Spanish at LarsonPrecalculus.com.*

Write the expansion of $(5 + y^2)^3$.

Sometimes you will need to find a specific term in a binomial expansion. Instead of writing the entire expansion, use the fact that, from the Binomial Theorem, the $(r + 1)$th term is $_nC_r\, x^{n-r}y^r$.

EXAMPLE 7   **Finding a Term or Coefficient**

a. Find the sixth term of $(a + 2b)^8$.

b. Find the coefficient of the term a^6b^5 in the expansion of $(3a - 2b)^{11}$.

Solution

a. Remember that the formula is for the $(r + 1)$th term, so r is one less than the number of the term you need. So, to find the sixth term in this binomial expansion, use $r = 5$, $n = 8$, $x = a$, and $y = 2b$, as shown.

$$_8C_5\, a^{8-5}(2b)^5 = 56 \cdot a^3 \cdot (2b)^5$$

$$= 56(2^5)a^3b^5$$

$$= 1792a^3b^5$$

b. In this case, $n = 11$, $r = 5$, $x = 3a$, and $y = -2b$. Substitute these values to obtain

$$_nC_r\, x^{n-r}y^r = {_{11}C_5}(3a)^6(-2b)^5$$

$$= (462)(729a^6)(-32b^5)$$

$$= -10{,}777{,}536a^6b^5.$$

So, the coefficient is $-10{,}777{,}536$.

✓ **Checkpoint** *Audio-video solution in English & Spanish at LarsonPrecalculus.com.*

a. Find the fifth term of $(a + 2b)^8$.

b. Find the coefficient of the term a^4b^7 in the expansion of $(3a - 2b)^{11}$.

Summarize   (Section 9.5)

1. State the Binomial Theorem *(page 473)*. For examples of using the Binomial Theorem to calculate binomial coefficients, see Examples 1 and 2.

2. Explain how to use Pascal's Triangle to calculate binomial coefficients *(page 475)*. For an example of using Pascal's Triangle to calculate binomial coefficients, see Example 3.

3. Explain how to use binomial coefficients to write a binomial expansion *(page 476)*. For examples of using binomial coefficients to write binomial expansions, see Examples 4–6.

9.6   Counting Principles

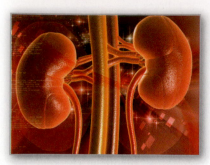

Counting principles can help you solve counting problems that occur in real life, such as determining the numbers of possible orders there are for best match, second-best match, and third-best match kidney donors.

■ Solve simple counting problems.
■ Use the Fundamental Counting Principle to solve counting problems.
■ Use permutations to solve counting problems.
■ Use combinations to solve counting problems.

Simple Counting Problems

This section and Section 9.7 present a brief introduction to some of the basic counting principles and their applications to probability. In Section 9.7, you will see that much of probability has to do with counting the number of ways an event can occur. The following two examples describe simple counting problems.

EXAMPLE 1   Selecting Pairs of Numbers at Random

You place eight pieces of paper, numbered from 1 to 8, in a box. You draw one piece of paper at random from the box, record its number, and *replace* the paper in the box. Then, you draw a second piece of paper at random from the box and record its number. Finally, you add the two numbers. How many different ways can you obtain a sum of 12?

Solution   To solve this problem, count the different ways to obtain a sum of 12 using two numbers from 1 to 8.

First number	4	5	6	7	8
Second number	8	7	6	5	4

So, a sum of 12 can occur in five different ways.

✓ *Checkpoint* *Audio-video solution in English & Spanish at LarsonPrecalculus.com.*

In Example 1, how many different ways can you obtain a sum of 14?

EXAMPLE 2   Selecting Pairs of Numbers at Random

You place eight pieces of paper, numbered from 1 to 8, in a box. You draw one piece of paper at random from the box, record its number, and *do not* replace the paper in the box. Then, you draw a second piece of paper at random from the box and record its number. Finally, you add the two numbers. How many different ways can you obtain a sum of 12?

Solution   To solve this problem, count the different ways to obtain a sum of 12 *using two different numbers* from 1 to 8.

First number	4	5	7	8
Second number	8	7	5	4

So, a sum of 12 can occur in four different ways.

✓ *Checkpoint* *Audio-video solution in English & Spanish at LarsonPrecalculus.com.*

Repeat Example 2 for drawing *three* pieces of paper.

   The difference between the counting problems in Examples 1 and 2 can be described by saying that the random selection in Example 1 occurs **with replacement,** whereas the random selection in Example 2 occurs **without replacement,** which eliminates the possibility of choosing two 6's.

The Fundamental Counting Principle

Examples 1 and 2 describe simple counting problems in which you can *list* each possible way that an event can occur. When it is possible, this is always the best way to solve a counting problem. However, some events can occur in so many different ways that it is not feasible to write out the entire list. In such cases, you must rely on formulas and counting principles. The most important of these is the **Fundamental Counting Principle.**

Fundamental Counting Principle

Let E_1 and E_2 be two events. The first event E_1 can occur in m_1 different ways. After E_1 has occurred, E_2 can occur in m_2 different ways. The number of ways that the two events can occur is $m_1 \cdot m_2$.

The Fundamental Counting Principle can be extended to three or more events. For instance, the number of ways that three events E_1, E_2, and E_3 can occur is $m_1 \cdot m_2 \cdot m_3$.

EXAMPLE 3   **Using the Fundamental Counting Principle**

How many different pairs of letters from the English alphabet are possible?

Solution   There are two events in this situation. The first event is the choice of the first letter, and the second event is the choice of the second letter. Because the English alphabet contains 26 letters, it follows that the number of two-letter pairs is $26 \cdot 26 = 676$.

✓ *Checkpoint* ◀))) *Audio-video solution in English & Spanish at LarsonPrecalculus.com.*

A combination lock will open when you select the right choice of three numbers (from 1 to 30, inclusive). How many different lock combinations are possible?

EXAMPLE 4   **Using the Fundamental Counting Principle**

Telephone numbers in the United States have 10 digits. The first three digits are the *area code* and the next seven digits are the *local telephone number.* How many different telephone numbers are possible within each area code? (Note that a local telephone number cannot begin with 0 or 1.)

Solution   Because the first digit of a local telephone number cannot be 0 or 1, there are only eight choices for the first digit. For each of the other six digits, there are 10 choices.

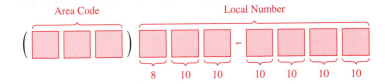

So, the number of telephone numbers that are possible within each area code is

$$8 \cdot 10 \cdot 10 \cdot 10 \cdot 10 \cdot 10 \cdot 10 = 8{,}000{,}000.$$

✓ *Checkpoint* ◀))) *Audio-video solution in English & Spanish at LarsonPrecalculus.com.*

A product's catalog number is made up of one letter from the English alphabet followed by a five-digit number. How many different catalog numbers are possible? ■

Permutations

One important application of the Fundamental Counting Principle is in determining the number of ways that n elements can be arranged (in order). An ordering of n elements is called a **permutation** of the elements.

Definition of Permutation

A **permutation** of n different elements is an ordering of the elements such that one element is first, one is second, one is third, and so on.

EXAMPLE 5 **Finding the Number of Permutations**

How many permutations of the letters

A, B, C, D, E, and F

are possible?

Solution   Consider the following reasoning.

   First position: Any of the *six* letters
   Second position: Any of the remaining *five* letters
   Third position: Any of the remaining *four* letters
   Fourth position: Any of the remaining *three* letters
   Fifth position: Either of the remaining *two* letters
   Sixth position: The *one* remaining letter

So, the numbers of choices for the six positions are as follows.

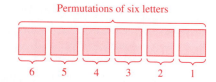

Permutations of six letters

The total number of permutations of the six letters is

$$6! = 6 \cdot 5 \cdot 4 \cdot 3 \cdot 2 \cdot 1$$
$$= 720.$$

✓ **Checkpoint** 🔊))) *Audio-video solution in English & Spanish at LarsonPrecalculus.com.*

How many permutations of the letters

W, X, Y, and Z

are possible?

Number of Permutations of n Elements

The number of permutations of n elements is

$$n \cdot (n - 1) \cdots 4 \cdot 3 \cdot 2 \cdot 1 = n!.$$

In other words, there are $n!$ different ways that n elements can be ordered.

Eleven thoroughbred racehorses hold the title of Triple Crown winner for winning the Kentucky Derby, the Preakness, and the Belmont Stakes in the same year. Forty-nine horses have won two out of the three races.

EXAMPLE 6 **Counting Horse Race Finishes**

Eight horses are running in a race. In how many different ways can these horses come in first, second, and third? (Assume that there are no ties.)

Solution Here are the different possibilities.

Win (first position): *Eight* choices

Place (second position): *Seven* choices

Show (third position): *Six* choices

Using the Fundamental Counting Principle, multiply these three numbers to obtain the following.

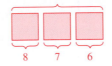

Different orders of horses

So, there are

$$8 \cdot 7 \cdot 6 = 336$$

different orders.

✓ *Checkpoint*  *Audio-video solution in English & Spanish at LarsonPrecalculus.com.*

A coin club has five members. In how many ways can there be a president and a vice-president?

It is useful, on occasion, to order a *subset* of a collection of elements rather than the entire collection. For example, you might want to order r elements out of a collection of n elements. Such an ordering is called a **permutation of n elements taken r at a time.**

▷ **TECHNOLOGY**

Most graphing calculators have the capability of evaluating $_nP_r$. If yours does, then evaluate $_8P_5$. You should get an answer of 6720.

Permutations of n Elements Taken r at a Time

The number of permutations of n elements taken r at a time is

$$_nP_r = \frac{n!}{(n-r)!} = n(n-1)(n-2) \cdots (n-r+1).$$

Using this formula, you can rework Example 6 to find that the number of permutations of eight horses taken three at a time is

$$_8P_3 = \frac{8!}{(8-3)!}$$

$$= \frac{8!}{5!}$$

$$= \frac{8 \cdot 7 \cdot 6 \cdot 5!}{5!}$$

$$= 336$$

which is the same answer obtained in the example.

Mikhail Pogosov/Shutterstock.com

Remember that for permutations, order is important. So, to find the possible permutations of the letters A, B, C, and D taken three at a time, you count the permutations (A, B, D) and (B, A, D) as different because the *order* of the elements is different.

Suppose, however, that you want to find the possible permutations of the letters A, A, B, and C. The total number of permutations of the four letters would be $_4P_4 = 4!$. However, not all of these arrangements would be *distinguishable* because there are two A's in the list. To find the number of distinguishable permutations, you can use the following formula.

Distinguishable Permutations

Suppose a set of n objects has n_1 of one kind of object, n_2 of a second kind, n_3 of a third kind, and so on, with

$$n = n_1 + n_2 + n_3 + \cdots + n_k.$$

Then the number of **distinguishable permutations** of the n objects is

$$\frac{n!}{n_1! \cdot n_2! \cdot n_3! \cdots \cdots n_k!}.$$

EXAMPLE 7   **Distinguishable Permutations**

In how many distinguishable ways can the letters in BANANA be written?

Solution   This word has six letters, of which three are A's, two are N's, and one is a B. So, the number of distinguishable ways the letters can be written is

$$\frac{n!}{n_1! \cdot n_2! \cdot n_3!} = \frac{6!}{3! \cdot 2! \cdot 1!}$$

$$= \frac{6 \cdot 5 \cdot 4 \cdot 3!}{3! \cdot 2!}$$

$$= 60.$$

The 60 different distinguishable permutations are as follows.

AAABNN	AAANBN	AAANNB	AABANN
AABNAN	AABNNA	AANABN	AANANB
AANBAN	AANBNA	AANNAB	AANNBA
ABAANN	ABANAN	ABANNA	ABNAAN
ABNANA	ABNNAA	ANAABN	ANAANB
ANABAN	ANABNA	ANANAB	ANANBA
ANBAAN	ANBANA	ANBNAA	ANNAAB
ANNABA	ANNBAA	BAAANN	BAANAN
BAANNA	BANAAN	BANANA	BANNAA
BNAAAN	BNAANA	BNANAA	BNNAAA
NAAABN	NAAANB	NAABAN	NAABNA
NAANAB	NAANBA	NABAAN	NABANA
NABNAA	NANAAB	NANABA	NANBAA
NBAAAN	NBAANA	NBANAA	NBNAAA
NNAAAB	NNAABA	NNABAA	NNBAAA

✓ *Checkpoint* 🔊)) *Audio-video solution in English & Spanish at LarsonPrecalculus.com.*

In how many distinguishable ways can the letters in MITOSIS be written? ■

Combinations

When you count the number of possible permutations of a set of elements, order is important. As a final topic in this section, you will look at a method of selecting subsets of a larger set in which order is *not* important. Such subsets are called **combinations of *n* elements taken *r* at a time.** For instance, the combinations

$$\{A, B, C\} \quad \text{and} \quad \{B, A, C\}$$

are equivalent because both sets contain the same three elements, and the order in which the elements are listed is not important. So, you would count only one of the two sets. A common example of how a combination occurs is in a card game in which players are free to reorder the cards after they have been dealt.

EXAMPLE 8   Combinations of *n* Elements Taken *r* at a Time

In how many different ways can three letters be chosen from the letters

A,   B,   C,   D,   and   E?

(The order of the three letters is not important.)

Solution   The following subsets represent the different combinations of three letters that can be chosen from the five letters.

$\{A, B, C\}$	$\{A, B, D\}$
$\{A, B, E\}$	$\{A, C, D\}$
$\{A, C, E\}$	$\{A, D, E\}$
$\{B, C, D\}$	$\{B, C, E\}$
$\{B, D, E\}$	$\{C, D, E\}$

From this list, you can conclude that there are 10 different ways that three letters can be chosen from five letters.

✓ **Checkpoint**  *Audio-video solution in English & Spanish at LarsonPrecalculus.com.*

In how many different ways can two letters be chosen from the letters A, B, C, D, E, F, and G? (The order of the two letters is not important.) ■

Combinations of *n* Elements Taken *r* at a Time

The number of combinations of *n* elements taken *r* at a time is

$$_nC_r = \frac{n!}{(n-r)!r!}$$

which is equivalent to $_nC_r = \dfrac{_nP_r}{r!}.$

Note that the formula for $_nC_r$ is the same one given for binomial coefficients. To see how to use this formula, solve the counting problem in Example 8. In that problem, you want to find the number of combinations of five elements taken three at a time. So, $n = 5, r = 3$, and the number of combinations is

$$_5C_3 = \frac{5!}{2!3!} = \frac{5 \cdot \overset{2}{\cancel{4}} \cdot \cancel{3!}}{2 \cdot 1 \cdot \cancel{3!}} = 10$$

which is the same answer obtained in Example 8.

A ♥	A ♦	A ♣	A ♠
2 ♥	2 ♦	2 ♣	2 ♠
3 ♥	3 ♦	3 ♣	3 ♠
4 ♥	4 ♦	4 ♣	4 ♠
5 ♥	5 ♦	5 ♣	5 ♠
6 ♥	6 ♦	6 ♣	6 ♠
7 ♥	7 ♦	7 ♣	7 ♠
8 ♥	8 ♦	8 ♣	8 ♠
9 ♥	9 ♦	9 ♣	9 ♠
10 ♥	10 ♦	10 ♣	10 ♠
J ♥	J ♦	J ♣	J ♠
Q ♥	Q ♦	Q ♣	Q ♠
K ♥	K ♦	K ♣	K ♠

Ranks and suits in a standard deck of playing cards
Figure 9.2

EXAMPLE 9   **Counting Card Hands**

A standard poker hand consists of five cards dealt from a deck of 52 (see Figure 9.2). How many different poker hands are possible? (After the cards are dealt, the player may reorder them, and so order is not important.)

Solution   To find the number of different poker hands, use the formula for the number of combinations of 52 elements taken five at a time, as follows.

$$_{52}C_5 = \frac{52!}{(52 - 5)!5!}$$

$$= \frac{52!}{47!5!}$$

$$= \frac{52 \cdot 51 \cdot 50 \cdot 49 \cdot 48 \cdot \cancel{47!}}{5 \cdot 4 \cdot 3 \cdot 2 \cdot 1 \cdot \cancel{47!}}$$

$$= 2{,}598{,}960$$

✓ *Checkpoint* Audio-video solution in English & Spanish at LarsonPrecalculus.com.

In three-card poker, each player is dealt three cards from a deck of 52. How many different three-card poker hands are possible? (Order is not important.)

EXAMPLE 10   **Forming a Team**

You are forming a 12-member swim team from 10 girls and 15 boys. The team must consist of five girls and seven boys. How many different 12-member teams are possible?

Solution   There are $_{10}C_5$ ways of choosing five girls. There are $_{15}C_7$ ways of choosing seven boys. By the Fundamental Counting Principle, there are $_{10}C_5 \cdot {}_{15}C_7$ ways of choosing five girls and seven boys.

$$_{10}C_5 \cdot {}_{15}C_7 = \frac{10!}{5! \cdot 5!} \cdot \frac{15!}{8! \cdot 7!} = 252 \cdot 6435 = 1{,}621{,}620$$

So, there are 1,621,620 12-member swim teams possible.

✓ *Checkpoint* Audio-video solution in English & Spanish at LarsonPrecalculus.com.

In Example 10, the team must consist of six boys and six girls. How many different 12-member teams are possible?

REMARK   When solving problems involving counting principles, you need to be able to distinguish among the various counting principles in order to determine which is necessary to solve the problem correctly. To do this, ask yourself the following questions.

1. Is the order of the elements important? *Permutation*

2. Are the chosen elements a subset of a larger set in which order is not important? *Combination*

3. Does the problem involve two or more separate events? *Fundamental Counting Principle*

Summarize   (Section 9.6)

1. Explain how to solve a simple counting problem (*page 478*). For examples of solving simple counting problems, see Examples 1 and 2.

2. State the Fundamental Counting Principle (*page 479*). For examples of using the Fundamental Counting Principle to solve counting problems, see Examples 3 and 4.

3. Explain how to find the number of permutations of *n* elements (*page 480*), the number of permutations of *n* elements taken *r* at a time (*page 481*), and the number of distinguishable permutations (*page 482*). For examples of using permutations to solve counting problems, see Examples 5–7.

4. Explain how to find the number of combinations of *n* elements taken *r* at a time (*page 483*). For examples of using combinations to solve counting problems, see Examples 8–10.

9.7 Probability

Probability applies to many real-life applications, such as the reliability of a communication network and an independent backup system for a space vehicle.

■ Find the probabilities of events.
■ Find the probabilities of mutually exclusive events.
■ Find the probabilities of independent events.
■ Find the probability of the complement of an event.

The Probability of an Event

Any happening for which the result is uncertain is called an **experiment.** The possible results of the experiment are **outcomes,** the set of all possible outcomes of the experiment is the **sample space** of the experiment, and any subcollection of a sample space is an **event.**

For instance, when you toss a six-sided die, the numbers 1 through 6 can represent the sample space. For this experiment, each of the outcomes is *equally likely.*

To describe sample spaces in such a way that each outcome is equally likely, you must sometimes distinguish between or among various outcomes in ways that appear artificial. Example 1 illustrates such a situation.

EXAMPLE 1 **Finding a Sample Space**

Find the sample space for each of the following experiments.

a. You toss one coin.

b. You toss two coins.

c. You toss three coins.

Solution

a. Because the coin will land either heads up (denoted by H) or tails up (denoted by T), the sample space is

$S = \{H, T\}.$

b. Because either coin can land heads up or tails up, the possible outcomes are as follows.

HH = heads up on both coins

HT = heads up on first coin and tails up on second coin

TH = tails up on first coin and heads up on second coin

TT = tails up on both coins

So, the sample space is

$S = \{HH, HT, TH, TT\}.$

Note that this list distinguishes between the two cases HT and TH, even though these two outcomes appear to be similar.

c. Following the notation of part (b), the sample space is

$S = \{HHH, HHT, HTH, HTT, THH, THT, TTH, TTT\}.$

Note that this list distinguishes among the cases HHT, HTH, and THH, and among the cases HTT, THT, and TTH.

✓ *Checkpoint* ◄))) *Audio-video solution in English & Spanish at LarsonPrecalculus.com.*

Find the sample space for the following experiment.

You toss a coin twice and a six-sided die once.

1971yes/Shutterstock.com

To calculate the probability of an event, count the number of outcomes in the event and in the sample space. The *number of outcomes* in event E is denoted by $n(E)$, and the number of outcomes in the sample space S is denoted by $n(S)$. The probability that event E will occur is given by $n(E)/n(S)$.

The Probability of an Event

If an event E has $n(E)$ equally likely outcomes and its sample space S has $n(S)$ equally likely outcomes, then the **probability** of event E is

$$P(E) = \frac{n(E)}{n(S)}.$$

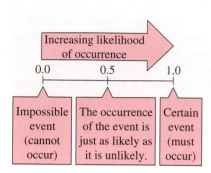

Figure 9.3

Because the number of outcomes in an event must be less than or equal to the number of outcomes in the sample space, the probability of an event must be a number between 0 and 1. That is,

$$0 \leq P(E) \leq 1$$

as indicated in Figure 9.3. If $P(E) = 0$, then event E *cannot occur*, and E is called an **impossible event.** If $P(E) = 1$, then event E *must occur,* and E is called a **certain event.**

EXAMPLE 2   Finding the Probability of an Event

a. You toss two coins. What is the probability that both land heads up?

b. You draw one card at random from a standard deck of playing cards. What is the probability that it is an ace?

Solution

a. Following the procedure in Example 1(b), let

$$E = \{HH\}$$

and

$$S = \{HH, HT, TH, TT\}.$$

The probability of getting two heads is

$$P(E) = \frac{n(E)}{n(S)} = \frac{1}{4}.$$

b. Because there are 52 cards in a standard deck of playing cards and there are four aces (one in each suit), the probability of drawing an ace is

$$P(E) = \frac{n(E)}{n(S)}$$

$$= \frac{4}{52}$$

$$= \frac{1}{13}.$$

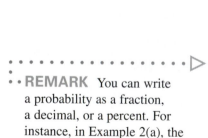

• • REMARK You can write a probability as a fraction, a decimal, or a percent. For instance, in Example 2(a), the probability of getting two heads can be written as $\frac{1}{4}$, 0.25, or 25%.

✓ *Checkpoint* 🔊))) *Audio-video solution in English & Spanish at LarsonPrecalculus.com.*

a. You toss three coins. What is the probability that all three land tails up?

b. You draw one card at random from a standard deck of playing cards. What is the probability that it is a diamond?

As shown in Example 3, when you toss two six-sided dice, the probability that the total of the two dice is 7 is $\frac{1}{6}$.

| EXAMPLE 3 | Finding the Probability of an Event |

You toss two six-sided dice. What is the probability that the total of the two dice is 7?

Solution   Because there are six possible outcomes on each die, use the Fundamental Counting Principle to conclude that there are 6 · 6 or 36 different outcomes when you toss two dice. To find the probability of rolling a total of 7, you must first count the number of ways in which this can occur.

First Die	Second Die
1	6
2	5
3	4
4	3
5	2
6	1

So, a total of 7 can be rolled in six ways, which means that the probability of rolling a 7 is

$$P(E) = \frac{n(E)}{n(S)} = \frac{6}{36} = \frac{1}{6}.$$

✓ **Checkpoint** *Audio-video solution in English & Spanish at LarsonPrecalculus.com.*

You toss two six-sided dice. What is the probability that the total of the two dice is 5?

| EXAMPLE 4 | Finding the Probability of an Event |

Twelve-sided dice, as shown in Figure 9.4, can be constructed (in the shape of regular dodecahedrons) such that each of the numbers from 1 to 6 appears twice on each die. Prove that these dice can be used in any game requiring ordinary six-sided dice without changing the probabilities of the various events.

Solution   For an ordinary six-sided die, each of the numbers

1, 2, 3, 4, 5, and 6

occurs only once, so the probability of rolling any particular number is

$$P(E) = \frac{n(E)}{n(S)} = \frac{1}{6}.$$

For one of the 12-sided dice, each number occurs twice, so the probability of rolling any particular number is

$$P(E) = \frac{n(E)}{n(S)} = \frac{2}{12} = \frac{1}{6}.$$

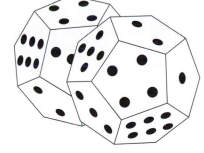

Figure 9.4

✓ **Checkpoint** *Audio-video solution in English & Spanish at LarsonPrecalculus.com.*

Show that the probability of drawing a club at random from a standard deck of playing cards is the same as the probability of drawing the ace of hearts at random from a set consisting of the aces of hearts, diamonds, clubs, and spades.

<table>
<tr><td>EXAMPLE 5</td><td>**Random Selection**</td></tr>
</table>

The following figure shows the numbers of colleges and universities in various regions of the United States in 2010. What is the probability that an institution selected at random is in one of the three southern regions? *(Source: National Center for Education Statistics)*

Solution From the figure, the total number of colleges and universities is 4490. Because there are $797 + 294 + 438 = 1529$ colleges and universities in the three southern regions, the probability that the institution is in one of these regions is

$$P(E) = \frac{n(E)}{n(S)} = \frac{1529}{4490} \approx 0.341.$$

 Checkpoint 🔊)) *Audio-video solution in English & Spanish at LarsonPrecalculus.com.*

In Example 5, what is the probability that the randomly selected institution is in the Pacific region?

<table>
<tr><td>EXAMPLE 6</td><td>**The Probability of Winning a Lottery**</td></tr>
</table>

In Arizona's The Pick game, a player chooses six different numbers from 1 to 44. If these six numbers match the six numbers drawn (in any order) by the lottery commission, then the player wins (or shares) the top prize. What is the probability of winning the top prize when the player buys one ticket?

Solution To find the number of elements in the sample space, use the formula for the number of combinations of 44 elements taken six at a time.

$$n(S) = {}_{44}C_6$$

$$= \frac{44 \cdot 43 \cdot 42 \cdot 41 \cdot 40 \cdot 39}{6 \cdot 5 \cdot 4 \cdot 3 \cdot 2 \cdot 1}$$

$$= 7{,}059{,}052$$

When a player buys one ticket, the probability of winning is

$$P(E) = \frac{1}{7{,}059{,}052}.$$

 Checkpoint 🔊)) *Audio-video solution in English & Spanish at LarsonPrecalculus.com.*

In Pennsylvania's Cash 5 game, a player chooses five different numbers from 1 to 43. If these five numbers match the five numbers drawn (in any order) by the lottery commission, then the player wins (or shares) the top prize. What is the probability of winning the top prize when the player buys one ticket?

Mutually Exclusive Events

Two events A and B (from the same sample space) are **mutually exclusive** when A and B have no outcomes in common. In the terminology of sets, the intersection of A and B is the empty set, which implies that

$$P(A \cap B) = 0.$$

For instance, when you toss two dice, the event A of rolling a total of 6 and the event B of rolling a total of 9 are mutually exclusive. To find the probability that one or the other of two mutually exclusive events will occur, *add* their individual probabilities.

Probability of the Union of Two Events

If A and B are events in the same sample space, then the probability of A or B occurring is given by

$$P(A \cup B) = P(A) + P(B) - P(A \cap B).$$

If A and B are mutually exclusive, then

$$P(A \cup B) = P(A) + P(B).$$

EXAMPLE 7   The Probability of a Union of Events

You draw one card at random from a standard deck of 52 playing cards. What is the probability that the card is either a heart or a face card?

Solution Because the deck has 13 hearts, the probability of drawing a heart (event A) is

$$P(A) = \frac{13}{52}.$$

Similarly, because the deck has 12 face cards, the probability of drawing a face card (event B) is

$$P(B) = \frac{12}{52}.$$

Because three of the cards are hearts *and* face cards (see Figure 9.5), it follows that

$$P(A \cap B) = \frac{3}{52}.$$

Finally, applying the formula for the probability of the union of two events, the probability of drawing a heart or a face card is

$$P(A \cup B) = P(A) + P(B) - P(A \cap B)$$

$$= \frac{13}{52} + \frac{12}{52} - \frac{3}{52}$$

$$= \frac{22}{52}$$

$$\approx 0.423.$$

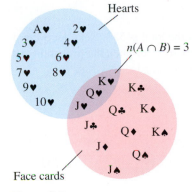

Hearts

$n(A \cap B) = 3$

Face cards

Figure 9.5

✓ **Checkpoint** ◀))) *Audio-video solution in English & Spanish at LarsonPrecalculus.com.*

You draw one card at random from a standard deck of 52 playing cards. What is the probability that the card is either an ace or a spade?

EXAMPLE 8 **Probability of Mutually Exclusive Events**

The human resources department of a company has compiled data showing the number of years of service for each employee. The table shows the results.

Years of Service	Number of Employees
0–4	157
5–9	89
10–14	74
15–19	63
20–24	42
25–29	38
30–34	35
35–39	21
40–44	8
45 or more	2

Spreadsheet at LarsonPrecalculus.com

a. What is the probability that an employee chosen at random has 4 or fewer years of service?

b. What is the probability that an employee chosen at random has 9 or fewer years of service?

Solution

a. To begin, add the number of employees to find that the total is 529. Next, let event A represent choosing an employee with 0 to 4 years of service. Then the probability of choosing an employee who has 4 or fewer years of service is

$$P(A) = \frac{157}{529}$$

$$\approx 0.297.$$

b. Let event B represent choosing an employee with 5 to 9 years of service. Then

$$P(B) = \frac{89}{529}.$$

Because event A from part (a) and event B have no outcomes in common, these two events are mutually exclusive and

$$P(A \cup B) = P(A) + P(B)$$

$$= \frac{157}{529} + \frac{89}{529}$$

$$= \frac{246}{529}$$

$$\approx 0.465.$$

So, the probability of choosing an employee who has 9 or fewer years of service is about 0.465.

 **Checkpoint**  *Audio-video solution in English & Spanish at LarsonPrecalculus.com.*

In Example 8, what is the probability that an employee chosen at random has 30 or more years of service?

Independent Events

Two events are **independent** when the occurrence of one has no effect on the occurrence of the other. For instance, rolling a total of 12 with two six-sided dice has no effect on the outcome of future rolls of the dice. To find the probability that two independent events will occur, *multiply* the probabilities of each.

Probability of Independent Events

If A and B are independent events, then the probability that both A and B will occur is

$$P(A \text{ and } B) = P(A) \cdot P(B).$$

EXAMPLE 9 **Probability of Independent Events**

A random number generator on a computer selects three integers from 1 to 20. What is the probability that all three numbers are less than or equal to 5?

Solution The probability of selecting a number from 1 to 5 is

$$P(A) = \frac{5}{20}$$

$$= \frac{1}{4}.$$

So, the probability that all three numbers are less than or equal to 5 is

$$P(A) \cdot P(A) \cdot P(A) = \left(\frac{1}{4}\right)\left(\frac{1}{4}\right)\left(\frac{1}{4}\right)$$

$$= \frac{1}{64}.$$

✓ *Checkpoint* 🔊))) *Audio-video solution in English & Spanish at LarsonPrecalculus.com.*

A random number generator on a computer selects two integers from 1 to 30. What is the probability that both numbers are less than 12?

EXAMPLE 10 **Probability of Independent Events**

In 2012, approximately 27% of the population of the United States got their news from mobile devices. In a survey, researchers selected 10 people at random from the population. What is the probability that all 10 got their news from mobile devices? (*Source: Pew Research Center*)

Solution Let A represent choosing a person who gets his or her news from a mobile device. The probability of choosing a person who got his or her news from a mobile device is 0.27, the probability of choosing a second person who got his or her news from a mobile device is 0.27, and so on. Because these events are independent, the probability that all 10 people got their news from mobile devices is

$$[P(A)]^{10} = (0.27)^{10} \approx 0.000002059.$$

✓ *Checkpoint* 🔊))) *Audio-video solution in English & Spanish at LarsonPrecalculus.com.*

In Example 10, researchers selected five people at random from the population. What is the probability that all five got their news from mobile devices?

The Complement of an Event

The **complement of an event** A is the collection of all outcomes in the sample space that are *not* in A. The complement of event A is denoted by A'. Because $P(A \text{ or } A') = 1$ and because A and A' are mutually exclusive, it follows that $P(A) + P(A') = 1$. So, the probability of A' is

$$P(A') = 1 - P(A).$$

For instance, if the probability of *winning* a certain game is $P(A) = \frac{1}{4}$, then the probability of *losing* the game is $P(A') = 1 - \frac{1}{4} = \frac{3}{4}$.

Probability of a Complement

Let A be an event and let A' be its complement. If the probability of A is $P(A)$, then the probability of the complement is

$$P(A') = 1 - P(A).$$

EXAMPLE 11   **Finding the Probability of a Complement**

A manufacturer has determined that a machine averages one faulty unit for every 1000 it produces. What is the probability that an order of 200 units will have one or more faulty units?

Solution   To solve this problem as stated, you would need to find the probabilities of having exactly one faulty unit, exactly two faulty units, exactly three faulty units, and so on. However, using complements, you can find the probability that all units are perfect and then subtract this value from 1. Because the probability that any given unit is perfect is $999/1000$, the probability that all 200 units are perfect is

$$P(A) = \left(\frac{999}{1000}\right)^{200} \approx 0.819.$$

So, the probability that at least one unit is faulty is

$$P(A') = 1 - P(A) \approx 1 - 0.819 = 0.181.$$

✓ **Checkpoint** *Audio-video solution in English & Spanish at LarsonPrecalculus.com.*

A manufacturer has determined that a machine averages one faulty unit for every 500 it produces. What is the probability that an order of 300 units will have one or more faulty units? ■

Summarize   (Section 9.7)

1. State the definition of the probability of an event *(page 486)*. For examples of finding the probabilities of events, see Examples 2–6.

2. State the definition of mutually exclusive events and explain how to find the probability of the union of two events *(page 489)*. For examples of finding the probabilities of the unions of two events, see Examples 7 and 8.

3. State the definition of, and explain how to find the probability of, independent events *(page 491)*. For examples of finding the probabilities of independent events, see Examples 9 and 10.

4. State the definition of, and explain how to find the probability of, the complement of an event *(page 492)*. For an example of finding the probability of the complement of an event, see Example 11.

Chapter Summary

What Did You Learn?	Explanation/Examples	Review Exercises
Section 9.1		
Use sequence notation to write the terms of sequences (p. 446).	$a_n = 7n - 4$; $a_1 = 7(1) - 4 = 3$, $a_2 = 7(2) - 4 = 10$, $a_3 = 7(3) - 4 = 17$, $a_4 = 7(4) - 4 = 24$	1–8
Use factorial notation (p. 448).	If n is a positive integer, then $n! = 1 \cdot 2 \cdot 3 \cdot 4 \cdots (n - 1) \cdot n$.	9–12
Use summation notation to write sums (p. 450).	The sum of the first n terms of a sequence is represented by $$\sum_{i=1}^{n} a_i = a_1 + a_2 + a_3 + a_4 + \cdots + a_n.$$	13–16
Find the sums of series (p. 451).	$$\sum_{i=1}^{\infty} \frac{8}{10^i} = \frac{8}{10^1} + \frac{8}{10^2} + \frac{8}{10^3} + \frac{8}{10^4} + \frac{8}{10^5} + \cdots$$ $$= 0.8 + 0.08 + 0.008 + 0.0008 + 0.00008 + \cdots$$ $$= 0.88888 \ldots = \frac{8}{9}$$	17, 18
Use sequences and series to model and solve real-life problems (p. 452).	A sequence can help you model the balance in an account that earns compound interest. (See Example 10.)	19, 20
Section 9.2		
Recognize, write, and find the nth terms of arithmetic sequences (p. 453).	$a_n = 9n + 5$; $a_1 = 9(1) + 5 = 14$, $a_2 = 9(2) + 5 = 23$, $a_3 = 9(3) + 5 = 32$, $a_4 = 9(4) + 5 = 41$	21–30
Find nth partial sums of arithmetic sequences (p. 456).	The sum of a finite arithmetic sequence with n terms is $S_n = (n/2)(a_1 + a_n)$.	31–36
Use arithmetic sequences to model and solve real-life problems (p. 458).	An arithmetic sequence can help you find the total sales of a small business. (See Example 9.)	37, 38
Section 9.3		
Recognize, write, and find the nth terms of geometric sequences (p. 459).	$a_n = 3(4^n)$; $a_1 = 3(4^1) = 12$, $a_2 = 3(4^2) = 48$, $a_3 = 3(4^3) = 192$, $a_4 = 3(4^4) = 768$	39–50
Find the sum of a finite geometric sequence (p. 462).	The sum of the finite geometric sequence $a_1, a_1 r, a_1 r^2, \ldots, a_1 r^{n-1}$ with common ratio $r \neq 1$ is given by $S_n = \sum_{i=1}^{n} a_1 r^{i-1} = a_1 \left(\dfrac{1 - r^n}{1 - r} \right).$	51–58
Find the sum of an infinite geometric series (p. 463).	If $|r| < 1$, then the infinite geometric series $a_1 + a_1 r + a_1 r^2 + \cdots + a_1 r^{n-1} + \cdots$ has the sum $S = \sum_{i=0}^{\infty} a_1 r^i = \dfrac{a_1}{1 - r}.$	59–62
Use geometric sequences to model and solve real-life problems (p. 464).	A finite geometric sequence can help you find the balance in an annuity at the end of two years. (See Example 8.)	63, 64

What Did You Learn?	Explanation/Examples	Review Exercises
Section 9.4 Use mathematical induction to prove statements involving a positive integer n (p. 465).	Let P_n be a statement involving the positive integer n. If (1) P_1 is true, and (2) for every positive integer k, the truth of P_k implies the truth of P_{k+1}, then the statement P_n must be true for all positive integers n.	65–68
Use pattern recognition and mathematical induction to write the nth term of a sequence (p. 470).	To find a formula for the nth term of a sequence, (1) calculate the first several terms of the sequence, (2) try to find a pattern for the terms and write a formula (hypothesis) for the nth term of the sequence, and (3) use mathematical induction to prove your hypothesis.	69–72
Find the sums of powers of integers (p. 471).	$\sum_{i=1}^{8} i^2 = \dfrac{n(n+1)(2n+1)}{6} = \dfrac{8(8+1)(16+1)}{6} = 204$	73, 74
Find finite differences of sequences (p. 472).	The first differences of a sequence are found by subtracting consecutive terms. The second differences are found by subtracting consecutive first differences.	75, 76
Section 9.5 Use the Binomial Theorem to calculate binomial coefficients (p. 473).	**The Binomial Theorem:** In the expansion of $(x+y)^n = x^n + nx^{n-1}y + \cdots + {}_nC_r x^{n-r}y^r + \cdots + nxy^{n-1} + y^n$, the coefficient of $x^{n-r}y^r$ is ${}_nC_r = \dfrac{n!}{(n-r)!r!}$.	77, 78
Use Pascal's Triangle to calculate binomial coefficients (p. 475).	First several rows of Pascal's Triangle: $\begin{matrix}&&&&1\\&&&1&&1\\&&1&&2&&1\\&1&&3&&3&&1\\1&&4&&6&&4&&1\end{matrix}$	79, 80
Use binomial coefficients to write binomial expansions (p. 476).	$(x+1)^3 = x^3 + 3x^2 + 3x + 1$ $(x-1)^4 = x^4 - 4x^3 + 6x^2 - 4x + 1$	81–84
Section 9.6 Solve simple counting problems (p. 478).	A computer randomly generates an integer from 1 through 15. The computer can generate an integer that is divisible by 3 in 5 ways (3, 6, 9, 12, and 15).	85, 86
Use the Fundamental Counting Principle to solve counting problems (p. 479).	**Fundamental Counting Principle:** Let E_1 and E_2 be two events. The first event E_1 can occur in m_1 different ways. After E_1 has occurred, E_2 can occur in m_2 different ways. The number of ways that the two events can occur is $m_1 \cdot m_2$.	87, 88
Use permutations (p. 480) and combinations (p. 483) to solve counting problems.	The number of permutations of n elements taken r at a time is ${}_nP_r = n!/(n-r)!$. The number of combinations of n elements taken r at a time is ${}_nC_r = n!/[(n-r)!r!]$, or ${}_nC_r = {}_nP_r/r!$.	89–92
Section 9.7 Find the probabilities of events (p. 485).	If an event E has $n(E)$ equally likely outcomes and its sample space S has $n(S)$ equally likely outcomes, then the probability of event E is $P(E) = n(E)/n(S)$.	93, 94
Find the probabilities of mutually exclusive events (p. 489) and independent events (p. 491), and find the probability of the complement of an event (p. 492).	If A and B are events in the same sample space, then the probability of A or B occurring is $P(A \cup B) = P(A) + P(B) - P(A \cap B)$. If A and B are mutually exclusive, then $P(A \cup B) = P(A) + P(B)$. If A and B are independent, then $P(A \text{ and } B) = P(A) \cdot P(B)$. The probability of the complement is $P(A') = 1 - P(A)$.	95–100

9.1 **Writing the Terms of a Sequence** In Exercises 1–4, write the first five terms of the sequence. (Assume that n begins with 1.)

1. $a_n = 2 + \dfrac{6}{n}$

2. $a_n = \dfrac{(-1)^n 5n}{2n - 1}$

3. $a_n = \dfrac{72}{n!}$

4. $a_n = n(n - 1)$

Finding the nth Term of a Sequence In Exercises 5–8, write an expression for the apparent nth term (a_n) of the sequence. (Assume that n begins with 1.)

5. $-2, 2, -2, 2, -2, \ldots$

6. $-1, 2, 7, 14, 23, \ldots$

7. $4, 2, \frac{4}{3}, 1, \frac{4}{5}, \ldots$

8. $1, -\frac{1}{2}, \frac{1}{3}, -\frac{1}{4}, \frac{1}{5}, \ldots$

Simplifying a Factorial Expression In Exercises 9–12, simplify the factorial expression.

9. $9!$

10. $4! \cdot 0!$

11. $\dfrac{3! \cdot 5!}{6!}$

12. $\dfrac{7! \cdot 6!}{6! \cdot 8!}$

Finding a Sum In Exercises 13 and 14, find the sum.

13. $\displaystyle\sum_{j=1}^{4} \dfrac{6}{j^2}$

14. $\displaystyle\sum_{k=1}^{10} 2k^3$

Using Sigma Notation to Write a Sum In Exercises 15 and 16, use sigma notation to write the sum.

15. $\dfrac{1}{2(1)} + \dfrac{1}{2(2)} + \dfrac{1}{2(3)} + \cdots + \dfrac{1}{2(20)}$

16. $\dfrac{1}{2} + \dfrac{2}{3} + \dfrac{3}{4} + \cdots + \dfrac{9}{10}$

Finding the Sum of an Infinite Series In Exercises 17 and 18, find the sum of the infinite series.

17. $\displaystyle\sum_{i=1}^{\infty} \dfrac{4}{10^i}$

18. $\displaystyle\sum_{k=1}^{\infty} 2\left(\dfrac{1}{100}\right)^k$

19. Compound Interest An investor deposits $10,000 in an account that earns 2.25% interest compounded monthly. The balance in the account after n months is given by

$$A_n = 10,000\left(1 + \dfrac{0.0225}{12}\right)^n, \quad n = 1, 2, 3, \ldots$$

(a) Write the first 10 terms of the sequence.

(b) Find the balance in the account after 10 years by computing the 120th term of the sequence.

20. Lottery Ticket Sales The total sales a_n (in billions of dollars) of lottery tickets in the United States from 2001 through 2010 can be approximated by the model

$$a_n = -0.18n^2 + 3.7n + 35, \quad n = 1, 2, \ldots, 10$$

where n is the year, with $n = 1$ corresponding to 2001. Write the terms of this finite sequence. Use a graphing utility to construct a bar graph that represents the sequence. *(Source: TLF Publications, Inc.)*

9.2 **Determining Whether a Sequence Is Arithmetic** In Exercises 21–24, determine whether the sequence is arithmetic. If so, then find the common difference.

21. $6, -1, -8, -15, -22, \ldots$

22. $0, 1, 3, 6, 10, \ldots$

23. $\frac{1}{2}, 1, \frac{3}{2}, 2, \frac{5}{2}, \ldots$

24. $1, \frac{15}{16}, \frac{7}{8}, \frac{13}{16}, \frac{3}{4}, \ldots$

Finding the nth Term In Exercises 25–28, find a formula for a_n for the arithmetic sequence.

25. $a_1 = 7, d = 12$

26. $a_1 = 28, d = -5$

27. $a_2 = 93, a_6 = 65$

28. $a_7 = 8, a_{13} = 6$

Writing the Terms of an Arithmetic Sequence In Exercises 29 and 30, write the first five terms of the arithmetic sequence.

29. $a_1 = 3, d = 11$

30. $a_1 = 25, a_{k+1} = a_k + 3$

31. Sum of a Finite Arithmetic Sequence Find the sum of the first 100 positive multiples of 7.

32. Sum of a Finite Arithmetic Sequence Find the sum of the integers from 40 to 90 (inclusive).

Finding a Partial Sum In Exercises 33–36, find the partial sum.

33. $\displaystyle\sum_{j=1}^{10} (2j - 3)$

34. $\displaystyle\sum_{j=1}^{8} (20 - 3j)$

35. $\displaystyle\sum_{k=1}^{11} \left(\dfrac{2}{3}k + 4\right)$

36. $\displaystyle\sum_{k=1}^{25} \left(\dfrac{3k + 1}{4}\right)$

37. Job Offer The starting salary for a job is $43,800 with a guaranteed increase of $1950 per year. Determine (a) the salary during the fifth year and (b) the total compensation through five full years of employment.

38. Baling Hay In the first two trips baling hay around a large field, a farmer obtains 123 bales and 112 bales, respectively. Because each round gets shorter, the farmer estimates that the same pattern will continue. Estimate the total number of bales made after the farmer takes another six trips around the field.

9.3 Determining Whether a Sequence Is Geometric   In Exercises 39–42, determine whether the sequence is geometric. If so, then find the common ratio.

39. 6, 12, 24, 48, . . .

40. 54, $-18, 6, -2, . . .$

41. $\frac{1}{5}, -\frac{3}{5}, \frac{9}{5}, -\frac{27}{5}, . . .$

42. $\frac{1}{4}, \frac{2}{5}, \frac{3}{6}, \frac{4}{7}, . . .$

Writing the Terms of a Geometric Sequence   In Exercises 43–46, write the first five terms of the geometric sequence.

43. $a_1 = 2, r = 15$

44. $a_1 = 4, r = -\frac{1}{4}$

45. $a_1 = 9, a_3 = 4$

46. $a_1 = 2, a_3 = 12$

Finding a Term of a Geometric Sequence   In Exercises 47–50, write an expression for the nth term of the geometric sequence. Then find the 10th term of the sequence.

47. $a_1 = 18, a_2 = -9$

48. $a_3 = 6, a_4 = 1$

49. $a_1 = 100, r = 1.05$

50. $a_1 = 5, r = 0.2$

Sum of a Finite Geometric Sequence   In Exercises 51–56, find the sum of the finite geometric sequence.

51. $\sum_{i=1}^{7} 2^{i-1}$

52. $\sum_{i=1}^{5} 3^{i-1}$

53. $\sum_{i=1}^{4} \left(\frac{1}{2}\right)^{i}$

54. $\sum_{i=1}^{6} \left(\frac{1}{3}\right)^{i-1}$

55. $\sum_{i=1}^{5} (2)^{i-1}$

56. $\sum_{i=1}^{4} 6(3)^{i}$

$\curvearrowright$ Sum of a Finite Geometric Sequence   In Exercises 57 and 58, use a graphing utility to find the sum of the finite geometric sequence.

57. $\sum_{i=1}^{10} 10\left(\frac{3}{5}\right)^{i-1}$

58. $\sum_{i=1}^{15} 20(0.2)^{i-1}$

Sum of an Infinite Geometric Sequence   In Exercises 59–62, find the sum of the infinite geometric series.

59. $\sum_{i=0}^{\infty} \left(\frac{7}{8}\right)^{i}$

60. $\sum_{i=0}^{\infty} (0.5)^{i}$

61. $\sum_{k=1}^{\infty} 4\left(\frac{2}{3}\right)^{k-1}$

62. $\sum_{k=1}^{\infty} 1.3\left(\frac{1}{10}\right)^{k-1}$

63. Depreciation   A paper manufacturer buys a machine for $120,000. During the next 5 years, it will depreciate at a rate of 30% per year. (In other words, at the end of each year the depreciated value will be 70% of what it was at the beginning of the year.)

(a) Find the formula for the nth term of a geometric sequence that gives the value of the machine t full years after it was purchased.

(b) Find the depreciated value of the machine after 5 full years.

64. Annuity   An investor deposits $800 in an account on the first day of each month for 10 years. The account pays 3%, compounded monthly. What will the balance be at the end of 10 years?

9.4 Using Mathematical Induction   In Exercises 65–68, use mathematical induction to prove the formula for every positive integer n.

65. $3 + 5 + 7 + \cdot\cdot\cdot + (2n + 1) = n(n + 2)$

66. $1 + \frac{3}{2} + 2 + \frac{5}{2} + \cdot\cdot\cdot + \frac{1}{2}(n + 1) = \frac{n}{4}(n + 3)$

67. $\sum_{i=0}^{n-1} ar^{i} = \frac{a(1 - r^{n})}{1 - r}$

68. $\sum_{k=0}^{n-1} (a + kd) = \frac{n}{2}[2a + (n - 1)d]$

Finding a Formula for a Sum   In Exercises 69–72, use mathematical induction to find a formula for the sum of the first n terms of the sequence.

69. 9, 13, 17, 21, . . .

70. 68, 60, 52, 44, . . .

71. $1, \frac{3}{5}, \frac{9}{25}, \frac{27}{125}, . . .$

72. $12, -1, \frac{1}{12}, -\frac{1}{144}, . . .$

Finding a Sum   In Exercises 73 and 74, find the sum using the formulas for the sums of powers of integers.

73. $\sum_{n=1}^{75} n$

74. $\sum_{n=1}^{6} (n^{5} - n^{2})$

Linear Model, Quadratic Model, or Neither?   In Exercises 75 and 76, write the first five terms of the sequence beginning with the given term. Then calculate the first and second differences of the sequence. State whether the sequence has a perfect linear model, a perfect quadratic model, or neither.

75. $a_1 = 5$
$a_n = a_{n-1} + 5$

76. $a_1 = -3$
$a_n = a_{n-1} - 2n$

9.5 Finding a Binomial Coefficient   In Exercises 77 and 78, find the binomial coefficient.

77. $_6C_4$

78. $_{12}C_3$

Using Pascal's Triangle   In Exercises 79 and 80, evaluate using Pascal's Triangle.

79. $\binom{7}{2}$

80. $\binom{10}{4}$

Expanding an Expression   In Exercises 81–84, use the Binomial Theorem to expand and simplify the expression. (Remember that $i = \sqrt{-1}$.)

81. $(x + 4)^{4}$

82. $(a - 3b)^{5}$

83. $(5 + 2i)^{4}$

84. $(4 - 5i)^{3}$

9.6

85. Numbers in a Hat You place slips of paper numbered 1 through 14 in a hat. In how many ways can you draw two numbers at random with replacement that total 12?

86. Shopping A customer in an electronics store can choose one of six speaker systems, one of five DVD players, and one of six flat screen televisions to design a home theater system. How many systems can the customer design?

87. Telephone Numbers All of the land line telephone numbers in a small town use the same three-digit prefix. How many different telephone numbers are possible by changing only the last four digits?

88. Course Schedule A college student is preparing a course schedule for the next semester. The student may select one of three mathematics courses, one of four science courses, and one of six history courses. How many schedules are possible?

89. Genetics A geneticist is using gel electrophoresis to analyze five DNA samples. The geneticist treats each sample with a different restriction enzyme and then injects it into one of five wells formed in a bed of gel. In how many orders can the geneticist inject the five samples into the wells?

90. Race There are 10 bicyclists entered in a race. In how many different ways could the top 3 places be decided?

91. Jury Selection A group of potential jurors has been narrowed down to 32 people. In how many ways can a jury of 12 people be selected?

92. Menu Choices A local sub shop offers five different breads, four different meats, three different cheeses, and six different vegetables. You can choose one bread and any number of the other items. Find the total number of combinations of sandwiches possible.

9.7

93. Apparel A man has five pairs of socks, of which no two pairs are the same color. He randomly selects two socks from a drawer. What is the probability that he gets a matched pair?

94. Bookshelf Order A child returns a five-volume set of books to a bookshelf. The child is not able to read, and so cannot distinguish one volume from another. What is the probability that the child shelves the books in the correct order?

95. Students by Class At a university, 31% of the students are freshmen, 26% are sophomores, 25% are juniors, and 18% are seniors. One student receives a cash scholarship randomly by lottery. What is the probability that the scholarship winner is

(a) a junior or senior?

(b) a freshman, sophomore, or junior?

96. Data Analysis Interviewers asked a sample of college students, faculty members, and administrators whether they favored a proposed increase in the annual activity fee to enhance student life on campus. The table lists the results.

	Students	Faculty	Admin.	Total
Favor	237	37	18	292
Oppose	163	38	7	208
Total	400	75	25	500

Find the probability that a person selected at random from the sample is as described.

(a) Not in favor of the proposal

(b) A student

(c) A faculty member in favor of the proposal

97. Tossing a Die You toss a six-sided die four times. What is the probability of getting a 5 on each roll?

98. Tossing a Die You toss a six-sided die six times. What is the probability of getting each number exactly once?

99. Drawing a Card You randomly draw a card from a standard deck of 52 playing cards. What is the probability that the card is not a club?

100. Tossing a Coin Find the probability of obtaining at least one tail when you toss a coin five times.

Exploration

True or False? In Exercises 101–104, determine whether the statement is true or false. Justify your answer.

101. $\dfrac{(n+2)!}{n!} = \dfrac{n+2}{n}$

102. $\displaystyle\sum_{i=1}^{5}(i^3 + 2i) = \sum_{i=1}^{5}i^3 + \sum_{i=1}^{5}2i$

103. $\displaystyle\sum_{k=1}^{8}3k = 3\sum_{k=1}^{8}k$

104. $\displaystyle\sum_{j=1}^{6}2^j = \sum_{j=3}^{8}2^{j-2}$

105. Think About It An infinite sequence is a function. What is the domain of the function?

106. Think About It How do the two sequences differ?

(a) $a_n = \dfrac{(-1)^n}{n}$ (b) $a_n = \dfrac{(-1)^{n+1}}{n}$

107. Writing Explain what is meant by a recursion formula.

108. Writing Write a brief paragraph explaining how to identify the graph of an arithmetic sequence and the graph of a geometric sequence.

Chapter Test

See **CalcChat.com** for tutorial help and worked-out solutions to odd-numbered exercises.

Take this test as you would take a test in class. When you are finished, check your work against the answers given in the back of the book.

1. Write the first five terms of the sequence $a_n = \dfrac{(-1)^n}{3n + 2}$. (Assume that n begins with 1.)

2. Write an expression for the apparent nth term (a_n) of the sequence.

    $$\frac{3}{1!}, \frac{4}{2!}, \frac{5}{3!}, \frac{6}{4!}, \frac{7}{5!}, \cdots$$

3. Write the next three terms of the series. Then find the sixth partial sum of the series.

    $8 + 21 + 34 + 47 + \cdots$

4. The fifth term of an arithmetic sequence is 5.4, and the 12th term is 11.0. Find the nth term.

5. The second term of a geometric sequence is 28, and the sixth term is 7168. Find the nth term.

6. Write the first five terms of the geometric sequence $a_n = 5(2)^{n-1}$. (Assume that n begins with 1.)

In Exercises 7–9, find the sum.

7. $\displaystyle\sum_{i=1}^{50} (2i^2 + 5)$

8. $\displaystyle\sum_{n=1}^{9} (12n - 7)$

9. $\displaystyle\sum_{i=1}^{\infty} 4\left(\tfrac{1}{2}\right)^i$

10. Use mathematical induction to prove the formula for every positive integer n.

    $$5 + 10 + 15 + \cdots + 5n = \frac{5n(n + 1)}{2}$$

11. Use the Binomial Theorem to expand and simplify (a) $(x + 6y)^4$ and (b) $3(x - 2)^5 + 4(x - 2)^3$.

12. Find the coefficient of the term $a^4 b^3$ in the expansion of $(3a - 2b)^7$.

In Exercises 13 and 14, evaluate each expression.

13. (a) $_9 P_2$ (b) $_{70} P_3$

14. (a) $_{11} C_4$ (b) $_{66} C_4$

15. How many distinct license plates can be issued consisting of one letter followed by a three-digit number?

16. Eight people are going for a ride in a boat that seats eight people. One person will drive, and only three of the remaining people are willing to ride in the two bow seats. How many seating arrangements are possible?

17. You attend a karaoke night and hope to hear your favorite song. The karaoke song book has 300 different songs (your favorite song is among them). Assuming that the singers are equally likely to pick any song and no song repeats, what is the probability that your favorite song is one of the 20 that you hear that night?

18. You are with three of your friends at a party. Names of all of the 30 guests are placed in a hat and drawn randomly to award four door prizes. Each guest can win only one prize. What is the probability that you and your friends win all four prizes?

19. The weather report calls for a 90% chance of snow. According to this report, what is the probability that it will *not* snow?

Cumulative Test for Chapters 7–9

Take this test as you would take a test in class. When you are finished, check your work against the answers given in the back of the book.

In Exercises 1–4, solve the system by the specified method.

1. Substitution
$$\begin{cases} y = 3 - x^2 \\ 2(y - 2) = x - 1 \end{cases}$$

2. Elimination
$$\begin{cases} x + 3y = -6 \\ 2x + 4y = -10 \end{cases}$$

3. Elimination
$$\begin{cases} -2x + 4y - z = -16 \\ x - 2y + 2z = 5 \\ x - 3y - z = 13 \end{cases}$$

4. Gauss-Jordan Elimination
$$\begin{cases} x + 3y - 2z = -7 \\ -2x + y - z = -5 \\ 4x + y + z = 3 \end{cases}$$

5. A custom-blend bird seed is to be mixed from seed mixtures costing $0.75 per pound and $1.25 per pound. How many pounds of each seed mixture are used to make 200 pounds of custom-blend bird seed costing $0.95 per pound?

6. Find the equation of the parabola $y = ax^2 + bx + c$ passing through the points $(0, 6)$, $(2, 3)$, and $(4, 2)$.

In Exercises 7 and 8, sketch the graph (and label the vertices) of the solution set of the system of inequalities.

7. $\begin{cases} 2x + y \geq -3 \\ x - 3y \leq 2 \end{cases}$

8. $\begin{cases} x - y > 6 \\ 5x + 2y < 10 \end{cases}$

9. Sketch the region determined by the constraints. Then find the minimum and maximum values of the objective function $z = 3x + 2y$ and where they occur, subject to the indicated constraints.

$$\begin{aligned} x + 4y &\leq 20 \\ 2x + y &\leq 12 \\ x &\geq 0 \\ y &\geq 0 \end{aligned}$$

$$\begin{cases} -x + 2y - z = 9 \\ 2x - y + 2z = -9 \\ 3x + 3y - 4z = 7 \end{cases}$$

System for 10 and 11

In Exercises 10 and 11, use the system of linear equations shown at the left.

10. Write the augmented matrix for the system.

11. Solve the system using the matrix found in Exercise 10 and Gauss-Jordan elimination.

In Exercises 12–17, perform the operations using the following matrices.

$$A = \begin{bmatrix} 3 & 0 \\ -1 & 4 \end{bmatrix}, \qquad B = \begin{bmatrix} -2 & 5 \\ 0 & -1 \end{bmatrix}$$

12. $A + B$

13. $-8B$

14. $2A - 5B$

15. AB

16. A^2

17. $BA - B^2$

$$\begin{bmatrix} 7 & 1 & 0 \\ -2 & 4 & -1 \\ 3 & 8 & 5 \end{bmatrix}$$

Matrix for 19

18. Find the inverse of the matrix (if it exists): $\begin{bmatrix} 1 & 2 & -1 \\ 3 & 7 & -10 \\ -5 & -7 & -15 \end{bmatrix}$.

19. Find the determinant of the matrix shown at the left.

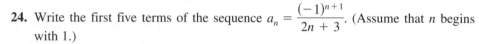

	Gym shoes	Jogging shoes	Walking shoes
14–17	0.079	0.064	0.029
18–24	0.050	0.060	0.022
25–34	0.103	0.259	0.085

Age group

Matrix for 20

20. The percents (in decimal form) of the total amounts spent on three types of footwear in a recent year are shown in the matrix at the left. The total amounts (in millions of dollars) spent by the age groups on the three types of footwear were $479.88 (14–17 age group), $365.88 (18–24 age group), and $1248.89 (25–34 age group). How many dollars worth of gym shoes, jogging shoes, and walking shoes were sold that year? *(Source: National Sporting Goods Association)*

In Exercises 21 and 22, use Cramer's Rule to solve the system of equations.

21. $\begin{cases} 8x - 3y = -52 \\ 3x + 5y = 5 \end{cases}$

22. $\begin{cases} 5x + 4y + 3z = 7 \\ -3x - 8y + 7z = -9 \\ 7x - 5y - 6z = -53 \end{cases}$

23. Use a determinant to find the area of the triangle shown at the left.

24. Write the first five terms of the sequence $a_n = \dfrac{(-1)^{n+1}}{2n + 3}$. (Assume that n begins with 1.)

25. Write an expression for the apparent nth term (a_n) of the sequence.

$$\frac{2!}{4}, \frac{3!}{5}, \frac{4!}{6}, \frac{5!}{7}, \frac{6!}{8}, \ldots$$

26. Find the 16th partial sum of the arithmetic sequence 6, 18, 30, 42, ...

27. The sixth term of an arithmetic sequence is 20.6, and the ninth term is 30.2.

    (a) Find the 20th term.

    (b) Find the nth term.

28. Write the first five terms of the sequence $a_n = 3(2)^{n-1}$. (Assume that n begins with 1.)

29. Find the sum: $\displaystyle\sum_{i=0}^{\infty} 1.3\left(\tfrac{1}{10}\right)^{i-1}$.

30. Use mathematical induction to prove the inequality

$$(n + 1)! > 2^n, \quad n \geq 2.$$

31. Use the Binomial Theorem to expand and simplify $(w - 9)^4$.

In Exercises 32–35, evaluate the expression.

32. $_{14}P_3$ **33.** $_{25}P_2$

34. $\dbinom{8}{4}$ **35.** $_{11}C_6$

In Exercises 36 and 37, find the number of distinguishable permutations of the group of letters.

36. B, A, S, K, E, T, B, A, L, L

37. A, N, T, A, R, C, T, I, C, A

38. A personnel manager at a department store has 10 applicants to fill three different sales positions. In how many ways can this be done, assuming that all the applicants are qualified for any of the three positions?

39. On a game show, contestants must arrange the digits 3, 4, and 5 in the proper order to form the price of an appliance. When the contestant arranges the digits correctly, he or she wins the appliance. What is the probability of winning when the contestant knows that the price is at least $400?

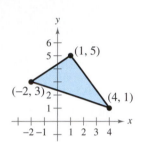

Figure for 23

Proofs in Mathematics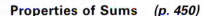

> **Properties of Sums**   *(p. 450)*
>
> **1.** $\displaystyle\sum_{i=1}^{n} c = cn,$ c is a constant.
>
> **2.** $\displaystyle\sum_{i=1}^{n} ca_i = c\sum_{i=1}^{n} a_i,$ c is a constant.
>
> **3.** $\displaystyle\sum_{i=1}^{n} (a_i + b_i) = \sum_{i=1}^{n} a_i + \sum_{i=1}^{n} b_i$
>
> **4.** $\displaystyle\sum_{i=1}^{n} (a_i - b_i) = \sum_{i=1}^{n} a_i - \sum_{i=1}^{n} b_i$

INFINITE SERIES

People considered the study of infinite series a novelty in the fourteenth century. Logician Richard Suiseth, whose nickname was Calculator, solved this problem.

If throughout the first half of a given time interval a variation continues at a certain intensity; throughout the next quarter of the interval at double the intensity; throughout the following eighth at triple the intensity and so ad infinitum; The average intensity for the whole interval will be the intensity of the variation during the second subinterval (or double the intensity).

This is the same as saying that the sum of the infinite series

$$\frac{1}{2} + \frac{2}{4} + \frac{3}{8} + \cdots$$

$$+ \frac{n}{2^n} + \cdots$$

is 2.

Proof

Each of these properties follows directly from the properties of real numbers.

1. $\displaystyle\sum_{i=1}^{n} c = c + c + c + \cdots + c = cn$ *n terms*

The proof of Property 2 uses the Distributive Property.

2. $\displaystyle\sum_{i=1}^{n} ca_i = ca_1 + ca_2 + ca_3 + \cdots + ca_n$

$$= c(a_1 + a_2 + a_3 + \cdots + a_n)$$

$$= c\sum_{i=1}^{n} a_i$$

The proof of Property 3 uses the Commutative and Associative Properties of Addition.

3. $\displaystyle\sum_{i=1}^{n} (a_i + b_i) = (a_1 + b_1) + (a_2 + b_2) + (a_3 + b_3) + \cdots + (a_n + b_n)$

$$= (a_1 + a_2 + a_3 + \cdots + a_n) + (b_1 + b_2 + b_3 + \cdots + b_n)$$

$$= \sum_{i=1}^{n} a_i + \sum_{i=1}^{n} b_i$$

The proof of Property 4 uses the Commutative and Associative Properties of Addition and the Distributive Property.

4. $\displaystyle\sum_{i=1}^{n} (a_i - b_i) = (a_1 - b_1) + (a_2 - b_2) + (a_3 - b_3) + \cdots + (a_n - b_n)$

$$= (a_1 + a_2 + a_3 + \cdots + a_n) + (-b_1 - b_2 - b_3 - \cdots - b_n)$$

$$= (a_1 + a_2 + a_3 + \cdots + a_n) - (b_1 + b_2 + b_3 + \cdots + b_n)$$

$$= \sum_{i=1}^{n} a_i - \sum_{i=1}^{n} b_i$$

The Sum of a Finite Arithmetic Sequence   (p. 456)

The sum of a finite arithmetic sequence with n terms is

$$S_n = \frac{n}{2}(a_1 + a_n).$$

Proof

Begin by generating the terms of the arithmetic sequence in two ways. In the first way, repeatedly add d to the first term to obtain

$$S_n = a_1 + a_2 + a_3 + \cdots + a_{n-2} + a_{n-1} + a_n$$
$$= a_1 + [a_1 + d] + [a_1 + 2d] + \cdots + [a_1 + (n-1)d].$$

In the second way, repeatedly subtract d from the nth term to obtain

$$S_n = a_n + a_{n-1} + a_{n-2} + \cdots + a_3 + a_2 + a_1$$
$$= a_n + [a_n - d] + [a_n - 2d] + \cdots + [a_n - (n-1)d].$$

If you add these two versions of S_n, then the multiples of d sum to zero and you obtain

$$2S_n = (a_1 + a_n) + (a_1 + a_n) + (a_1 + a_n) + \cdots + (a_1 + a_n) \qquad \textcolor{red}{n \text{ terms}}$$
$$2S_n = n(a_1 + a_n)$$
$$S_n = \frac{n}{2}(a_1 + a_n).$$

The Sum of a Finite Geometric Sequence   (p. 462)

The sum of the finite geometric sequence

$$a_1, \ a_1r, \ a_1r^2, \ a_1r^3, \ a_1r^4, \ \ldots, \ a_1r^{n-1}$$

with common ratio $r \neq 1$ is given by $S_n = \displaystyle\sum_{i=1}^{n} a_1 r^{i-1} = a_1\left(\dfrac{1 - r^n}{1 - r}\right)$.

Proof

$$S_n = a_1 + a_1r + a_1r^2 + \cdots + a_1r^{n-2} + a_1r^{n-1}$$
$$rS_n = a_1r + a_1r^2 + a_1r^3 + \cdots + a_1r^{n-1} + a_1r^n \qquad \textcolor{red}{\text{Multiply by } r.}$$

Subtracting the second equation from the first yields

$$S_n - rS_n = a_1 - a_1r^n.$$

So, $S_n(1 - r) = a_1(1 - r^n)$, and, because $r \neq 1$, you have $S_n = a_1\left(\dfrac{1 - r^n}{1 - r}\right)$.

The Binomial Theorem   *(p. 473)*

In the expansion of $(x + y)^n$

$$(x + y)^n = x^n + nx^{n-1}y + \cdots + {}_nC_r\, x^{n-r}y^r + \cdots + nxy^{n-1} + y^n$$

the coefficient of $x^{n-r}y^r$ is

$${}_nC_r = \frac{n!}{(n - r)!r!}.$$

Proof

The Binomial Theorem can be proved quite nicely using mathematical induction. The steps are straightforward but look a little messy, so only an outline of the proof follows.

1. For $n = 1$, you have $(x + y)^1 = x^1 + y^1 = {}_1C_0 x + {}_1C_1 y$, and the formula is valid.

2. Assuming that the formula is true for $n = k$, the coefficient of $x^{k-r}y^r$ is

$${}_kC_r = \frac{k!}{(k - r)!r!} = \frac{k(k - 1)(k - 2) \cdots (k - r + 1)}{r!}.$$

To show that the formula is true for $n = k + 1$, look at the coefficient of $x^{k+1-r}y^r$ in the expansion of

$$(x + y)^{k+1} = (x + y)^k(x + y).$$

On the right-hand side, the term involving $x^{k+1-r}y^r$ is the sum of two products.

$$({}_kC_r x^{k-r}y^r)(x) + ({}_kC_{r-1}x^{k+1-r}y^{r-1})(y)$$

$$= \left[\frac{k!}{(k - r)!r!} + \frac{k!}{(k + 1 - r)!(r - 1)!}\right]x^{k+1-r}y^r$$

$$= \left[\frac{(k + 1 - r)k!}{(k + 1 - r)!r!} + \frac{k!r}{(k + 1 - r)!r!}\right]x^{k+1-r}y^r$$

$$= \left[\frac{k!(k + 1 - r + r)}{(k + 1 - r)!r!}\right]x^{k+1-r}y^r$$

$$= \left[\frac{(k + 1)!}{(k + 1 - r)!r!}\right]x^{k+1-r}y^r$$

$$= {}_{k+1}C_r x^{k+1-r}y^r$$

So, by mathematical induction, the Binomial Theorem is valid for all positive integers n.

P.S. Problem Solving ▪ ▪ ▪ ▪ ▪ ▪ ▪ ▪ ▪ ▪ ▪ ▪ ▪ ▪ ▪

 1. Decreasing Sequence Consider the sequence

$$a_n = \frac{n+1}{n^2+1}.$$

(a) Use a graphing utility to graph the first 10 terms of the sequence.

(b) Use the graph from part (a) to estimate the value of a_n as n approaches infinity.

(c) Complete the table.

n	1	10	100	1000	10,000
a_n					

(d) Use the table from part (c) to determine (if possible) the value of a_n as n approaches infinity.

 2. Alternating Sequence Consider the sequence

$$a_n = 3 + (-1)^n.$$

(a) Use a graphing utility to graph the first 10 terms of the sequence.

(b) Use the graph from part (a) to describe the behavior of the graph of the sequence.

(c) Complete the table.

n	1	10	101	1000	10,001
a_n					

(d) Use the table from part (c) to determine (if possible) the value of a_n as n approaches infinity.

3. Greek Mythology Can the Greek hero Achilles, running at 20 feet per second, ever catch a tortoise, starting 20 feet ahead of Achilles and running at 10 feet per second? The Greek mathematician Zeno said no. When Achilles runs 20 feet, the tortoise will be 10 feet ahead. Then, when Achilles runs 10 feet, the tortoise will be 5 feet ahead. Achilles will keep cutting the distance in half but will never catch the tortoise. The table shows Zeno's reasoning. From the table, you can see that both the distances and the times required to achieve them form infinite geometric series. Using the table, show that both series have finite sums. What do these sums represent?

DATA	Distance (in feet)	Time (in seconds)
	20	1
	10	0.5
	5	0.25
	2.5	0.125
	1.25	0.0625
	0.625	0.03125

Spreadsheet at LarsonPrecalculus.com

 4. Conjecture Let $x_0 = 1$ and consider the sequence x_n given by

$$x_n = \frac{1}{2}x_{n-1} + \frac{1}{x_{n-1}}, \quad n = 1, 2, \ldots$$

Use a graphing utility to compute the first 10 terms of the sequence and make a conjecture about the value of x_n as n approaches infinity.

5. Operations on an Arithmetic Sequence The following operations are performed on each term of an arithmetic sequence. Determine whether the resulting sequence is arithmetic. If so, then state the common difference.

(a) A constant C is added to each term.

(b) Each term is multiplied by a nonzero constant C.

(c) Each term is squared.

6. Sequences of Powers The following sequence of perfect squares is not arithmetic.

1, 4, 9, 16, 25, 36, 49, 64, 81, . . .

However, you can form a related sequence that is arithmetic by finding the differences of consecutive terms.

(a) Write the first eight terms of the related arithmetic sequence described above. What is the nth term of this sequence?

(b) Describe how you can find an arithmetic sequence that is related to the following sequence of perfect cubes.

1, 8, 27, 64, 125, 216, 343, 512, 729, . . .

(c) Write the first seven terms of the related sequence in part (b) and find the nth term of the sequence.

(d) Describe how you can find the arithmetic sequence that is related to the following sequence of perfect fourth powers.

1, 16, 81, 256, 625, 1296, 2401, 4096, 6561, . . .

(e) Write the first six terms of the related sequence in part (d) and find the nth term of the sequence.

7. Piecewise-Defined Sequence You can define a sequence using a piecewise formula. The following is an example of a piecewise-defined sequence.

$$a_1 = 7, \quad a_n = \begin{cases} \dfrac{a_{n-1}}{2}, & \text{when } a_{n-1} \text{ is even.} \\ 3a_{n-1} + 1, & \text{when } a_{n-1} \text{ is odd.} \end{cases}$$

(a) Write the first 20 terms of the sequence.

(b) Write the first 10 terms of the sequences for which $a_1 = 4$, $a_1 = 5$, and $a_1 = 12$ (using a_n as defined above). What conclusion can you make about the behavior of each sequence?

8. Fibonacci Sequence Let $f_1, f_2, \ldots, f_n, \ldots$ be the Fibonacci sequence.

(a) Use mathematical induction to prove that

$$f_1 + f_2 + \cdots + f_n = f_{n+2} - 1.$$

(b) Find the sum of the first 20 terms of the Fibonacci sequence.

9. Pentagonal Numbers The numbers 1, 5, 12, 22, 35, 51, . . . are called pentagonal numbers because they represent the numbers of dots used to make pentagons, as shown below. Use mathematical induction to prove that the nth pentagonal number P_n is given by

$$P_n = \frac{n(3n - 1)}{2}.$$

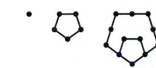 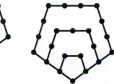

10. Think About It What conclusion can be drawn from the following information about the sequence of statements P_n?

(a) P_3 is true and P_k implies P_{k+1}.

(b) $P_1, P_2, P_3, \ldots, P_{50}$ are all true.

(c) $P_1, P_2,$ and P_3 are all true, but the truth of P_k does not imply that P_{k+1} is true.

(d) P_2 is true and P_{2k} implies P_{2k+2}.

11. Sierpinski Triangle Recall that a *fractal* is a geometric figure that consists of a pattern that is repeated infinitely on a smaller and smaller scale. A well-known fractal is called the *Sierpinski Triangle*. In the first stage, the midpoints of the three sides are used to create the vertices of a new triangle, which is then removed, leaving three triangles. The figure below shows the first two stages. Note that each remaining triangle is similar to the original triangle. Assume that the length of each side of the original triangle is one unit. Write a formula that describes the side length of the triangles generated in the nth stage. Write a formula for the area of the triangles generated in the nth stage.

12. Job Offer You work for a company that pays $0.01 the first day, $0.02 the second day, $0.04 the third day, and so on. If the daily wage keeps doubling, then what would your total income be for working 30 days?

13. Multiple Choice A multiple choice question has five possible answers. You know that the answer is not B or D, but you are not sure about answers A, C, and E. What is the probability that you will get the right answer when you take a guess?

14. Throwing a Dart You throw a dart at the circular target shown below. The dart is equally likely to hit any point inside the target. What is the probability that it hits the region outside the triangle?

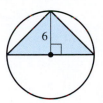

15. Odds The odds in favor of an event occurring are the ratio of the probability that the event will occur to the probability that the event will not occur. The reciprocal of this ratio represents the odds against the event occurring.

(a) Six of the marbles in a bag are red. The odds against choosing a red marble are 4 to 1. How many marbles are in the bag?

(b) A bag contains three blue marbles and seven yellow marbles. What are the odds in favor of choosing a blue marble? What are the odds against choosing a blue marble?

(c) Write a formula for converting the odds in favor of an event to the probability of the event.

(d) Write a formula for converting the probability of an event to the odds in favor of the event.

16. Expected Value An event A has n possible outcomes, which have the values $x_1, x_2, \ldots, x_n$. The probabilities of the n outcomes occurring are $p_1, p_2, \ldots, p_n$. The **expected value** V of an event A is the sum of the products of the outcomes' probabilities and their values,

$$V = p_1 x_1 + p_2 x_2 + \cdots + p_n x_n.$$

(a) To win California's Super Lotto Plus game, you must match five different numbers chosen from the numbers 1 to 47, plus one MEGA number chosen from the numbers 1 to 27. You purchase a ticket for $1. If the jackpot for the next drawing is $12,000,000, then what is the expected value of the ticket?

(b) You are playing a dice game in which you need to score 60 points to win. On each turn, you toss two six-sided dice. Your score for the turn is 0 when the dice do not show the same number. Your score for the turn is the product of the numbers on the dice when they do show the same number. What is the expected value of each turn? How many turns will it take on average to score 60 points?

10 Topics in Analytic Geometry

Satellite Orbit

Microphone Pickup Pattern

Nuclear Cooling Towers

Halley's Comet

Radio Telescopes

507

Clockwise from top left, nikkytok/Shutterstock.com; Cristi Matei/Shutterstock.com;
Digital Vision/Getty Images; John A Davis/Shutterstock.com; Malyshev Maksim/Shutterstock.com

You can use the inclination of a line to measure heights indirectly. For instance, the inclination of a line can be used to determine the change in elevation from the base to the top of the Falls Incline Railway in Niagara Falls, Ontario, Canada.

■ Find the inclination of a line.
■ Find the angle between two lines.
■ Find the distance between a point and a line.

Inclination of a Line

In Section 1.3, the slope of a line was described as the ratio of the change in y to the change in x. In this section, you will look at the slope of a line in terms of the angle of inclination of the line.

Every nonhorizontal line must intersect the x-axis. The angle formed by such an intersection determines the **inclination** of the line, as specified in the following definition.

Definition of Inclination

The **inclination** of a nonhorizontal line is the positive angle θ (less than π) measured counterclockwise from the x-axis to the line. (See figures below.)

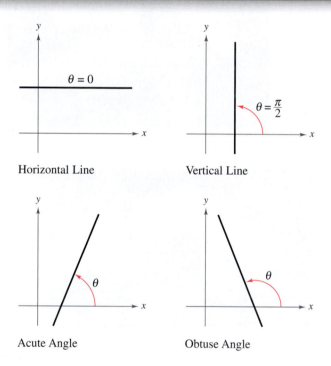

Horizontal Line

Vertical Line

Acute Angle

Obtuse Angle

The inclination of a line is related to its slope in the following manner.

Inclination and Slope

If a nonvertical line has inclination θ and slope m, then

$$m = \tan \theta.$$

For a proof of this relation between inclination and slope, see Proofs in Mathematics on page 562.

Note that if $m \geq 0$, then $\theta = \arctan m$ because $0 \leq \theta < \pi/2$. On the other hand, if $m < 0$, then $\theta = \pi + \arctan m$ because $\pi/2 < \theta < \pi$.

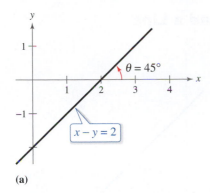

(a)

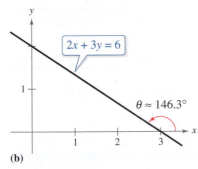

(b)

Figure 10.1

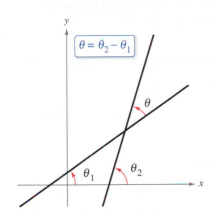

Figure 10.2

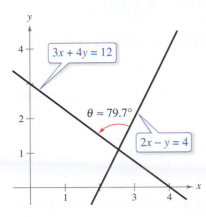

Figure 10.3

EXAMPLE 1   **Finding the Inclination of a Line**

Find the inclination of (a) $x - y = 2$ and (b) $2x + 3y = 6$.

Solution

a. The slope of this line is $m = 1$. So, its inclination is determined from $\tan \theta = 1$. Note that $m \geq 0$. This means that

$$\theta = \arctan 1 = \pi/4 \text{ radian} = 45°$$

as shown in Figure 10.1(a).

b. The slope of this line is $m = -\frac{2}{3}$. So, its inclination is determined from $\tan \theta = -\frac{2}{3}$. Note that $m < 0$. This means that

$$\theta = \pi + \arctan\left(-\frac{2}{3}\right) \approx \pi + (-0.5880) \approx 2.5536 \text{ radians} \approx 146.3°$$

as shown in Figure 10.1(b).

✓ **Checkpoint** 🔊))) *Audio-video solution in English & Spanish at LarsonPrecalculus.com.*

Find the inclination of $4x - 5y = 7$.

The Angle Between Two Lines

When two distinct lines intersect and are nonperpendicular, their intersection forms two pairs of opposite angles. One pair is acute and the other pair is obtuse. The smaller of these angles is the **angle between the two lines.** If two lines have inclinations θ_1 and θ_2, where $\theta_1 < \theta_2$ and $\theta_2 - \theta_1 < \pi/2$, then the angle between the two lines is $\theta = \theta_2 - \theta_1$, as shown in Figure 10.2. You can use the formula for the tangent of the difference of two angles

$$\tan \theta = \tan(\theta_2 - \theta_1) = \frac{\tan \theta_2 - \tan \theta_1}{1 + \tan \theta_1 \tan \theta_2}$$

to obtain the formula for the angle between two lines.

Angle Between Two Lines

If two nonperpendicular lines have slopes m_1 and m_2, then the tangent of the angle between the two lines is

$$\tan \theta = \left| \frac{m_2 - m_1}{1 + m_1 m_2} \right|.$$

EXAMPLE 2   **Finding the Angle Between Two Lines**

Find the angle between $2x - y = 4$ and $3x + 4y = 12$.

Solution   The two lines have slopes of $m_1 = 2$ and $m_2 = -\frac{3}{4}$, respectively. So, the tangent of the angle between the two lines is

$$\tan \theta = \left| \frac{m_2 - m_1}{1 + m_1 m_2} \right| = \left| \frac{(-3/4) - 2}{1 + (2)(-3/4)} \right| = \left| \frac{-11/4}{-2/4} \right| = \frac{11}{2}.$$

Finally, you can conclude that the angle is $\theta = \arctan \frac{11}{2} \approx 1.3909 \text{ radians} \approx 79.7°$ as shown in Figure 10.3.

✓ **Checkpoint** 🔊))) *Audio-video solution in English & Spanish at LarsonPrecalculus.com.*

Find the angle between $4x - 5y + 10 = 0$ and $3x + 2y + 5 = 0$.

The Distance Between a Point and a Line

Finding the distance between a line and a point not on the line is an application of perpendicular lines. This distance is defined as the length of the perpendicular line segment joining the point and the line, as shown at the right.

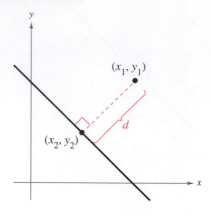

Distance Between a Point and a Line

The distance between the point (x_1, y_1) and the line $Ax + By + C = 0$ is

$$d = \frac{|Ax_1 + By_1 + C|}{\sqrt{A^2 + B^2}}.$$

Remember that the values of A, B, and C in this distance formula correspond to the general equation of a line, $Ax + By + C = 0$. For a proof of this formula for the distance between a point and a line, see Proofs in Mathematics on page 562.

EXAMPLE 3   **Finding the Distance Between a Point and a Line**

Find the distance between the point $(4, 1)$ and the line $y = 2x + 1$.

Solution   The general form of the equation is $-2x + y - 1 = 0$. So, the distance between the point and the line is

$$d = \frac{|-2(4) + 1(1) + (-1)|}{\sqrt{(-2)^2 + 1^2}} = \frac{8}{\sqrt{5}} \approx 3.58 \text{ units.}$$

The line and the point are shown in Figure 10.4.

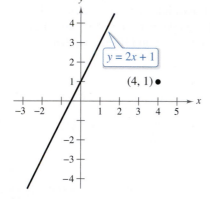

Figure 10.4

✓ **Checkpoint**  Audio-video solution in English & Spanish at LarsonPrecalculus.com.

Find the distance between the point $(5, -1)$ and the line $y = -3x + 2$.

EXAMPLE 4   **Finding the Distance Between a Point and a Line**

Find the distance between the point $(2, -1)$ and the line $-5x + 4y = 8$.

Solution   The general form of the equation is $-5x + 4y - 8 = 0$. So, the distance between the point and the line is

$$d = \frac{|-5(2) + 4(-1) + (-8)|}{\sqrt{(-5)^2 + 4^2}} = \frac{22}{\sqrt{41}} \approx 3.44 \text{ units.}$$

The line and the point are shown in Figure 10.5.

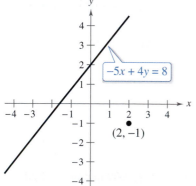

Figure 10.5

✓ **Checkpoint**  Audio-video solution in English & Spanish at LarsonPrecalculus.com.

Find the distance between the point $(3, 2)$ and the line $-3x + 5y = -2$.

EXAMPLE 5 **An Application of Two Distance Formulas**

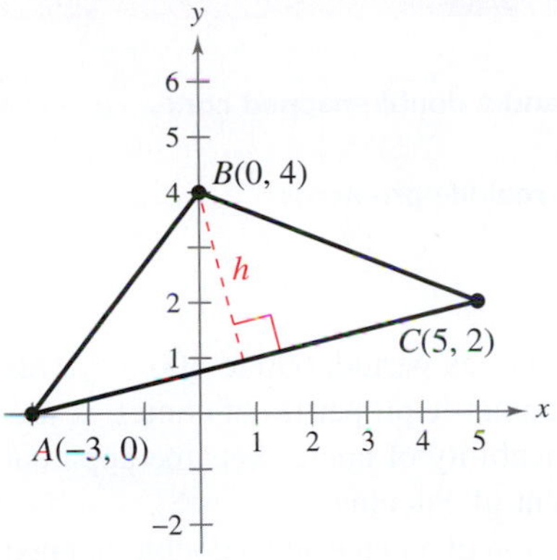

Figure 10.6 shows a triangle with vertices $A(-3, 0)$, $B(0, 4)$, and $C(5, 2)$.

a. Find the altitude h from vertex B to side AC.

b. Find the area of the triangle.

Solution

a. To find the altitude, use the formula for the distance between line AC and the point $(0, 4)$. The equation of line AC is obtained as follows.

$$\text{Slope:}\quad m = \frac{2 - 0}{5 - (-3)} = \frac{2}{8} = \frac{1}{4}$$

$$\text{Equation:}\qquad y - 0 = \frac{1}{4}(x + 3) \qquad \text{Point-slope form}$$

$$4y = x + 3 \qquad \text{Multiply each side by 4.}$$

$$x - 4y + 3 = 0 \qquad \text{General form}$$

So, the distance between this line and the point $(0, 4)$ is

$$\text{Altitude} = h = \frac{|1(0) + (-4)(4) + 3|}{\sqrt{1^2 + (-4)^2}} = \frac{13}{\sqrt{17}} \text{ units.}$$

b. Using the formula for the distance between two points, you can find the length of the base AC to be

$$b = \sqrt{[5 - (-3)]^2 + (2 - 0)^2} \qquad \text{Distance Formula}$$

$$= \sqrt{8^2 + 2^2} \qquad \text{Simplify.}$$

$$= 2\sqrt{17} \text{ units.} \qquad \text{Simplify.}$$

Finally, the area of the triangle in Figure 10.6 is

$$A = \frac{1}{2}bh \qquad \text{Formula for the area of a triangle}$$

$$= \frac{1}{2}(2\sqrt{17})\left(\frac{13}{\sqrt{17}}\right) \qquad \text{Substitute for } b \text{ and } h.$$

$$= 13 \text{ square units.} \qquad \text{Simplify.}$$

✓ **Checkpoint**  *Audio-video solution in English & Spanish at LarsonPrecalculus.com.*

A triangle has vertices $A(-2, 0)$, $B(0, 5)$, and $C(4, 3)$.

a. Find the altitude from vertex B to side AC.

b. Find the area of the triangle.

Figure 10.6

Summarize (Section 10.1)

1. State the definition of the inclination of a line *(page 508)*. For an example of finding the inclinations of lines, see Example 1.

2. Describe how to find the angle between two lines *(page 509)*. For an example of finding the angle between two lines, see Example 2.

3. Describe how to find the distance between a point and a line *(page 510)*. For examples of finding the distances between points and lines, see Examples 3–5.

10.2 Introduction to Conics: Parabolas

You can use parabolas to model and solve many types of real-life problems. For instance, a parabola can be used to model the cables of the Golden Gate Bridge.

■ Recognize a conic as the intersection of a plane and a double-napped cone.
■ Write equations of parabolas in standard form.
■ Use the reflective property of parabolas to solve real-life problems.

Conics

Conic sections were discovered during the classical Greek period, 600 to 300 B.C. This early Greek study was largely concerned with the geometric properties of conics. It was not until the early 17th century that the broad applicability of conics became apparent and played a prominent role in the early development of calculus.

A **conic section** (or simply **conic**) is the intersection of a plane and a double-napped cone. Notice in Figure 10.7 that in the formation of the four basic conics, the intersecting plane does not pass through the vertex of the cone. When the plane does pass through the vertex, the resulting figure is a **degenerate conic,** as shown in Figure 10.8.

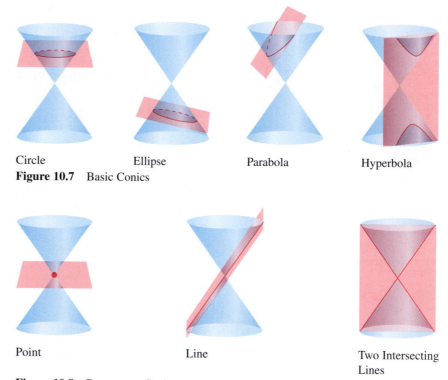

| Circle | Ellipse | Parabola | Hyperbola |

Figure 10.7 Basic Conics

| Point | Line | Two Intersecting Lines |

Figure 10.8 Degenerate Conics

There are several ways to approach the study of conics. You could begin by defining conics in terms of the intersections of planes and cones, as the Greeks did, or you could define them algebraically, in terms of the general second-degree equation

$$Ax^2 + Bxy + Cy^2 + Dx + Ey + F = 0.$$

However, you will study a third approach, in which each of the conics is defined as a **locus** (collection) of points satisfying a geometric property. For example, in Section 1.2, you saw how the definition of a circle as the set of all points (x, y) that are equidistant from a fixed point (h, k) led to the standard form of the equation of a circle

$$(x - h)^2 + (y - k)^2 = r^2. \qquad \text{Equation of circle}$$

Parabolas

In Section 2.1, you learned that the graph of the quadratic function

$$f(x) = ax^2 + bx + c$$

is a parabola that opens upward or downward. The following definition of a parabola is more general in the sense that it is independent of the orientation of the parabola.

Definition of Parabola

A **parabola** is the set of all points (x, y) in a plane that are equidistant from a fixed line, called the **directrix,** and a fixed point, called the **focus,** not on the line. (See figure.) The **vertex** is the midpoint between the focus and the directrix. The **axis** of the parabola is the line passing through the focus and the vertex.

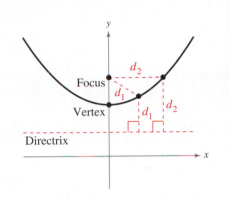

Note that a parabola is symmetric with respect to its axis. Using the definition of a parabola, you can derive the following **standard form of the equation of a parabola** whose directrix is parallel to the x-axis or to the y-axis.

Standard Equation of a Parabola

The **standard form of the equation of a parabola** with vertex at (h, k) is as follows.

$$(x - h)^2 = 4p(y - k), \; p \neq 0 \qquad \text{Vertical axis; directrix: } y = k - p$$

$$(y - k)^2 = 4p(x - h), \; p \neq 0 \qquad \text{Horizontal axis; directrix: } x = h - p$$

The focus lies on the axis p units (*directed distance*) from the vertex. If the vertex is at the origin, then the equation takes one of the following forms.

$$x^2 = 4py \qquad \text{Vertical axis}$$

$$y^2 = 4px \qquad \text{Horizontal axis}$$

See the figures below.

For a proof of the standard form of the equation of a parabola, see Proofs in Mathematics on page 562.

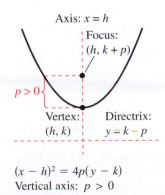

$(x - h)^2 = 4p(y - k)$
Vertical axis: $p > 0$

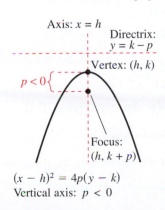

$(x - h)^2 = 4p(y - k)$
Vertical axis: $p < 0$

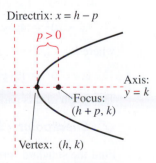

$(y - k)^2 = 4p(x - h)$
Horizontal axis: $p > 0$

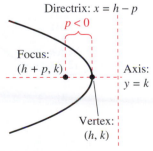

$(y - k)^2 = 4p(x - h)$
Horizontal axis: $p < 0$

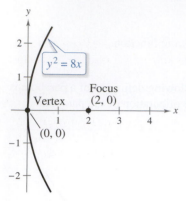

Figure 10.9

▷ **ALGEBRA HELP** The technique of completing the square is used to write the equation in Example 2 in standard form. You can review completing the square in Appendix A.5.

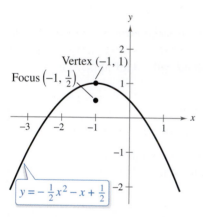

Figure 10.10

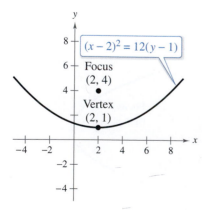

Figure 10.11

EXAMPLE 1   **Finding the Standard Equation of a Parabola**

Find the standard form of the equation of the parabola with vertex at the origin and focus $(2, 0)$.

Solution   The axis of the parabola is horizontal, passing through $(0, 0)$ and $(2, 0)$, as shown in Figure 10.9. The standard form is $y^2 = 4px$, where $p = 2$. So, the equation is $y^2 = 8x$. You can use a graphing utility to confirm this equation. To do this, let $y_1 = \sqrt{8x}$ to graph the upper portion and let $y_2 = -\sqrt{8x}$ to graph the lower portion of the parabola.

✓ *Checkpoint* ◀))) Audio-video solution in English & Spanish at LarsonPrecalculus.com.

Find the standard form of the equation of the parabola with vertex at the origin and focus $\left(0, \frac{3}{8}\right)$.

EXAMPLE 2   **Finding the Focus of a Parabola**

Find the focus of the parabola $y = -\frac{1}{2}x^2 - x + \frac{1}{2}$.

Solution   Convert to standard form by completing the square.

$$
\begin{aligned}
y &= -\tfrac{1}{2}x^2 - x + \tfrac{1}{2} & \text{Write original equation.}\\
-2y &= x^2 + 2x - 1 & \text{Multiply each side by } -2.\\
1 - 2y &= x^2 + 2x & \text{Add 1 to each side.}\\
1 + 1 - 2y &= x^2 + 2x + 1 & \text{Complete the square.}\\
2 - 2y &= x^2 + 2x + 1 & \text{Combine like terms.}\\
-2(y - 1) &= (x + 1)^2 & \text{Standard form}
\end{aligned}
$$

Comparing this equation with

$$(x - h)^2 = 4p(y - k)$$

you can conclude that $h = -1$, $k = 1$, and $p = -\frac{1}{2}$. Because p is negative, the parabola opens downward, as shown in Figure 10.10. So, the focus of the parabola is $(h, k + p) = \left(-1, \frac{1}{2}\right)$.

✓ *Checkpoint* ◀))) Audio-video solution in English & Spanish at LarsonPrecalculus.com.

Find the focus of the parabola $x = \frac{1}{4}y^2 + \frac{3}{2}y + \frac{13}{4}$.

EXAMPLE 3   **Finding the Standard Equation of a Parabola**

Find the standard form of the equation of the parabola with vertex $(2, 1)$ and focus $(2, 4)$.

Solution   Because the axis of the parabola is vertical, passing through $(2, 1)$ and $(2, 4)$, consider the equation

$$(x - h)^2 = 4p(y - k)$$

where $h = 2$, $k = 1$, and $p = 4 - 1 = 3$. So, the standard form is

$$(x - 2)^2 = 12(y - 1).$$

The graph of this parabola is shown in Figure 10.11.

✓ *Checkpoint* ◀))) Audio-video solution in English & Spanish at LarsonPrecalculus.com.

Find the standard form of the equation of the parabola with vertex $(2, -3)$ and focus $(4, -3)$.

One important application of parabolas is in astronomy. Radio telescopes use parabolic dishes to collect radio waves from space.

Application

A line segment that passes through the focus of a parabola and has endpoints on the parabola is called a **focal chord.** The specific focal chord perpendicular to the axis of the parabola is called the **latus rectum.**

Parabolas occur in a wide variety of applications. For instance, a parabolic reflector can be formed by revolving a parabola about its axis. The resulting surface has the property that all incoming rays parallel to the axis are reflected through the focus of the parabola. This is the principle behind the construction of the parabolic mirrors used in reflecting telescopes. Conversely, the light rays emanating from the focus of a parabolic reflector used in a flashlight are all parallel to one another, as shown below.

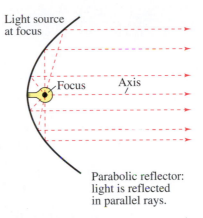

Parabolic reflector: light is reflected in parallel rays.

A line is **tangent** to a parabola at a point on the parabola when the line intersects, but does not cross, the parabola at the point. Tangent lines to parabolas have special properties related to the use of parabolas in constructing reflective surfaces.

Reflective Property of a Parabola

The tangent line to a parabola at a point P makes equal angles with the following two lines (see figure below).

1. The line passing through P and the focus
2. The axis of the parabola

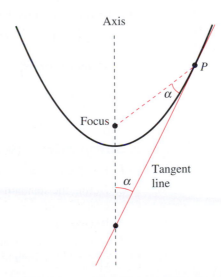

Finding the Tangent Line at a Point on a Parabola

Find the equation of the tangent line to the parabola $y = x^2$ at the point $(1, 1)$.

Solution   For this parabola, $p = \frac{1}{4}$ and the focus is $\left(0, \frac{1}{4}\right)$, as shown in the figure below.

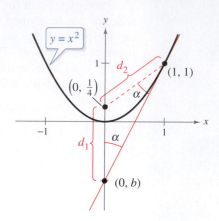

You can find the y-intercept $(0, b)$ of the tangent line by equating the lengths of the two sides of the isosceles triangle shown in the figure:

$$d_1 = \tfrac{1}{4} - b$$

and

$$d_2 = \sqrt{(1 - 0)^2 + \left(1 - \tfrac{1}{4}\right)^2}$$
$$= \tfrac{5}{4}.$$

> **TECHNOLOGY** Use a graphing utility to confirm the result of Example 4. By graphing
>
> $$y_1 = x^2 \quad \text{and} \quad y_2 = 2x - 1$$
>
> in the same viewing window, you should be able to see that the line touches the parabola at the point $(1, 1)$.

Note that $d_1 = \frac{1}{4} - b$ rather than $b - \frac{1}{4}$. The order of subtraction for the distance is important because the distance must be positive. Setting $d_1 = d_2$ produces

$$\tfrac{1}{4} - b = \tfrac{5}{4}$$
$$b = -1.$$

So, the slope of the tangent line is

$$m = \frac{1 - (-1)}{1 - 0}$$
$$= 2$$

> **ALGEBRA HELP** You can review techniques for writing linear equations in Section 1.3.

and the equation of the tangent line in slope-intercept form is

$$y = 2x - 1.$$

✓ **Checkpoint** ◀))) *Audio-video solution in English & Spanish at LarsonPrecalculus.com.*

Find the equation of the tangent line to the parabola $y = 3x^2$ at the point $(1, 3)$.   ∎

Summarize   (Section 10.2)

1. List the four basic conic sections *(page 512)*.
2. Define the standard form of the equation of a parabola *(page 513)*. For examples of writing the equations of parabolas in standard form, see Examples 1–3.
3. Describe an application involving the reflective property of parabolas *(page 516, Example 4)*.

10.3 Ellipses

Ellipses have many real-life applications. For instance, the focal properties of an ellipse are used by a lithotripter machine to break up kidney stones.

- Write equations of ellipses in standard form and graph ellipses.
- Use properties of ellipses to model and solve real-life problems.
- Find eccentricities of ellipses.

Introduction

The second type of conic is called an **ellipse.** It is defined as follows.

> **Definition of Ellipse**
>
> An **ellipse** is the set of all points (x, y) in a plane, the sum of whose distances from two distinct fixed points, called **foci,** is constant. See Figure 10.12.

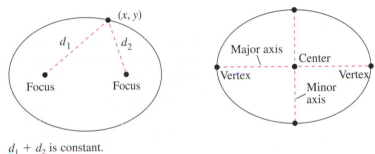

$d_1 + d_2$ is constant.

Figure 10.12

Figure 10.13

The line through the foci intersects the ellipse at two points called **vertices.** The chord joining the vertices is the **major axis,** and its midpoint is the **center** of the ellipse. The chord perpendicular to the major axis at the center is the **minor axis** of the ellipse. See Figure 10.13.

You can visualize the definition of an ellipse by imagining two thumbtacks placed at the foci, as shown below. If the ends of a fixed length of string are fastened to the thumbtacks and the string is drawn taut with a pencil, then the path traced by the pencil will be an ellipse.

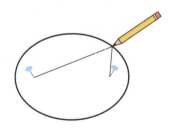

To derive the standard form of the equation of an ellipse, consider the ellipse in Figure 10.14 with the following points.

Center: (h, k) Vertices: $(h \pm a, k)$ Foci: $(h \pm c, k)$

Note that the center is the midpoint of the segment joining the foci.

The sum of the distances from any point on the ellipse to the two foci is constant. Using a vertex point, this constant sum is

$$(a + c) + (a - c) = 2a \qquad \text{Length of major axis}$$

or simply the length of the major axis.

$$2\sqrt{b^2 + c^2} = 2a$$
$$b^2 + c^2 = a^2$$

Figure 10.14

Now, if you let (x, y) be *any* point on the ellipse, then the sum of the distances between (x, y) and the two foci must also be $2a$. That is,

$$\sqrt{[x - (h - c)]^2 + (y - k)^2} + \sqrt{[x - (h + c)]^2 + (y - k)^2} = 2a$$

which, after expanding and regrouping, reduces to

$$(a^2 - c^2)(x - h)^2 + a^2(y - k)^2 = a^2(a^2 - c^2).$$

Finally, in Figure 10.14, you can see that

$$b^2 = a^2 - c^2$$

which implies that the equation of the ellipse is

$$b^2(x - h)^2 + a^2(y - k)^2 = a^2 b^2$$

$$\frac{(x - h)^2}{a^2} + \frac{(y - k)^2}{b^2} = 1.$$

You would obtain a similar equation in the derivation by starting with a vertical major axis. Both results are summarized as follows.

> **•• REMARK** Consider the equation of the ellipse
>
> $$\frac{(x - h)^2}{a^2} + \frac{(y - k)^2}{b^2} = 1.$$
>
> If you let $a = b$, then the equation can be rewritten as
>
> $$(x - h)^2 + (y - k)^2 = a^2$$
>
> which is the standard form of the equation of a circle with radius $r = a$ (see Section 1.2). Geometrically, when $a = b$ for an ellipse, the major and minor axes are of equal length, and so the graph is a circle.

Standard Equation of an Ellipse

The **standard form of the equation of an ellipse** with center (h, k) and major and minor axes of lengths $2a$ and $2b$, respectively, where $0 < b < a$, is

$$\frac{(x - h)^2}{a^2} + \frac{(y - k)^2}{b^2} = 1 \qquad \text{Major axis is horizontal.}$$

$$\frac{(x - h)^2}{b^2} + \frac{(y - k)^2}{a^2} = 1. \qquad \text{Major axis is vertical.}$$

The foci lie on the major axis, c units from the center, with

$$c^2 = a^2 - b^2.$$

If the center is at the origin, then the equation takes one of the following forms.

$$\frac{x^2}{a^2} + \frac{y^2}{b^2} = 1 \qquad \text{Major axis is horizontal.}$$

$$\frac{x^2}{b^2} + \frac{y^2}{a^2} = 1 \qquad \text{Major axis is vertical.}$$

Both the horizontal and vertical orientations for an ellipse are shown below.

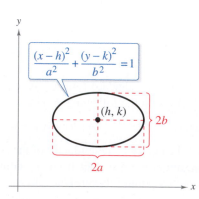

Major axis is horizontal.

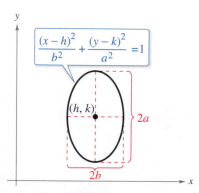

Major axis is vertical.

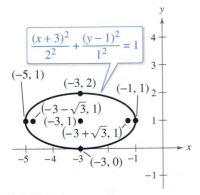

Figure 10.15

EXAMPLE 1 **Finding the Standard Equation of an Ellipse**

Find the standard form of the equation of the ellipse having foci $(0, 1)$ and $(4, 1)$ and a major axis of length 6, as shown in Figure 10.15.

Solution Because the foci occur at $(0, 1)$ and $(4, 1)$, the center of the ellipse is $(2, 1)$ and the distance from the center to one of the foci is $c = 2$. Because $2a = 6$, you know that $a = 3$. Now, from $c^2 = a^2 - b^2$, you have

$$b = \sqrt{a^2 - c^2} = \sqrt{3^2 - 2^2} = \sqrt{5}.$$

Because the major axis is horizontal, the standard equation is

$$\frac{(x - 2)^2}{3^2} + \frac{(y - 1)^2}{(\sqrt{5})^2} = 1.$$

This equation simplifies to

$$\frac{(x - 2)^2}{9} + \frac{(y - 1)^2}{5} = 1.$$

✓ **Checkpoint** 🔊 *Audio-video solution in English & Spanish at LarsonPrecalculus.com.*

Find the standard form of the equation of the ellipse having foci $(2, 0)$ and $(2, 6)$ and a major axis of length 8.

EXAMPLE 2 **Sketching an Ellipse**

Find the center, vertices, and foci of the ellipse $x^2 + 4y^2 + 6x - 8y + 9 = 0$. Then sketch the ellipse.

Solution Begin by writing the original equation in standard form. In the fourth step, note that 9 and 4 are added to *both* sides of the equation when completing the squares.

$$x^2 + 4y^2 + 6x - 8y + 9 = 0 \qquad \text{Write original equation.}$$

$$\left(x^2 + 6x + \boxed{}\right) + \left(4y^2 - 8y + \boxed{}\right) = -9 \qquad \text{Group terms.}$$

$$\left(x^2 + 6x + \boxed{}\right) + 4\left(y^2 - 2y + \boxed{}\right) = -9 \qquad \text{Factor 4 out of } y\text{-terms.}$$

$$(x^2 + 6x + 9) + 4(y^2 - 2y + 1) = -9 + 9 + 4(1)$$

$$(x + 3)^2 + 4(y - 1)^2 = 4 \qquad \text{Write in completed square form.}$$

$$\frac{(x + 3)^2}{4} + \frac{(y - 1)^2}{1} = 1 \qquad \text{Divide each side by 4.}$$

$$\frac{(x + 3)^2}{2^2} + \frac{(y - 1)^2}{1^2} = 1 \qquad \text{Write in standard form.}$$

From this standard form, it follows that the center is $(h, k) = (-3, 1)$. Because the denominator of the x-term is $a^2 = 2^2$, the endpoints of the major axis lie two units to the right and left of the center. So, the vertices are $(-5, 1)$ and $(-1, 1)$. Similarly, because the denominator of the y-term is $b^2 = 1^2$, the endpoints of the minor axis lie one unit up and down from the center. Now, from $c^2 = a^2 - b^2$, you have $c = \sqrt{2^2 - 1^2} = \sqrt{3}$. So, the foci of the ellipse are $\left(-3 - \sqrt{3}, 1\right)$ and $\left(-3 + \sqrt{3}, 1\right)$. The ellipse is shown in Figure 10.16.

Figure 10.16

✓ **Checkpoint** 🔊 *Audio-video solution in English & Spanish at LarsonPrecalculus.com.*

Find the center, vertices, and foci of the ellipse $9x^2 + 4y^2 + 36x - 8y + 4 = 0$. Then sketch the ellipse.

EXAMPLE 3   **Sketching an Ellipse**

Find the center, vertices, and foci of the ellipse $4x^2 + y^2 - 8x + 4y - 8 = 0$. Then sketch the ellipse.

Solution   By completing the square, you can write the original equation in standard form.

$$4x^2 + y^2 - 8x + 4y - 8 = 0 \qquad \text{Write original equation.}$$

$$\left(4x^2 - 8x + \quad\right) + \left(y^2 + 4y + \quad\right) = 8 \qquad \text{Group terms.}$$

$$4\left(x^2 - 2x + \quad\right) + \left(y^2 + 4y + \quad\right) = 8 \qquad \text{Factor 4 out of } x\text{-terms.}$$

$$4(x^2 - 2x + 1) + (y^2 + 4y + 4) = 8 + 4(1) + 4$$

$$4(x - 1)^2 + (y + 2)^2 = 16 \qquad \text{Write in completed square form.}$$

$$\frac{(x - 1)^2}{4} + \frac{(y + 2)^2}{16} = 1 \qquad \text{Divide each side by 16.}$$

$$\frac{(x - 1)^2}{2^2} + \frac{(y + 2)^2}{4^2} = 1 \qquad \text{Write in standard form.}$$

The major axis is vertical, where $h = 1$, $k = -2$, $a = 4$, $b = 2$, and

$$c = \sqrt{a^2 - b^2} = \sqrt{16 - 4} = \sqrt{12} = 2\sqrt{3}.$$

So, you have the following.

Center: $(1, -2)$ Vertices: $(1, -6)$ Foci: $\left(1, -2 - 2\sqrt{3}\right)$

$\qquad\qquad\qquad\qquad\qquad\quad (1, 2)$ $\qquad\qquad \left(1, -2 + 2\sqrt{3}\right)$

The ellipse is shown in Figure 10.17.

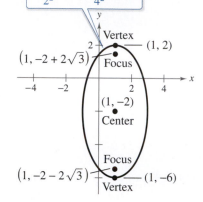

$$\frac{(x - 1)^2}{2^2} + \frac{(y + 2)^2}{4^2} = 1$$

Figure 10.17

✓ **Checkpoint** 🔊))) *Audio-video solution in English & Spanish at LarsonPrecalculus.com.*

Find the center, vertices, and foci of the ellipse $5x^2 + 9y^2 + 10x - 54y + 41 = 0$. Then sketch the ellipse. ◼

▷ **TECHNOLOGY** You can use a graphing utility to graph an ellipse by graphing the upper and lower portions in the same viewing window. For instance, to graph the ellipse in Example 3, first solve for y to get

$$y_1 = -2 + 4\sqrt{1 - \frac{(x - 1)^2}{4}}$$

and

$$y_2 = -2 - 4\sqrt{1 - \frac{(x - 1)^2}{4}}.$$

Use a viewing window in which $-6 \le x \le 9$ and $-7 \le y \le 3$. You should obtain the graph shown below.

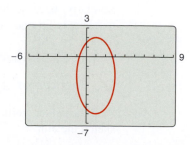

Application

Ellipses have many practical and aesthetic uses. For instance, machine gears, supporting arches, and acoustic designs often involve elliptical shapes. The orbits of satellites and planets are also ellipses. Example 4 investigates the elliptical orbit of the moon about Earth.

Halley's comet has an elliptical orbit about the sun and is visible from Earth approximately every 75 years. The comet's latest appearance was in 1986.

▷

REMARK Note in Example 4 and in the figure that Earth is *not* the center of the moon's orbit.

EXAMPLE 4 An Application Involving an Elliptical Orbit

The moon travels about Earth in an elliptical orbit with Earth at one focus, as shown below. The major and minor axes of the orbit have lengths of 768,800 kilometers and 767,640 kilometers, respectively. Find the greatest (*apogee*) and least (*perigee*) distances from Earth's center to the moon's center.

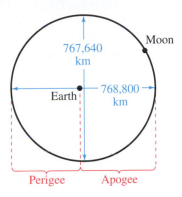

Solution   Because

$$2a = 768,800 \quad \text{and} \quad 2b = 767,640$$

you have

$$a = 384,400 \quad \text{and} \quad b = 383,820$$

which implies that

$$c = \sqrt{a^2 - b^2}$$
$$= \sqrt{384,400^2 - 383,820^2}$$
$$\approx 21,108.$$

So, the greatest distance between the center of Earth and the center of the moon is

$$a + c \approx 384,400 + 21,108$$
$$= 405,508 \text{ kilometers}$$

and the least distance is

$$a - c \approx 384,400 - 21,108$$
$$= 363,292 \text{ kilometers}.$$

✓ *Checkpoint* *Audio-video solution in English & Spanish at LarsonPrecalculus.com.*

Encke's comet travels about the sun in an elliptical orbit with the sun at one focus. The major and minor axes of the orbit have lengths of approximately 4.429 astronomical units and 2.345 astronomical units, respectively. (An astronomical unit is about 93 million miles.) Find the greatest (*aphelion*) and least (*perihelion*) distances from the sun's center to the comet's center.

Eccentricity

One of the reasons it was difficult for early astronomers to detect that the orbits of the planets are ellipses is that the foci of the planetary orbits are relatively close to their centers, and so the orbits are nearly circular. To measure the ovalness of an ellipse, you can use the concept of **eccentricity.**

Definition of Eccentricity

The eccentricity e of an ellipse is given by the ratio $e = \dfrac{c}{a}$.

Note that $0 < e < 1$ for *every* ellipse.

To see how this ratio is used to describe the shape of an ellipse, note that because the foci of an ellipse are located along the major axis between the vertices and the center, it follows that

$$0 < c < a.$$

For an ellipse that is nearly circular, the foci are close to the center and the ratio c/a is close to 0, as shown in Figure 10.18. On the other hand, for an elongated ellipse, the foci are close to the vertices and the ratio c/a is close to 1, as shown in Figure 10.19.

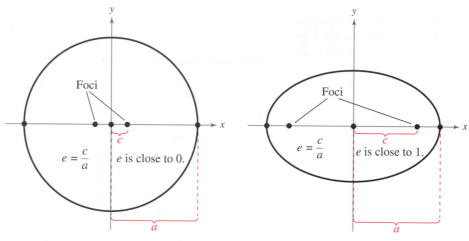

Figure 10.18   **Figure 10.19**

The orbit of the moon has an eccentricity of $e \approx 0.0549$, and the eccentricities of the eight planetary orbits are as follows.

Mercury:	$e \approx 0.2056$	Jupiter:	$e \approx 0.0484$
Venus:	$e \approx 0.0068$	Saturn:	$e \approx 0.0542$
Earth:	$e \approx 0.0167$	Uranus:	$e \approx 0.0472$
Mars:	$e \approx 0.0934$	Neptune:	$e \approx 0.0086$

The time it takes Saturn to orbit the sun is about 29.4 Earth years.

Summarize   (Section 10.3)

1. State the definition of an ellipse *(page 517)*. For examples involving the equations and graphs of ellipses, see Examples 1–3.

2. Describe a real-life application of an ellipse *(page 520, Example 4)*.

3. State the definition of the eccentricity of an ellipse *(page 521)*.

10.4   Hyperbolas

You can use hyperbolas to model and solve many types of real-life problems, such as long distance radio navigation for aircraft and ships.

- ■ Write equations of hyperbolas in standard form.
- ■ Find asymptotes of and graph hyperbolas.
- ■ Use properties of hyperbolas to solve real-life problems.
- ■ Classify conics from their general equations.

Introduction

The definition of a **hyperbola** is similar to that of an ellipse. For an ellipse, the *sum* of the distances between the foci and a point on the ellipse is fixed. For a hyperbola, the absolute value of the *difference* of the distances between the foci and a point on the hyperbola is fixed.

Definition of Hyperbola

A **hyperbola** is the set of all points (x, y) in a plane for which the absolute value of the difference of the distances from two distinct fixed points, called **foci,** is constant. See Figure 10.20.

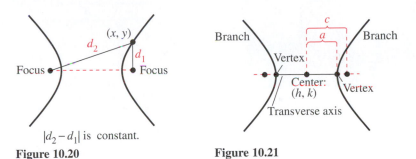

$|d_2 - d_1|$ is constant.
Figure 10.20

Figure 10.21

The graph of a hyperbola has two disconnected parts called **branches.** The line through the two foci intersects the hyperbola at two points called the **vertices.** The line segment connecting the vertices is the **transverse axis,** and the midpoint of the transverse axis is the **center** of the hyperbola.

Consider the hyperbola in Figure 10.21 with the following points.

Center: (h, k) Vertices: $(h \pm a, k)$ Foci: $(h \pm c, k)$

Note that the center is the midpoint of the segment joining the foci.

The absolute value of the difference of the distances from *any* point on the hyperbola to the two foci is constant. Using a vertex, this constant value is

$$|[2a + (c - a)] - (c - a)| = |2a|$$

$$= 2a \qquad \text{Length of transverse axis}$$

or simply the length of the transverse axis. Now, if you let (x, y) be *any* point on the hyperbola, then

$$|d_2 - d_1| = 2a.$$

You would obtain the same result for a hyperbola with a vertical transverse axis.

The development of the standard form of the equation of a hyperbola is similar to that of an ellipse. Note in the definition on the next page that a, b, and c are related differently for hyperbolas than for ellipses.

Standard Equation of a Hyperbola

The **standard form of the equation of a hyperbola** with center (h, k) is

$$\frac{(x-h)^2}{a^2} - \frac{(y-k)^2}{b^2} = 1 \qquad \text{Transverse axis is horizontal.}$$

$$\frac{(y-k)^2}{a^2} - \frac{(x-h)^2}{b^2} = 1. \qquad \text{Transverse axis is vertical.}$$

The vertices are a units from the center, and the foci are c units from the center. Moreover, $c^2 = a^2 + b^2$. If the center of the hyperbola is at the origin, then the equation takes one of the following forms.

$$\frac{x^2}{a^2} - \frac{y^2}{b^2} = 1 \qquad \text{Transverse axis is horizontal.}$$

$$\frac{y^2}{a^2} - \frac{x^2}{b^2} = 1 \qquad \text{Transverse axis is vertical.}$$

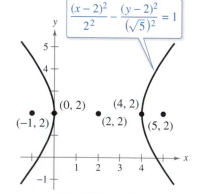

Nuclear cooling towers such as the one shown above are hyperboloids. This means that they have hyperbolic vertical cross sections.

Both the horizontal and vertical orientations for a hyperbola are shown below.

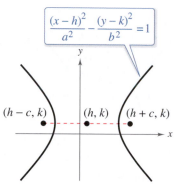

Transverse axis is horizontal.

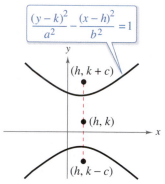

Transverse axis is vertical.

EXAMPLE 1 **Finding the Standard Equation of a Hyperbola**

Find the standard form of the equation of the hyperbola with foci $(-1, 2)$ and $(5, 2)$ and vertices $(0, 2)$ and $(4, 2)$.

Solution By the Midpoint Formula, the center of the hyperbola occurs at the point $(2, 2)$. Furthermore, $c = 5 - 2 = 3$ and $a = 4 - 2 = 2$, and it follows that

$$b = \sqrt{c^2 - a^2} = \sqrt{3^2 - 2^2} = \sqrt{9 - 4} = \sqrt{5}.$$

So, the hyperbola has a horizontal transverse axis and the standard form of the equation is

$$\frac{(x-2)^2}{2^2} - \frac{(y-2)^2}{(\sqrt{5})^2} = 1. \qquad \text{See Figure 10.22.}$$

This equation simplifies to

$$\frac{(x-2)^2}{4} - \frac{(y-2)^2}{5} = 1.$$

Figure 10.22

✓ **Checkpoint** 🔊)) *Audio-video solution in English & Spanish at LarsonPrecalculus.com.*

Find the standard form of the equation of the hyperbola with foci $(2, -5)$ and $(2, 3)$ and vertices $(2, -4)$ and $(2, 2)$.

Malyshev Maksim/Shutterstock.com

Asymptotes of a Hyperbola

Each hyperbola has two **asymptotes** that intersect at the center of the hyperbola, as shown in Figure 10.23. The asymptotes pass through the vertices of a rectangle of dimensions $2a$ by $2b$, with its center at (h, k). The line segment of length $2b$ joining $(h, k + b)$ and $(h, k - b)$ [or $(h + b, k)$ and $(h - b, k)$] is the **conjugate axis** of the hyperbola.

Conjugate axis

$(h, k + b)$

Asymptote

(h, k)

$(h, k - b)$

Asymptote

$(h - a, k)$ $(h + a, k)$

Figure 10.23

Asymptotes of a Hyperbola

The equations of the asymptotes of a hyperbola are

$$y = k \pm \frac{b}{a}(x - h) \qquad \text{Transverse axis is horizontal.}$$

$$y = k \pm \frac{a}{b}(x - h). \qquad \text{Transverse axis is vertical.}$$

EXAMPLE 2   **Using Asymptotes to Sketch a Hyperbola**

Sketch the hyperbola $4x^2 - y^2 = 16$.

Algebraic Solution

Divide each side of the original equation by 16, and write the equation in standard form.

$$\frac{x^2}{2^2} - \frac{y^2}{4^2} = 1 \qquad \text{Write in standard form.}$$

From this, you can conclude that $a = 2$, $b = 4$, and the transverse axis is horizontal. So, the vertices occur at $(-2, 0)$ and $(2, 0)$, and the endpoints of the conjugate axis occur at $(0, -4)$ and $(0, 4)$. Using these four points, you are able to sketch the rectangle shown in Figure 10.24. Now, from $c^2 = a^2 + b^2$, you have $c = \sqrt{2^2 + 4^2} = \sqrt{20} = 2\sqrt{5}$. So, the foci of the hyperbola are $\left(-2\sqrt{5}, 0\right)$ and $\left(2\sqrt{5}, 0\right)$. Finally, by drawing the asymptotes through the corners of this rectangle, you can complete the sketch shown in Figure 10.25. Note that the asymptotes are $y = 2x$ and $y = -2x$.

Graphical Solution

Solve the equation of the hyperbola for y, as follows.

$$4x^2 - y^2 = 16$$
$$4x^2 - 16 = y^2$$
$$\pm\sqrt{4x^2 - 16} = y$$

Then use a graphing utility to graph

$$y_1 = \sqrt{4x^2 - 16} \quad \text{and} \quad y_2 = -\sqrt{4x^2 - 16}$$

in the same viewing window, as shown below. Be sure to use a square setting.

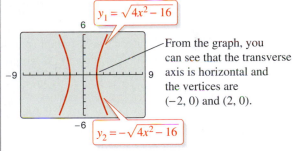

$y_1 = \sqrt{4x^2 - 16}$

From the graph, you can see that the transverse axis is horizontal and the vertices are $(-2, 0)$ and $(2, 0)$.

$y_2 = -\sqrt{4x^2 - 16}$

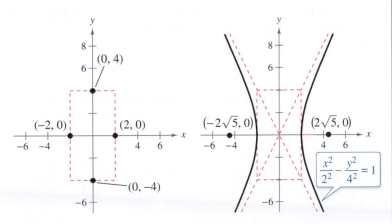

Figure 10.24   **Figure 10.25**

✔ **Checkpoint** 🔊 *Audio-video solution in English & Spanish at LarsonPrecalculus.com.*

Sketch the hyperbola $4y^2 - 9x^2 = 36$.

EXAMPLE 3   **Finding the Asymptotes of a Hyperbola**

Sketch the hyperbola $4x^2 - 3y^2 + 8x + 16 = 0$, find the equations of its asymptotes, and find the foci.

Solution

$4x^2 - 3y^2 + 8x + 16 = 0$	Write original equation.
$(4x^2 + 8x) - 3y^2 = -16$	Group terms.
$4(x^2 + 2x) - 3y^2 = -16$	Factor 4 out of x-terms.
$4(x^2 + 2x + 1) - 3y^2 = -16 + 4(1)$	Complete the square.
$4(x + 1)^2 - 3y^2 = -12$	Write in completed square form.
$-\dfrac{(x + 1)^2}{3} + \dfrac{y^2}{4} = 1$	Divide each side by -12.
$\dfrac{y^2}{2^2} - \dfrac{(x + 1)^2}{(\sqrt{3})^2} = 1$	Write in standard form.

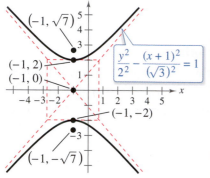

$$\frac{y^2}{2^2} - \frac{(x+1)^2}{(\sqrt{3})^2} = 1$$

Figure 10.26

From this equation you can conclude that the hyperbola has a vertical transverse axis, is centered at $(-1, 0)$, has vertices $(-1, 2)$ and $(-1, -2)$, and has a conjugate axis with endpoints $\left(-1 - \sqrt{3}, 0\right)$ and $\left(-1 + \sqrt{3}, 0\right)$. To sketch the hyperbola, draw a rectangle through these four points. The asymptotes are the lines passing through the corners of the rectangle. Using $a = 2$ and $b = \sqrt{3}$, you can conclude that the equations of the asymptotes are

$$y = \frac{2}{\sqrt{3}}(x + 1) \quad \text{and} \quad y = -\frac{2}{\sqrt{3}}(x + 1).$$

Finally, you can determine the foci by using the equation $c^2 = a^2 + b^2$. So, you have $c = \sqrt{2^2 + \left(\sqrt{3}\right)^2} = \sqrt{7}$, and the foci are $\left(-1, \sqrt{7}\right)$ and $\left(-1, -\sqrt{7}\right)$. The hyperbola is shown in Figure 10.26.

✔ **Checkpoint** ◀))) Audio-video solution in English & Spanish at LarsonPrecalculus.com.

Sketch the hyperbola $9x^2 - 4y^2 + 8y - 40 = 0$, find the equations of its asymptotes, and find the foci.

▷ **TECHNOLOGY** You can use a graphing utility to graph a hyperbola by graphing the upper and lower portions in the same viewing window. For instance, to graph the hyperbola in Example 3, first solve for y to get

$$y_1 = 2\sqrt{1 + \frac{(x + 1)^2}{3}} \quad \text{and} \quad y_2 = -2\sqrt{1 + \frac{(x + 1)^2}{3}}.$$

Use a viewing window in which $-9 \le x \le 9$ and $-6 \le y \le 6$. You should obtain the graph shown below. Notice that the graphing utility does not draw the asymptotes. However, when you trace along the branches, you can see that the values of the hyperbola approach the asymptotes.

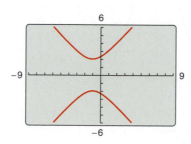

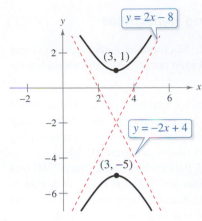

Figure 10.27

EXAMPLE 4   **Using Asymptotes to Find the Standard Equation**

Find the standard form of the equation of the hyperbola having vertices $(3, -5)$ and $(3, 1)$ and having asymptotes

$$y = 2x - 8 \quad \text{and} \quad y = -2x + 4$$

as shown in Figure 10.27.

Solution   By the Midpoint Formula, the center of the hyperbola is $(3, -2)$. Furthermore, the hyperbola has a vertical transverse axis with $a = 3$. From the original equations, you can determine the slopes of the asymptotes to be

$$m_1 = 2 = \frac{a}{b} \quad \text{and} \quad m_2 = -2 = -\frac{a}{b}$$

and, because $a = 3$, you can conclude

$$2 = \frac{a}{b} \quad \Longrightarrow \quad 2 = \frac{3}{b} \quad \Longrightarrow \quad b = \frac{3}{2}.$$

So, the standard form of the equation of the hyperbola is

$$\frac{(y + 2)^2}{3^2} - \frac{(x - 3)^2}{\left(\frac{3}{2}\right)^2} = 1.$$

✓ **Checkpoint** 🔊)) *Audio-video solution in English & Spanish at LarsonPrecalculus.com.*

Find the standard form of the equation of the hyperbola having vertices $(3, 2)$ and $(9, 2)$ and having asymptotes

$$y = -2 + \frac{2}{3}x \quad \text{and} \quad y = 6 - \frac{2}{3}x.$$

As with ellipses, the *eccentricity* of a hyperbola is

$$e = \frac{c}{a} \qquad \text{Eccentricity}$$

and because $c > a$, it follows that $e > 1$. When the eccentricity is large, the branches of the hyperbola are nearly flat, as shown in Figure 10.28. When the eccentricity is close to 1, the branches of the hyperbola are more narrow, as shown in Figure 10.29.

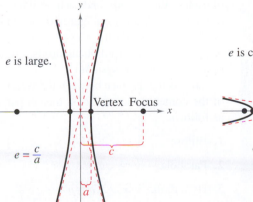

Figure 10.28

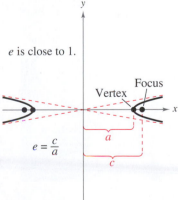

Figure 10.29

Chapter 10

Applications

EXAMPLE 5 An Application Involving Hyperbolas

Two microphones, 1 mile apart, record an explosion. Microphone A receives the sound 2 seconds before microphone B. Where did the explosion occur? (Assume sound travels at 1100 feet per second.)

Solution Begin by representing the situation in a coordinate plane. The distance between the microphones is 1 mile, or 5280 feet. So, position the point representing microphone A 2640 units to the right of the origin and the point representing microphone B 2640 units to the left of the origin, as shown in Figure 10.30.

Assuming sound travels at 1100 feet per second, the explosion took place 2200 feet farther from B than from A. The locus of all points that are 2200 feet closer to A than to B is one branch of a hyperbola with foci at A and B. Because the hyperbola is centered at the origin and has a horizontal transverse axis, the standard form of its equation is

$$\frac{x^2}{a^2} - \frac{y^2}{b^2} = 1.$$

Because the foci are 2640 units from the center, $c = 2640$. Let d_A and d_B be the distances of any point on the hyperbola from the foci at A and B, respectively. From page 523, you have

$$|d_B - d_A| = 2a$$

$$|2200| = 2a \qquad \text{The points are 2200 feet closer to A than to B.}$$

$$1100 = a. \qquad \text{Divide each side by 2.}$$

So, $b^2 = c^2 - a^2 = 2640^2 - 1100^2 = 5,759,600$, and you can conclude that the explosion occurred somewhere on the right branch of the hyperbola

$$\frac{x^2}{1,210,000} - \frac{y^2}{5,759,600} = 1.$$

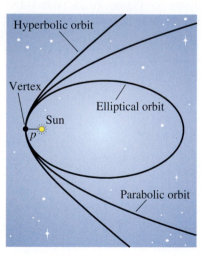

Figure 10.30

✓ **Checkpoint** 🔊)) Audio-video solution in English & Spanish at LarsonPrecalculus.com.

Repeat Example 5 when microphone A receives the sound 4 seconds before microphone B.

Another interesting application of conic sections involves the orbits of comets in our solar system. Of the 610 comets identified prior to 1970, 245 have elliptical orbits, 295 have parabolic orbits, and 70 have hyperbolic orbits. The center of the sun is a focus of each of these orbits, and each orbit has a vertex at the point where the comet is closest to the sun, as shown in Figure 10.31. Undoubtedly, there have been many comets with parabolic or hyperbolic orbits that were not identified. We only get to see such comets *once*. Comets with elliptical orbits, such as Halley's comet, are the only ones that remain in our solar system.

If p is the distance between the vertex and the focus (in meters), and v is the velocity of the comet at the vertex (in meters per second), then the type of orbit is determined as follows.

1. Ellipse: $v < \sqrt{2GM/p}$

2. Parabola: $v = \sqrt{2GM/p}$

3. Hyperbola: $v > \sqrt{2GM/p}$

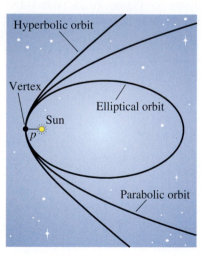

Figure 10.31

In each of these relations, $M = 1.989 \times 10^{30}$ kilograms (the mass of the sun) and $G \approx 6.67 \times 10^{-11}$ cubic meter per kilogram-second squared (the universal gravitational constant).

General Equations of Conics

> ### Classifying a Conic from Its General Equation
>
> The graph of $Ax^2 + Cy^2 + Dx + Ey + F = 0$ is one of the following.
>
> **1.** *Circle:* $A = C$ $A \neq 0$
>
> **2.** *Parabola:* $AC = 0$ $A = 0$ or $C = 0$, but not both.
>
> **3.** *Ellipse:* $AC > 0$ A and C have like signs.
>
> **4.** *Hyperbola:* $AC < 0$ A and C have unlike signs.

The test above is valid when the graph is a conic. The test does not apply to equations such as $x^2 + y^2 = -1$, whose graph is not a conic.

EXAMPLE 6   **Classifying Conics from General Equations**

a. For the equation $4x^2 - 9x + y - 5 = 0$, you have

$$AC = 4(0) = 0. \qquad \text{Parabola}$$

So, the graph is a parabola.

b. For the equation $4x^2 - y^2 + 8x - 6y + 4 = 0$, you have

$$AC = 4(-1) < 0. \qquad \text{Hyperbola}$$

So, the graph is a hyperbola.

c. For the equation $2x^2 + 4y^2 - 4x + 12y = 0$, you have

$$AC = 2(4) > 0. \qquad \text{Ellipse}$$

So, the graph is an ellipse.

d. For the equation $2x^2 + 2y^2 - 8x + 12y + 2 = 0$, you have

$$A = C = 2. \qquad \text{Circle}$$

So, the graph is a circle.

✓ *Checkpoint*  *Audio-video solution in English & Spanish at LarsonPrecalculus.com.*

Classify the graph of each equation.

a. $3x^2 + 3y^2 - 6x + 6y + 5 = 0$ **b.** $2x^2 - 4y^2 + 4x + 8y - 3 = 0$

c. $3x^2 + y^2 + 6x - 2y + 3 = 0$ **d.** $2x^2 + 4x + y - 2 = 0$

Caroline Herschel (1750–1848) was the first woman to be credited with detecting a new comet. During her long life, this English astronomer discovered a total of eight new comets.

> ## Summarize   (Section 10.4)
>
> **1.** State the definition of a hyperbola *(page 523)*. For an example of finding the standard form of the equation of a hyperbola, see Example 1.
>
> **2.** Write the equations of the asymptotes of hyperbolas with horizontal and vertical transverse axes *(page 525)*. For examples involving the asymptotes of hyperbolas, see Examples 2–4.
>
> **3.** Describe a real-life application of a hyperbola *(page 528, Example 5)*.
>
> **4.** Describe how to classify a conic from its general equation *(page 529)*. For an example of classifying conics, see Example 6.

10.5 Rotation of Conics

Rotated conics can help you model and solve real-life problems. For instance, a rotated parabola can be used to model the cross section of a satellite dish.

■ Rotate the coordinate axes to eliminate the *xy*-term in equations of conics.
■ Use the discriminant to classify conics.

Rotation

The equation of a conic with axes parallel to one of the coordinate axes has a standard form that can be written in the general form

$$Ax^2 + Cy^2 + Dx + Ey + F = 0. \qquad \text{\color{red}Horizontal or vertical axis}$$

Conics whose axes are rotated so that they are not parallel to either the *x*-axis or the *y*-axis have general equations that contain an *xy*-term.

$$Ax^2 + Bxy + Cy^2 + Dx + Ey + F = 0 \qquad \text{\color{red}Equation in } xy\text{-plane}$$

To eliminate this *xy*-term, you can use a procedure called **rotation of axes.** The objective is to rotate the *x*- and *y*-axes until they are parallel to the axes of the conic. The rotated axes are denoted as the *x′*-axis and the *y′*-axis, as shown in Figure 10.32. After the rotation, the equation of the conic in the new *x′y′*-plane will have the form

$$A'(x')^2 + C'(y')^2 + D'x' + E'y' + F' = 0. \qquad \text{\color{red}Equation in } x'y'\text{-plane}$$

Because this equation has no *x′y′*-term, you can obtain a standard form by completing the square. The following theorem identifies how much to rotate the axes to eliminate the *xy*-term and also the equations for determining the new coefficients *A′*, *C′*, *D′*, *E′*, and *F′*.

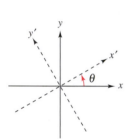

Figure 10.32

Rotation of Axes to Eliminate an *xy*-Term

The general second-degree equation

$$Ax^2 + Bxy + Cy^2 + Dx + Ey + F = 0$$

can be rewritten as

$$A'(x')^2 + C'(y')^2 + D'x' + E'y' + F' = 0$$

by rotating the coordinate axes through an angle θ, where

$$\cot 2\theta = \frac{A - C}{B}.$$

The coefficients of the new equation are obtained by making the substitutions

$$x = x' \cos \theta - y' \sin \theta$$

and

$$y = x' \sin \theta + y' \cos \theta.$$

Remember that the substitutions

$$x = x' \cos \theta - y' \sin \theta$$

and

$$y = x' \sin \theta + y' \cos \theta$$

were developed to eliminate the *x′y′*-term in the rotated system. You can use this as a check of your work. In other words, when your final equation contains an *x′y′*-term, you know that you have made a mistake.

EXAMPLE 1   **Rotation of Axes for a Hyperbola**

Rotate the axes to eliminate the xy-term in the equation $xy - 1 = 0$. Then write the equation in standard form and sketch its graph.

Solution   Because $A = 0$, $B = 1$, and $C = 0$, you have

$$\cot 2\theta = \frac{A - C}{B} = 0 \quad \Longrightarrow \quad 2\theta = \frac{\pi}{2} \quad \Longrightarrow \quad \theta = \frac{\pi}{4}$$

which implies that

$$x = x' \cos \frac{\pi}{4} - y' \sin \frac{\pi}{4}$$

$$= x'\left(\frac{1}{\sqrt{2}}\right) - y'\left(\frac{1}{\sqrt{2}}\right)$$

$$= \frac{x' - y'}{\sqrt{2}}$$

and

$$y = x' \sin \frac{\pi}{4} + y' \cos \frac{\pi}{4}$$

$$= x'\left(\frac{1}{\sqrt{2}}\right) + y'\left(\frac{1}{\sqrt{2}}\right)$$

$$= \frac{x' + y'}{\sqrt{2}}.$$

The equation in the $x'y'$-system is obtained by substituting these expressions in the original equation.

$$xy - 1 = 0$$

$$\left(\frac{x' - y'}{\sqrt{2}}\right)\left(\frac{x' + y'}{\sqrt{2}}\right) - 1 = 0$$

$$\frac{(x')^2 - (y')^2}{2} - 1 = 0$$

$$\frac{(x')^2}{(\sqrt{2})^2} - \frac{(y')^2}{(\sqrt{2})^2} = 1 \qquad \textcolor{red}{\text{Write in standard form.}}$$

In the $x'y'$-system, this is a hyperbola centered at the origin with vertices at $\left(\pm\sqrt{2}, 0\right)$, as shown in Figure 10.33. To find the coordinates of the vertices in the xy-system, substitute the coordinates $\left(\pm\sqrt{2}, 0\right)$ in the equations

$$x = \frac{x' - y'}{\sqrt{2}} \quad \text{and} \quad y = \frac{x' + y'}{\sqrt{2}}.$$

This substitution yields the vertices $(1, 1)$ and $(-1, -1)$ in the xy-system. Note also that the asymptotes of the hyperbola have equations

$$y' = \pm x'$$

which correspond to the original x- and y-axes.

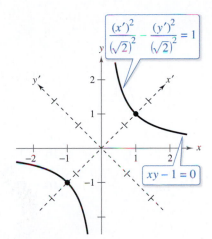

$$\frac{(x')^2}{(\sqrt{2})^2} - \frac{(y')^2}{(\sqrt{2})^2} = 1$$

$$xy - 1 = 0$$

Vertices:
In $x'y'$-system: $\left(\sqrt{2}, 0\right), \left(-\sqrt{2}, 0\right)$
In xy-system: $(1, 1), (-1, -1)$
Figure 10.33

✓ **Checkpoint**   *Audio-video solution in English & Spanish at LarsonPrecalculus.com.*

Rotate the axes to eliminate the xy-term in the equation $xy + 6 = 0$. Then write the equation in standard form and sketch its graph.

> **EXAMPLE 2** **Rotation of Axes for an Ellipse**

Rotate the axes to eliminate the xy-term in the equation

$$7x^2 - 6\sqrt{3}xy + 13y^2 - 16 = 0.$$

Then write the equation in standard form and sketch its graph.

Solution Because $A = 7$, $B = -6\sqrt{3}$, and $C = 13$, you have

$$\cot 2\theta = \frac{A - C}{B} = \frac{7 - 13}{-6\sqrt{3}} = \frac{1}{\sqrt{3}}$$

which implies that $\theta = \pi/6$. The equation in the $x'y'$-system is obtained by making the substitutions

$$x = x' \cos \frac{\pi}{6} - y' \sin \frac{\pi}{6}$$

$$= x'\left(\frac{\sqrt{3}}{2}\right) - y'\left(\frac{1}{2}\right)$$

$$= \frac{\sqrt{3}x' - y'}{2}$$

and

$$y = x' \sin \frac{\pi}{6} + y' \cos \frac{\pi}{6}$$

$$= x'\left(\frac{1}{2}\right) + y'\left(\frac{\sqrt{3}}{2}\right)$$

$$= \frac{x' + \sqrt{3}y'}{2}$$

in the original equation. So, you have

$$7x^2 - 6\sqrt{3}xy + 13y^2 - 16 = 0$$

$$7\left(\frac{\sqrt{3}x' - y'}{2}\right)^2 - 6\sqrt{3}\left(\frac{\sqrt{3}x' - y'}{2}\right)\left(\frac{x' + \sqrt{3}y'}{2}\right) + 13\left(\frac{x' + \sqrt{3}y'}{2}\right)^2 - 16 = 0$$

which simplifies to

$$4(x')^2 + 16(y')^2 - 16 = 0$$

$$4(x')^2 + 16(y')^2 = 16$$

$$\frac{(x')^2}{4} + \frac{(y')^2}{1} = 1$$

$$\frac{(x')^2}{2^2} + \frac{(y')^2}{1^2} = 1. \qquad \text{Write in standard form.}$$

This is the equation of an ellipse centered at the origin with vertices $(\pm 2, 0)$ in the $x'y'$-system, as shown in Figure 10.34.

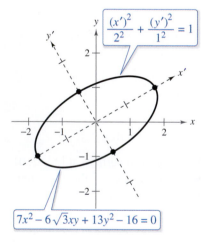

$$\frac{(x')^2}{2^2} + \frac{(y')^2}{1^2} = 1$$

$$7x^2 - 6\sqrt{3}xy + 13y^2 - 16 = 0$$

Vertices:
In $x'y'$-system: $(\pm 2, 0)$
In xy-system: $\left(\sqrt{3}, 1\right), \left(-\sqrt{3}, -1\right)$
Figure 10.34

✓ **Checkpoint** ◀))) *Audio-video solution in English & Spanish at LarsonPrecalculus.com.*

Rotate the axes to eliminate the xy-term in the equation

$$12x^2 + 16\sqrt{3}xy + 28y^2 - 36 = 0.$$

Then write the equation in standard form and sketch its graph.

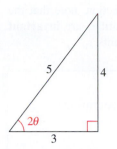

Figure 10.35

| EXAMPLE 3 | **Rotation of Axes for a Parabola** |

Rotate the axes to eliminate the xy-term in the equation $x^2 - 4xy + 4y^2 + 5\sqrt{5}y + 1 = 0$.
Then write the equation in standard form and sketch its graph.

Solution   Because $A = 1$, $B = -4$, and $C = 4$, you have

$$\cot 2\theta = \frac{A - C}{B} = \frac{1 - 4}{-4} = \frac{3}{4}.$$

Using this information, draw a right triangle as shown in Figure 10.35. From the figure,
you can see that $\cos 2\theta = \frac{3}{5}$. To find the values of $\sin \theta$ and $\cos \theta$, you can use the
half-angle formulas in the forms

$$\sin \theta = \sqrt{\frac{1 - \cos 2\theta}{2}} \quad \text{and} \quad \cos \theta = \sqrt{\frac{1 + \cos 2\theta}{2}}.$$

So,

$$\sin \theta = \sqrt{\frac{1 - \cos 2\theta}{2}} = \sqrt{\frac{1 - \frac{3}{5}}{2}} = \sqrt{\frac{1}{5}} = \frac{1}{\sqrt{5}}$$

$$\cos \theta = \sqrt{\frac{1 + \cos 2\theta}{2}} = \sqrt{\frac{1 + \frac{3}{5}}{2}} = \sqrt{\frac{4}{5}} = \frac{2}{\sqrt{5}}.$$

Consequently, you use the substitutions

$$x = x' \cos \theta - y' \sin \theta = x'\left(\frac{2}{\sqrt{5}}\right) - y'\left(\frac{1}{\sqrt{5}}\right) = \frac{2x' - y'}{\sqrt{5}}$$

and

$$y = x' \sin \theta + y' \cos \theta = x'\left(\frac{1}{\sqrt{5}}\right) + y'\left(\frac{2}{\sqrt{5}}\right) = \frac{x' + 2y'}{\sqrt{5}}.$$

Substituting these expressions in the original equation, you have

$$x^2 - 4xy + 4y^2 + 5\sqrt{5}y + 1 = 0$$

$$\left(\frac{2x' - y'}{\sqrt{5}}\right)^2 - 4\left(\frac{2x' - y'}{\sqrt{5}}\right)\left(\frac{x' + 2y'}{\sqrt{5}}\right) + 4\left(\frac{x' + 2y'}{\sqrt{5}}\right)^2 + 5\sqrt{5}\left(\frac{x' + 2y'}{\sqrt{5}}\right) + 1 = 0$$

which simplifies as follows.

$$5(y')^2 + 5x' + 10y' + 1 = 0$$

$$5[(y')^2 + 2y'] = -5x' - 1 \qquad \text{\color{red}Group terms.}$$

$$5[(y')^2 + 2y' + 1] = -5x' - 1 + 5(1) \qquad \text{\color{red}Complete the square.}$$

$$5(y' + 1)^2 = -5x' + 4 \qquad \text{\color{red}Write in completed square form.}$$

$$(y' + 1)^2 = (-1)\left(x' - \frac{4}{5}\right) \qquad \text{\color{red}Write in standard form.}$$

The graph of this equation is a parabola with vertex $\left(\frac{4}{5}, -1\right)$ in the $x'y'$-system. Its axis
is parallel to the x'-axis in the $x'y'$-system, and because $\sin \theta = 1/\sqrt{5}$, $\theta \approx 26.6°$, as
shown in Figure 10.36.

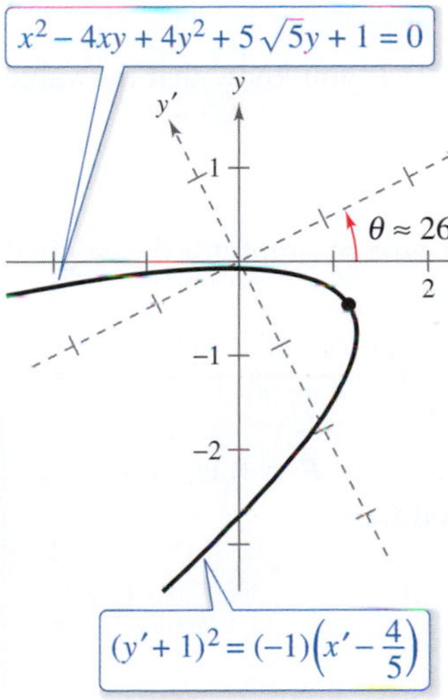

$x^2 - 4xy + 4y^2 + 5\sqrt{5}y + 1 = 0$

$\theta \approx 26.6°$

$(y' + 1)^2 = (-1)\left(x' - \frac{4}{5}\right)$

Vertex:

In $x'y'$-system: $\left(\frac{4}{5}, -1\right)$

In xy-system: $\left(\frac{13}{5\sqrt{5}}, -\frac{6}{5\sqrt{5}}\right)$

Figure 10.36

✓ **Checkpoint**))) *Audio-video solution in English & Spanish at LarsonPrecalculus.com.*

Rotate the axes to eliminate the xy-term in the equation

$$4x^2 + 4xy + y^2 - 2\sqrt{5}x + 4\sqrt{5}y - 30 = 0.$$

Then write the equation in standard form and sketch its graph.

Invariants Under Rotation

In the rotation of axes theorem listed at the beginning of this section, note that the constant term is the same in both equations, $F' = F$. Such quantities are **invariant under rotation**. The next theorem lists some other rotation invariants.

Rotation Invariants

The rotation of the coordinate axes through an angle θ that transforms the equation $Ax^2 + Bxy + Cy^2 + Dx + Ey + F = 0$ into the form

$$A'(x')^2 + C'(y')^2 + D'x' + E'y' + F' = 0$$

has the following rotation invariants.

1. $F = F'$

2. $A + C = A' + C'$

3. $B^2 - 4AC = (B')^2 - 4A'C'$

•• REMARK When there is an xy-term in the equation of a conic, you should realize that the conic is rotated. Before rotating the axes, you should use the discriminant to classify the conic.

You can use the results of this theorem to classify the graph of a second-degree equation *with* an xy-term in much the same way you do for a second-degree equation *without* an xy-term. Note that because $B' = 0$, the invariant $B^2 - 4AC$ reduces to

$$B^2 - 4AC = -4A'C'. \qquad \text{Discriminant}$$

This quantity is called the **discriminant** of the equation

$$Ax^2 + Bxy + Cy^2 + Dx + Ey + F = 0.$$

Now, from the classification procedure given in Section 10.4, you know that the value of $A'C'$ determines the type of graph for the equation

$$A'(x')^2 + C'(y')^2 + D'x' + E'y' + F' = 0.$$

Consequently, the value of $B^2 - 4AC$ will determine the type of graph for the original equation, as given in the following classification.

Classification of Conics by the Discriminant

The graph of the equation $Ax^2 + Bxy + Cy^2 + Dx + Ey + F = 0$ is, except in degenerate cases, determined by its discriminant as follows.

1. *Ellipse or circle:* $B^2 - 4AC < 0$

2. *Parabola:* $B^2 - 4AC = 0$

3. *Hyperbola:* $B^2 - 4AC > 0$

For example, in the general equation

$$3x^2 + 7xy + 5y^2 - 6x - 7y + 15 = 0$$

you have $A = 3$, $B = 7$, and $C = 5$. So, the discriminant is

$$B^2 - 4AC = 7^2 - 4(3)(5)$$
$$= 49 - 60$$
$$= -11.$$

Because $-11 < 0$, the graph of the equation is an ellipse or a circle.

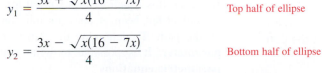

EXAMPLE 4 **Rotation and Graphing Utilities**

For each equation, classify the graph of the equation, use the Quadratic Formula to solve for y, and then use a graphing utility to graph the equation.

a. $2x^2 - 3xy + 2y^2 - 2x = 0$ **b.** $x^2 - 6xy + 9y^2 - 2y + 1 = 0$

c. $3x^2 + 8xy + 4y^2 - 7 = 0$

Solution

a. Because $B^2 - 4AC = 9 - 16 < 0$, the graph is a circle or an ellipse.

$$2x^2 - 3xy + 2y^2 - 2x = 0 \qquad \text{Write original equation.}$$

$$2y^2 - 3xy + (2x^2 - 2x) = 0 \qquad \text{Quadratic form } ay^2 + by + c = 0$$

$$y = \frac{-(-3x) \pm \sqrt{(-3x)^2 - 4(2)(2x^2 - 2x)}}{2(2)}$$

$$y = \frac{3x \pm \sqrt{x(16 - 7x)}}{4}$$

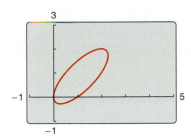

Figure 10.37

Graph both of the equations to obtain the ellipse shown in Figure 10.37.

$$y_1 = \frac{3x + \sqrt{x(16 - 7x)}}{4} \qquad \text{Top half of ellipse}$$

$$y_2 = \frac{3x - \sqrt{x(16 - 7x)}}{4} \qquad \text{Bottom half of ellipse}$$

b. Because $B^2 - 4AC = 36 - 36 = 0$, the graph is a parabola.

$$x^2 - 6xy + 9y^2 - 2y + 1 = 0 \qquad \text{Write original equation.}$$

$$9y^2 - (6x + 2)y + (x^2 + 1) = 0 \qquad \text{Quadratic form } ay^2 + by + c = 0$$

$$y = \frac{(6x + 2) \pm \sqrt{(6x + 2)^2 - 4(9)(x^2 + 1)}}{2(9)}$$

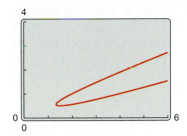

Figure 10.38

Graph both of the equations to obtain the parabola shown in Figure 10.38.

c. Because $B^2 - 4AC = 64 - 48 > 0$, the graph is a hyperbola.

$$3x^2 + 8xy + 4y^2 - 7 = 0 \qquad \text{Write original equation.}$$

$$4y^2 + 8xy + (3x^2 - 7) = 0 \qquad \text{Quadratic form } ay^2 + by + c = 0$$

$$y = \frac{-8x \pm \sqrt{(8x)^2 - 4(4)(3x^2 - 7)}}{2(4)}$$

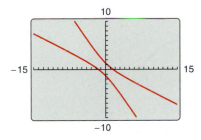

Figure 10.39

Graph both of the equations to obtain the hyperbola shown in Figure 10.39.

✓ **Checkpoint** 🔊 *Audio-video solution in English & Spanish at LarsonPrecalculus.com.*

Classify the graph of the equation $2x^2 - 8xy + 8y^2 + 3x + 5 = 0$, use the Quadratic Formula to solve for y, and then use a graphing utility to graph the equation. ◼

Summarize **(Section 10.5)**

1. Describe how to rotate coordinate axes to eliminate the xy-term in equations of conics *(page 530)*. For examples of rotating coordinate axes to eliminate the xy-term in equations of conics, see Examples 1–3.

2. Describe how to use the discriminant to classify conics *(page 534)*. For an example of using the discriminant to classify conics, see Example 4.

10.6 Parametric Equations

- ■ Evaluate sets of parametric equations for given values of the parameter.
- ■ Sketch curves that are represented by sets of parametric equations.
- ■ Rewrite sets of parametric equations as single rectangular equations by eliminating the parameter.
- ■ Find sets of parametric equations for graphs.

Plane Curves

Parametric equations are useful for modeling the path of an object, such as the path of a baseball.

Up to this point, you have been representing a graph by a single equation involving *two* variables such as x and y. In this section, you will study situations in which it is useful to introduce a *third* variable to represent a curve in the plane.

To see the usefulness of this procedure, consider the path of an object that is propelled into the air at an angle of 45°. When the initial velocity of the object is 48 feet per second, it can be shown that the object follows the parabolic path

$$y = -\frac{x^2}{72} + x. \qquad \text{Rectangular equation}$$

However, this equation does not tell the whole story. Although it does tell you *where* the object has been, it does not tell you *when* the object was at a given point (x, y) on the path. To determine this time, you can introduce a third variable t, called a **parameter.** It is possible to write both x and y as functions of t to obtain the **parametric equations**

$$x = 24\sqrt{2}\,t \qquad \text{Parametric equation for } x$$

$$y = -16t^2 + 24\sqrt{2}\,t. \qquad \text{Parametric equation for } y$$

From this set of equations you can determine that at time $t = 0$, the object is at the point $(0, 0)$. Similarly, at time $t = 1$, the object is at the point $\left(24\sqrt{2}, 24\sqrt{2} - 16\right)$, and so on, as shown below.

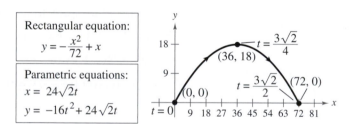

Curvilinear Motion: Two Variables for Position, One Variable for Time

For this particular motion problem, x and y are continuous functions of t, and the resulting path is a **plane curve.** (Recall that a *continuous function* is one whose graph has no breaks, holes, or gaps.)

Definition of Plane Curve

If f and g are continuous functions of t on an interval I, then the set of ordered pairs $(f(t), g(t))$ is a **plane curve** C. The equations

$$x = f(t) \quad \text{and} \quad y = g(t)$$

are **parametric equations** for C, and t is the **parameter.**

Sketching a Plane Curve

When sketching a curve represented by a pair of parametric equations, you still plot points in the *xy*-plane. Each set of coordinates (x, y) is determined from a value chosen for the parameter t. Plotting the resulting points in the order of *increasing* values of t traces the curve in a specific direction. This is called the **orientation** of the curve.

EXAMPLE 1   Sketching a Curve

Sketch the curve given by the parametric equations

$$x = t^2 - 4 \quad \text{and} \quad y = \frac{t}{2}, \quad -2 \le t \le 3.$$

Solution   Using values of t in the specified interval, the parametric equations yield the points (x, y) shown in the table.

REMARK   When using a value of t to find x, be sure to use the same value of t to find the corresponding value of y. Organizing your results in a table, as shown in Example 1, can be helpful.

t	x	y
-2	0	-1
-1	-3	$-\frac{1}{2}$
0	-4	0
1	-3	$\frac{1}{2}$
2	0	1
3	5	$\frac{3}{2}$

By plotting these points in the order of increasing t, you obtain the curve shown in Figure 10.40. The arrows on the curve indicate its orientation as t increases from -2 to 3. So, when a particle moves on this curve, it starts at $(0, -1)$ and then moves along the curve to the point $\left(5, \frac{3}{2}\right)$.

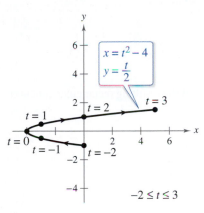

Figure 10.40

Checkpoint   Audio-video solution in English & Spanish at LarsonPrecalculus.com.

Sketch the curve given by the parametric equations

$$x = 2t \quad \text{and} \quad y = 4t^2 + 2, \quad -2 \le t \le 2.$$

Note that the graph shown in Figure 10.40 does not define y as a function of x. This points out one benefit of parametric equations—they can be used to represent graphs that are more general than graphs of functions.

Two different sets of parametric equations can have the same graph. For example, the set of parametric equations

$$x = 4t^2 - 4 \quad \text{and} \quad y = t, \quad -1 \le t \le \frac{3}{2}$$

has the same graph as the set of parametric equations given in Example 1. However, by comparing the values of t in Figures 10.40 and 10.41, you can see that the second graph is traced out more *rapidly* (considering t as time) than the first graph. So, in applications, different parametric representations can be used to represent various *speeds* at which objects travel along a given path.

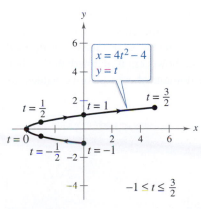

Figure 10.41

Eliminating the Parameter

Example 1 uses simple point plotting to sketch the curve. This tedious process can sometimes be simplified by finding a rectangular equation (in x and y) that has the same graph. This process is called **eliminating the parameter.**

Parametric equations	⇒	Solve for t in one equation.	⇒	Substitute into other equation.	⇒	Rectangular equation

$$x = t^2 - 4 \qquad t = 2y \qquad x = (2y)^2 - 4 \qquad x = 4y^2 - 4$$
$$y = \frac{t}{2}$$

Now you can recognize that the equation $x = 4y^2 - 4$ represents a parabola with a horizontal axis and vertex at $(-4, 0)$.

When converting equations from parametric to rectangular form, you may need to alter the domain of the rectangular equation so that its graph matches the graph of the parametric equations. Such a situation is demonstrated in Example 2.

EXAMPLE 2 **Eliminating the Parameter**

Sketch the curve represented by the equations

$$x = \frac{1}{\sqrt{t+1}} \quad \text{and} \quad y = \frac{t}{t+1}$$

by eliminating the parameter and adjusting the domain of the resulting rectangular equation.

Solution Solving for t in the equation for x produces

$$x = \frac{1}{\sqrt{t+1}} \quad \Longrightarrow \quad x^2 = \frac{1}{t+1}$$

which implies that

$$t = \frac{1 - x^2}{x^2}.$$

Now, substituting in the equation for y, you obtain the rectangular equation

$$y = \frac{t}{t+1} = \frac{\dfrac{1-x^2}{x^2}}{\dfrac{1-x^2}{x^2}+1} = \frac{\dfrac{1-x^2}{x^2}}{\dfrac{1-x^2}{x^2}+1} \cdot \frac{x^2}{x^2} = 1 - x^2.$$

From this rectangular equation, you can recognize that the curve is a parabola that opens downward and has its vertex at $(0, 1)$. Also, this rectangular equation is defined for all values of x. The parametric equation for x, however, is defined only when $t > -1$. This implies that you should restrict the domain of x to positive values, as shown in Figure 10.42.

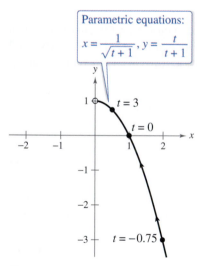

Figure 10.42

Parametric equations:
$$x = \frac{1}{\sqrt{t+1}}, \ y = \frac{t}{t+1}$$

$t = 3$

$t = 0$

$t = -0.75$

✓ **Checkpoint**))) *Audio-video solution in English & Spanish at LarsonPrecalculus.com.*

Sketch the curve represented by the equations

$$x = \frac{1}{\sqrt{t-1}} \quad \text{and} \quad y = \frac{t+1}{t-1}$$

by eliminating the parameter and adjusting the domain of the resulting rectangular equation.

It is not necessary for the parameter in a set of parametric equations to represent time. The next example uses an *angle* as the parameter.

EXAMPLE 3   Eliminating an Angle Parameter

Sketch the curve represented by

$$x = 3 \cos \theta \quad \text{and} \quad y = 4 \sin \theta, \quad 0 \le \theta < 2\pi$$

by eliminating the parameter.

Solution   Begin by solving for $\cos \theta$ and $\sin \theta$ in the equations.

$$\cos \theta = \frac{x}{3} \quad \text{and} \quad \sin \theta = \frac{y}{4} \qquad \text{Solve for } \cos \theta \text{ and } \sin \theta.$$

Use the identity $\sin^2 \theta + \cos^2 \theta = 1$ to form an equation involving only x and y.

$$\cos^2 \theta + \sin^2 \theta = 1 \qquad \text{Pythagorean identity}$$

$$\left(\frac{x}{3}\right)^2 + \left(\frac{y}{4}\right)^2 = 1 \qquad \text{Substitute } \frac{x}{3} \text{ for } \cos \theta \text{ and } \frac{y}{4} \text{ for } \sin \theta.$$

$$\frac{x^2}{9} + \frac{y^2}{16} = 1 \qquad \text{Rectangular equation}$$

From this rectangular equation, you can see that the graph is an ellipse centered at $(0, 0)$, with vertices $(0, 4)$ and $(0, -4)$ and minor axis of length $2b = 6$, as shown below. Note that the elliptic curve is traced out *counterclockwise* as θ increases on the interval $[0, 2\pi)$.

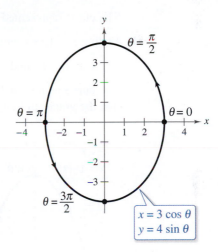

✓ *Checkpoint* 🔊)) *Audio-video solution in English & Spanish at LarsonPrecalculus.com.*

Sketch the curve represented by

$$x = 5 \cos \theta \quad \text{and} \quad y = 3 \sin \theta, \quad 0 \le \theta < 2\pi$$

by eliminating the parameter. ∎

In Examples 2 and 3, it is important to realize that eliminating the parameter is primarily an *aid to curve sketching*. When the parametric equations represent the path of a moving object, the graph alone is not sufficient to describe the object's motion. You still need the parametric equations to tell you the *position*, *direction*, and *speed* at a given time.

Finding Parametric Equations for a Graph

You have been studying techniques for sketching the graph represented by a set of parametric equations. Now consider the *reverse* problem—that is, how can you find a set of parametric equations for a given graph or a given physical description? From the discussion following Example 1, you know that such a representation is not unique. That is, the equations

$$x = 4t^2 - 4 \quad \text{and} \quad y = t, \quad -1 \le t \le \frac{3}{2}$$

produced the same graph as the equations

$$x = t^2 - 4 \quad \text{and} \quad y = \frac{t}{2}, \quad -2 \le t \le 3.$$

This is further demonstrated in Example 4.

EXAMPLE 4 **Finding Parametric Equations for a Graph**

Find a set of parametric equations to represent the graph of $y = 1 - x^2$, using the following parameters.

a. $t = x$ **b.** $t = 1 - x$

Solution

a. Letting $t = x$, you obtain the parametric equations

$$x = t \quad \text{and} \quad y = 1 - x^2 = 1 - t^2.$$

The curve represented by the parametric equations is shown in Figure 10.43.

b. Letting $t = 1 - x$, you obtain the parametric equations

$$x = 1 - t \quad \text{and} \quad y = 1 - x^2 = 1 - (1 - t)^2 = 2t - t^2.$$

The curve represented by the parametric equations is shown in Figure 10.44. Note that the graphs in Figures 10.43 and 10.44 have opposite orientations.

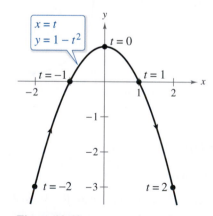

Figure 10.43

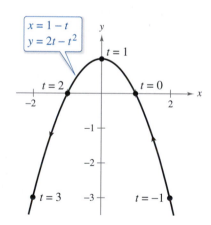

Figure 10.44

✓ *Checkpoint* 🔊)) *Audio-video solution in English & Spanish at LarsonPrecalculus.com.*

Find a set of parametric equations to represent the graph of $y = x^2 + 2$, using the following parameters.

a. $t = x$ **b.** $t = 2 - x$

A **cycloid** is a curve traced out by a point P on a circle as the circle rolls along a straight line in a plane.

EXAMPLE 5   **Parametric Equations for a Cycloid**

Write parametric equations for a cycloid traced out by a point P on a circle of radius a as the circle rolls along the x-axis given that P is at a minimum when $x = 0$.

Solution   As the parameter, let θ be the measure of the circle's rotation, and let the point $P(x, y)$ begin at the origin. When $\theta = 0$, P is at the origin; when $\theta = \pi$, P is at a maximum point $(\pi a, 2a)$; and when $\theta = 2\pi$, P is back on the x-axis at $(2\pi a, 0)$. From the figure below, you can see that $\angle APC = \pi - \theta$. So, you have

$$\sin\theta = \sin(\pi - \theta) = \sin(\angle APC) = \frac{AC}{a} = \frac{BD}{a}$$

•• **REMARK**   In Example 5, $\overparen{PD}$ represents the arc of the circle between points P and D. ▷

$$\cos\theta = -\cos(\pi - \theta) = -\cos(\angle APC) = -\frac{AP}{a}$$

which implies that $BD = a\sin\theta$ and $AP = -a\cos\theta$. Because the circle rolls along the x-axis, you know that $OD = \overparen{PD} = a\theta$. Furthermore, because $BA = DC = a$, you have

$$x = OD - BD = a\theta - a\sin\theta$$

and

$$y = BA + AP = a - a\cos\theta.$$

So, the parametric equations are $x = a(\theta - \sin\theta)$ and $y = a(1 - \cos\theta)$.

▷ **TECHNOLOGY**   You can use a graphing utility in *parametric* mode to obtain a graph similar to the one in Example 5 by graphing the following equations.

$$X_{1T} = T - \sin T$$

$$Y_{1T} = 1 - \cos T$$

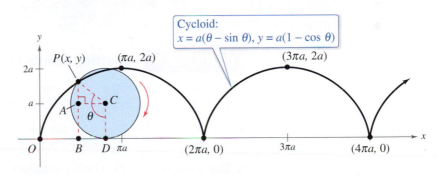

Cycloid:
$x = a(\theta - \sin\theta), \ y = a(1 - \cos\theta)$

✔ **Checkpoint** 🔊))) *Audio-video solution in English & Spanish at LarsonPrecalculus.com.*

Write parametric equations for a cycloid traced out by a point P on a circle of radius a as the circle rolls along the x-axis given that P is at a maximum when $x = 0$.   ■

Summarize   (Section 10.6)

1. Describe how to sketch a curve by evaluating a set of parametric equations for given values of the parameter (*page 537*). For an example of sketching a curve given by parametric equations, see Example 1.

2. Describe the process of eliminating the parameter (*page 538*). For examples of sketching curves by eliminating the parameter, see Examples 2 and 3.

3. Describe how to find a set of parametric equations for a graph (*page 540*). For examples of finding sets of parametric equations for graphs, see Examples 4 and 5.

10.7 Polar Coordinates

You can use polar coordinates in mathematical modeling. For instance, polar coordinates can be used to model the path of a passenger car on a Ferris wheel.

- Plot points in the polar coordinate system.
- Convert points from rectangular to polar form and vice versa.
- Convert equations from rectangular to polar form and vice versa.

Introduction

So far, you have been representing graphs of equations as collections of points (x, y) in the rectangular coordinate system, where x and y represent the directed distances from the coordinate axes to the point (x, y). In this section, you will study a different system called the **polar coordinate system.**

To form the polar coordinate system in the plane, fix a point O, called the **pole** (or **origin**), and construct from O an initial ray called the **polar axis,** as shown below. Then each point P in the plane can be assigned **polar coordinates** (r, θ) as follows.

1. $r =$ *directed distance* from O to P

2. $\theta =$ *directed angle,* counterclockwise from polar axis to segment $\overline{OP}$

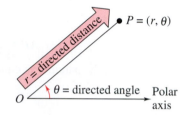

EXAMPLE 1   Plotting Points in the Polar Coordinate System

a. The point $(r, \theta) = (2, \pi/3)$ lies two units from the pole on the terminal side of the angle $\theta = \pi/3$, as shown in Figure 10.45.

b. The point $(r, \theta) = (3, -\pi/6)$ lies three units from the pole on the terminal side of the angle $\theta = -\pi/6$, as shown in Figure 10.46.

c. The point $(r, \theta) = (3, 11\pi/6)$ coincides with the point $(3, -\pi/6)$, as shown in Figure 10.47.

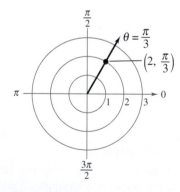

Figure 10.45

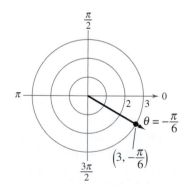

Figure 10.46

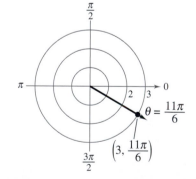

Figure 10.47

✓ *Checkpoint* ◀))) *Audio-video solution in English & Spanish at LarsonPrecalculus.com.*

Plot each point given in polar coordinates.

a. $(3, \pi/4)$ **b.** $(2, -\pi/3)$ **c.** $(2, 5\pi/3)$

In rectangular coordinates, each point (x, y) has a unique representation. This is not true for polar coordinates. For instance, the coordinates

$$(r, \theta) \quad \text{and} \quad (r, \theta + 2\pi)$$

represent the same point, as illustrated in Example 1. Another way to obtain multiple representations of a point is to use negative values for r. Because r is a *directed distance*, the coordinates

$$(r, \theta) \quad \text{and} \quad (-r, \theta + \pi)$$

represent the same point. In general, the point (r, θ) can be represented as

$$(r, \theta) = (r, \theta \pm 2n\pi) \quad \text{or} \quad (r, \theta) = (-r, \theta \pm (2n + 1)\pi)$$

where n is any integer. Moreover, the pole is represented by $(0, \theta)$, where θ is any angle.

EXAMPLE 2   **Multiple Representations of Points**

Plot the point

$$\left(3, -\frac{3\pi}{4}\right)$$

and find three additional polar representations of this point, using

$$-2\pi < \theta < 2\pi.$$

Solution   The point is shown below. Three other representations are as follows.

$$\left(3, -\frac{3\pi}{4} + 2\pi\right) = \left(3, \frac{5\pi}{4}\right) \qquad \text{Add } 2\pi \text{ to } \theta.$$

$$\left(-3, -\frac{3\pi}{4} - \pi\right) = \left(-3, -\frac{7\pi}{4}\right) \qquad \text{Replace } r \text{ by } -r; \text{ subtract } \pi \text{ from } \theta.$$

$$\left(-3, -\frac{3\pi}{4} + \pi\right) = \left(-3, \frac{\pi}{4}\right) \qquad \text{Replace } r \text{ by } -r; \text{ add } \pi \text{ to } \theta.$$

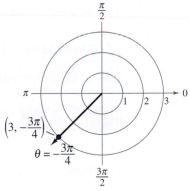

$$\left(3, -\tfrac{3\pi}{4}\right) = \left(3, \tfrac{5\pi}{4}\right) = \left(-3, -\tfrac{7\pi}{4}\right) = \left(-3, \tfrac{\pi}{4}\right) = \cdots$$

✓ *Checkpoint* 🔊 *Audio-video solution in English & Spanish at LarsonPrecalculus.com.*

Plot each point and find three additional polar representations of the point, using $-2\pi < \theta < 2\pi.$

a. $\left(3, \dfrac{4\pi}{3}\right)$ **b.** $\left(2, -\dfrac{5\pi}{6}\right)$ **c.** $\left(-1, \dfrac{3\pi}{4}\right)$

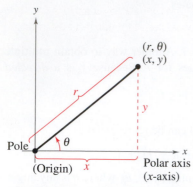

Figure 10.48

Coordinate Conversion

To establish the relationship between polar and rectangular coordinates, let the polar axis coincide with the positive x-axis and the pole with the origin, as shown in Figure 10.48. Because (x, y) lies on a circle of radius r, it follows that $r^2 = x^2 + y^2$. Moreover, for $r > 0$, the definitions of the trigonometric functions imply that

$$\tan \theta = \frac{y}{x}, \quad \cos \theta = \frac{x}{r}, \quad \text{and} \quad \sin \theta = \frac{y}{r}.$$

You can show that the same relationships hold for $r < 0$.

Coordinate Conversion

The polar coordinates (r, θ) are related to the rectangular coordinates (x, y) as follows.

Polar-to-Rectangular	**Rectangular-to-Polar**
$x = r \cos \theta$	$\tan \theta = \dfrac{y}{x}$
$y = r \sin \theta$	$r^2 = x^2 + y^2$

EXAMPLE 3 **Polar-to-Rectangular Conversion**

Convert $\left(\sqrt{3}, \dfrac{\pi}{6}\right)$ to rectangular coordinates.

Solution For the point $(r, \theta) = \left(\sqrt{3}, \pi/6\right)$, you have the following.

$$x = r \cos \theta = \sqrt{3} \cos \frac{\pi}{6} = \sqrt{3}\left(\frac{\sqrt{3}}{2}\right) = \frac{3}{2}$$

$$y = r \sin \theta = \sqrt{3} \sin \frac{\pi}{6} = \sqrt{3}\left(\frac{1}{2}\right) = \frac{\sqrt{3}}{2}$$

The rectangular coordinates are $(x, y) = \left(\dfrac{3}{2}, \dfrac{\sqrt{3}}{2}\right)$. (See Figure 10.49.)

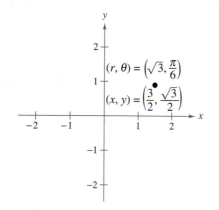

Figure 10.49

✓ **Checkpoint** ◀))) Audio-video solution in English & Spanish at LarsonPrecalculus.com.

Convert $(2, \pi)$ to rectangular coordinates.

EXAMPLE 4 **Rectangular-to-Polar Conversion**

Convert $(-1, 1)$ to polar coordinates.

Solution For the second-quadrant point $(x, y) = (-1, 1)$, you have

$$\tan \theta = \frac{y}{x} = \frac{1}{-1} = -1 \quad \Longrightarrow \quad \theta = \pi + \arctan(-1) = \frac{3\pi}{4}.$$

Because θ lies in the same quadrant as (x, y), use positive r.

$$r = \sqrt{x^2 + y^2} = \sqrt{(-1)^2 + (1)^2} = \sqrt{2}$$

So, *one* set of polar coordinates is $(r, \theta) = \left(\sqrt{2}, 3\pi/4\right)$, as shown in Figure 10.50.

✓ **Checkpoint** ◀))) Audio-video solution in English & Spanish at LarsonPrecalculus.com.

Convert $(0, 2)$ to polar coordinates.

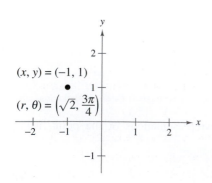

Figure 10.50

Equation Conversion

To convert a rectangular equation to polar form, replace x by $r\cos\theta$ and y by $r\sin\theta$. For instance, the rectangular equation $y = x^2$ can be written in polar form as follows.

$$y = x^2 \qquad \text{Rectangular equation}$$

$$r\sin\theta = (r\cos\theta)^2 \qquad \text{Polar equation}$$

$$r = \sec\theta\tan\theta \qquad \text{Simplest form}$$

Converting a polar equation to rectangular form requires considerable ingenuity. Example 5 demonstrates several polar-to-rectangular conversions that enable you to sketch the graphs of some polar equations.

> **EXAMPLE 5**   **Converting Polar Equations to Rectangular Form**

Convert each polar equation to a rectangular equation.

a. $r = 2$ **b.** $\theta = \pi/3$ **c.** $r = \sec\theta$

Solution

a. The graph of the polar equation $r = 2$ consists of all points that are two units from the pole. In other words, this graph is a circle centered at the origin with a radius of 2, as shown in Figure 10.51. You can confirm this by converting to rectangular form, using the relationship $r^2 = x^2 + y^2$.

$$r = 2 \implies r^2 = 2^2 \implies x^2 + y^2 = 2^2$$

Polar equation   Rectangular equation

b. The graph of the polar equation $\theta = \pi/3$ consists of all points on the line that makes an angle of $\pi/3$ with the positive polar axis, as shown in Figure 10.52. To convert to rectangular form, make use of the relationship $\tan\theta = y/x$.

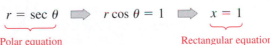

$$\theta = \pi/3 \implies \tan\theta = \sqrt{3} \implies y = \sqrt{3}x$$

Polar equation   Rectangular equation

c. The graph of the polar equation $r = \sec\theta$ is not evident by simple inspection, so convert to rectangular form by using the relationship $r\cos\theta = x$.

$$r = \sec\theta \implies r\cos\theta = 1 \implies x = 1$$

Polar equation   Rectangular equation

Now you see that the graph is a vertical line, as shown in Figure 10.53.

✓ **Checkpoint** ◀))) *Audio-video solution in English & Spanish at LarsonPrecalculus.com.*

Convert $r = 6\sin\theta$ to a rectangular equation.   ■

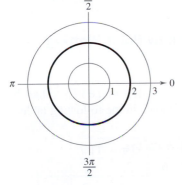

Figure 10.51

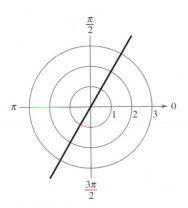

Figure 10.52

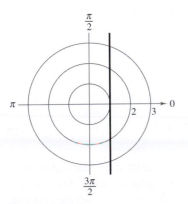

Figure 10.53

> **Summarize**   **(Section 10.7)**
>
> 1. Describe how to plot the point (r, θ) in the polar coordinate system *(page 542)*. For examples of plotting points in the polar coordinate system, see Examples 1 and 2.
>
> 2. Describe how to convert points from rectangular to polar form and vice versa *(page 544)*. For examples of converting between forms, see Examples 3 and 4.
>
> 3. Describe how to convert equations from rectangular to polar form and vice versa *(page 545)*. For an example of converting polar equations to rectangular form, see Example 5.

10.8 Graphs of Polar Equations

- Graph polar equations by point plotting.
- Use symmetry, zeros, and maximum *r*-values to sketch graphs of polar equations.
- Recognize special polar graphs.

Introduction

In previous chapters, you learned how to sketch graphs in rectangular coordinate systems. You began with the basic point-plotting method. Then you used sketching aids such as symmetry, intercepts, asymptotes, periods, and shifts to further investigate the natures of graphs. This section approaches curve sketching in the polar coordinate system similarly, beginning with a demonstration of point plotting.

You can graph the pickup pattern of a microphone in a polar coordinate system.

EXAMPLE 1 **Graphing a Polar Equation by Point Plotting**

Sketch the graph of the polar equation $r = 4 \sin \theta$.

Solution The sine function is periodic, so you can get a full range of *r*-values by considering values of θ in the interval $0 \le \theta \le 2\pi$, as shown in the following table.

θ	0	$\frac{\pi}{6}$	$\frac{\pi}{3}$	$\frac{\pi}{2}$	$\frac{2\pi}{3}$	$\frac{5\pi}{6}$	π	$\frac{7\pi}{6}$	$\frac{3\pi}{2}$	$\frac{11\pi}{6}$	2π
r	0	2	$2\sqrt{3}$	4	$2\sqrt{3}$	2	0	-2	-4	-2	0

By plotting these points, as shown in Figure 10.54, it appears that the graph is a circle of radius 2 whose center is at the point $(x, y) = (0, 2)$.

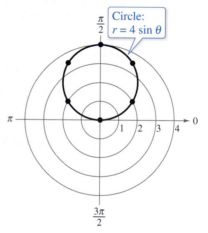

Figure 10.54

✓ *Checkpoint* ◀))) *Audio-video solution in English & Spanish at LarsonPrecalculus.com.*

Sketch the graph of the polar equation $r = 6 \cos \theta$. ■

You can confirm the graph in Figure 10.54 by converting the polar equation to rectangular form and then sketching the graph of the rectangular equation. You can also use a graphing utility set to *polar* mode and graph the polar equation or set the graphing utility to *parametric* mode and graph a parametric representation.

nikkytok/Shutterstock.com

Symmetry, Zeros, and Maximum *r*-Values

In Figure 10.54 on the preceding page, note that as θ increases from 0 to 2π the graph is traced out twice. Moreover, note that the graph is *symmetric with respect to the line* $\theta = \pi/2$. Had you known about this symmetry and retracing ahead of time, you could have used fewer points. The three important types of symmetry to consider in polar curve sketching are shown below.

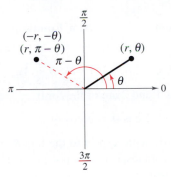

Symmetry with Respect to the Line $\theta = \dfrac{\pi}{2}$

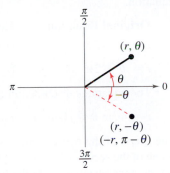

Symmetry with Respect to the Polar Axis

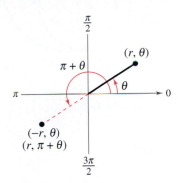

Symmetry with Respect to the Pole

Tests for Symmetry in Polar Coordinates

The graph of a polar equation is symmetric with respect to the following when the given substitution yields an equivalent equation.

1. The line $\theta = \pi/2$: Replace (r, θ) by $(r, \pi - \theta)$ or $(-r, -\theta)$.

2. The polar axis: Replace (r, θ) by $(r, -\theta)$ or $(-r, \pi - \theta)$.

3. The pole: Replace (r, θ) by $(r, \pi + \theta)$ or $(-r, \theta)$.

EXAMPLE 2   **Using Symmetry to Sketch a Polar Graph**

Use symmetry to sketch the graph of $r = 3 + 2 \cos \theta$.

Solution Replacing (r, θ) by $(r, -\theta)$ produces

$$r = 3 + 2\cos(-\theta) = 3 + 2\cos\theta. \qquad \cos(-\theta) = \cos\theta$$

So, you can conclude that the curve is symmetric with respect to the polar axis. Plotting the points in the table and using polar axis symmetry, you obtain the graph shown in Figure 10.55. This graph is called a **limaçon**.

θ	0	$\dfrac{\pi}{6}$	$\dfrac{\pi}{3}$	$\dfrac{\pi}{2}$	$\dfrac{2\pi}{3}$	$\dfrac{5\pi}{6}$	π
r	5	$3 + \sqrt{3}$	4	3	2	$3 - \sqrt{3}$	1

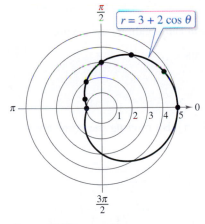

$r = 3 + 2\cos\theta$

Figure 10.55

✓ *Checkpoint* ◀))) *Audio-video solution in English & Spanish at LarsonPrecalculus.com.*

Use symmetry to sketch the graph of $r = 3 + 2 \sin \theta$.

Note in Example 2 that $\cos(-\theta) = \cos\theta$. This is because the cosine function is *even*. Recall from Section 4.2 that the cosine function is even and the sine function is odd. That is, $\sin(-\theta) = -\sin\theta$.

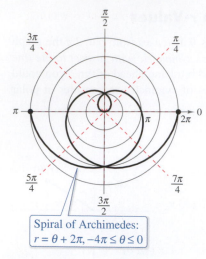

Spiral of Archimedes:
$r = \theta + 2\pi, -4\pi \le \theta \le 0$

Figure 10.56

The three tests for symmetry in polar coordinates listed on page 547 are sufficient to guarantee symmetry, but they are not necessary. For instance, Figure 10.56 shows the graph of

$$r = \theta + 2\pi \qquad \text{Spiral of Archimedes}$$

to be symmetric with respect to the line $\theta = \pi/2$, and yet the tests on page 547 fail to indicate symmetry because neither of the following replacements yields an equivalent equation.

Original Equation	Replacement	New Equation
$r = \theta + 2\pi$	(r, θ) by $(-r, -\theta)$	$-r = -\theta + 2\pi$
$r = \theta + 2\pi$	(r, θ) by $(r, \pi - \theta)$	$r = -\theta + 3\pi$

The equations discussed in Examples 1 and 2 are of the form

$$r = 4 \sin \theta = f(\sin \theta) \quad \text{and} \quad r = 3 + 2 \cos \theta = g(\cos \theta).$$

The graph of the first equation is symmetric with respect to the line $\theta = \pi/2$, and the graph of the second equation is symmetric with respect to the polar axis. This observation can be generalized to yield the following tests.

Quick Tests for Symmetry in Polar Coordinates

1. The graph of $r = f(\sin \theta)$ is symmetric with respect to the line $\theta = \dfrac{\pi}{2}$.

2. The graph of $r = g(\cos \theta)$ is symmetric with respect to the polar axis.

Two additional aids to sketching graphs of polar equations involve knowing the θ-values for which $|r|$ is maximum and knowing the θ-values for which $r = 0$. For instance, in Example 1, the maximum value of $|r|$ for $r = 4 \sin \theta$ is $|r| = 4$, and this occurs when $\theta = \pi/2$, as shown in Figure 10.54. Moreover, $r = 0$ when $\theta = 0$.

EXAMPLE 3 Sketching a Polar Graph

Sketch the graph of $r = 1 - 2 \cos \theta$.

Solution From the equation $r = 1 - 2 \cos \theta$, you can obtain the following.

Symmetry: With respect to the polar axis

Maximum value of $|r|$: $r = 3$ when $\theta = \pi$

Zero of r: $r = 0$ when $\theta = \pi/3$

The table shows several θ-values in the interval $[0, \pi]$. By plotting the corresponding points, you can sketch the graph shown in Figure 10.57.

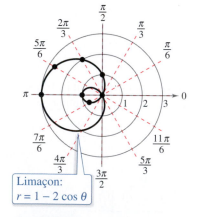

Limaçon:
$r = 1 - 2 \cos \theta$

Figure 10.57

θ	0	$\dfrac{\pi}{6}$	$\dfrac{\pi}{3}$	$\dfrac{\pi}{2}$	$\dfrac{2\pi}{3}$	$\dfrac{5\pi}{6}$	π
r	-1	$1 - \sqrt{3}$	0	1	2	$1 + \sqrt{3}$	3

Note how the negative r-values determine the *inner loop* of the graph in Figure 10.57. This graph, like the one in Figure 10.55, is a limaçon.

✓ *Checkpoint*  *Audio-video solution in English & Spanish at LarsonPrecalculus.com.*

Sketch the graph of $r = 1 + 2 \sin \theta$.

Some curves reach their zeros and maximum *r*-values at more than one point, as shown in Example 4.

EXAMPLE 4   **Sketching a Polar Graph**

Sketch the graph of

$$r = 2 \cos 3\theta.$$

Solution

Symmetry: With respect to the polar axis

Maximum value of $|r|$: $|r| = 2$ when $3\theta = 0, \pi, 2\pi, 3\pi$ or $\theta = 0, \dfrac{\pi}{3}, \dfrac{2\pi}{3}, \pi$

Zeros of r: $r = 0$ when $3\theta = \dfrac{\pi}{2}, \dfrac{3\pi}{2}, \dfrac{5\pi}{2}$ or $\theta = \dfrac{\pi}{6}, \dfrac{\pi}{2}, \dfrac{5\pi}{6}$

θ	0	$\dfrac{\pi}{12}$	$\dfrac{\pi}{6}$	$\dfrac{\pi}{4}$	$\dfrac{\pi}{3}$	$\dfrac{5\pi}{12}$	$\dfrac{\pi}{2}$
r	2	$\sqrt{2}$	0	$-\sqrt{2}$	-2	$-\sqrt{2}$	0

By plotting these points and using the specified symmetry, zeros, and maximum values, you can obtain the graph, as shown below. This graph is called a **rose curve,** and each loop on the graph is called a *petal*. Note how the entire curve is generated as θ increases from 0 to π.

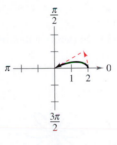

$0 \le \theta \le \dfrac{\pi}{6}$

$0 \le \theta \le \dfrac{\pi}{3}$

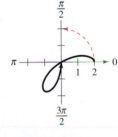

$0 \le \theta \le \dfrac{\pi}{2}$

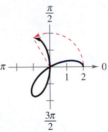

$0 \le \theta \le \dfrac{2\pi}{3}$

$0 \le \theta \le \dfrac{5\pi}{6}$

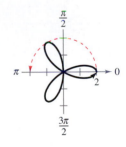

$0 \le \theta \le \pi$

▷ **TECHNOLOGY** Use a graphing utility in *polar* mode to verify the graph of $r = 2 \cos 3\theta$ shown in Example 4.

✓ *Checkpoint* ◖)) *Audio-video solution in English & Spanish at LarsonPrecalculus.com.*

Sketch the graph of

$$r = 2 \sin 3\theta.$$

Special Polar Graphs

Several important types of graphs have equations that are simpler in polar form than in rectangular form. For example, the circle

$$r = 4 \sin \theta$$

in Example 1 has the more complicated rectangular equation

$$x^2 + (y - 2)^2 = 4.$$

Several other types of graphs that have simple polar equations are shown below.

Limaçons

$$r = a \pm b \cos \theta, \, r = a \pm b \sin \theta \quad (a > 0, b > 0)$$

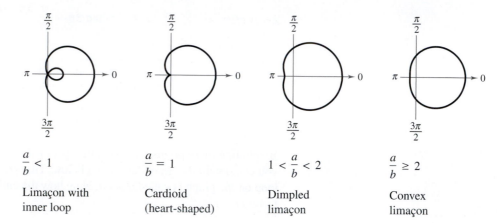

$$\frac{a}{b} < 1$$

Limaçon with
inner loop

$$\frac{a}{b} = 1$$

Cardioid
(heart-shaped)

$$1 < \frac{a}{b} < 2$$

Dimpled
limaçon

$$\frac{a}{b} \geq 2$$

Convex
limaçon

Rose Curves

n petals when n is odd, $2n$ petals when n is even $(n \geq 2)$

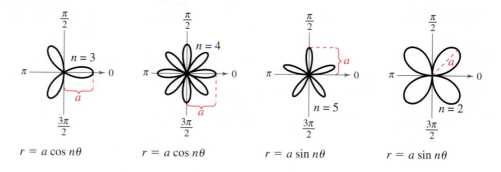

$$r = a \cos n\theta$$

$$r = a \cos n\theta$$

$$r = a \sin n\theta$$

$$r = a \sin n\theta$$

Circles and Lemniscates

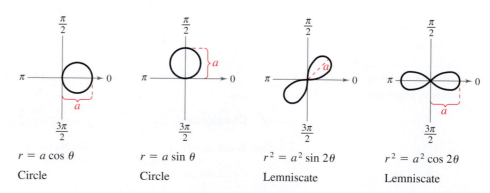

$$r = a \cos \theta$$

Circle

$$r = a \sin \theta$$

Circle

$$r^2 = a^2 \sin 2\theta$$

Lemniscate

$$r^2 = a^2 \cos 2\theta$$

Lemniscate

θ	r
0	3
$\dfrac{\pi}{6}$	$\dfrac{3}{2}$
$\dfrac{\pi}{4}$	0
$\dfrac{\pi}{3}$	$-\dfrac{3}{2}$

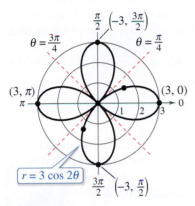

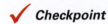

Figure 10.58

θ	$r = \pm 3\sqrt{\sin 2\theta}$
0	0
$\dfrac{\pi}{12}$	$\pm\dfrac{3}{\sqrt{2}}$
$\dfrac{\pi}{4}$	± 3
$\dfrac{5\pi}{12}$	$\pm\dfrac{3}{\sqrt{2}}$
$\dfrac{\pi}{2}$	0

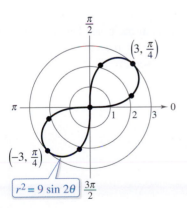

Figure 10.59

EXAMPLE 5   **Sketching a Rose Curve**

Sketch the graph of $r = 3\cos 2\theta$.

Solution

Type of curve: Rose curve with $2n = 4$ petals

Symmetry: With respect to the polar axis, the line $\theta = \pi/2$, and the pole

Maximum value of $|r|$: $|r| = 3$ when $\theta = 0,\ \pi/2,\ \pi,\ 3\pi/2$

Zeros of r: $r = 0$ when $\theta = \pi/4,\ 3\pi/4$

Using this information together with the additional points shown in the table at the left, you obtain the graph shown in Figure 10.58.

✓ *Checkpoint*   ◀))) *Audio-video solution in English & Spanish at LarsonPrecalculus.com.*

Sketch the graph of $r = 3\cos 3\theta$.

EXAMPLE 6   **Sketching a Lemniscate**

Sketch the graph of $r^2 = 9\sin 2\theta$.

Solution

Type of curve: Lemniscate

Symmetry: With respect to the pole

Maximum value of $|r|$: $|r| = 3$ when $\theta = \pi/4$

Zeros of r: $r = 0$ when $\theta = 0,\ \pi/2$

When $\sin 2\theta < 0$, this equation has no solution points. So, you restrict the values of θ to those for which $\sin 2\theta \geq 0$.

$$0 \leq \theta \leq \frac{\pi}{2} \quad \text{or} \quad \pi \leq \theta \leq \frac{3\pi}{2}$$

Using symmetry, you need to consider only the first of these two intervals. By finding a few additional points (see table at the left), you obtain the graph shown in Figure 10.59.

✓ *Checkpoint*   ◀))) *Audio-video solution in English & Spanish at LarsonPrecalculus.com.*

Sketch the graph of $r^2 = 4\cos 2\theta$.   ◼

Summarize   **(Section 10.8)**

1. Describe how to graph polar equations by point plotting *(page 546)*. For an example of graphing a polar equation by point plotting, see Example 1.

2. State the tests for symmetry in polar coordinates *(page 547)*. For an example of using symmetry to sketch the graph of a polar equation, see Example 2.

3. Describe how to use zeros and maximum r-values to sketch graphs of polar equations *(pages 548 and 549)*. For examples of using zeros and maximum r-values to sketch graphs of polar equations, see Examples 3 and 4.

4. State and give an example of a special polar graph covered in this lesson *(page 550)*. For examples of sketching special polar graphs, see Examples 5 and 6.

10.9 Polar Equations of Conics

- Define conics in terms of eccentricity, and write and graph equations of conics in polar form.
- Use equations of conics in polar form to model real-life problems.

Alternative Definition and Polar Equations of Conics

In Sections 10.3 and 10.4, you learned that the rectangular equations of ellipses and hyperbolas take simple forms when the origin lies at their *centers*. As it happens, there are many important applications of conics in which it is more convenient to use one of the *foci* as the origin. In this section, you will learn that polar equations of conics take simple forms when one of the foci lies at the pole.

To begin, consider the following alternative definition of a conic that uses the concept of eccentricity.

You can model the orbits of planets and satellites with polar equations.

> ### Alternative Definition of a Conic
>
> The locus of a point in the plane that moves such that its distance from a fixed point (focus) is in a constant ratio to its distance from a fixed line (directrix) is a **conic**. The constant ratio is the eccentricity of the conic and is denoted by e. Moreover, the conic is an **ellipse** when $0 < e < 1$, a **parabola** when $e = 1$, and a **hyperbola** when $e > 1$. (See the figures below.)

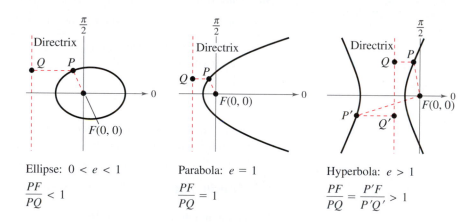

Ellipse: $0 < e < 1$

$$\frac{PF}{PQ} < 1$$

Parabola: $e = 1$

$$\frac{PF}{PQ} = 1$$

Hyperbola: $e > 1$

$$\frac{PF}{PQ} = \frac{P'F}{P'Q'} > 1$$

In the figures, note that for each type of conic, the focus is at the pole. The benefit of locating a focus of a conic at the pole is that the equation of the conic takes on a simpler form. For a proof of the polar equations of conics, see Proofs in Mathematics on page 564.

> ### Polar Equations of Conics
>
> The graph of a polar equation of the form
>
> **1.** $r = \dfrac{ep}{1 \pm e \cos \theta}$ or   **2.** $r = \dfrac{ep}{1 \pm e \sin \theta}$
>
> is a conic, where $e > 0$ is the eccentricity and $|p|$ is the distance between the focus (pole) and the directrix.

An equation of the form

$$r = \frac{ep}{1 \pm e \cos \theta}$$ Vertical directrix

corresponds to a conic with a vertical directrix and symmetry with respect to the polar axis. An equation of the form

$$r = \frac{ep}{1 \pm e \sin \theta}$$ Horizontal directrix

corresponds to a conic with a horizontal directrix and symmetry with respect to the line $\theta = \pi/2$. Moreover, the converse is also true—that is, any conic with a focus at the pole and having a horizontal or vertical directrix can be represented by one of these equations.

EXAMPLE 1   **Identifying a Conic from Its Equation**

Identify the type of conic represented by the equation $r = \dfrac{15}{3 - 2 \cos \theta}$.

Algebraic Solution

To identify the type of conic, rewrite the equation in the form

$$r = \frac{ep}{1 \pm e \cos \theta}.$$

$$r = \frac{15}{3 - 2 \cos \theta}$$ Write original equation.

$$= \frac{5}{1 - (2/3) \cos \theta}$$ Divide numerator and denominator by 3.

Because $e = \frac{2}{3} < 1$, you can conclude that the graph is an ellipse.

Graphical Solution

Use a graphing utility in *polar* mode and be sure to use a square setting, as shown below.

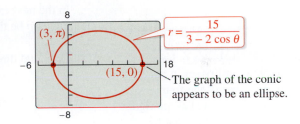

The graph of the conic appears to be an ellipse.

✔ **Checkpoint** ◀))) *Audio-video solution in English & Spanish at LarsonPrecalculus.com.*

Identify the type of conic represented by the equation $r = \dfrac{8}{2 - 3 \sin \theta}$.

For the ellipse in Example 1, the major axis is horizontal and the vertices lie at $(15, 0)$ and $(3, \pi)$. So, the length of the *major* axis is $2a = 18$. To find the length of the *minor* axis, you can use the equations $e = c/a$ and $b^2 = a^2 - c^2$ to conclude that

$$b^2 = a^2 - c^2$$

$$= a^2 - (ea)^2$$

$$= a^2(1 - e^2).$$ Ellipse

Because $e = \frac{2}{3}$, you have

$$b^2 = 9^2 \left[1 - \left(\frac{2}{3} \right)^2 \right] = 45$$

which implies that $b = \sqrt{45} = 3\sqrt{5}$. So, the length of the minor axis is $2b = 6\sqrt{5}$. A similar analysis for hyperbolas yields

$$b^2 = c^2 - a^2$$

$$= (ea)^2 - a^2$$

$$= a^2(e^2 - 1).$$ Hyperbola

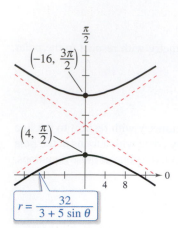

$r = \dfrac{32}{3 + 5\sin\theta}$

Figure 10.60

EXAMPLE 2 **Sketching a Conic from Its Polar Equation**

Identify the conic $r = \dfrac{32}{3 + 5\sin\theta}$ and sketch its graph.

Solution Dividing the numerator and denominator by 3, you have

$$r = \frac{32/3}{1 + (5/3)\sin\theta}.$$

Because $e = \frac{5}{3} > 1$, the graph is a hyperbola. The transverse axis of the hyperbola lies on the line $\theta = \pi/2$, and the vertices occur at $(4, \pi/2)$ and $(-16, 3\pi/2)$. Because the length of the transverse axis is 12, you can see that $a = 6$. To find b, write

$$b^2 = a^2(e^2 - 1) = 6^2\left[\left(\tfrac{5}{3}\right)^2 - 1\right] = 64.$$

So, $b = 8$. You can use a and b to determine that the asymptotes of the hyperbola are $y = 10 \pm \frac{3}{4}x$. The graph is shown in Figure 10.60.

✓ **Checkpoint** 🔊)) *Audio-video solution in English & Spanish at LarsonPrecalculus.com.*

Identify the conic $r = \dfrac{3}{2 - 4\sin\theta}$ and sketch its graph. ∎

In the next example, you are asked to find a polar equation of a specified conic. To do this, let p be the distance between the pole and the directrix.

▷ **TECHNOLOGY** Use a graphing utility set in *polar* mode to verify the four orientations listed at the right. Remember that e must be positive, but p can be positive or negative.

1. Horizontal directrix above the pole: $r = \dfrac{ep}{1 + e\sin\theta}$

2. Horizontal directrix below the pole: $r = \dfrac{ep}{1 - e\sin\theta}$

3. Vertical directrix to the right of the pole: $r = \dfrac{ep}{1 + e\cos\theta}$

4. Vertical directrix to the left of the pole: $r = \dfrac{ep}{1 - e\cos\theta}$

EXAMPLE 3 **Finding the Polar Equation of a Conic**

Find a polar equation of the parabola whose focus is the pole and whose directrix is the line $y = 3$.

Solution Because the directrix is horizontal and above the pole, use an equation of the form

$$r = \frac{ep}{1 + e\sin\theta}.$$

Moreover, because the eccentricity of a parabola is $e = 1$ and the distance between the pole and the directrix is $p = 3$, you have the equation

$$r = \frac{3}{1 + \sin\theta}.$$

The parabola is shown in Figure 10.61.

✓ **Checkpoint** 🔊)) *Audio-video solution in English & Spanish at LarsonPrecalculus.com.*

Find a polar equation of the parabola whose focus is the pole and whose directrix is the line $x = -2$. ∎

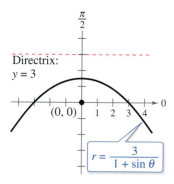

Directrix: $y = 3$

$r = \dfrac{3}{1 + \sin\theta}$

Figure 10.61

Application

Kepler's Laws (listed below), named after the German astronomer Johannes Kepler (1571–1630), can be used to describe the orbits of the planets about the sun.

1. Each planet moves in an elliptical orbit with the sun at one focus.

2. A ray from the sun to a planet sweeps out equal areas in equal times.

3. The square of the period (the time it takes for a planet to orbit the sun) is proportional to the cube of the mean distance between the planet and the sun.

Although Kepler stated these laws on the basis of observation, they were later validated by Isaac Newton (1642–1727). In fact, Newton showed that these laws apply to the orbits of all heavenly bodies, including comets and satellites. This is illustrated in the next example, which involves the comet named after the English mathematician and physicist Edmund Halley (1656–1742).

If you use Earth as a reference with a period of 1 year and a distance of 1 astronomical unit (about 93 million miles), then the proportionality constant in Kepler's third law is 1. For example, because Mars has a mean distance to the sun of $d \approx 1.524$ astronomical units, its period P is given by $d^3 = P^2$. So, the period of Mars is $P \approx 1.88$ years.

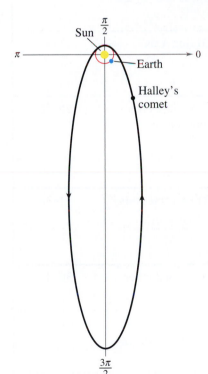

Figure 10.62

EXAMPLE 4   Halley's Comet

Halley's comet has an elliptical orbit with an eccentricity of $e \approx 0.967$. The length of the major axis of the orbit is approximately 35.88 astronomical units. Find a polar equation for the orbit. How close does Halley's comet come to the sun?

Solution   Using a vertical major axis, as shown in Figure 10.62, choose an equation of the form $r = ep/(1 + e \sin \theta)$. Because the vertices of the ellipse occur when $\theta = \pi/2$ and $\theta = 3\pi/2$, you can determine the length of the major axis to be the sum of the r-values of the vertices. That is,

$$2a = \frac{0.967p}{1 + 0.967} + \frac{0.967p}{1 - 0.967} \approx 29.79p \approx 35.88.$$

So, $p \approx 1.204$ and $ep \approx (0.967)(1.204) \approx 1.164$. Using this value of ep in the equation, you have

$$r = \frac{1.164}{1 + 0.967 \sin \theta}$$

where r is measured in astronomical units. To find the closest point to the sun (the focus), substitute $\theta = \pi/2$ into this equation to obtain

$$r = \frac{1.164}{1 + 0.967 \sin(\pi/2)} \approx 0.59 \text{ astronomical unit} \approx 55{,}000{,}000 \text{ miles.}$$

✓ **Checkpoint**  *Audio-video solution in English & Spanish at LarsonPrecalculus.com.*

Encke's comet has an elliptical orbit with an eccentricity of $e \approx 0.848$. The length of the major axis of the orbit is approximately 4.429 astronomical units. Find a polar equation for the orbit. How close does Encke's comet come to the sun? ■

Summarize   (Section 10.9)

1. State the definition of a conic in terms of eccentricity *(page 552)*. For examples of identifying and sketching conics in polar form, see Examples 1 and 2.

2. Describe a real-life problem that can be modeled by an equation of a conic in polar form *(page 555, Example 4)*.

Chapter Summary

	What Did You Learn?	Explanation/Examples	Review Exercises
Section 10.1	Find the inclination of a line *(p. 508)*.	If a nonvertical line has inclination θ and slope m, then $m = \tan \theta$.	1–4
	Find the angle between two lines *(p. 509)*.	If two nonperpendicular lines have slopes m_1 and m_2, then the tangent of the angle between the lines is $\tan \theta = |(m_2 - m_1)/(1 + m_1 m_2)|$.	5–8
	Find the distance between a point and a line *(p. 510)*.	The distance between the point (x_1, y_1) and the line $Ax + By + C = 0$ is $d = |Ax_1 + By_1 + C|/\sqrt{A^2 + B^2}$.	9, 10
Section 10.2	Recognize a conic as the intersection of a plane and a double-napped cone *(p. 512)*.	In the formation of the four basic conics, the intersecting plane does not pass through the vertex of the cone. (See Figure 10.7.)	11, 12
	Write equations of parabolas in standard form *(p. 513)*.	**Horizontal Axis** $(y - k)^2 = 4p(x - h)$, $p \neq 0$ **Vertical Axis** $(x - h)^2 = 4p(y - k)$, $p \neq 0$	13–16
	Use the reflective property of parabolas to solve real-life problems *(p. 515)*.	The tangent line to a parabola at a point P makes equal angles with (1) the line passing through P and the focus and (2) the axis of the parabola.	17–20
Section 10.3	Write equations of ellipses in standard form and graph ellipses *(p. 518)*.	**Horizontal Major Axis** $\dfrac{(x-h)^2}{a^2} + \dfrac{(y-k)^2}{b^2} = 1$ **Vertical Major Axis** $\dfrac{(x-h)^2}{b^2} + \dfrac{(y-k)^2}{a^2} = 1$	21–24, 27–30
	Use properties of ellipses to model and solve real-life problems *(p. 521)*.	You can use the properties of ellipses to find distances from Earth's center to the moon's center in the moon's orbit. (See Example 4.)	25, 26
	Find eccentricities *(p. 522)*.	The eccentricity e of an ellipse is given by $e = c/a$.	27–30
Section 10.4	Write equations of hyperbolas in standard form *(p. 524)*, and find asymptotes of and graph hyperbolas *(p. 525)*.	**Horizontal Transverse Axis** $\dfrac{(x-h)^2}{a^2} - \dfrac{(y-k)^2}{b^2} = 1$ **Vertical Transverse Axis** $\dfrac{(y-k)^2}{a^2} - \dfrac{(x-h)^2}{b^2} = 1$	31–38
	Use properties of hyperbolas to solve real-life problems *(p. 528)*.	You can use the properties of hyperbolas in radar and other detection systems. (See Example 5.)	39, 40
	Classify conics from their general equations *(p. 529)*.	The graph of $Ax^2 + Cy^2 + Dx + Ey + F = 0$ is, except in degenerate cases, a circle $(A = C)$, a parabola $(AC = 0)$, an ellipse $(AC > 0)$, or a hyperbola $(AC < 0)$.	41–44
Section 10.5	Rotate the coordinate axes to eliminate the xy-term in equations of conics *(p. 530)*.	The equation $Ax^2 + Bxy + Cy^2 + Dx + Ey + F = 0$ can be rewritten as $A'(x')^2 + C'(y')^2 + D'x' + E'y' + F' = 0$ by rotating the coordinate axes through an angle θ, where $\cot 2\theta = (A - C)/B$.	45–48
	Use the discriminant to classify conics *(p. 534)*.	The graph of $Ax^2 + Bxy + Cy^2 + Dx + Ey + F = 0$ is, except in degenerate cases, an ellipse or a circle $\left(B^2 - 4AC < 0\right)$, a parabola $\left(B^2 - 4AC = 0\right)$, or a hyperbola $\left(B^2 - 4AC > 0\right)$.	49–52

What Did You Learn?	**Explanation/Examples**	**Review Exercises**
Section 10.6 Evaluate sets of parametric equations for given values of the parameter *(p. 536)*.	If f and g are continuous functions of t on an interval I, then the set of ordered pairs $(f(t), g(t))$ is a plane curve C. The equations $x = f(t)$ and $y = g(t)$ are parametric equations for C, and t is the parameter.	53, 54
Sketch curves that are represented by sets of parametric equations *(p. 537)*.	Sketching a curve represented by parametric equations requires plotting points in the *xy*-plane. Each set of coordinates (x, y) is determined from a value chosen for t.	55–60
Rewrite sets of parametric equations as single rectangular equations by eliminating the parameter *(p. 538)*.	To eliminate the parameter in a pair of parametric equations, solve for t in one equation and substitute the value of t into the other equation. The result is the corresponding rectangular equation.	55–60
Find sets of parametric equations for graphs *(p. 540)*.	When finding a set of parametric equations for a given graph, remember that the parametric equations are not unique.	61–66
Section 10.7 Plot points in the polar coordinate system *(p. 542)*.		67–70
Convert points *(p. 544)* and equations *(p. 545)* from rectangular to polar form and vice versa.	**Polar Coordinates (r, θ) and Rectangular Coordinates (x, y)** Polar-to-Rectangular: $x = r \cos \theta$, $y = r \sin \theta$ Rectangular-to-Polar: $\tan \theta = y/x$, $r^2 = x^2 + y^2$	71–90
Section 10.8 Graph polar equations by point plotting *(p. 546)*.	Graphing a polar equation by point plotting is similar to graphing a rectangular equation.	91–100
Use symmetry, zeros, and maximum r-values to sketch graphs of polar equations *(p. 547)*.	The graph of a polar equation is symmetric with respect to the following when the given substitution yields an equivalent equation. **1.** Line $\theta = \pi/2$: Replace (r, θ) by $(r, \pi - \theta)$ or $(-r, -\theta)$. **2.** Polar axis: Replace (r, θ) by $(r, -\theta)$ or $(-r, \pi - \theta)$. **3.** Pole: Replace (r, θ) by $(r, \pi + \theta)$ or $(-r, \theta)$. Other aids to graphing polar equations are the θ-values for which $|r|$ is maximum and the θ-values for which $r = 0$.	91–100
Recognize special polar graphs *(p. 550)*.	Several types of graphs, such as limaçons, rose curves, circles, and lemniscates, have equations that are simpler in polar form than in rectangular form. (See page 550.)	101–104
Section 10.9 Define conics in terms of eccentricity, and write and graph equations of conics in polar form *(p. 552)*.	The eccentricity of a conic is denoted by e. **ellipse:** $0 < e < 1$ **parabola:** $e = 1$ **hyperbola:** $e > 1$ The graph of a polar equation of the form (1) $r = (ep)/(1 \pm e \cos \theta)$ or (2) $r = (ep)/(1 \pm e \sin \theta)$ is a conic, where $e > 0$ is the eccentricity and $|p|$ is the distance between the focus (pole) and the directrix.	105–112
Use equations of conics in polar form to model real-life problems *(p. 555)*.	You can use the equation of a conic in polar form to model the orbit of Halley's comet. (See Example 4.)	113, 114

Review Exercises   See **CalcChat.com** for tutorial help and worked-out solutions to odd-numbered exercises.

10.1 **Finding the Inclination of a Line** In Exercises 1–4, find the inclination θ (in radians and degrees) of the line with the given characteristics.

1. Passes through the points $(-1, 2)$ and $(2, 5)$
2. Passes through the points $(3, 4)$ and $(-2, 7)$
3. Equation: $y = 2x + 4$
4. Equation: $x - 5y = 7$

Finding the Angle Between Two Lines In Exercises 5–8, find the angle θ (in radians and degrees) between the lines.

5. $4x + y = 2$
   $-5x + y = -1$

6. $-5x + 3y = 3$
   $-2x + 3y = 1$

7. $2x - 7y = 8$
   $\frac{2}{5}x + y = 0$

8. $0.02x + 0.07y = 0.18$
   $0.09x - 0.04y = 0.17$

Finding the Distance Between a Point and a Line In Exercises 9 and 10, find the distance between the point and the line.

Point	Line
9. $(5, 3)$	$x - y = 10$
10. $(0, 4)$	$x + 2y = 2$

10.2 **Forming a Conic Section** In Exercises 11 and 12, state what type of conic is formed by the intersection of the plane and the double-napped cone.

11.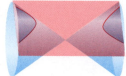
12.

Finding the Standard Equation of a Parabola In Exercises 13–16, find the standard form of the equation of the parabola with the given characteristics. Then sketch the parabola.

13. Vertex: $(0, 0)$
    Focus: $(4, 0)$

14. Vertex: $(2, 0)$
    Focus: $(0, 0)$

15. Vertex: $(0, 2)$
    Directrix: $x = -3$

16. Vertex: $(-3, -3)$
    Directrix: $y = 0$

Finding the Tangent Line at a Point on a Parabola In Exercises 17 and 18, find the equation of the tangent line to the parabola at the given point.

17. $y = 2x^2$, $(-1, 2)$
18. $x^2 = -2y$, $(-4, -8)$

19. **Architecture** A parabolic archway is 10 meters high at the vertex. At a height of 8 meters, the width of the archway is 6 meters (see figure). How wide is the archway at ground level?

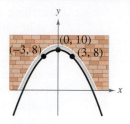

Figure for 19

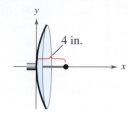

Figure for 20

20. **Parabolic Microphone** The receiver of a parabolic microphone is at the focus of its parabolic reflector, 4 inches from the vertex (see figure). Write an equation of a cross section of the reflector with its focus on the positive x-axis and its vertex at the origin.

10.3 **Finding the Standard Equation of an Ellipse** In Exercises 21–24, find the standard form of the equation of the ellipse with the given characteristics.

21. Vertices: $(-2, 0)$, $(8, 0)$; foci: $(0, 0)$, $(6, 0)$
22. Vertices: $(4, 3)$, $(4, 7)$; foci: $(4, 4)$, $(4, 6)$
23. Vertices: $(0, 1)$, $(4, 1)$; endpoints of the minor axis: $(2, 0)$, $(2, 2)$
24. Vertices: $(-4, -1)$, $(-4, 11)$; endpoints of the minor axis: $(-6, 5)$, $(-2, 5)$

25. **Architecture** A contractor plans to construct a semielliptical arch 10 feet wide and 4 feet high. Where should the foci be placed in order to sketch the arch?

26. **Wading Pool** You are building a wading pool that is in the shape of an ellipse. Your plans give an equation for the elliptical shape of the pool measured in feet as

$$\frac{x^2}{324} + \frac{y^2}{196} = 1.$$

Find the longest distance across the pool, the shortest distance, and the distance between the foci.

Sketching an Ellipse In Exercises 27–30, find the center, vertices, foci, and eccentricity of the ellipse. Then sketch the ellipse.

27. $\dfrac{(x + 1)^2}{25} + \dfrac{(y - 2)^2}{49} = 1$

28. $\dfrac{(x - 5)^2}{1} + \dfrac{(y + 3)^2}{36} = 1$

29. $16x^2 + 9y^2 - 32x + 72y + 16 = 0$

30. $4x^2 + 25y^2 + 16x - 150y + 141 = 0$

10.4 **Finding the Standard Equation of a Hyperbola** In Exercises 31–34, find the standard form of the equation of the hyperbola with the given characteristics.

31. Vertices: $(0, \pm 1)$; foci: $(0, \pm 2)$

32. Vertices: $(3, 3), (-3, 3)$; foci: $(4, 3), (-4, 3)$

33. Foci: $(\pm 5, 0)$; asymptotes: $y = \pm \frac{3}{4}x$

34. Foci: $(0, \pm 13)$; asymptotes: $y = \pm \frac{5}{12}x$

Sketching a Hyperbola In Exercises 35–38, find the center, vertices, foci, and the equations of the asymptotes of the hyperbola. Then sketch the hyperbola using the asymptotes as an aid.

35. $\dfrac{(x-5)^2}{36} - \dfrac{(y+3)^2}{16} = 1$ 36. $\dfrac{(y-1)^2}{4} - x^2 = 1$

37. $9x^2 - 16y^2 - 18x - 32y - 151 = 0$

38. $-4x^2 + 25y^2 - 8x + 150y + 121 = 0$

39. **Navigation** Radio transmitting station A is located 200 miles east of transmitting station B. A ship is in an area to the north and 40 miles west of station A. Synchronized radio pulses transmitted at 186,000 miles per second by the two stations are received 0.0005 second sooner from station A than from station B. How far north is the ship?

40. **Locating an Explosion** Two of your friends live 4 miles apart and on the same "east-west" street, and you live halfway between them. You are having a three-way phone conversation when you hear an explosion. Six seconds later, your friend to the east hears the explosion, and your friend to the west hears it 8 seconds after you do. Find equations of two hyperbolas that would locate the explosion. (Assume that the coordinate system is measured in feet and that sound travels at 1100 feet per second.)

Classifying a Conic from a General Equation In Exercises 41–44, classify the graph of the equation as a circle, a parabola, an ellipse, or a hyperbola.

41. $5x^2 - 2y^2 + 10x - 4y + 17 = 0$

42. $-4y^2 + 5x + 3y + 7 = 0$

43. $3x^2 + 2y^2 - 12x + 12y + 29 = 0$

44. $4x^2 + 4y^2 - 4x + 8y - 11 = 0$

10.5 **Rotation of Axes** In Exercises 45–48, rotate the axes to eliminate the xy-term in the equation. Then write the equation in standard form. Sketch the graph of the resulting equation, showing both sets of axes.

45. $xy + 3 = 0$

46. $x^2 - 4xy + y^2 + 9 = 0$

47. $5x^2 - 2xy + 5y^2 - 12 = 0$

48. $4x^2 + 8xy + 4y^2 + 7\sqrt{2}x + 9\sqrt{2}y = 0$

Rotation and Graphing Utilities In Exercises 49–52, (a) use the discriminant to classify the graph of the equation, (b) use the Quadratic Formula to solve for y, and (c) use a graphing utility to graph the equation.

49. $16x^2 - 24xy + 9y^2 - 30x - 40y = 0$

50. $13x^2 - 8xy + 7y^2 - 45 = 0$

51. $x^2 + y^2 + 2xy + 2\sqrt{2}x - 2\sqrt{2}y + 2 = 0$

52. $x^2 - 10xy + y^2 + 1 = 0$

10.6 **Sketching a Curve** In Exercises 53 and 54, (a) create a table of x- and y-values for the parametric equations using $t = -2, -1, 0, 1,$ and 2, and (b) plot the points (x, y) generated in part (a) and sketch a graph of the parametric equations.

53. $x = 3t - 2$ and $y = 7 - 4t$

54. $x = \dfrac{1}{4}t$ and $y = \dfrac{6}{t+3}$

Sketching a Curve In Exercises 55–60, (a) sketch the curve represented by the parametric equations (indicate the orientation of the curve) and (b) eliminate the parameter and write the resulting rectangular equation whose graph represents the curve. Adjust the domain of the rectangular equation, if necessary. (c) Verify your result with a graphing utility.

55. $x = 2t$ 56. $x = 1 + 4t$
    $y = 4t$ $y = 2 - 3t$

57. $x = t^2$ 58. $x = t + 4$
    $y = \sqrt{t}$ $y = t^2$

59. $x = 3\cos\theta$ 60. $x = 3 + 3\cos\theta$
    $y = 3\sin\theta$ $y = 2 + 5\sin\theta$

Finding Parametric Equations for a Graph In Exercises 61–64, find a set of parametric equations to represent the graph of the rectangular equation using (a) $t = x$, (b) $t = x + 1$, and (c) $t = 3 - x$.

61. $y = 2x + 3$

62. $y = 4 - 3x$

63. $y = x^2 + 3$

64. $y = 2 - x^2$

65. $y = 2x^2 + 2$

66. $y = 1 - 4x^2$

10.7 **Plotting Points in the Polar Coordinate System** In Exercises 67–70, plot the point given in polar coordinates and find two additional polar representations of the point, using $-2\pi < \theta < 2\pi$.

67. $\left(2, \dfrac{\pi}{4}\right)$ 68. $\left(-5, -\dfrac{\pi}{3}\right)$

69. $(-7, 4.19)$

70. $\left(\sqrt{3}, 2.62\right)$

Polar-to-Rectangular Conversion In Exercises 71–74, a point in polar coordinates is given. Convert the point to rectangular coordinates.

71. $\left(-1, \dfrac{\pi}{3}\right)$ **72.** $\left(2, \dfrac{5\pi}{4}\right)$

73. $\left(3, \dfrac{3\pi}{4}\right)$ **74.** $\left(0, \dfrac{\pi}{2}\right)$

Rectangular-to-Polar Conversion In Exercises 75–78, a point in rectangular coordinates is given. Convert the point to polar coordinates.

75. $(0, 1)$ **76.** $\left(-\sqrt{5}, \sqrt{5}\right)$

77. $(4, 6)$ **78.** $(3, -4)$

Converting a Rectangular Equation to Polar Form In Exercises 79–84, convert the rectangular equation to polar form.

79. $x^2 + y^2 = 81$ **80.** $x^2 + y^2 = 48$

81. $x^2 + y^2 - 6y = 0$ **82.** $x^2 + y^2 - 4x = 0$

83. $xy = 5$ **84.** $xy = -2$

Converting a Polar Equation to Rectangular Form In Exercises 85–90, convert the polar equation to rectangular form.

85. $r = 5$ **86.** $r = 12$

87. $r = 3\cos\theta$ **88.** $r = 8\sin\theta$

89. $r^2 = \sin\theta$ **90.** $r^2 = 4\cos 2\theta$

10.8 **Sketching the Graph of a Polar Equation In Exercises 91–100, sketch the graph of the polar equation using symmetry, zeros, maximum r-values, and any other additional points.**

91. $r = 6$ **92.** $r = 11$

93. $r = 4\sin 2\theta$ **94.** $r = \cos 5\theta$

95. $r = -2(1 + \cos\theta)$ **96.** $r = 1 - 4\cos\theta$

97. $r = 2 + 6\sin\theta$ **98.** $r = 5 - 5\cos\theta$

99. $r = -3\cos 2\theta$ **100.** $r^2 = \cos 2\theta$

Identifying Types of Polar Graphs In Exercises 101–104, identify the type of polar graph and use a graphing utility to graph the equation.

101. $r = 3(2 - \cos\theta)$ **102.** $r = 5(1 - 2\cos\theta)$

103. $r = 8\cos 3\theta$ **104.** $r^2 = 2\sin 2\theta$

10.9 **Sketching a Conic In Exercises 105–108, identify the conic and sketch its graph.**

105. $r = \dfrac{1}{1 + 2\sin\theta}$ **106.** $r = \dfrac{6}{1 + \sin\theta}$

107. $r = \dfrac{4}{5 - 3\cos\theta}$ **108.** $r = \dfrac{16}{4 + 5\cos\theta}$

Finding the Polar Equation of a Conic In Exercises 109–112, find a polar equation of the conic with its focus at the pole.

109. Parabola   Vertex: $(2, \pi)$

110. Parabola   Vertex: $(2, \pi/2)$

111. Ellipse   Vertices: $(5, 0)$, $(1, \pi)$

112. Hyperbola   Vertices: $(1, 0)$, $(7, 0)$

113. Explorer 18 On November 27, 1963, the United States launched Explorer 18. Its low and high points above the surface of Earth were 110 miles and 122,800 miles, respectively. The center of Earth was at one focus of the orbit (see figure). Find the polar equation of the orbit and find the distance between the surface of Earth (assume Earth has a radius of 4000 miles) and the satellite when $\theta = \pi/3$.

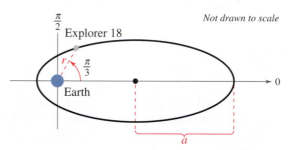

114. Asteroid An asteroid takes a parabolic path with Earth as its focus. It is about 6,000,000 miles from Earth at its closest approach. Write the polar equation of the path of the asteroid with its vertex at $\theta = \pi/2$. Find the distance between the asteroid and Earth when $\theta = -\pi/3$.

Exploration

True or False? In Exercises 115–117, determine whether the statement is true or false. Justify your answer.

115. The graph of $\frac{1}{4}x^2 - y^4 = 1$ is a hyperbola.

116. Only one set of parametric equations can represent the line $y = 3 - 2x$.

117. There is a unique polar coordinate representation of each point in the plane.

118. Think About It Consider an ellipse with the major axis horizontal and 10 units in length. The number b in the standard form of the equation of the ellipse must be less than what real number? Explain the change in the shape of the ellipse as b approaches this number.

119. Think About It What is the relationship between the graphs of the rectangular and polar equations?

(a) $x^2 + y^2 = 25$, $r = 5$

(b) $x - y = 0$, $\theta = \dfrac{\pi}{4}$

Chapter Test

See CalcChat.com for tutorial help and worked-out solutions to odd-numbered exercises.

Take this test as you would take a test in class. When you are finished, check your work against the answers given in the back of the book.

1. Find the inclination of the line $2x - 5y + 5 = 0$.
2. Find the angle between the lines $3x + 2y = 4$ and $4x - y = -6$.
3. Find the distance between the point $(7, 5)$ and the line $y = 5 - x$.

In Exercises 4–7, identify the conic and write the equation in standard form. Find the center, vertices, foci, and the equations of the asymptotes (if applicable). Then sketch the conic.

4. $y^2 - 2x + 2 = 0$
5. $x^2 - 4y^2 - 4x = 0$
6. $9x^2 + 16y^2 + 54x - 32y - 47 = 0$
7. $2x^2 + 2y^2 - 8x - 4y + 9 = 0$

8. Find the standard form of the equation of the parabola with vertex $(2, -3)$ and a vertical axis that passes through the point $(4, 0)$.
9. Find the standard form of the equation of the hyperbola with foci $(0, \pm 2)$ and asymptotes $y = \pm \frac{1}{9}x$.
10. (a) Determine the number of degrees the axes must be rotated to eliminate the xy-term of the conic $x^2 + 6xy + y^2 - 6 = 0$.
    (b) Sketch the conic from part (a) and use a graphing utility to confirm your result.
11. Sketch the curve represented by the parametric equations $x = 2 + 3 \cos \theta$ and $y = 2 \sin \theta$. Eliminate the parameter and write the resulting rectangular equation.
12. Find a set of parametric equations to represent the graph of the rectangular equation $y = 3 - x^2$ using (a) $t = x$ and (b) $t = x + 2$.
13. Convert the polar coordinates $\left(-2, \frac{5\pi}{6}\right)$ to rectangular form.
14. Convert the rectangular coordinates $(2, -2)$ to polar form and find two additional polar representations of the point, using $-2\pi < \theta < 2\pi$.
15. Convert the rectangular equation $x^2 + y^2 - 3x = 0$ to polar form.

In Exercises 16–19, sketch the graph of the polar equation. Identify the type of graph.

16. $r = \dfrac{4}{1 + \cos \theta}$
17. $r = \dfrac{4}{2 + \sin \theta}$
18. $r = 2 + 3 \sin \theta$
19. $r = 2 \sin 4\theta$

20. Find a polar equation of the ellipse with focus at the pole, eccentricity $e = \frac{1}{4}$, and directrix $y = 4$.
21. A straight road rises with an inclination of 0.15 radian from the horizontal. Find the slope of the road and the change in elevation over a one-mile stretch of the road.
22. A baseball is hit at a point 3 feet above the ground toward the left field fence. The fence is 10 feet high and 375 feet from home plate. The path of the baseball can be modeled by the parametric equations $x = (115 \cos \theta)t$ and $y = 3 + (115 \sin \theta)t - 16t^2$. Will the baseball go over the fence when it is hit at an angle of $\theta = 30°$? Will the baseball go over the fence when $\theta = 35°$?

Proofs in Mathematics ■ ■ ■ ■ ■ ■ ■ ■ ■ ■ ■ ■ ■ ■

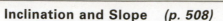

Inclination and Slope   *(p. 508)*

If a nonvertical line has inclination θ and slope m, then $m = \tan \theta$.

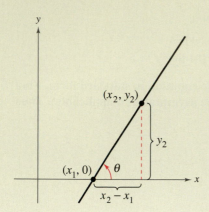

Proof

If $m = 0$, then the line is horizontal and $\theta = 0$. So, the result is true for horizontal lines because $m = 0 = \tan 0$.

If the line has a positive slope, then it will intersect the x-axis. Label this point $(x_1, 0)$, as shown in the figure. If (x_2, y_2) is a second point on the line, then the slope is

$$m = \frac{y_2 - 0}{x_2 - x_1} = \frac{y_2}{x_2 - x_1} = \tan \theta.$$

The case in which the line has a negative slope can be proved in a similar manner. ■

Distance Between a Point and a Line   *(p. 510)*

The distance between the point (x_1, y_1) and the line $Ax + By + C = 0$ is

$$d = \frac{|Ax_1 + By_1 + C|}{\sqrt{A^2 + B^2}}.$$

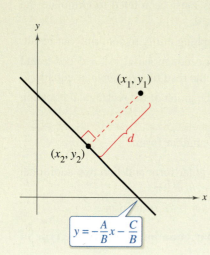

Proof

For simplicity, assume that the given line is neither horizontal nor vertical (see figure). By writing the equation $Ax + By + C = 0$ in slope-intercept form

$$y = -\frac{A}{B}x - \frac{C}{B}$$

you can see that the line has a slope of $m = -A/B$. So, the slope of the line passing through (x_1, y_1) and perpendicular to the given line is B/A, and its equation is $y - y_1 = (B/A)(x - x_1)$. These two lines intersect at the point (x_2, y_2), where

$$x_2 = \frac{B(Bx_1 - Ay_1) - AC}{A^2 + B^2} \quad \text{and} \quad y_2 = \frac{A(-Bx_1 + Ay_1) - BC}{A^2 + B^2}.$$

Finally, the distance between (x_1, y_1) and (x_2, y_2) is

$$d = \sqrt{(x_2 - x_1)^2 + (y_2 - y_1)^2}$$

$$= \sqrt{\left(\frac{B^2x_1 - ABy_1 - AC}{A^2 + B^2} - x_1\right)^2 + \left(\frac{-ABx_1 + A^2y_1 - BC}{A^2 + B^2} - y_1\right)^2}$$

$$= \sqrt{\frac{A^2(Ax_1 + By_1 + C)^2 + B^2(Ax_1 + By_1 + C)^2}{(A^2 + B^2)^2}}$$

$$= \frac{|Ax_1 + By_1 + C|}{\sqrt{A^2 + B^2}}.$$ ■

PARABOLIC PATHS

There are many natural occurrences of parabolas in real life. For instance, the famous astronomer Galileo discovered in the 17th century that an object that is projected upward and obliquely to the pull of gravity travels in a parabolic path. Examples of this are the center of gravity of a jumping dolphin and the path of water molecules in a drinking fountain.

Standard Equation of a Parabola *(p. 513)*

The standard form of the equation of a parabola with vertex at (h, k) is as follows.

$$(x - h)^2 = 4p(y - k), \quad p \neq 0 \qquad \text{Vertical axis; directrix: } y = k - p$$

$$(y - k)^2 = 4p(x - h), \quad p \neq 0 \qquad \text{Horizontal axis; directrix: } x = h - p$$

The focus lies on the axis p units *(directed distance)* from the vertex. If the vertex is at the origin, then the equation takes one of the following forms.

$$x^2 = 4py \qquad \text{Vertical axis}$$

$$y^2 = 4px \qquad \text{Horizontal axis}$$

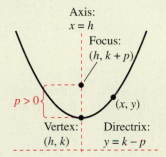

Axis:
$x = h$

Focus:
$(h, k + p)$

$p > 0$

(x, y)

Vertex:
(h, k)

Directrix:
$y = k - p$

Parabola with vertical axis

Proof

For the case in which the directrix is parallel to the x-axis and the focus lies above the vertex, as shown in the top figure, if (x, y) is any point on the parabola, then, by definition, it is equidistant from the focus

$$(h, k + p)$$

and the directrix

$$y = k - p.$$

So, you have

$$\sqrt{(x - h)^2 + [y - (k + p)]^2} = y - (k - p)$$

$$(x - h)^2 + [y - (k + p)]^2 = [y - (k - p)]^2$$

$$(x - h)^2 + y^2 - 2y(k + p) + (k + p)^2 = y^2 - 2y(k - p) + (k - p)^2$$

$$(x - h)^2 + y^2 - 2ky - 2py + k^2 + 2pk + p^2 = y^2 - 2ky + 2py + k^2 - 2pk + p^2$$

$$(x - h)^2 - 2py + 2pk = 2py - 2pk$$

$$(x - h)^2 = 4p(y - k).$$

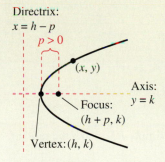

Directrix:
$x = h - p$

$p > 0$

(x, y)

Axis:
$y = k$

Focus:
$(h + p, k)$

Vertex: (h, k)

Parabola with horizontal axis

For the case in which the directrix is parallel to the y-axis and the focus lies to the right of the vertex, as shown in the bottom figure, if (x, y) is any point on the parabola, then, by definition, it is equidistant from the focus

$$(h + p, k)$$

and the directrix

$$x = h - p.$$

So, you have

$$\sqrt{[x - (h + p)]^2 + (y - k)^2} = x - (h - p)$$

$$[x - (h + p)]^2 + (y - k)^2 = [x - (h - p)]^2$$

$$x^2 - 2x(h + p) + (h + p)^2 + (y - k)^2 = x^2 - 2x(h - p) + (h - p)^2$$

$$x^2 - 2hx - 2px + h^2 + 2ph + p^2 + (y - k)^2 = x^2 - 2hx + 2px + h^2 - 2ph + p^2$$

$$-2px + 2ph + (y - k)^2 = 2px - 2ph$$

$$(y - k)^2 = 4p(x - h).$$

Note that if a parabola is centered at the origin, then the two equations above would simplify to $x^2 = 4py$ and $y^2 = 4px$, respectively. ■

Polar Equations of Conics   *(p. 552)*

The graph of a polar equation of the form

1. $r = \dfrac{ep}{1 \pm e \cos \theta}$

or

2. $r = \dfrac{ep}{1 \pm e \sin \theta}$

is a conic, where $e > 0$ is the eccentricity and $|p|$ is the distance between the focus (pole) and the directrix.

Proof

A proof for

$$r = \frac{ep}{1 + e \cos \theta}$$

with $p > 0$ is shown here. The proofs of the other cases are similar. In the figure, consider a vertical directrix, p units to the right of the focus $F(0, 0)$. If $P(r, \theta)$ is a point on the graph of

$$r = \frac{ep}{1 + e \cos \theta}$$

then the distance between P and the directrix is

$$
\begin{aligned}
PQ &= |p - x| \\
&= |p - r \cos \theta| \\
&= \left| p - \left(\frac{ep}{1 + e \cos \theta} \right) \cos \theta \right| \\
&= \left| p \left(1 - \frac{e \cos \theta}{1 + e \cos \theta} \right) \right| \\
&= \left| \frac{p}{1 + e \cos \theta} \right| \\
&= \left| \frac{r}{e} \right|.
\end{aligned}
$$

Moreover, because the distance between P and the pole is simply $PF = |r|$, the ratio of PF to PQ is

$$
\begin{aligned}
\frac{PF}{PQ} &= \frac{|r|}{\left| \dfrac{r}{e} \right|} \\
&= |e| \\
&= e
\end{aligned}
$$

and, by definition, the graph of the equation must be a conic.   ■

P.S. Problem Solving ■ ■ ■ ■ ■ ■ ■ ■ ■ ■ ■ ■ ■

1. **Mountain Climbing** Several mountain climbers are located in a mountain pass between two peaks. The angles of elevation to the two peaks are 0.84 radian and 1.10 radians. A range finder shows that the distances to the peaks are 3250 feet and 6700 feet, respectively (see figure).

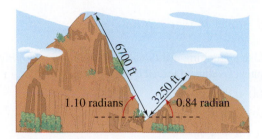

(a) Find the angle between the two lines.

(b) Approximate the amount of vertical climb that is necessary to reach the summit of each peak.

2. **Finding the Equation of a Parabola** Find the general equation of a parabola that has the x-axis as the axis of symmetry and the focus at the origin.

3. **Area** Find the area of the square inscribed in the ellipse, as shown below.

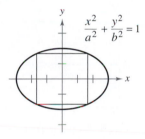

$$\frac{x^2}{a^2} + \frac{y^2}{b^2} = 1$$

4. **Involute** The *involute* of a circle is described by the endpoint P of a string that is held taut as it is unwound from a spool (see figure). The spool does not rotate. Show that

$$x = r(\cos \theta + \theta \sin \theta)$$

and

$$y = r(\sin \theta - \theta \cos \theta)$$

is a parametric representation of the involute of a circle.

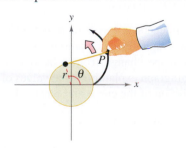

5. **Tour Boat** A tour boat travels between two islands that are 12 miles apart (see figure). There is enough fuel for a 20-mile trip.

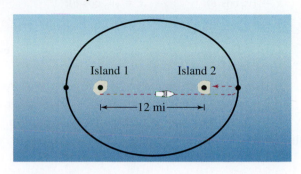

(a) Explain why the region in which the boat can travel is bounded by an ellipse.

(b) Let $(0, 0)$ represent the center of the ellipse. Find the coordinates of each island.

(c) The boat travels from Island 1, past Island 2 to a vertex of the ellipse, and then to Island 2. How many miles does the boat travel? Use your answer to find the coordinates of the vertex.

(d) Use the results from parts (b) and (c) to write an equation of the ellipse that bounds the region in which the boat can travel.

6. **Finding the Equation of a Hyperbola** Find an equation of the hyperbola such that for any point on the hyperbola, the absolute value of the difference of its distances from the points $(2, 2)$ and $(10, 2)$ is 6.

7. **Proof** Prove that the graph of the equation

$$Ax^2 + Cy^2 + Dx + Ey + F = 0$$

is one of the following (except in degenerate cases).

Conic	Condition
(a) Circle	$A = C$
(b) Parabola	$A = 0$ or $C = 0$ (but not both)
(c) Ellipse	$AC > 0$
(d) Hyperbola	$AC < 0$

8. **Proof** Prove that

$$c^2 = a^2 + b^2$$

for the equation of the hyperbola

$$\frac{x^2}{a^2} - \frac{y^2}{b^2} = 1$$

where the distance from the center of the hyperbola $(0, 0)$ to a focus is c.

565

9. Projectile Motion The following sets of parametric equations model projectile motion.

$$x = (v_0 \cos \theta)t \qquad x = (v_0 \cos \theta)t$$
$$y = (v_0 \sin \theta)t \qquad y = h + (v_0 \sin \theta)t - 16t^2$$

(a) Under what circumstances would you use each model?

(b) Eliminate the parameter for each set of equations.

(c) In which case is the path of the moving object not affected by a change in the velocity v? Explain.

10. Orientation of an Ellipse As t increases, the ellipse given by the parametric equations

$$x = \cos t$$

and

$$y = 2 \sin t$$

is traced out *counterclockwise*. Find a parametric representation for which the same ellipse is traced out *clockwise*.

11. Rose Curves The rose curves described in this chapter are of the form

$$r = a \cos n\theta$$

or

$$r = a \sin n\theta$$

where n is a positive integer that is greater than or equal to 2. Use a graphing utility to graph $r = a \cos n\theta$ and $r = a \sin n\theta$ for some noninteger values of n. Describe the graphs.

12. Strophoid The curve given by the parametric equations

$$x = \frac{1 - t^2}{1 + t^2}$$

and

$$y = \frac{t(1 - t^2)}{1 + t^2}$$

is called a **strophoid.**

(a) Find a rectangular equation of the strophoid.

(b) Find a polar equation of the strophoid.

(c) Use a graphing utility to graph the strophoid.

13. Hypocycloid A **hypocycloid** has the parametric equations

$$x = (a - b) \cos t + b \cos\left(\frac{a - b}{b}t\right)$$

and

$$y = (a - b) \sin t - b \sin\left(\frac{a - b}{b}t\right).$$

Use a graphing utility to graph the hypocycloid for each pair of values. Describe each graph.

(a) $a = 2, b = 1$

(b) $a = 3, b = 1$

(c) $a = 4, b = 1$

(d) $a = 10, b = 1$

(e) $a = 3, b = 2$

(f) $a = 4, b = 3$

14. Butterfly Curve The graph of the polar equation

$$r = e^{\cos \theta} - 2 \cos 4\theta + \sin^5\left(\frac{\theta}{12}\right)$$

is called the *butterfly curve*, as shown in the figure.

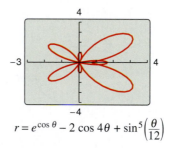

$$r = e^{\cos \theta} - 2 \cos 4\theta + \sin^5\left(\frac{\theta}{12}\right)$$

(a) The graph shown was produced using $0 \le \theta \le 2\pi$. Does this show the entire graph? Explain your reasoning.

(b) Approximate the maximum r-value of the graph. Does this value change when you use $0 \le \theta \le 4\pi$ instead of $0 \le \theta \le 2\pi$? Explain.

15. Writing Use a graphing utility to graph the polar equation

$$r = \cos 5\theta + n \cos \theta$$

for the integers $n = -5$ to $n = 5$ using $0 \le \theta \le \pi$. As you graph these equations, you should see the graph's shape change from a heart to a bell. Write a short paragraph explaining what values of n produce the heart portion of the curve and what values of n produce the bell portion.

A.1 Real Numbers and Their Properties

Real numbers can represent many real-life quantities, such as the federal deficit.

- Represent and classify real numbers.
- Order real numbers and use inequalities.
- Find the absolute values of real numbers and find the distance between two real numbers.
- Evaluate algebraic expressions.
- Use the basic rules and properties of algebra.

Real Numbers

Real numbers can describe quantities in everyday life such as age, miles per gallon, and population. Symbols such as

$$-5, 9, 0, \tfrac{4}{3}, 0.666\ldots, 28.21, \sqrt{2}, \pi, \text{ and } \sqrt[3]{-32}$$

represent real numbers. Here are some important **subsets** (each member of a subset B is also a member of a set A) of the real numbers. The three dots, called *ellipsis points,* indicate that the pattern continues indefinitely.

$$\{1, 2, 3, 4, \ldots\} \qquad \text{Set of natural numbers}$$

$$\{0, 1, 2, 3, 4, \ldots\} \qquad \text{Set of whole numbers}$$

$$\{\ldots, -3, -2, -1, 0, 1, 2, 3, \ldots\} \qquad \text{Set of integers}$$

A real number is **rational** when it can be written as the ratio p/q of two integers, where $q \neq 0$. For instance, the numbers

$$\tfrac{1}{3} = 0.3333\ldots = 0.\overline{3}, \tfrac{1}{8} = 0.125, \text{ and } \tfrac{125}{111} = 1.126126\ldots = 1.\overline{126}$$

are rational. The decimal representation of a rational number either repeats $\left(\text{as in } \tfrac{173}{55} = 3.1\overline{45}\right)$ or terminates $\left(\text{as in } \tfrac{1}{2} = 0.5\right)$. A real number that cannot be written as the ratio of two integers is called **irrational.** Irrational numbers have infinite nonrepeating decimal representations. For instance, the numbers

$$\sqrt{2} = 1.4142135\ldots \approx 1.41 \quad \text{and} \quad \pi = 3.1415926\ldots \approx 3.14$$

are irrational. (The symbol $\approx$ means "is approximately equal to.") Figure A.1 shows subsets of real numbers and their relationships to each other.

EXAMPLE 1 Classifying Real Numbers

Determine which numbers in the set $\left\{-13, -\sqrt{5}, -1, -\tfrac{1}{3}, 0, \tfrac{5}{8}, \sqrt{2}, \pi, 7\right\}$ are (a) natural numbers, (b) whole numbers, (c) integers, (d) rational numbers, and (e) irrational numbers.

Solution

a. Natural numbers: $\{7\}$

b. Whole numbers: $\{0, 7\}$

c. Integers: $\{-13, -1, 0, 7\}$

d. Rational numbers: $\left\{-13, -1, -\dfrac{1}{3}, 0, \dfrac{5}{8}, 7\right\}$

e. Irrational numbers: $\left\{-\sqrt{5}, \sqrt{2}, \pi\right\}$

✓ **Checkpoint** 🔊) *Audio-video solution in English & Spanish at LarsonPrecalculus.com.*

Repeat Example 1 for the set $\left\{-\pi, -\tfrac{1}{4}, \tfrac{6}{3}, \tfrac{1}{2}\sqrt{2}, -7.5, -1, 8, -22\right\}$.

Real numbers

Irrational numbers — Rational numbers

Integers — Noninteger fractions (positive and negative)

Negative integers — Whole numbers

Natural numbers — Zero

Subsets of real numbers
Figure A.1

Real numbers are represented graphically on the **real number line.** When you draw a point on the real number line that corresponds to a real number, you are **plotting** the real number. The point 0 on the real number line is the **origin.** Numbers to the right of 0 are positive, and numbers to the left of 0 are negative, as shown below. The term **nonnegative** describes a number that is either positive or zero.

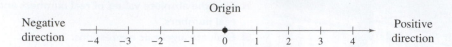

As illustrated below, there is a *one-to-one correspondence* between real numbers and points on the real number line.

Every real number corresponds to exactly one point on the real number line.

Every point on the real number line corresponds to exactly one real number.

EXAMPLE 2 **Plotting Points on the Real Number Line**

Plot the real numbers on the real number line.

a. $-\dfrac{7}{4}$

b. 2.3

c. $\dfrac{2}{3}$

d. -1.8

Solution The following figure shows all four points.

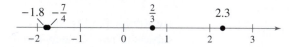

a. The point representing the real number $-\frac{7}{4} = -1.75$ lies between -2 and -1, but closer to -2, on the real number line.

b. The point representing the real number 2.3 lies between 2 and 3, but closer to 2, on the real number line.

c. The point representing the real number $\frac{2}{3} = 0.666\ldots$ lies between 0 and 1, but closer to 1, on the real number line.

d. The point representing the real number -1.8 lies between -2 and -1, but closer to -2, on the real number line. Note that the point representing -1.8 lies slightly to the left of the point representing $-\frac{7}{4}$.

✓ *Checkpoint*  *Audio-video solution in English & Spanish at LarsonPrecalculus.com.*

Plot the real numbers on the real number line.

a. $\dfrac{5}{2}$ b. -1.6

c. $-\dfrac{3}{4}$ d. 0.7

Ordering Real Numbers

One important property of real numbers is that they are *ordered*.

> ### Definition of Order on the Real Number Line
>
> If a and b are real numbers, then a is less than b when $b - a$ is positive. The **inequality** $a < b$ denotes the **order** of a and b. This relationship can also be described by saying that b is *greater than a* and writing $b > a$. The inequality $a \le b$ means that a is *less than or equal to b*, and the inequality $b \ge a$ means that b is *greater than or equal to a*. The symbols $<$, $>$, $\le$, and $\ge$ are *inequality symbols*.

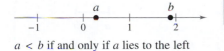

$a < b$ if and only if a lies to the left of b.

Figure A.2

Geometrically, this definition implies that $a < b$ if and only if a lies to the *left* of b on the real number line, as shown in Figure A.2.

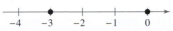

Figure A.3

Figure A.4

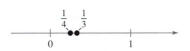

Figure A.5

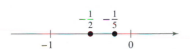

Figure A.6

EXAMPLE 3   Ordering Real Numbers

Place the appropriate inequality symbol ($<$ or $>$) between the pair of real numbers.

a. $-3, 0$ **b.** $-2, -4$ **c.** $\frac{1}{4}, \frac{1}{3}$ **d.** $-\frac{1}{5}, -\frac{1}{2}$

Solution

a. Because -3 lies to the left of 0 on the real number line, as shown in Figure A.3, you can say that -3 is *less than* 0, and write $-3 < 0$.

b. Because -2 lies to the right of -4 on the real number line, as shown in Figure A.4, you can say that -2 is *greater than* -4, and write $-2 > -4$.

c. Because $\frac{1}{4}$ lies to the left of $\frac{1}{3}$ on the real number line, as shown in Figure A.5, you can say that $\frac{1}{4}$ is *less than* $\frac{1}{3}$, and write $\frac{1}{4} < \frac{1}{3}$.

d. Because $-\frac{1}{5}$ lies to the right of $-\frac{1}{2}$ on the real number line, as shown in Figure A.6, you can say that $-\frac{1}{5}$ is *greater than* $-\frac{1}{2}$, and write $-\frac{1}{5} > -\frac{1}{2}$.

✓ *Checkpoint* *Audio-video solution in English & Spanish at LarsonPrecalculus.com.*

Place the appropriate inequality symbol ($<$ or $>$) between the pair of real numbers.

a. $1, -5$ **b.** $\frac{3}{2}, 7$ **c.** $-\frac{2}{3}, -\frac{3}{4}$ **d.** $-3.5, 1$

EXAMPLE 4   Interpreting Inequalities

Describe the subset of real numbers that the inequality represents.

a. $x \le 2$ **b.** $-2 \le x < 3$

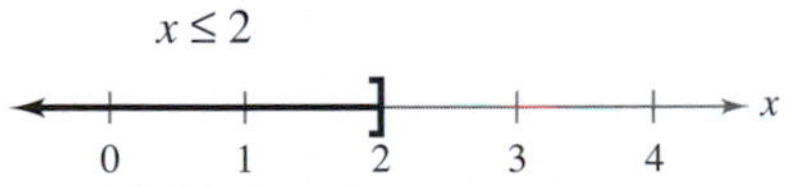

Figure A.7

Solution

a. The inequality $x \le 2$ denotes all real numbers less than or equal to 2, as shown in Figure A.7.

b. The inequality $-2 \le x < 3$ means that $x \ge -2$ *and* $x < 3$. This "double inequality" denotes all real numbers between -2 and 3, including -2 but not including 3, as shown in Figure A.8.

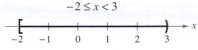

Figure A.8

✓ *Checkpoint*  *Audio-video solution in English & Spanish at LarsonPrecalculus.com.*

Describe the subset of real numbers that the inequality represents.

a. $x > -3$ **b.** $0 < x \le 4$

Inequalities can describe subsets of real numbers called **intervals.** In the bounded intervals below, the real numbers a and b are the **endpoints** of each interval. The endpoints of a closed interval are included in the interval, whereas the endpoints of an open interval are not included in the interval.

• •REMARK The reason that the four types of intervals at the right are called *bounded* is that each has a finite length. An interval that does not have a finite length is *unbounded* (see below).

Bounded Intervals on the Real Number Line

Notation	Interval Type	Inequality	Graph
$[a, b]$	Closed	$a \leq x \leq b$	
(a, b)	Open	$a < x < b$	
$[a, b)$		$a \leq x < b$	
$(a, b]$		$a < x \leq b$	

The symbols ∞, **positive infinity,** and $-\infty$, **negative infinity,** do not represent real numbers. They are simply convenient symbols used to describe the unboundedness of an interval such as $(1, \infty)$ or $(-\infty, 3]$.

• •REMARK Whenever you write an interval containing ∞ or $-\infty$, always use a parenthesis and never a bracket next to these symbols. This is because ∞ and $-\infty$ are never an endpoint of an interval and therefore are not included in the interval.

Unbounded Intervals on the Real Number Line

Notation	Interval Type	Inequality	Graph
$[a, \infty)$		$x \geq a$	
(a, ∞)	Open	$x > a$	
$(-\infty, b]$		$x \leq b$	
$(-\infty, b)$	Open	$x < b$	
$(-\infty, \infty)$	Entire real line	$-\infty < x < \infty$	

EXAMPLE 5 **Interpreting Intervals**

a. The interval $(-1, 0)$ consists of all real numbers greater than -1 and less than 0.

b. The interval $[2, \infty)$ consists of all real numbers greater than or equal to 2.

✓ *Checkpoint* ◄))) *Audio-video solution in English & Spanish at LarsonPrecalculus.com.*

Give a verbal description of the interval $[-2, 5)$.

EXAMPLE 6 **Using Inequalities to Represent Intervals**

a. The inequality $c \leq 2$ can represent the statement "c is at most 2."

b. The inequality $-3 < x \leq 5$ can represent "all x in the interval $(-3, 5]$."

✓ *Checkpoint* ◄))) *Audio-video solution in English & Spanish at LarsonPrecalculus.com.*

Use inequality notation to represent the statement "x is greater than -2 and at most 4."

Absolute Value and Distance

The **absolute value** of a real number is its *magnitude,* or the distance between the origin and the point representing the real number on the real number line.

> **Definition of Absolute Value**
>
> If a is a real number, then the absolute value of a is
>
> $$|a| = \begin{cases} a, & \text{if } a \geq 0 \\ -a, & \text{if } a < 0 \end{cases}.$$

Notice in this definition that the absolute value of a real number is never negative. For instance, if $a = -5$, then $|-5| = -(-5) = 5$. The absolute value of a real number is either positive or zero. Moreover, 0 is the only real number whose absolute value is 0. So, $|0| = 0$.

EXAMPLE 7   **Finding Absolute Values**

a. $|-15| = 15$

b. $\left| \dfrac{2}{3} \right| = \dfrac{2}{3}$

c. $|-4.3| = 4.3$

d. $-|-6| = -(6) = -6$

✓ **Checkpoint**  *Audio-video solution in English & Spanish at LarsonPrecalculus.com.*

Evaluate each expression.

a. $|1|$

b. $-\left| \dfrac{3}{4} \right|$

c. $\dfrac{2}{|-3|}$

d. $-|0.7|$

EXAMPLE 8   **Evaluating the Absolute Value of a Number**

Evaluate $\dfrac{|x|}{x}$ for (a) $x > 0$ and (b) $x < 0$.

Solution

a. If $x > 0$, then $|x| = x$ and $\dfrac{|x|}{x} = \dfrac{x}{x} = 1$.

b. If $x < 0$, then $|x| = -x$ and $\dfrac{|x|}{x} = \dfrac{-x}{x} = -1$.

✓ **Checkpoint**  *Audio-video solution in English & Spanish at LarsonPrecalculus.com.*

Evaluate $\dfrac{|x + 3|}{x + 3}$ for (a) $x > -3$ and (b) $x < -3$.

The **Law of Trichotomy** states that for any two real numbers a and b, *precisely* one of three relationships is possible:

$a = b$, $a < b$,   or $a > b$.   Law of Trichotomy

EXAMPLE 9   **Comparing Real Numbers**

Place the appropriate symbol ($<$, $>$, or $=$) between the pair of real numbers.

a. $|-4|$ ▨ $|3|$ **b.** $|-10|$ ▨ $|10|$ **c.** $-|-7|$ ▨ $|-7|$

Solution

a. $|-4| > |3|$ because $|-4| = 4$ and $|3| = 3$, and 4 is greater than 3.

b. $|-10| = |10|$ because $|-10| = 10$ and $|10| = 10$.

c. $-|-7| < |-7|$ because $-|-7| = -7$ and $|-7| = 7$, and -7 is less than 7.

✓ **Checkpoint** ◀⧽⧽) *Audio-video solution in English & Spanish at LarsonPrecalculus.com.*

Place the appropriate symbol ($<$, $>$, or $=$) between the pair of real numbers.

a. $|-3|$ ▨ $|4|$ **b.** $-|-4|$ ▨ $-|4|$ **c.** $|-3|$ ▨ $-|-3|$ ■

Properties of Absolute Values

1. $|a| \geq 0$ **2.** $|-a| = |a|$

3. $|ab| = |a||b|$ **4.** $\left|\dfrac{a}{b}\right| = \dfrac{|a|}{|b|}, \quad b \neq 0$

Absolute value can be used to define the distance between two points on the real number line. For instance, the distance between -3 and 4 is

$$|-3 - 4| = |-7|$$
$$= 7$$

as shown in Figure A.9.

The distance between -3 and 4 is 7.

Figure A.9

Distance Between Two Points on the Real Number Line

Let a and b be real numbers. The **distance between a and b** is

$$d(a, b) = |b - a| = |a - b|.$$

EXAMPLE 10   **Finding a Distance**

Find the distance between -25 and 13.

Solution

The distance between -25 and 13 is

$$|-25 - 13| = |-38| = 38. \qquad \textcolor{red}{\text{Distance between } -25 \text{ and } 13}$$

The distance can also be found as follows.

$$|13 - (-25)| = |38| = 38 \qquad \textcolor{red}{\text{Distance between } -25 \text{ and } 13}$$

✓ **Checkpoint** *Audio-video solution in English & Spanish at LarsonPrecalculus.com.*

a. Find the distance between 35 and -23.

b. Find the distance between -35 and -23.

c. Find the distance between 35 and 23. ■

Algebraic Expressions

One characteristic of algebra is the use of letters to represent numbers. The letters are **variables,** and combinations of letters and numbers are **algebraic expressions.** Here are a few examples of algebraic expressions.

$$5x, \qquad 2x - 3, \qquad \frac{4}{x^2 + 2}, \qquad 7x + y$$

> ### Definition of an Algebraic Expression
>
> An **algebraic expression** is a collection of letters (**variables**) and real numbers (**constants**) combined using the operations of addition, subtraction, multiplication, division, and exponentiation.

The **terms** of an algebraic expression are those parts that are separated by *addition.* For example, $x^2 - 5x + 8 = x^2 + (-5x) + 8$ has three terms: x^2 and $-5x$ are the **variable terms** and 8 is the **constant term.** The numerical factor of a term is called the **coefficient.** For instance, the coefficient of $-5x$ is -5, and the coefficient of x^2 is 1.

EXAMPLE 11 Identifying Terms and Coefficients

Algebraic Expression	Terms	Coefficients
a. $5x - \dfrac{1}{7}$	$5x, -\dfrac{1}{7}$	$5, -\dfrac{1}{7}$
b. $2x^2 - 6x + 9$	$2x^2, -6x, 9$	$2, -6, 9$
c. $\dfrac{3}{x} + \dfrac{1}{2}x^4 - y$	$\dfrac{3}{x}, \dfrac{1}{2}x^4, -y$	$3, \dfrac{1}{2}, -1$

✔ *Checkpoint* *Audio-video solution in English & Spanish at LarsonPrecalculus.com.*

Identify the terms and coefficients of $-2x + 4$.

To **evaluate** an algebraic expression, substitute numerical values for each of the variables in the expression, as shown in the next example.

EXAMPLE 12 Evaluating Algebraic Expressions

Expression	Value of Variable	Substitute.	Value of Expression
a. $-3x + 5$	$x = 3$	$-3(3) + 5$	$-9 + 5 = -4$
b. $3x^2 + 2x - 1$	$x = -1$	$3(-1)^2 + 2(-1) - 1$	$3 - 2 - 1 = 0$
c. $\dfrac{2x}{x + 1}$	$x = -3$	$\dfrac{2(-3)}{-3 + 1}$	$\dfrac{-6}{-2} = 3$

Note that you must substitute the value for *each* occurrence of the variable.

✔ *Checkpoint* 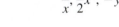 *Audio-video solution in English & Spanish at LarsonPrecalculus.com.*

Evaluate $4x - 5$ when $x = 0$.

Use the **Substitution Principle** to evaluate algebraic expressions. It states that "If $a = b$, then b can replace a in any expression involving a." In Example 12(a), for instance, 3 is *substituted* for x in the expression $-3x + 5$.

Basic Rules of Algebra

There are four arithmetic operations with real numbers: *addition, multiplication, subtraction,* and *division,* denoted by the symbols $+$, $\times$ or $\cdot$, $-$, and $\div$ or $/$, respectively. Of these, addition and multiplication are the two primary operations. Subtraction and division are the inverse operations of addition and multiplication, respectively.

Definitions of Subtraction and Division

Subtraction: Add the opposite.   **Division:** Multiply by the reciprocal.

$$a - b = a + (-b) \qquad\qquad \text{If } b \neq 0, \text{ then } a/b = a\left(\frac{1}{b}\right) = \frac{a}{b}.$$

In these definitions, $-b$ is the **additive inverse** (or opposite) of b, and $1/b$ is the **multiplicative inverse** (or reciprocal) of b. In the fractional form a/b, a is the **numerator** of the fraction and b is the **denominator**.

Because the properties of real numbers below are true for variables and algebraic expressions as well as for real numbers, they are often called the **Basic Rules of Algebra.** Try to formulate a verbal description of each property. For instance, the first property states that *the order in which two real numbers are added does not affect their sum.*

Basic Rules of Algebra

Let a, b, and c be real numbers, variables, or algebraic expressions.

Property		**Example**
Commutative Property of Addition:	$a + b = b + a$	$4x + x^2 = x^2 + 4x$
Commutative Property of Multiplication:	$ab = ba$	$(4 - x)x^2 = x^2(4 - x)$
Associative Property of Addition:	$(a + b) + c = a + (b + c)$	$(x + 5) + x^2 = x + (5 + x^2)$
Associative Property of Multiplication:	$(ab)c = a(bc)$	$(2x \cdot 3y)(8) = (2x)(3y \cdot 8)$
Distributive Properties:	$a(b + c) = ab + ac$	$3x(5 + 2x) = 3x \cdot 5 + 3x \cdot 2x$
	$(a + b)c = ac + bc$	$(y + 8)y = y \cdot y + 8 \cdot y$
Additive Identity Property:	$a + 0 = a$	$5y^2 + 0 = 5y^2$
Multiplicative Identity Property:	$a \cdot 1 = a$	$(4x^2)(1) = 4x^2$
Additive Inverse Property:	$a + (-a) = 0$	$5x^3 + (-5x^3) = 0$
Multiplicative Inverse Property:	$a \cdot \dfrac{1}{a} = 1, \quad a \neq 0$	$(x^2 + 4)\left(\dfrac{1}{x^2 + 4}\right) = 1$

Because subtraction is defined as "adding the opposite," the Distributive Properties are also true for subtraction. For instance, the "subtraction form" of $a(b + c) = ab + ac$ is $a(b - c) = ab - ac$. Note that the operations of subtraction and division are neither commutative nor associative. The examples

$$7 - 3 \neq 3 - 7 \quad \text{and} \quad 20 \div 4 \neq 4 \div 20$$

show that subtraction and division are not commutative. Similarly

$$5 - (3 - 2) \neq (5 - 3) - 2 \quad \text{and} \quad 16 \div (4 \div 2) \neq (16 \div 4) \div 2$$

demonstrate that subtraction and division are not associative.

EXAMPLE 13   **Identifying Rules of Algebra**

Identify the rule of algebra illustrated by the statement.

a. $(5x^3)2 = 2(5x^3)$

b. $(4x + 3) - (4x + 3) = 0$

c. $7x \cdot \dfrac{1}{7x} = 1, \quad x \neq 0$

d. $(2 + 5x^2) + x^2 = 2 + (5x^2 + x^2)$

Solution

a. This statement illustrates the Commutative Property of Multiplication. In other words, you obtain the same result whether you multiply $5x^3$ by 2, or 2 by $5x^3$.

b. This statement illustrates the Additive Inverse Property. In terms of subtraction, this property states that when any expression is subtracted from itself the result is 0.

c. This statement illustrates the Multiplicative Inverse Property. Note that x must be a nonzero number. The reciprocal of x is undefined when x is 0.

d. This statement illustrates the Associative Property of Addition. In other words, to form the sum $2 + 5x^2 + x^2$, it does not matter whether 2 and $5x^2$, or $5x^2$ and x^2 are added first.

✓ **Checkpoint** *Audio-video solution in English & Spanish at LarsonPrecalculus.com.*

Identify the rule of algebra illustrated by the statement.

a. $x + 9 = 9 + x$ **b.** $5(x^3 \cdot 2) = (5x^3)2$ **c.** $(2 + 5x^2)y^2 = 2 \cdot y^2 + 5x^2 \cdot y^2$

••**REMARK** Notice the difference between the *opposite of a number* and a *negative number*. If a is negative, then its opposite, $-a$, is positive. For instance, if $a = -5$, then

$$-a = -(-5) = 5.$$

Properties of Negation and Equality

Let a, b, and c be real numbers, variables, or algebraic expressions.

Property	**Example**
1. $(-1)a = -a$	$(-1)7 = -7$
2. $-(-a) = a$	$-(-6) = 6$
3. $(-a)b = -(ab) = a(-b)$	$(-5)3 = -(5 \cdot 3) = 5(-3)$
4. $(-a)(-b) = ab$	$(-2)(-x) = 2x$
5. $-(a + b) = (-a) + (-b)$	$-(x + 8) = (-x) + (-8)$
	$= -x - 8$
6. If $a = b$, then $a \pm c = b \pm c$.	$\frac{1}{2} + 3 = 0.5 + 3$
7. If $a = b$, then $ac = bc$.	$4^2 \cdot 2 = 16 \cdot 2$
8. If $a \pm c = b \pm c$, then $a = b$.	$1.4 - 1 = \frac{7}{5} - 1 \implies 1.4 = \frac{7}{5}$
9. If $ac = bc$ and $c \neq 0$, then $a = b$.	$3x = 3 \cdot 4 \implies x = 4$

••**REMARK** The "or" in the Zero-Factor Property includes the possibility that either or both factors may be zero. This is an *inclusive or,* and it is generally the way the word "or" is used in mathematics.

Properties of Zero

Let a and b be real numbers, variables, or algebraic expressions.

1. $a + 0 = a$ and $a - 0 = a$ **2.** $a \cdot 0 = 0$

3. $\dfrac{0}{a} = 0, \quad a \neq 0$ **4.** $\dfrac{a}{0}$ is undefined.

5. Zero-Factor Property: If $ab = 0$, then $a = 0$ or $b = 0$.

▷

REMARK In Property 1 of fractions, the phrase "if and only if" implies two statements. One statement is: If $a/b = c/d$, then $ad = bc$. The other statement is: If $ad = bc$, where $b \neq 0$ and $d \neq 0$, then $a/b = c/d$.

Properties and Operations of Fractions

Let a, b, c, and d be real numbers, variables, or algebraic expressions such that $b \neq 0$ and $d \neq 0$.

1. **Equivalent Fractions:** $\dfrac{a}{b} = \dfrac{c}{d}$ if and only if $ad = bc$.

2. **Rules of Signs:** $-\dfrac{a}{b} = \dfrac{-a}{b} = \dfrac{a}{-b}$ and $\dfrac{-a}{-b} = \dfrac{a}{b}$

3. **Generate Equivalent Fractions:** $\dfrac{a}{b} = \dfrac{ac}{bc}$, $c \neq 0$

4. **Add or Subtract with Like Denominators:** $\dfrac{a}{b} \pm \dfrac{c}{b} = \dfrac{a \pm c}{b}$

5. **Add or Subtract with Unlike Denominators:** $\dfrac{a}{b} \pm \dfrac{c}{d} = \dfrac{ad \pm bc}{bd}$

6. **Multiply Fractions:** $\dfrac{a}{b} \cdot \dfrac{c}{d} = \dfrac{ac}{bd}$

7. **Divide Fractions:** $\dfrac{a}{b} \div \dfrac{c}{d} = \dfrac{a}{b} \cdot \dfrac{d}{c} = \dfrac{ad}{bc}$, $c \neq 0$

EXAMPLE 14 **Properties and Operations of Fractions**

a. Equivalent fractions: $\dfrac{x}{5} = \dfrac{3 \cdot x}{3 \cdot 5} = \dfrac{3x}{15}$ **b.** Divide fractions: $\dfrac{7}{x} \div \dfrac{3}{2} = \dfrac{7}{x} \cdot \dfrac{2}{3} = \dfrac{14}{3x}$

✓ **Checkpoint** 🔊 *Audio-video solution in English & Spanish at LarsonPrecalculus.com.*

a. Multiply fractions: $\dfrac{3}{5} \cdot \dfrac{x}{6}$ **b.** Add fractions: $\dfrac{x}{10} + \dfrac{2x}{5}$ ■

REMARK The number 1 is neither prime nor composite.

▷

If a, b, and c are integers such that $ab = c$, then a and b are **factors** or **divisors** of c. A **prime number** is an integer that has exactly two positive factors—itself and 1—such as 2, 3, 5, 7, and 11. The numbers 4, 6, 8, 9, and 10 are **composite** because each can be written as the product of two or more prime numbers. The **Fundamental Theorem of Arithmetic** states that every positive integer greater than 1 is a prime number or can be written as the product of prime numbers in precisely one way (disregarding order). For instance, the *prime factorization* of 24 is $24 = 2 \cdot 2 \cdot 2 \cdot 3$.

Summarize (Appendix A.1)

1. Describe how to represent and classify real numbers (*pages A1 and A2*). For examples of representing and classifying real numbers, see Examples 1 and 2.

2. Describe how to order real numbers and use inequalities (*pages A3 and A4*). For examples of ordering real numbers and using inequalities, see Examples 3–6.

3. State the absolute value of a real number (*page A5*). For examples of using absolute value, see Examples 7–10.

4. Explain how to evaluate an algebraic expression (*page A7*). For examples involving algebraic expressions, see Examples 11 and 12.

5. State the basic rules and properties of algebra (*pages A8–A10*). For examples involving the basic rules and properties of algebra, see Examples 13 and 14.

A.2 Exponents and Radicals

Real numbers and algebraic expressions are often written with exponents and radicals. For instance, an expression involving rational exponents can be used to find the times required for a funnel to empty for different water heights.

■ Use properties of exponents.
■ Use scientific notation to represent real numbers.
■ Use properties of radicals.
■ Simplify and combine radicals.
■ Rationalize denominators and numerators.
■ Use properties of rational exponents.

Integer Exponents

Repeated *multiplication* can be written in **exponential form.**

Repeated Multiplication	Exponential Form
$a \cdot a \cdot a \cdot a \cdot a$	a^5
$(-4)(-4)(-4)$	$(-4)^3$
$(2x)(2x)(2x)(2x)$	$(2x)^4$

Exponential Notation

If a is a real number and n is a positive integer, then

$$a^n = \underbrace{a \cdot a \cdot a \cdots a}_{n \text{ factors}}$$

where n is the **exponent** and a is the **base.** You read a^n as "a to the nth **power.**"

An exponent can also be negative. In Property 3 below, be sure you see how to use a negative exponent.

Properties of Exponents

Let a and b be real numbers, variables, or algebraic expressions, and let m and n be integers. (All denominators and bases are nonzero.)

Property	Example
1. $a^m a^n = a^{m+n}$	$3^2 \cdot 3^4 = 3^{2+4} = 3^6 = 729$
2. $\dfrac{a^m}{a^n} = a^{m-n}$	$\dfrac{x^7}{x^4} = x^{7-4} = x^3$
3. $a^{-n} = \dfrac{1}{a^n} = \left(\dfrac{1}{a}\right)^n$	$y^{-4} = \dfrac{1}{y^4} = \left(\dfrac{1}{y}\right)^4$
4. $a^0 = 1, \quad a \neq 0$	$(x^2 + 1)^0 = 1$
5. $(ab)^m = a^m b^m$	$(5x)^3 = 5^3 x^3 = 125x^3$
6. $(a^m)^n = a^{mn}$	$(y^3)^{-4} = y^{3(-4)} = y^{-12} = \dfrac{1}{y^{12}}$
7. $\left(\dfrac{a}{b}\right)^m = \dfrac{a^m}{b^m}$	$\left(\dfrac{2}{x}\right)^3 = \dfrac{2^3}{x^3} = \dfrac{8}{x^3}$
8. $|a^2| = |a|^2 = a^2$	$|(-2)^2| = |-2|^2 = (2)^2 = 4$

▷ **TECHNOLOGY** You can use a calculator to evaluate exponential expressions. When doing so, it is important to know when to use parentheses because the calculator follows the order of operations. For instance, you would evaluate $(-2)^4$ on a graphing calculator as follows.

((−) 2) ^ 4 [ENTER]

The display will be 16. If you omit the parentheses, then the display will be -16.

It is important to recognize the difference between expressions such as $(-2)^4$ and -2^4. In $(-2)^4$, the parentheses indicate that the exponent applies to the negative sign as well as to the 2, but in $-2^4 = -(2^4)$, the exponent applies only to the 2. So, $(-2)^4 = 16$ and $-2^4 = -16$.

The properties of exponents listed on the preceding page apply to *all* integers m and n, not just to positive integers, as shown in the examples below.

EXAMPLE 1   **Evaluating Exponential Expressions**

a. $(-5)^2 = (-5)(-5) = 25$ Negative sign is part of the base.

b. $-5^2 = -(5)(5) = -25$ Negative sign is *not* part of the base.

c. $2 \cdot 2^4 = 2^{1+4} = 2^5 = 32$ Property 1

d. $\dfrac{4^4}{4^6} = 4^{4-6} = 4^{-2} = \dfrac{1}{4^2} = \dfrac{1}{16}$ Properties 2 and 3

✓ *Checkpoint*  *Audio-video solution in English & Spanish at LarsonPrecalculus.com.*

Evaluate each expression.

a. -3^4 **b.** $(-3)^4$

c. $3^2 \cdot 3$ **d.** $\dfrac{3^5}{3^8}$

EXAMPLE 2   **Evaluating Algebraic Expressions**

Evaluate each algebraic expression when $x = 3$.

a. $5x^{-2}$ **b.** $\dfrac{1}{3}(-x)^3$

Solution

a. When $x = 3$, the expression $5x^{-2}$ has a value of

$$5x^{-2} = 5(3)^{-2} = \dfrac{5}{3^2} = \dfrac{5}{9}.$$

b. When $x = 3$, the expression $\dfrac{1}{3}(-x)^3$ has a value of

$$\dfrac{1}{3}(-x)^3 = \dfrac{1}{3}(-3)^3 = \dfrac{1}{3}(-27) = -9.$$

✓ *Checkpoint*  *Audio-video solution in English & Spanish at LarsonPrecalculus.com.*

Evaluate each algebraic expression when $x = 4$.

a. $-x^{-2}$ **b.** $\dfrac{1}{4}(-x)^4$

EXAMPLE 3   **Using Properties of Exponents**

Use the properties of exponents to simplify each expression.

a. $(-3ab^4)(4ab^{-3})$ **b.** $(2xy^2)^3$ **c.** $3a(-4a^2)^0$ **d.** $\left(\dfrac{5x^3}{y}\right)^2$

Solution

a. $(-3ab^4)(4ab^{-3}) = (-3)(4)(a)(a)(b^4)(b^{-3}) = -12a^2b$

b. $(2xy^2)^3 = 2^3(x)^3(y^2)^3 = 8x^3y^6$

c. $3a(-4a^2)^0 = 3a(1) = 3a, \quad a \neq 0$

d. $\left(\dfrac{5x^3}{y}\right)^2 = \dfrac{5^2(x^3)^2}{y^2} = \dfrac{25x^6}{y^2}$

✓ **Checkpoint** *Audio-video solution in English & Spanish at LarsonPrecalculus.com.*

Use the properties of exponents to simplify each expression.

a. $(2x^{-2}y^3)(-x^4y)$ **b.** $(4a^2b^3)^0$ **c.** $(-5z)^3(z^2)$ **d.** $\left(\dfrac{3x^4}{x^2y^2}\right)^2$

EXAMPLE 4   **Rewriting with Positive Exponents**

a. $x^{-1} = \dfrac{1}{x}$ Property 3

b. $\dfrac{1}{3x^{-2}} = \dfrac{1(x^2)}{3}$ The exponent -2 does not apply to 3.

$\qquad = \dfrac{x^2}{3}$ Simplify.

c. $\dfrac{12a^3b^{-4}}{4a^{-2}b} = \dfrac{12a^3 \cdot a^2}{4b \cdot b^4}$ Property 3

$\qquad = \dfrac{3a^5}{b^5}$ Property 1

d. $\left(\dfrac{3x^2}{y}\right)^{-2} = \dfrac{3^{-2}(x^2)^{-2}}{y^{-2}}$ Properties 5 and 7

$\qquad = \dfrac{3^{-2}x^{-4}}{y^{-2}}$ Property 6

$\qquad = \dfrac{y^2}{3^2x^4}$ Property 3

$\qquad = \dfrac{y^2}{9x^4}$ Simplify.

> •• **REMARK** Rarely in algebra is there only one way to solve a problem. Do not be concerned when the steps you use to solve a problem are not exactly the same as the steps presented in this text. It is important to use steps that you understand and, of course, steps that are justified by the rules of algebra. For instance, you might prefer the following steps for Example 4(d).
>
> $$\left(\dfrac{3x^2}{y}\right)^{-2} = \left(\dfrac{y}{3x^2}\right)^2 = \dfrac{y^2}{9x^4}$$
>
> Note how the first step of this solution uses Property 3. The fractional form of this property is
>
> $$\left(\dfrac{a}{b}\right)^{-m} = \left(\dfrac{b}{a}\right)^m.$$

✓ **Checkpoint** *Audio-video solution in English & Spanish at LarsonPrecalculus.com.*

Rewrite each expression with positive exponents.

a. $2a^{-2}$ **b.** $\dfrac{3a^{-3}b^4}{15ab^{-1}}$

c. $\left(\dfrac{x}{10}\right)^{-1}$ **d.** $(-2x^2)^3(4x^3)^{-1}$

Scientific Notation

Exponents provide an efficient way of writing and computing with very large (or very small) numbers. For instance, there are about 359 billion billion gallons of water on Earth—that is, 359 followed by 18 zeros.

359,000,000,000,000,000,000

It is convenient to write such numbers in **scientific notation.** This notation has the form $\pm c \times 10^n$, where $1 \le c < 10$ and n is an integer. So, the number of gallons of water on Earth, written in scientific notation, is

$3.59 \times 100,000,000,000,000,000,000 = 3.59 \times 10^{20}.$

The *positive* exponent 20 indicates that the number is *large* (10 or more) and that the decimal point has been moved 20 places. A *negative* exponent indicates that the number is *small* (less than 1). For instance, the mass (in grams) of one electron is approximately

$9.0 \times 10^{-28} = 0.0000000000000000000000000009.$

28 decimal places

EXAMPLE 5 **Scientific Notation**

a. $0.0000782 = 7.82 \times 10^{-5}$

b. $836,100,000 = 8.361 \times 10^8$

✓ *Checkpoint* ◀)) *Audio-video solution in English & Spanish at LarsonPrecalculus.com.*

Write 45,850 in scientific notation.

EXAMPLE 6 **Decimal Notation**

a. $-9.36 \times 10^{-6} = -0.00000936$

b. $1.345 \times 10^2 = 134.5$

✓ *Checkpoint* ◀)) *Audio-video solution in English & Spanish at LarsonPrecalculus.com.*

Write -2.718×10^{-3} in decimal notation.

▷ **TECHNOLOGY** Most calculators automatically switch to scientific notation when they are showing large (or small) numbers that exceed the display range.

To enter numbers in scientific notation, your calculator should have an exponential entry key labeled

$\boxed{\text{EE}}$ or $\boxed{\text{EXP}}$.

Consult the user's guide for instructions on keystrokes and how your calculator displays numbers in scientific notation.

EXAMPLE 7 **Using Scientific Notation**

Evaluate $\dfrac{(2,400,000,000)(0.0000045)}{(0.00003)(1500)}$.

Solution Begin by rewriting each number in scientific notation and simplifying.

$$\frac{(2,400,000,000)(0.0000045)}{(0.00003)(1500)} = \frac{(2.4 \times 10^9)(4.5 \times 10^{-6})}{(3.0 \times 10^{-5})(1.5 \times 10^3)}$$

$$= \frac{(2.4)(4.5)(10^3)}{(4.5)(10^{-2})}$$

$$= (2.4)(10^5)$$

$$= 240,000$$

✓ *Checkpoint* ◀)) *Audio-video solution in English & Spanish at LarsonPrecalculus.com.*

Evaluate $(24,000,000,000)(0.00000012)(300,000)$.

Radicals and Their Properties

A **square root** of a number is one of its two equal factors. For example, 5 is a square root of 25 because 5 is one of the two equal factors of 25. In a similar way, a **cube root** of a number is one of its three equal factors, as in $125 = 5^3$.

Definition of *n*th Root of a Number

Let a and b be real numbers and let $n \geq 2$ be a positive integer. If

$a = b^n$

then b is an ***n*th root of a.** When $n = 2$, the root is a **square root.** When $n = 3$, the root is a **cube root.**

Some numbers have more than one nth root. For example, both 5 and -5 are square roots of 25. The *principal square* root of 25, written as $\sqrt{25}$, is the positive root, 5. The **principal nth root** of a number is defined as follows.

Principal *n*th Root of a Number

Let a be a real number that has at least one nth root. The **principal nth root of a** is the nth root that has the same sign as a. It is denoted by a **radical symbol**

$\sqrt[n]{a}.$ Principal nth root

The positive integer $n \geq 2$ is the **index** of the radical, and the number a is the **radicand.** When $n = 2$, omit the index and write $\sqrt{a}$ rather than $\sqrt[2]{a}$. (The plural of index is *indices*.)

A common misunderstanding is that the square root sign implies both negative and positive roots. This is not correct. The square root sign implies only a positive root. When a negative root is needed, you must use the negative sign with the square root sign.

$$\textit{Incorrect: } \cancel{\sqrt{4} = \pm 2} \qquad \textit{Correct: } -\sqrt{4} = -2 \quad \textit{and} \quad \sqrt{4} = 2$$

EXAMPLE 8   **Evaluating Expressions Involving Radicals**

a. $\sqrt{36} = 6$ because $6^2 = 36$.

b. $-\sqrt{36} = -6$ because $-\left(\sqrt{36}\right) = -\left(\sqrt{6^2}\right) = -(6) = -6$.

c. $\sqrt[3]{\dfrac{125}{64}} = \dfrac{5}{4}$ because $\left(\dfrac{5}{4}\right)^3 = \dfrac{5^3}{4^3} = \dfrac{125}{64}$.

d. $\sqrt[5]{-32} = -2$ because $(-2)^5 = -32$.

e. $\sqrt[4]{-81}$ is not a real number because no real number raised to the fourth power produces -81.

✓ *Checkpoint* ◀))) *Audio-video solution in English & Spanish at LarsonPrecalculus.com.*

Evaluate each expression (if possible).

a. $-\sqrt{144}$ **b.** $\sqrt{-144}$

c. $\sqrt{\dfrac{25}{64}}$ **d.** $-\sqrt[3]{\dfrac{8}{27}}$

Here are some generalizations about the *n*th roots of real numbers.

Generalizations About *n*th Roots of Real Numbers			
Real Number *a*	Integer *n* > 0	Root(s) of *a*	Example
a > 0	*n* is even.	$\sqrt[n]{a}, \ -\sqrt[n]{a}$	$\sqrt[4]{81} = 3, \ -\sqrt[4]{81} = -3$
a > 0 or *a* < 0	*n* is odd.	$\sqrt[n]{a}$	$\sqrt[3]{-8} = -2$
a < 0	*n* is even.	No real roots	$\sqrt{-4}$ is not a real number.
a = 0	*n* is even or odd.	$\sqrt[n]{0} = 0$	$\sqrt[5]{0} = 0$

Integers such as 1, 4, 9, 16, 25, and 36 are called **perfect squares** because they have integer square roots. Similarly, integers such as 1, 8, 27, 64, and 125 are called **perfect cubes** because they have integer cube roots.

Properties of Radicals

Let *a* and *b* be real numbers, variables, or algebraic expressions such that the indicated roots are real numbers, and let *m* and *n* be positive integers.

Property

1. $\sqrt[n]{a^m} = \left(\sqrt[n]{a}\right)^m$

2. $\sqrt[n]{a} \cdot \sqrt[n]{b} = \sqrt[n]{ab}$

3. $\dfrac{\sqrt[n]{a}}{\sqrt[n]{b}} = \sqrt[n]{\dfrac{a}{b}}, \quad b \neq 0$

4. $\sqrt[m]{\sqrt[n]{a}} = \sqrt[mn]{a}$

5. $\left(\sqrt[n]{a}\right)^n = a$

6. For *n* even, $\sqrt[n]{a^n} = |a|$.

   For *n* odd, $\sqrt[n]{a^n} = a$.

Example

$\sqrt[3]{8^2} = \left(\sqrt[3]{8}\right)^2 = (2)^2 = 4$

$\sqrt{5} \cdot \sqrt{7} = \sqrt{5 \cdot 7} = \sqrt{35}$

$\dfrac{\sqrt[4]{27}}{\sqrt[4]{9}} = \sqrt[4]{\dfrac{27}{9}} = \sqrt[4]{3}$

$\sqrt[3]{\sqrt{10}} = \sqrt[6]{10}$

$\left(\sqrt{3}\right)^2 = 3$

$\sqrt{(-12)^2} = |-12| = 12$

$\sqrt[3]{(-12)^3} = -12$

A common use of Property 6 is $\sqrt{a^2} = |a|$.

EXAMPLE 9 **Using Properties of Radicals**

Use the properties of radicals to simplify each expression.

a. $\sqrt{8} \cdot \sqrt{2}$ **b.** $\left(\sqrt[3]{5}\right)^3$

c. $\sqrt[3]{x^3}$ **d.** $\sqrt[6]{y^6}$

Solution

a. $\sqrt{8} \cdot \sqrt{2} = \sqrt{8 \cdot 2} = \sqrt{16} = 4$ **b.** $\left(\sqrt[3]{5}\right)^3 = 5$

c. $\sqrt[3]{x^3} = x$ **d.** $\sqrt[6]{y^6} = |y|$

✓ *Checkpoint*  *Audio-video solution in English & Spanish at LarsonPrecalculus.com.*

Use the properties of radicals to simplify each expression.

a. $\dfrac{\sqrt{125}}{\sqrt{5}}$ **b.** $\sqrt[3]{125^2}$ **c.** $\sqrt[3]{x^2} \cdot \sqrt[3]{x}$ **d.** $\sqrt{\sqrt{x}}$

Simplifying Radicals

An expression involving radicals is in **simplest form** when the following conditions are satisfied.

1. All possible factors have been removed from the radical.
2. All fractions have radical-free denominators (a process called *rationalizing the denominator* accomplishes this).
3. The index of the radical is reduced.

   To simplify a radical, factor the radicand into factors whose exponents are multiples of the index. Write the roots of these factors outside the radical. The "leftover" factors make up the new radicand.

REMARK When you simplify a radical, it is important that both expressions are defined for the same values of the variable. For instance, in Example 10(c), $\sqrt{75x^3}$ and $5x\sqrt{3x}$ are both defined only for nonnegative values of x. Similarly, in Example 10(e), $\sqrt[4]{(5x)^4}$ and $5|x|$ are both defined for all real values of x.

EXAMPLE 10 **Simplifying Radicals**

Perfect cube · Leftover factor

a. $\sqrt[3]{24} = \sqrt[3]{8 \cdot 3} = \sqrt[3]{2^3 \cdot 3} = 2\sqrt[3]{3}$

Perfect 4th power · Leftover factor

b. $\sqrt[4]{48} = \sqrt[4]{16 \cdot 3} = \sqrt[4]{2^4 \cdot 3} = 2\sqrt[4]{3}$

c. $\sqrt{75x^3} = \sqrt{25x^2 \cdot 3x} = \sqrt{(5x)^2 \cdot 3x} = 5x\sqrt{3x}$

d. $\sqrt[3]{24a^4} = \sqrt[3]{8a^3 \cdot 3a} = \sqrt[3]{(2a)^3 \cdot 3a} = 2a\sqrt[3]{3a}$

e. $\sqrt[4]{(5x)^4} = |5x| = 5|x|$

✓ *Checkpoint* ◄))) Audio-video solution in English & Spanish at LarsonPrecalculus.com.

Simplify each radical expression.

a. $\sqrt{32}$ **b.** $\sqrt[3]{250}$ **c.** $\sqrt{24a^5}$ **d.** $\sqrt[3]{-135x^3}$ ◼

Radical expressions can be combined (added or subtracted) when they are **like radicals**—that is, when they have the same index and radicand. For instance, $\sqrt{2}$, $3\sqrt{2}$, and $\frac{1}{2}\sqrt{2}$ are like radicals, but $\sqrt{3}$ and $\sqrt{2}$ are unlike radicals. To determine whether two radicals can be combined, you should first simplify each radical.

EXAMPLE 11 **Combining Radicals**

a. $2\sqrt{48} - 3\sqrt{27} = 2\sqrt{16 \cdot 3} - 3\sqrt{9 \cdot 3}$ — Find square factors.

$\qquad\qquad\qquad = 8\sqrt{3} - 9\sqrt{3}$ — Find square roots and multiply by coefficients.

$\qquad\qquad\qquad = (8 - 9)\sqrt{3}$ — Combine like radicals.

$\qquad\qquad\qquad = -\sqrt{3}$ — Simplify.

b. $\sqrt[3]{16x} - \sqrt[3]{54x^4} = \sqrt[3]{8 \cdot 2x} - \sqrt[3]{27 \cdot x^3 \cdot 2x}$ — Find cube factors.

$\qquad\qquad\qquad = 2\sqrt[3]{2x} - 3x\sqrt[3]{2x}$ — Find cube roots.

$\qquad\qquad\qquad = (2 - 3x)\sqrt[3]{2x}$ — Combine like radicals.

✓ *Checkpoint* ◄)))  Audio-video solution in English & Spanish at LarsonPrecalculus.com.

Simplify each radical expression.

a. $3\sqrt{8} + \sqrt{18}$ **b.** $\sqrt[3]{81x^5} - \sqrt[3]{24x^2}$ ◼

Rationalizing Denominators and Numerators

To rationalize a denominator or numerator of the form $a - b\sqrt{m}$ or $a + b\sqrt{m}$, multiply both numerator and denominator by a **conjugate**: $a + b\sqrt{m}$ and $a - b\sqrt{m}$ are conjugates of each other. If $a = 0$, then the rationalizing factor for $\sqrt{m}$ is itself, $\sqrt{m}$. For cube roots, choose a rationalizing factor that generates a perfect cube.

EXAMPLE 12 Rationalizing Single-Term Denominators

Rationalize the denominator of each expression.

a. $\dfrac{5}{2\sqrt{3}}$

b. $\dfrac{2}{\sqrt[3]{5}}$

Solution

a. $\dfrac{5}{2\sqrt{3}} = \dfrac{5}{2\sqrt{3}} \cdot \dfrac{\sqrt{3}}{\sqrt{3}}$ $\sqrt{3}$ is rationalizing factor.

$= \dfrac{5\sqrt{3}}{2(3)}$ Multiply.

$= \dfrac{5\sqrt{3}}{6}$ Simplify.

b. $\dfrac{2}{\sqrt[3]{5}} = \dfrac{2}{\sqrt[3]{5}} \cdot \dfrac{\sqrt[3]{5^2}}{\sqrt[3]{5^2}}$ $\sqrt[3]{5^2}$ is rationalizing factor.

$= \dfrac{2\sqrt[3]{5^2}}{\sqrt[3]{5^3}}$ Multiply.

$= \dfrac{2\sqrt[3]{25}}{5}$ Simplify.

✓ **Checkpoint**))) *Audio-video solution in English & Spanish at LarsonPrecalculus.com.*

Rationalize the denominator of each expression.

a. $\dfrac{5}{3\sqrt{2}}$ b. $\dfrac{1}{\sqrt[3]{25}}$

EXAMPLE 13 Rationalizing a Denominator with Two Terms

$\dfrac{2}{3 + \sqrt{7}} = \dfrac{2}{3 + \sqrt{7}} \cdot \dfrac{3 - \sqrt{7}}{3 - \sqrt{7}}$ Multiply numerator and denominator by conjugate of denominator.

$= \dfrac{2(3 - \sqrt{7})}{3(3) + 3(-\sqrt{7}) + \sqrt{7}(3) - (\sqrt{7})(\sqrt{7})}$ Use Distributive Property.

$= \dfrac{2(3 - \sqrt{7})}{(3)^2 - (\sqrt{7})^2}$ Simplify.

$= \dfrac{2(3 - \sqrt{7})}{2} = 3 - \sqrt{7}$ Simplify.

✓ **Checkpoint**))) *Audio-video solution in English & Spanish at LarsonPrecalculus.com.*

Rationalize the denominator: $\dfrac{8}{\sqrt{6} - \sqrt{2}}$.

Sometimes it is necessary to rationalize the numerator of an expression. For instance, in Appendix A.4 you will use the technique shown in the next example to rationalize the numerator of an expression from calculus.

EXAMPLE 14   **Rationalizing a Numerator**

$$\frac{\sqrt{5} - \sqrt{7}}{2} = \frac{\sqrt{5} - \sqrt{7}}{2} \cdot \frac{\sqrt{5} + \sqrt{7}}{\sqrt{5} + \sqrt{7}} \qquad \text{Multiply numerator and denominator by conjugate of denominator.}$$

$$= \frac{\left(\sqrt{5}\right)^2 - \left(\sqrt{7}\right)^2}{2\left(\sqrt{5} + \sqrt{7}\right)} \qquad \text{Simplify.}$$

$$= \frac{5 - 7}{2\left(\sqrt{5} + \sqrt{7}\right)} \qquad \text{Square terms of numerator.}$$

$$= \frac{-2}{2\left(\sqrt{5} + \sqrt{7}\right)} = \frac{-1}{\sqrt{5} + \sqrt{7}} \qquad \text{Simplify.}$$

✓ **Checkpoint**   *Audio-video solution in English & Spanish at LarsonPrecalculus.com.*

Rationalize the numerator: $\dfrac{2 - \sqrt{2}}{3}$.

> •• **REMARK** Do not confuse the expression $\sqrt{5} + \sqrt{7}$ with the expression $\sqrt{5 + 7}$. In general, $\sqrt{x + y}$ does not equal $\sqrt{x} + \sqrt{y}$. Similarly, $\sqrt{x^2 + y^2}$ does not equal $x + y$.

Rational Exponents

> **Definition of Rational Exponents**
>
> If a is a real number and n is a positive integer such that the principal nth root of a exists, then $a^{1/n}$ is defined as
>
> $$a^{1/n} = \sqrt[n]{a}, \text{ where } 1/n \text{ is the } \textbf{rational exponent} \text{ of } a.$$
>
> Moreover, if m is a positive integer that has no common factor with n, then
>
> $$a^{m/n} = (a^{1/n})^m = \left(\sqrt[n]{a}\right)^m \quad \text{and} \quad a^{m/n} = (a^m)^{1/n} = \sqrt[n]{a^m}.$$

> •• **REMARK** You must remember that the expression $b^{m/n}$ is not defined unless $\sqrt[n]{b}$ is a real number. This restriction produces some unusual results. For instance, the number $(-8)^{1/3}$ is defined because $\sqrt[3]{-8} = -2$, but the number $(-8)^{2/6}$ is undefined because $\sqrt[6]{-8}$ is not a real number.

The numerator of a rational exponent denotes the *power* to which the base is raised, and the denominator denotes the *index* or the *root* to be taken.

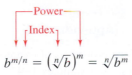

$$b^{m/n} = \left(\sqrt[n]{b}\right)^m = \sqrt[n]{b^m}$$

When you are working with rational exponents, the properties of integer exponents still apply. For instance, $2^{1/2} 2^{1/3} = 2^{(1/2)+(1/3)} = 2^{5/6}$.

EXAMPLE 15   **Changing From Radical to Exponential Form**

a. $\sqrt{3} = 3^{1/2}$ **b.** $\sqrt{(3xy)^5} = \sqrt[2]{(3xy)^5} = (3xy)^{5/2}$

c. $2x\sqrt[4]{x^3} = (2x)(x^{3/4}) = 2x^{1+(3/4)} = 2x^{7/4}$

✓ **Checkpoint**    *Audio-video solution in English & Spanish at LarsonPrecalculus.com.*

Write each expression in exponential form.

a. $\sqrt[3]{27}$ **b.** $\sqrt{x^3 y^5 z}$ **c.** $3x\sqrt[3]{x^2}$

▷ **TECHNOLOGY** There are four methods of evaluating radicals on most graphing calculators. For square roots, you can use the *square root key* ☑. For cube roots, you can use the *cube root key* ∛☐. For other roots, you can first convert the radical to exponential form and then use the *exponential key* ∧, or you can use the *x*th root key ☒ (or menu choice). Consult the user's guide for your calculator for specific keystrokes.

EXAMPLE 16 **Changing From Exponential to Radical Form**

a. $(x^2 + y^2)^{3/2} = \left(\sqrt{x^2+y^2}\right)^3 = \sqrt{(x^2+y^2)^3}$

b. $2y^{3/4}z^{1/4} = 2(y^3z)^{1/4} = 2\sqrt[4]{y^3z}$

c. $a^{-3/2} = \dfrac{1}{a^{3/2}} = \dfrac{1}{\sqrt{a^3}}$

d. $x^{0.2} = x^{1/5} = \sqrt[5]{x}$

✓ **Checkpoint** ◀))) *Audio-video solution in English & Spanish at LarsonPrecalculus.com.*

Write each expression in radical form.

a. $(x^2 - 7)^{-1/2}$ b. $-3b^{1/3}c^{2/3}$

c. $a^{0.75}$ d. $(x^2)^{2/5}$

Rational exponents are useful for evaluating roots of numbers on a calculator, for reducing the index of a radical, and for simplifying expressions in calculus.

EXAMPLE 17 **Simplifying with Rational Exponents**

a. $(-32)^{-4/5} = \left(\sqrt[5]{-32}\right)^{-4} = (-2)^{-4} = \dfrac{1}{(-2)^4} = \dfrac{1}{16}$

b. $(-5x^{5/3})(3x^{-3/4}) = -15x^{(5/3)-(3/4)} = -15x^{11/12}, \quad x \neq 0$

c. $\sqrt[9]{a^3} = a^{3/9} = a^{1/3} = \sqrt[3]{a}$ Reduce index.

d. $\sqrt[3]{\sqrt{125}} = \sqrt[6]{125} = \sqrt[6]{(5)^3} = 5^{3/6} = 5^{1/2} = \sqrt{5}$

e. $(2x-1)^{4/3}(2x-1)^{-1/3} = (2x-1)^{(4/3)-(1/3)} = 2x - 1, \quad x \neq \tfrac{1}{2}$

REMARK The expression in Example 17(e) is not defined when $x = \tfrac{1}{2}$ because

$$\left(2 \cdot \tfrac{1}{2} - 1\right)^{-1/3} = (0)^{-1/3}$$

is not a real number.

✓ **Checkpoint** ◀))) *Audio-video solution in English & Spanish at LarsonPrecalculus.com.*

Simplify each expression.

a. $(-125)^{-2/3}$ b. $(4x^2y^{3/2})(-3x^{-1/3}y^{-3/5})$

c. $\sqrt[3]{\sqrt[4]{27}}$ d. $(3x+2)^{5/2}(3x+2)^{-1/2}$

Summarize (Appendix A.2)

1. Make a list of the properties of exponents (*page A11*). For examples that use these properties, see Examples 1–4.

2. State the definition of scientific notation (*page A14*). For examples of scientific notation, see Examples 5–7.

3. Make a list of the properties of radicals (*page A16*). For an example that uses these properties, see Example 9.

4. Explain how to simplify a radical expression (*page A17*). For examples of simplifying radical expressions, see Examples 10 and 11.

5. Explain how to rationalize a denominator or a numerator (*page A18*). For examples of rationalizing denominators and numerators, see Examples 12–14.

6. State the definition of a rational exponent (*page A19*). For an example of simplifying expressions with rational exponents, see Example 17.

A.3 Polynomials and Factoring

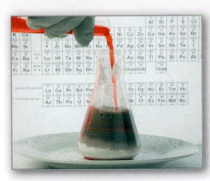

Polynomial factoring can help you solve real-life problems, such as developing an alternative model for the rate of change of an autocatalytic chemical reaction.

- ■ Write polynomials in standard form.
- ■ Add, subtract, and multiply polynomials.
- ■ Use special products to multiply polynomials.
- ■ Remove common factors from polynomials.
- ■ Factor special polynomial forms.
- ■ Factor trinomials as the product of two binomials.
- ■ Factor polynomials by grouping.

Polynomials

One of the most common types of algebraic expressions is the **polynomial.** Some examples are $2x + 5$, $3x^4 - 7x^2 + 2x + 4$, and $5x^2y^2 - xy + 3$. The first two are *polynomials in x and* the third is a *polynomial in x and y.* The terms of a polynomial in x have the form ax^k, where a is the **coefficient** and k is the **degree** of the term. For instance, the polynomial $2x^3 - 5x^2 + 1 = 2x^3 + (-5)x^2 + (0)x + 1$ has coefficients 2, -5, 0, and 1.

▷

REMARK Expressions are not polynomials when a variable is underneath a radical or when a polynomial expression (with degree greater than 0) is in the denominator of a term. For example, the expressions $x^3 - \sqrt{3x} = x^3 - (3x)^{1/2}$ and $x^2 + (5/x) = x^2 + 5x^{-1}$ are not polynomials.

> **Definition of a Polynomial in x**
>
> Let $a_0, a_1, a_2, \ldots, a_n$ be real numbers and let n be a nonnegative integer. A polynomial in x is an expression of the form
>
> $$a_n x^n + a_{n-1}x^{n-1} + \cdots + a_1 x + a_0$$
>
> where $a_n \neq 0$. The polynomial is of **degree** n, a_n is the **leading coefficient,** and a_0 is the **constant term.**

Polynomials with one, two, and three terms are called **monomials, binomials,** and **trinomials,** respectively. A polynomial written with descending powers of x is in **standard form.**

EXAMPLE 1   **Writing Polynomials in Standard Form**

Polynomial	Standard Form	Degree	Leading Coefficient
a. $4x^2 - 5x^7 - 2 + 3x$	$-5x^7 + 4x^2 + 3x - 2$	7	-5
b. $4 - 9x^2$	$-9x^2 + 4$	2	-9
c. 8	$8 \ (8 = 8x^0)$	0	8

✓ *Checkpoint*   *Audio-video solution in English & Spanish at LarsonPrecalculus.com.*

Write the polynomial $6 - 7x^3 + 2x$ in standard form. Then identify the degree and leading coefficient of the polynomial. ■

A polynomial that has all zero coefficients is called the **zero polynomial,** denoted by 0. No degree is assigned to the zero polynomial. For polynomials in more than one variable, the degree of a *term* is the sum of the exponents of the variables in the term. The degree of the *polynomial* is the highest degree of its terms. For instance, the degree of the polynomial $-2x^3y^6 + 4xy - x^7y^4$ is 11 because the sum of the exponents in the last term is the greatest. The leading coefficient of the polynomial is the coefficient of the highest-degree term.

Operations with Polynomials

You can add and subtract polynomials in much the same way you add and subtract real numbers. Add or subtract the *like terms* (terms having the same variables to the same powers) by adding or subtracting their coefficients. For instance, $-3xy^2$ and $5xy^2$ are like terms and their sum is

$$-3xy^2 + 5xy^2 = (-3 + 5)xy^2 = 2xy^2.$$

EXAMPLE 2   Sums and Differences of Polynomials

a. $(5x^3 - 7x^2 - 3) + (x^3 + 2x^2 - x + 8)$

$= (5x^3 + x^3) + (-7x^2 + 2x^2) - x + (-3 + 8)$ Group like terms.

$= 6x^3 - 5x^2 - x + 5$ Combine like terms.

b. $(7x^4 - x^2 - 4x + 2) - (3x^4 - 4x^2 + 3x)$

$= 7x^4 - x^2 - 4x + 2 - 3x^4 + 4x^2 - 3x$ Distributive Property

$= (7x^4 - 3x^4) + (-x^2 + 4x^2) + (-4x - 3x) + 2$ Group like terms.

$= 4x^4 + 3x^2 - 7x + 2$ Combine like terms.

✓ *Checkpoint* 🔊)) *Audio-video solution in English & Spanish at LarsonPrecalculus.com.*

Find the difference $(2x^3 - x + 3) - (x^2 - 2x - 3)$ and write the resulting polynomial in standard form. ◼

> **REMARK** When a negative sign precedes an expression inside parentheses, remember to distribute the negative sign to each term inside the parentheses, as shown.
>
> $-(3x^4 - 4x^2 + 3x)$
>
> $= -3x^4 + 4x^2 - 3x$

To find the *product* of two polynomials, use the left and right Distributive Properties. For example, if you treat $5x + 7$ as a single quantity, then you can multiply $3x - 2$ by $5x + 7$ as follows.

$(3x - 2)(5x + 7) = 3x(5x + 7) - 2(5x + 7)$

$= (3x)(5x) + (3x)(7) - (2)(5x) - (2)(7)$

$= 15x^2 + 21x - 10x - 14$

Product of First terms	Product of Outer terms	Product of Inner terms	Product of Last terms

$= 15x^2 + 11x - 14$

Note in this **FOIL Method** (which can only be used to multiply two binomials) that the outer (O) and inner (I) terms are like terms and can be combined.

EXAMPLE 3   Finding a Product by the FOIL Method

Use the FOIL Method to find the product of $2x - 4$ and $x + 5$.

Solution

$$\overset{\text{F}\qquad\text{O}\qquad\text{I}\qquad\text{L}}{(2x - 4)(x + 5) = 2x^2 + 10x - 4x - 20}$$

$$= 2x^2 + 6x - 20$$

✓ *Checkpoint* 🔊)) *Audio-video solution in English & Spanish at LarsonPrecalculus.com.*

Use the FOIL Method to find the product of $3x - 1$ and $x - 5$. ◼

Special Products

Some binomial products have special forms that occur frequently in algebra. You do not need to memorize these formulas because you can use the Distributive Property to multiply. However, becoming familiar with these formulas will enable you to manipulate the algebra more quickly.

Special Products

Let u and v be real numbers, variables, or algebraic expressions.

Special Product   **Example**

Sum and Difference of Same Terms

$$(u + v)(u - v) = u^2 - v^2$$ $$(x + 4)(x - 4) = x^2 - 4^2$$
$$= x^2 - 16$$

Square of a Binomial

$$(u + v)^2 = u^2 + 2uv + v^2$$ $$(x + 3)^2 = x^2 + 2(x)(3) + 3^2$$
$$= x^2 + 6x + 9$$

$$(u - v)^2 = u^2 - 2uv + v^2$$ $$(3x - 2)^2 = (3x)^2 - 2(3x)(2) + 2^2$$
$$= 9x^2 - 12x + 4$$

Cube of a Binomial

$$(u + v)^3 = u^3 + 3u^2v + 3uv^2 + v^3$$ $$(x + 2)^3 = x^3 + 3x^2(2) + 3x(2^2) + 2^3$$
$$= x^3 + 6x^2 + 12x + 8$$

$$(u - v)^3 = u^3 - 3u^2v + 3uv^2 - v^3$$ $$(x - 1)^3 = x^3 - 3x^2(1) + 3x(1^2) - 1^3$$
$$= x^3 - 3x^2 + 3x - 1$$

EXAMPLE 4   **Sum and Difference of Same Terms**

Find each product.

a. $(5x + 9)(5x - 9)$ **b.** $(x + y - 2)(x + y + 2)$

Solution

a. The product of a sum and a difference of the *same* two terms has no middle term and takes the form $(u + v)(u - v) = u^2 - v^2$.

$$(5x + 9)(5x - 9) = (5x)^2 - 9^2 = 25x^2 - 81$$

b. By grouping $x + y$ in parentheses, you can write the product of the trinomials as a special product.

Difference Sum

$$(x + y - 2)(x + y + 2) = [(x + y) - 2][(x + y) + 2]$$
$$= (x + y)^2 - 2^2 \qquad \text{Sum and difference of same terms}$$
$$= x^2 + 2xy + y^2 - 4$$

✓ ***Checkpoint***  *Audio-video solution in English & Spanish at LarsonPrecalculus.com.*

Find the product: $(x - 2 + 3y)(x - 2 - 3y)$.

Polynomials with Common Factors

The process of writing a polynomial as a product is called **factoring.** It is an important tool for solving equations and for simplifying rational expressions.

Unless noted otherwise, when you are asked to factor a polynomial, assume that you are looking for factors that have integer coefficients. If a polynomial does not factor using integer coefficients, then it is **prime** or **irreducible over the integers.** For instance, the polynomial

$$x^2 - 3$$

is irreducible over the integers. Over the *real numbers,* this polynomial factors as

$$x^2 - 3 = (x + \sqrt{3})(x - \sqrt{3}).$$

A polynomial is **completely factored** when each of its factors is prime. For instance,

$$x^3 - x^2 + 4x - 4 = (x - 1)(x^2 + 4) \qquad \text{Completely factored}$$

is completely factored, but

$$x^3 - x^2 - 4x + 4 = (x - 1)(x^2 - 4) \qquad \text{Not completely factored}$$

is not completely factored. Its complete factorization is

$$x^3 - x^2 - 4x + 4 = (x - 1)(x + 2)(x - 2).$$

The simplest type of factoring involves a polynomial that can be written as the product of a monomial and another polynomial. The technique used here is the Distributive Property, $a(b + c) = ab + ac$, in the *reverse* direction.

$$ab + ac = a(b + c) \qquad \text{a is a common factor.}$$

Removing (factoring out) any common factors is the first step in completely factoring a polynomial.

EXAMPLE 5   Removing Common Factors

Factor each expression.

a. $6x^3 - 4x$

b. $-4x^2 + 12x - 16$

c. $(x - 2)(2x) + (x - 2)(3)$

Solution

a. $6x^3 - 4x = 2x(3x^2) - 2x(2) \qquad \text{$2x$ is a common factor.}$

$\qquad\qquad = 2x(3x^2 - 2)$

b. $-4x^2 + 12x - 16 = -4(x^2) + (-4)(-3x) + (-4)4 \qquad \text{-4 is a common factor.}$

$\qquad\qquad\qquad\qquad = -4(x^2 - 3x + 4)$

c. $(x - 2)(2x) + (x - 2)(3) = (x - 2)(2x + 3) \qquad \text{$(x - 2)$ is a common factor.}$

✓ **Checkpoint** ◖))) *Audio-video solution in English & Spanish at LarsonPrecalculus.com.*

Factor each expression.

a. $5x^3 - 15x^2$

b. $-3 + 6x - 12x^3$

c. $(x + 1)(x^2) - (x + 1)(2)$

Factoring Special Polynomial Forms

Some polynomials have special forms that arise from the special product forms on page A23. You should learn to recognize these forms.

Factoring Special Polynomial Forms

Factored Form	Example

Difference of Two Squares

$u^2 - v^2 = (u + v)(u - v)$ $9x^2 - 4 = (3x)^2 - 2^2 = (3x + 2)(3x - 2)$

Perfect Square Trinomial

$u^2 + 2uv + v^2 = (u + v)^2$ $x^2 + 6x + 9 = x^2 + 2(x)(3) + 3^2 = (x + 3)^2$

$u^2 - 2uv + v^2 = (u - v)^2$ $x^2 - 6x + 9 = x^2 - 2(x)(3) + 3^2 = (x - 3)^2$

Sum or Difference of Two Cubes

$u^3 + v^3 = (u + v)(u^2 - uv + v^2)$ $x^3 + 8 = x^3 + 2^3 = (x + 2)(x^2 - 2x + 4)$

$u^3 - v^3 = (u - v)(u^2 + uv + v^2)$ $27x^3 - 1 = (3x)^3 - 1^3 = (3x - 1)(9x^2 + 3x + 1)$

The factored form of a difference of two squares is always a set of *conjugate pairs*.

$$u^2 - v^2 = (u + v)(u - v)$$ Conjugate pairs

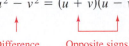

Difference   Opposite signs

To recognize perfect square terms, look for coefficients that are squares of integers and variables raised to *even powers*.

EXAMPLE 6   **Removing a Common Factor First**

$$3 - 12x^2 = 3(1 - 4x^2)$$ 3 is a common factor.

$$= 3[1^2 - (2x)^2]$$

$$= 3(1 + 2x)(1 - 2x)$$ Difference of two squares

 ✓ *Checkpoint*  Audio-video solution in English & Spanish at LarsonPrecalculus.com.

Factor $100 - 4y^2$.

· · REMARK In Example 6, note that the first step in factoring a polynomial is to check for any common factors. Once you have removed any common factors, it is often possible to recognize patterns that were not immediately obvious.

EXAMPLE 7   **Factoring the Difference of Two Squares**

a. $(x + 2)^2 - y^2 = [(x + 2) + y][(x + 2) - y]$

$$= (x + 2 + y)(x + 2 - y)$$

b. $16x^4 - 81 = (4x^2)^2 - 9^2$

$$= (4x^2 + 9)(4x^2 - 9)$$ Difference of two squares

$$= (4x^2 + 9)[(2x)^2 - 3^2]$$

$$= (4x^2 + 9)(2x + 3)(2x - 3)$$ Difference of two squares

 ✓ *Checkpoint*  Audio-video solution in English & Spanish at LarsonPrecalculus.com.

Factor $(x - 1)^2 - 9y^4$.

A **perfect square trinomial** is the square of a binomial, and it has the form

$$u^2 + 2uv + v^2 = (u + v)^2 \quad \text{or} \quad u^2 - 2uv + v^2 = (u - v)^2.$$

Note that the first and last terms are squares and the middle term is twice the product of u and v.

EXAMPLE 8   **Factoring Perfect Square Trinomials**

Factor each trinomial.

a. $x^2 - 10x + 25$ **b.** $16x^2 + 24x + 9$

Solution

a. $x^2 - 10x + 25 = x^2 - 2(x)(5) + 5^2 = (x - 5)^2$

b. $16x^2 + 24x + 9 = (4x)^2 + 2(4x)(3) + 3^2 = (4x + 3)^2$

✓ *Checkpoint* �))) *Audio-video solution in English & Spanish at LarsonPrecalculus.com.*

Factor $9x^2 - 30x + 25$.

The next two formulas show the sums and differences of cubes. Pay special attention to the signs of the terms.

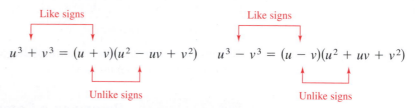

EXAMPLE 9   **Factoring the Difference of Cubes**

$x^3 - 27 = x^3 - 3^3$ Rewrite 27 as 3^3.

$\qquad\quad = (x - 3)(x^2 + 3x + 9)$ Factor.

✓ *Checkpoint* �))) *Audio-video solution in English & Spanish at LarsonPrecalculus.com.*

Factor $64x^3 - 1$.

EXAMPLE 10   **Factoring the Sum of Cubes**

a. $y^3 + 8 = y^3 + 2^3$ Rewrite 8 as 2^3.

$\qquad\quad = (y + 2)(y^2 - 2y + 4)$ Factor.

b. $3x^3 + 192 = 3(x^3 + 64)$ 3 is a common factor.

$\qquad\qquad = 3(x^3 + 4^3)$ Rewrite 64 as 4^3.

$\qquad\qquad = 3(x + 4)(x^2 - 4x + 16)$ Factor.

✓ *Checkpoint* �))) *Audio-video solution in English & Spanish at LarsonPrecalculus.com.*

Factor each expression.

a. $x^3 + 216$ **b.** $5y^3 + 135$

Trinomials with Binomial Factors

To factor a trinomial of the form $ax^2 + bx + c$, use the following pattern.

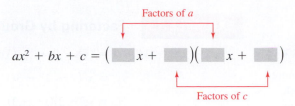

Factors of a

$$ax^2 + bx + c = (\boxed{}x + \boxed{})(\boxed{}x + \boxed{})$$

Factors of c

The goal is to find a combination of factors of a and c such that the outer and inner products add up to the middle term bx. For instance, for the trinomial $6x^2 + 17x + 5$, you can write all possible factorizations and determine which one has outer and inner products that add up to $17x$.

$$(6x + 5)(x + 1), \ (6x + 1)(x + 5), \ (2x + 1)(3x + 5), \ (2x + 5)(3x + 1)$$

You can see that $(2x + 5)(3x + 1)$ is the correct factorization because the outer (O) and inner (I) products add up to $17x$.

$$\begin{array}{ccccc} \text{F} & \text{O} & \text{I} & \text{L} & \text{O + I} \\ \downarrow & \downarrow & \downarrow & \downarrow & \downarrow \end{array}$$

$$(2x + 5)(3x + 1) = 6x^2 + 2x + 15x + 5 = 6x^2 + 17x + 5$$

EXAMPLE 11   Factoring a Trinomial: Leading Coefficient Is 1

Factor $x^2 - 7x + 12$.

Solution   For this trinomial, you have $a = 1$, $b = -7$, and $c = 12$. Because b is negative and c is positive, both factors of 12 must be negative. So, the possible factorizations of $x^2 - 7x + 12$ are

$$(x - 2)(x - 6), \quad (x - 1)(x - 12), \quad \text{and} \quad (x - 3)(x - 4).$$

Testing the middle term, you will find the correct factorization to be

$$x^2 - 7x + 12 = (x - 3)(x - 4).$$

✓ **Checkpoint** ◀))) Audio-video solution in English & Spanish at LarsonPrecalculus.com.

Factor $x^2 + x - 6$.

REMARK Factoring a trinomial can involve trial and error. However, once you have produced the factored form, it is relatively easy to check your answer. For instance, verify the factorization in Example 11 by multiplying $(x - 3)$ by $(x - 4)$ to see that you obtain the original trinomial $x^2 - 7x + 12$.

EXAMPLE 12   Factoring a Trinomial: Leading Coefficient Is Not 1

Factor $2x^2 + x - 15$.

Solution   For this trinomial, you have $a = 2$ and $c = -15$, which means that the factors of -15 must have unlike signs. The eight possible factorizations are as follows.

$$(2x - 1)(x + 15) \quad (2x + 1)(x - 15) \quad (2x - 3)(x + 5) \quad (2x + 3)(x - 5)$$

$$(2x - 5)(x + 3) \quad (2x + 5)(x - 3) \quad (2x - 15)(x + 1) \quad (2x + 15)(x - 1)$$

Testing the middle term, you will find the correct factorization to be

$$2x^2 + x - 15 = (2x - 5)(x + 3). \qquad \text{O + I} = 6x - 5x = x$$

✓ **Checkpoint** ◀))) Audio-video solution in English & Spanish at LarsonPrecalculus.com.

Factor each trinomial.

a. $2x^2 - 5x + 3$ **b.** $12x^2 + 7x + 1$

Factoring by Grouping

Sometimes, polynomials with more than three terms can be **factored by grouping.**

EXAMPLE 13 Factoring by Grouping

$$x^3 - 2x^2 - 3x + 6 = (x^3 - 2x^2) - (3x - 6) \qquad \text{Group terms.}$$
$$= x^2(x - 2) - 3(x - 2) \qquad \text{Factor each group.}$$
$$= (x - 2)(x^2 - 3) \qquad \text{Distributive Property}$$

✓ *Checkpoint*  *Audio-video solution in English & Spanish at LarsonPrecalculus.com.*

Factor $x^3 + x^2 - 5x - 5$. ■

Factoring by grouping can save you some of the trial and error involved in factoring a trinomial. To factor a trinomial of the form $ax^2 + bx + c$ by grouping, rewrite the middle term using the sum of two factors of the product ac that add up to b. Example 14 illustrates this technique.

EXAMPLE 14 Factoring a Trinomial by Grouping

In the trinomial $2x^2 + 5x - 3$, $a = 2$ and $c = -3$, so the product ac is -6. Now, -6 factors as $(6)(-1)$ and $6 - 1 = 5 = b$. So, rewrite the middle term as $5x = 6x - x$.

$$2x^2 + 5x - 3 = 2x^2 + 6x - x - 3 \qquad \text{Rewrite middle term.}$$
$$= (2x^2 + 6x) - (x + 3) \qquad \text{Group terms.}$$
$$= 2x(x + 3) - (x + 3) \qquad \text{Factor groups.}$$
$$= (x + 3)(2x - 1) \qquad \text{Distributive Property}$$

✓ *Checkpoint*  *Audio-video solution in English & Spanish at LarsonPrecalculus.com.*

Use factoring by grouping to factor $2x^2 + 5x - 12$. ■

> **REMARK** Sometimes, more than one grouping will work. For instance, another way to factor the polynomial in Example 13 is as follows.
>
> $$x^3 - 2x^2 - 3x + 6$$
> $$= (x^3 - 3x) - (2x^2 - 6)$$
> $$= x(x^2 - 3) - 2(x^2 - 3)$$
> $$= (x^2 - 3)(x - 2)$$
>
> As you can see, you obtain the same result as in Example 13.

Summarize (Appendix A.3)

1. State the definition of a polynomial in x and explain what is meant by the standard form of a polynomial *(page A21)*. For an example of writing polynomials in standard form, see Example 1.

2. Describe how to add and subtract polynomials *(page A22)*. For an example of adding and subtracting polynomials, see Example 2.

3. Describe the FOIL Method *(page A22)*. For an example of finding a product using the FOIL Method, see Example 3.

4. Explain how to find binomial products that have special forms *(page A23)*. For an example of binomial products that have special forms, see Example 4.

5. Describe what it means to completely factor a polynomial *(page A24)*. For an example of removing common factors, see Example 5.

6. Make a list of the special polynomial forms of factoring *(page A25)*. For examples of factoring these special forms, see Examples 6–10.

7. Describe how to factor a trinomial of the form $ax^2 + bx + c$ *(page A27)*. For examples of factoring trinomials of this form, see Examples 11 and 12.

8. Explain how to factor a polynomial by grouping *(page A28)*. For examples of factoring by grouping, see Examples 13 and 14.

A.4 Rational Expressions

- Find domains of algebraic expressions.
- Simplify rational expressions.
- Add, subtract, multiply, and divide rational expressions.
- Simplify complex fractions and rewrite difference quotients.

Domain of an Algebraic Expression

The set of real numbers for which an algebraic expression is defined is the **domain** of the expression. Two algebraic expressions are **equivalent** when they have the same domain and yield the same values for all numbers in their domain. For instance,

$$(x + 1) + (x + 2) \quad \text{and} \quad 2x + 3$$

are equivalent because

$$
\begin{aligned}
(x + 1) + (x + 2) &= x + 1 + x + 2 \\
&= x + x + 1 + 2 \\
&= 2x + 3.
\end{aligned}
$$

Rational expressions can help you solve real-life problems. For instance, a rational expression can be used to model the temperature of food in a refrigerator.

EXAMPLE 1   **Finding the Domain of an Algebraic Expression**

a. The domain of the polynomial

$$2x^3 + 3x + 4$$

is the set of all real numbers. In fact, the domain of any polynomial is the set of all real numbers, unless the domain is specifically restricted.

b. The domain of the radical expression

$$\sqrt{x - 2}$$

is the set of real numbers greater than or equal to 2, because the square root of a negative number is not a real number.

c. The domain of the expression

$$\frac{x + 2}{x - 3}$$

is the set of all real numbers except $x = 3$, which would result in division by zero, which is undefined.

✓ *Checkpoint* *Audio-video solution in English & Spanish at LarsonPrecalculus.com.*

Find the domain of each expression.

a. $4x^3 + 3, \quad x \geq 0$ **b.** $\sqrt{x + 7}$ **c.** $\dfrac{1 - x}{x}$ ■

The quotient of two algebraic expressions is a *fractional expression*. Moreover, the quotient of two *polynomials* such as

$$\frac{1}{x}, \quad \frac{2x - 1}{x + 1}, \quad \text{or} \quad \frac{x^2 - 1}{x^2 + 1}$$

is a **rational expression**.

Simplifying Rational Expressions

Recall that a fraction is in simplest form when its numerator and denominator have no factors in common aside from ±1. To write a fraction in simplest form, divide out common factors.

$$\frac{a \cdot \not c}{b \cdot \not c} = \frac{a}{b}, \quad c \neq 0$$

The key to success in simplifying rational expressions lies in your ability to *factor* polynomials. When simplifying rational expressions, factor each polynomial completely to determine whether the numerator and denominator have factors in common.

EXAMPLE 2 **Simplifying a Rational Expression**

$$\frac{x^2 + 4x - 12}{3x - 6} = \frac{(x + 6)(x - 2)}{3(x - 2)} \qquad \text{Factor completely.}$$

$$= \frac{x + 6}{3}, \quad x \neq 2 \qquad \text{Divide out common factor.}$$

Note that the original expression is undefined when $x = 2$ (because division by zero is undefined). To make the simplified expression *equivalent* to the original expression, you must list the domain restriction $x \neq 2$ with the simplified expression.

✓ **Checkpoint**  *Audio-video solution in English & Spanish at LarsonPrecalculus.com.*

Write $\dfrac{4x + 12}{x^2 - 3x - 18}$ in simplest form. ■

Sometimes it may be necessary to change the sign of a factor by factoring out (-1) to simplify a rational expression, as shown in Example 3.

EXAMPLE 3 **Simplifying a Rational Expression**

$$\frac{12 + x - x^2}{2x^2 - 9x + 4} = \frac{(4 - x)(3 + x)}{(2x - 1)(x - 4)} \qquad \text{Factor completely.}$$

$$= \frac{-(x - 4)(3 + x)}{(2x - 1)(x - 4)} \qquad (4 - x) = -(x - 4)$$

$$= -\frac{3 + x}{2x - 1}, \quad x \neq 4 \qquad \text{Divide out common factor.}$$

✓ **Checkpoint** *Audio-video solution in English & Spanish at LarsonPrecalculus.com.*

Write $\dfrac{3x^2 - x - 2}{5 - 4x - x^2}$ in simplest form. ■

In this text, when writing a rational expression, the domain is usually not listed with the expression. It is *implied* that the real numbers that make the denominator zero are excluded from the expression. Also, when performing operations with rational expressions, this text follows the convention of listing *by the simplified expression* all values of x that must be specifically excluded from the domain in order to make the domains of the simplified and original expressions agree. Example 3, for instance, lists the restriction $x \neq 4$ with the simplified expression to make the two domains agree. Note that the value $x = \frac{1}{2}$ is excluded from *both* domains, so it is not necessary to list this value.

• • • • • • • • • • • • • • • ▷

•• **REMARK** In Example 2, do not make the mistake of trying to simplify further by dividing out terms.

$$\frac{x + 6}{3} = \frac{x + \cancel{6}}{\cancel{3}} = x + 2$$

Remember that to simplify fractions, divide out common *factors,* not terms.

Operations with Rational Expressions

To multiply or divide rational expressions, use the properties of fractions discussed in Appendix A.1. Recall that to divide fractions, you invert the divisor and multiply.

EXAMPLE 4   **Multiplying Rational Expressions**

$$\frac{2x^2 + x - 6}{x^2 + 4x - 5} \cdot \frac{x^3 - 3x^2 + 2x}{4x^2 - 6x} = \frac{(2x - 3)(x + 2)}{(x + 5)(x - 1)} \cdot \frac{x(x - 2)(x - 1)}{2x(2x - 3)}$$

$$= \frac{(x + 2)(x - 2)}{2(x + 5)}, \quad x \neq 0, x \neq 1, x \neq \frac{3}{2}$$

> ▷ **REMARK** Note that Example 4 lists the restrictions $x \neq 0$, $x \neq 1$, and $x \neq \frac{3}{2}$ with the simplified expression in order to make the two domains agree. Also note that the value $x = -5$ is excluded from both domains, so it is not necessary to list this value.

✓ **Checkpoint** ◀))) Audio-video solution in English & Spanish at LarsonPrecalculus.com.

Multiply and simplify: $\dfrac{15x^2 + 5x}{x^3 - 3x^2 - 18x} \cdot \dfrac{x^2 - 2x - 15}{3x^2 - 8x - 3}$.

EXAMPLE 5   **Dividing Rational Expressions**

$$\frac{x^3 - 8}{x^2 - 4} \div \frac{x^2 + 2x + 4}{x^3 + 8} = \frac{x^3 - 8}{x^2 - 4} \cdot \frac{x^3 + 8}{x^2 + 2x + 4} \qquad \text{Invert and multiply.}$$

$$= \frac{(x - 2)(x^2 + 2x + 4)}{(x + 2)(x - 2)} \cdot \frac{(x + 2)(x^2 - 2x + 4)}{(x^2 + 2x + 4)}$$

$$= x^2 - 2x + 4, \quad x \neq \pm 2 \qquad \text{Divide out common factors.}$$

✓ **Checkpoint** ◀))) Audio-video solution in English & Spanish at LarsonPrecalculus.com.

Divide and simplify: $\dfrac{x^3 - 1}{x^2 - 1} \div \dfrac{x^2 + x + 1}{x^2 + 2x + 1}$.

To add or subtract rational expressions, use the LCD (least common denominator) method or the *basic definition*

$$\frac{a}{b} \pm \frac{c}{d} = \frac{ad \pm bc}{bd}, \quad b \neq 0, d \neq 0. \qquad \text{Basic definition}$$

This definition provides an efficient way of adding or subtracting *two* fractions that have no common factors in their denominators.

EXAMPLE 6   **Subtracting Rational Expressions**

$$\frac{x}{x - 3} - \frac{2}{3x + 4} = \frac{x(3x + 4) - 2(x - 3)}{(x - 3)(3x + 4)} \qquad \text{Basic definition}$$

$$= \frac{3x^2 + 4x - 2x + 6}{(x - 3)(3x + 4)} \qquad \text{Distributive Property}$$

$$= \frac{3x^2 + 2x + 6}{(x - 3)(3x + 4)} \qquad \text{Combine like terms.}$$

> ▷ **REMARK** When subtracting rational expressions, remember to distribute the negative sign to all the terms in the quantity that is being subtracted.

✓ **Checkpoint** ◀))) Audio-video solution in English & Spanish at LarsonPrecalculus.com.

Subtract: $\dfrac{x}{2x - 1} - \dfrac{1}{x + 2}$.

For three or more fractions, or for fractions with a repeated factor in the denominators, the LCD method works well. Recall that the least common denominator of several fractions consists of the product of all prime factors in the denominators, with each factor given the highest power of its occurrence in any denominator. Here is a numerical example.

$$\frac{1}{6} + \frac{3}{4} - \frac{2}{3} = \frac{1 \cdot 2}{6 \cdot 2} + \frac{3 \cdot 3}{4 \cdot 3} - \frac{2 \cdot 4}{3 \cdot 4} \qquad \text{The LCD is 12.}$$

$$= \frac{2}{12} + \frac{9}{12} - \frac{8}{12}$$

$$= \frac{3}{12}$$

$$= \frac{1}{4}$$

Sometimes, the numerator of the answer has a factor in common with the denominator. In such cases, simplify the answer. For instance, in the example above, $\frac{3}{12}$ simplifies to $\frac{1}{4}$.

EXAMPLE 7   Combining Rational Expressions: The LCD Method

Perform the operations and simplify.

$$\frac{3}{x - 1} - \frac{2}{x} + \frac{x + 3}{x^2 - 1}$$

Solution   Using the factored denominators

$$(x - 1), \quad x, \quad \text{and} \quad (x + 1)(x - 1)$$

you can see that the LCD is $x(x + 1)(x - 1)$.

$$\frac{3}{x - 1} - \frac{2}{x} + \frac{x + 3}{(x + 1)(x - 1)}$$

$$= \frac{3(x)(x + 1)}{x(x + 1)(x - 1)} - \frac{2(x + 1)(x - 1)}{x(x + 1)(x - 1)} + \frac{(x + 3)(x)}{x(x + 1)(x - 1)}$$

$$= \frac{3(x)(x + 1) - 2(x + 1)(x - 1) + (x + 3)(x)}{x(x + 1)(x - 1)}$$

$$= \frac{3x^2 + 3x - 2x^2 + 2 + x^2 + 3x}{x(x + 1)(x - 1)} \qquad \text{Distributive Property}$$

$$= \frac{3x^2 - 2x^2 + x^2 + 3x + 3x + 2}{x(x + 1)(x - 1)} \qquad \text{Group like terms.}$$

$$= \frac{2x^2 + 6x + 2}{x(x + 1)(x - 1)} \qquad \text{Combine like terms.}$$

$$= \frac{2(x^2 + 3x + 1)}{x(x + 1)(x - 1)} \qquad \text{Factor.}$$

✓ **Checkpoint** 🔊))) *Audio-video solution in English & Spanish at LarsonPrecalculus.com.*

Perform the operations and simplify.

$$\frac{4}{x} - \frac{x + 5}{x^2 - 4} + \frac{4}{x + 2}$$

Complex Fractions and the Difference Quotient

Fractional expressions with separate fractions in the numerator, denominator, or both are called **complex fractions.** Here are two examples.

$$\frac{\left(\dfrac{1}{x}\right)}{x^2 + 1} \quad \text{and} \quad \frac{\left(\dfrac{1}{x}\right)}{\left(\dfrac{1}{x^2 + 1}\right)}$$

To simplify a complex fraction, combine the fractions in the numerator into a single fraction and then combine the fractions in the denominator into a single fraction. Then invert the denominator and multiply.

EXAMPLE 8   **Simplifying a Complex Fraction**

$$\frac{\left(\dfrac{2}{x} - 3\right)}{\left(1 - \dfrac{1}{x - 1}\right)} = \frac{\left[\dfrac{2 - 3(x)}{x}\right]}{\left[\dfrac{1(x - 1) - 1}{x - 1}\right]} \qquad \text{Combine fractions.}$$

$$= \frac{\left(\dfrac{2 - 3x}{x}\right)}{\left(\dfrac{x - 2}{x - 1}\right)} \qquad \text{Simplify.}$$

$$= \frac{2 - 3x}{x} \cdot \frac{x - 1}{x - 2} \qquad \text{Invert and multiply.}$$

$$= \frac{(2 - 3x)(x - 1)}{x(x - 2)}, \quad x \neq 1$$

✓ **Checkpoint** 🔊))) *Audio-video solution in English & Spanish at LarsonPrecalculus.com.*

Simplify the complex fraction $\dfrac{\left(\dfrac{1}{x + 2} + 1\right)}{\left(\dfrac{x}{3} - 1\right)}$.   ■

Another way to simplify a complex fraction is to multiply its numerator and denominator by the LCD of all fractions in its numerator and denominator. This method is applied to the fraction in Example 8 as follows.

$$\frac{\left(\dfrac{2}{x} - 3\right)}{\left(1 - \dfrac{1}{x - 1}\right)} = \frac{\left(\dfrac{2}{x} - 3\right)}{\left(1 - \dfrac{1}{x - 1}\right)} \cdot \frac{x(x - 1)}{x(x - 1)} \qquad \text{LCD is } x(x - 1).$$

$$= \frac{\left(\dfrac{2 - 3x}{x}\right) \cdot x(x - 1)}{\left(\dfrac{x - 2}{x - 1}\right) \cdot x(x - 1)}$$

$$= \frac{(2 - 3x)(x - 1)}{x(x - 2)}, \quad x \neq 1$$

The next three examples illustrate some methods for simplifying rational expressions involving negative exponents and radicals. These types of expressions occur frequently in calculus.

To simplify an expression with negative exponents, one method is to begin by factoring out the common factor with the *lesser* exponent. Remember that when factoring, you *subtract* exponents. For instance, in $3x^{-5/2} + 2x^{-3/2}$, the lesser exponent is $-\frac{5}{2}$ and the common factor is $x^{-5/2}$.

$$3x^{-5/2} + 2x^{-3/2} = x^{-5/2}[3(1) + 2x^{-3/2-(-5/2)}]$$

$$= x^{-5/2}(3 + 2x^1)$$

$$= \frac{3 + 2x}{x^{5/2}}$$

EXAMPLE 9 **Simplifying an Expression**

Simplify the following expression containing negative exponents.

$$x(1 - 2x)^{-3/2} + (1 - 2x)^{-1/2}$$

Solution Begin by factoring out the common factor with the *lesser exponent.*

$$x(1 - 2x)^{-3/2} + (1 - 2x)^{-1/2} = (1 - 2x)^{-3/2}[x + (1 - 2x)^{(-1/2)-(-3/2)}]$$

$$= (1 - 2x)^{-3/2}[x + (1 - 2x)^1]$$

$$= \frac{1 - x}{(1 - 2x)^{3/2}}$$

✓ *Checkpoint* 🔊))) *Audio-video solution in English & Spanish at LarsonPrecalculus.com.*

Simplify $(x - 1)^{-1/3} - x(x - 1)^{-4/3}$.

The next example shows a second method for simplifying an expression with negative exponents.

EXAMPLE 10 **Simplifying an Expression**

$$\frac{(4 - x^2)^{1/2} + x^2(4 - x^2)^{-1/2}}{4 - x^2}$$

$$= \frac{(4 - x^2)^{1/2} + x^2(4 - x^2)^{-1/2}}{4 - x^2} \cdot \frac{(4 - x^2)^{1/2}}{(4 - x^2)^{1/2}}$$

$$= \frac{(4 - x^2)^1 + x^2(4 - x^2)^0}{(4 - x^2)^{3/2}}$$

$$= \frac{4 - x^2 + x^2}{(4 - x^2)^{3/2}}$$

$$= \frac{4}{(4 - x^2)^{3/2}}$$

✓ *Checkpoint* 🔊))) *Audio-video solution in English & Spanish at LarsonPrecalculus.com.*

Simplify

$$\frac{x^2(x^2 - 2)^{-1/2} + (x^2 - 2)^{1/2}}{x^2 - 2}.$$

EXAMPLE 11 **Rewriting a Difference Quotient**

The following expression from calculus is an example of a *difference quotient*.

$$\frac{\sqrt{x+h} - \sqrt{x}}{h}$$

Rewrite this expression by rationalizing its numerator.

Solution

$$\frac{\sqrt{x+h} - \sqrt{x}}{h} = \frac{\sqrt{x+h} - \sqrt{x}}{h} \cdot \frac{\sqrt{x+h} + \sqrt{x}}{\sqrt{x+h} + \sqrt{x}}$$

> ▷ **ALGEBRA HELP** You can review the techniques for rationalizing a numerator in Appendix A.2.

$$= \frac{\left(\sqrt{x+h}\right)^2 - \left(\sqrt{x}\right)^2}{h\left(\sqrt{x+h} + \sqrt{x}\right)}$$

$$= \frac{x + h - x}{h\left(\sqrt{x+h} + \sqrt{x}\right)}$$

$$= \frac{h}{h\left(\sqrt{x+h} + \sqrt{x}\right)}$$

$$= \frac{1}{\sqrt{x+h} + \sqrt{x}}, \quad h \neq 0$$

Notice that the original expression is undefined when $h = 0$. So, you must exclude $h = 0$ from the domain of the simplified expression so that the expressions are equivalent.

✓ **Checkpoint** ◀)) *Audio-video solution in English & Spanish at LarsonPrecalculus.com.*

Rewrite the difference quotient

$$\frac{\sqrt{9+h} - 3}{h}$$

by rationalizing its numerator.

Difference quotients, such as that in Example 11, occur frequently in calculus. Often, they need to be rewritten in an equivalent form that can be evaluated when $h = 0$.

Summarize **(Appendix A.4)**

1. State the definition of the domain of an algebraic expression *(page A29)*. For an example of finding the domains of algebraic expressions, see Example 1.

2. State the definition of a rational expression and describe how to simplify a rational expression *(pages A29 and A30)*. For examples of simplifying rational expressions, see Examples 2 and 3.

3. Describe how to multiply, divide, add, and subtract rational expressions *(page A31)*. For examples of operations with rational expressions, see Examples 4–7.

4. State the definition of a complex fraction *(page A33)*. For an example of simplifying a complex fraction, see Example 8.

5. Describe how to rewrite a difference quotient and why you would want to do so *(page A35)*. For an example of rewriting a difference quotient, see Example 11.

A.5 Solving Equations

You can use linear equations in many real-life applications. For example, you can use linear equations in forensics to determine height from femur length.

- Identify different types of equations.
- Solve linear equations in one variable and rational equations that lead to linear equations.
- Solve quadratic equations by factoring, extracting square roots, completing the square, and using the Quadratic Formula.
- Solve polynomial equations of degree three or greater.
- Solve radical equations.
- Solve absolute value equations.
- Use common formulas to solve real-life problems.

Equations and Solutions of Equations

An **equation** in x is a statement that two algebraic expressions are equal. For example,

$$3x - 5 = 7, \quad x^2 - x - 6 = 0, \quad \text{and} \quad \sqrt{2x} = 4$$

are equations. To **solve** an equation in x means to find all values of x for which the equation is true. Such values are **solutions.** For instance, $x = 4$ is a solution of the equation $3x - 5 = 7$ because $3(4) - 5 = 7$ is a true statement.

The solutions of an equation depend on the kinds of numbers being considered. For instance, in the set of rational numbers, $x^2 = 10$ has no solution because there is no rational number whose square is 10. However, in the set of real numbers, the equation has the two solutions $x = \sqrt{10}$ and $x = -\sqrt{10}$.

An equation that is true for *every* real number in the domain of the variable is called an **identity.** For example,

$$x^2 - 9 = (x + 3)(x - 3) \qquad \text{Identity}$$

is an identity because it is a true statement for any real value of x. The equation

$$\frac{x}{3x^2} = \frac{1}{3x} \qquad \text{Identity}$$

where $x \neq 0$, is an identity because it is true for any nonzero real value of x.

An equation that is true for just *some* (but not all) of the real numbers in the domain of the variable is called a **conditional equation.** For example, the equation

$$x^2 - 9 = 0 \qquad \text{Conditional equation}$$

is conditional because $x = 3$ and $x = -3$ are the only values in the domain that satisfy the equation.

A **contradiction** is an equation that is false for *every* real number in the domain of the variable. For example, the equation

$$2x - 4 = 2x + 1 \qquad \text{Contradiction}$$

is a contradiction because there are no real values of x for which the equation is true.

Linear and Rational Equations

Definition of Linear Equation in One Variable

A **linear equation in one variable** x is an equation that can be written in the standard form

$$ax + b = 0$$

where a and b are real numbers with $a \neq 0$.

A linear equation has exactly one solution. To see this, consider the following steps. (Remember that $a \neq 0$.)

$$ax + b = 0 \qquad \text{Write original equation.}$$

$$ax = -b \qquad \text{Subtract } b \text{ from each side.}$$

$$x = -\frac{b}{a} \qquad \text{Divide each side by } a.$$

To solve a conditional equation in x, isolate x on one side of the equation by a sequence of **equivalent equations,** each having the same solution(s) as the original equation. The operations that yield equivalent equations come from the properties of equality reviewed in Appendix A.1.

Generating Equivalent Equations

An equation can be transformed into an *equivalent equation* by one or more of the following steps.

		Given Equation	Equivalent Equation
1.	Remove symbols of grouping, combine like terms, or simplify fractions on one or both sides of the equation.	$2x - x = 4$	$x = 4$
2.	Add (or subtract) the same quantity to (from) *each* side of the equation.	$x + 1 = 6$	$x = 5$
3.	Multiply (or divide) *each* side of the equation by the same *nonzero* quantity.	$2x = 6$	$x = 3$
4.	Interchange the two sides of the equation.	$2 = x$	$x = 2$

The following example shows the steps for solving a linear equation in one variable x.

EXAMPLE 1 **Solving a Linear Equation**

............▷

••REMARK After solving an equation, you should check each solution in the original equation. For instance, you can check the solution of Example 1(a) as follows.

$3x - 6 = 0$ Write original equation.

$3(2) - 6 \overset{?}{=} 0$ Substitute 2 for x.

$0 = 0$ Solution checks. ✓

Try checking the solution of Example 1(b).

a. $3x - 6 = 0$ Original equation

     $3x = 6$ Add 6 to each side.

     $x = 2$ Divide each side by 3.

b. $5x + 4 = 3x - 8$ Original equation

     $2x + 4 = -8$ Subtract $3x$ from each side.

     $2x = -12$ Subtract 4 from each side.

     $x = -6$ Divide each side by 2.

✓ ***Checkpoint*** 🔊)) *Audio-video solution in English & Spanish at LarsonPrecalculus.com.*

Solve each equation.

a. $7 - 2x = 15$

b. $7x - 9 = 5x + 7$

A **rational equation** is an equation that involves one or more fractional expressions. To solve a rational equation, find the least common denominator (LCD) of all terms and multiply every term by the LCD. This process will clear the original equation of fractions and produce a simpler equation to work with.

EXAMPLE 2 **Solving a Rational Equation**

Solve $\dfrac{x}{3} + \dfrac{3x}{4} = 2.$

Solution

$$\frac{x}{3} + \frac{3x}{4} = 2 \qquad \text{Original equation}$$

$$(12)\frac{x}{3} + (12)\frac{3x}{4} = (12)2 \qquad \text{Multiply each term by the LCD.}$$

$$4x + 9x = 24 \qquad \text{Simplify.}$$

$$13x = 24 \qquad \text{Combine like terms.}$$

$$x = \frac{24}{13} \qquad \text{Divide each side by 13.}$$

The solution is $x = \frac{24}{13}$. Check this in the original equation.

✓ *Checkpoint* ◀))) *Audio-video solution in English & Spanish at LarsonPrecalculus.com.*

Solve $\dfrac{4x}{9} - \dfrac{1}{3} = x + \dfrac{5}{3}.$ ∎

When multiplying or dividing an equation by a *variable expression,* it is possible to introduce an **extraneous solution** that does not satisfy the original equation.

EXAMPLE 3 **An Equation with an Extraneous Solution**

Solve $\dfrac{1}{x-2} = \dfrac{3}{x+2} - \dfrac{6x}{x^2-4}.$

Solution The LCD is $x^2 - 4 = (x+2)(x-2)$. Multiply each term by this LCD.

$$\frac{1}{x-2}(x+2)(x-2) = \frac{3}{x+2}(x+2)(x-2) - \frac{6x}{x^2-4}(x+2)(x-2)$$

$$x + 2 = 3(x-2) - 6x, \quad x \neq \pm 2$$

$$x + 2 = 3x - 6 - 6x$$

$$x + 2 = -3x - 6$$

$$4x = -8$$

$$x = -2 \qquad \text{Extraneous solution}$$

In the original equation, $x = -2$ yields a denominator of zero. So, $x = -2$ is an extraneous solution, and the original equation has *no solution.*

✓ *Checkpoint* ◀))) *Audio-video solution in English & Spanish at LarsonPrecalculus.com.*

Solve $\dfrac{3x}{x-4} = 5 + \dfrac{12}{x-4}.$ ∎

Quadratic Equations

A **quadratic equation** in x is an equation that can be written in the general form

$$ax^2 + bx + c = 0$$

where a, b, and c are real numbers with $a \neq 0$. A quadratic equation in x is also called a **second-degree polynomial equation** in x.

You should be familiar with the following four methods of solving quadratic equations.

Solving a Quadratic Equation

Factoring

If $ab = 0$, then $a = 0$ or $b = 0$.

Example:
$$x^2 - x - 6 = 0$$
$$(x - 3)(x + 2) = 0$$
$$x - 3 = 0 \implies x = 3$$
$$x + 2 = 0 \implies x = -2$$

Square Root Principle

If $u^2 = c$, where $c > 0$, then $u = \pm\sqrt{c}$.

Example:
$$(x + 3)^2 = 16$$
$$x + 3 = \pm 4$$
$$x = -3 \pm 4$$
$$x = 1 \quad \text{or} \quad x = -7$$

Completing the Square

If $x^2 + bx = c$, then

$$x^2 + bx + \left(\frac{b}{2}\right)^2 = c + \left(\frac{b}{2}\right)^2 \qquad \text{Add } \left(\frac{b}{2}\right)^2 \text{ to each side.}$$

$$\left(x + \frac{b}{2}\right)^2 = c + \frac{b^2}{4}.$$

Example:
$$x^2 + 6x = 5$$
$$x^2 + 6x + 3^2 = 5 + 3^2 \qquad \text{Add } \left(\frac{6}{2}\right)^2 \text{ to each side.}$$
$$(x + 3)^2 = 14$$
$$x + 3 = \pm\sqrt{14}$$
$$x = -3 \pm \sqrt{14}$$

Quadratic Formula

If $ax^2 + bx + c = 0$, then $x = \dfrac{-b \pm \sqrt{b^2 - 4ac}}{2a}$.

Example:
$$2x^2 + 3x - 1 = 0$$
$$x = \frac{-3 \pm \sqrt{3^2 - 4(2)(-1)}}{2(2)}$$
$$= \frac{-3 \pm \sqrt{17}}{4}$$

REMARK The Square Root Principle is also referred to as *extracting square roots.*

REMARK You can solve every quadratic equation by completing the square or using the Quadratic Formula.

EXAMPLE 4   **Solving a Quadratic Equation by Factoring**

a.
$$2x^2 + 9x + 7 = 3$$ — Original equation

$$2x^2 + 9x + 4 = 0$$ — Write in general form.

$$(2x + 1)(x + 4) = 0$$ — Factor.

$$2x + 1 = 0 \implies x = -\tfrac{1}{2}$$ — Set 1st factor equal to 0.

$$x + 4 = 0 \implies x = -4$$ — Set 2nd factor equal to 0.

The solutions are $x = -\tfrac{1}{2}$ and $x = -4$. Check these in the original equation.

b.
$$6x^2 - 3x = 0$$ — Original equation

$$3x(2x - 1) = 0$$ — Factor.

$$3x = 0 \implies x = 0$$ — Set 1st factor equal to 0.

$$2x - 1 = 0 \implies x = \tfrac{1}{2}$$ — Set 2nd factor equal to 0.

The solutions are $x = 0$ and $x = \tfrac{1}{2}$. Check these in the original equation.

✓ *Checkpoint*  **Audio-video solution in English & Spanish at LarsonPrecalculus.com.**

Solve $2x^2 - 3x + 1 = 6$ by factoring.

Note that the method of solution in Example 4 is based on the Zero-Factor Property from Appendix A.1. This property applies *only* to equations written in general form (in which the right side of the equation is zero). So, all terms must be collected on one side *before* factoring. For instance, in the equation $(x - 5)(x + 2) = 8$, it is *incorrect* to set each factor equal to 8. Try to solve this equation correctly.

EXAMPLE 5   **Extracting Square Roots**

Solve each equation by extracting square roots.

a. $4x^2 = 12$

b. $(x - 3)^2 = 7$

Solution

a.
$$4x^2 = 12$$ — Write original equation.

$$x^2 = 3$$ — Divide each side by 4.

$$x = \pm\sqrt{3}$$ — Extract square roots.

The solutions are $x = \sqrt{3}$ and $x = -\sqrt{3}$. Check these in the original equation.

b.
$$(x - 3)^2 = 7$$ — Write original equation.

$$x - 3 = \pm\sqrt{7}$$ — Extract square roots.

$$x = 3 \pm \sqrt{7}$$ — Add 3 to each side.

The solutions are $x = 3 \pm \sqrt{7}$. Check these in the original equation.

✓ *Checkpoint*  **Audio-video solution in English & Spanish at LarsonPrecalculus.com.**

Solve each equation by extracting square roots.

a. $3x^2 = 36$

b. $(x - 1)^2 = 10$

When solving quadratic equations by completing the square, you must add $(b/2)^2$ to *each side* in order to maintain equality. When the leading coefficient is *not* 1, you must divide each side of the equation by the leading coefficient *before* completing the square, as shown in Example 7.

EXAMPLE 6   **Completing the Square: Leading Coefficient Is 1**

Solve $x^2 + 2x - 6 = 0$ by completing the square.

Solution

$x^2 + 2x - 6 = 0$	Write original equation.
$x^2 + 2x = 6$	Add 6 to each side.
$x^2 + 2x + 1^2 = 6 + 1^2$	Add 1^2 to each side.

$$\text{(half of 2)}^2$$

$(x + 1)^2 = 7$	Simplify.
$x + 1 = \pm\sqrt{7}$	Extract square roots.
$x = -1 \pm \sqrt{7}$	Subtract 1 from each side.

The solutions are

$$x = -1 \pm \sqrt{7}.$$

Check these in the original equation.

✓ *Checkpoint*  *Audio-video solution in English & Spanish at LarsonPrecalculus.com.*

Solve $x^2 - 4x - 1 = 0$ by completing the square.

EXAMPLE 7   **Completing the Square: Leading Coefficient Is Not 1**

Solve $3x^2 - 4x - 5 = 0$ by completing the square.

Solution

$3x^2 - 4x - 5 = 0$	Write original equation.
$3x^2 - 4x = 5$	Add 5 to each side.
$x^2 - \dfrac{4}{3}x = \dfrac{5}{3}$	Divide each side by 3.
$x^2 - \dfrac{4}{3}x + \left(-\dfrac{2}{3}\right)^2 = \dfrac{5}{3} + \left(-\dfrac{2}{3}\right)^2$	Add $\left(-\dfrac{2}{3}\right)^2$ to each side.

$$\left(\text{half of } -\tfrac{4}{3}\right)^2$$

$\left(x - \dfrac{2}{3}\right)^2 = \dfrac{19}{9}$	Simplify.
$x - \dfrac{2}{3} = \pm\dfrac{\sqrt{19}}{3}$	Extract square roots.
$x = \dfrac{2}{3} \pm \dfrac{\sqrt{19}}{3}$	Add $\tfrac{2}{3}$ to each side.

✓ *Checkpoint*  *Audio-video solution in English & Spanish at LarsonPrecalculus.com.*

Solve $3x^2 - 10x - 2 = 0$ by completing the square.

The Quadratic Formula: Two Distinct Solutions

Use the Quadratic Formula to solve $x^2 + 3x = 9$.

Solution

$$x^2 + 3x = 9$$ Write original equation.

$$x^2 + 3x - 9 = 0$$ Write in general form.

$$x = \frac{-b \pm \sqrt{b^2 - 4ac}}{2a}$$ Quadratic Formula

$$x = \frac{-3 \pm \sqrt{(3)^2 - 4(1)(-9)}}{2(1)}$$ Substitute $a = 1$, $b = 3$, and $c = -9$.

$$x = \frac{-3 \pm \sqrt{45}}{2}$$ Simplify.

$$x = \frac{-3 \pm 3\sqrt{5}}{2}$$ Simplify.

The two solutions are

$$x = \frac{-3 + 3\sqrt{5}}{2} \quad \text{and} \quad x = \frac{-3 - 3\sqrt{5}}{2}.$$

Check these in the original equation.

✓ **Checkpoint**  Audio-video solution in English & Spanish at LarsonPrecalculus.com.

Use the Quadratic Formula to solve $3x^2 + 2x - 10 = 0$.

> **REMARK** When using the Quadratic Formula, remember that *before* applying the formula, you must first write the quadratic equation in general form.

The Quadratic Formula: One Solution

Use the Quadratic Formula to solve $8x^2 - 24x + 18 = 0$.

Solution

$$8x^2 - 24x + 18 = 0$$ Write original equation.

$$4x^2 - 12x + 9 = 0$$ Divide out common factor of 2.

$$x = \frac{-b \pm \sqrt{b^2 - 4ac}}{2a}$$ Quadratic Formula

$$x = \frac{-(-12) \pm \sqrt{(-12)^2 - 4(4)(9)}}{2(4)}$$ Substitute $a = 4$, $b = -12$, and $c = 9$.

$$x = \frac{12 \pm \sqrt{0}}{8} = \frac{3}{2}$$ Simplify.

This quadratic equation has only one solution:

$$x = \tfrac{3}{2}.$$

Check this in the original equation.

✓ **Checkpoint**  Audio-video solution in English & Spanish at LarsonPrecalculus.com.

Use the Quadratic Formula to solve $18x^2 - 48x + 32 = 0$.   ■

Note that you could have solved Example 9 without first dividing out a common factor of 2. Substituting $a = 8$, $b = -24$, and $c = 18$ into the Quadratic Formula produces the same result.

Polynomial Equations of Higher Degree

The methods used to solve quadratic equations can sometimes be extended to solve polynomial equations of higher degrees.

. . . . ▷

·· REMARK A common mistake in solving an equation such as that in Example 10 is to divide each side of the equation by the variable factor x^2. This loses the solution $x = 0$. When solving an equation, always write the equation in general form, then factor the equation and set each factor equal to zero. Do not divide each side of an equation by a variable factor in an attempt to simplify the equation.

EXAMPLE 10   **Solving a Polynomial Equation by Factoring**

Solve $3x^4 = 48x^2$.

Solution   First write the polynomial equation in general form with zero on one side. Then factor the other side, set each factor equal to zero, and solve.

$3x^4 = 48x^2$	Write original equation.
$3x^4 - 48x^2 = 0$	Write in general form.
$3x^2(x^2 - 16) = 0$	Factor out common factor.
$3x^2(x + 4)(x - 4) = 0$	Write in factored form.
$3x^2 = 0$ ⟹ $x = 0$	Set 1st factor equal to 0.
$x + 4 = 0$ ⟹ $x = -4$	Set 2nd factor equal to 0.
$x - 4 = 0$ ⟹ $x = 4$	Set 3rd factor equal to 0.

You can check these solutions by substituting in the original equation, as follows.

Check

$3(0)^4 = 48(0)^2$	0 checks. ✔
$3(-4)^4 = 48(-4)^2$	-4 checks. ✔
	4 checks. ✔

So, the solutions are

$$x = 0, \quad x = -4, \quad \text{and} \quad x = 4.$$

✔ **Checkpoint** ◀))) *Audio-video solution in English & Spanish at LarsonPrecalculus.com.*

Solve $9x^4 - 12x^2 = 0$.

EXAMPLE 11   **Solving a Polynomial Equation by Factoring**

Solve $x^3 - 3x^2 - 3x + 9 = 0$.

Solution

$x^3 - 3x^2 - 3x + 9 = 0$	Write original equation.
$x^2(x - 3) - 3(x - 3) = 0$	Factor by grouping.
$(x - 3)(x^2 - 3) = 0$	Distributive Property
$x - 3 = 0$ ⟹ $x = 3$	Set 1st factor equal to 0.
$x^2 - 3 = 0$ ⟹ $x = \pm\sqrt{3}$	Set 2nd factor equal to 0.

The solutions are $x = 3$, $x = \sqrt{3}$, and $x = -\sqrt{3}$. Check these in the original equation.

✔ **Checkpoint** ◀))) *Audio-video solution in English & Spanish at LarsonPrecalculus.com.*

Solve each equation.

a. $x^3 - 5x^2 - 2x + 10 = 0$

b. $6x^3 - 27x^2 - 54x = 0$

Radical Equations

A **radical equation** is an equation that involves one or more radical expressions.

EXAMPLE 12   Solving Radical Equations

a.

$\sqrt{2x + 7} - x = 2$	Original equation
$\sqrt{2x + 7} = x + 2$	Isolate radical.
$2x + 7 = x^2 + 4x + 4$	Square each side.
$0 = x^2 + 2x - 3$	Write in general form.
$0 = (x + 3)(x - 1)$	Factor.
$x + 3 = 0 \implies x = -3$	Set 1st factor equal to 0.
$x - 1 = 0 \implies x = 1$	Set 2nd factor equal to 0.

By checking these values, you can determine that the only solution is $x = 1$.

b.

$\sqrt{2x - 5} - \sqrt{x - 3} = 1$	Original equation
$\sqrt{2x - 5} = \sqrt{x - 3} + 1$	Isolate $\sqrt{2x - 5}$.
$2x - 5 = x - 3 + 2\sqrt{x - 3} + 1$	Square each side.
$2x - 5 = x - 2 + 2\sqrt{x - 3}$	Combine like terms.
$x - 3 = 2\sqrt{x - 3}$	Isolate $2\sqrt{x - 3}$.
$x^2 - 6x + 9 = 4(x - 3)$	Square each side.
$x^2 - 10x + 21 = 0$	Write in general form.
$(x - 3)(x - 7) = 0$	Factor.
$x - 3 = 0 \implies x = 3$	Set 1st factor equal to 0.
$x - 7 = 0 \implies x = 7$	Set 2nd factor equal to 0.

The solutions are $x = 3$ and $x = 7$. Check these in the original equation.

✓ *Checkpoint*))) *Audio-video solution in English & Spanish at LarsonPrecalculus.com.*

Solve $-\sqrt{40 - 9x} + 2 = x$.

EXAMPLE 13   Solving an Equation Involving a Rational Exponent

Solve $(x - 4)^{2/3} = 25$.

Solution

$(x - 4)^{2/3} = 25$	Write original equation.
$\sqrt[3]{(x - 4)^2} = 25$	Rewrite in radical form.
$(x - 4)^2 = 15{,}625$	Cube each side.
$x - 4 = \pm 125$	Extract square roots.
$x = 129, \quad x = -121$	Add 4 to each side.

✓ *Checkpoint*))) *Audio-video solution in English & Spanish at LarsonPrecalculus.com.*

Solve $(x - 5)^{2/3} = 16$.

Absolute Value Equations

An **absolute value equation** is an equation that involves one or more absolute value expressions. To solve an absolute value equation, remember that the expression inside the absolute value bars can be positive or negative. This results in *two* separate equations, each of which must be solved. For instance, the equation

$$|x - 2| = 3$$

results in the two equations

$$x - 2 = 3 \quad \text{and} \quad -(x - 2) = 3$$

which implies that the equation has two solutions: $x = 5$ and $x = -1$.

EXAMPLE 14   **Solving an Absolute Value Equation**

Solve $|x^2 - 3x| = -4x + 6$.

Solution   Because the variable expression inside the absolute value bars can be positive or negative, you must solve the following two equations.

First Equation

$x^2 - 3x = -4x + 6$	Use positive expression.
$x^2 + x - 6 = 0$	Write in general form.
$(x + 3)(x - 2) = 0$	Factor.
$x + 3 = 0 \implies x = -3$	Set 1st factor equal to 0.
$x - 2 = 0 \implies x = 2$	Set 2nd factor equal to 0.

Second Equation

$-(x^2 - 3x) = -4x + 6$	Use negative expression.
$x^2 - 7x + 6 = 0$	Write in general form.
$(x - 1)(x - 6) = 0$	Factor.
$x - 1 = 0 \implies x = 1$	Set 1st factor equal to 0.
$x - 6 = 0 \implies x = 6$	Set 2nd factor equal to 0.

Check

$|(-3)^2 - 3(-3)| \overset{?}{=} -4(-3) + 6$	Substitute -3 for x.
$18 = 18$	-3 checks. ✓
$|(2)^2 - 3(2)| \overset{?}{=} -4(2) + 6$	Substitute 2 for x.
$2 \neq -2$	2 does not check.
$|(1)^2 - 3(1)| \overset{?}{=} -4(1) + 6$	Substitute 1 for x.
$2 = 2$	1 checks. ✓
$|(6)^2 - 3(6)| \overset{?}{=} -4(6) + 6$	Substitute 6 for x.
$18 \neq -18$	6 does not check.

The solutions are $x = -3$ and $x = 1$.

✓ **Checkpoint** ◀))) *Audio-video solution in English & Spanish at LarsonPrecalculus.com.*

Solve $|x^2 + 4x| = 5x + 12$.

Common Formulas

The following geometric formulas are used at various times throughout this course. For your convenience, some of these formulas along with several others are also provided on the inside cover of this text.

Common Formulas for Area A, Perimeter P, Circumference C, and Volume V

Rectangle	Circle	Rectangular Solid	Circular Cylinder	Sphere
$A = lw$	$A = \pi r^2$	$V = lwh$	$V = \pi r^2 h$	$V = \frac{4}{3}\pi r^3$
$P = 2l + 2w$	$C = 2\pi r$			

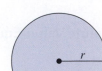
↤—4 cm—↦

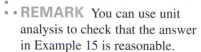
h

Figure A.10

REMARK You can use unit analysis to check that the answer in Example 15 is reasonable.

$$\frac{200 \text{ cm}^3}{16\pi \text{ cm}^2} = \frac{200 \text{ cm} \cdot \text{ cm} \cdot \text{ cm}}{16\pi \text{ cm} \cdot \text{ cm}}$$

$$= \frac{200}{16\pi} \text{ cm}$$

$$\approx 3.98 \text{ cm}$$

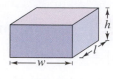 **EXAMPLE 15**   **Using a Geometric Formula**

The cylindrical can shown in Figure A.10 has a volume of 200 cubic centimeters (cm^3). Find the height of the can.

Solution   The formula for the *volume of a cylinder* is $V = \pi r^2 h$. To find the height of the can, solve for h. Then, using $V = 200$ and $r = 4$, find the height.

$$V = \pi r^2 h \implies h = \frac{V}{\pi r^2} = \frac{200}{\pi (4)^2} = \frac{200}{16\pi} \approx 3.98$$

So, the height of the can is about 3.98 centimeters.

✓ **Checkpoint** 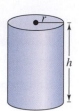 *Audio-video solution in English & Spanish at LarsonPrecalculus.com.*

A cylindrical container has a volume of 84 cubic inches and a radius of 3 inches. Find the height of the container.   ■

Summarize   **(Appendix A.5)**

1. State the definition of an identity, a conditional equation, and a contradiction *(page A36)*.

2. State the definition of a linear equation *(page A36)*. For examples of solving linear equations and rational equations that lead to linear equations, see Examples 1–3.

3. List the four methods of solving quadratic equations discussed in this section *(page A39)*. For examples of solving quadratic equations, see Examples 4–9.

4. Describe how to solve a polynomial equation by factoring *(page A43)*. For examples of solving polynomial equations by factoring, see Examples 10 and 11.

5. Describe how to solve a radical equation *(page A44)*. For an example of solving radical equations, see Example 12.

6. Describe how to solve an absolute value equation *(page A45)*. For an example of solving an absolute value equation, see Example 14.

7. Describe real-life problems that can be solved using common geometric formulas *(page A46)*. For an example that uses a volume formula, see Example 15.

A.6 Linear Inequalities in One Variable

- Represent solutions of linear inequalities in one variable.
- Use properties of inequalities to create equivalent inequalities.
- Solve linear inequalities in one variable.
- Solve absolute value inequalities.
- Use inequalities to model and solve real-life problems.

You can use inequalities to model and solve real-life problems. For instance, an absolute value inequality can be used to analyze the cover layer thickness of a Blu-ray Disc™.

Introduction

Simple inequalities were discussed in Appendix A.1. There, you used the inequality symbols $<$, $\leq$, $>$, and $\geq$ to compare two numbers and to denote subsets of real numbers. For instance, the simple inequality

$$x \geq 3$$

denotes all real numbers x that are greater than or equal to 3.

Now, you will expand your work with inequalities to include more involved statements such as

$$5x - 7 < 3x + 9 \quad \text{and} \quad -3 \leq 6x - 1 < 3.$$

As with an equation, you **solve an inequality** in the variable x by finding all values of x for which the inequality is true. Such values are **solutions** and are said to **satisfy** the inequality. The set of all real numbers that are solutions of an inequality is the **solution set** of the inequality. For instance, the solution set of

$$x + 1 < 4$$

is all real numbers that are less than 3.

The set of all points on the real number line that represents the solution set is the **graph of the inequality.** Graphs of many types of inequalities consist of intervals on the real number line. See Appendix A.1 to review the nine basic types of intervals on the real number line. Note that each type of interval can be classified as *bounded* or *unbounded*.

EXAMPLE 1   Intervals and Inequalities

Write an inequality to represent each interval. Then state whether the interval is bounded or unbounded.

a. $(-3, 5]$ **b.** $(-3, \infty)$

c. $[0, 2]$ **d.** $(-\infty, \infty)$

Solution

a. $(-3, 5]$ corresponds to $-3 < x \leq 5$.   Bounded

b. $(-3, \infty)$ corresponds to $x > -3$.   Unbounded

c. $[0, 2]$ corresponds to $0 \leq x \leq 2$.   Bounded

d. $(-\infty, \infty)$ corresponds to $-\infty < x < \infty$.   Unbounded

✓ **Checkpoint** 🔊)) *Audio-video solution in English & Spanish at LarsonPrecalculus.com.*

Write an inequality to represent each interval. Then state whether the interval is bounded or unbounded.

a. $[-1, 3]$ **b.** $(-1, 6)$

c. $(-\infty, 4)$ **d.** $[0, \infty)$

Properties of Inequalities

The procedures for solving linear inequalities in one variable are much like those for solving linear equations. To isolate the variable, you can make use of the **properties of inequalities.** These properties are similar to the properties of equality, but there are two important exceptions. When each side of an inequality is multiplied or divided by a negative number, the direction of the inequality symbol must be reversed. Here is an example.

$$-2 < 5 \qquad \text{Original inequality}$$

$$(-3)(-2) > (-3)(5) \qquad \text{Multiply each side by } -3 \text{ and reverse inequality symbol.}$$

$$6 > -15 \qquad \text{Simplify.}$$

Notice that when you do not reverse the inequality symbol in the example above, you obtain the false statement

$$6 < -15. \qquad \text{False statement}$$

Two inequalities that have the same solution set are **equivalent.** For instance, the inequalities

$$x + 2 < 5$$

and

$$x < 3$$

are equivalent. To obtain the second inequality from the first, you can subtract 2 from each side of the inequality. The following list describes the operations that can be used to create equivalent inequalities.

Properties of Inequalities

Let a, b, c, and d be real numbers.

1. **Transitive Property**

$$a < b \text{ and } b < c \implies a < c$$

2. **Addition of Inequalities**

$$a < b \text{ and } c < d \implies a + c < b + d$$

3. **Addition of a Constant**

$$a < b \implies a + c < b + c$$

4. **Multiplication by a Constant**

For $c > 0$, $a < b \implies ac < bc$

For $c < 0$, $a < b \implies ac > bc$ Reverse the inequality symbol.

Each of the properties above is true when the symbol $<$ is replaced by $\leq$ and the symbol $>$ is replaced by $\geq$. For instance, another form of the multiplication property is shown below.

For $c > 0$, $a \leq b \implies ac \leq bc$

For $c < 0$, $a \leq b \implies ac \geq bc$

On your own, try to verify each of the properties of inequalities by using several examples with real numbers.

Solving a Linear Inequality in One Variable

The simplest type of inequality is a **linear inequality** in one variable. For instance,

$$2x + 3 > 4$$

is a linear inequality in x.

EXAMPLE 2   Solving a Linear Inequality

Solve $5x - 7 > 3x + 9$. Then graph the solution set.

Solution

$5x - 7 > 3x + 9$	Write original inequality.
$2x - 7 > 9$	Subtract $3x$ from each side.
$2x > 16$	Add 7 to each side.
$x > 8$	Divide each side by 2.

The solution set is all real numbers that are greater than 8, which is denoted by $(8, \infty)$. The graph of this solution set is shown below. Note that a parenthesis at 8 on the real number line indicates that 8 *is not* part of the solution set.

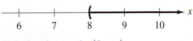

Solution interval: $(8, \infty)$

REMARK Checking the solution set of an inequality is not as simple as checking the solutions of an equation. You can, however, get an indication of the validity of a solution set by substituting a few convenient values of x. For instance, in Example 2, try substituting $x = 5$ and $x = 10$ into the original inequality.

✓ **Checkpoint** 🔊 *Audio-video solution in English & Spanish at LarsonPrecalculus.com.*

Solve $7x - 3 \leq 2x + 7$. Then graph the solution set.

EXAMPLE 3   Solving a Linear Inequality

Solve $1 - \frac{3}{2}x \geq x - 4$.

Algebraic Solution

$1 - \dfrac{3x}{2} \geq x - 4$	Write original inequality.
$2 - 3x \geq 2x - 8$	Multiply each side by 2.
$2 - 5x \geq -8$	Subtract $2x$ from each side.
$-5x \geq -10$	Subtract 2 from each side.
$x \leq 2$	Divide each side by -5 and reverse the inequality symbol.

The solution set is all real numbers that are less than or equal to 2, which is denoted by $(-\infty, 2]$. The graph of this solution set is shown below. Note that a bracket at 2 on the real number line indicates that 2 *is* part of the solution set.

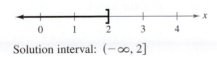

Solution interval: $(-\infty, 2]$

Graphical Solution

Use a graphing utility to graph $y_1 = 1 - \frac{3}{2}x$ and $y_2 = x - 4$ in the same viewing window, as shown below.

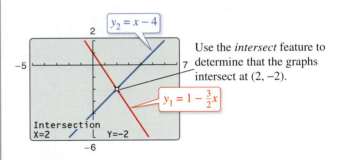

Use the *intersect* feature to determine that the graphs intersect at $(2, -2)$.

The graph of y_1 lies above the graph of y_2 to the left of their point of intersection, which implies that $y_1 \geq y_2$ for all $x \leq 2$.

✓ **Checkpoint** 🔊 *Audio-video solution in English & Spanish at LarsonPrecalculus.com.*

Solve $2 - \frac{5}{3}x > x - 6$ (a) algebraically and (b) graphically.

Sometimes it is possible to write two inequalities as a **double inequality.** For instance, you can write the two inequalities

$$-4 \le 5x - 2$$

and

$$5x - 2 < 7$$

more simply as

$$-4 \le 5x - 2 < 7. \qquad \text{Double inequality}$$

This form allows you to solve the two inequalities together, as demonstrated in Example 4.

EXAMPLE 4   **Solving a Double Inequality**

Solve $-3 \le 6x - 1 < 3$. Then graph the solution set.

Solution   To solve a double inequality, you can isolate x as the middle term.

$$-3 \le 6x - 1 < 3 \qquad \text{Write original inequality.}$$

$$-3 + 1 \le 6x - 1 + 1 < 3 + 1 \qquad \text{Add 1 to each part.}$$

$$-2 \le 6x < 4 \qquad \text{Simplify.}$$

$$\frac{-2}{6} \le \frac{6x}{6} < \frac{4}{6} \qquad \text{Divide each part by 6.}$$

$$-\frac{1}{3} \le x < \frac{2}{3} \qquad \text{Simplify.}$$

The solution set is all real numbers that are greater than or equal to $-\frac{1}{3}$ and less than $\frac{2}{3}$, which is denoted by $\left[-\frac{1}{3}, \frac{2}{3}\right)$. The graph of this solution set is shown below.

Solution interval: $\left[-\frac{1}{3}, \frac{2}{3}\right)$

✓ **Checkpoint**))) *Audio-video solution in English & Spanish at LarsonPrecalculus.com.*

Solve $1 < 2x + 7 < 11$. Then graph the solution set.

You can solve the double inequality in Example 4 in two parts, as follows.

$$-3 \le 6x - 1 \quad \text{and} \quad 6x - 1 < 3$$

$$-2 \le 6x \qquad\qquad\qquad 6x < 4$$

$$-\frac{1}{3} \le x \qquad\qquad\qquad x < \frac{2}{3}$$

The solution set consists of all real numbers that satisfy *both* inequalities. In other words, the solution set is the set of all values of x for which

$$-\frac{1}{3} \le x < \frac{2}{3}.$$

When combining two inequalities to form a double inequality, be sure that the inequalities satisfy the Transitive Property. For instance, it is *incorrect* to combine the inequalities $3 < x$ and $x \le -1$ as $3 < x \le -1$. This "inequality" is wrong because 3 is not less than -1.

Absolute Value Inequalities

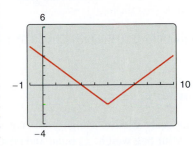
> **Solving an Absolute Value Inequality**
>
> Let *u* be an algebraic expression and let *a* be a real number such that $a > 0$.
>
> 1. $|u| < a$ if and only if $-a < u < a$.
>
> 2. $|u| \le a$ if and only if $-a \le u \le a$.
>
> 3. $|u| > a$ if and only if $u < -a$ or $u > a$.
>
> 4. $|u| \ge a$ if and only if $u \le -a$ or $u \ge a$.

EXAMPLE 5   **Solving an Absolute Value Inequality**

Solve each inequality. Then graph the solution set.

a. $|x - 5| < 2$ **b.** $|x + 3| \ge 7$

Solution

a. $\quad |x - 5| < 2$ Write original inequality.

$\quad\quad -2 < x - 5 < 2$ Write equivalent inequalities.

$\quad -2 + 5 < x - 5 + 5 < 2 + 5$ Add 5 to each part.

$\quad\quad\quad 3 < x < 7$ Simplify.

The solution set is all real numbers that are greater than 3 and less than 7, which is denoted by $(3, 7)$. The graph of this solution set is shown below. Note that the graph of the inequality can be described as all real numbers less than two units from 5.

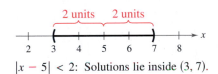

$|x - 5| < 2$: Solutions lie inside $(3, 7)$.

b. $\quad |x + 3| \ge 7$ Write original inequality.

$\quad\quad x + 3 \le -7 \quad \text{or} \quad x + 3 \ge 7$ Write equivalent inequalities.

$\quad x + 3 - 3 \le -7 - 3 \quad\quad x + 3 - 3 \ge 7 - 3$ Subtract 3 from each side.

$\quad\quad\quad x \le -10 \quad\quad\quad\quad\quad x \ge 4$ Simplify.

The solution set is all real numbers that are less than or equal to -10 *or* greater than or equal to 4, which is denoted by $(-\infty, -10] \cup [4, \infty)$. The symbol $\cup$ is the *union* symbol, which denotes the combining of two sets. The graph of this solution set is shown below. Note that the graph of the inequality can be described as all real numbers at least seven units from -3.

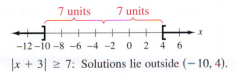

$|x + 3| \ge 7$: Solutions lie outside $(-10, 4)$.

✓ *Checkpoint* 🔊 *Audio-video solution in English & Spanish at LarsonPrecalculus.com.*

Solve $|x - 20| \le 4$. Then graph the solution set. ■

Applications

EXAMPLE 6 **Comparative Shopping**

Cell Phone Plans
Plan A:
$49.99 per month for 500 minutes plus $0.40 for each additional minute
Plan B:
$45.99 per month for 500 minutes plus $0.45 for each additional minute

Figure A.11

Consider the two cell phone plans shown in Figure A.11. How many *additional* minutes must you use in one month for plan B to cost more than plan A?

Solution Let m represent your additional minutes in one month. Write and solve an inequality.

$$0.45m + 45.99 > 0.40m + 49.99$$

$$0.05m > 4$$

$$m > 80$$

Plan B costs more when you use more than 80 additional minutes in one month.

✓ *Checkpoint*  *Audio-video solution in English & Spanish at LarsonPrecalculus.com.*

Rework Example 6 when plan A costs $54.99 per month for 500 minutes plus $0.35 for each additional minute.

EXAMPLE 7 **Accuracy of a Measurement**

You buy chocolates that cost $9.89 per pound. The scale used to weigh your bag is accurate to within $\frac{1}{32}$ pound. According to the scale, your bag weighs $\frac{1}{2}$ pound and costs $4.95. How much might you have been undercharged or overcharged?

Solution Let x represent the actual weight of your bag. The difference of the actual weight and the weight shown on the scale is at most $\frac{1}{32}$ pound. That is, $\left| x - \frac{1}{2} \right| \leq \frac{1}{32}$. You can solve this inequality as follows.

$$-\frac{1}{32} \leq x - \frac{1}{2} \leq \frac{1}{32}$$

$$\frac{15}{32} \leq x \leq \frac{17}{32}$$

The least your bag can weigh is $\frac{15}{32}$ pound, which would have cost $4.64. The most the bag can weigh is $\frac{17}{32}$ pound, which would have cost $5.25. So, you might have been overcharged by as much as $0.31 or undercharged by as much as $0.30.

✓ *Checkpoint*  *Audio-video solution in English & Spanish at LarsonPrecalculus.com.*

Rework Example 7 when the scale is accurate to within $\frac{1}{64}$ pound.

Summarize **(Appendix A.6)**

1. Describe how to use inequalities to represent intervals (*page A47, Example 1*).

2. State the properties of inequalities (*page A48*).

3. Describe how to solve a linear inequality (*pages A49 and A50, Examples 2–4*).

4. Describe how to solve an absolute value inequality (*page A51, Example 5*).

5. Describe a real-life example that uses an inequality (*page A52, Examples 6 and 7*).

A.7 Errors and the Algebra of Calculus

- Avoid common algebraic errors.
- Recognize and use algebraic techniques that are common in calculus.

Algebraic Errors to Avoid

This section contains five lists of common algebraic errors: errors involving parentheses, errors involving fractions, errors involving exponents, errors involving radicals, and errors involving dividing out. Many of these errors are made because they seem to be the *easiest* things to do. For instance, the operations of subtraction and division are often believed to be commutative and associative. The following examples illustrate the fact that subtraction and division are neither commutative nor associative.

Not commutative	Not associative
$4 - 3 \neq 3 - 4$	$8 - (6 - 2) \neq (8 - 6) - 2$
$15 \div 5 \neq 5 \div 15$	$20 \div (4 \div 2) \neq (20 \div 4) \div 2$

Errors Involving Parentheses

Potential Error	Correct Form	Comment
$a - (x - b) = a - x - b$	$a - (x - b) = a - x + b$	Distribute to each term in parentheses.
$(a + b)^2 = a^2 + b^2$	$(a + b)^2 = a^2 + 2ab + b^2$	Remember the middle term when squaring binomials.
$\left(\frac{1}{2}a\right)\left(\frac{1}{2}b\right) = \frac{1}{2}(ab)$	$\left(\frac{1}{2}a\right)\left(\frac{1}{2}b\right) = \frac{1}{4}(ab) = \frac{ab}{4}$	$\frac{1}{2}$ occurs twice as a factor.
$(3x + 6)^2 = 3(x + 2)^2$	$(3x + 6)^2 = [3(x + 2)]^2$ $= 3^2(x + 2)^2$	When factoring, raise all factors to the power.

Errors Involving Fractions

Potential Error	Correct Form	Comment
$\dfrac{2}{x + 4} = \dfrac{2}{x} + \dfrac{2}{4}$	Leave as $\dfrac{2}{x + 4}$.	The fraction is already in simplest form.
	$\dfrac{\left(\frac{x}{a}\right)}{b} = \left(\frac{x}{a}\right)\left(\frac{1}{b}\right) = \dfrac{x}{ab}$	Multiply by the reciprocal when dividing fractions.
	$\dfrac{1}{a} + \dfrac{1}{b} = \dfrac{b + a}{ab}$	Use the property for adding fractions.
	$\dfrac{1}{3x} = \dfrac{1}{3} \cdot \dfrac{1}{x}$	Use the property for multiplying fractions.
$(1/3)x = \dfrac{1}{3x}$	$(1/3)x = \dfrac{1}{3} \cdot x = \dfrac{x}{3}$	Be careful when expressing fractions in the form $1/a$.
$(1/x) + 2 = \dfrac{1}{x + 2}$	$(1/x) + 2 = \dfrac{1}{x} + 2 = \dfrac{1 + 2x}{x}$	Be careful when expressing fractions in the form $1/a$. Be sure to find a common denominator before adding fractions.

Errors Involving Exponents

Potential Error	Correct Form	Comment
$(x^2)^3 = x^5$	$(x^2)^3 = x^{2 \cdot 3} = x^6$	Multiply exponents when raising a power to a power.
$x^2 \cdot x^3 = x^6$	$x^2 \cdot x^3 = x^{2+3} = x^5$	Add exponents when multiplying powers with like bases.
$(2x)^3 = 2x^3$	$(2x)^3 = 2^3 x^3 = 8x^3$	Raise each factor to the power.
$\dfrac{1}{x^2 - x^3} = x^{-2} - x^{-3}$	Leave as $\dfrac{1}{x^2 - x^3}$.	Do not move term-by-term from denominator to numerator.

Errors Involving Radicals

Potential Error	Correct Form	Comment
$\sqrt{5x} = 5\sqrt{x}$	$\sqrt{5x} = \sqrt{5}\sqrt{x}$	Radicals apply to every factor inside the radical.
$\sqrt{x^2 + a^2} = x + a$	Leave as $\sqrt{x^2 + a^2}$.	Do not apply radicals term-by-term when adding or subtracting terms.
$\sqrt{-x + a} = -\sqrt{x - a}$	Leave as $\sqrt{-x + a}$.	Do not factor minus signs out of square roots.

Errors Involving Dividing Out

Potential Error	Correct Form	Comment
$\dfrac{a + bx}{a} = 1 + bx$	$\dfrac{a + bx}{a} = \dfrac{a}{a} + \dfrac{bx}{a} = 1 + \dfrac{b}{a}x$	Divide out common factors, not common terms.
$\dfrac{a + ax}{a} = a + x$	$\dfrac{a + ax}{a} = \dfrac{a(1 + x)}{a} = 1 + x$	Factor before dividing out common factors.
$1 + \dfrac{x}{2x} = 1 + \dfrac{1}{x}$	$1 + \dfrac{x}{2x} = 1 + \dfrac{1}{2} = \dfrac{3}{2}$	Divide out common factors.

A good way to avoid errors is to *work slowly, write neatly,* and *talk to yourself.* Each time you write a step, ask yourself why the step is algebraically legitimate. You can justify the step below because *dividing the numerator and denominator by the same nonzero number produces an equivalent fraction.*

$$\frac{2x}{6} = \frac{2 \cdot x}{2 \cdot 3} = \frac{x}{3}$$

EXAMPLE 1   **Describing and Correcting an Error**

Describe and correct the error. $\dfrac{1}{2x} + \dfrac{1}{3x} = \dfrac{1}{5x}$

Solution   Use the property for adding fractions: $\dfrac{1}{a} + \dfrac{1}{b} = \dfrac{b + a}{ab}$.

$$\frac{1}{2x} + \frac{1}{3x} = \frac{3x + 2x}{6x^2} = \frac{5x}{6x^2} = \frac{5}{6x}$$

✓ *Checkpoint*   *Audio-video solution in English & Spanish at LarsonPrecalculus.com.*

Describe and correct the error. $\sqrt{x^2 + 4} = x + 2$

Some Algebra of Calculus

In calculus it is often necessary to take a simplified algebraic expression and rewrite it. See the following lists, taken from a standard calculus text.

Unusual Factoring

Expression	Useful Calculus Form	Comment
$\dfrac{5x^4}{8}$	$\dfrac{5}{8}x^4$	Write with fractional coefficient.
$\dfrac{x^2 + 3x}{-6}$	$-\dfrac{1}{6}(x^2 + 3x)$	Write with fractional coefficient.
$2x^2 - x - 3$	$2\left(x^2 - \dfrac{x}{2} - \dfrac{3}{2}\right)$	Factor out the leading coefficient.
$\dfrac{x}{2}(x + 1)^{-1/2} + (x + 1)^{1/2}$	$\dfrac{(x + 1)^{-1/2}}{2}[x + 2(x + 1)]$	Factor out the variable expression with the lesser exponent.

Writing with Negative Exponents

Expression	Useful Calculus Form	Comment
$\dfrac{9}{5x^3}$	$\dfrac{9}{5}x^{-3}$	Move the factor to the numerator and change the sign of the exponent.
$\dfrac{7}{\sqrt{2x - 3}}$	$7(2x - 3)^{-1/2}$	Move the factor to the numerator and change the sign of the exponent.

Writing a Fraction as a Sum

Expression	Useful Calculus Form	Comment
$\dfrac{x + 2x^2 + 1}{\sqrt{x}}$	$x^{1/2} + 2x^{3/2} + x^{-1/2}$	Divide each term of the numerator by $x^{1/2}$.
$\dfrac{1 + x}{x^2 + 1}$	$\dfrac{1}{x^2 + 1} + \dfrac{x}{x^2 + 1}$	Rewrite the fraction as a sum of fractions.
$\dfrac{2x}{x^2 + 2x + 1}$	$\dfrac{2x + 2 - 2}{x^2 + 2x + 1}$	Add and subtract the same term.
	$= \dfrac{2x + 2}{x^2 + 2x + 1} - \dfrac{2}{(x + 1)^2}$	Rewrite the fraction as a difference of fractions.
$\dfrac{x^2 - 2}{x + 1}$	$x - 1 - \dfrac{1}{x + 1}$	Use long division. (See Section 2.3.)
$\dfrac{x + 7}{x^2 - x - 6}$	$\dfrac{2}{x - 3} - \dfrac{1}{x + 2}$	Use the method of partial fractions. (See Section 7.4.)

Inserting Factors and Terms

Expression	Useful Calculus Form	Comment
$(2x - 1)^3$	$\dfrac{1}{2}(2x - 1)^3(2)$	Multiply and divide by 2.
$7x^2(4x^3 - 5)^{1/2}$	$\dfrac{7}{12}(4x^3 - 5)^{1/2}(12x^2)$	Multiply and divide by 12.
$\dfrac{4x^2}{9} - 4y^2 = 1$	$\dfrac{x^2}{9/4} - \dfrac{y^2}{1/4} = 1$	Write with fractional denominators.
$\dfrac{x}{x + 1}$	$\dfrac{x + 1 - 1}{x + 1} = 1 - \dfrac{1}{x + 1}$	Add and subtract the same term.

The next five examples demonstrate many of the steps in the preceding lists.

EXAMPLE 2 **Factors Involving Negative Exponents**

Factor $x(x + 1)^{-1/2} + (x + 1)^{1/2}$.

Solution When multiplying powers with like bases, you add exponents. When factoring, you are undoing multiplication, and so you *subtract* exponents.

$$x(x + 1)^{-1/2} + (x + 1)^{1/2} = (x + 1)^{-1/2}[x(x + 1)^0 + (x + 1)^1]$$
$$= (x + 1)^{-1/2}[x + (x + 1)]$$
$$= (x + 1)^{-1/2}(2x + 1)$$

✓ *Checkpoint*  Audio-video solution in English & Spanish at LarsonPrecalculus.com.

Factor $x(x - 2)^{-1/2} + 3(x - 2)^{1/2}$.

Another way to simplify the expression in Example 2 is to multiply the expression by a fractional form of 1 and then use the Distributive Property.

$$x(x + 1)^{-1/2} + (x + 1)^{1/2} = [x(x + 1)^{-1/2} + (x + 1)^{1/2}] \cdot \dfrac{(x + 1)^{1/2}}{(x + 1)^{1/2}}$$
$$= \dfrac{x(x + 1)^0 + (x + 1)^1}{(x + 1)^{1/2}} = \dfrac{2x + 1}{\sqrt{x + 1}}$$

EXAMPLE 3 **Inserting Factors in an Expression**

Insert the required factor: $\dfrac{x + 2}{(x^2 + 4x - 3)^2} = (\ \ )\dfrac{1}{(x^2 + 4x - 3)^2}(2x + 4)$.

Solution The expression on the right side of the equation is twice the expression on the left side. To make both sides equal, insert a factor of $\frac{1}{2}$.

$$\dfrac{x + 2}{(x^2 + 4x - 3)^2} = \left(\dfrac{1}{2}\right)\dfrac{1}{(x^2 + 4x - 3)^2}(2x + 4)$$ Right side is multiplied and divided by 2.

✓ *Checkpoint* Audio-video solution in English & Spanish at LarsonPrecalculus.com.

Insert the required factor: $\dfrac{6x - 3}{(x^2 - x + 4)^2} = (\ \ )\dfrac{1}{(x^2 - x + 4)^2}(2x - 1)$.

EXAMPLE 4   Rewriting Fractions

Explain why the two expressions are equivalent.

$$\frac{4x^2}{9} - 4y^2 = \frac{x^2}{9/4} - \frac{y^2}{1/4}$$

Solution   To write the expression on the left side of the equation in the form given on the right side, multiply the numerator and denominator of each term by $\frac{1}{4}$.

$$\frac{4x^2}{9} - 4y^2 = \frac{4x^2}{9}\left(\frac{1/4}{1/4}\right) - 4y^2\left(\frac{1/4}{1/4}\right) = \frac{x^2}{9/4} - \frac{y^2}{1/4}$$

✓ **Checkpoint**  *Audio-video solution in English & Spanish at LarsonPrecalculus.com.*

Explain why the two expressions are equivalent.

$$\frac{9x^2}{16} + 25y^2 = \frac{x^2}{16/9} + \frac{y^2}{1/25}$$

EXAMPLE 5   Rewriting with Negative Exponents

Rewrite each expression using negative exponents.

a. $\dfrac{-4x}{(1 - 2x^2)^2}$ **b.** $\dfrac{2}{5x^3} - \dfrac{1}{\sqrt{x}} + \dfrac{3}{5(4x)^2}$

Solution

a. $\dfrac{-4x}{(1 - 2x^2)^2} = -4x(1 - 2x^2)^{-2}$

b. $\dfrac{2}{5x^3} - \dfrac{1}{\sqrt{x}} + \dfrac{3}{5(4x)^2} = \dfrac{2}{5x^3} - \dfrac{1}{x^{1/2}} + \dfrac{3}{5(4x)^2}$

$$= \frac{2}{5}x^{-3} - x^{-1/2} + \frac{3}{5}(4x)^{-2}$$

✓ **Checkpoint** *Audio-video solution in English & Spanish at LarsonPrecalculus.com.*

Rewrite $\dfrac{-6x}{(1 - 3x^2)^2} + \dfrac{1}{\sqrt[3]{x}}$ using negative exponents.

EXAMPLE 6   Rewriting a Fraction as a Sum of Terms

Rewrite each fraction as the sum of three terms.

a. $\dfrac{x^2 - 4x + 8}{2x}$ **b.** $\dfrac{x + 2x^2 + 1}{\sqrt{x}}$

Solution

a. $\dfrac{x^2 - 4x + 8}{2x} = \dfrac{x^2}{2x} - \dfrac{4x}{2x} + \dfrac{8}{2x}$ **b.** $\dfrac{x + 2x^2 + 1}{\sqrt{x}} = \dfrac{x}{x^{1/2}} + \dfrac{2x^2}{x^{1/2}} + \dfrac{1}{x^{1/2}}$

$$= \frac{x}{2} - 2 + \frac{4}{x}$$ $$= x^{1/2} + 2x^{3/2} + x^{-1/2}$$

✓ **Checkpoint**  *Audio-video solution in English & Spanish at LarsonPrecalculus.com.*

Rewrite $\dfrac{x^4 - 2x^3 + 5}{x^3}$ as the sum of three terms.

Answers to Selected Exercises and Chapter Tests

Chapter 1

Review Exercises   *(page 71)*

1.

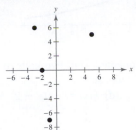

3. Quadrant IV

5.

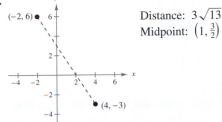

Distance: $3\sqrt{13}$
Midpoint: $\left(1, \frac{3}{2}\right)$

7.

x	-2	-1	0	1	2
y	-11	-8	-5	-2	1

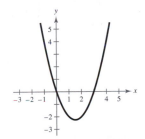

9.

x	-1	0	1	2	3	4
y	4	0	-2	-2	0	4

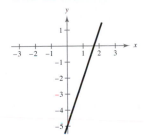

11. x-intercept: $\left(-\frac{7}{2}, 0\right)$
y-intercept: $(0, 7)$

13. x-intercepts: $(1, 0), (5, 0)$
y-intercept: $(0, 5)$

15. x-intercept: $\left(\frac{1}{4}, 0\right)$
y-intercept: $(0, 1)$
No symmetry

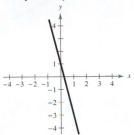

17. x-intercepts: $\left(\pm\sqrt{5}, 0\right)$
y-intercept: $(0, 5)$
y-axis symmetry

19. x-intercept: $\left(\sqrt[3]{-3}, 0\right)$
y-intercept: $(0, 3)$
No symmetry

21. x-intercept: $(-5, 0)$
y-intercept: $\left(0, \sqrt{5}\right)$
No symmetry

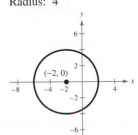

23. Center: $(0, 0)$
Radius: 3

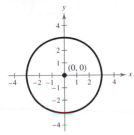

25. Center: $(-2, 0)$
Radius: 4

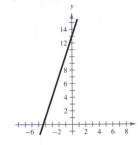

27. $(x - 2)^2 + (y + 3)^2 = 13$

29. slope: 3
y-intercept: 13

31. slope: 0
y-intercept: 6

33.

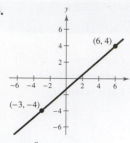

$m = \frac{8}{9}$

35. $y = -\frac{1}{2}x + 2$

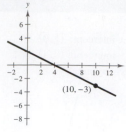

37. $y = \frac{2}{7}x + \frac{2}{7}$ **39.** (a) $y = \frac{5}{4}x - \frac{23}{4}$ (b) $y = -\frac{4}{5}x + \frac{2}{5}$

41. $S = 0.80L$ **43.** No   **45.** Yes

47. (a) 5   (b) 17   (c) $t^4 + 1$ (d) $t^2 + 2t + 2$

49. All real numbers x such that $-5 \le x \le 5$

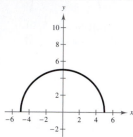

51. 16 ft/sec   **53.** $4x + 2h + 3, \quad h \ne 0$

55. Function   **57.** $\frac{7}{3}, 3$ **59.** $-\frac{3}{8}$

61.

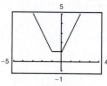

63.

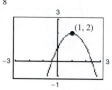

Increasing on $(0, \infty)$
Decreasing on $(-\infty, -1)$
Constant on $(-1, 0)$

65. 4   **67.** Even; y-axis symmetry

69. (a) $f(x) = -3x$ **71.**

(b)

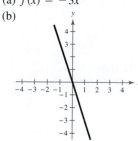

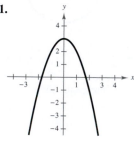

73.

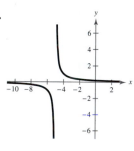

75. (a) $f(x) = x^2$
(b) Vertical shift nine units down
(c)

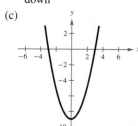

(d) $h(x) = f(x) - 9$

77. (a) $f(x) = \sqrt{x}$ (c)
(b) Reflection in the
x-axis and vertical
shift four units up
(d) $h(x) = -f(x) + 4$

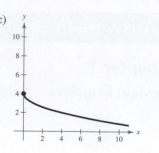

79. (a) $f(x) = x^2$
(b) Reflection in the x-axis, horizontal shift two units to the left, and vertical shift three units up
(c)   (d) $h(x) = -f(x + 2) + 3$

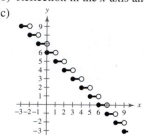

81. (a) $f(x) = [\![x]\!]$
(b) Reflection in the x-axis and vertical shift six units up
(c)

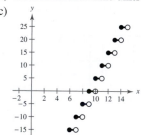

(d) $h(x) = -f(x) + 6$

83. (a) $f(x) = [\![x]\!]$
(b) Horizontal shift nine units to the right and vertical stretch
(c)

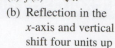

(d) $h(x) = 5f(x - 9)$

85. (a) $x^2 + 2x + 2$
(b) $x^2 - 2x + 4$
(c) $2x^3 - x^2 + 6x - 3$
(d) $\dfrac{x^2 + 3}{2x - 1}$; all real numbers x except $x = \dfrac{1}{2}$

87. (a) $x - \frac{8}{3}$
(b) $x - 8$
Domains of $f, g, f \circ g$, and $g \circ f$: all real numbers x

89. $N(T(t)) = 100t^2 + 275$
The composition function $N(T(t))$ represents the number of bacteria in the food as a function of time.

91. $f^{-1}(x) = \dfrac{x-8}{3}$

$f(f^{-1}(x)) = 3\left(\dfrac{x-8}{3}\right) + 8 = x$

$f^{-1}(f(x)) = \dfrac{3x+8-8}{3} = x$

93.

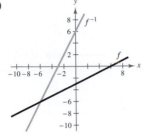

The function does not have an inverse function.

95. (a) $f^{-1}(x) = 2x + 6$

(b)

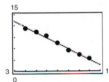

(c) The graphs are reflections of each other in the line $y = x$.

(d) Both f and f^{-1} have domains and ranges that are all real numbers.

97. $x > 4; f^{-1}(x) = \sqrt{\dfrac{x}{2} + 4}, \ x \neq 0$

99. (a) and (b)

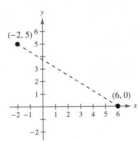

$V = -1.453t + 17.58$

The model is a good fit for the actual data.

101. \$44.80

103. True. If $f(x) = x^3$ and $g(x) = \sqrt[3]{x}$, then the domain of g is all real numbers, which is equal to the range of f, and vice versa.

Chapter Test   (page 74)

1.

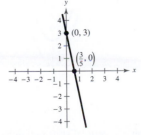

2. $V(h) = 16\pi h$

Midpoint: $\left(2, \tfrac{5}{2}\right)$; Distance: $\sqrt{89}$

3. No symmetry

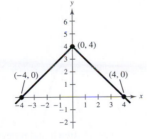

4. y-axis symmetry

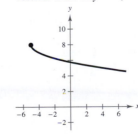

5. y-axis symmetry

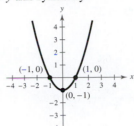

6. $(x-1)^2 + (y-3)^2 = 16$

7. $y = -2x + 1$ **8.** $y = -1.7x + 5.9$

9. (a) $5x + 2y - 8 = 0$ (b) $-2x + 5y - 20 = 0$

10. (a) $-\dfrac{1}{8}$ (b) $-\dfrac{1}{28}$ (c) $\dfrac{\sqrt{x}}{x^2 - 18x}$ **11.** $x \leq 3$

12. (a) $0, \pm 0.4314$

(b)

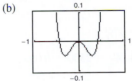

(c) Increasing on $(-0.31, 0), (0.31, \infty)$

Decreasing on $(-\infty, -0.31), (0, 0.31)$

(d) Even

13. (a) $0, 3$

(b)

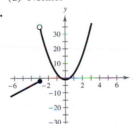

(c) Increasing on $(-\infty, 2)$

Decreasing on $(2, 3)$

(d) Neither

14. (a) -5

(b)

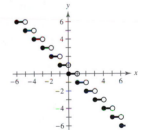

(c) Increasing on $(-5, \infty)$

Decreasing on $(-\infty, -5)$

(d) Neither

15.

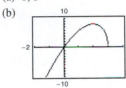

16. Reflection in the x-axis of $y = [\![x]\!]$

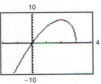

17. Reflection in the x-axis, horizontal shift, and vertical shift of $y = \sqrt{x}$

18. Reflection in the x-axis, vertical stretch, horizontal shift, and vertical shift of $y = x^3$

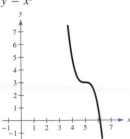

19. (a) $2x^2 - 4x - 2$ (b) $4x^2 + 4x - 12$
(c) $-3x^4 - 12x^3 + 22x^2 + 28x - 35$
(d) $\dfrac{3x^2 - 7}{-x^2 - 4x + 5}$, $x \neq -5, 1$
(e) $3x^4 + 24x^3 + 18x^2 - 120x + 68$
(f) $-9x^4 + 30x^2 - 16$

20. (a) $\dfrac{1 + 2x^{3/2}}{x}$, $x > 0$ (b) $\dfrac{1 - 2x^{3/2}}{x}$, $x > 0$
(c) $\dfrac{2\sqrt{x}}{x}$, $x > 0$ (d) $\dfrac{1}{2x^{3/2}}$, $x > 0$
(e) $\dfrac{\sqrt{x}}{2x}$, $x > 0$ (f) $\dfrac{2\sqrt{x}}{x}$, $x > 0$

21. $f^{-1}(x) = \sqrt[3]{x - 8}$ **22.** No inverse
23. $f^{-1}(x) = \left(\tfrac{1}{3}x\right)^{2/3}$, $x \geq 0$ **24.** $v = 6\sqrt{s}$
25. $A = \dfrac{25}{6}xy$ **26.** $b = \dfrac{48}{a}$

Problem Solving   (page 76)

1. (a) $W_1 = 2000 + 0.07S$ (b) $W_2 = 2300 + 0.05S$
(c)
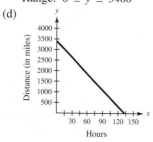

Both jobs pay the same monthly salary when sales equal
$15,000.
(d) No. Job 1 would pay $3400 and job 2 would pay $3300.
3. (a) The function will be even.   (b) The function will be odd.
(c) The function will be neither even nor odd.
5. $f(x) = a_{2n}x^{2n} + a_{2n-2}x^{2n-2} + \cdots + a_2x^2 + a_0$
$f(-x) = a_{2n}(-x)^{2n} + a_{2n-2}(-x)^{2n-2}$
$\qquad\qquad + \cdots + a_2(-x)^2 + a_0$
$\qquad = f(x)$
7. (a) $81\tfrac{2}{3}$ h   (b) $25\tfrac{5}{7}$ mi/h
(c) $y = -\tfrac{180}{7}x + 3400$
Domain: $0 \leq x \leq \tfrac{1190}{9}$
Range: $0 \leq y \leq 3400$
(d)
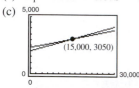

9. (a) $(f \circ g)(x) = 4x + 24$ (b) $(f \circ g)^{-1}(x) = \tfrac{1}{4}x - 6$
(c) $f^{-1}(x) = \tfrac{1}{4}x$; $g^{-1}(x) = x - 6$
(d) $(g^{-1} \circ f^{-1})(x) = \tfrac{1}{4}x - 6$; They are the same.
(e) $(f \circ g)(x) = 8x^3 + 1$; $(f \circ g)^{-1}(x) = \tfrac{1}{2}\sqrt[3]{x - 1}$;
$\quad f^{-1}(x) = \sqrt[3]{x - 1}$; $g^{-1}(x) = \tfrac{1}{2}x$;
$\quad (g^{-1} \circ f^{-1})(x) = \tfrac{1}{2}\sqrt[3]{x - 1}$
(f) Answers will vary.
(g) $(f \circ g)^{-1}(x) = (g^{-1} \circ f^{-1})(x)$

11. (a)   (b)
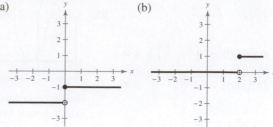

(c)   (d)
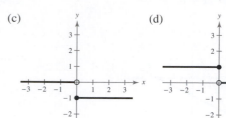

(e)   (f)
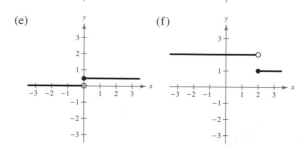

13. Proof
15. (a)

x	-4	-2	0	4
$f(f^{-1}(x))$	-4	-2	0	4

(b)

x	-3	-2	0	1
$(f + f^{-1})(x)$	5	1	-3	-5

(c)

x	-3	-2	0	1
$(f \cdot f^{-1})(x)$	4	0	2	6

(d)

x	-4	-3	0	4
$|f^{-1}(x)|$	2	1	1	3

Chapter 2

Review Exercises   (page 134)

1. (a)   (b)
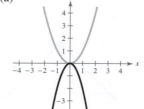

Vertical stretch and reflection   Vertical shift two units up
in the x-axis

3. $g(x) = (x - 1)^2 - 1$

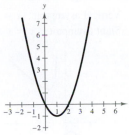

Vertex: $(1, -1)$
Axis of symmetry: $x = 1$
x-intercepts: $(0, 0), (2, 0)$

5. $h(x) = -(x - 2)^2 + 7$

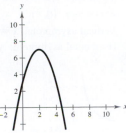

Vertex: $(2, 7)$
Axis of symmetry: $x = 2$
x-intercepts: $(2 \pm \sqrt{7}, 0)$

7. $h(x) = 4\left(x + \frac{1}{2}\right)^2 + 12$

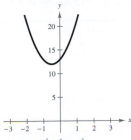

Vertex: $\left(-\frac{1}{2}, 12\right)$
Axis of symmetry: $x = -\frac{1}{2}$
No x-intercept

9. (a) $y = 500 - x$
$A(x) = 500x - x^2$
(b) $x = 250, y = 250$

11.

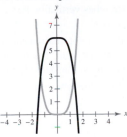

13. Falls to the left, falls to the right

15. Rises to the left, rises to the right

17. (a) Falls to the left, rises to the right
(b) $-2, 0$
(c) Answers will vary.
(d)

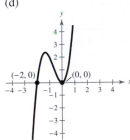

19. (a) Rises to the left, falls to the right
(b) -1
(c) Answers will vary.
(d)

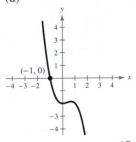

21. (a) $[-1, 0]$　(b) About -0.900

23. $6x + 3 + \dfrac{17}{5x - 3}$

25. (a) Yes　(b) Yes　(c) Yes　(d) No

27. (a) Answers will vary.　(b) $(2x + 5), (x - 3)$
(c) $f(x) = (2x + 5)(x - 3)(x + 6)$　(d) $-\frac{5}{2}, 3, -6$
(e)

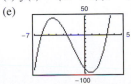

29. $8 + 10i$　　**31.** $3 + 7i$　　**33.** $63 + 77i$　　**35.** $\frac{23}{17} + \frac{10}{17}i$

37. $1 \pm 3i$　　**39.** 2　　**41.** $-6, -2, 5$

43. $g(x) = (x + 2)(x - 3)(x - 6); -2, 3, 6$

45. One or three positive zeros, two or no negative zeros

47. Domain: all real numbers x except $x = -10$
Vertical asymptote: $x = -10$
Horizontal asymptote: $y = 3$

49. (a) Domain: all real numbers x except $x = 0$
(b) No intercepts
(c) Vertical asymptote: $x = 0$
Horizontal asymptote: $y = 0$
(d)

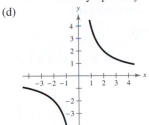

51. (a) Domain: all real numbers x　(b) Intercept: $(0, 0)$
(c) Horizontal asymptote: $y = 0$
(d)

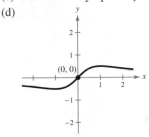

53. (a) Domain: all real numbers x except $x = 0, \frac{1}{3}$
(b) x-intercept: $\left(\frac{3}{2}, 0\right)$
(c) Vertical asymptote: $x = 0$
Horizontal asymptote: $y = 2$
(d)

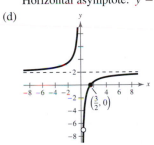

55. (a) Domain: all real numbers x
(b) Intercept: $(0, 0)$　(c) Slant asymptote: $y = 2x$
(d)

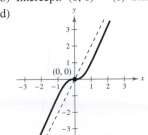

CHAPTER 2

57. (a)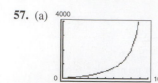

(b) $176 million;
$528 million;
$1584 million
(or $1.584 billion)

(c) No

59. $\left(-\frac{2}{3}, \frac{1}{4}\right)$

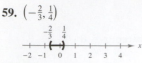

61. $[-5, -1) \cup (1, \infty)$

63. 9 days

65. False. A fourth-degree polynomial can have at most four zeros, and complex zeros occur in conjugate pairs.

Chapter Test   *(page 136)*

1. (a) Reflection in the x-axis followed by a vertical shift two units up   (b) Horizontal shift $\frac{3}{2}$ units to the right

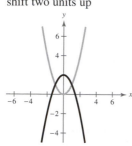 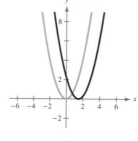

2. $y = (x - 3)^2 - 6$

3. (a) 50 ft

(b) 5. Yes, changing the constant term results in a vertical translation of the graph and therefore changes the maximum height.

4. Rises to the left, falls to the right

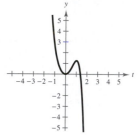

5. $3x + \dfrac{x - 1}{x^2 + 1}$ **6.** $2x^3 + 4x^2 + 3x + 6 + \dfrac{9}{x - 2}$

7. $(2x - 5)(x + \sqrt{3})(x - \sqrt{3})$; Zeros: $\frac{5}{2}, \pm\sqrt{3}$

8. (a) $-3 + 5i$ (b) 7   **9.** $2 - i$

10. $f(x) = x^4 - 7x^3 + 17x^2 - 15x$ **11.** $f(x) = 4x^2 - 16x + 16$

12. $-5, -\frac{2}{3}, 1$ **13.** $-2, 4, -1 \pm \sqrt{2}i$

14. x-intercepts: $(-2, 0), (2, 0)$
Vertical asymptote: $x = 0$
Horizontal asymptote: $y = -1$

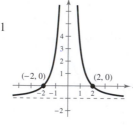

15. x-intercept: $\left(-\frac{3}{2}, 0\right)$
y-intercept: $\left(0, \frac{3}{4}\right)$
Vertical asymptote: $x = -4$
Horizontal asymptote: $y = 2$

16. y-intercept: $(0, -2)$
Vertical asymptote: $x = 1$
Slant asymptote: $y = x + 1$

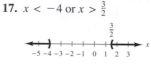 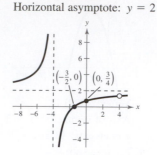

17. $x < -4$ or $x > \frac{3}{2}$

18. $x \le -12$ or $-6 < x < 0$

Problem Solving   *(page 139)*

1. Answers will vary.   **3.** 2 in. $\times$ 2 in. $\times$ 5 in.

5. (a) and (b) $y = -x^2 + 5x - 4$

7. (a) $f(x) = (x - 2)x^2 + 5 = x^3 - 2x^2 + 5$
(b) $f(x) = -(x + 3)x^2 + 1 = -x^3 - 3x^2 + 1$

9. $(a + bi)(a - bi) = a^2 + abi - abi - b^2i^2$
$= a^2 + b^2$

11. (a) As $|a|$ increases, the graph stretches vertically. For $a < 0$, the graph is reflected in the x-axis.

(b) As $|b|$ increases, the vertical asymptote is translated. For $b > 0$, the graph is translated to the right. For $b < 0$, the graph is reflected in the x-axis and is translated to the left.

13. No. Complex zeros always occur in conjugate pairs.

Chapter 3

Review Exercises   *(page 177)*

1. 0.164   **3.** 0.337   **5.** 1456.529

7. Shift the graph of f one unit up.

9. Reflect f in the x-axis and shift one unit up.

11.

x	-1	0	1	2	3
$f(x)$	8	5	4.25	4.063	4.016

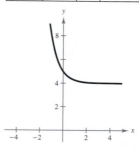

13.

x	-1	0	1	2	3
$f(x)$	4.008	4.04	4.2	5	9

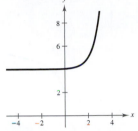

15.

x	-2	-1	0	1	2
$f(x)$	3.25	3.5	4	5	7

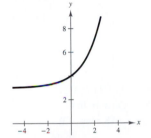

17. $x = 1$ **19.** $x = 4$ **21.** 2980.958   **23.** 0.183

25.

x	-2	-1	0	1	2
$h(x)$	2.72	1.65	1	0.61	0.37

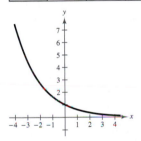

27.

x	-3	-2	-1	0	1
$f(x)$	0.37	1	2.72	7.39	20.09

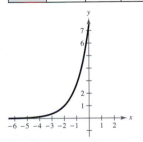

29. (a) 0.154   (b) 0.487   (c) 0.811

31.

n	1	2	4	12
A	\$6719.58	\$6734.28	\$6741.74	\$6746.77

n	365	Continuous
A	\$6749.21	\$6749.29

33. $\log_3 27 = 3$ **35.** $\ln 2.2255\ldots = 0.8$ **37.** 3
39. -2 **41.** $x = 7$ **43.** $x = -5$
45. Domain: $(0, \infty)$ **47.** Domain: $(-5, \infty)$
x-intercept: $(1, 0)$ x-intercept: $(9995, 0)$
Vertical asymptote: $x = 0$ Vertical asymptote: $x = -5$

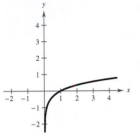   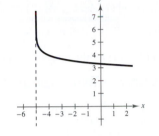

49. 3.118   **51.** 0.25
53. Domain: $(0, \infty)$ **55.** Domain: $(-\infty, 0), (0, \infty)$
x-intercept: $(e^{-3}, 0)$ x-intercepts: $(\pm 1, 0)$
Vertical asymptote: $x = 0$ Vertical asymptote: $x = 0$

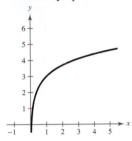   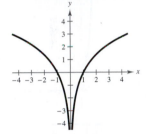

57. 53.4 in.   **59.** (a) and (b) 2.585   **61.** (a) and (b) -2.322
63. $\log 2 + 2 \log 3 \approx 1.255$ **65.** $2 \ln 2 + \ln 5 \approx 2.996$
67. $1 + 2 \log_5 x$ **69.** $2 - \frac{1}{2}\log_3 x$ **71.** $2 \ln x + 2 \ln y + \ln z$
73. $\log_2 5x$ **75.** $\ln \dfrac{x}{\sqrt[4]{y}}$ **77.** $\log_3 \dfrac{\sqrt{x}}{(y + 8)^2}$
79. (a) $0 \le h < 18{,}000$
(b) 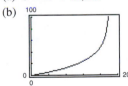   Vertical asymptote:
                                    $h = 18{,}000$

(c) The plane is climbing at a slower rate, so the time required increases.
(d) 5.46 min
81. 3   **83.** $\ln 3 \approx 1.099$ **85.** $e^4 \approx 54.598$ **87.** 1, 3
89. $\dfrac{\ln 32}{\ln 2} = 5$
91. 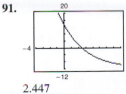   **93.** $\frac{1}{3}e^{8.2} \approx 1213.650$

2.447

95. $3e^2 \approx 22.167$ **97.** No solution   **99.** 0.900

101.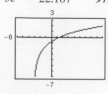

1.482

103.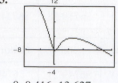

0, 0.416, 13.627

105. 73.2 yr   **107.** e   **108.** b   **109.** f

110. d   **111.** a   **112.** c   **113.** $y = 2e^{0.1014x}$

115.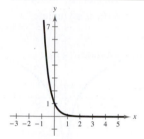

71

117. (a) 10^{-6} W/m²
(b) $10\sqrt{10}$ W/m²
(c) 1.259×10^{-12} W/m²

119. True by the inverse properties.

Chapter Test   *(page 180)*

1. 2.366   **2.** 687.291   **3.** 0.497   **4.** 22.198

5.

x	-1	$-\frac{1}{2}$	0	$\frac{1}{2}$	1
$f(x)$	10	3.162	1	0.316	0.1

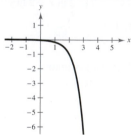

6.

x	-1	0	1	2	3
$f(x)$	-0.005	-0.028	-0.167	-1	-6

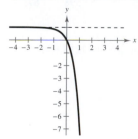

7.

x	-1	$-\frac{1}{2}$	0	$\frac{1}{2}$	1
$f(x)$	0.865	0.632	0	-1.718	-6.389

8. (a) -0.89 (b) 9.2

9. Domain: $(0, \infty)$
x-intercept: $(10^{-6}, 0)$
Vertical asymptote: $x = 0$

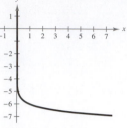

10. Domain: $(4, \infty)$
x-intercept: $(5, 0)$
Vertical asymptote: $x = 4$

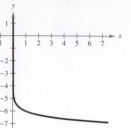

11. Domain: $(-6, \infty)$
x-intercept: $(e^{-1} - 6, 0)$
Vertical asymptote: $x = -6$

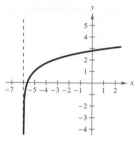

12. 1.945   **13.** -0.167 **14.** -11.047

15. $\log_2 3 + 4 \log_2 a$ **16.** $\ln 5 + \frac{1}{2} \ln x - \ln 6$

17. $3 \log(x - 1) - 2 \log y - \log z$ **18.** $\log_3 13y$

19. $\ln \dfrac{x^4}{y^4}$ **20.** $\ln \dfrac{x^3 y^2}{x + 3}$ **21.** -2 **22.** $\dfrac{\ln 44}{-5} \approx -0.757$

23. $\dfrac{\ln 197}{4} \approx 1.321$ **24.** $e^{1/2} \approx 1.649$ **25.** $e^{-11/4} \approx 0.0639$

26. 20   **27.** $y = 2745e^{0.1570t}$ **28.** 55%

29. (a)

x	$\frac{1}{4}$	1	2	4	5	6
H	58.720	75.332	86.828	103.43	110.59	117.38

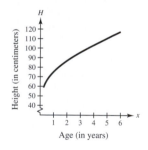

(b) 103 cm; 103.43 cm

Cumulative Test for Chapters 1–3   *(page 181)*

1.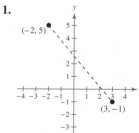

2.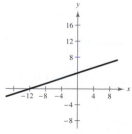

Midpoint: $\left(\frac{1}{2}, 2\right)$
Distance: $\sqrt{61}$

3.

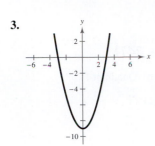

4.

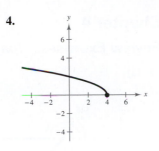

5. $y = 2x + 2$

6. For some values of x there correspond two values of y.

7. (a) $\dfrac{3}{2}$ (b) Division by 0 is undefined.   (c) $\dfrac{s+2}{s}$

8. (a) Vertical shrink by $\frac{1}{2}$ (b) Vertical shift two units up
(c) Horizontal shift two units to the left

9. (a) $5x - 2$ (b) $-3x - 4$ (c) $4x^2 - 11x - 3$
(d) $\dfrac{x-3}{4x+1}$; Domain: all real numbers x except $x = -\dfrac{1}{4}$

10. (a) $\sqrt{x-1} + x^2 + 1$ (b) $\sqrt{x-1} - x^2 - 1$
(c) $x^2\sqrt{x-1} + \sqrt{x-1}$
(d) $\dfrac{\sqrt{x-1}}{x^2+1}$; Domain: all real numbers x such that $x \geq 1$

11. (a) $2x + 12$ (b) $\sqrt{2x^2 + 6}$
Domain of $f \circ g$: all real numbers x such that $x \geq -6$
Domain of $g \circ f$: all real numbers

12. (a) $|x| - 2$ (b) $|x - 2|$
Domain of $f \circ g$ and $g \circ f$: all real numbers

13. Yes; $h^{-1}(x) = -\frac{1}{5}(x - 3)$ **14.** 2438.65 kW

15. $y = -\frac{3}{4}(x + 8)^2 + 5$

16.

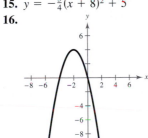

17.

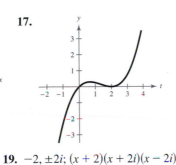

18.

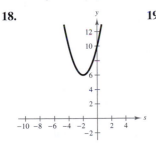

19. $-2, \pm 2i; (x + 2)(x + 2i)(x - 2i)$

20. $-7, 0, 3; x(x)(x - 3)(x + 7)$

21. $4, -\frac{1}{2}, 1 \pm 3i; (x - 4)(2x + 1)(x - 1 + 3i)(x - 1 - 3i)$

22. $3x - 2 - \dfrac{3x - 2}{2x^2 + 1}$ **23.** $3x^3 + 6x^2 + 14x + 23 + \dfrac{49}{x - 2}$

24.

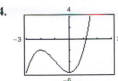

Interval: $[1, 2]$; 1.20

25. Intercept: $(0, 0)$
Vertical asymptotes:
$x = -3, x = 1$
Horizontal asymptote: $y = 0$

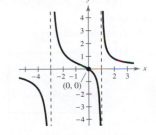

26. y-intercept: $(0, 2)$
x-intercept: $(2, 0)$
Vertical asymptote: $x = 1$
Horizontal asymptote:
$y = 1$

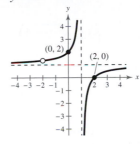

27. y-intercept: $(0, 6)$
x-intercepts: $(2, 0), (3, 0)$
Vertical asymptote: $x = -1$
Slant asymptote:
$y = x - 6$

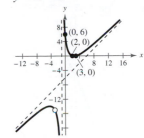

28. $x \leq -3$ or $0 \leq x \leq 3$

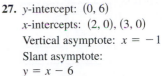

29. All real numbers x such that
$x < -5$ or $x > -1$

30. Reflect f in the x-axis and y-axis, and shift three units to the right.

31. Reflect f in the x-axis, and shift four units up.

32. 1.991   **33.** -0.067 **34.** 1.717   **35.** 0.281

36. $\ln(x + 4) + \ln(x - 4) - 4 \ln x, x > 4$

37. $\ln \dfrac{x^2}{\sqrt{x + 5}}, x > 0$ **38.** $x = \dfrac{\ln 12}{2} \approx 1.242$

39. $\ln 6 \approx 1.792$ or $\ln 7 \approx 1.946$ **40.** $e^6 - 2 \approx 401.429$

41. (a) and (c)

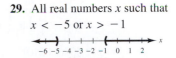

The model is a good fit for the data.
(b) $S = -0.0172t^3 + 0.119t^2 + 2.22t + 36.8$
(d) \$15.0 billion; No; The trend appears to show sales increasing.

42. 6.3 h

Problem Solving   *(page 184)*

1.

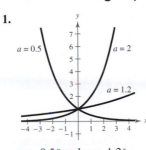

$y = 0.5^x$ and $y = 1.2^x$
$0 < a \leq e^{1/e}$

3. As $x \to \infty$, the graph of e^x increases at a greater rate than the graph of x^n.

5. Answers will vary.

7. (a) 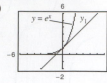 (b)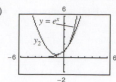

(c)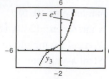

9.

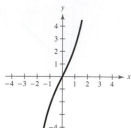

$$f^{-1}(x) = \ln\left(\frac{x + \sqrt{x^2 + 4}}{2}\right)$$

11. c   **13.** $t = \dfrac{\ln c_1 - \ln c_2}{\left(\dfrac{1}{k_2} - \dfrac{1}{k_1}\right)\ln\dfrac{1}{2}}$

15. (a) $y_1 = 252{,}606(1.0310)^t$

(b) $y_2 = 400.88t^2 - 1464.6t + 291{,}782$

(c)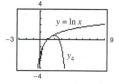

(d) The exponential model is a better fit. No, because the model is rapidly approaching infinity.

17. $1, e^2$

19. $y_4 = (x - 1) - \frac{1}{2}(x - 1)^2 + \frac{1}{3}(x - 1)^3 - \frac{1}{4}(x - 1)^4$

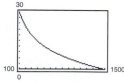

The pattern implies that
$\ln x = (x - 1) - \frac{1}{2}(x - 1)^2 + \frac{1}{3}(x - 1)^3 - \cdots$.

21.

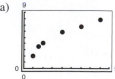

17.7 ft³/min

23. (a)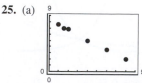

25. (a)

(b)–(e) Answers will vary.   (b)–(e) Answers will vary.

Chapter 4

Review Exercises   *(page 241)*

1. (a) 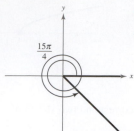   (b) Quadrant IV

(c) $\dfrac{23\pi}{4}, -\dfrac{\pi}{4}$

3. (a) 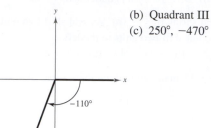   (b) Quadrant III

(c) $250°, -470°$

5. 7.854   **7.** -0.589 **9.** 54.000°   **11.** $-200.535°$

13. 198° 24′   **15.** 48.17 in.   **17.** Area $\approx$ 339.29 in.²

19. $\left(-\dfrac{1}{2}, \dfrac{\sqrt{3}}{2}\right)$ **21.** $\left(-\dfrac{\sqrt{3}}{2}, -\dfrac{1}{2}\right)$

23. $\sin\dfrac{3\pi}{4} = \dfrac{\sqrt{2}}{2}$ $\csc\dfrac{3\pi}{4} = \sqrt{2}$

$\cos\dfrac{3\pi}{4} = -\dfrac{\sqrt{2}}{2}$ $\sec\dfrac{3\pi}{4} = -\sqrt{2}$

$\tan\dfrac{3\pi}{4} = -1$ $\cot\dfrac{3\pi}{4} = -1$

25. $\sin\dfrac{11\pi}{4} = \sin\dfrac{3\pi}{4} = \dfrac{\sqrt{2}}{2}$

27. $\sin\left(-\dfrac{17\pi}{6}\right) = \sin\dfrac{7\pi}{6} = -\dfrac{1}{2}$ **29.** -75.3130 **31.** 3.2361

33. $\sin\theta = \dfrac{4\sqrt{41}}{41}$ $\csc\theta = \dfrac{\sqrt{41}}{4}$

$\cos\theta = \dfrac{5\sqrt{41}}{41}$ $\sec\theta = \dfrac{\sqrt{41}}{5}$

$\tan\theta = \dfrac{4}{5}$ $\cot\theta = \dfrac{5}{4}$

35. 0.6494   **37.** 3.6722

39. (a) 3   (b) $\dfrac{2\sqrt{2}}{3}$ (c) $\dfrac{3\sqrt{2}}{4}$ (d) $\dfrac{\sqrt{2}}{4}$ **41.** 71.3 m

43. $\sin\theta = \frac{4}{5}$ $\csc\theta = \frac{5}{4}$ **45.** $\sin\theta = \frac{4}{5}$ $\csc\theta = \frac{5}{4}$

$\cos\theta = \frac{3}{5}$ $\sec\theta = \frac{5}{3}$ $\cos\theta = \frac{3}{5}$ $\sec\theta = \frac{5}{3}$

$\tan\theta = \frac{4}{3}$ $\cot\theta = \frac{3}{4}$ $\tan\theta = \frac{4}{3}$ $\cot\theta = \frac{3}{4}$

47. $\sin\theta = -\dfrac{\sqrt{11}}{6}$ **49.** $\sin\theta = \dfrac{\sqrt{21}}{5}$

$\cos\theta = \dfrac{5}{6}$ $\tan\theta = -\dfrac{\sqrt{21}}{2}$

$\tan\theta = -\dfrac{\sqrt{11}}{5}$ $\csc\theta = \dfrac{5\sqrt{21}}{21}$

$\csc\theta = -\dfrac{6\sqrt{11}}{11}$ $\sec\theta = -\dfrac{5}{2}$

$\cot\theta = -\dfrac{5\sqrt{11}}{11}$ $\cot\theta = -\dfrac{2\sqrt{21}}{21}$

51. $\theta' = 84°$

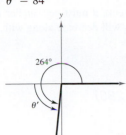

53. $\theta' = \dfrac{\pi}{5}$

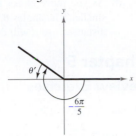

55. $\sin \dfrac{\pi}{3} = \dfrac{\sqrt{3}}{2}$; $\cos \dfrac{\pi}{3} = \dfrac{1}{2}$; $\tan \dfrac{\pi}{3} = \sqrt{3}$

57. $\sin(-150°) = -\dfrac{1}{2}$; $\cos(-150°) = -\dfrac{\sqrt{3}}{2}$; $\tan(-150°) = \dfrac{\sqrt{3}}{3}$

59. -0.7568 **61.** 0.9511

63.

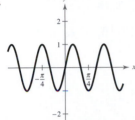

65.

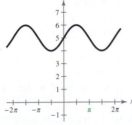

67.

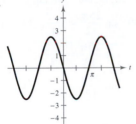

69. (a) $y = 2 \sin 528\pi x$
(b) 264 cycles/sec

71.

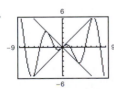

73.

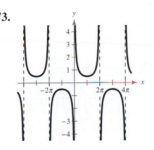

75.

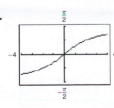

As $x \to +\infty$, $f(x)$ oscillates.

81. -0.98 **83.** 0.09

85.

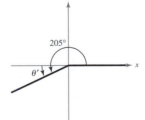

87. $\dfrac{4}{5}$ **89.** $\dfrac{4\sqrt{15}}{15}$

91. $\dfrac{\sqrt{4 - x^2}}{x}$ **93.**

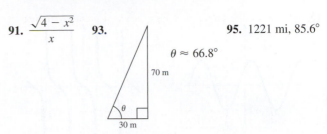

$\theta \approx 66.8°$

70 m

30 m

95. 1221 mi, $85.6°$

97. False. For each θ there corresponds exactly one value of y.
99. The function is undefined because $\sec \theta = 1/\cos \theta$.
101. The ranges of the other four trigonometric functions are $(-\infty, \infty)$ or $(-\infty, -1] \cup [1, \infty)$.

Chapter Test   *(page 244)*

1. (a)

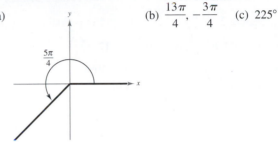

(b) $\dfrac{13\pi}{4}, -\dfrac{3\pi}{4}$ (c) $225°$

2. 3500 rad/min   **3.** About 709.04 ft^2

4. $\sin \theta = \dfrac{3\sqrt{10}}{10}$ $\csc \theta = \dfrac{\sqrt{10}}{3}$

$\cos \theta = -\dfrac{\sqrt{10}}{10}$ $\sec \theta = -\sqrt{10}$

$\tan \theta = -3$ $\cot \theta = -\dfrac{1}{3}$

5. For $0 \le \theta < \dfrac{\pi}{2}$:   For $\pi \le \theta < \dfrac{3\pi}{2}$:

$\sin \theta = \dfrac{3\sqrt{13}}{13}$ $\sin \theta = -\dfrac{3\sqrt{13}}{13}$

$\cos \theta = \dfrac{2\sqrt{13}}{13}$ $\cos \theta = -\dfrac{2\sqrt{13}}{13}$

$\csc \theta = \dfrac{\sqrt{13}}{3}$ $\csc \theta = -\dfrac{\sqrt{13}}{3}$

$\sec \theta = \dfrac{\sqrt{13}}{2}$ $\sec \theta = -\dfrac{\sqrt{13}}{2}$

$\cot \theta = \dfrac{2}{3}$ $\cot \theta = \dfrac{2}{3}$

6. $\theta' = 25°$ **7.** Quadrant III

8. $150°, 210°$
9. $1.33, 1.81$

77. $-\dfrac{\pi}{2}$ **79.** $\dfrac{\pi}{6}$

10. $\sin \theta = -\dfrac{4}{5}$ **11.** $\sin \theta = \dfrac{21}{29}$

$\tan \theta = -\dfrac{4}{3}$ $\cos \theta = -\dfrac{20}{29}$

$\csc \theta = -\dfrac{5}{4}$ $\tan \theta = -\dfrac{21}{20}$

$\sec \theta = \dfrac{5}{3}$ $\csc \theta = \dfrac{29}{21}$

$\cot \theta = -\dfrac{3}{4}$ $\cot \theta = -\dfrac{20}{21}$

CHAPTER 4

12.

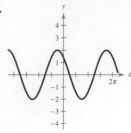

13.

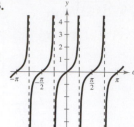

14.

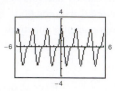

Period: 2

15.

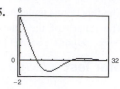

Not periodic

16. $a = -2, b = \dfrac{1}{2}, c = -\dfrac{\pi}{4}$

17. $\dfrac{\sqrt{55}}{3}$

18.

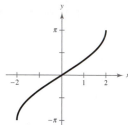

19. 309.3°

20. $d = -6 \cos \pi t$

Problem Solving   *(page 246)*

1. (a) $\dfrac{11\pi}{2}$ rad or 990°   (b) About 816.42 ft

3. (a) 4767 ft   (b) 3705 ft

(c) $w \approx 2183$ ft, $\tan 63° = \dfrac{w + 3705}{3000}$

5. (a)

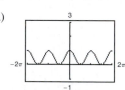

(b)

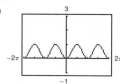

Even

Even

7. $h = 51 - 50 \sin\left(8\pi t + \dfrac{\pi}{2}\right)$

9. (a)

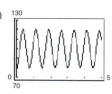

(b) Period $= \dfrac{3}{4}$ sec;

Answers will vary.

(c) 20 mm; Answers will vary.

(d) 80 beats/min

(e) Period $= \dfrac{15}{16}$ sec; $\dfrac{32\pi}{15}$

11. (a)

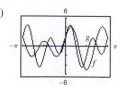

(b) Period of f: 2π; Period of g: π

(c) Yes, because the sine and cosine functions are periodic.

13. (a) 40.5°   (b) $x \approx 1.71$ ft; $y \approx 3.46$ ft   (c) About 1.75 ft

(d) As you move closer to the rock, d must get smaller and smaller. The angles θ_1 and θ_2 will decrease along with the distance y, so d will decrease.

Chapter 5

Review Exercises   *(page 280)*

1. $\tan x$ **3.** $\cot x$

5. $\sin \theta = \dfrac{2\sqrt{13}}{13}$ $\cos \theta = \dfrac{3\sqrt{13}}{13}$

$\csc \theta = \dfrac{\sqrt{13}}{2}$ $\cot \theta = \dfrac{3}{2}$

7. $\sin^2 x$ **9.** 1   **11.** $\tan u \sec u$ **13.** $\cot^2 x$

15. $-2\tan^2 \theta$ **17.** $5 \cos \theta$ **19–25.** Answers will vary.

27. $\dfrac{\pi}{3} + 2n\pi, \dfrac{2\pi}{3} + 2n\pi$ **29.** $\dfrac{\pi}{6} + n\pi$

31. $\dfrac{\pi}{3} + n\pi, \dfrac{2\pi}{3} + n\pi$ **33.** $0, \dfrac{2\pi}{3}, \dfrac{4\pi}{3}$ **35.** $0, \dfrac{\pi}{2}, \pi$

37. $\dfrac{\pi}{8}, \dfrac{3\pi}{8}, \dfrac{9\pi}{8}, \dfrac{11\pi}{8}$ **39.** $\dfrac{\pi}{2}$

41. $0, \dfrac{\pi}{8}, \dfrac{3\pi}{8}, \dfrac{5\pi}{8}, \dfrac{7\pi}{8}, \dfrac{9\pi}{8}, \dfrac{11\pi}{8}, \dfrac{13\pi}{8}, \dfrac{15\pi}{8}$

43. $0, \pi, \arctan 2, \arctan 2 + \pi$

45. $\arctan(-3) + \pi, \arctan(-3) + 2\pi, \arctan 2, \arctan 2 + \pi$

47. $\sin 285° = -\dfrac{\sqrt{2}}{4}(\sqrt{3} + 1)$ **49.** $\sin \dfrac{25\pi}{12} = \dfrac{\sqrt{2}}{4}(\sqrt{3} - 1)$

$\cos 285° = \dfrac{\sqrt{2}}{4}(\sqrt{3} - 1)$ $\cos \dfrac{25\pi}{12} = \dfrac{\sqrt{2}}{4}(\sqrt{3} + 1)$

$\tan 285° = -2 - \sqrt{3}$ $\tan \dfrac{25\pi}{12} = 2 - \sqrt{3}$

51. $\sin 15°$ **53.** $-\dfrac{24}{25}$ **55.** -1

57–59. Answers will vary.   **61.** $\dfrac{\pi}{4}, \dfrac{7\pi}{4}$ **63.** $\sin 2u = \dfrac{24}{25}$

$\cos 2u = -\dfrac{7}{25}$

$\tan 2u = -\dfrac{24}{7}$

65.

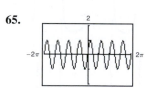

67. $\dfrac{1 - \cos 4x}{1 + \cos 4x}$

69. $\sin(-75°) = -\dfrac{1}{2}\sqrt{2 + \sqrt{3}}$

$\cos(-75°) = \dfrac{1}{2}\sqrt{2 - \sqrt{3}}$

$\tan(-75°) = -2 - \sqrt{3}$

71. (a) Quadrant II

(b) $\sin \dfrac{u}{2} = \dfrac{2\sqrt{5}}{5}$, $\cos \dfrac{u}{2} = -\dfrac{\sqrt{5}}{5}$, $\tan \dfrac{u}{2} = -2$

73. $-|\cos 5x|$ **75.** $\dfrac{1}{2}[\sin 10\theta - \sin(-2\theta)]$

77. $2 \cos \dfrac{11\theta}{2} \cos \dfrac{\theta}{2}$ **79.** $\theta = 15°$ or $\dfrac{\pi}{12}$

81. False. If $(\pi/2) < \theta < \pi$, then $\cos(\theta/2) > 0$. The sign of $\cos(\theta/2)$ depends on the quadrant in which $\theta/2$ lies.

83. True. $4 \sin(-x) \cos(-x) = 4(-\sin x) \cos x$

$= -4 \sin x \cos x$

$= -2(2 \sin x \cos x)$

$= -2 \sin 2x$

85. No. For an equation to be an identity, the equation must be true for all real numbers x. $\sin \theta = \frac{1}{2}$ has an infinite number of solutions but is not an identity.

Chapter Test *(page 282)*

1. $\sin \theta = -\dfrac{6\sqrt{61}}{61}$ $\csc \theta = -\dfrac{\sqrt{61}}{6}$

$\cos \theta = -\dfrac{5\sqrt{61}}{61}$ $\sec \theta = -\dfrac{\sqrt{61}}{5}$

$\tan \theta = \dfrac{6}{5}$ $\cot \theta = \dfrac{5}{6}$

2. 1   **3.** 1   **4.** $\csc \theta \sec \theta$

5. $\theta = 0,\ \pi/2 < \theta \le \pi,\ 3\pi/2 < \theta < 2\pi$

6.

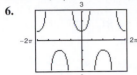

$y_1 = y_2$

7–12. Answers will vary.

13. $\frac{1}{8}(3 - 4\cos x + \cos 2x)$

14. $\tan 2\theta$ **15.** $2(\sin 5\theta + \sin \theta)$ **16.** $-2\sin 2\theta \sin \theta$

17. $0, \dfrac{3\pi}{4}, \pi, \dfrac{7\pi}{4}$ **18.** $\dfrac{\pi}{6}, \dfrac{\pi}{2}, \dfrac{5\pi}{6}, \dfrac{3\pi}{2}$ **19.** $\dfrac{\pi}{6}, \dfrac{5\pi}{6}, \dfrac{7\pi}{6}, \dfrac{11\pi}{6}$

20. $\dfrac{\pi}{6}, \dfrac{5\pi}{6}, \dfrac{3\pi}{2}$ **21.** $0, 2.596$ **22.** $\dfrac{\sqrt{2} - \sqrt{6}}{4}$

23. $\sin 2u = -\frac{20}{29},\ \cos 2u = -\frac{21}{29},\ \tan 2u = \frac{20}{21}$

24. Day 123 to day 223

25. $t = 0.26, 0.58, 0.89, 1.20, 1.52, 1.83$

Problem Solving *(page 286)*

1. $\sin \theta = \pm\sqrt{1 - \cos^2 \theta}$

$\csc \theta = \pm\dfrac{1}{\sqrt{1 - \cos^2 \theta}}$ $\sec \theta = \dfrac{1}{\cos \theta}$

$\tan \theta = \pm\dfrac{\sqrt{1 - \cos^2 \theta}}{\cos \theta}$ $\cot \theta = \pm\dfrac{\cos \theta}{\sqrt{1 - \cos^2 \theta}}$

3. Answers will vary.   **5.** $u + v = w$

7. (a) $A = 100 \sin \dfrac{\theta}{2} \cos \dfrac{\theta}{2}$ (b) $A = 50 \sin \theta;\ \theta = \dfrac{\pi}{2}$

9. (a) $F = \dfrac{0.6W \cos \theta}{\sin 12°}$ (b)

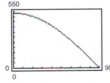

(c) Maximum: $\theta = 0°$
    Minimum: $\theta = 90°$

11. (a) High tides:
        6:12 A.M., 6:36 P.M.
        Low tides:
        12:00 A.M., 12:24 P.M.

(b) The water depth never
    falls below 7 feet.

(c)

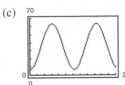

13. (a) $n = \dfrac{1}{2}\left(\cot \dfrac{\theta}{2} + \sqrt{3}\right)$ (b) $\theta \approx 76.5°$

15. (a) $\dfrac{\pi}{6} \le x \le \dfrac{5\pi}{6}$ (b) $\dfrac{2\pi}{3} \le x \le \dfrac{4\pi}{3}$

(c) $\dfrac{\pi}{2} < x < \pi, \dfrac{3\pi}{2} < x < 2\pi$

(d) $0 \le x \le \dfrac{\pi}{4}, \dfrac{5\pi}{4} \le x < 2\pi$

Chapter 6

Review Exercises *(page 325)*

1. $C = 72°, b \approx 12.21, c \approx 12.36$

3. $A = 26°, a \approx 24.89, c \approx 56.23$

5. $C = 66°, a \approx 2.53, b \approx 9.11$

7. $B = 108°, a \approx 11.76, c \approx 21.49$

9. $A \approx 20.41°, C \approx 9.59°, a \approx 20.92$

11. $B \approx 39.48°, C \approx 65.52°, c \approx 48.24$

13. 19.06   **15.** 47.23

17. 31.1 m   **19.** 31.01 ft

21. $A \approx 27.81°, B \approx 54.75°, C \approx 97.44°$

23. $A \approx 16.99°, B \approx 26.00°, C \approx 137.01°$

25. $A \approx 29.92°, B \approx 86.18°, C \approx 63.90°$

27. $A = 36°, C = 36°, b \approx 17.80$

29. $A \approx 45.76°, B \approx 91.24°, c \approx 21.42$

31. Law of Sines; $A \approx 77.52°, B \approx 38.48°, a \approx 14.12$

33. Law of Cosines; $A \approx 28.62°, B \approx 33.56°, C \approx 117.82°$

35. About 4.3 ft, about 12.6 ft

37. 615.1 m   **39.** 7.64

41. 8.36   **43.** $\|u\| = \|v\| = \sqrt{61},\ \text{slope}_u = \text{slope}_v = \frac{5}{6}$

45. $\langle 7, -5 \rangle$ **47.** $\langle 7, -7 \rangle$ **49.** $\langle -4, 4\sqrt{3} \rangle$

51. (a) $\langle -4, 3 \rangle$

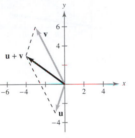

(b) $\langle 2, -9 \rangle$

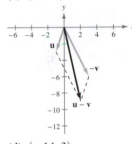

(c) $\langle -4, -12 \rangle$

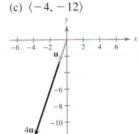

(d) $\langle -14, 3 \rangle$

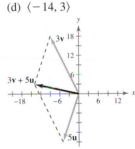

53. (a) $\langle -1, 6 \rangle$

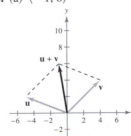

(b) $\langle -9, -2 \rangle$

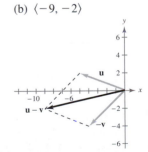

(c) $\langle -20, 8 \rangle$

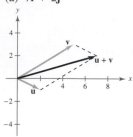

(d) $\langle -13, 22 \rangle$

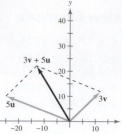

63. $\langle -10, -37 \rangle$

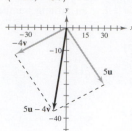

55. (a) $7\mathbf{i} + 2\mathbf{j}$

(b) $-3\mathbf{i} - 4\mathbf{j}$

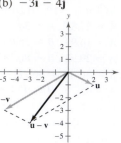

65. $-\mathbf{i} + 5\mathbf{j}$ **67.** $6\mathbf{i} + 4\mathbf{j}$ **69.** $\|\mathbf{v}\| = 7;\ \theta = 60°$
71. $\|\mathbf{v}\| = \sqrt{41};\ \theta = 38.7°$ **73.** $\|\mathbf{v}\| = 3\sqrt{2};\ \theta = 225°$
75. The resultant force is 133.92 pounds and 5.6° from the 85-pound force.
77. 422.30 mi/h; 130.4°   **79.** 45   **81.** -2
83. 40; scalar   **85.** $4 - 2\sqrt{5}$; scalar

87. $\langle 72, -36 \rangle$; vector   **89.** 38; scalar   **91.** $\dfrac{11\pi}{12}$

(c) $8\mathbf{i} - 4\mathbf{j}$

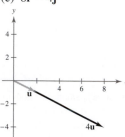

(d) $25\mathbf{i} + 4\mathbf{j}$

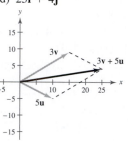

93. 160.5°   **95.** Orthogonal   **97.** Not orthogonal
99. $-\dfrac{13}{17}\langle 4, 1 \rangle,\ \dfrac{16}{17}\langle -1, 4 \rangle$ **101.** $\dfrac{5}{2}\langle -1, 1 \rangle, \dfrac{9}{2}\langle 1, 1 \rangle$ **103.** 48
105. 72,000 ft-lb
107.

109.

7 $\sqrt{34}$

111.

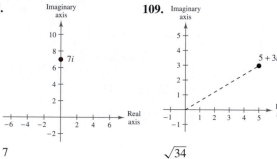

2

57. (a) $3\mathbf{i} + 6\mathbf{j}$

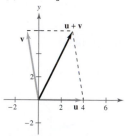

(b) $5\mathbf{i} - 6\mathbf{j}$

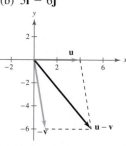

113. $4\left(\cos\dfrac{\pi}{2} + i\sin\dfrac{\pi}{2}\right)$ **115.** $7\sqrt{2}\left(\cos\dfrac{7\pi}{4} + i\sin\dfrac{7\pi}{4}\right)$
117. $13(\cos 4.32 + i\sin 4.32)$

(c) $16\mathbf{i}$

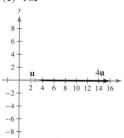

(d) $17\mathbf{i} + 18\mathbf{j}$

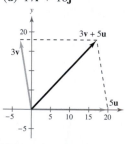

119. (a) $z_1 = 4\left(\cos\dfrac{11\pi}{6} + i\sin\dfrac{11\pi}{6}\right)$

$z_2 = 10\left(\cos\dfrac{3\pi}{2} + i\sin\dfrac{3\pi}{2}\right)$

(b) $z_1 z_2 = 40\left(\cos\dfrac{10\pi}{3} + i\sin\dfrac{10\pi}{3}\right)$

$\dfrac{z_1}{z_2} = \dfrac{2}{5}\left(\cos\dfrac{\pi}{3} + i\sin\dfrac{\pi}{3}\right)$

59. $\langle 30, 9 \rangle$

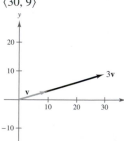

61. $\langle 22, -7 \rangle$

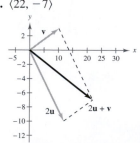

121. $\dfrac{625}{2} + \dfrac{625\sqrt{3}}{2}i$ **123.** $2035 - 828i$

125. (a) $3\left(\cos\dfrac{\pi}{4} + i\sin\dfrac{\pi}{4}\right)$ (b)

$3\left(\cos\dfrac{7\pi}{12} + i\sin\dfrac{7\pi}{12}\right)$

$3\left(\cos\dfrac{11\pi}{12} + i\sin\dfrac{11\pi}{12}\right)$

$3\left(\cos\dfrac{5\pi}{4} + i\sin\dfrac{5\pi}{4}\right)$

$3\left(\cos\dfrac{19\pi}{12} + i\sin\dfrac{19\pi}{12}\right)$

$3\left(\cos\dfrac{23\pi}{12} + i\sin\dfrac{23\pi}{12}\right)$

(c) $\dfrac{3\sqrt{2}}{2} + \dfrac{3\sqrt{2}}{2}i,\ -0.776 + 2.898i,$

$-2.898 + 0.776i,\ -\dfrac{3\sqrt{2}}{2} - \dfrac{3\sqrt{2}}{2}i,$

$0.776 - 2.898i,\ 2.898 - 0.776i$

127. (a) $2(\cos 0 + i\sin 0)$ (b)

$2\left(\cos\dfrac{2\pi}{3} + i\sin\dfrac{2\pi}{3}\right)$

$2\left(\cos\dfrac{4\pi}{3} + i\sin\dfrac{4\pi}{3}\right)$

(c) $2,\ -1 + \sqrt{3}i,$
$-1 - \sqrt{3}i$

129. $3\left(\cos\dfrac{\pi}{4} + i\sin\dfrac{\pi}{4}\right) = \dfrac{3\sqrt{2}}{2} + \dfrac{3\sqrt{2}}{2}i$

$3\left(\cos\dfrac{3\pi}{4} + i\sin\dfrac{3\pi}{4}\right) = -\dfrac{3\sqrt{2}}{2} + \dfrac{3\sqrt{2}}{2}i$

$3\left(\cos\dfrac{5\pi}{4} + i\sin\dfrac{5\pi}{4}\right) = -\dfrac{3\sqrt{2}}{2} - \dfrac{3\sqrt{2}}{2}i$

$3\left(\cos\dfrac{7\pi}{4} + i\sin\dfrac{7\pi}{4}\right) = \dfrac{3\sqrt{2}}{2} - \dfrac{3\sqrt{2}}{2}i$

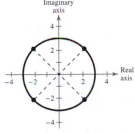

131. $2\left(\cos\dfrac{\pi}{2} + i\sin\dfrac{\pi}{2}\right) = 2i$

$2\left(\cos\dfrac{7\pi}{6} + i\sin\dfrac{7\pi}{6}\right) = -\sqrt{3} - i$

$2\left(\cos\dfrac{11\pi}{6} + i\sin\dfrac{11\pi}{6}\right) = \sqrt{3} - i$

133. $\cos 0 + i\sin 0 = 1$

$\cos\dfrac{\pi}{2} + i\sin\dfrac{\pi}{2} = i$

$\cos\dfrac{2\pi}{3} + i\sin\dfrac{2\pi}{3} = -\dfrac{1}{2} + \dfrac{\sqrt{3}}{2}i$

$\cos\dfrac{4\pi}{3} + i\sin\dfrac{4\pi}{3} = -\dfrac{1}{2} - \dfrac{\sqrt{3}}{2}i$

$\cos\dfrac{3\pi}{2} + i\sin\dfrac{3\pi}{2} = -i$

135. True. Sin 90° is defined in the Law of Sines.

137. False. The solutions to $x^2 - 8i = 0$ are $x = 2 + 2i$ and $x = -2 - 2i$.

139. $a^2 = b^2 + c^2 - 2bc\cos A,\ b^2 = a^2 + c^2 - 2ac\cos B,$
$c^2 = a^2 + b^2 - 2ab\cos C$

141. **A** and **C**

143. For $k > 0$, the direction is the same and the magnitude is k times as great.

For $k < 0$, the result is a vector in the opposite direction and the magnitude is $|k|$ times as great.

145. (a) $4(\cos 60° + i\sin 60°)$ (b) -64
$4(\cos 180° + i\sin 180°)$
$4(\cos 300° + i\sin 300°)$

147. (a) $2(\cos 30° + i\sin 30°)$ (b) $8i$
$2(\cos 150° + i\sin 150°)$
$2(\cos 270° + i\sin 270°)$

149. $z_1 z_2 = -4$

$\dfrac{z_1}{z_2} = \cos(2\theta - \pi) + i\sin(2\theta - \pi)$

$= -\cos 2\theta - i\sin 2\theta$

Chapter Test *(page 329)*

1. Law of Sines; $C = 88°, b \approx 27.81, c \approx 29.98$

2. Law of Sines; $A = 42°, b \approx 21.91, c \approx 10.95$

3. Law of Sines
Two solutions: $B \approx 29.12°, C \approx 126.88°, c \approx 22.03$
$\quad\quad\quad\quad\quad\quad B \approx 150.88°, C \approx 5.12°, c \approx 2.46$

4. Law of Cosines; No solution

5. Law of Sines; $A \approx 39.96°, C \approx 40.04°, c \approx 15.02$

6. Law of Cosines; $A \approx 21.90°, B \approx 37.10°, c \approx 78.15$

7. 2052.5 m² **8.** 606.3 mi; 29.1° **9.** $\langle 14, -23\rangle$

10. $\left\langle \dfrac{18\sqrt{34}}{17}, -\dfrac{30\sqrt{34}}{17}\right\rangle$

CHAPTER 6

11. $\langle -4, 12 \rangle$

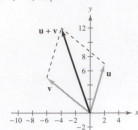

12. $\langle 8, 2 \rangle$

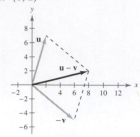

2. $-83.1°$ **3.** $\frac{20}{29}$

4.

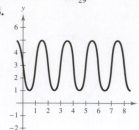

5.

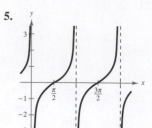

13. $\langle 28, 20 \rangle$

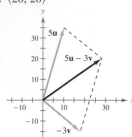

14. $\langle -4, 38 \rangle$

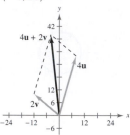

6.

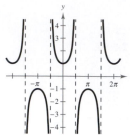

7. $a = -3, b = \pi, c = 0$

15. $\left\langle \frac{24}{25}, -\frac{7}{25} \right\rangle$ **16.** 14.9°; 250.15 lb   **17.** 135°

18. Yes   **19.** $\frac{37}{26}\langle 5, 1 \rangle$; $\frac{29}{26}\langle -1, 5 \rangle$ **20.** About 104 lb

21. $4\sqrt{2}\left(\cos \frac{7\pi}{4} + i \sin \frac{7\pi}{4} \right)$ **22.** $-3 + 3\sqrt{3}i$

23. $-\frac{6561}{2} - \frac{6561\sqrt{3}}{2}i$ **24.** 5832i

25. $4\sqrt[4]{2}\left(\cos \frac{\pi}{12} + i \sin \frac{\pi}{12} \right)$ **26.** $3\left(\cos \frac{\pi}{6} + i \sin \frac{\pi}{6} \right)$

$4\sqrt[4]{2}\left(\cos \frac{7\pi}{12} + i \sin \frac{7\pi}{12} \right)$ $3\left(\cos \frac{5\pi}{6} + i \sin \frac{5\pi}{6} \right)$

$4\sqrt[4]{2}\left(\cos \frac{13\pi}{12} + i \sin \frac{13\pi}{12} \right)$ $3\left(\cos \frac{3\pi}{2} + i \sin \frac{3\pi}{2} \right)$

$4\sqrt[4]{2}\left(\cos \frac{19\pi}{12} + i \sin \frac{19\pi}{12} \right)$

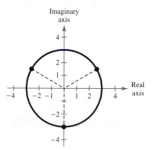

8.

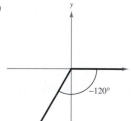

9. 4.9

10. $\frac{3}{4}$

11. $\sqrt{1 - 4x^2}$

12. 1

13. $2 \tan \theta$

14–16. Answers will vary.

17. $\frac{\pi}{3}, \frac{\pi}{2}, \frac{3\pi}{2}, \frac{5\pi}{3}$

18. $\frac{\pi}{6}, \frac{5\pi}{6}, \frac{7\pi}{6}, \frac{11\pi}{6}$ **19.** $\frac{3\pi}{2}$ **20.** $\frac{16}{63}$ **21.** $\frac{4}{3}$

22. $\frac{\sqrt{5}}{5}, \frac{2\sqrt{5}}{5}$ **23.** $\frac{5}{2}\left(\sin \frac{5\pi}{2} - \sin \pi \right)$ **24.** $-2 \sin 8x \sin x$

25. Law of Sines; $B \approx 26.39°, C \approx 123.61°, c \approx 14.99$

26. Law of Cosines; $B \approx 52.48°, C \approx 97.52°, a \approx 5.04$

27. Law of Sines; $B = 60°, a \approx 5.77, c \approx 11.55$

28. Law of Cosines; $A \approx 26.28°, B \approx 49.74°, C \approx 103.98°$

29. Law of Sines; $C = 109°, a \approx 14.96, b \approx 9.27$

30. Law of Cosines; $A \approx 6.75°, B \approx 93.25°, c \approx 9.86$

31. 41.48 in.^{2} **32.** 599.09 m^2 **33.** $7i + 8j$

34. $\left\langle \frac{\sqrt{2}}{2}, \frac{\sqrt{2}}{2} \right\rangle$ **35.** -5 **36.** $-\frac{1}{13}\langle 1, 5 \rangle$; $\frac{21}{13}\langle 5, -1 \rangle$

37. $2\sqrt{2}\left(\cos \frac{3\pi}{4} + i \sin \frac{3\pi}{4} \right)$ **38.** $-12\sqrt{3} + 12i$

39. $\cos 0 + i \sin 0 = 1$

$\cos \frac{2\pi}{3} + i \sin \frac{2\pi}{3} = -\frac{1}{2} + \frac{\sqrt{3}}{2}i$

$\cos \frac{4\pi}{3} + i \sin \frac{4\pi}{3} = -\frac{1}{2} - \frac{\sqrt{3}}{2}i$

Cumulative Test for Chapters 4–6   *(page 330)*

1. (a)

(b) 240°

(c) $-\frac{2\pi}{3}$

(d) 60°

(e) $\sin(-120°) = -\frac{\sqrt{3}}{2}$ $\csc(-120°) = -\frac{2\sqrt{3}}{3}$

$\cos(-120°) = -\frac{1}{2}$ $\sec(-120°) = -2$

$\tan(-120°) = \sqrt{3}$ $\cot(-120°) = \frac{\sqrt{3}}{3}$

40. $3\left(\cos \frac{\pi}{5} + i \sin \frac{\pi}{5} \right)$ **41.** About 395.8 rad/min; about 8312.7 in./min

$3\left(\cos \frac{3\pi}{5} + i \sin \frac{3\pi}{5} \right)$

$3(\cos \pi + i \sin \pi)$

$3\left(\cos \frac{7\pi}{5} + i \sin \frac{7\pi}{5} \right)$

$3\left(\cos \frac{9\pi}{5} + i \sin \frac{9\pi}{5} \right)$

42. 42π yd$^2 \approx 131.95$ yd^2 **43.** 5 ft   **44.** 22.6°

45. $d = 4\cos\dfrac{\pi}{4}t$ **46.** 32.6°; 543.9 km/h   **47.** 425 ft-lb

Problem Solving   *(page 336)*

1. 2.01 ft

3. (a) 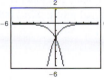   (b) Station A: 27.45 mi;
                                        Station B: 53.03 mi
                                        (c) 11.03 mi; S 21.7° E

5. (a) (i) $\sqrt{2}$ (ii) $\sqrt{5}$ (iii) 1
      (iv) 1   (v) 1   (vi) 1
  (b) (i) 1   (ii) $3\sqrt{2}$ (iii) $\sqrt{13}$
      (iv) 1   (v) 1   (vi) 1
  (c) (i) $\dfrac{\sqrt{5}}{2}$ (ii) $\sqrt{13}$ (iii) $\dfrac{\sqrt{85}}{2}$
      (iv) 1   (v) 1   (vi) 1
  (d) (i) $2\sqrt{5}$ (ii) $5\sqrt{2}$ (iii) $5\sqrt{2}$
      (iv) 1   (v) 1   (vi) 1

7. $\mathbf{u} \cdot \mathbf{v} = 0$ and $\mathbf{u} \cdot \mathbf{w} = 0$
$$\mathbf{u} \cdot (c\mathbf{v} + d\mathbf{w}) = \mathbf{u} \cdot c\mathbf{v} + \mathbf{u} \cdot d\mathbf{w}$$
$$= c(\mathbf{u} \cdot \mathbf{v}) + d(\mathbf{u} \cdot \mathbf{w})$$
$$= 0$$

9. (a) $\mathbf{u} = \langle 0, -120 \rangle$, $\mathbf{v} = \langle 40, 0 \rangle$
  (b)   (c) 126.5 miles per hour;
                  The magnitude gives
                  the actual rate of the
                  skydiver's fall.
               (d) 71.57°

  (e)   123.7 mi/h

Chapter 7

Review Exercises   *(page 384)*

1. $(1, 1)$ **3.** $\left(\dfrac{3}{2}, 5\right)$ **5.** $(0.25, 0.625)$ **7.** $(5, 4)$

9. $(0, 0), (2, 8), (-2, 8)$ **11.** $(4, -2)$

13. $(1.41, -0.66), (-1.41, 10.66)$

15.

17.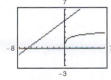

$(0, -2)$ No solution

19.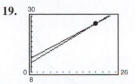

The BMI for males exceeds the BMI for females after age 20.

21. 16 ft × 18 ft   **23.** $\left(\dfrac{5}{2}, 3\right)$ **25.** $(0, 0)$

27. $\left(\dfrac{8}{5}a + \dfrac{14}{5}, a\right)$ **29.** d, one solution, consistent

30. c, infinitely many solutions, consistent

31. b, no solution, inconsistent

32. a, one solution, consistent   **33.** $(100{,}000, 23)$

35. $(2, -4, -5)$ **37.** $(-6, 7, 10)$ **39.** $\left(\dfrac{24}{5}, \dfrac{22}{5}, -\dfrac{8}{5}\right)$

41. $\left(-\dfrac{3}{4}, 0, -\dfrac{5}{4}\right)$ **43.** $(a - 4, a - 3, a)$

45. $y = 2x^2 + x - 5$ **47.** $x^2 + y^2 - 4x + 4y - 1 = 0$

49. (a) $y = -5.950t^2 + 104.45t - 312.9$
  (b)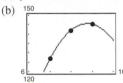

The model is a good fit for the data.
  (c) $-\$84.9$ billion; no

51. $\$16{,}000$ at 7%   **53.** $s = -16t^2 + 150$
   $\$13{,}000$ at 9%
   $\$11{,}000$ at 11%

55. $\dfrac{A}{x} + \dfrac{B}{x + 20}$ **57.** $\dfrac{A}{x} + \dfrac{B}{x^2} + \dfrac{C}{x - 5}$

59. $\dfrac{3}{x + 2} - \dfrac{4}{x + 4}$ **61.** $1 - \dfrac{25}{8(x + 5)} + \dfrac{9}{8(x - 3)}$

63. $\dfrac{1}{2}\left(\dfrac{3}{x - 1} - \dfrac{x - 3}{x^2 + 1}\right)$ **65.** $\dfrac{3}{x^2 + 1} + \dfrac{4x - 3}{(x^2 + 1)^2}$

67.   **69.**
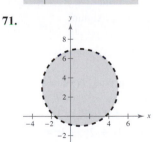   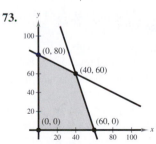

71.   **73.**

75.

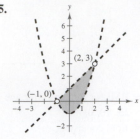

77.

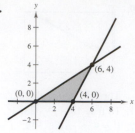

Chapter Test *(page 388)*

1. $(-4, -5)$ **2.** $(0, -1), (1, 0), (2, 1)$ **3.** $(8, 4), (2, -2)$

4.

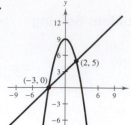

$\left(3, \frac{3}{2}\right)$

5.

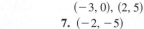

$(-3, 0), (2, 5)$

79. (a)

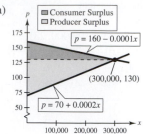

(b) Consumer surplus:
$4,500,000

Producer surplus:
$9,000,000

6.

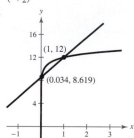

$(1, 12), (0.034, 8.619)$

7. $(-2, -5)$ **8.** $(10, -3)$

9. $(2, -3, 1)$ **10.** No solution

11. $-\dfrac{1}{x+1} + \dfrac{3}{x-2}$ **12.** $\dfrac{2}{x^2} + \dfrac{3}{2-x}$

13. $-\dfrac{5}{x} + \dfrac{3}{x+1} + \dfrac{3}{x-1}$ **14.** $-\dfrac{2}{x} + \dfrac{3x}{x^2+2}$

81. $\begin{cases} x \geq 3 \\ x \leq 7 \\ y \geq 1 \\ y \leq 10 \end{cases}$

83. $\begin{cases} 20x + 30y \leq 24{,}000 \\ 12x + 8y \leq 12{,}400 \\ x \qquad\;\; \geq 0 \\ \qquad y \geq 0 \end{cases}$

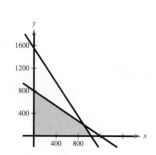

15.

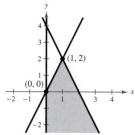

16.

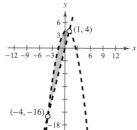

85.

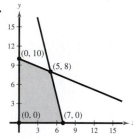

Minimum at $(0, 0)$: 0
Maximum at $(5, 8)$: 47

87.

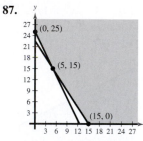

Minimum at $(15, 0)$: 26.25
No maximum

17.

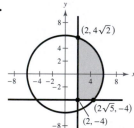

89. 72 haircuts
0 permanents
Optimal revenue: $1800

91. True. The nonparallel sides of the trapezoid are equal in length.

93. $\begin{cases} 4x + y = -22 \\ \frac{1}{2}x + y = 6 \end{cases}$ **95.** $\begin{cases} 3x + y = 7 \\ -6x + 3y = 1 \end{cases}$

97. $\begin{cases} x + y + z = 6 \\ x + y - z = 0 \\ x - y - z = 2 \end{cases}$ **99.** $\begin{cases} 2x + 2y - 3z = 7 \\ x - 2y + z = 4 \\ -x + 4y - z = -1 \end{cases}$

101. An inconsistent system of linear equations has no solution.

18. Minimum at $(0, 0)$: 0
Maximum at $(12, 0)$: 240

20. $y = -\frac{1}{2}x^2 + x + 6$

19. $24,000 in 4% fund
$26,000 in 5.5% fund

21. 0 units of model I
5300 units of model II
Optimal profit: $212,000

Problem Solving (page 390)

1.

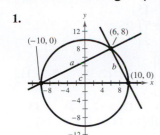

$a = 8\sqrt{5}, b = 4\sqrt{5}, c = 20$
$(8\sqrt{5})^2 + (4\sqrt{5})^2 = 20^2$
Therefore, the triangle is a right triangle.

3. $ad \neq bc$

5. (a)

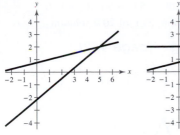 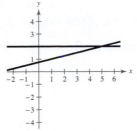

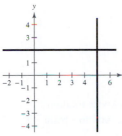

$(5, 2)$
Answers will vary.

(b)

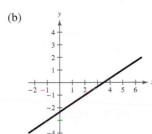

$\left(\frac{3}{2}a + \frac{7}{2}, a\right)$
Answers will vary.

7. 10.1 ft; About 252.7 ft   **9.** $12.00

11. (a) $(3, -4)$ (b) $\left(\dfrac{2}{-a+5}, \dfrac{1}{4a-1}, \dfrac{1}{a}\right)$

13. (a) $\left(\dfrac{-5a+16}{6}, \dfrac{5a-16}{6}, a\right)$

(b) $\left(\dfrac{-11a+36}{14}, \dfrac{13a-40}{14}, a\right)$

(c) $(-a+3, a-3, a)$ (d) Infinitely many

15. $\begin{cases} a + \quad t \leq 32 \\ 0.15a \qquad \geq 1.9 \\ 193a + 772t \geq 11{,}000 \end{cases}$

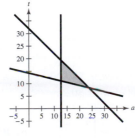

17. (a) $\begin{cases} 0 < y \leq 130 \\ x \geq 40 \\ x + y \leq 200 \end{cases}$

(b)

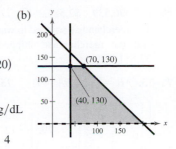

(c) No. The point $(90, 120)$ is not in the solution region.

(d) LDL/VLDL: 135 mg/dL
    HDL: 65 mg/dL

(e) $(75, 90)$; $\frac{165}{75} = 2.2 < 4$

Chapter 8

Review Exercises (page 435)

1. 3×1 **3.** 1×1 **5.** $\begin{bmatrix} 3 & -10 & \vdots & 15 \\ 5 & 4 & \vdots & 22 \end{bmatrix}$

7. $\begin{cases} 5x + y + 7z = -9 \\ 4x + 2y \quad = 10 \\ 9x + 4y + 2z = 3 \end{cases}$ **9.** $\begin{bmatrix} 1 & 2 & 3 \\ 0 & 1 & 1 \\ 0 & 0 & 1 \end{bmatrix}$

11. $\begin{cases} x + 2y + 3z = 9 \\ y - 2z = 2 \\ z = 0 \end{cases}$ **13.** $\begin{cases} x - 5y + 4z = 1 \\ y + 2z = 3 \\ z = 4 \end{cases}$

$(5, 2, 0)$ $(-40, -5, 4)$

15. $(10, -12)$ **17.** $\left(-\frac{1}{5}, \frac{7}{10}\right)$ **19.** Inconsistent

21. $(1, -2, 2)$ **23.** $\left(-2a + \frac{3}{2}, 2a + 1, a\right)$ **25.** $(5, 2, -6)$

27. $(1, 0, 4, 3)$ **29.** $(1, 2, 2)$ **31.** $(2, -3, 3)$

33. $(2, 3, -1)$ **35.** $(2, 6, -10, -3)$ **37.** $x = 12, y = -7$

39. $x = 1, y = 11$

41. (a) $\begin{bmatrix} -1 & 8 \\ 15 & 13 \end{bmatrix}$ (b) $\begin{bmatrix} 5 & -12 \\ -9 & -3 \end{bmatrix}$

(c) $\begin{bmatrix} 8 & -8 \\ 12 & 20 \end{bmatrix}$ (d) $\begin{bmatrix} -7 & 28 \\ 39 & 29 \end{bmatrix}$

43. (a) $\begin{bmatrix} 5 & 7 \\ -3 & 14 \\ 31 & 42 \end{bmatrix}$ (b) $\begin{bmatrix} 5 & 1 \\ -11 & -10 \\ -9 & -38 \end{bmatrix}$

(c) $\begin{bmatrix} 20 & 16 \\ -28 & 8 \\ 44 & 8 \end{bmatrix}$ (d) $\begin{bmatrix} 5 & 13 \\ 5 & 38 \\ 71 & 122 \end{bmatrix}$

45. $\begin{bmatrix} 17 & -17 \\ 13 & 2 \end{bmatrix}$ **47.** $\begin{bmatrix} 54 & 4 \\ -2 & 24 \\ -4 & 32 \end{bmatrix}$ **49.** $\begin{bmatrix} 48 & -18 & -3 \\ 15 & 51 & 33 \end{bmatrix}$

51. $\begin{bmatrix} -11 & -6 \\ 8 & -13 \\ -18 & -8 \end{bmatrix}$ **53.** $\begin{bmatrix} 3 & \frac{2}{3} \\ -\frac{4}{3} & \frac{11}{3} \\ \frac{10}{3} & 0 \end{bmatrix}$ **55.** $\begin{bmatrix} -30 & 4 \\ 51 & 70 \end{bmatrix}$

57. $\begin{bmatrix} 100 & 220 \\ 12 & -4 \\ 84 & 212 \end{bmatrix}$ **59.** $\begin{bmatrix} 14 & -2 & 8 \\ 14 & -10 & 40 \\ 36 & -12 & 48 \end{bmatrix}$

61. $\begin{bmatrix} 44 & 4 \\ 20 & 8 \end{bmatrix}$ **63.** $\begin{bmatrix} 14 & -22 & 22 \\ 19 & -41 & 80 \\ 42 & -66 & 66 \end{bmatrix}$

65. Not possible. The number of columns of the first matrix does not equal the number of rows of the second matrix.

67. $\begin{bmatrix} 76 & 114 & 133 \\ 38 & 95 & 76 \end{bmatrix}$

69. [$2,396,539 $2,581,388]
The merchandise shipped to warehouse 1 is worth $2,396,539 and the merchandise shipped to warehouse 2 is worth $2,581,388.

71–73. $AB = I$ and $BA = I$ **75.** $\begin{bmatrix} 4 & -5 \\ 5 & -6 \end{bmatrix}$

77. $\begin{bmatrix} \frac{1}{2} & -1 & -\frac{1}{2} \\ \frac{1}{2} & -\frac{2}{3} & -\frac{5}{6} \\ 0 & \frac{2}{3} & \frac{1}{3} \end{bmatrix}$ **79.** $\begin{bmatrix} 13 & 6 & -4 \\ -12 & -5 & 3 \\ 5 & 2 & -1 \end{bmatrix}$

81. $\begin{bmatrix} -3 & 6 & -5.5 & 3.5 \\ 1 & -2 & 2 & -1 \\ 7 & -15 & 14.5 & -9.5 \\ -1 & 2.5 & -2.5 & 1.5 \end{bmatrix}$ **83.** $\begin{bmatrix} 1 & -1 \\ 4 & -\frac{7}{2} \end{bmatrix}$

85. Does not exist   **87.** $\begin{bmatrix} 2 & \frac{20}{3} \\ \frac{1}{10} & \frac{1}{6} \end{bmatrix}$ **89.** $\begin{bmatrix} \frac{20}{9} & \frac{5}{9} \\ -\frac{10}{9} & -\frac{25}{9} \end{bmatrix}$

91. $(36, 11)$ **93.** $(-6, -1)$ **95.** $(2, 3)$ **97.** $(-8, 18)$
99. $(2, -1, -2)$ **101.** $(6, 1, -1)$ **103.** $(-3, 1)$
105. $\left(\frac{1}{6}, -\frac{7}{4}\right)$ **107.** $(1, 1, -2)$ **109.** -42 **111.** 550
113. (a) $M_{11} = 4, M_{12} = 7, M_{21} = -1, M_{22} = 2$
  (b) $C_{11} = 4, C_{12} = -7, C_{21} = 1, C_{22} = 2$
115. (a) $M_{11} = 30, M_{12} = -12, M_{13} = -21, M_{21} = 20,$
   $M_{22} = 19, M_{23} = 22, M_{31} = 5, M_{32} = -2, M_{33} = 19$
  (b) $C_{11} = 30, C_{12} = 12, C_{13} = -21, C_{21} = -20,$
   $C_{22} = 19, C_{23} = -22, C_{31} = 5, C_{32} = 2, C_{33} = 19$
117. -6 **119.** 15   **121.** 130   **123.** -8 **125.** 279
127. $(4, 7)$ **129.** $(-1, 4, 5)$ **131.** 16   **133.** 10
135. Collinear   **137.** $x - 2y + 4 = 0$
139. $2x + 6y - 13 = 0$
141. (a) Uncoded: $[12\ \ 15\ \ 15], [11\ \ 0\ \ 15], [21\ \ 20\ \ 0],$
   $[2\ \ 5\ \ 12], [15\ \ 23\ \ 0]$
  (b) Encoded: $-21\ \ 6\ \ 0\ \ -68\ \ 8\ \ 45\ \ 102\ \ -42$
   $-60\ \ -53\ \ 20\ \ 21\ \ 99\ \ -30\ \ -69$
143. SEE YOU FRIDAY
145. False. The matrix must be square.
147. An error message appears because $1(6) - (-2)(-3) = 0$.
149. If A is a square matrix, then the cofactor C_{ij} of the entry a_{ij} is $(-1)^{i+j} M_{ij}$, where M_{ij} is the determinant obtained by deleting the ith row and jth column of A. The determinant of A is the sum of the entries of any row or column of A multiplied by their respective cofactors.
151. The part of the matrix corresponding to the coefficients of the system reduces to a matrix in which the number of rows with nonzero entries is the same as the number of variables.

Chapter Test   *(page 440)*

1. $\begin{bmatrix} 1 & 0 & 0 \\ 0 & 1 & 0 \\ 0 & 0 & 1 \end{bmatrix}$

2. $\begin{bmatrix} 1 & 0 & -1 & 2 \\ 0 & 1 & 0 & -1 \\ 0 & 0 & 0 & 0 \\ 0 & 0 & 0 & 0 \end{bmatrix}$

3. $\begin{bmatrix} 4 & 3 & -2 & \vdots & 14 \\ -1 & -1 & 2 & \vdots & -5 \\ 3 & 1 & -4 & \vdots & 8 \end{bmatrix}, \left(1, 3, -\frac{1}{2}\right)$

4. (a) $\begin{bmatrix} 1 & 5 \\ 0 & -4 \end{bmatrix}$ (b) $\begin{bmatrix} 18 & 15 \\ -15 & -15 \end{bmatrix}$
  (c) $\begin{bmatrix} 8 & 15 \\ -5 & -13 \end{bmatrix}$ (d) $\begin{bmatrix} 5 & -5 \\ 0 & 5 \end{bmatrix}$

5. $\begin{bmatrix} \frac{2}{7} & \frac{3}{7} \\ \frac{5}{7} & \frac{4}{7} \end{bmatrix}$ **6.** $\begin{bmatrix} -\frac{5}{2} & 4 & -3 \\ 5 & -7 & 6 \\ 4 & -6 & 5 \end{bmatrix}$ **7.** $(12, 18)$ **8.** -112

9. 29   **10.** 43   **11.** $(-3, 5)$ **12.** $(-2, 4, 6)$ **13.** 7
14. Uncoded: $[11\ \ 14\ \ 15], [3\ \ 11\ \ 0], [15\ \ 14\ \ 0], [23\ \ 15\ \ 15],$
   $[4\ \ 0\ \ 0]$
Encoded: $115\ \ -41\ \ -59\ \ 14\ \ -3\ \ -11\ \ 29\ \ -15$
   $-14\ \ 128\ \ -53\ \ -60\ \ 4\ \ -4\ \ 0$
15. 75 L of 60% solution, 25 L of 20% solution

Problem Solving   *(page 442)*

1. (a) $AT = \begin{bmatrix} -1 & -4 & -2 \\ 1 & 2 & 3 \end{bmatrix}, AAT = \begin{bmatrix} -1 & -2 & -3 \\ -1 & -4 & -2 \end{bmatrix}$

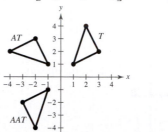

  A represents a counterclockwise rotation.
  (b) AAT is rotated clockwise 90° to obtain AT. AT is then rotated clockwise 90° to obtain T.

3. (a) Yes   (b) No   (c) No   (d) No
5. (a) Gold Satellite System: 28,750 subscribers
   Galaxy Satellite Network: 35,750 subscribers
   Nonsubscribers: 35,500
   Answers will vary.
  (b) Gold Satellite System: 30,813 subscribers
   Galaxy Satellite Network: 39,675 subscribers
   Nonsubscribers: 29,513
   Answers will vary.
  (c) Gold Satellite System: 31,947 subscribers
   Galaxy Satellite Network: 42,329 subscribers
   Nonsubscribers: 25,724
   Answers will vary.
  (d) Satellite companies are increasing the number of subscribers, while the nonsubscribers are decreasing.

7. $x = 4$ **9.** Answers will vary. For example: $A = \begin{bmatrix} 1 & 1 \\ 0 & 0 \end{bmatrix}$

11. Answers will vary.   **13.** $\begin{vmatrix} x & 0 & 0 & d \\ -1 & x & 0 & c \\ 0 & -1 & x & b \\ 0 & 0 & -1 & a \end{vmatrix}$

15. Transformer: $10.00
Foot of wire: $0.20
Light: $1.00

17. (a) $A^{-1} = \begin{bmatrix} 1 & -2 \\ 1 & -3 \end{bmatrix}$ (b) JOHN RETURN TO BASE

19. $|A| = 0$

Chapter 9

Review Exercises   *(page 495)*

1. $8, 5, 4, \frac{7}{2}, \frac{16}{5}$ **3.** $72, 36, 12, 3, \frac{3}{5}$ **5.** $a_n = 2(-1)^n$

7. $a_n = \dfrac{4}{n}$ **9.** 362,880   **11.** 1   **13.** $\dfrac{205}{24}$

15. $\displaystyle\sum_{k=1}^{20} \dfrac{1}{2k}$ **17.** $\dfrac{4}{9}$

19. (a) $A_1 = \$10,018.75$
$A_2 \approx \$10,037.54$
$A_3 \approx \$10,056.36$
$A_4 \approx \$10,075.21$
$A_5 \approx \$10,094.10$
$A_6 \approx \$10,113.03$
$A_7 \approx \$10,131.99$
$A_8 \approx \$10,150.99$
$A_9 \approx \$10,170.02$
$A_{10} \approx \$10,189.09$
(b) $12,520.59

21. Arithmetic sequence, $d = -7$

23. Arithmetic sequence, $d = \frac{1}{2}$ **25.** $a_n = 12n - 5$

27. $a_n = -7n + 107$ **29.** 3, 14, 25, 36, 47   **31.** 35,350

33. 80   **35.** 88   **37.** (a) $51,600   (b) $238,500

39. Geometric sequence, $r = 2$

41. Geometric sequence, $r = -3$

43. 2, 30, 450, 6750, 101,250

45. $9, 6, 4, \frac{8}{3}, \frac{16}{9}$ or $9, -6, 4, -\frac{8}{3}, \frac{16}{9}$ **47.** $a_n = 18\left(-\frac{1}{2}\right)^{n-1}; \frac{9}{512}$

49. $a_n = 100(1.05)^{n-1}$; About 155.133   **51.** 127   **53.** $\frac{15}{16}$

55. 31   **57.** 24.85   **59.** 8   **61.** 12

63. (a) $a_n = 120,000(0.7)^n$ (b) $20,168.40

65–67. Proofs   **69.** $S_n = n(2n + 7)$

71. $S_n = \frac{5}{2}\left[1 - \left(\frac{3}{5}\right)^n\right]$ **73.** 2850

75. 5, 10, 15, 20, 25
First differences: 5, 5, 5, 5
Second differences: 0, 0, 0
Linear

77. 15   **79.** 21   **81.** $x^4 + 16x^3 + 96x^2 + 256x + 256$

83. $41 + 840i$ **85.** 11   **87.** 10,000   **89.** 120

91. 225,792,840   **93.** $\frac{1}{9}$ **95.** (a) 43%   (b) 82%

97. $\frac{1}{1296}$ **99.** $\frac{3}{4}$

101. False. $\dfrac{(n + 2)!}{n!} = \dfrac{(n + 2)(n + 1)n!}{n!} = (n + 2)(n + 1)$

103. True by the Properties of Sums.

105. The set of natural numbers

107. Each term of the sequence is defined in terms of preceding terms.

Chapter Test   *(page 498)*

1. $-\dfrac{1}{5}, \dfrac{1}{8}, -\dfrac{1}{11}, \dfrac{1}{14}, -\dfrac{1}{17}$ **2.** $a_n = \dfrac{n + 2}{n!}$

3. 60, 73, 86; 243   **4.** $a_n = 0.8n + 1.4$

5. $a_n = 7(4)^{n-1}$ **6.** 5, 10, 20, 40, 80   **7.** 86,100

8. 477   **9.** 4   **10.** Proof

11. (a) $x^4 + 24x^3y + 216x^2y^2 + 864xy^3 + 1296y^4$
(b) $3x^5 - 30x^4 + 124x^3 - 264x^2 + 288x - 128$

12. $-22,680$ **13.** (a) 72   (b) 328,440
14. (a) 330   (b) 720,720   **15.** 26,000   **16.** 720

17. $\dfrac{1}{15}$ **18.** $\dfrac{1}{27,405}$ **19.** 10%

Cumulative Test for Chapters 7–9   *(page 499)*

1. $(1, 2), \left(-\frac{3}{2}, \frac{3}{4}\right)$ **2.** $(-3, -1)$ **3.** $(5, -2, -2)$

4. $(1, -2, 1)$

5. $0.75 mixture: 120 lb; $1.25 mixture: 80 lb

6. $y = \frac{1}{4}x^2 - 2x + 6$

7.
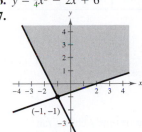

8.
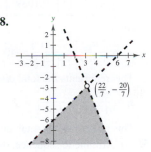

9.
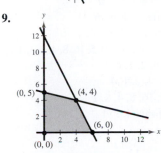

Maximum at $(4, 4)$: $z = 20$
Minimum at $(0, 0)$: $z = 0$

10. $\begin{bmatrix} -1 & 2 & -1 & \vdots & 9 \\ 2 & -1 & 2 & \vdots & -9 \\ 3 & 3 & -4 & \vdots & 7 \end{bmatrix}$

11. $(-2, 3, -1)$ **12.** $\begin{bmatrix} 1 & 5 \\ -1 & 3 \end{bmatrix}$ **13.** $\begin{bmatrix} 16 & -40 \\ 0 & 8 \end{bmatrix}$

14. $\begin{bmatrix} 16 & -25 \\ -2 & 13 \end{bmatrix}$ **15.** $\begin{bmatrix} -6 & 15 \\ 2 & -9 \end{bmatrix}$ **16.** $\begin{bmatrix} 9 & 0 \\ -7 & 16 \end{bmatrix}$

17. $\begin{bmatrix} -15 & 35 \\ 1 & -5 \end{bmatrix}$ **18.** $\begin{bmatrix} -175 & 37 & -13 \\ 95 & -20 & 7 \\ 14 & -3 & 1 \end{bmatrix}$ **19.** 203

20. Gym shoes: $2539 million
Jogging shoes: $2362 million
Walking shoes: $4418 million

21. $(-5, 4)$ **22.** $(-3, 4, 2)$ **23.** 9

24. $\frac{1}{5}, -\frac{1}{7}, \frac{1}{9}, -\frac{1}{11}, \frac{1}{13}$

25. $a_n = \dfrac{(n + 1)!}{n + 3}$ **26.** 1536

27. (a) 65.4   (b) $a_n = 3.2n + 1.4$

28. 3, 6, 12, 24, 48   **29.** $\frac{130}{9}$ **30.** Proof

31. $w^4 - 36w^3 + 486w^2 - 2916w + 6561$

32. 2184   **33.** 600   **34.** 70

35. 462   **36.** 453,600

37. 151,200   **38.** 720   **39.** $\frac{1}{4}$

Problem Solving   *(page 504)*

1. (a) 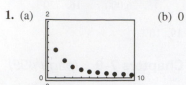   (b) 0

(c)

n	1	10	100	1000	10,000
a_n	1	0.1089	0.0101	0.0010	0.0001

(d) 0

3. $s_d = \dfrac{a_1}{1 - r} = \dfrac{20}{1 - \frac{1}{2}} = 40$

This represents the total distance Achilles ran.

$s_t = \dfrac{a_1}{1 - r} = \dfrac{1}{1 - \frac{1}{2}} = 2$

This represents the total amount of time Achilles ran.

5. (a) Arithmetic sequence, difference $= d$

(b) Arithmetic sequence, difference $= dC$

(c) Not an arithmetic sequence

7. (a) 7, 22, 11, 34, 17, 52, 26, 13, 40, 20, 10, 5, 16, 8, 4, 2, 1, 4, 2, 1

(b) $a_1 = 4$: 4, 2, 1, 4, 2, 1, 4, 2, 1, 4

$a_1 = 5$: 5, 16, 8, 4, 2, 1, 4, 2, 1, 4

$a_1 = 12$: 12, 6, 3, 10, 5, 16, 8, 4, 2, 1

Eventually, the terms repeat: 4, 2, 1

9. Proof

11. $S_n = \left(\dfrac{1}{2}\right)^{n-1}$; $A_n = \dfrac{\sqrt{3}}{4}S_n^2$ **13.** $\dfrac{1}{3}$

15. (a) 30 marbles   (b) 3 to 7; 7 to 3

(c) $P(E) = \dfrac{\text{odds in favor of } E}{\text{odds in favor of } E + 1}$

(d) Odds in favor of event $E = \dfrac{P(E)}{P(E')}$

Chapter 10

Review Exercises   *(page 558)*

1. $\dfrac{\pi}{4}$ rad, 45°   **3.** 1.1071 rad, 63.43°   **5.** 0.4424 rad, 25.35°

7. 0.6588 rad, 37.75°   **9.** $4\sqrt{2}$ **11.** Hyperbola

13. $y^2 = 16x$ **15.** $(y - 2)^2 = 12x$

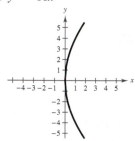   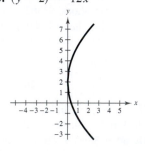

17. $y = -4x - 2$ **19.** $6\sqrt{5}$ m

21. $\dfrac{(x - 3)^2}{25} + \dfrac{y^2}{16} = 1$ **23.** $\dfrac{(x - 2)^2}{4} + \dfrac{(y - 1)^2}{1} = 1$

25. The foci occur 3 feet from the center of the arch.

27. Center: $(-1, 2)$

Vertices:

$(-1, 9), (-1, -5)$

Foci: $\left(-1, 2 \pm 2\sqrt{6}\right)$

Eccentricity: $\dfrac{2\sqrt{6}}{7}$

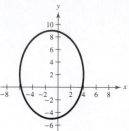

29. Center: $(1, -4)$

Vertices:

$(1, 0), (1, -8)$

Foci: $\left(1, -4 \pm \sqrt{7}\right)$

Eccentricity: $\dfrac{\sqrt{7}}{4}$

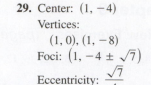

31. $\dfrac{y^2}{1} - \dfrac{x^2}{3} = 1$ **33.** $\dfrac{x^2}{16} - \dfrac{y^2}{9} = 1$

35. Center: $(5, -3)$

Vertices:

$(11, -3), (-1, -3)$

Foci: $\left(5 \pm 2\sqrt{13}, -3\right)$

Asymptotes:

$y = -3 \pm \frac{2}{3}(x - 5)$

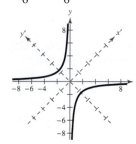

37. Center: $(1, -1)$

Vertices:

$(5, -1), (-3, -1)$

Foci: $(6, -1), (-4, -1)$

Asymptotes:

$y = -1 \pm \frac{3}{4}(x - 1)$

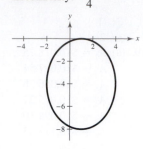

39. 72 mi   **41.** Hyperbola   **43.** Ellipse

45. $\dfrac{(y')^2}{6} - \dfrac{(x')^2}{6} = 1$ **47.** $\dfrac{(x')^2}{3} + \dfrac{(y')^2}{2} = 1$

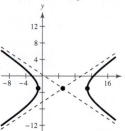   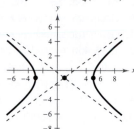

49. (a) Parabola

(b) $y = \dfrac{24x + 40 \pm \sqrt{(24x + 40)^2 - 36(16x^2 - 30x)}}{18}$

(c)

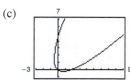

51. (a) Parabola

(b) $y = \dfrac{-\left(2x - 2\sqrt{2}\right) \pm \sqrt{\left(2x - 2\sqrt{2}\right)^2 - 4\left(x^2 + 2\sqrt{2}x + 2\right)}}{2}$

(c)

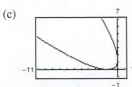

53. (a)

t	-2	-1	0	1	2
x	-8	-5	-2	1	4
y	15	11	7	3	-1

(b)

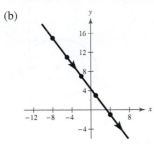

55. (a)

57. (a)

(b) $y = 2x$

(b) $y = \sqrt[4]{x}$

59. (a)

61. (a) $x = t,\ y = 2t + 3$
(b) $x = t - 1,\ y = 2t + 1$
(c) $x = 3 - t,\ y = 9 - 2t$

(b) $x^2 + y^2 = 9$

63. (a) $x = t,\ y = t^2 + 3$ (b) $x = t - 1,\ y = t^2 - 2t + 4$
(c) $x = 3 - t,\ y = t^2 - 6t + 12$

65. (a) $x = t,\ y = 2t^2 + 2$ (b) $x = t - 1,\ y = 2t^2 - 4t + 4$
(c) $x = 3 - t,\ y = 2t^2 - 12t + 20$

67.   **69.**

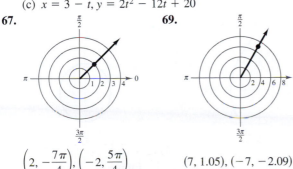

$\left(2, -\dfrac{7\pi}{4}\right), \left(-2, \dfrac{5\pi}{4}\right)$ $(7, 1.05), (-7, -2.09)$

71. $\left(-\dfrac{1}{2}, -\dfrac{\sqrt{3}}{2}\right)$ **73.** $\left(-\dfrac{3\sqrt{2}}{2}, \dfrac{3\sqrt{2}}{2}\right)$ **75.** $\left(1, \dfrac{\pi}{2}\right)$

77. $\left(2\sqrt{13}, 0.9828\right)$ **79.** $r = 9$ **81.** $r = 6 \sin \theta$
83. $r^2 = 10 \csc 2\theta$ **85.** $x^2 + y^2 = 25$
87. $x^2 + y^2 = 3x$ **89.** $x^2 + y^2 = y^{2/3}$

91. Symmetry: $\theta = \dfrac{\pi}{2}$, polar axis, pole

Maximum value of $|r|$: $|r| = 6$ for all values of θ
No zeros of r

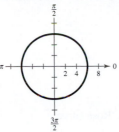

93. Symmetry: $\theta = \dfrac{\pi}{2}$, polar axis, pole

Maximum value of $|r|$: $|r| = 4$ when $\theta = \dfrac{\pi}{4}, \dfrac{3\pi}{4}, \dfrac{5\pi}{4}, \dfrac{7\pi}{4}$

Zeros of r: $r = 0$ when $\theta = 0, \dfrac{\pi}{2}, \pi, \dfrac{3\pi}{2}$

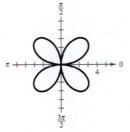

95. Symmetry: polar axis
Maximum value of $|r|$: $|r| = 4$ when $\theta = 0$
Zeros of r: $r = 0$ when $\theta = \pi$

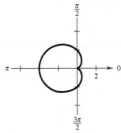

97. Symmetry: $\theta = \dfrac{\pi}{2}$

Maximum value of $|r|$: $|r| = 8$ when $\theta = \dfrac{\pi}{2}$

Zeros of r: $r = 0$ when $\theta = 3.4814, 5.9433$

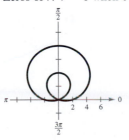

99. Symmetry: $\theta = \dfrac{\pi}{2}$, polar axis, pole

Maximum value of $|r|$: $|r| = 3$ when $\theta = 0, \dfrac{\pi}{2}, \pi, \dfrac{3\pi}{2}$

Zeros of r: $r = 0$ when $\theta = \dfrac{\pi}{4}, \dfrac{3\pi}{4}, \dfrac{5\pi}{4}, \dfrac{7\pi}{4}$

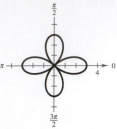

101. Limaçon

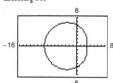

103. Rose curve

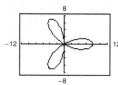

105. Hyperbola

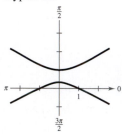

107. Ellipse

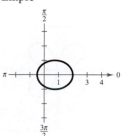

109. $r = \dfrac{4}{1 - \cos \theta}$ **111.** $r = \dfrac{5}{3 - 2 \cos \theta}$

113. $r = \dfrac{7961.93}{1 - 0.937 \cos \theta}$; 10,980.11 mi

115. False. The equation of a hyperbola is a second-degree equation.

117. False. $(2, \pi/4)$, $(-2, 5\pi/4)$, and $(2, 9\pi/4)$ all represent the same point.

119. (a) The graphs are the same.   (b) The graphs are the same.

Chapter Test *(page 561)*

1. 0.3805 rad, 21.8°   **2.** 0.8330 rad, 47.7°   **3.** $\dfrac{7\sqrt{2}}{2}$

4. Parabola: $y^2 = 2(x - 1)$

Vertex: $(1, 0)$

Focus: $\left(\dfrac{3}{2}, 0\right)$

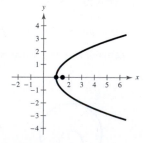

5. Hyperbola: $\dfrac{(x - 2)^2}{4} - y^2 = 1$

Center: $(2, 0)$

Vertices: $(0, 0), (4, 0)$

Foci: $\left(2 \pm \sqrt{5}, 0\right)$

Asymptotes: $y = \pm\dfrac{1}{2}(x - 2)$

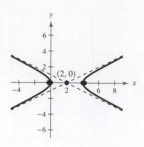

6. Ellipse: $\dfrac{(x + 3)^2}{16} + \dfrac{(y - 1)^2}{9} = 1$

Center: $(-3, 1)$

Vertices: $(1, 1), (-7, 1)$

Foci: $\left(-3 \pm \sqrt{7}, 1\right)$

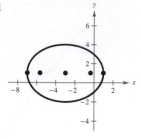

7. Circle: $(x - 2)^2 + (y - 1)^2 = \dfrac{1}{2}$

Center: $(2, 1)$

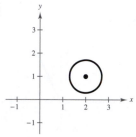

8. $(x - 2)^2 = \dfrac{4}{3}(y + 3)$ **9.** $\dfrac{y^2}{2/5} - \dfrac{x^2}{18/5} = 1$

10. (a) 45°

(b)

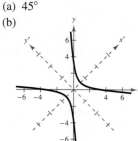

11.

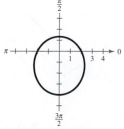

$\dfrac{(x - 2)^2}{9} + \dfrac{y^2}{4} = 1$

12. (a) $x = t, y = 3 - t^2$ (b) $x = t - 2, y = -t^2 + 4t - 1$

13. $\left(\sqrt{3}, -1\right)$

14. $\left(2\sqrt{2}, \dfrac{7\pi}{4}\right), \left(-2\sqrt{2}, \dfrac{3\pi}{4}\right), \left(2\sqrt{2}, -\dfrac{\pi}{4}\right)$

15. $r = 3 \cos \theta$

16.

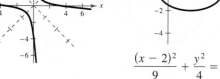

Parabola

17.

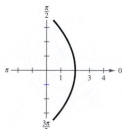

Ellipse

18.

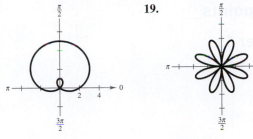

Limaçon with inner loop

19.

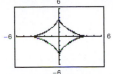

Rose curve

20. Answers will vary. For example: $r = \dfrac{1}{1 + 0.25 \sin \theta}$

21. Slope: 0.1511; Change in elevation: 789 ft

22. No; Yes

Problem Solving   *(page 565)*

1. (a) 1.2016 rad   (b) 2420 ft, 5971 ft   **3.** $A = \dfrac{4a^2b^2}{a^2 + b^2}$

5. (a) Because $d_1 + d_2 \le 20$, by definition, the outer bound that the boat can travel is an ellipse. The islands are the foci.

(b) Island 1: $(-6, 0)$;
Island 2: $(6, 0)$

(c) 20 mi; Vertex: $(10, 0)$

(d) $\dfrac{x^2}{100} + \dfrac{y^2}{64} = 1$

7. Proof

9. (a) The first set of parametric equations models projectile motion along a straight line. The second set of parametric equations models projectile motion of an object launched at a height of h units above the ground that will eventually fall back to the ground.

(b) $y = (\tan \theta)x$; $y = h + x \tan \theta - \dfrac{16x^2 \sec^2 \theta}{v_0{}^2}$

(c) In the first case, the path of the moving object is not affected by a change in the velocity because eliminating the parameter removes v_0.

11.

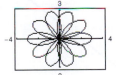

$r = 3 \sin\left(\dfrac{5\theta}{2}\right)$ $r = -\cos(\sqrt{2}\theta),\ -2\pi \le \theta \le 2\pi$

Sample answer: If n is a rational number, then the curve has a finite number of petals. If n is an irrational number, then the curve has an infinite number of petals.

13. (a)

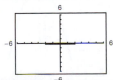

The graph is a line between -2 and 2 on the x-axis.

(b)

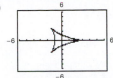

The graph is a three-sided figure with counterclockwise orientation.

(c)

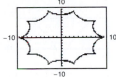

The graph is a four-sided figure with counterclockwise orientation.

(d)

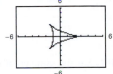

The graph is a 10-sided figure with counterclockwise orientation.

(e)

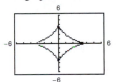

The graph is a three-sided figure with clockwise orientation.

(f)

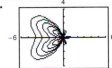

The graph is a four-sided figure with clockwise orientation.

15.

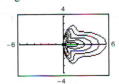

For $n \ge 1$, a bell is produced.
For $n \le -1$, a heart is produced.
For $n = 0$, a rose curve is produced.

Technology

Chapter 1   *(page 3)*

Sample answer:

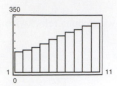

(page 21)

The lines appear perpendicular on the square setting.

(page 30)

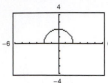   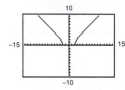

Domain: $[-2, 2]$ Domain: $(-\infty, -2] \cup [2, \infty)$
    Yes, for -2 and 2.

Chapter 3   *(page 169)*

$S = 9.30(1.1195)^t$
The model given by the graphing utility has the same coefficient as the model in Example 1. However, the model in Example 1 contains the natural exponential function.

Chapter 4   *(page 214)*

No graph is visible. Try $-\pi \le x \le \pi$ and $-0.5 \le y \le 0.5$ as a viewing window.

Chapter 7   *(page 342)*

The point of intersection (7000, 5000) agrees with the solution.

(page 344)

$(2, 0)$; $(0, -1)$, $(2, 1)$; None

Chapter 8   *(page 406)*

$$\begin{bmatrix} 1 & 1 \\ 1 & -5 \end{bmatrix}$$

Chapter 10   *(page 541)*

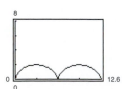

Checkpoints

Chapter 1

Section 1.1

1.

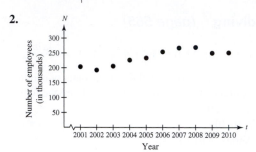

2.

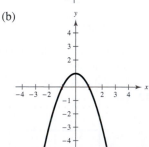

3. $\sqrt{37} \approx 6.08$

4. $d_1 = \sqrt{45}$, $d_2 = \sqrt{20}$, $d_3 = \sqrt{65}$
    $\left(\sqrt{45}\right)^2 + \left(\sqrt{20}\right)^2 = \left(\sqrt{65}\right)^2$

5. $(1, -1)$ **6.** $\sqrt{709} \approx 27$ yd   **7.** \$6.75 billion

8. $(1, 2)$, $(1, -2)$, $(-1, 0)$, $(-1, -4)$

Section 1.2

1. (a) No   (b) Yes

2. (a)   (b)

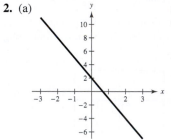   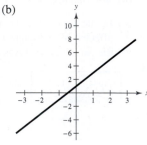

3. (a)   (b)

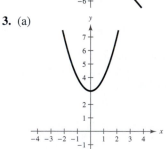

4. x-intercepts: $(0, 0)$, $(-5, 0)$ **5.** x-axis symmetry
    y-intercept: $(0, 0)$

6.

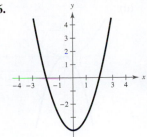

7.

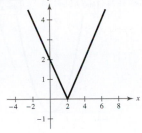

8. $(x + 3)^2 + (y + 5)^2 = 25$ **9.** About 175 lb

Section 1.3

1. (a)

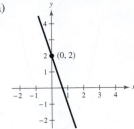

(b)

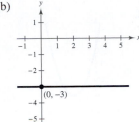

(c)

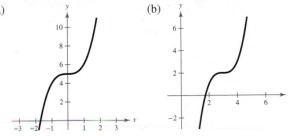

2. (a) 2   (b) $-\frac{3}{2}$ (c) Undefined   (d) 0
3. (a) $y = 2x - 13$ (b) $y = -\frac{2}{3}x + \frac{5}{3}$ (c) $y = 1$
4. (a) $y = \frac{5}{3}x + \frac{23}{3}$ (b) $y = -\frac{3}{5}x - \frac{7}{5}$ **5.** Yes
6. The y-intercept, $(0, 1500)$, tells you that the initial value of a copier at the time it is purchased is $1500. The slope, $m = -300$, tells you that the value of the copier decreases by $300 each year after it is purchased.
7. $y = -4125x + 24{,}750$ **8.** $y = -2t + 76.6$; $50.6 billion

Section 1.4

1. (a) Not a function   (b) Function
2. (a) Not a function   (b) Function
3. (a) -2 (b) -38 (c) $-3x^2 + 6x + 7$
4. $f(-2) = 5, f(2) = 1, f(3) = 2$ **5.** ± 4 **6.** $-4, 3$
7. (a) $\{-2, -1, 0, 1, 2\}$ (b) All real numbers x except $x = 3$
(c) All real numbers r such that $r > 0$
(d) All real numbers x such that $x \geq 16$
8. (a) $S(r) = 10\pi r^2$ (b) $S(h) = \frac{5}{8}\pi h^2$ **9.** No
10. 2003: 178.75   2004: 212.40   2005: 246.05
2006: 297.90   2007: 368.65   2008: 439.40
2009: 510.15
11. $2x + h + 2, h \neq 0$

Section 1.5

1. (a) All real numbers x except $x = -3$
(b) $f(0) = 3; f(3) = -6$ (c) $(-\infty, 3]$
2. Function

3. (a) $x = -8, x = \frac{3}{2}$ (b) $t = 25$ (c) $x = \pm\sqrt{2}$
4.

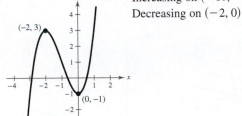

Increasing on $(-\infty, -2)$ and $(0, \infty)$
Decreasing on $(-2, 0)$

5. $(-0.88, 6.06)$ **6.** (a) -3 (b) 0
7. (a) 20 ft/sec   (b) $\frac{140}{3}$ ft/sec
8. (a) Neither; No symmetry   (b) Even; y-axis symmetry
(c) Odd; Origin symmetry

Section 1.6

1. $f(x) = -\frac{5}{2}x + 1$ **2.** $f\left(-\frac{3}{2}\right) = 0, f(1) = 3, f\left(-\frac{5}{2}\right) = -1$
3.

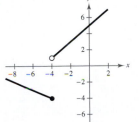

Section 1.7

1. (a)

(b)

2. $j(x) = -(x + 3)^4$
3. (a) The graph of g is a reflection in the x-axis of the graph of f.
(b) The graph of h is a reflection in the y-axis of the graph of f.
4. (a) The graph of g is a vertical stretch of the graph of f by a factor of 4.
(b) The graph of h is a vertical shrink of the graph of $f(x)$ by a factor of $\frac{1}{4}$.
5. (a) The graph of g is a horizontal shrink of the graph of f by a factor of 2.
(b) The graph of h is a horizontal stretch of the graph of f by a factor of $\frac{1}{2}$.

Section 1.8

1. $x^2 - x + 1; 3$ **2.** $x^2 + x - 1; 11$ **3.** $x^2 - x^3; -18$
4. $\left(\frac{f}{g}\right)(x) = \frac{\sqrt{x - 3}}{\sqrt{16 - x^2}}$; $\left(\frac{g}{f}\right)(x) = \frac{\sqrt{16 - x^2}}{\sqrt{x - 3}}$;
Domain: $[3, 4)$ Domain: $(3, 4]$
5. (a) $8x^2 + 7$ (b) $16x^2 + 80x + 101$ (c) 9
6. All real numbers x **7.** $f(x) = \frac{1}{5}\sqrt[3]{x}, g(x) = 8 - x$
8. (a) $(N \circ T)(t) = 32t^2 + 36t + 204$ (b) About 4.5 h

Section 1.9

1. $f^{-1}(x) = 5x$, $f(f^{-1}(x)) = \frac{1}{5}(5x) = x$, $f^{-1}(f(x)) = 5\left(\frac{1}{5}x\right) = x$

2. $g(x) = 7x + 4$

3. 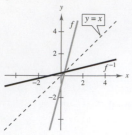   **4.**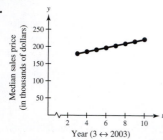

5. (a) Yes   (b) No   **6.** $f^{-1}(x) = \dfrac{5 - 2x}{x + 3}$

7. $f^{-1}(x) = x^3 - 10$

Section 1.10

1.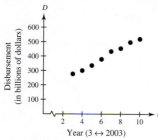

The model is a good fit for the data.

2. 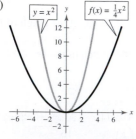   **3.** $I = 0.075P$

$D = 36.85t + 161.2$

4. 576 ft   **5.** 508 units   **6.** 506 ft   **7.** 14,000 joules

Chapter 2

Section 2.1

1. (a) (b)

The graph of $f(x) = \frac{1}{4}x^2$ is broader than the graph of $y = x^2$.

The graph of $g(x) = -\frac{1}{6}x^2$ is a reflection in the x-axis and is broader than the graph of $y = x^2$.

(c) 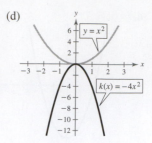 (d)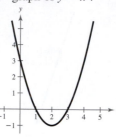

The graph of $h(x) = \frac{5}{2}x^2$ is narrower than the graph of $y = x^2$.

The graph of $k(x) = -4x^2$ is a reflection in the x-axis and is narrower than the graph of $y = x^2$.

2. 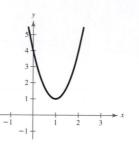   **3.**

Vertex: $(1, 1)$
Axis: $x = 1$

Vertex: $(2, -1)$
x-intercepts: $(1, 0)$, $(3, 0)$

4. $y = (x + 4)^2 + 11$ **5.** About 39.7 ft

Section 2.2

1. (a) 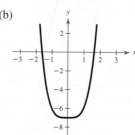 (b)

(c) 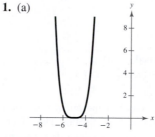 (d)

2. (a) Falls to the left, rises to the right
(b) Rises to the left, falls to the right.

3. Real zeros: $x = 0$, $x = 6$; Turning points: 2

4. 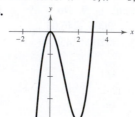   **5.**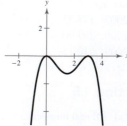

6. $x \approx 3.196$

Section 2.3

1. $(x + 4)(3x + 7)$ 2. $x^2 + x + 3, x \neq 3$

3. $2x^3 - x^2 + 3 - \dfrac{6}{3x + 1}$ 4. $5x^2 - 2x + 3$

5. (a) 1   (b) 396   (c) $-\dfrac{13}{2}$ (d) -17

6. $f(-3) = 0, f(x) = (x + 3)(x - 5)(x + 2)$

Section 2.4

1. (a) $12 - i$ (b) $-2 + 7i$ (c) i (d) 0

2. (a) $18 - 6i$ (b) 41   (c) $12 + 6i$ 3. (a) 45   (b) 29

4. $\dfrac{3}{5} + \dfrac{4}{5}i$ 5. $-2\sqrt{7}$ 6. $-\dfrac{7}{8} \pm \dfrac{\sqrt{23}}{8}i$

Section 2.5

1. 4   2. (a) $-1, 2, 4$ (b) No rational zeros   (c) 3

3. 5   4. $-3, \dfrac{1}{2}, 2, 5$ 5. $-6, -1, 3$

6. (a) $f(x) = x^4 + 45x^2 - 196$

   (b) $f(x) = x^4 - 12x^3 + 52x^2 - 92x + 51$

   (c) $f(x) = x^4 - 7x^3 + 14x^2 + 2x - 20$

7. $\dfrac{2}{3}, \pm 4i$

8. (a) $f(x) = (x - 1)(x + 1)(x - 3i)(x + 3i); \pm 1, \pm 3i$

   (b) $g(x) = (x - 1)(x - 1 - 2i)(x - 1 + 2i); 1, 1 \pm 2i$

   (c) $h(x) = (x - 3)(x - 4 - i)(x - 4 + i); 3, 4 \pm i$

9. Three or one positive real zeros, no negative real zeros

10. $\dfrac{1}{2}$ 11. 7 in. × 7 in. × 9 in.

Section 2.6

1. Domain: all real numbers such that $x \neq 1$

   $f(x)$ decreases without bound as x approaches 1 from the left and increases without bound as x approaches 1 from the right.

2. Vertical asymptote: $x = -1$
   Horizontal asymptote: $y = 3$

3.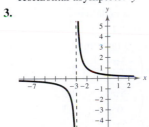

Domain: all real numbers x except $x = -3$

4.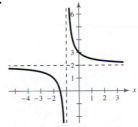

Domain: all real numbers x except $x = -1$

5.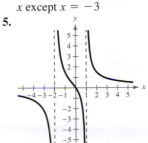

Domain: all real numbers x except $x = -2, 1$

6.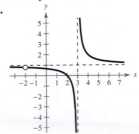

Domain: all real numbers x except $x = -2, 3$

7.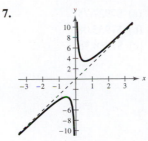

Domain: all real numbers x except $x = 0$

8. (a) \$63.75 million; \$208.64 million; \$1020 million

   (b) No. The function is undefined at $p = 100$.

9. 12.9 in. by 6.5 in.

Section 2.7

1. $(-4, 5)$ 2. $\left(-2, \dfrac{1}{3}\right) \cup (2, \infty)$

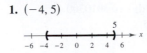   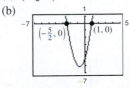

3. (a) $\left(-\dfrac{5}{2}, 1\right)$

   (b)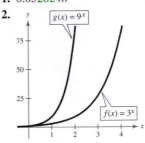

   The graph is below the x-axis when x is greater than $-\dfrac{5}{2}$ and less than 1. So, the solution set is $\left(-\dfrac{5}{2}, 1\right)$.

4. (a) The solution set is empty.

   (b) The solution set consists of the single real number $\{-2\}$.

   (c) The solution set consists of all real numbers except $x = 3$.

   (d) The solution set is all real numbers.

5. (a) $\left(-\infty, \dfrac{11}{4}\right] \cup (3, \infty)$ (b) $(-\infty, -17) \cup (6, \infty)$

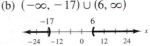

6. $180{,}000 \le x \le 300{,}000$ 7. $(-\infty, 2] \cup [5, \infty)$

Chapter 3

Section 3.1

1. 0.0528248

2.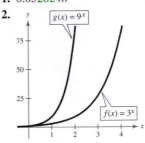

3.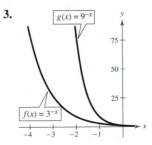

4. (a) 2   (b) 3

5. (a) Shift the graph of f two units to the right.

   (b) Shift the graph of f three units up.

   (c) Reflect the graph of f in the y-axis and shift three units down.

6. (a) 1.3498588   (b) 0.3011942   (c) 492.7490411

7.

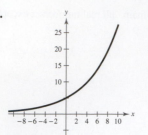

8. (a) $7927.75
   (b) $7935.08
   (c) $7938.78

9. About 9.970 lb; about 0.275 lb

Section 3.2

1. (a) 0   (b) -3 (c) 4
2. (a) 2.4393327   (b) -0.5606673
   (c) Error or complex number   (d) -0.3010300
3. (a) 1   (b) 3   (c) 0   **4.** ± 3
5.

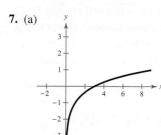

6.

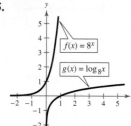

Vertical asymptote: $x = 0$

7. (a)

(b)

8. (a) -4.6051702 (b) 1.3862944   (c) 1.3169579
   (d) Error or complex number
9. (a) $\frac{1}{3}$ (b) 0   (c) $\frac{3}{4}$ (d) 7   **10.** $(-3, \infty)$
11. (a) 70.84   (b) 61.18   (c) 59.61

Section 3.3

1. 3.5850   **2.** 3.5850
3. (a) $\log 3 + 2 \log 5$ (b) $2 \log 3 - 3 \log 5$ **4.** 4
5. $\log_3 4 + 2 \log_3 x - \frac{1}{2} \log_3 y$ **6.** $\log \frac{(x+3)^2}{(x-2)^4}$
7. $\ln y = \frac{2}{3} \ln x$

Section 3.4

1. (a) 9   (b) 216   (c) $\ln 5$ (d) $-\frac{1}{2}$
2. (a) $4, -2$ (b) 1.723   **3.** 3.401   **4.** -4.778
5. 1.099, 1.386   **6.** (a) $e^{2/3}$ (b) 7   (c) 12
7. 0.513   **8.** $\frac{32}{3}$ **9.** 10
10. About 13.2 years; It takes longer for your money to double at a lower interest rate.
11. 2004

Section 3.5

1. 2021   **2.** 400 bacteria   **3.** About 38,000 yr

4.

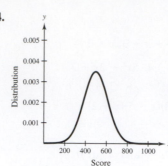

5. About 7 days
6. (a) 1,000,000
   (b) About 80,000,000

497

Chapter 4
Section 4.1

1. (a) $\frac{\pi}{4}, -\frac{7\pi}{4}$ (b) $\frac{5\pi}{3}, -\frac{7\pi}{3}$
2. (a) Complement: $\frac{\pi}{3}$; Supplement: $\frac{5\pi}{6}$
   (b) Complement: Not possible; Supplement: $\frac{\pi}{6}$
3. (a) $\frac{\pi}{3}$ (b) $\frac{16\pi}{9}$ **4.** (a) $30°$ (b) $300°$
5. 24π in. ≈ 75.40 in.   **6.** 0.84 cm/sec
7. (a) 4800π rad/min   (b) 60,319 in./min   **8.** 1117 ft²

Section 4.2

1. (a) $\sin \frac{\pi}{2} = 1$ $\csc \frac{\pi}{2} = 1$

   $\cos \frac{\pi}{2} = 0$ $\sec \frac{\pi}{2}$ is undefined.

   $\tan \frac{\pi}{2}$ is undefined. $\cot \frac{\pi}{2} = 0$

   (b) $\sin 0 = 0$ $\csc 0$ is undefined.
   $\cos 0 = 1$ $\sec 0 = 1$
   $\tan 0 = 0$ $\cot 0$ is undefined.

   (c) $\sin\left(-\frac{5\pi}{6}\right) = -\frac{1}{2}$ $\csc\left(-\frac{5\pi}{6}\right) = -2$

   $\cos\left(-\frac{5\pi}{6}\right) = -\frac{\sqrt{3}}{2}$ $\sec\left(-\frac{5\pi}{6}\right) = -\frac{2\sqrt{3}}{3}$

   $\tan\left(-\frac{5\pi}{6}\right) = \frac{\sqrt{3}}{3}$ $\cot\left(-\frac{5\pi}{6}\right) = \sqrt{3}$

   (d) $\sin\left(-\frac{3\pi}{4}\right) = -\frac{\sqrt{2}}{2}$ $\csc\left(-\frac{3\pi}{4}\right) = -\sqrt{2}$

   $\cos\left(-\frac{3\pi}{4}\right) = -\frac{\sqrt{2}}{2}$ $\sec\left(-\frac{3\pi}{4}\right) = -\sqrt{2}$

   $\tan\left(-\frac{3\pi}{4}\right) = 1$ $\cot\left(-\frac{3\pi}{4}\right) = 1$

2. (a) 0   (b) $-\frac{\sqrt{3}}{2}$ (c) 0.3
3. (a) 0.78183148   (b) 1.0997502

Section 4.3

1. $\sin \theta = \frac{1}{2}$ $\cos \theta = \frac{\sqrt{3}}{2}$ $\tan \theta = \frac{\sqrt{3}}{3}$

   $\csc \theta = 2$ $\sec \theta = \frac{2\sqrt{3}}{3}$ $\cot \theta = \sqrt{3}$

2. $\sqrt{2}$ **3.** $\tan 60° = \sqrt{3}$, $\tan 30° = \dfrac{\sqrt{3}}{3}$ **4.** 1.7650691

5. (a) $\dfrac{\sqrt{15}}{4}$ (b) $\sqrt{15}$ **6.** (a) $\frac{1}{2}$ (b) $\sqrt{5}$

7. Answers will vary.   **8.** 40 ft   **9.** 60°   **10.** 17.6 ft; 17.2 ft

Section 4.4

1. $\sin \theta = \dfrac{3\sqrt{13}}{13}$, $\cos \theta = -\dfrac{2\sqrt{13}}{13}$, $\tan \theta = -\dfrac{3}{2}$ **2.** $-\dfrac{3}{5}$

3. $-1; 0$ **4.** (a) 33°   (b) $\dfrac{4\pi}{9}$ (c) $\dfrac{\pi}{5}$

5. (a) $-\dfrac{\sqrt{2}}{2}$ (b) $-\dfrac{1}{2}$ (c) $-\dfrac{\sqrt{3}}{3}$ **6.** (a) $-\dfrac{3}{5}$ (b) $\dfrac{4}{3}$

7. (a) -1.8040478 (b) -1.0428352 (c) 0.8090170

Section 4.5

1.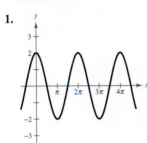

2.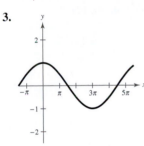

3.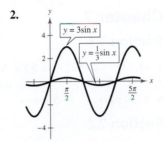

4. Period: 2π
Amplitude: 2
$\left[\dfrac{\pi}{2}, \dfrac{5\pi}{2}\right]$ corresponds to one cycle.
Key points: $\left(\dfrac{\pi}{2}, 2\right)$, $(\pi, 0)$, $\left(\dfrac{3\pi}{2}, -2\right)$, $(2\pi, 0)$, $\left(\dfrac{5\pi}{2}, 2\right)$

5.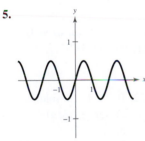

6.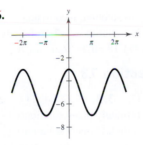

7. $y = 5.6 \sin(0.524t - 0.525) + 5.7$

Section 4.6

1.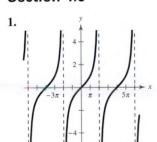

2.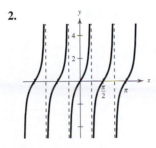

3.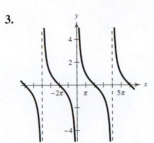

4.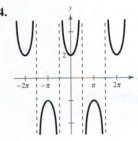

5.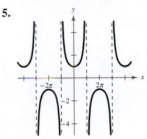

6.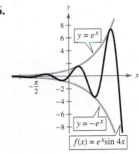

Section 4.7

1. (a) $\dfrac{\pi}{2}$
(b) Not possible

2.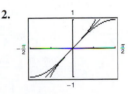

3. π **4.** (a) 1.3670516   (b) Not possible   (c) 1.9273001

5. (a) -14 (b) $-\dfrac{\pi}{4}$ (c) 0.54   **6.** $\dfrac{4}{5}$ **7.** $\sqrt{x^2 + 1}$

Section 4.8

1. $a \approx 5.46$, $c \approx 15.96$, $B = 70°$ **2.** 15.8 ft   **3.** 15.1 ft

4. 3.58°   **5.** Bearing: N 53° W, Distance: 3.2 nm

6. $d = 6 \sin \dfrac{2\pi}{3}t$; Frequency: $\dfrac{1}{3}$

7. (a) 4   (b) 3 cycles per unit of time   (c) 4   (d) $\frac{1}{12}$

Chapter 5

Section 5.1

1. $\sin x = -\dfrac{\sqrt{10}}{10}$, $\cos x = -\dfrac{3\sqrt{10}}{10}$, $\tan x = \dfrac{1}{3}$, $\csc x = -\sqrt{10}$
$\sec x = -\dfrac{\sqrt{10}}{3}$, $\cot x = 3$

2. $-\sin x$
3. (a) $(1 + \cos \theta)(1 - \cos \theta)$ (b) $(2 \csc \theta - 3)(\csc \theta - 2)$
4. $(\tan x + 1)(\tan x + 2)$ **5.** $\sin x$ **6.** $2 \sec^2 \theta$
7. $1 + \sin \theta$ **8.** $3 \cos \theta$ **9.** $\ln|\tan x|$

Section 5.2

1–7. Answers will vary.

Section 5.3

1. $\dfrac{\pi}{4} + 2n\pi$, $\dfrac{3\pi}{4} + 2n\pi$

2. $\dfrac{\pi}{3} + 2n\pi$, $\dfrac{2\pi}{3} + 2n\pi$, $\dfrac{4\pi}{3} + 2n\pi$, $\dfrac{5\pi}{3} + 2n\pi$ **3.** $n\pi$

4. $\dfrac{\pi}{6}, \dfrac{\pi}{2}, \dfrac{5\pi}{6}$ **5.** $\dfrac{\pi}{4} + n\pi, \dfrac{3\pi}{4} + n\pi$ **6.** $0, \dfrac{3\pi}{2}$

7. $\dfrac{\pi}{6} + n\pi, \dfrac{\pi}{3} + n\pi$ **8.** $\dfrac{\pi}{2} + 2n\pi$

9. $\arctan \frac{3}{4} + n\pi, \arctan(-2) + n\pi$ **10.** $49.9°, 59.9°$

Section 5.4

1. $\dfrac{\sqrt{2} + \sqrt{6}}{4}$ **2.** $\dfrac{\sqrt{2} + \sqrt{6}}{4}$ **3.** $-\dfrac{63}{65}$

4. $\dfrac{x + \sqrt{1 - x^2}}{\sqrt{2}}$ **5.** Answers will vary.

6. (a) $-\cos\theta$ (b) $\dfrac{\tan\theta - 1}{\tan\theta + 1}$ **7.** $\dfrac{\pi}{3}, \dfrac{5\pi}{3}$

8. Answers will vary.

Section 5.5

1. $\dfrac{\pi}{3} + 2n\pi, \pi + 2n\pi, \dfrac{5\pi}{3} + 2n\pi$

2. $\sin 2\theta = \frac{24}{25}, \cos 2\theta = \frac{7}{25}, \tan 2\theta = \frac{24}{7}$

3. $4\cos^3 x - 3\cos x$ **4.** $\dfrac{\cos 4x - 4\cos 2x + 3}{\cos 4x + 4\cos 2x + 3}$

5. $\dfrac{-\sqrt{2 - \sqrt{3}}}{2}$ **6.** $\dfrac{\pi}{3}, \pi, \dfrac{5\pi}{3}$ **7.** $\dfrac{1}{2}\sin 8x + \dfrac{1}{2}\sin 2x$

8. $\dfrac{\sqrt{2}}{2}$ **9.** $\dfrac{\pi}{6} + \dfrac{2n\pi}{3}, \dfrac{\pi}{2} + \dfrac{2n\pi}{3}, n\pi$ **10.** $45°$

Chapter 6

Section 6.1

1. $C = 105°, b \approx 45.25, c \approx 61.82$ **2.** 13.29 m
3. $B \approx 12.39°, C \approx 136.61°, c \approx 16.01$ **4.** $\sin B \approx 3.0311 > 1$
5. Two solutions:
   $B \approx 70.4°, C \approx 51.6°, c \approx 4.16$ ft
   $B \approx 109.6°, C \approx 12.4°, c \approx 1.14$ ft
6. 213 in.^{2} **7.** 1856.59 m

Section 6.2

1. $A \approx 26.38°, B \approx 36.34°, C \approx 117.28°$
2. $B \approx 59.6°, C \approx 40.4°, a \approx 18.26$ **3.** 202 ft
4. N 15.37° E   **5.** 19.90

Section 6.3

1. $\|\overrightarrow{PQ}\| = \|\overrightarrow{RS}\| = \sqrt{10}$, slope$_{\overrightarrow{PQ}}$ = slope$_{\overrightarrow{RS}} = \frac{1}{3}$
$\overrightarrow{PQ}$ and $\overrightarrow{RS}$ have the same magnitude and direction, so they are equal.
2. $\mathbf{v} = \langle -5, 6 \rangle, \|\mathbf{v}\| = \sqrt{61}$
3. (a) $\langle 4, 6 \rangle$ (b) $\langle -2, 2 \rangle$ (c) $\langle -7, 2 \rangle$
4. $\left\langle \dfrac{6}{\sqrt{37}}, -\dfrac{1}{\sqrt{37}} \right\rangle$ **5.** $-6\mathbf{i} - 3\mathbf{j}$ **6.** $11\mathbf{i} - 14\mathbf{j}$
7. (a) $135°$ (b) $209.74°$ **8.** $\langle -70.71, -70.71 \rangle$
9. 2405 lb   **10.** $\|\mathbf{v}\| \approx 451.8$ mi/h, $\theta \approx 305.1°$

Section 6.4

1. -6 **2.** (a) $\langle -36, 108 \rangle$ (b) 43   (c) $2\sqrt{10}$ **3.** $45°$
4. Yes   **5.** $\frac{1}{17}\langle 64, 16 \rangle; \frac{1}{17}\langle -13, 52 \rangle$ **6.** 38.8 lb
7. 1212 ft-lb

Section 6.5

1.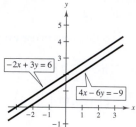

2. $6\sqrt{2}\left(\cos \dfrac{7\pi}{4} + i \sin \dfrac{7\pi}{4}\right)$

3. $-4 + 4\sqrt{3}i$ **4.** 10   **5.** $12i$ **6.** $\dfrac{\sqrt{3}}{2} + \dfrac{1}{2}i$

7. -4 **8.** $1, i, -1, -i$
9. $\sqrt[3]{3} + \sqrt[3]{3}i, -1.9701 + 0.5279i, 0.5279 - 1.9701i$

Chapter 7

Section 7.1

1. $(3, 3)$ **2.** \$6250 at 6.5%, \$18,750 at 8.5%
3. $(-3, -1), (2, 9)$ **4.** No solution   **5.** $(1, 3)$
6. 6250 pairs   **7.** 5 weeks

Section 7.2

1. $\left(\frac{3}{4}, \frac{5}{2}\right)$ **2.** $(4, 3)$ **3.** $(3, -1)$ **4.** $(9, 12)$
5.

No solution; inconsistent
6. No solution   **7.** Infinitely many solutions: $(a, 4a + 3)$
8. 471.18 mi/h; 16.63 mi/h   **9.** $(1,500,000, 537)$

Section 7.3

1. $(2, -3, 3)$ **2.** $(1, 1)$ **3.** $(1, 2, 3)$ **4.** No solution
5. Infinitely many solutions: $(-23a + 22, 15a - 13, a)$
6. Infinitely many solutions: $\left(\frac{1}{4}a, \frac{17}{4}a - 3, a\right)$
7. $s = -16t^2 + 20t + 100$; The object was thrown upward at a velocity of 20 feet per second from a height of 100 feet.
8. $y = \frac{1}{3}x^2 - 2x$

Section 7.4

1. $-\dfrac{3}{2x + 1} + \dfrac{2}{x - 1}$ **2.** $x - \dfrac{3}{x} + \dfrac{4}{x^2} + \dfrac{3}{x + 1}$

3. $-\dfrac{5}{x} + \dfrac{7x}{x^2 + 1}$ **4.** $\dfrac{x + 3}{x^2 + 4} - \dfrac{6x + 5}{(x^2 + 4)^2}$

5. $\dfrac{1}{x} - \dfrac{2}{x^2} + \dfrac{2 - x}{x^2 + 2} + \dfrac{4 - 2x}{(x^2 + 2)^2}$

Section 7.5

1.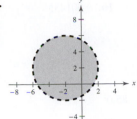

2.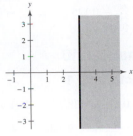

3.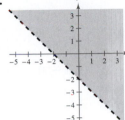

4.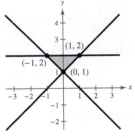

5.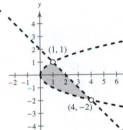

6.

No solution

7.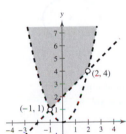

8. Consumer surplus: $22,500,000
Producer surplus: $33,750,000

9. $\begin{cases} 8x + 2y \geq 16 & \text{Nutrient A} \\ x + y \geq 5 & \text{Nutrient B} \\ 2x + 7y \geq 20 & \text{Nutrient C} \\ x \geq 0 \\ y \geq 0 \end{cases}$

Section 7.6

1. Maximum at $(0, 6)$: 30   **2.** Minimum at $(0, 0)$: 0

3. Maximum at $(60, 20)$: 880   **4.** Minimum at $(10, 0)$: 30

5. $2925; 1050 boxes of chocolate-covered creams, 150 boxes of chocolate-covered nuts.

6. 3 bags of brand X, 2 bags of brand Y

Chapter 8

Section 8.1

1. 2×3 **2.** $\begin{bmatrix} 1 & 1 & 1 & \vdots & 2 \\ 2 & -1 & 3 & \vdots & -1 \\ -1 & 2 & -1 & \vdots & 4 \end{bmatrix}$

3. Add -3 times Row 1 to Row 2.

4. Answers will vary. Solution: $(-1, 0, 1)$

5. Reduced row-echelon form   **6.** $(4, -2, 1)$

7. No solution   **8.** $(7, 4, -3)$ **9.** $(3a + 8, 2a - 5, a)$

Section 8.2

1. $a_{11} = 6, a_{12} = 3, a_{21} = -2, a_{22} = 4$ **2.** $\begin{bmatrix} 6 & -2 \\ 2 & 3 \end{bmatrix}$

3. (a) $\begin{bmatrix} 4 & 3 \\ -1 & 7 \\ -2 & 15 \end{bmatrix}$ (b) $\begin{bmatrix} 4 & -5 \\ 1 & 1 \\ -4 & 1 \end{bmatrix}$

(c) $\begin{bmatrix} 12 & -3 \\ 0 & 12 \\ -9 & 24 \end{bmatrix}$ (d) $\begin{bmatrix} 12 & -11 \\ 2 & 6 \\ -11 & 10 \end{bmatrix}$

4. $\begin{bmatrix} 1 & 2 \\ 10 & -4 \end{bmatrix}$ **5.** $\begin{bmatrix} -6 & 6 \\ -10 & 6 \end{bmatrix}$ **6.** $\begin{bmatrix} 5 & 0 \\ -1 & 4 \end{bmatrix}$

7. $\begin{bmatrix} -1 & 30 \\ 2 & -4 \\ 1 & 12 \end{bmatrix}$ **8.** $\begin{bmatrix} -3 & -22 \\ 3 & 10 \\ -5 & 10 \end{bmatrix}$ **9.** Not possible

10. $AB = [6], BA = \begin{bmatrix} 3 & -1 \\ -9 & 3 \end{bmatrix}$ **11.** $\begin{bmatrix} 7 & 0 \\ 0 & 7 \end{bmatrix}$

12. (a) $\begin{bmatrix} -2 & -3 \\ 6 & 1 \end{bmatrix}\begin{bmatrix} x_1 \\ x_2 \end{bmatrix} = \begin{bmatrix} -4 \\ -36 \end{bmatrix}$ (b) $\begin{bmatrix} -7 \\ 6 \end{bmatrix}$

13. Total cost for women's team: $2490
Total cost for men's team: $2871

Section 8.3

1. $AB = I$ and $BA = I$ **2.** $\begin{bmatrix} 3 & 2 \\ 1 & 1 \end{bmatrix}$

3. $\begin{bmatrix} -4 & -2 & 5 \\ -2 & -1 & 2 \\ -1 & 0 & 1 \end{bmatrix}$ **4.** $\begin{bmatrix} \frac{4}{23} & \frac{1}{23} \\ -\frac{3}{23} & \frac{5}{23} \end{bmatrix}$ **5.** $(2, -1, -2)$

Section 8.4

1. (a) -7 (b) 10   (c) 0

2. $M_{11} = -9, M_{12} = -10, M_{13} = 2, M_{21} = 5, M_{22} = -2,$
$M_{23} = -3, M_{31} = 13, M_{32} = 5, M_{33} = -1$
$C_{11} = -9, C_{12} = 10, C_{13} = 2, C_{21} = -5, C_{22} = -2,$
$C_{23} = 3, C_{31} = 13, C_{32} = -5, C_{33} = -1$

3. -31 **4.** 704

Section 8.5

1. $(3, -2)$ **2.** $(2, -3, 1)$ **3.** 9 square units

4. Collinear   **5.** $x - y + 2 = 0$

6. $\begin{bmatrix} 15 & 23 & 12 \\ 14 & 15 & 3 \end{bmatrix}\begin{bmatrix} 19 & 0 & 1 \\ 20 & 21 & 18 \end{bmatrix}\begin{bmatrix} 18 & 5 & 0 \\ 14 & 1 & 12 \end{bmatrix}$

7. 110 -39 -59 25 -21 -3 23 -18 -5 47 -20
-24 149 -56 -75 87 -38 -37

8. OWLS ARE NOCTURNAL

Chapter 9

Section 9.1

1. 3, 5, 7, 9   **2.** $1, \frac{3}{2}, \frac{1}{3}, \frac{3}{4}$

3. (a) $a_n = 4n - 3$ (b) $a_n = (-1)^{n+1}(2n)$

4. 6, 7, 8, 9, 10   **5.** 1, 3, 4, 7, 11   **6.** 2, 4, 5, $\frac{14}{3}, \frac{41}{12}$
7. $4(n + 1)$ **8.** 44   **9.** (a) 0.5555   (b) $\frac{5}{9}$
10. (a) $1000, $1002.50, $1005.01   (b) $1127.33

Section 9.2

1. 2, 5, 8, 11; $d = 3$ **2.** $a_n = 5n - 6$
3. -3, 1, 5, 9, 13, 17, 21, 25, 29, 33, 37   **4.** 79   **5.** 217
6. (a) 630   (b) $N(1 + 2N)$ **7.** 43,560   **8.** 1470
9. $2,500,000

Section 9.3

1. -12, 24, -48, 96; $r = -2$
2. 2, 8, 32, 128, 512

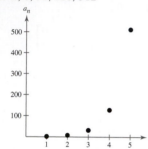

3. 104.02   **4.** 195,312,500   **5.** $\frac{2187}{32}$ **6.** 2.667
7. (a) 10   (b) 6.25   **8.** $3500.85

Section 9.4

1. (a) $\dfrac{6}{(k + 1)(k + 4)}$ (b) $k + 3 \le 3k^2$
    (c) $2^{4k + 2} + 1 > 5k + 5$
2–5. Proofs   **6.** $S_k = k(2k + 1)$; Proof
7. (a) 210   (b) 785   **8.** $a_n = n^2 - n - 2$

Section 9.5

1. (a) 462   (b) 36   (c) 1   (d) 1
2. (a) 21   (b) 21   (c) 14   (d) 14
3. 1, 9, 36, 84, 126, 126, 84, 36, 9, 1
4. $x^4 + 8x^3 + 24x^2 + 32x + 16$
5. (a) $y^4 - 8y^3 + 24y^2 - 32y + 16$
    (b) $32x^5 - 80x^4y + 80x^3y^2 - 40x^2y^3 + 10xy^4 - y^5$
6. $125 + 75y^2 + 15y^4 + y^6$
7. (a) $1120a^4b^4$ (b) $-3,421,440$

Section 9.6

1. 3 ways   **2.** 36 ways   **3.** 27,000 combinations
4. 2,600,000 numbers   **5.** 24 permutations   **6.** 20 ways
7. 1260 ways   **8.** 21 ways
9. 22,100 three-card poker hands   **10.** 1,051,050 teams

Section 9.7

1. {$HH1, HH2, HH3, HH4, HH5, HH6, HT1, HT2, HT3, HT4,$
$HT5, HT6, TH1, TH2, TH3, TH4, TH5, TH6, TT1, TT2, TT3$
$TT4, TT5, TT6$}
2. (a) $\frac{1}{8}$ (b) $\frac{1}{4}$ **3.** $\frac{1}{9}$ **4.** Answers will vary.

5. $\dfrac{302}{2245} \approx 0.135$ **6.** $\dfrac{1}{962,598}$ **7.** $\dfrac{4}{13} \approx 0.308$
8. $\frac{66}{529} \approx 0.125$ **9.** $\frac{121}{900} \approx 0.134$ **10.** 0.001435
11. 0.452

Chapter 10

Section 10.1

1. $0.6747 \approx 38.7°$ **2.** $1.4841 \approx 85.0°$
3. $\dfrac{12}{\sqrt{10}} \approx 3.79$ units   **4.** $\dfrac{3}{\sqrt{34}} \approx 0.51$ unit
5. (a) $\dfrac{8}{\sqrt{5}} \approx 3.58$ units   (b) 12 square units

Section 10.2

1. $x^2 = \frac{3}{2}y$ **2.** $(2, -3)$ **3.** $(y + 3)^2 = 8(x - 2)$
4. $y = 6x - 3$

Section 10.3

1. $\dfrac{(x - 2)^2}{7} + \dfrac{(y - 3)^2}{16} = 1$
2. Center: $(-2, 1)$ **3.** Center: $(-1, 3)$
    Vertices: Vertices:
    $(-2, -2), (-2, 4)$ $(-4, 3), (2, 3)$
    Foci: $\left(-2, 1 \pm \sqrt{5}\right)$ Foci: $(-3, 3), (1, 3)$

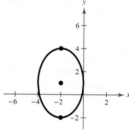 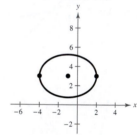

4. Greatest distance: 4.093 astronomical units
    Least distance: 0.336 astronomical units

Section 10.4

1. $\dfrac{(y + 1)^2}{9} - \dfrac{(x - 2)^2}{7} = 1$
2.   **3.** Foci: $\left(\pm \sqrt{13}, 1\right)$
                              Asymptotes: $y = 1 \pm \frac{3}{2}x$

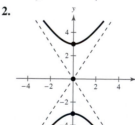

4. $\dfrac{(x - 6)^2}{9} - \dfrac{(y - 2)^2}{4} = 1$
5. The explosion occurred somewhere on the right branch of the hyperbola
$$\dfrac{x^2}{4,840,000} - \dfrac{y^2}{2,129,600} = 1.$$

6. (a) Circle   (b) Hyperbola   (c) Ellipse   (d) Parabola

Section 10.5

1. $\dfrac{(y')^2}{12} - \dfrac{(x')^2}{12} = 1$

2. $\dfrac{(x')^2}{1} + \dfrac{(y')^2}{9} = 1$

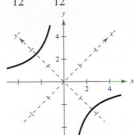

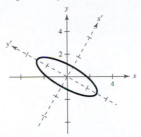

3. $(x')^2 = -2(y' - 3)$

4. Parabola:
$$y = \dfrac{2x \pm \sqrt{-6x - 10}}{4}$$

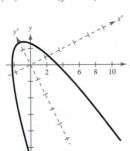

Section 10.6

1.

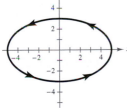

2.

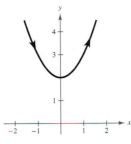

3.

4. (a) $x = t,\ y = t^2 + 2$
   (b) $x = 2 - t,\ y = t^2 - 4t + 6$

5. $x = a(\theta + \sin \theta),\ y = a(1 + \cos \theta)$

Section 10.7

1. (a)

(b)

(c)

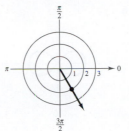

2. (a)

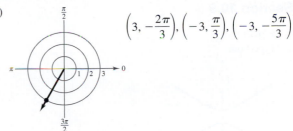

$\left(3, -\dfrac{2\pi}{3}\right), \left(-3, \dfrac{\pi}{3}\right), \left(-3, -\dfrac{5\pi}{3}\right)$

(b)

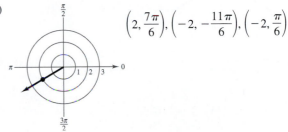

$\left(2, \dfrac{7\pi}{6}\right), \left(-2, -\dfrac{11\pi}{6}\right), \left(-2, \dfrac{\pi}{6}\right)$

(c)

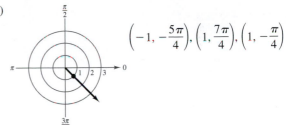

$\left(-1, -\dfrac{5\pi}{4}\right), \left(1, \dfrac{7\pi}{4}\right), \left(1, -\dfrac{\pi}{4}\right)$

3. $(-2, 0)$ **4.** $\left(2, \dfrac{\pi}{2}\right)$ **5.** $x^2 + y^2 - 6y = 0$

Section 10.8

1.

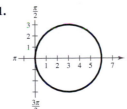

2.

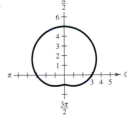

3.

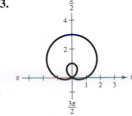

4.

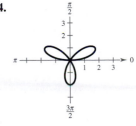

5.

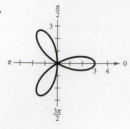

6.

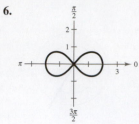

Section 10.9

1. Hyperbola

2. Hyperbola

3. $r = \dfrac{2}{1 - \cos \theta}$ **4.** $r = \dfrac{0.622}{1 + 0.848 \sin \theta}$; about 31,000,000 mi

Index

GRAPHS OF PARENT FUNCTIONS

Linear Function

$$f(x) = mx + b$$

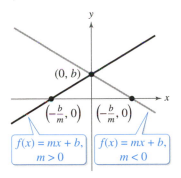

Domain: $(-\infty, \infty)$
Range: $(-\infty, \infty)$
x-intercept: $(-b/m, 0)$
y-intercept: $(0, b)$
Increasing when $m > 0$
Decreasing when $m < 0$

Absolute Value Function

$$f(x) = |x| = \begin{cases} x, & x \ge 0 \\ -x, & x < 0 \end{cases}$$

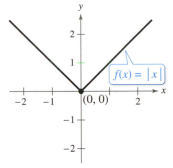

Domain: $(-\infty, \infty)$
Range: $[0, \infty)$
Intercept: $(0, 0)$
Decreasing on $(-\infty, 0)$
Increasing on $(0, \infty)$
Even function
y-axis symmetry

Square Root Function

$$f(x) = \sqrt{x}$$

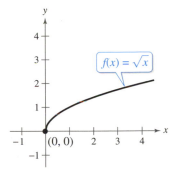

Domain: $[0, \infty)$
Range: $[0, \infty)$
Intercept: $(0, 0)$
Increasing on $(0, \infty)$

Greatest Integer Function

$$f(x) = [\![x]\!]$$

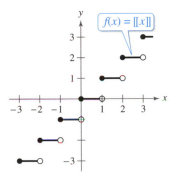

Domain: $(-\infty, \infty)$
Range: the set of integers
x-intercepts: in the interval $[0, 1)$
y-intercept: $(0, 0)$
Constant between each pair of
consecutive integers
Jumps vertically one unit at
each integer value

Quadratic (Squaring) Function

$$f(x) = ax^2$$

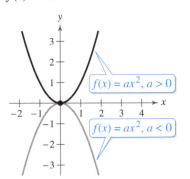

Domain: $(-\infty, \infty)$
Range $(a > 0)$: $[0, \infty)$
Range $(a < 0)$: $(-\infty, 0]$
Intercept: $(0, 0)$
Decreasing on $(-\infty, 0)$ for $a > 0$
Increasing on $(0, \infty)$ for $a > 0$
Increasing on $(-\infty, 0)$ for $a < 0$
Decreasing on $(0, \infty)$ for $a < 0$
Even function
y-axis symmetry
Relative minimum $(a > 0)$,
relative maximum $(a < 0)$,
or vertex: $(0, 0)$

Cubic Function

$$f(x) = x^3$$

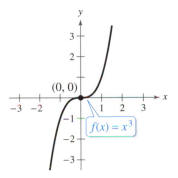

Domain: $(-\infty, \infty)$
Range: $(-\infty, \infty)$
Intercept: $(0, 0)$
Increasing on $(-\infty, \infty)$
Odd function
Origin symmetry

Rational (Reciprocal) Function

$f(x) = \dfrac{1}{x}$

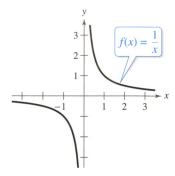

Domain: $(-\infty, 0) \cup (0, \infty)$
Range: $(-\infty, 0) \cup (0, \infty)$
No intercepts
Decreasing on $(-\infty, 0)$ and $(0, \infty)$
Odd function
Origin symmetry
Vertical asymptote: y-axis
Horizontal asymptote: x-axis

Exponential Function

$f(x) = a^x, \ a > 1$

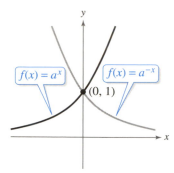

Domain: $(-\infty, \infty)$
Range: $(0, \infty)$
Intercept: $(0, 1)$
Increasing on $(-\infty, \infty)$
  for $f(x) = a^x$
Decreasing on $(-\infty, \infty)$
  for $f(x) = a^{-x}$
Horizontal asymptote: x-axis
Continuous

Logarithmic Function

$f(x) = \log_a x, \ a > 1$

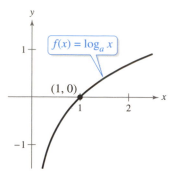

Domain: $(0, \infty)$
Range: $(-\infty, \infty)$
Intercept: $(1, 0)$
Increasing on $(0, \infty)$
Vertical asymptote: y-axis
Continuous
Reflection of graph of $f(x) = a^x$
  in the line $y = x$

Sine Function

$f(x) = \sin x$

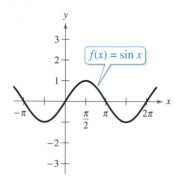

Domain: $(-\infty, \infty)$
Range: $[-1, 1]$
Period: 2π
x-intercepts: $(n\pi, 0)$
y-intercept: $(0, 0)$
Odd function
Origin symmetry

Cosine Function

$f(x) = \cos x$

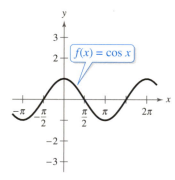

Domain: $(-\infty, \infty)$
Range: $[-1, 1]$
Period: 2π
x-intercepts: $\left(\dfrac{\pi}{2} + n\pi, 0\right)$
y-intercept: $(0, 1)$
Even function
y-axis symmetry

Tangent Function

$f(x) = \tan x$

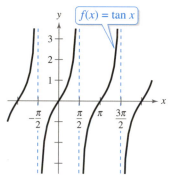

Domain: all $x \neq \dfrac{\pi}{2} + n\pi$

Range: $(-\infty, \infty)$
Period: π
x-intercepts: $(n\pi, 0)$
y-intercept: $(0, 0)$
Vertical asymptotes:

$$x = \dfrac{\pi}{2} + n\pi$$

Odd function
Origin symmetry

Cosecant Function

$f(x) = \csc x$

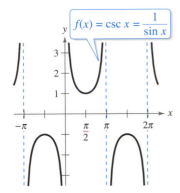

Domain: all $x \neq n\pi$

Range: $(-\infty, -1] \cup [1, \infty)$

Period: 2π

No intercepts

Vertical asymptotes: $x = n\pi$

Odd function

Origin symmetry

Secant Function

$f(x) = \sec x$

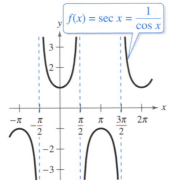

Domain: all $x \neq \dfrac{\pi}{2} + n\pi$

Range: $(-\infty, -1] \cup [1, \infty)$

Period: 2π

y-intercept: $(0, 1)$

Vertical asymptotes:

$$x = \frac{\pi}{2} + n\pi$$

Even function

y-axis symmetry

Cotangent Function

$f(x) = \cot x$

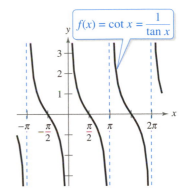

Domain: all $x \neq n\pi$

Range: $(-\infty, \infty)$

Period: π

x-intercepts: $\left(\dfrac{\pi}{2} + n\pi, 0\right)$

Vertical asymptotes: $x = n\pi$

Odd function

Origin symmetry

Inverse Sine Function

$f(x) = \arcsin x$

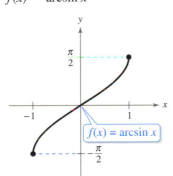

Domain: $[-1, 1]$

Range: $\left[-\dfrac{\pi}{2}, \dfrac{\pi}{2}\right]$

Intercept: $(0, 0)$

Odd function

Origin symmetry

Inverse Cosine Function

$f(x) = \arccos x$

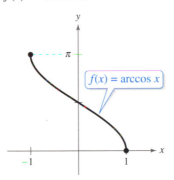

Domain: $[-1, 1]$

Range: $[0, \pi]$

y-intercept: $\left(0, \dfrac{\pi}{2}\right)$

Inverse Tangent Function

$f(x) = \arctan x$

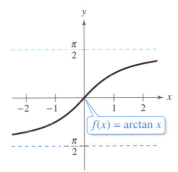

Domain: $(-\infty, \infty)$

Range: $\left(-\dfrac{\pi}{2}, \dfrac{\pi}{2}\right)$

Intercept: $(0, 0)$

Horizontal asymptotes:

$$y = \pm\frac{\pi}{2}$$

Odd function

Origin symmetry

Definition of the Six Trigonometric Functions

Right triangle definitions, where $0 < \theta < \pi/2$

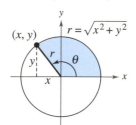

$$\sin \theta = \frac{\text{opp}}{\text{hyp}} \qquad \csc \theta = \frac{\text{hyp}}{\text{opp}}$$

$$\cos \theta = \frac{\text{adj}}{\text{hyp}} \qquad \sec \theta = \frac{\text{hyp}}{\text{adj}}$$

$$\tan \theta = \frac{\text{opp}}{\text{adj}} \qquad \cot \theta = \frac{\text{adj}}{\text{opp}}$$

Circular function definitions, where θ is any angle

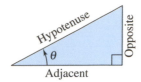

$$\sin \theta = \frac{y}{r} \qquad \csc \theta = \frac{r}{y}$$

$$\cos \theta = \frac{x}{r} \qquad \sec \theta = \frac{r}{x}$$

$$\tan \theta = \frac{y}{x} \qquad \cot \theta = \frac{x}{y}$$

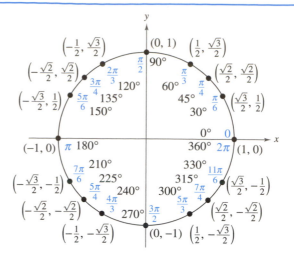

Reciprocal Identities

$$\sin u = \frac{1}{\csc u} \qquad \cos u = \frac{1}{\sec u} \qquad \tan u = \frac{1}{\cot u}$$

$$\csc u = \frac{1}{\sin u} \qquad \sec u = \frac{1}{\cos u} \qquad \cot u = \frac{1}{\tan u}$$

Quotient Identities

$$\tan u = \frac{\sin u}{\cos u} \qquad \cot u = \frac{\cos u}{\sin u}$$

Pythagorean Identities

$$\sin^2 u + \cos^2 u = 1$$
$$1 + \tan^2 u = \sec^2 u \qquad 1 + \cot^2 u = \csc^2 u$$

Cofunction Identities

$$\sin\left(\frac{\pi}{2} - u\right) = \cos u \qquad \cot\left(\frac{\pi}{2} - u\right) = \tan u$$

$$\cos\left(\frac{\pi}{2} - u\right) = \sin u \qquad \sec\left(\frac{\pi}{2} - u\right) = \csc u$$

$$\tan\left(\frac{\pi}{2} - u\right) = \cot u \qquad \csc\left(\frac{\pi}{2} - u\right) = \sec u$$

Even/Odd Identities

$$\sin(-u) = -\sin u \qquad \cot(-u) = -\cot u$$
$$\cos(-u) = \cos u \qquad \sec(-u) = \sec u$$
$$\tan(-u) = -\tan u \qquad \csc(-u) = -\csc u$$

Sum and Difference Formulas

$$\sin(u \pm v) = \sin u \cos v \pm \cos u \sin v$$
$$\cos(u \pm v) = \cos u \cos v \mp \sin u \sin v$$
$$\tan(u \pm v) = \frac{\tan u \pm \tan v}{1 \mp \tan u \tan v}$$

Double-Angle Formulas

$$\sin 2u = 2 \sin u \cos u$$
$$\cos 2u = \cos^2 u - \sin^2 u = 2 \cos^2 u - 1 = 1 - 2 \sin^2 u$$
$$\tan 2u = \frac{2 \tan u}{1 - \tan^2 u}$$

Power-Reducing Formulas

$$\sin^2 u = \frac{1 - \cos 2u}{2}$$

$$\cos^2 u = \frac{1 + \cos 2u}{2}$$

$$\tan^2 u = \frac{1 - \cos 2u}{1 + \cos 2u}$$

Sum-to-Product Formulas

$$\sin u + \sin v = 2 \sin\left(\frac{u + v}{2}\right) \cos\left(\frac{u - v}{2}\right)$$

$$\sin u - \sin v = 2 \cos\left(\frac{u + v}{2}\right) \sin\left(\frac{u - v}{2}\right)$$

$$\cos u + \cos v = 2 \cos\left(\frac{u + v}{2}\right) \cos\left(\frac{u - v}{2}\right)$$

$$\cos u - \cos v = -2 \sin\left(\frac{u + v}{2}\right) \sin\left(\frac{u - v}{2}\right)$$

Product-to-Sum Formulas

$$\sin u \sin v = \frac{1}{2}[\cos(u - v) - \cos(u + v)]$$

$$\cos u \cos v = \frac{1}{2}[\cos(u - v) + \cos(u + v)]$$

$$\sin u \cos v = \frac{1}{2}[\sin(u + v) + \sin(u - v)]$$

$$\cos u \sin v = \frac{1}{2}[\sin(u + v) - \sin(u - v)]$$

FORMULAS FROM GEOMETRY

Triangle:

$h = a \sin \theta$

$\text{Area} = \dfrac{1}{2}bh$

$c^2 = a^2 + b^2 - 2ab \cos \theta$ (Law of Cosines)

Right Triangle:

Pythagorean Theorem
$c^2 = a^2 + b^2$

Equilateral Triangle:

$h = \dfrac{\sqrt{3}s}{2}$

$\text{Area} = \dfrac{\sqrt{3}s^2}{4}$

Parallelogram:

$\text{Area} = bh$

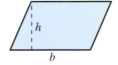

Trapezoid:

$\text{Area} = \dfrac{h}{2}(a + b)$

Circle:

$\text{Area} = \pi r^2$

$\text{Circumference} = 2\pi r$

Sector of Circle:

$\text{Area} = \dfrac{\theta r^2}{2}$

$s = r\theta$

θ in radians

Circular Ring:

$\text{Area} = \pi(R^2 - r^2)$

$\quad\quad\;\; = 2\pi pw$

$p = \text{average radius,}$

$w = \text{width of ring}$

Sector of Circular Ring:

$\text{Area} = \theta pw$

$p = \text{average radius,}$

$w = \text{width of ring,}$

θ in radians

Ellipse:

$\text{Area} = \pi ab$

$\text{Circumference} \approx 2\pi \sqrt{\dfrac{a^2 + b^2}{2}}$

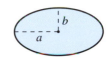

Cone:

$\text{Volume} = \dfrac{Ah}{3}$

$A = \text{area of base}$

Right Circular Cone:

$\text{Volume} = \dfrac{\pi r^2 h}{3}$

$\text{Lateral Surface Area} = \pi r \sqrt{r^2 + h^2}$

Frustum of Right Circular Cone:

$\text{Volume} = \dfrac{\pi(r^2 + rR + R^2)h}{3}$

$\text{Lateral Surface Area} = \pi s(R + r)$

Right Circular Cylinder:

$\text{Volume} = \pi r^2 h$

$\text{Lateral Surface Area} = 2\pi rh$

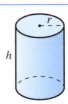

Sphere:

$\text{Volume} = \dfrac{4}{3}\pi r^3$

$\text{Surface Area} = 4\pi r^2$

Wedge:

$A = B \sec \theta$

$A = \text{area of upper face,}$

$B = \text{area of base}$

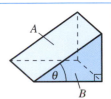

ALGEBRA

Factors and Zeros of Polynomials:

Given the polynomial $p(x) = a_n x^n + a_{n-1} x^{n-1} + \cdots + a_1 x + a_0$. If $p(b) = 0$, then b is a *zero* of the polynomial and a *solution* of the equation $p(x) = 0$. Furthermore, $(x - b)$ is a *factor* of the polynomial.

Fundamental Theorem of Algebra:
An nth degree polynomial has n (not necessarily distinct) zeros.

Quadratic Formula:
If $p(x) = ax^2 + bx + c$, $a \neq 0$ and $b^2 - 4ac \geq 0$, then the real zeros of p are $x = \left(-b \pm \sqrt{b^2 - 4ac}\right)/2a$.

Special Factors:

$x^2 - a^2 = (x - a)(x + a)$

$x^3 - a^3 = (x - a)(x^2 + ax + a^2)$

$x^3 + a^3 = (x + a)(x^2 - ax + a^2)$

$x^4 - a^4 = (x - a)(x + a)(x^2 + a^2)$

$x^4 + a^4 = \left(x^2 + \sqrt{2}\,ax + a^2\right)\left(x^2 - \sqrt{2}\,ax + a^2\right)$

$x^n - a^n = (x - a)(x^{n-1} + ax^{n-2} + \cdots + a^{n-1})$, for n odd

$x^n + a^n = (x + a)(x^{n-1} - ax^{n-2} + \cdots + a^{n-1})$, for n odd

$x^{2n} - a^{2n} = (x^n - a^n)(x^n + a^n)$

Examples

$x^2 - 9 = (x - 3)(x + 3)$

$x^3 - 8 = (x - 2)(x^2 + 2x + 4)$

$x^3 + 4 = \left(x + \sqrt[3]{4}\right)\left(x^2 - \sqrt[3]{4}x + \sqrt[3]{16}\right)$

$x^4 - 4 = \left(x - \sqrt{2}\right)\left(x + \sqrt{2}\right)(x^2 + 2)$

$x^4 + 4 = (x^2 + 2x + 2)(x^2 - 2x + 2)$

$x^5 - 1 = (x - 1)(x^4 + x^3 + x^2 + x + 1)$

$x^7 + 1 = (x + 1)(x^6 - x^5 + x^4 - x^3 + x^2 - x + 1)$

$x^6 - 1 = (x^3 - 1)(x^3 + 1)$

Binomial Theorem:

$(x + a)^2 = x^2 + 2ax + a^2$

$(x - a)^2 = x^2 - 2ax + a^2$

$(x + a)^3 = x^3 + 3ax^2 + 3a^2 x + a^3$

$(x - a)^3 = x^3 - 3ax^2 + 3a^2 x - a^3$

$(x + a)^4 = x^4 + 4ax^3 + 6a^2 x^2 + 4a^3 + a^4$

$(x - a)^4 = x^4 - 4ax^3 + 6a^2 x^2 - 4a^3 x + a^4$

$(x + a)^n = x^n + nax^{n-1} + \dfrac{n(n-1)}{2!} a^2 x^{n-2} + \cdots + na^{n-1}x + a^n$

$(x - a)^n = x^n - nax^{n-1} + \dfrac{n(n-1)}{2!} a^2 x^{n-2} - \cdots \pm na^{n-1}x \mp a^n$

Examples

$(x + 3)^2 = x^2 + 6x + 9$

$(x^2 - 5)^2 = x^4 - 10x^2 + 25$

$(x + 2)^3 = x^3 + 6x^2 + 12x + 8$

$(x - 1)^3 = x^3 - 3x^2 + 3x - 1$

$\left(x + \sqrt{2}\right)^4 = x^4 + 4\sqrt{2}x^3 + 12x^2 + 8\sqrt{2}x + 4$

$(x - 4)^4 = x^4 - 16x^3 + 96x^2 - 256x + 256$

$(x + 1)^5 = x^5 + 5x^4 + 10x^3 + 10x^2 + 5x + 1$

$(x - 1)^6 = x^6 - 6x^5 + 15x^4 - 20x^3 + 15x^2 - 6x + 1$

Rational Zero Test:
If $p(x) = a_n x^n + a_{n-1} x^{n-1} + \cdots + a_1 x + a_0$ has integer coefficients, then every *rational* zero of $p(x) = 0$ is of the form $x = r/s$, where r is a factor of a_0 and s is a factor of a_n.

Exponents and Radicals:

$a^0 = 1$, $a \neq 0$

$a^{-x} = \dfrac{1}{a^x}$

$a^x a^y = a^{x+y}$

$\dfrac{a^x}{a^y} = a^{x-y}$

$(a^x)^y = a^{xy}$

$(ab)^x = a^x b^x$

$\left(\dfrac{a}{b}\right)^x = \dfrac{a^x}{b^x}$

$\sqrt{a} = a^{1/2}$

$\sqrt[n]{a} = a^{1/n}$

$\sqrt[n]{a^m} = a^{m/n} = \left(\sqrt[n]{a}\right)^m$

$\sqrt[n]{ab} = \sqrt[n]{a}\,\sqrt[n]{b}$

$\sqrt[n]{\left(\dfrac{a}{b}\right)} = \dfrac{\sqrt[n]{a}}{\sqrt[n]{b}}$

Conversion Table:

1 centimeter $\approx$ 0.394 inch

1 meter $\approx$ 39.370 inches

    $\approx$ 3.281 feet

1 kilometer $\approx$ 0.621 mile

1 liter $\approx$ 0.264 gallon

1 newton $\approx$ 0.225 pound

1 joule $\approx$ 0.738 foot-pound

1 gram $\approx$ 0.035 ounce

1 kilogram $\approx$ 2.205 pounds

1 inch = 2.54 centimeters

1 foot = 30.48 centimeters

    $\approx$ 0.305 meter

1 mile $\approx$ 1.609 kilometers

1 gallon $\approx$ 3.785 liters

1 pound $\approx$ 4.448 newtons

1 foot-lb $\approx$ 1.356 joules

1 ounce $\approx$ 28.350 grams

1 pound $\approx$ 0.454 kilogram

Using the Graphing Utility in Enhanced WebAssign

The Enhanced WebAssign Graphing Utility lets you graph one or more mathematical elements directly on a set of coordinate axes. Your graph is then scored automatically when you submit the assignment for grading.

The Graphing Utility currently supports points, rays, segments, lines, circles, and parabolas. Inequalities can also be indicated by filling one or more areas.

■ Graphing Utility Interface Overview

The middle of Graphing Utility is the drawing area. It contains labeled coordinate axes, which may have different axis scales and extents depending on the nature of the question you are working on.

On the left side of Graphing Utility is the list of Tools that let you create graph objects and select objects to edit.

The bottom of the Graphing Utility is the Object Properties Toolbar, which becomes active when you have a graph element selected. This toolbar shows you all the details about the selected graph object, and also lets you edit properties of that object.

On the right side of Graphing Utility is the list of Actions that lets you create fills and delete objects from your graph.

Drawing Tools	
Points: (point tool icon)	Click the Point Tool, and then click where you want the point to appear.
Lines: (line tool icons)	Click the Line Tool, and then place two points (which are on the desired line). The arrows on the end of the line indicate that the line goes off to infinity on both ends.
Rays: (ray tool icons)	Click the Ray Tool, place the endpoint, and then place another point on the ray.
Line Segments: (line segment tool icons)	Click the Line Segment Tool, and then place the two endpoints of the line segment.
Circles: (circle tool icon)	Click the Circle Tool, place the point at the center first, and then place a point on the radius of the circle.
Parabolas: (parabola tool icons)	Click the Parabola Tool (either vertical or horizontal), place the vertex first, and then place another point on the parabola.
No Solution: (No Solution tool icon)	If the question has no solution, simply click the No Solution Tool.

Note: Don't worry if you don't place the points exactly where you want them initially; you can move these points around before submitting for grading.

■ Selecting Graph Objects

To edit a graph object, that object must be "selected" as the active object. (When you first draw an object, it is created in the selected state.) When a graph element is "selected," the color of the line changes and two "handles" are visible. The handles are the two square points you clicked to create the object. To select an object, click the select tool, and then click on the object. To deselect the object, click on a blank area on the drawing area or on a drawing tool.

Not Selected   Selected

Once an object is selected, you can modify it by using your mouse or the keyboard. As you move it, you'll notice that you cannot move the handles off the drawing area. To move an object with the mouse, click and drag the object's line. Or, click and drag one of the handles to move just that handle. On the keyboard, the arrow keys also move the selected object around by one unit.

As you move the object or handle you'll see that the Object Properties toolbar changes to remain up to date.

Object Properties Toolbar	
Coordinate Fields: ● Point 1 (-13 , -9) ■ Point 2 (15 , -6)	You can use the Coordinate Fields to edit the coordinates of the handles directly. Use this method to enter decimal or fractional coordinates.
Endpoint Controls: Endpoint ■ ◉ Endpoint ◉ ■	If the selected object is a segment or ray, the Endpoint Controls are used to define the endpoint as closed or open. As a shortcut, you can also define an endpoint by clicking on the endpoint when the ray or segment is in the unselected state.
Solid/Dash Controls: Solid Dash	For any selected object other than a point, the Solid/Dash Controls are used to define the object as solid or dashed. To change graph objects to solid or dashed, select the object and then click the solid or dash button.

■ Using Fractions or Decimals as Coordinates

To draw an object with handle coordinates that are fractions or decimals, you must use the Object Properties Toolbar. Draw the desired object anywhere on the drawing area, then use the Coordinate Fields to change the endpoint(s) to the desired value. For example, to enter a fraction, just type "3/4."

Note: The points and lines you draw must be exactly correct when you submit for grading. This means you should not round any of your values—if you want a point at 11/3, you should enter 11/3 in the coordinate box rather than 3.667. Also, mixed fractions are not acceptable entries. This means that 3 2/3 is an incorrect entry.

Actions	
Fill Tool: Fill	To graph an inequality, you must specify a region on the graph. To do this, first draw the line(s), circle(s), or other object(s) that will define the region you want to represent your answer. Be sure to specify the objects as either solid or dashed, according to the inequality you are graphing! Then choose the Fill Tool and click inside the region that you want filled.
	If you decide you wanted the fill in a different area, click the filled region that you want to unfill, and then click the region that you do want to fill.
Delete Tool: Delete	To erase a single graph object, first select that element in the drawing area then click the Delete Tool or press the Delete key on your keyboard.
Clear All Tool: Clear All	The Clear All Tool will erase all of your graph objects. (If the drawing area is already empty, the Clear All Tool is disabled.)

▶ **EXAMPLE:**

Let's suppose you're asked to graph the inequality $y > 5x + \frac{1}{5}$, and you want to use the points $(0, \frac{1}{5})$ and $(1, 5\frac{1}{5})$. You would first place any line on the drawing area.

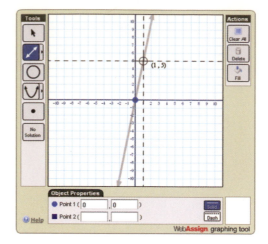

Then, adjust the points using the Coordinate Fields. Remember, you need to enter 5 1/5 as a decimal (5.2) or an improper fraction (26/5).

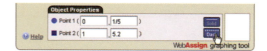

Next, you would define the line as dashed since the inequality does not include the values on the line.

Finally, you would select the Fill Tool and click on the desired region to complete the graph.

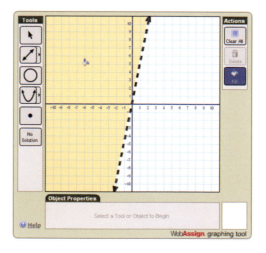